# HANDBOOK OF MATHEMATICAL MODELS IN COMPUTER VISION

# HANDBOOK OF MATHEMATICAL MODELS IN COMPUTER VISION

*Edited by*

**Nikos Paragios**
*Ecole Nationale des Ponts et Chaussees*

**Yunmei Chen**
*University of Florida*

**Olivier Faugeras**
*INRIA*

Library of Congress Cataloging-in-Publication Data

A C.I.P. Catalogue record for this book is available from the Library of Congress.

***Handbook of Mathematical Models in Computer Vision,*** Edited by Nikos Paragios, Yunmei Chen and Olivier Faugeras

p.cm.

ISBN-13: 978-1-4419-3885-5

ISBN-10: (eBook) 0-387-28831-7
ISBN-13: (eBook) 978-0387-28831-4

Printed on acid-free paper.



Printed in the United States of America.

springeronline.com

# Contents

# Preface

Abstract

Biological vision is a rather fascinating domain of research. Scientists of various origins like biology, medicine, neurophysiology, engineering, mathematics, etc. aim to understand the processes leading to visual perception process and at reproducing such systems. Understanding the environment is most of the time done through visual perception which appears to be one of the most fundamental sensory abilities in humans and therefore a significant amount of research effort has been dedicated towards modelling and reproducing human visual abilities. Mathematical methods play a central role in this endeavour.

## Introduction

David Marr's theory was a pioneering step towards understanding visual perception. In his view human vision was based on a complete surface reconstruction of the environment that was then used to address visual subtasks. This approach was proven to be insufficient by neuro-biologists and complementary ideas from statistical pattern recognition and artificial intelligence were introduced to better address the visual perception problem. In this framework visual perception is represented by a set of actions and rules connecting these actions. The emerging concept of active vision consists of a selective visual perception paradigm that is basically equivalent to recovering from the environment the minimal piece information required to address a particular task of interest.

Mathematical methods are an alternative to tackle visual perception. The central idea behind these methods is to reformulate the visual perception components as optimization problems where the minima of a specifically designed objective function "solve" the task under consideration. The definition of such functions is often an ill-posed problem since the number of variables to be recovered is much larger than the number of constraints. Furthermore, often the optimization process itself is ill-posed due the non-convexity of the designed function inducing the presence of local minima. Variational, statistical and combinatorial methods are three distinct and important categories of mathematical methods in computational vision.

Variational techniques are either based on the optimization of cost functions through the calculus of variations or on the design of partial differential equations whose steady state corresponds to the solution of the visual perception task. Such techniques have gained significant attention over the past decade and have been used to address image restoration and enhancement, image segmentation, tracking and stereo reconstruction among other problems. The possibility to use the calculus of variations in the optimization process is the most important strength of these methods combined with the fact that one can integrate many terms and build quite complicated objective functions at the expense of converging toward local minima.

Statistical methods often consist of two stages, a learning and an execution one. Complex conditional, multi-dimensional distributions are used to describe visual perception tasks that are learnt through a training procedure. Visual perception is then formulated as an inference problem, conditional to the observations (images). One can claim that such methods are suitable to address constrained optimization problems, in particular when the subset of solutions can be well described through a conditional parametric density function. They suffer from the curse of dimensionality, e.g. in the Bayesian case when very-high dimensional integrals have to be computed.

Discrete optimization is an alternative to the continuous case often addressed through statistical and variational methods. To this end, visual perception is often redefined as a labelling procedure at the image element level according to a predefined set of plausible classes. Such a simplification often reduces the dimensionality of the problem and makes possible the design of efficient optimization algorithms. On the other hand such methods can have limited performance because of the discretization of the solution space, in particular when the solution lives in a rather continuous in-homogeneous space. One can refer to graph-based methods for addressing such tasks.

The choice of the most appropriate technique to address visual perception is rather task-driven and one cannot claim the existence of a universal solution to most of the visual perception problems. In this edited volume, our intention is to present the most promising and representative mathematical models to address visual perception through variational, statistical and combinatorial methods. In order to be faithful to the current state of the art in visual perception, a rather complete set of computational vision components has been considered starting from low level vision tasks like image enhancement and restoration and ending at complete reconstruction of scene's geometry from images.

The volume is organized in six thematic areas and thirty-three chapters presenting an overview of existing mathematical methodologies to address an important number of visual perception tasks.

## Contributions & Contributors

Image reconstruction from either destroyed or incomplete data is a crucial low level task of visual perception. Local filter operators, diffusion methods as well as variational methods are among the most studied methods in the domain. The book starts with three tutorial chapters in this thematic area. The total variation method and diffusion filters as well as image decomposition in orthogonal bases, two of the most instrumental methods to address image reconstruction are presented in the first chapter. Image inpainting/completion is a more advanced problem consisting of restoring missing information in images ; it belongs to the same family and is covered in chapter 2. In the third chapter of this thematic area, an introduction to the problem as well as the most prominent techniques from the area of variational methods are presented.

Image segmentation and object extraction are of particular interest with applications in numerous domains. In its simplest instantiation the problem consists of creating an image partition with respect to some feature space, the regions being assumed to have uniform visual structure in this space. Such a problem can be solved in many ways. Labelling is an example where the objective is to assign to the local image element the most likely hypothesis given the observation. Two chapters explore such a concept in this thematic area, the watershed transformation is one of them and combinatorial optimization through the graph-cuts paradigm is another. Evolution of curves and surfaces is an alternative method to address the same problem. Classes are represented through moving interfaces that are deforming in order to capture image regions with consistent visual properties. The snake model - a pioneering framework - is the predecessor of the methods presented. First, an overview for finding multiple contours for contour completion from points or curves in 2D or 3D images is presented using the concept of minimal paths. Then in order a method that integrate region statistics is presented within deformable models leading to a new class of deformable shape and texture models. Use of prior knowledge is important within the segmentation process and therefore in the next chapter the design of shape priors for variational region-based segmentation is presented. Segmentation through the propagation of curves through the level set method is an established technique to grouping and object extraction Therefore, methods to address model-free as well as model-based segmentation are part of this thematic area. Last, but not least, a stochastic snake model based the theory of interacting particle systems and hydrodynamic limits is presented as a new way of evolving curves as a possible alternative to level set methods.

Representing and understanding structures is an essential component of biological vision, often used as a basis for high level vision tasks. Therefore, a thematic area dedicated to shape modelling and registration is present in this volume. Shape representations of various form are explored while at the same time the notions of establishing correspondences between different structures representing the same object are presented as well as methods recovering correspondences between shapes and images.

Motion analysis is a fundamental area of computational vision and mostly consists of two problems, estimating correspondences between images and being able to track objects of interest in a sequence of images. Optical flow estimation can be addressed in different ways. In this thematic area we explore the use of parametric motion models as well as the estimation of dense correspondences between images. Furthermore, we present a compendium of existing methods to detect and track objects in a consistent fashion within several frames as well as variational formulations to segment images and track objects in several frames. Understanding the real 3D motion is a far more complicated task of computational vision in particular when considering objects that do exhibit a number of articulations. Human motion capture is an example that is presented in this thematic area. We conclude with methods going beyond objects that are able to account, describe and reproduce the dynamics of structured scenes.

Stereo reconstruction is one of the best studied tasks in high level vision. Understanding and reproducing the 3D geometry of a scene is a fundamental component of biological vision. In this thematic area the shape from shading problem i.e. that of recovering the structure of the scene from one single image is first addressed. Different methods exploring the use of multiple cameras to recover 3D from images are then presented, based on differential geometry, variational formulations and combinatorial optimization. The notion of time and dynamic behaviour of scenes is also addressed where the objective is to create 3D temporal models of the evolving geometry.

Medical image analysis is one of the most prominent application domains of computer vision and in such a constrained solution space one can develop methods that can better capture the expected form of the structures of interest. Regularization, segmentation, object extraction and registration are the tasks presented in this thematic area. Model-free combinatorial methods that aim to recover organs of particular interest, statistical methods that aim to capture the variation of anatomical structures, and variational methods that aim to recover and segment smooth vectorial images are presented. Last, but not least a comprehensive review of statistical methods to image registration is presented, a problem that consists of recovering correspondences between different modalities measuring the same anatomical structure.

In order to capture the spectrum of the different methods and present an overview of mathematical methodologies in computational vision a notable number of contributors was invited to complete such an effort. Eighty-three contributors from the academic and the industrial world, from nine different countries and thirty-eight institutions have participated in this effort. The final outcome consists of 6 thematic areas, 33 chapters, 625 pages and 929 references.

N. Paragios, Y. Chen & O. Faugeras

# List of Contributors

**Agrawal, Motilal**
Artificial Intelligence Center
SRI International, Menlo Park, USA
mailto:agrawal@ai.sri.com
http://www.ai.sri.com/people/agrawal/

**van Assen, Hans**
Division of Image Processing, Department of Radiology
Leiden University, Leiden, Netherlands
mailto:H.C.van_ Assen@lumc.nl

**Aubert, Gilles**
Department of Mathematics
Universite de Nice/Sophia Antipolis, France
mailto:gaubert@math.unice.fr
http://math1.unice.fr/~gaubert/

**Barlaud, Michel**
Laboratoire I3S
CNRS-Universite de Nice/Sophia Antipolis, France
mailto:barlaud@i3s.unice.fr
http://www.i3s.unice.fr/~barlaud/

**Barron, Carlos**
Department of Computer Science
University of Houston, Houston, USA
mailto:cbarron@uh.edu

**Bertalmío, Marcelo**
Departament de Tecnologia
Universitat Pompeu Fabra, Barcelona, Spain
mailto:marcelo.bertalmio@upf.edu

http://www.tecn.upf.es/~mbertalmio

**Bergtholdt, Martin**
Department of Mathematics & Computer Science
University of Mannheim, Germany
mailto:bergtholdt@uni-mannheim.de
http://www.cvgpr.uni-mannheim.de/

**Blake, Andrew**
Microsoft Research Cambridge, UK
mailto:ablake@microsoft.com
http://www.research.microsoft.com/~ablake

**Boykov, Yuri**
Departament of Computer Science
University of Western Ontario, Canada
mailto:yuri@csd.uwo.ca
http://www.csd.uwo.ca/faculty/yuri/

**Brox, Thomas**
Faculty of Mathematics and Computer Science
Saarland University, 66041 Saarbrücken, Germany
mailto:brox@mia.uni-saarland.de
http://www.mia.uni-saarland.de/brox/

**Bruckstein, Alfred M.**
Computer Science Department
Technion, Haifa, Israel
mailto:freddy@cs.technion.ac.il

**Caselles, Vicent**
Departament de Tecnologia
Universitat Pompeu Fabra, Barcelona, Spain
mailto:vicent.caselles@upf.edu
http://www.iua.upf.es/~vcaselles/

**Chan, Tony**
Department of Mathematics
University of California at Los Angeles, USA
mailto:chan@math.ucla.edu
http://www.math.ucla.edu/~chan

**Chen, Ting**
Department of Radiology
NYU Medical School, New York, USA

mailto:ting.chen@med.nyu.edu

**Chen, Yunmei**
Department of Mathematics
University of Florida, Gainesville, USA
mailto:yun@math.ufl.edu
http://www.math.ufl.edu/~yun/

**Cohen, Laurent**
CEREMADE
Universite Paris IX Dauphine, Paris, France
mailto:cohen@ceremade.dauphine.fr
http://www.ceremade.dauphine.fr/~cohen

**Cremers, Daniel**
Imaging & Visualization Department
Siemens Corporate Research, Princeton, NJ, USA
mailto:daniel.cremers@scr.siemens.com
http://www.cs.ucla.edu/~cremers

**Davis, Larry**
Department of Computer Science
University of Maryland, College Park, USA
mailto:lsd@cs.umd.edu
http://cvl.umiacs.umd.edu/users/lsd/

**Deriche, Rachid**
I.N.R.I.A. Sophia Antipolis, France
mailto:Rachid.Deriche@inria.fr
http://www-sop.inria.fr/odyssee/team/Rachid.Deriche/

**Doretto, Gianfranco**
Computer Science Department
University of California at Los Angeles, USA
mailto:doretto@cs.ucla.edu
http://www.cs.ucla.edu/~doretto/

**Esedoglu, Selim**
Department of Mathematics
University of California at Los Angeles, USA
mailto:esedoglu@math.ucla.edu
http://www.math.ucla.edu/~esedoglu

**Faugeras, Olivier**
I.N.R.I.A. Sophia Antipolis, France

mailto:Olivier.Faugeras@inria.fr
http://www-sop.inria.fr/odyssee/team/Olivier.Faugeras

**Fisher III, John**
Computer Science and Artificial Intelligence Laboratory
Massachusetts Institute of Technology, Cambridge, USA
mailto:fisher@ai.mit.edu
http://www.ai.mit.edu/people/fisher/

**Fleet, David**
Department of Computer Science
University of Toronto, Toronto, Canada
mailto:fleet@cs.toronto.edu
http://www.cs.toronto.edu/~fleet/

**Frangi, Alejandro**
Department of Technology
Pompeu Fabra University, Barcelona, Spain
mailto:alejandro.frangi@upf.edu
http://www.tecn.upf.es/~afrangi/

**Grady, Leo**
Imaging and Visualization Department
Siemens Corporate Research, Princeton, USA
mailto:leo.grady@siemens.com

**Guo, Hongyu**
Department of Computer, Information Science and Engineering
University of Florida, Gainesville, USA
mailto:hguo@cise.ufl.edu
http://www.cise.ufl.edu/~hguo

**Haro, Gloria**
Departament de Tecnologia
Universitat Pompeu Fabra, Barcelona, Spain
mailto:gloria.haro@upf.edu
http://www.tecn.upf.es/~gharo

**Herbulot, Ariane**
Laboratoire I3S
CNRS-Universite de Nice/Sophia Antipolis, France
mailto:herbulot@i3s.unice.fr
http://www.i3s.unice.fr/~herbulot/

**Huang, Xiaolei**

Division of Computer and Information Sciences
Rutgers, the State University of New Jersey, New Brunswick, USA
mailto:xiaolei@paul.rutgers.edu
http://www.research.rutgers.edu/~xiaolei/

**Jehan-Besson, Stephanie**
Laboratoire GREYC-Image
Ecole Nationale Supérieure d'Ingénieurs de Caen, France
mailto:stephanie.jehan@greyc.ensicaen.fr
http://www.greyc.ensicaen.fr/~jehan

**Joshi, Sarang**
Department of Radiation Oncology and Biomedical Engineering
University of North Carolina, Chapel Hill, USA
mailto:sjoshi@unc.edu
http://www.cs.unc.edu/~joshi

**Joshi, Shantanu**
Department of Electrical Engineering
Florida State University, Tallahassee, USA
mailto:joshi@eng.fsu.edu

**Kakadiaris, Ioannis**
Department of Computer Science
University of Houston, Houston, USA
mailto:ikakadia@central.uh.edu
http://www.vcl.uh.edu/~ioannisk/

**Kaziska, Dave**
Department of Statistics
Florida State University, Tallahassee, USA
mailto:kaziska@stat.fsu.edu

**Keriven, Renaud**
Département d'Informatique
Ecole Normale Supérieure, Paris, France
mailto:Renaud.Keriven@ens.fr
http://cermics.enpc.fr/~keriven/home.html

**Kolmogorov, Vladimir**
Microsoft Research Cambridge, UK
mailto:vnk@microsoft.com
http://www.research.microsoft.com/~vnk

**Lenglet, Christophe**

I.N.R.I.A. Sophia Antipolis, France
mailto:clenglet@sophia.inria.fr
http://www-sop.inria.fr/odyssee/team/Christophe.Lenglet/

**Lelieveldt, Boudewijn**
Division of Image Processing, Department of Radiology
Leiden University Medical Center, Leiden, Netherlands
mailto:B.Lelieveldt@lumc.nl

**Markussen, Bo**
Department of Computer Science
University of Copenhagen, Denmark
mailto:boma@diku.dk
http://www.bomar.dk/

**Metaxas, Dimitris**
Division of Computer and Information Sciences
Rutgers, the State University of New Jersey, New Brunswick, USA
mailto:dnm@cs.rutgers.edu
http://www.cs.rutgers.edu/~dnm/

**Meyer, Fernand**
Centre de Morphologie Mathématique
Ecole des Mines de Paris, Paris, France
mailto:fernand.meyer@cmm.ensmp.fr
http://cmm.ensmp.fr

**Mitchell, Steven**
The University of Iowa, Iowa City, USA
mailto:steve@componica.com

**Mittal, Anurag**
Real-time Vision and Modeling Department
Siemens Corporate Research, Princeton, USA
mailto:anurag.mittal@siemens.com
http://www.umiacs.umd.edu/~anurag

**Mrázek, Pavel**
Upek, Husinecká 7, Praha 3, Czech Republic
mailto:pavel.mrazek@upek.com

**Nain, Delphine**
Departments of Electrical and Computer and Biomedical Engineering
Georgia Institute of Technology, Atlanta, USA
mailto:delfin@cc.gatech.edu

http://www.bme.gatech.edu/groups/bil/

**Nielsen, Mads**
Department of Innovation
IT University of Copenhagen, Denmark
mailto:malte@itu.dk
http://www.itu.dk/people/malte

**Ordas, Sebastian**
Department of Technology
Pompeu Fabra University, Barcelona, Spain
mailto:sebastian.ordas@upf.edu

**Paragios, Nikos**
C.E.R.T.I.S.
Ecole Nationale des Ponts et Chaussées, Champs sur Marne, France
mailto:nikos.paragios@certis.enpc.fr
http://www.enpc.fr/certis/people/paragios.html

**Park, Frederick**
Department of Mathematics
University of California at Los Angeles, USA
mailto:fpark@math.ucla.edu
http://www.math.ucla.edu/~fpark

**Pollefeys, Marc**
Department of Computer Science
University of North Carolina, Chapel Hill, USA
mailto:marc@cs.unc.edu
http://www.cs.unc.edu/~marc/

**Pons, Jean-Philippe**
C.E.R.T.I.S.
Ecole Nationale des Ponts et Chaussées, Champs sur Marne, France
mailto:Jean-Philippe.Pons@certis.enpc.fr
http://www.enpc.fr/certis/people/pons.html

**Prados, Emmanuel**
I.N.R.I.A. Sophia Antipolis, France
mailto:Emmanuel.Prados@sophia.inria.fr
http://www-sop.inria.fr/odyssee/team/Emmanuel.Prados/

**Rangarajan, Anand**
Department of Computer, Information Science and Engineering
University of Florida, Gainesville, USA

mailto:anand@cise.ufl.edu
http://www.cise.ufl.edu/~anand

**Reiber, Johan H.C.**
Department of Radiology
Leiden University Medical Center, Leiden, the Netherlands
mailto:J.H.C.Reiber@lumc.nl
http://www.lkeb.nl

**Rousson, Mikael**
Imaging and Visualization Department
Siemens Corporate Research, Princeton, USA
mailto:mikael.rousson@scr.siemens.com

**Sapiro, Guillermo**
Department of Electrical and Computer Engineering
University of Minnesota, Minneapolis, USA
mailto:guille@ece.umn.edu
http://www.ece.umn.edu/users/guille/

**Schnörr, Christoph**
Department of Mathematics & Computer Science
University of Mannheim, Germany
mailto:schnoerr@uni-mannheim.de
http://www.cvgpr.uni-mannheim.de/

**Soatto, Stefano**
Computer Science Department
University of California at Los Angeles, USA
mailto:soatto@cs.ucla.edu
http://www.cs.ucla.edu/~soatto/

**Sonka, Milan**
Dept. of Electrical and Computer Engineering
The University of Iowa, Iowa City, USA
mailto:milan-sonka@uiowa.edu
http://www.engineering.uiowa.edu/~sonka/

**Srivastava, Anuj**
Department of Statistics
Florida State University, Tallahassee, USA
mailto:anuj@stat.fsu.edu
http://stat.fsu.edu/~anuj

**Steidl, Gabriele**

Faculty of Mathematics and Computer Science
University of Mannheim, Mannheim, Germany
mailto:steidl@math.uni-mannheim.de
http://kiwi.math.uni-mannheim.de/

**Stewart, Charles**
Department of Computer Science
Rensselaer Polytechnic Institute, Troy, USA
mailto:stewart@cs.rpi.edu
http://www.cs.rpi.edu/~stewart

**Sun, Yiyong**
Imaging and Visualization Department
Siemens Corporate Research, Princeton, USA
mailto:yiyong.sun@siemens.com

**Szeliski, Richard**
Microsoft Research, Redmond, USA
mailto:szeliski@microsoft.com
http://www.research.microsoft.com/~szeliski/

**Tannenbaum, Allen**
Departments of Electrical and Computer and Biomedical Engineering
Georgia Institute of Technology, Atlanta, USA
mailto:tannenba@bme.gatech.edu
http://www.bme.gatech.edu/groups/bil/

**Tschumperlé, David**
GREYC - UMR CNRS 6072
Centre National de la Recherche Scientifique (CNRS), Caen, France
mailto:David.Tschumperle@greyc.ensicaen.fr
http://www.greyc.ensicaen.fr/~dtschump/

**Unal, Gozde**
Intelligent Vision and Reasoning
Siemens Corporate Research, Princeton, USA
mailto:gozde.unal@siemens.com

**Veksler, Olga**
Departament of Computer Science
University of Western Ontario, Canada
mailto:olga@csd.uwo.ca
http://www.csd.uwo.ca/faculty/olga/

**Vemuri, Baba**

Department of Computer, Information Science and Engineering
Univiversity of Florida, Gainesville, USA
mailto:vemuri@cise.ufl.edu
http://www.cise.ufl.edu/~vemuri

**Wang, Zhizhou**
Imaging and Visualization Department
Siemens Corporate Research, Princeton, USA
mailto:zhizhou.wang@siemens.com

**Weickert, Joachim**
Faculty of Mathematics and Computer Science
Saarland University, Saarbrücken, Germany
mailto:weickert@mia.uni-saarland.de
http://www.mia.uni-saarland.de/weickert/

**Welk, Martin**
Faculty of Mathematics and Computer Science
Saarland University, Saarbrücken, Germany
mailto:welk@mia.uni-saarland.de
http://www.mia.uni-saarland.de/welk/

**Weiss, Yair**
School of Computer Science and Engineering
The Hebrew University of Jerusalem, Jerusalem, Israel
mailto:yweiss@cs.huji.ac.il
http://www.cs.huji.ac.il/~yweiss/

**Wells III, William**
Department of Radiology
Harvard Medical School and Brigham and Women's Hospital, Boston, USA
mailto:sw@bwh.harvard.edu
http://splweb.bwh.harvard.edu:8000/pages/ppl/sw/homepage.html

**Williams, James**
Imaging and Visualization Department
Siemens Corporate Research, Princeton, USA
mailto:jimwilliams@siemens.com

**Wilson, Dave**
Department of Mathematics
University of Florida, Gainesville, USA
mailto:dcw@math.ufl.edu

http://www.math.ufl.edu/~dcw/

**Yezzi, Anthony**
Departments of Electrical and Computer and Biomedical Engineering
Georgia Institute of Technology, Atlanta, USA
mailto:ayezzi@ece.gatech.edu
http://www.ece.gatech.edu/profiles/ayezzi/

**Yip, Andy**
Department of Mathematics
University of California at Los Angeles, USA
mailto:mhyip@math.ucla.edu
http://www.math.ucla.edu/~mhyip

**Zabih, Ramin**
Department of Computer Science
Cornell University, Ithaca, USA
mailto:rdz@cs.cornell.edu
http://www.cs.cornell.edu/~rdz

**Zeitouni, Ofer**
School of Mathematics
University of Minnesota, Minneapolis, USA
mailto:zeitouni@math.umn.edu

**Zöllei, Lilla**
Computer Science and Artificial Intelligence Laboratory
Massachusetts Institute of Technology, Cambridge, USA
mailto:lzollei@csail.mit.edu
http://people.csail.mit.edu/people/lzollei/

**Zucker, Steven**
Department of Computer Science and Program in Applied Mathematics
Yale University, New Haven, USA
mailto:steven.zucker@yale.edu
http://www.cs.yale.edu/homes/vision/zucker/steve.html

# Part I

# Image Reconstruction

# Chapter 1

## Diffusion Filters and Wavelets: What Can They Learn from Each Other?

**J. Weickert, G. Steidl, P. Mrázek, M. Welk, and T. Brox**

Abstract

Nonlinear diffusion filtering and wavelet shrinkage are two methods that serve the same purpose, namely discontinuity-preserving denoising. In this chapter we give a survey on relations between both paradigms when space-discrete or fully discrete versions of nonlinear diffusion filters are considered. For the case of space-discrete diffusion, we show equivalence between soft Haar wavelet shrinkage and total variation (TV) diffusion for 2-pixel signals. For the general case of $N$-pixel signals, this leads us to a numerical scheme for TV diffusion with many favourable properties. Both considerations are then extended to 2-D images, where an analytical solution for $2 \times 2$ pixel images serves as building block for a wavelet-inspired numerical scheme for TV diffusion. When replacing space-discrete diffusion by fully discrete one with an explicit time discretisation, we obtain a general relation between the shrinkage function of a shift-invariant Haar wavelet shrinkage on a single scale and the diffusivity of a nonlinear diffusion filter. This allows to study novel, diffusion-inspired shrinkage functions with competitive performance, to suggest new shrinkage rules for 2-D images with better rotation invariance, and to propose coupled shrinkage rules for colour images where a desynchronisation of the colour channels is avoided. Finally we present a new result which shows that one is not restricted to shrinkage with Haar wavelets: By using wavelets with a higher number of vanishing moments, equivalences to higher-order diffusion-like PDEs are discovered.

## 1.1 Introduction

Signal and image denoising is a field where one often is interested in removing noise without sacrificing important structures such as discontinuities. To this end, a large variety of nonlinear strategies has been proposed in the literature including

Figure 1.1. **(a) Left:** Original image with additive Gaussian noise. **(b) Middle:** Result after shift invariant soft wavelet shrinkage. **(c) Right:** Result after nonlinear diffusion filtering with total variation diffusivity.

wavelet shrinkage [275] and nonlinear diffusion filtering [642]; see Figure 1.1. The goal of this chapter is to survey a number of connections between these two techniques and to outline how they can benefit from each other.

While many publications on the connections between wavelet shrinkage and PDE-based evolutions (as well as related variational methods) focus on the analysis in the *continuous* setting (see e.g. [49, 114, 161, 163, 568]), significantly less investigations have been carried out in the *discrete* setting [214]. In this chapter we give a survey on our contributions that are based on discrete considerations. Due to the lack of space we can only present the main ideas and refer the reader to the original papers [584, 585, 586, 760, 882] for more details.

This chapter is organised as follows: In Section 1.2 we start with briefly sketching the main ideas behind wavelet shrinkage and nonlinear diffusion filtering. Afterwards in Section 1.3 we focus on relations between both worlds, when we restrict ourselves to space-discrete nonlinear diffusion with a total variation (TV) diffusivity and to soft Haar wavelet shrinkage. Section 1.4 presents additional relations that arise from considering fully discrete nonlinear diffusion with arbitrary diffusivities, and Haar wavelet shrinkage with arbitrary shrinkage functions. In Section 1.5 we present a new result that generalises these considerations to higher-order diffusion-like PDEs and shrinkage with wavelets having a higher number of vanishing moments. The chapter is concluded with a summary in Section 1.6.

## 1.2 Basic Methods

### *1.2.1 Wavelet Shrinkage*

Wavelet shrinkage has been made popular by a series of papers by Donoho and Johnstone (see e.g. [274, 275]). Assume we are given some discrete 1-D signal $f = (f_i)_{i\in\mathbb{Z}}$ that we may also interpret as a piecewise constant function. Then the *discrete wavelet transform* represents $f$ in terms of shifted versions of a dilated scaling function $\varphi$, and shifted and dilated versions of a wavelet function $\psi$. In case of orthonormal wavelets, this gives

$$f = \sum_{i\in\mathbb{Z}} \langle f, \varphi_i^n \rangle\, \varphi_i^n + \sum_{j=-\infty}^{n} \sum_{i\in\mathbb{Z}} \langle f, \psi_i^j \rangle\, \psi_i^j, \tag{1.1}$$

where $\psi_i^j(s) := 2^{-j/2}\psi(2^{-j}s - i)$ and where $\langle\cdot,\cdot\rangle$ denotes the inner product in $L_2(\mathbb{R})$. If the measurement $f$ is corrupted by moderate white Gaussian noise, then this noise is contained to a small amount in all wavelet coefficients $\langle f, \psi_i^j\rangle$, while the original signal is in general determined by a few significant wavelet coefficients [540]. Therefore, wavelet shrinkage attempts to eliminate noise from the wavelet coefficients by the following three-step procedure:

1. *Analysis*: Transform the noisy data $f$ to the wavelet coefficients $d_i^j = \langle f, \psi_i^j\rangle$ and scaling function coefficients $c_i^n = \langle f, \varphi_i^n\rangle$ according to (1.1).
2. *Shrinkage*: Apply a shrinkage function $S_\theta$ with a threshold parameter $\theta$ to the wavelet coefficients, i.e., $S_\theta(d_i^j) = S_\theta(\langle f, \psi_i^j\rangle)$.
3. *Synthesis*: Reconstruct the denoised version $u$ of $f$ from the shrunken wavelet coefficients:

$$u := \sum_{i\in\mathbb{Z}} \langle f, \varphi_i^n \rangle\, \varphi_i^n + \sum_{j=-\infty}^{n} \sum_{i\in\mathbb{Z}} S_\theta(\langle f, \psi_i^j \rangle)\, \psi_i^j. \tag{1.2}$$

In this paper we pay particular attention to *Haar wavelets*, well suited for piecewise constant signals with discontinuities. The Haar wavelet and Haar scaling functions are given respectively by

$$\psi(x) = \mathbf{1}_{[0,\frac{1}{2})} - \mathbf{1}_{[\frac{1}{2},1)}, \tag{1.3}$$

$$\varphi(x) = \mathbf{1}_{[0,1)} \tag{1.4}$$

where $\mathbf{1}_{[a,b)}$ denotes the characteristic function, equal to 1 on $[a, b)$ and zero everywhere else. In the case of the so-called *soft wavelet shrinkage* [274], one uses the shrinkage function

$$S_\theta(s) := \begin{cases} s - \theta \operatorname{sgn} s & \text{if } |s| > \theta, \\ 0 & \text{if } |s| \le \theta. \end{cases} \tag{1.5}$$

### 1.2.2 Nonlinear Diffusion Filtering

The basic idea behind nonlinear diffusion filtering [642, 870] in the 1-D case is to obtain a family $u(x, t)$ of filtered versions of a continuous signal $f(x)$ as the solution of a suitable diffusion process

$$u_t = (g(|u_x|)\, u_x)_x \tag{1.6}$$

with $f$ as initial condition,

$$u(x, 0) = f(x)$$

and reflecting boundary conditions. Here subscripts denote partial derivatives, and the diffusion time $t$ is a simplification parameter: Larger values correspond to more pronounced filtering.

The diffusivity $g(|u_x|)$ is a nonnegative function that controls the amount of diffusion. Usually, it is decreasing in $|u_x|$. This ensures that strong edges are less blurred by the diffusion filter than low-contrast details. In this chapter, the *total variation (TV)* diffusivity

$$g(|s|) = \frac{1}{|s|} \tag{1.7}$$

plays an important role, since the resulting TV diffusion [27, 272] does not require to specify additional contrast parameters, leads to scale invariant filters, has finite extinction time, interesting shape-preserving qualities, and is equivalent to TV regularisation [695] in the 1-D setting; see the references in [882] for more details.

Unfortunately, TV diffusion is not unproblematic in practice: In corresponding numerical algorithms the unbounded diffusivity requires infinitesimally small time steps or creates very ill-conditioned linear systems. Therefore, TV diffusion is often approximated by a model with bounded diffusivity:

$$u_t = \left( \frac{1}{\sqrt{\varepsilon^2 + u_x^2}} \, u_x \right)_x \tag{1.8}$$

This regularisation, however, may introduce undesirable blurring effects and destroy some of the favourable properties of unregularised TV diffusion.

## 1.3 Relations for Space-Discrete Diffusion

In this section we study connections between soft Haar wavelet shrinkage and nonlinear diffusion with TV diffusivity in the space-discrete case. This allows us to find analytical solutions for simple scenarios. They are used as building blocks for numerical schemes for TV diffusion.

### *1.3.1 Equivalence for Two-Pixel Signals*

We start by considering wavelet shrinkage of a two-pixel signal $(f_0, f_1)$ in the Haar basis [760]. Its coefficients with respect to the scaling function $\varphi = (\frac{1}{\sqrt{2}}, \frac{1}{\sqrt{2}})$ and the wavelet $\psi = (\frac{1}{\sqrt{2}}, \frac{-1}{\sqrt{2}})$ are given by

$$c = \frac{f_0 + f_1}{\sqrt{2}}, \qquad d = \frac{f_0 - f_1}{\sqrt{2}}. \tag{1.9}$$

Soft thresholding of the wavelet coefficient yields

$$S_\theta(d) = \begin{cases} d - \theta \operatorname{sgn} d & \text{if } |d| > \theta, \\ 0 & \text{if } |d| \le \theta, \end{cases} \tag{1.10}$$

leading to the filtered signal $(u_0, u_1)$ with

$$u_0(\theta) = \begin{cases} f_0 + \frac{\theta}{\sqrt{2}} \operatorname{sgn}(f_1 - f_0) & \text{if } \theta < |f_1 - f_0|/\sqrt{2}, \\ (f_0 + f_1)/2 & \text{else,} \end{cases} \tag{1.11}$$

$$u_1(\theta) = \begin{cases} f_1 - \frac{\theta}{\sqrt{2}} \operatorname{sgn}(f_1 - f_0) & \text{if } \theta < |f_1 - f_0|/\sqrt{2}, \\ (f_0 + f_1)/2 & \text{else.} \end{cases} \tag{1.12}$$

On the other hand, space discrete TV diffusion of a two-pixel signal with reflecting boundary conditions and grid size 1 creates the dynamical system

$$\dot{u}_0 = \operatorname{sgn}(u_1 - u_0) \tag{1.13}$$

$$\dot{u}_1 = -\operatorname{sgn}(u_1 - u_0) \tag{1.14}$$

with initial conditions $u_0(0) = f_0$ and $u_1(0) = f_1$. The dot denotes differentiation with respect to time. It is easy to verify that this system with discontinuous right hand side has the unique analytical solution

$$u_0(t) = \begin{cases} f_0 + t \operatorname{sgn}(f_1 - f_0) & \text{if } t < |f_1 - f_0|/2, \\ (f_0 + f_1)/2 & \text{else,} \end{cases} \tag{1.15}$$

$$u_1(t) = \begin{cases} f_1 - t \operatorname{sgn}(f_1 - f_0) & \text{if } t < |f_1 - f_0|/2, \\ (f_0 + f_1)/2 & \text{else.} \end{cases} \tag{1.16}$$

Interestingly, this is equivalent to soft Haar wavelet shrinkage with threshold $\theta = \sqrt{2}t$. Moreover, we observe that a finite extinction time is obvious in the two-pixel model and that no problems with degenerated diffusivities appear [760].

### *1.3.2 A Wavelet-Inspired Scheme for TV Diffusion of Signals*

Let us now investigate if we can also benefit from the 2-pixel equivalences in the case of general discrete 1-D signals with $N$ pixels. To this end, we perform a wavelet decomposition on the finest scale only. Haar wavelets create natural two-pixel pairings, but unfortunately, their shrinkage is not shift invariant. As a remedy, Coifman and Donoho have proposed to apply *cycle spinning* [213]: On one hand, shrinkage is performed on the original signal. In parallel to this the signal is shifted by 1 pixel, shrinkage is performed, and then the result is shifted back. Averaging both filtered signals creates a process that is shift invariant by construction.

Interestingly this procedure does also inspire a novel numerical scheme for TV diffusion. It uses the analytical solution of the two-pixel model as a building block. With the two-pixel model, TV diffusion with time step size $2\tau$ is performed on all pixel pairs $(u_{2i}, u_{2i+1})$. In parallel we perform TV diffusion on all pixel pairs $(u_{2i-1}, u_{2i})$. Averaging both results leads to the following numerical scheme for

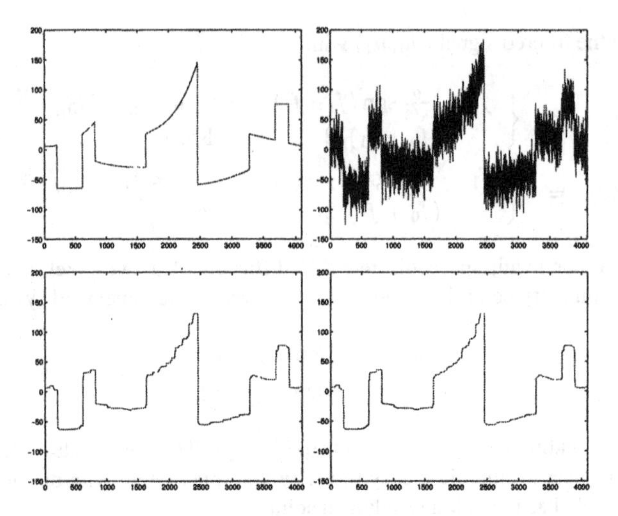


Figure 1.2. **(a) Top left:** Original signal without noise. **(b) Top right:** With additive Gaussian noise, SNR=8 dB. **(c) Bottom left:** Result with two-pixel scheme. SNR = 24.5 dB. **(d) Bottom right:** Result with classical regularised scheme. SNR = 24.6 dB. From [760].

TV diffusion [760]:

$$
\begin{aligned}
u_i^{k+1} \;=\; u_i^k \;&+\; \frac{\tau}{h}\,\mathrm{sgn}\,(u_{i+1}^k - u_i^k)\;\min\left(1,\; \frac{h}{4\tau}|u_{i+1}^k - u_i^k|\right) \\
&-\; \frac{\tau}{h}\,\mathrm{sgn}\,(u_i^k - u_{i-1}^k)\;\min\left(1,\; \frac{h}{4\tau}|u_i^k - u_{i-1}^k|\right), \quad (1.17)
\end{aligned}
$$

where the upper index $k$ denotes the time level $k\tau$, and $h$ is the spatial grid size. Although this scheme is explicit, it is even absolutely stably since it is based on a linear combination of analytical two-pixel interactions that satisfy a maximum–minimum principle. Moreover, it can be shown that the scheme is also conditionally consistent to the continuous TV diffusion [760]. It should be noted that it does not require any regularisation of the diffusivity such as (1.8), and hence does not suffer from corresponding dissipative artifacts at edges. In Figure 1.2 it is shown that it is a competitive alternative to conventional schemes based that approximate regularised TV diffusion.

### *1.3.3 Generalisations to Images*

Interestingly, the considerations in Subsections 1.3.1 and 1.3.2 can be generalised to the 2-D setting [882]. By considering an image with $2 \times 2$ pixels, one shows that soft Haar wavelet shrinkage and space-discrete TV diffusion are equivalent by deriving the same analytical solution for both processes. In order to use this 4-pixel solution as a building block for a numerical scheme for 2-D TV diffusion, we consider the four $2 \times 2$ cells containing some pixel $(i, j)$. By computing their

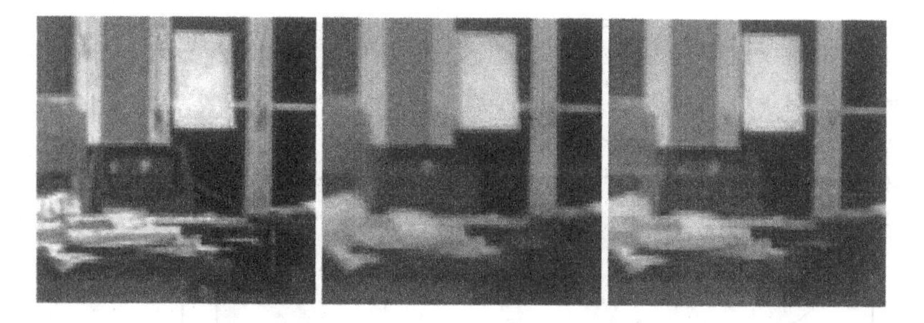

Figure 1.3. (a) Left: Original image, 93 × 93 pixels. (b) Middle: Standard explicit scheme for regularised TV diffusion ($\varepsilon = 0.01$, $\tau = 0.0025$, 10000 iterations). (c) Right: Same with four-pixel scheme without regularisation ($\tau = 0.1$, 250 iterations). Note that 40 times larger time steps are used. From [882].

analytical solutions and averaging the results, we obtain a wavelet-inspired numerical scheme for 2-D TV diffusion. In the same way as its 1-D counterpart, it is explicit, absolutely stable, conditionally consistent, and does not require any regularisation of the singular TV diffusion equation. Compared to classical explicit discretisations based on regularised TV diffusion, it creates sharper edges, even when significantly larger time step sizes are used; see Figure 1.3.

## 1.4 Relations for Fully Discrete Diffusion

The previous section focused on space-discrete TV diffusion and soft Haar wavelet shrinkage. This restriction allowed us to derive analytical solutions for both paradigms. In order to obtain additonal connections let us now investigate fully discrete nonlinear diffusion with arbitrary diffusivities and Haar wavelet shrinkage with general shrinkage functions.

### *1.4.1 Diffusion-Inspired Shrinkage Functions*

Let us consider a discrete signal $(f_i)_{i\in\mathbb{Z}}$. It is easily seen that one cycle of shift-invariant Haar wavelet shrinkage on a single level creates a filtered signal $(u_i)_{i\in\mathbb{Z}}$ with

$$\begin{aligned} u_i &= \frac{f_{i-1}+2f_i+f_{i+1}}{4} + \frac{\sqrt{2}}{4}\,S_\theta\left(\frac{f_i-f_{i+1}}{\sqrt{2}}\right) \\ &- \frac{\sqrt{2}}{4}\,S_\theta\left(\frac{f_{i-1}-f_i}{\sqrt{2}}\right). \end{aligned} \tag{1.18}$$

On the other hand, the first iteration of an explicit (Euler forward) scheme for a nonlinear diffusion filter with initial state $f$, time step size $\tau$ and spatial step size

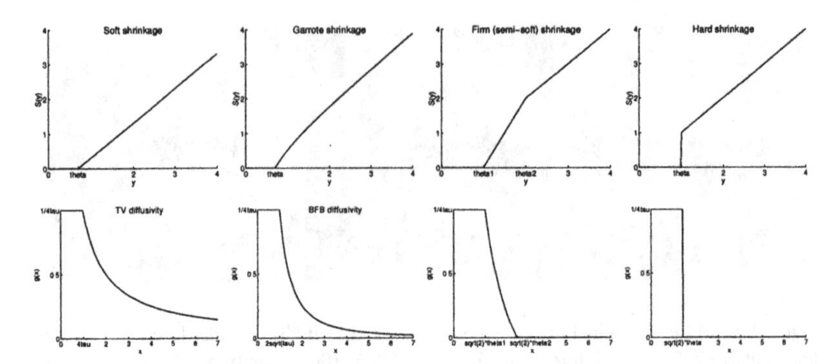


Figure 1.4. **(a) Top:** Four popular shrinkage functions: soft, garrote, firm, and hard shrinkage. **(b) Bottom:** Corresponding diffusivities. From [585].

1 leads to

$$\frac{u_i - f_i}{\tau} = g(|f_{i+1} - f_i|)\,(f_{i+1} - f_i) - g(|f_i^k - f_{i-1}^k|)\,(f_i^k - f_{i-1}^k), \quad (1.19)$$

which can be rewritten as

$$\begin{aligned} u_i &= \frac{f_{i-1} + 2f_i + f_{i+1}}{4} + (f_i - f_{i+1})\left(\frac{1}{4} - \tau\, g(|f_i - f_{i+1}|)\right) \\ &- (f_{i-1} - f_i)\left(\frac{1}{4} - \tau\, g(|f_{i-1} - f_i|)\right). \end{aligned} \quad (1.20)$$

Comparing (1.18) and (1.20) shows that both methods are equivalent if

$$\frac{\sqrt{2}}{4}\, S_\theta\left(\frac{s}{\sqrt{2}}\right) = s\left(\frac{1}{4} - \tau\, g(|s|)\right). \quad (1.21)$$

This formula states a general correspondence between a shrinkage function $S_\theta$ of a shift-invariant single scale Haar wavelet shrinkage and the diffusivity $g$ of an explicit nonlinear diffusion scheme [585]. It does not only allow to reinterpret a number of shrinkage strategies as nonlinear diffusion filters (Figure 1.4), it also leads to novel, diffusion-inspired shrinkage functions (Figure 1.5). Interestingly, some of these diffusion-inspired shrinkage functions turn out to belong to the ones with the best denoising capabilities [585]. A detailed analysis of this connection in terms of extremum principles, monotonicity preservation and sign stability can be found in [586].

### *1.4.2 Wavelet Shrinkage with Improved Rotation Invariance*

In order to extend our results from 1-D signals to 2-D greyscale images, we have to specify the 2-D Haar Wavelet transform first. It is based on a lowpass filter $L$ with coefficients $(\frac{1}{\sqrt{2}}, \frac{1}{\sqrt{2}})$ and a highpass filter $H$ with coefficients $(\frac{1}{\sqrt{2}}, -\frac{1}{\sqrt{2}})$ Applying the 1-D filters $L$ and $H$ alternatingly in $x$ and $y$ direction gives a 2-D

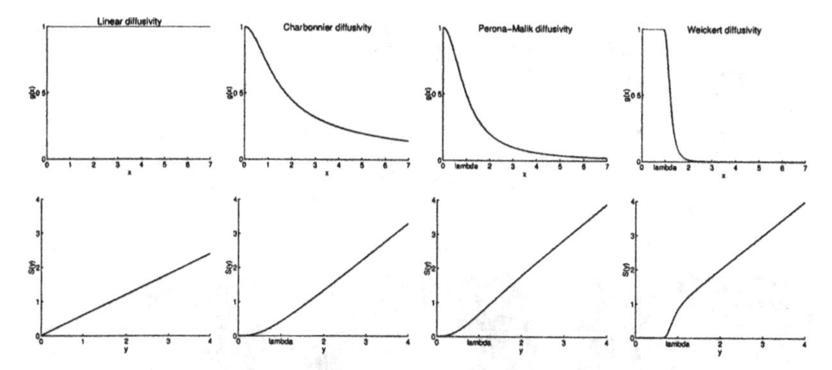


Figure 1.5. **(a) Top:** Four popular diffusivities: linear, Charbonnier, Perona–Malik, and Weickert diffusivity. **(b) Bottom:** Corresponding shrinkage functions. From [585].

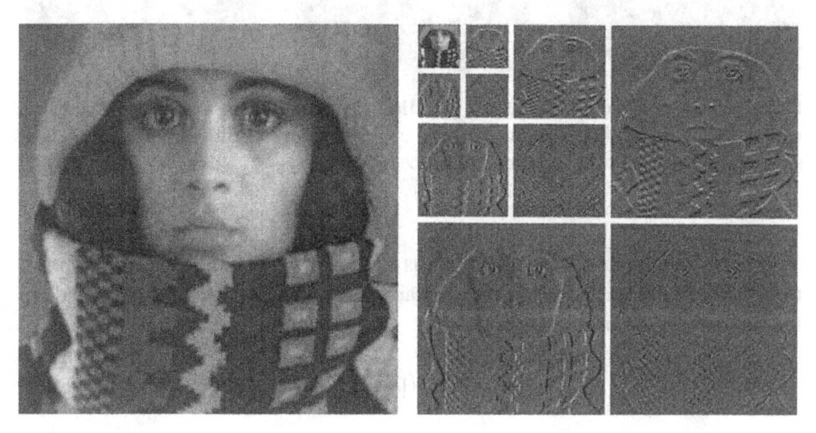

Figure 1.6. **(a) Left:** Original image. **(b) Right:** The first three levels of a 2-D Haar wavelet decomposition.

Haar wavelet decomposition with the following structure:

$$v^{l+1} = L(x) * L(y) * v^l, \tag{1.22}$$

$$w_y^{l+1} = L(x) * H(y) * v^l, \tag{1.23}$$

$$w_x^{l+1} = H(x) * L(y) * v^l, \tag{1.24}$$

$$w_{xy}^{l+1} = H(x) * H(y) * v^l \tag{1.25}$$

with $v^0 := f$. Figure 1.6 illustrates this principle.

The basic idea behind classical 2-D wavelet shrinkage is now to shrink all wavelet coefficients $w_y$, $w_x$ and $w_{xy}$ separately according to their magnitude. If shift invariance is required, one averages the results for the 4 shift possibilities. However, even in this case, one usually observes a severe lack of rotation invariance.

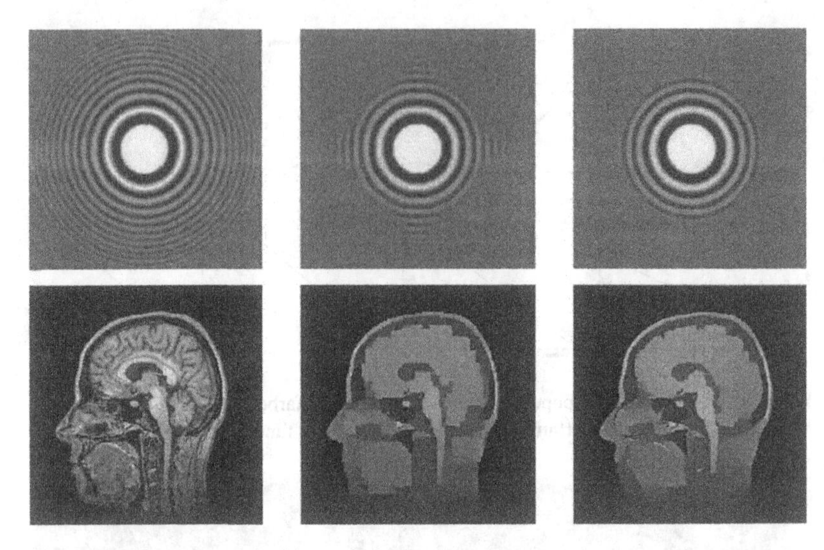

Figure 1.7. (a) **Left:** original images. (b) **Middle:** reconstruction after iterated shift invariant hard wavelet shrinkage. (c) **Right:** reconstruction by a diffusion-inspired wavelet shrinkage with much better rotation invariance. From [584].

In order to address this problem, let us investigate 2-D nonlinear diffusion filtering. In its isotropic variant with a scalar-valued diffusivity [642], it is based on the rotationally invariant equation

$$u_t = \text{div}(g(|\nabla u|)\, \nabla u) \tag{1.26}$$

In a similar way as in the 1-D situation, one can now consider explicit discretisations and relate the diffusivities to shrinkage functions for shift invariant Haar wavelet shrinkage. In contrast to classical shrinkage where the wavelet coefficients are shrunken separately, this leads to novel shrinkage rules where the wavelets are coupled [584], e.g.

$$S(w_x) = w_x \left(1 - 4\,\tau\, g\left(\sqrt{w_x^2 + w_y^2 + 2w_{xy}^2}\right)\right), \tag{1.27}$$

$$S(w_y) = w_y \left(1 - 4\,\tau\, g\left(\sqrt{w_x^2 + w_y^2 + 2w_{xy}^2}\right)\right), \tag{1.28}$$

$$S(w_{xy}) = w_{xy} \left(1 - 4\,\tau\, g\left(\sqrt{w_x^2 + w_y^2 + 2w_{xy}^2}\right)\right). \tag{1.29}$$

Because of the rotation invariance of the nonlinear diffusion equation, one can expect that these shrinkage rules lead to a significantly better realisiation of rotation invariance than classical 2-D wavelet shrinkage. These expectations are confirmed by the experiments in Figure 1.7.

  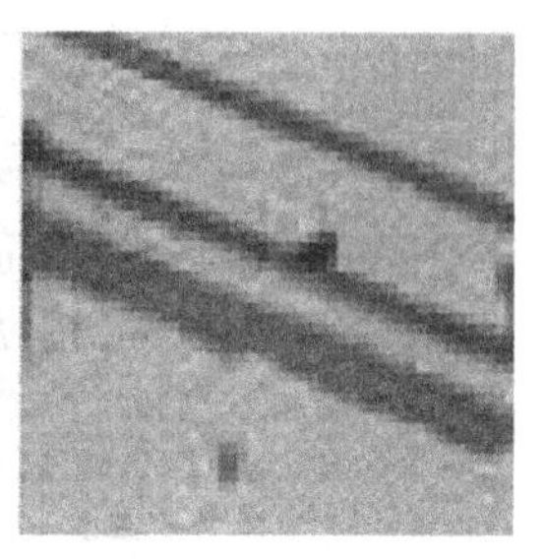

Figure 1.8. **(a) Left:** Zoom into an original image. **(b) Middle:** After classical wavelet shrinkage without coupling the RGB channels. **(c) Right:** Wavelet shrinkage with diffusion-inspired channel coupling.

### *1.4.3 Diffusion-Inspired Wavelet Shrinkage of Colour Images*

While we have investigated diffusion-inspired shrinkage of greyscale images in the previous section, let us now turn our attention to colour images. In this case wavelet shrinkage is frequently applied such that the different colour channels (e.g. RGB or YUV) are shrunken separately. This can result in a lack of synchronisation that creates artifacts at colour edges.

For nonlinear diffusion filtering of colour images, one often uses a process with a joint diffusivity that steers the evolution of all three channels [344]. In the continuous setting such an evolution has the structure

$$\partial_t u_i = \operatorname{div}\Big(g\Big(\Big(\sum_{j=1}^{3} |\nabla u_j|^2\Big)^{1/2}\Big)\,\nabla u_i\Big) \tag{1.30}$$

where the index $i$ specifies the colour channel. By considering an explicit discretisation and relating it to wavelet shrinkage, we end up with shrinkage rules where all channels are coupled. Figure 1.8 illustrates that this diffusion-inspired shrinkage of colour images leads to a more convincing behaviour at edges where all channels remain synchronised.

## 1.5 Wavelets with Higher Vanishing Moments

Up to now we have only considered relations between *Haar* wavelet shrinkage and nonlinear diffusion with diffusivities depending on *first* order derivatives. In this section, we will see that there exists also a relation between one step of translation invariant wavelet shrinkage with wavelets having $m \geq 1$ vanishing moments and explicit difference schemes of diffusion-like equations whose diffusivities include $m$-th order derivatives. To our knowledge these relations have not been considered in the literature before.

For the sake of simplicity, we restrict our attention to the periodic setting, i.e., in the following all indices are taken modulo $N$. We are concerned with wavelet filters $h^i := (h_0^i, \ldots, h_{M_i-1}^i)$, $i = 1, 2$ having the perfect reconstruction property

$$\frac{1}{2}\left(\sum_{k=0}^{M_0-1} h_k^0 h_{k-l}^0 + \sum_{k=0}^{M_1-1} h_k^1 h_{k-l}^1\right) = \delta_{0,l}. \tag{1.31}$$

Moreover, we assume that $h^1$ has $m \geq 1$ vanishing moments:

$$\sum_{k=0}^{M_1-1} k^r\, h_k^1 \;=\; 0, \qquad r = 0, \ldots, m-1, \tag{1.32}$$

$$\sum_{k=0}^{M_1-1} k^m\, h_k^1 \;=\; \gamma_m \neq 0. \tag{1.33}$$

Examples of such filters are for $m = 1$ the Haar filter pair

$$h^0 := \frac{1}{\sqrt{2}}(1,1), \qquad h^1 := \frac{1}{\sqrt{2}}(1,-1) \tag{1.34}$$

with $\gamma_1 = -1/\sqrt{2}$, and for $m = 2$ the Daubechies filter pair

$$h^0 \;:=\; \frac{1}{4\sqrt{2}}\left(1+\sqrt{3},\, 3+\sqrt{3},\, 3-\sqrt{3},\, 1-\sqrt{3}\right), \tag{1.35}$$

$$h^1 \;:=\; \frac{1}{4\sqrt{2}}\left(-1+\sqrt{3},\, 3-\sqrt{3},\, -3-\sqrt{3},\, 1+\sqrt{3}\right) \tag{1.36}$$

with $\gamma_2 = \sqrt{3}/\sqrt{2}$. Then the three steps of wavelet shrinkage applied to the signal $f := (f_0, \ldots, f_{N-1})$ read as follows:

- *Analysis step:* For $j = 0, \ldots, N-1$, we compute

$$c_j \;:=\; \sum_{k=0}^{M_0-1} h_k^0 f_{k+j} \;=\; \sum_{k=0}^{N-1} h_{k-j}^0 f_k, \tag{1.37}$$

$$d_j \;:=\; \sum_{k=0}^{M_1-1} h_k^1 f_{k+j} \;=\; \sum_{k=0}^{N-1} h_{k-j}^1 f_k, . \tag{1.38}$$

- *Shrinkage step:* For $j = 0, \ldots, N-1$ we shrink the highpass coefficients $d_j$ as $S_\theta(d_j)$, $j = 0, \ldots, N-1$.
- *Synthesis step:* For $j = 0, \ldots, N-1$, we compute

$$u_j := \frac{1}{2}\left(\sum_{k=0}^{M_0-1} h_k^0\, c_{j-k} + \sum_{k=0}^{M_1-1} h_k^1\, S_\theta(d_{j-k})\right). \tag{1.39}$$

Assume now that the samples $f_k := f(kh)$ with $h := 1/N$ were taken from a sufficiently smooth periodic function with period 1. Then we obtain by the Taylor expansion that

$$f_{k+l} = \sum_{r=0}^{m} \frac{(kh)^r}{r!} f^{(r)}(lh) + \mathcal{O}(h^{m+1}). \tag{1.40}$$

Since $h^1$ has $m$ vanishing moments, it follows with (1.38) that

$$\begin{aligned} d_l &= \sum_{r=0}^{m} \frac{h^r}{r!} f^{(r)}(lh) \sum_{k=0}^{M_1-1} k^r\, h_k^1 \;+\; \mathcal{O}(h^{m+1}) \\ &= \frac{h^m}{m!} f^{(m)}(lh)\gamma_m \;+\; \mathcal{O}(h^{m+1}). \end{aligned} \tag{1.41}$$

Thus,

$$f^{(m)}(lh) = \frac{m!}{\gamma_m h^m} d_l \;+\; \mathcal{O}(h). \tag{1.42}$$

Similarly, we conclude that

$$f^{(m)}(lh) = \frac{(-1)^m m!}{\gamma_m h^m} \sum_{k=0}^{M_1-1} h_k^1 f_{l-k} \;+\; \mathcal{O}(h). \tag{1.43}$$

Let us now consider a higher-order diffusion-like equation with periodic boundary conditions:

$$u_t = \left((g(|u^{(m)}|)u^{(m)}\right)^{(m)}, \tag{1.44}$$

$$u(x,0) = f(x), \tag{1.45}$$

$$u^{(r)}(0) = u^{(r)}(1), \qquad r = 0,\dots,2m-1. \tag{1.46}$$

We approximate the inner and outer $m$–th derivatives by (1.42) and (1.43), respectively. This results in

$$u_t(jh) \approx \frac{(-1)^m (m!)^2}{(\gamma_m h^m)^2} \sum_{k=0}^{M_1-1} h_k^1\, g\left(\left|\frac{m!}{\gamma_m h^m} d_{j-k}\right|\right) d_{j-k}. \tag{1.47}$$

Finally, the approximation of $u_t$ by a forward difference with time step $\tau$ leads to an iterative scheme whose first step reads

$$u_j^{(1)} := f_j + \tau \frac{(-1)^m (m!)^2}{(\gamma_m h^m)^2} \sum_{k=0}^{M_1-1} h_k^1\, g\left(\left|\frac{m!}{\gamma_m h^m} d_{j-k}\right|\right) d_{j-k}. \tag{1.48}$$

Since our filter pair has the perfect reconstruction property (1.31), we have with $S_\theta(s) = s$ in (1.39) that $u_j = f_j$. Thus, $u_j^{(1)}$ can be rewritten as

$$\begin{aligned} u_j^{(1)} = \frac{1}{2}\Bigg(&\sum_{k=0}^{M_0-1} h_k^0\, c_{j-k} + \sum_{k=0}^{M_1-1} h_k^1 d_{j-k} \cdot \\ &\cdot \left(1 + 2\tau \frac{(-1)^m (m!)^2}{(\gamma_m h^m)^2}\, g\left(\left|\frac{m!\, d_{j-k}}{\gamma_m h^m}\right|\right)\right)\Bigg). \end{aligned} \tag{1.49}$$

Comparing this equation with (1.39) we see that the signal obtained by wavelet shrinkage coincides with those of the first step of our iterative scheme if

$$S_\theta\left(\frac{\gamma_m h^m}{m!} s\right) = s\left(\frac{\gamma_m h^m}{m!} + 2\tau \frac{(-1)^m m!}{\gamma_m h^m}\, g\left(|s|\right)\right). \tag{1.50}$$

This fundamental relation generalises (1.21). It gives the connection between the shinkage function $S_\theta$ of single scale, shift-invariant wavelet shrinkage with $m$ vanishing moments and the "diffusivity" $g$ of the diffusion-like PDE (1.44) of order $2m$. For $m = 1$ it coincides with our result (1.21) for Haar wavelet shrinkage. For $m = 2$ we obtain

$$S_\theta\left(\frac{\sqrt{3}}{2\sqrt{2}}\,s\right) = s\left(\frac{\sqrt{3}}{2\sqrt{2}} + \tau\frac{4\sqrt{2}}{\sqrt{3}}\,g(|s|)\right). \tag{1.51}$$

## 1.6 Summary

The goal of this chapter was to give a survey on connections between two discontinuity-preserving paradigms for signal and image denoising: wavelet shrinkage and nonlinear diffusion filtering. Unlike most other researchers in this field we focused on discrete connections. It turned out that the wavelet and the diffusion community can indeed learn much from each other.

Focusing on soft Haar wavelet shrinkage and space-discrete TV diffusion, we showed that diffusion filters can benefit from wavelet shrinkage: It was possible to derive wavelet-inspired schemes for TV diffusion that are explicit, absolutely stable, do not require regularisations in order to cope with singularities, and perform favourably.

On the other hand, investigating fully discrete schemes for nonlinear diffusion filtering and its higher-order generalisations allowed us to find a general relation between its diffusivity and the shrinkage function of shift-invariant wavelet shrinkage on a single scale. This led to diffusion-inspired shrinkage functions with competitive performance, to shrinkage rules with improved rotation invariance, and to coupling strategies for wavelet shrinkage of colour images. Hence, also wavelet methods can benefit from diffusion methods.

These connections give rise to the question whether it is also possible to design hybrid methods that benefit from both worlds by attempting to combine the efficiency of wavelet strategies with the quality of diffusion methods. They can be either regarded as iterated shift-invariant wavelet shrinkage methods, or as multiscale diffusion filters. First experiments confirm that this is indeed an interesting class of methods [587]. Performing a theoretical analysis of the connections between single-step multiscale procedures and iterated single scale methods, however, still leads to a lot of challenging questions. They are a topic of our current research.

## Acknowledgements

Our research described in this chapter has been partly funded by the projects We 2602/2-1, We 2602/2-2, and We 2602/1-1 of the *Deutsche Forschungsgemeinschaft (DFG)*. This is gratefully acknowledged.

# Chapter2

# Total Variation Image Restoration: Overview and Recent Developments

**T. Chan, S. Esedoglu, F. Park and A. Yip**

Abstract

Since their introduction in a classic paper by Rudin, Osher and Fatemi [695], total variation minimizing models have become one of the most popular and successful methodology for image restoration. More recently, there has been a resurgence of interest and exciting new developments, some extending the applicabilities to inpainting, blind deconvolution and vector-valued images, while others offer improvements in better preservation of contrast, geometry and textures, in ameliorating the staircasing effect, and in exploiting the multiscale nature of the models. In addition, new computational methods have been proposed with improved computational speed and robustness. We shall review some of these recent developments.

## 2.1 Introduction

Variational models have been extremely successful in a wide variety of restoration problems, and remain one of the most active areas of research in mathematical image processing and computer vision. By now, their scope encompasses not only the fundamental problem of image denoising, but also other restoration tasks such as deblurring, blind deconvolution, and inpainting. Variational models exhibit the solution of these problems as minimizers of appropriately chosen functionals. The minimization technique of choice for such models routinely involves the solution of nonlinear partial differential equations (PDEs) derived as necessary optimality conditions.

Perhaps the most basic (fundamental) image restoration problem is denoising. It forms a significant preliminary step in many machine vision tasks, such as object detection and recognition. It is also one of the mathematically most intriguing problems in vision. A major concern in designing image denoising models is to preserve important image features, such as those most easily detected by the human visual system, while removing noise. One such important image feature are the edges; these are places in an image where there is a sharp change in image properties, which happens for instance at object boundaries. A great deal of research has gone into designing models for removing noise while preserving edges; recently there has also been a lot of effort in preserving other fine scale image features, such as texture. All successful denoising models take advantage of the fact that there is an inherent regularity found in natural images; this is how they attempt to tell apart noise and actual image information. Variational and PDE based models make it particularly easy to impose geometric regularity on the solutions obtained as denoised images, such as smoothness of boundaries. This is one of the main reasons behind their success.

Total variation based image restoration models were first introduced by Rudin, Osher, and Fatemi (ROF) in their pioneering work [695] on edge preserving image denoising. It is one of the earliest and best known examples of PDE based edge preserving denoising. It was designed with the explicit goal of preserving sharp discontinuities (edges) in images while removing noise and other unwanted fine scale detail. Being convex, the ROF model is one of the simplest variational models having this most desirable property. The revolutionary aspect of this model is its regularization term that allows for discontinuities but at the same time disfavors oscillations. It was originally formulated in [695] for grayscale imagery in the following form:

$$\inf_{\int_\Omega (u-f)^2\,dx=\sigma^2} \int_\Omega |\nabla u|. \tag{2.1}$$

Here, $\Omega$ denotes the image domain (for instance, the computer screen), and is usually a rectangle. The function $f(x) : \Omega \longrightarrow \mathbb{R}$ represents the given observed image, which is assumed to be corrupted by Gaussian noise of variance $\sigma^2$. The constraint of the optimization forces the minimization to take place over images that are consistent with this known noise level. The objective functional itself is called the *total variation* (TV) of the function $u(x)$; for smooth images it is equivalent to the $L^1$ norm of the derivative, and hence is some measure of the amount of oscillation found in the function $u(x)$. Optimization problem (2.1) is equivalent to the following *unconstrained* optimization, which was also first introduced in [695]:

$$\inf_{u\in L^2(\Omega)} \int_\Omega |\nabla u| + \lambda \int_\Omega (u-f)^2\,dx. \tag{2.2}$$

Here, $\lambda \geq 0$ is a Lagrange multiplier. The equivalence of problems (2.1) and (2.2) has been established in [162]. In the original ROF paper [695] there is an iterative numerical procedure given for choosing $\lambda$ so that the solution $u(x)$ obtained solves (2.1).

We point out that total variation based energies appear, and have been previously studied in, many different areas of pure and applied mathematics. For instance, the notion of total variation of a function and functions of bounded variation appear in the theory of minimal surfaces. In applied mathematics, total variation based models and analysis appear in more classical applications such as elasticity and fluid dynamics. Due to ROF, this notion has now become central also in image processing.

Over the years, the ROF model has been extended to many other image restoration tasks, and has been modified in a variety of ways to improve its performance. In this article, we will concentrate on some recent developments in total variation based image restoration research. Some of these developments have led to new algorithms, and others to new models and theory. While we try to be comprehensive, we are of course limited to those topics and works that are of interest to us, and that we are familiar with. In particular, we aim to provide highlights of a number of new ideas that include the use of different norms in measuring fidelity, applications to new image processing tasks such as inpainting, and so on. We also hope that this article can serve as a guide to recent literature on some of these developments.

## 2.2 Properties and Extensions

### *2.2.1 BV Space and Basic Properties*

The space of functions with bounded variation (BV) is an ideal choice for minimizers to the ROF model since BV provides regularity of solutions but also allows sharp discontinuities (edges). Many other spaces like the Sobolev space $W^{1,1}$ do not allow edges. Before defining the space BV, we formally state the definition of TV as:

$$\int_{\Omega} |\nabla f| = \sup\left\{ \int_{\Omega} f \nabla \cdot \mathbf{g} d\mathbf{x} \mid \mathbf{g} \in C_c^1(\Omega, \mathbb{R}^n), |\mathbf{g}(\mathbf{x})| \leq 1 \forall\, \mathbf{x} \in \Omega \right\} \tag{2.3}$$

where $f \in L^1(\Omega)$ and $\Omega \subseteq \mathbb{R}^n$ is a bounded open set. We can now define the space BV as $\{f \in L^1(\Omega) \mid \int_{\Omega} |\nabla f| < \infty\}$. Thus, BV functions amount to $L^1$ functions with bounded TV semi-norm. Moreover, through the TV semi-norm there is a natural link between BV and the ROF model.

Given the choice of $BV(\Omega)$ as the appropriate space for minimizers of the ROF model (2.2), there are the basic properties of existence and uniqueness to settle. The ROF model in unconstrained form (2.2) is a strictly convex functional, hence, admits a unique minimum. Moreover, it is shown in [162] that the equality constraint $\int_{\Omega} (u - f)^2 d\mathbf{x} = \sigma^2$ in the non-convex ROF model (2.1) is equivalent to the convex inequality constraint $\int_{\Omega} (u - f)^2 d\mathbf{x} \leq \sigma^2$. Hence, the non-convex minimization in (2.1) is equivalent to a convex minimization problem which under some additional assumptions is further equivalent to the above unconstrained minimization (2.2).

For BV functions there is a useful coarea formulation linking the total variation to the level sets giving some insight into the behavior of the TV norm. Given a function $f \in BV(\Omega)$ and $\gamma \in \mathbb{R}$, denote by $\{f = \gamma\}$ the set:

$\{\mathbf{x} \in \mathbb{R}^2 \mid f(\mathbf{x}) = \gamma\}$. Then, if $f$ is regular, the TV of $f$ can be given by:

$$\int_\Omega |\nabla f| = \int_{-\infty}^{\infty} \int_{\{f=\gamma\}} ds\, d\gamma. \tag{2.4}$$

Here, the term $\int_{\{f=\gamma\}} ds$ represents the length of the set $\{f = \gamma\}$. The formula states that the TV norm of $f$ can be obtained by integrating along all contours of $\{f = \gamma\}$ for all values of $\gamma$. Thus, one can view TV as controlling both the size of the jumps in an image and the geometry of the level sets.

### 2.2.2 *Multi-channel TV*

Total variation based models can be extended to vector valued images in various ways.

An interesting generalization of TV denoising to vector valued images was proposed by Sapiro and Ringach [704]. The idea is to think of the image $u : \mathbb{R}^2 \longrightarrow \mathbb{R}^m$ as a parametrized two dimensional surface in $\mathbb{R}^m$, and to use the difference between eigenvalues of the first fundamental form as a measure of edge strength. A variational model results from integrating the square root of the magnitude of this difference as the regularization term.

Blomgren and Chan [98] generalized total variation regularization to vectorial data as the Euclidean norm of the vector of (scalar) total variations of the components. This generalization has the benefit that vector valued images defined on the line whose components are monotone functions with identical boundary conditions all have the same energy, regardless of their smoothness. This implies good edge preserving properties.

Another interesting approach generalizing edge preserving variational denoising models to vector valued images is due to Kimmel, Malladi, and Sochen [473]. They regard the given image $u(x) : \mathbb{R}^2 \longrightarrow \mathbb{R}^m$ as a surface in $\mathbb{R}^{m+2}$, and propose an area minimizing flow (which they call Beltrami flow) as a means of denoising it.

### 2.2.3 *Scale*

The constant $\lambda$ that appears in the ROF model plays the role of a "scale parameter". By tweaking $\lambda$, a user can select the level of detail desired in the reconstructed image. In this sense, $\lambda$ in (2.2) is analogous to the time variable in scale space theories for nonlinear diffusion based denoising models. The geometric interpretation of the regularization term in (2.2) given by the co-area formula suggests that $\lambda$ determines which image features are kept based on, roughly speaking, their "perimeter to area" ratio.

The intuitive link between $\lambda$ and scale of image features can be exactly verified in the case of an image that consists of a white disk on a black background. Strong and Chan [770] determined the solution of the ROF functional for such a given image $f(x)$. It turns out to be $(1 - \frac{1}{\lambda r})f(x)$ for $\lambda > \frac{1}{r}$. In particular, there is always a loss of contrast in the reconstruction, no matter how large the fidelity constant $\lambda$ is. And when $\lambda \leq \frac{1}{r}$, the solution is identically 0, meaning that the model prefers to remove disks of radius less than $\frac{1}{\lambda}$. This simple but instructive example indicates how to relate the parameter $\lambda$ to the scale of objects we desire to preserve in reconstructions. Strong and Chan's observation has been generalized to other exact solutions of the ROF model in [69].

The parameter $\lambda$ can thus be used for performing multiscale decomposition of images: Image features at different scales are separated by minimizing the ROF energy using different values of $\lambda$. Recent research along these lines is described in section 2.5.3.

## 2.3 Caveats

While using TV-norm as regularization can reduce oscillations and regularize the geometry of level sets without penalizing discontinuities, it possesses some properties which may be undesirable under some circumstances.

*Loss of contrast.* The total variation of a function, defined on a bounded domain, is decreased if we re-scale it around its mean value in such a way that the difference between the maximum and minimum value (contrast) is reduced. In [770, 567], the authors showed that for any non-trivial regularization parameter, the solution to the ROF model has a contrast loss. The example of a white disk with radius $R$ over a black background discussed in 2.2.2 is a simple illustration. In this case, the contrast loss is inversely proportional to $f(x)/r$ before the disk merges with the background. In general, reduction of the contrast of a feature by $h > 0$ would induce a decrease in the regularization term of the ROF model by $O(h)$ and an increase in the fidelity term by $O(h^2)$ only. Such scalings of the regularization and fidelity terms favors the reduction of the contrast.

*Loss of geometry.* The co-area formula (2.4) reveals that, in addition to loss of contrast, the TV of a function may be decreased by reducing the length of each level set. In some cases, such a property of the TV-norm may lead to distortion of the geometry of level sets when applying the ROF model. In [770], Strong and Chan show that, for circular image features, their shape is preserved at least for a small change in the regularization parameter and their location is also preserved even they are corrupted by noise of moderate level. In [69], Bellettini et al. extend Strong and Chan's results and show that the set of all bounded connected shapes $C$ that are shape-invariant in the solution of the ROF model is precisely given by

$$\left\{ C \subset \mathbb{R}^N : C \text{ convex}, \partial C \in C^{1,1} \text{ and ess} \sup_{p \in \partial C} \kappa_{\partial C}(p) \leq |\partial C|/|C| \right\}.$$

Here, $|\partial C|$ is the perimeter of $C$, $|C|$ is the area of $C$ and $\kappa_{\partial C}(p)$ is the curvature of $\partial C$ at $p$. The downside of the above characterization is that the ROF model distorts the geometry of shapes that do not belong to the shape-invariant set. For instance, it has been shown in [567], if the input image is a rectangle $R$ over a background with a different intensity, then cutting a corner (an isosceles triangle) with height $h$ of the rectangle would induce a reduction in the TV-norm by $O(h)$ and an increment of the fitting term by $O(h^2)$, thus favoring cutting the corners.

*Staircasing.* This refers to the phenomenon that the denoised image may look blocky (piecewise constant). In the 1-D discrete case, there is a simple explanation to this — the preservation of monotonicity of neighboring values. Such a property requires that, for each $i$, if the input $f = \{f_i\}$ satisfies $f_i \leq f_{i+1}$ (resp. $\geq$), then the output must satisfy $u_i \leq u_{i+1}$ (resp. $\geq$) for any $\lambda$. In the case where $f$ satisfies $f_{i_0-1} < f_{i_0} > f_{i_0+1} < f_{i_0+2}$ for some $i_0$, which often happens when the true signal is monotonically increasing around $i_0$ and is corrupted by noise but $u$ satisfies $u_{i_0-1} < u_{i_0} = u_{i_0+1} < u_{i_0+2}$, then, visually, $u$ looks like a staircase at $i_0$ but a monotonically increasing signal is more desirable. In the 2-D case, the monotonicity preserving property is no longer true in general, for instance, near corners of image features. However, away from the corners where the curvature of the level sets is high, staircase is often observed.

*Loss of Texture.* Although highly effective for denoising, the TV norm cannot preserve delicate small scale features like texture. This can be accounted for from a combination of the above mentioned geometry and contrast loss caveats of the ROF model which have the tendency to affect small scale features most severely.

## 2.4 Variants

Total variation based image reconstruction models have been extended in a variety of ways. Many of these are modifications of the original ROF functional (2.2), addressing the above mentioned caveats.

### 2.4.1 *Iterated Refinement*

A very interesting and innovative new perspective on the standard ROF model has been recently proposed by Osher et al. [615]. The new framework involved can be generalized to many convex reconstruction models (inverse problems) beyond TV based denoising. When applied to the ROF model in particular, this new approach fixes a number of its caveats, such as loss of contrast, and promises even further improvements in other significant aspects of reconstruction, such as preservation of textures.

The key idea is to compensate for the loss of signal in reconstructed images by minimizing the ROF model repeatedly, each time adding back the signal removed in the previous iteration. Thus, starting with a given $f_0(x) := f(x)$, repeat for $j = 1, 2, 3, \ldots$:

1. Set $u_j(x) = \text{argmin}_u$ of (2.2) using $f_j(x)$ as the given image.

2. Set $f_{j+1}(x) = f_j(x) + \big(f - u_j(x)\big)$.

When applied to the characteristic function of a disk, this algorithm recovers it perfectly after a finite number of iterations without loss of contrast.

The algorithm can be generalized to inverse problems of the form $\inf_u J(u) + H(u, f)$. Here, $J$ is a convex regularization term, and $H(u, f)$ a fidelity term that is required to be convex in $u$ for every $f$. In this setting, the iterative procedure above becomes: Start with $u_0 = 0$, repeat for $j = 1, 2, 3, \ldots$

$$u_{j+1} = \arg\min_w H(w, f) + J(w) - J(u_j) - \langle D_u J(u_j), w - u_j \rangle. \qquad (2.5)$$

Here, $D_u J(u_j)$ denotes the derivative of the functional $J$ at the $j$-th iterate $u_j$, and $\langle \cdot, \cdot \rangle$ represents the duality pairing. If $J$ is non-differentiable (as in the ROF model), then $D_u J(u_j)$ needs to be understood as an element of the subgradient $\partial J(u_j)$ of $J$ at $u_j$. It is clear from formula (2.5) that the algorithm involves removing from the regularization term $J(u)$ its linearization at the current iterate $u_j$.

Formula (2.5) suggests the following definition: For $p \in \partial J(v)$, let

$$D^p(u, v) := J(u) - J(v) - \langle p, u - v \rangle$$

be the generalized *Bregman distance* associated with the functional $J$. It defines a notion of distance between two functions $u$ and $v$ because it satisfies the conditions $D^p(u, v) \geq 0$ for all $u, v$, and $D^p(u, u) = 0$. However, it is not a metric as it needs not be symmetric or satisfy a triangle inequality.

A number of important general theorems have been established in [615], including:

- As long as the distance of the reconstructed image $u_j$ to the given noisy $f(x)$ remains greater than $\sigma$ (the noise variance), the iteration decreases the Bregman distance of the iterates $u_j$ to the *true* (i.e. noise-free) image.
- $H(u_j, f)$ decreases monotonically and tends to 0 as $j \longrightarrow \infty$.

In [615], further results can be found about the convergence rate of the iterates $u_j$ to the given image $f$ under certain regularity assumptions on $f$.

### *2.4.2 $L^1$ Fitting*

A simple way to modify the ROF model in order to compensate for the loss of contrast is to replace the squared $L^2$ norm in the fidelity term in (2.2) by the $L^1$ norm instead. The resulting energy is

$$\int_\Omega |\nabla u| + \lambda \int_\Omega |u - f| \, dx. \qquad (2.6)$$

Discrete versions of this model were studied for one dimensional signals by Alliney [14], and in higher dimensions by Nikolova [602]. In particular, it has been shown to be more effective that the standard ROF model in the presence of certain types of noise, such as salt and pepper. Recently, it has been studied in the continuous setting by Chan and Esedoglu [165].

Although the modification involved in (2.6) seems minor, it has certain desirable consequences. First and foremost, the scaling between the two terms of (2.6) is different from the one in the original ROF model (2.2), and leads to contrast invariance: If $u(x)$ is the solution of (2.6) with $f(x)$ as the given image, then $cu(x)$ is the solution of (2.6) with $cf(x)$ as the given image. This property does not hold for (2.2). A related consequence is: If the given image $f(x)$ is the characteristic function of a set $\Omega$ with smooth boundary, then the image is perfectly recovered by model (2.6) for large enough choices of the parameter $\lambda$. This is in contrast to the behavior of the ROF model, which always prefers to remove some of the original signal from the reconstructed one, and preserves a very small class of shapes. This statement can be generalized beyond original images given by characteristic functions of sets to show that a wide class of regular images are left unmodified by model (2.6) for large enough choices of the parameter $\lambda$.

In addition to having better contrast preservation properties, model (2.6) also turns out to be useful for the denoising of *shapes*. A natural variational model for denoising a shape $S$, which we model as a subset of $\mathbb{R}^n$, is the following: $\min_{\Sigma\subset\mathbb{R}^n} \mathrm{Per}(\Sigma) + \lambda|S \vartriangle \Sigma|$, where the first term in the energy represents the perimeter of the set $\Sigma$, and the second represents the volume of the symmetric difference of the sets $S$ and $\Sigma$ weighted by the scale parameter $\lambda \geq 0$. This model is exactly the one we would get if the minimization in the standard ROF model (2.2) is restricted to functions of the form $u(x) = 1_\Sigma(x)$ and $f(x) = 1_S(x)$. Unlike the standard ROF problem, however, this minimization is non-convex. In particular, standard approaches for solving it run the risk of getting stuck in local minima. The total variation model with $L^1$ fidelity term (2.6) turns out to be a convex formulation of the shape denoising problem given above. Indeed, the following statement has been proved in [165]: Let $u(x)$ be a minimizer of (2.6) for $f(x) = 1_S(x)$. Then, for a.e. $\mu \in [0,1]$, the set $\Sigma(\mu) = \{x \in \mathbb{R}^N : u(x) \geq \mu\}$ is a minimizer of the shape denoising problem. Thus, in order to solve the *non-convex* shape denoising problem, it suffices to solve instead the *convex* problem (2.6) and then take (essentially) any level set of that solution.

### 2.4.3 *Anisotropic TV*

In [299], Esedoglu and Osher introduced and studied anisotropic versions of the ROF model (2.2). The motivation is to privilege certain edge directions so that they are preferred in reconstructions. This can be useful in applications in which there may be prior geometric information available about the shapes expected in the recovered image. In particular, it can be used to restore characteristic functions of convex regions having desired shapes.

The idea proposed in [299] is to replace the total variation penalty term in (2.2) with the following more general term:

$$\int_\Omega \phi(\nabla u) := \sup_{\substack{g\in C_c^1(\Omega;\mathbb{R}^n)\\ g(x)\in W_\phi \forall x\in\Omega}} \int_\Omega u(x)\mathrm{div}g(x)\,dx$$

where the function $\phi : \mathbb{R}^n \longrightarrow \mathbb{R}$ is a convex, positively one-homogeneous function that is 0 at the origin, and the set $W_\phi$ is defined as follows:

$$W_\phi := \{y \in \mathbb{R}^n : x \cdot y \leq \phi(x) \; \forall x \in \mathbb{R}^n\}.$$

For example, if $\phi(x) = |x|$, then the set $W_\phi$ turns out to be simply the unit ball $\{y \in \mathbb{R}^N : |y| \leq 1\}$, and the definition of $\int_\Omega \phi(\nabla u)$ given above reduces to the standard definition of total variation. Another simple example in two dimensions is $\phi(x, y) = |x| + |y|$, in which case the set $W_\phi$ is just the closed unit square.

The set $W_\phi$ defined above is the *Wulff shape* associated with the function $\phi$. It determines the shapes that are compatible with the anisotropy $\phi$. For example, it is proved in [299] that if $f(x)$ is the characteristic function of (a scaled or translated version of) the Wulff shape $W_\phi$, then the solution $u$ is a constant multiple of $f(x)$. This result generalizes that of Strong and Chan [770] and Meyer in [567] that concern the case of a disk for the standard ROF model.

If $W_\phi$ is a convex polygon in two dimensions, then its sides act as preferred edge directions for the reconstructions obtained by the anisotropic ROF model. Indeed, it is proved in [299] that if $u(x) = 1_\Sigma(x)$ is a solution to the anisotropic model, and if $\Sigma$ is known to be a set with piecewise smooth boundary $\partial\Sigma$, then $\partial\Sigma$ should include a line segment parallel to one of the sides of $W_\phi$ wherever its tangent becomes parallel to one of those sides. On the other hand, one can show that $\partial\Sigma$ can include corners that are different than the ones in $\partial W_\phi$.

In addition to being of interest for applications, the results of [299] are also of theoretical interest. Indeed, these anisotropic variants of total variation constitute an infinitude of equivalent regularizations (in the sense that the semi-norms they define are equivalent), yet the properties of their minimizers have been shown to be extremely different. That suggests that in general one should not expect an image restoration model to perform quite as well as the original ROF model just because its regularization term is equivalent to total variation.

### 2.4.4 $H^{1,p}$ Regularization and Inf Convolution

As discussed in Section 2.3, staircasing is one of the potential caveats to watch for when using total variation based regularization. It occurs even more severely in reconstructions by functionals that have a non-convex dependence on image gradients; one famous example is the Perona-Malik scheme, which can be thought of as gradient descent for such an energy functional whose dependence on image gradients grows sublinearly at infinity. The TV model is borderline convex: its dependence on image gradients is linear at infinity. This feature, which is responsible for its ability to reconstruct images with discontinuities, is also responsible for the staircasing effect.

A natural approach to overcoming the staircasing effect is to make the reconstruction model more convex in regions of moderate gradient (away from the edges). A functional designed to accomplish this was proposed by Blomgren,

Mulet, Chan, and Wong [99]. It has the form

$$\int_{\Omega} |\nabla u|^{P(|\nabla u|)}\, dx + \lambda \int_{\Omega} (u - f)^2\, dx. \tag{2.7}$$

Here, the function $P(\xi) : \mathbb{R}^+ \longrightarrow [0, 2]$ is to be chosen so that it monotonically decreases from 2 to 0. A simple example is $P(\xi) = \frac{2}{1+2\xi}$.

The idea behind (2.7) is that the model automatically adapts the gradient exponent to fit the data, so that near edges it behaves exactly like the ROF model, and away from the edges it may behave more like the Dirichlet energy. This leads to much smoother reconstructions in regions of moderate gradient and thus prevents staircasing. On the other hand, unlike the ROF model, (2.7) is non-convex and difficult to analyze.

Another approach to preventing staircasing is to introduce higher order derivatives into the energy; the cost of moderately high but constant gradient regions is zero for such terms. On the other hand, a functional that depends on higher order derivatives would not maintain edges in its reconstructions. It is therefore necessary to once again allow the model to decide for itself where to use the total variation norm and where to use higher order derivative norms. One of the earliest proposals of this kind was made by Chambolle and Lions in [162], where they introduced the notion of *inf convolution* between two convex functionals. In this approach, an image $u$ is decomposed into two parts: $u = u_1 + u_2$. The $u_1$ component is measured using the total variation norm, while the second component $u_2$ is measured using a higher order norm. The precise decomposition of $u$ into these two components is part of the minimization problem. More precisely, one solves the following variational problem that now involves two unknowns:

$$\inf_{u_1, u_2} \int_{\Omega} |\nabla u_1| + \alpha |D^2 u_2| + \lambda (u_1 + u_2 - f)^2\, dx.$$

Minimizing this energy requires the discontinuous component of the image to be allocated to the $u_1$ component, while regions that are well approximated by moderate but nearly constant slopes get allocated to the $u_2$ component at very little cost. This prevents staircasing to a remarkable degree in the one dimensional examples presented in [162]. Another method that utilizes total variation and higher order derivatives to suppress staircasing is by Chan, Marquina, and Mulet in [168].

Despite the important contributions listed above, staircasing remains one of the challenges of total variation based image reconstructions.

## 2.5 Further Applications to Image Reconstruction

### 2.5.1 Deconvolution

The TV norm can also be used to regularize image deblurring problems. The forward degradation model for a blurred and noisy image can be realized as: $f = k * u + \eta$, where $f$ is the observed (degraded) image, $k$ a given point spread function (PSF), $u$ the clean image, $\eta$ an additive noise (often Gaussian), and $*$ denoting the convolution operator.

The task of restoring an image $u$ under the above degradation is known as deconvolution if the PSF $k$ is known or blind deconvolution if there is little or no known a priori information on the PSF. If we replace the $u$ in the unconstrained ROF model (2.2) with the convolution $k * u$, then we arrive at the TV deconvolution model:

$$\min_{u \in BV} ||k * u - f||_2^2 + \lambda_u ||u||_{TV}. \tag{2.8}$$

Here, as in the ROF model (2.2), the regularization parameter $\lambda_u$ is related to the statistical signal to noise ratio (SNR).

Extending the work by You and Kaveh [911], Chan and Wong introduce in [176] the TV blind deconvolution model:

$$\min_{u,k \in BV} \|k * u - f\|_2^2 + \lambda_u \|u\|_{\text{TV}} + \lambda_k \|k\|_{\text{TV}}. \tag{2.9}$$

where the additional parameter $\lambda_k$ controls the spread of $k$. Moreover, solutions $\{u(\lambda_k)\}$ of (2.9) form a one parameter family corresponding to $\lambda_k$. The authors also propose an alternating minimization algorithm for minimizing the above energy (2.9) which we denote by $F(u,k)$. Here, given $u^n$ one solves for $k^{n+1} := \arg\min_k F(u^n, k)$, then given $k^{n+1}$, one solves for $u^{n+1} := \arg\min_u F(u, k^{n+1})$ alternatingly. Such an alternating procedure is shown to be convergent when the TV-norm is replaced by the $H^1$-norm.

A key advantage of using TV regularization for blind deconvolution is that the TV norm can recover sharp edges in the PSF (e.g. motion blur or out-of-focus blur) while not penalizing smooth transitions.

### *2.5.2 Inpainting*

Image inpainting refers to the filling-in of missing or occluded regions in an image based on information available on the observed regions. A common principle for inpainting is to complete isophotes (level sets) in a natural way — such a philosophy is also true for professional artists to restore damaged ancient paintings. To this end, several successful inpainting models have been proposed such as Masnou and Morel [553] and Bertalmio et al. [79]. We refer the reader to [171] and the references therein for other more recent models. Among these models, Chan and Shen proposed in [171] a *TV inpainting model* which uses variational methods in inpainting. The basic ingredient is to solve the boundary value problem:

$$\min_u \int_\Omega |\nabla u| \qquad \text{subject to} \qquad u = u_0 \qquad \text{in } \Omega \setminus D. \tag{2.10}$$

Here, $D$ is the missing region to be inpainted, $u_0$ is the observed image whose value in $D$ is missing. Thus, the TV inpainting method simply fills-in the missing region such that the TV in $\Omega$ is minimized. The use of TV-norm is desirable because it has the effect of extending level sets into $D$ without smearing discontinuities along the tangential direction of the boundary of $D$.

With a slight modification of (2.10), simultaneous inpainting (in $D$) and denoising (in $\Omega \setminus D$) may be done as follows:

$$\min_u \int_\Omega |\nabla u| + \lambda \int_{\Omega \setminus D} (u - u_0)^2 dx. \tag{2.11}$$

Define a spatial varying parameter $\lambda_e(x)$ which is 0 in $D$ and is $\lambda$ in $\Omega \setminus D$. Then the Euler-Lagrange equation for (2.11) can be written as

$$-\nabla \cdot \left( \frac{\nabla u}{|\nabla u|} \right) + 2\lambda_e(u - u_0) = 0$$

which has the same form as that for the ROF model, except the regularization is switching between 0 and $\lambda$ in different regions. Thus, it is easy to modify an implementation of the ROF model to the TV inpainting model. Finally, we remark that some variants of (2.11) such as curvature-driven diffusion [172] and Euler's Elastica [167] have been proposed which complete isophotes in a smoother way.

### 2.5.3 *Texture and Multiscale Decompositions*

Another way of looking at denoising problems is by separating a given noisy image $f$ into two components to form the decomposition: $f = u + v$, where $u$ is the denoised image and $v = f - u$ the noise. In [567], Meyer adopts this view for the purpose of texture extraction where $v$ captures not only noise but also texture. To do this, he proposed a new decomposition model:

$$\inf_u \left\{ E(u) = \int_\Omega |\nabla u| + \lambda \|v\|_*, f = u + v \right\} \tag{2.12}$$

where the $*$ norm is given by:

$$\|v\|_* = \inf_{\mathbf{g}=(g_1,g_2)} \left\{ \|\sqrt{{g_1}^2 + {g_2}^2}\|_{L^\infty} \mid v = \partial_x g_1 + \partial_y g_2 \right\} \tag{2.13}$$

and the $v$ component lies in what is essentially the dual space of BV, the $G$ space:

$$G = \left\{ v \mid v = \partial_x g_1 + \partial_y g_2 \,,\, g_1, g_2 \in L^\infty(\mathbb{R}^2) \right\}. \tag{2.14}$$

Here, $v$ is an oscillatory function representing texture and the $*$ norm is designed to give small value for these functions. Thus, the main idea in (2.12) is to try to pull out texture by controlling $\|v\|_*$. Experiments in [843, 619] (discussed below) visually show that the model (2.12) extracts texture better than the standard ROF model.

In practice, the model (2.12) is difficult to implement due to the nature of the $*$ norm. Vese and Osher [843] were the first to overcome this difficulty where they devise an $L^p$ approximation to the norm $\|\cdot\|_*$. In a later work [619], Osher et al. propose another $L^p$ approximation based on the $H^{-1}$ norm and introduce a resulting fourth order PDE. Both works numerically demonstrate the effectiveness of the model (2.12) for texture extraction and also give some further applications to denoising and deblurring.

In a related work, Aujol et al. [36] propose a decomposition algorithm based on Meyer's work [567] where they further decompose an image as $f = u+v+w$ where $u$, $v$, and $w$ are cartoon, texture, and noise respectively.

Given the scale properties of the ROF model seen in section 2.2.3, it is natural to consider a multiscale decomposition based on the ROF model. Multiscale decompositions are of particular interest since one may want to extract image features of many different scales (either coarse or fine). One such multiscale decomposition is Tadmor et al. [784] and proceeds in a hierarchical manner. After choosing an initial $\lambda_0 = \lambda$ to remove the smallest oscillation in a given image $f$, the regularization parameters $\{\lambda_j\}$, $\lambda_j = 2^j\lambda$ induce a sequence of dyadic scales for $j = 1, \ldots, k$. If we denote by $u_{\lambda_j}$ the solution to the ROF model (2.2) for parameter $\lambda_j$, then $f$ has the decomposition:

$$f = u_{\lambda_0} + u_{\lambda_1} + u_{\lambda_2} + \cdots + u_{\lambda_k} + v_{\lambda_k}.$$

with $v_{\lambda_k}$ denoting the $k$-th stage residual $v_{\lambda_k} = f - (u_{\lambda_0} + u_{\lambda_1} + u_{\lambda_2} + \cdots + u_{\lambda_k})$. Furthermore, the authors show that $\|v_{\lambda_k}\|_* \to 0$ as $k \to \infty$. Hence $\|f - \sum_{i=0}^{k} u_{\lambda_i}\|_* \to 0$ as $k \to \infty$ and the decomposition converges to $f$ in the $*$ norm. A related work based on merging dynamics of a monotonicity constrained TV model can be found in [169].

## 2.6 Numerical Methods

There have been numerous numerical algorithms proposed for minimizing the ROF objective. Most of them fall into the three main approaches, namely, direct optimization, solving the associated Euler-Lagrange equations and using the dual variable explicitly in the solution process to overcome some computational difficulties encountered in the primal problem. We will focus on the latter two approaches.

### *2.6.1 Artificial Time Marching and Fixed Point Iteration*

In their original paper [695], Rudin et al. proposed the use of artificial time marching to solve the Euler-Lagrange equations which is equivalent to the steepest descent of the energy function. More precisely, consider the image as a function of space and time and seek the steady state of the equation

$$\frac{\partial u}{\partial t} = \nabla \cdot \left( \frac{\nabla u}{|\nabla u|_\beta} \right) - 2\lambda(u - f). \tag{2.15}$$

Here, $|\nabla u|_\beta := \sqrt{|\nabla u| + \beta^2}$ is a regularized version of $|\nabla u|$ to reduce degeneracies in flat regions where $|\nabla u| \approx 0$. In numerical implementation, an explicit time marching scheme with time step $\Delta t$ and space step size $\Delta x$ is used. Under this method, the objective value of the ROF model is guaranteed to be decreasing and the solution will tend to the unique minimizer as time increases. However, the convergence is usually slow due to the Courant-Friedrichs-Lewy (CFL) condition, $\Delta t \leq c\Delta x^2|\nabla u|$ for some constant $c > 0$ (see [546]), imposed on the size of the time step, especially in flat regions where $|\nabla u| \approx 0$. To relax the CFL condition, Marquina and Osher use, in [546], a "preconditioning" technique to cancel singularities due to the degenerate diffusion coefficient $1/|\nabla u|$:

$$\frac{\partial u}{\partial t} = |\nabla u| \left[ \nabla \cdot \left( \frac{\nabla u}{|\nabla u|_\beta} \right) - 2\lambda(u - f) \right] \tag{2.16}$$

which can also be viewed as mean curvature motion with a forcing term $-2\lambda(u - f)$. Explicit schemes suggested in [546] for solving the above equation improve the CFL to $\Delta t \leq c\Delta x^2$ which is independent of $|\nabla u|$.

To completely get rid of CFL conditions, Vogel and Oman proposed in [849] a fixed point iteration scheme (FP) which solves the stationary Euler-Lagrange directly. The Euler-Lagrange equation is linearized by lagging the diffusion coefficient and thus the $(i + 1)$-th iterate is obtained by solving the sparse linear equation:

$$\nabla \cdot \left( \frac{\nabla u^{i+1}}{|\nabla u^i|_\beta} \right) - \lambda(u^{i+1} - f) = 0. \tag{2.17}$$

While this method converges only linearly, empirically, only a few iterations are needed to achieve visual accuracy. In practice, one typically employs specifically designed fast solvers to solve (2.17) in each iteration.

### 2.6.2 *Duality-based Methods*

The methods described in Section 2.5.1 are based on solving the primal Euler-Lagrange equation which is degenerate in regions where $\nabla u = 0$. Although regularization by $1/|\nabla u|_\beta$ avoids the coefficient of the parabolic term becoming arbitrarily large, the use of a large enough $\beta$ for effective regularization will reduce the ability of the ROF model to preserve edges.

Chan et al. in [166], Carter in [151] and Chambolle in [160] exploit the dual formulation of the ROF model By using the identity $\|\mathbf{x}\| \equiv \sup_{\|\mathbf{g}\| \leq 1} \mathbf{x} \cdot \mathbf{g}$ for vectors in Euclidean spaces and treating $\mathbf{g}$ as the dual variable, one arrives at the dual formulation:

$$\sup_{\mathbf{g} \in C_c^1(\Omega, B^2)} \int_\Omega f \nabla \cdot \mathbf{g} dx - \frac{1}{2\lambda} \int_\Omega (\nabla \cdot \mathbf{g})^2 dx \tag{2.18}$$

where $B^2$ is the unit disk in $\mathbb{R}^2$. Once $\mathbf{g}$ is obtained, the primal variable can be recovered by $u = f - \lambda^{-1} \nabla \cdot \mathbf{g}$. A promise of the dual formulation is that the objective function is differentiable in $\mathbf{g}$, unlike the primal problem which is badly behaved when $\nabla u = 0$. However, the optimization problem becomes a constrained one which requires additional complexity to solve.

The approach used in [166] solves for $u$ and g simultaneously. Its derivation starts by treating the term $\nabla u/|\nabla u|$ in the primal Euler-Lagrange equation as an independent variable g, leading to the system:

$$-\nabla \cdot \mathbf{g} + \lambda(u - f) = 0, \qquad \mathbf{g}|\nabla u|_\beta - \nabla u = 0.$$

The above system of nonlinear equations is solved by Newton's method and quadratic convergence rate is almost always achieved. In the Newton updates, one may combine the two equations to eliminate the need to update g, thus the cost per iteration is as cheap as the fixed point iteration (2.17). Empirically, this primal-dual method is much more robust than applying Newton's method directly to the primal problem in $u$ only.

In [160], Chambolle devised an efficient algorithm solely based on the dual formulation (2.18). By carefully looking at the Euler-Lagrange equation for (2.18) and eliminating the associated Lagrange multipliers, one arrives at solving $H(\mathbf{g}) - |H(\mathbf{g})| = 0$ where $H(\mathbf{g}) = -\nabla(f - \lambda^{-1}\nabla \cdot \mathbf{g})$ is the negative of the gradient of the primal variable $u$. The update formula for g used in [160] is a simple relaxation $\mathbf{g}^{n+1} = \frac{\mathbf{g}^n + \tau H(\mathbf{g}^n)}{1+\tau|H(\mathbf{g}^n)|}$ where $\tau > 0$ is chosen to be small enough so that the iteration converges.

# Chapter3

# PDE-Based Image and Surface Inpainting

**M. Bertalmío, V. Caselles, G. Haro, and G. Sapiro**

Abstract

Inpainting, the technique of modifying an image in an undetectable form, is as ancient as art itself. The goals and applications of inpainting are numerous, from the restoration of damaged paintings, photographs and films, to the removal of selected undesirable objects. This chapter is intended to present an overview of PDE based image inpainting algorithms, with emphasis in models developed by the authors. These models are based on the propagation of information along the image isophotes and on the minimization of an energy functional which follows a relaxation of the Elastica model. This last variational formulation can be easily extended to 3D to fill holes in surfaces, a problem closely related to image inpainting. Basic PDE-based approaches to inpainting share the shortcoming that they cannot restore texture, so combinations of these algorithms with texture synthesis techniques are also discussed. Some results are shown for applications such as removal of text, restoration of scratched photographs, removal of selected objects and reconstruction of 3D surfaces with holes. Other recent approaches to the image inpainting problem are also briefly reviewed.

## 3.1 Introduction

The modification of images in a way that is non-detectable for an observer who does not know the original image is a practice as old as artistic creation itself. Medieval artwork started to be restored as early as the Renaissance, the motives being often as much to bring medieval pictures "up to date" as to fill in any gaps [298, 852]. This practice is called *retouching* or *inpainting*. The object of inpainting is to reconstitute the missing or damaged portions of the work, in order to make it more legible and to restore its unity [298].

The need to retouch the image in an unobtrusive way extended naturally from paintings to photography and film. The purposes remain the same: to revert deterioration (e.g., cracks in photographs or scratches and dust spots in film), or to add or remove elements (e.g., removal of stamped date and red-eye from photographs, the infamous "airbrushing" of political enemies [475]).

Digital techniques are starting to be a widespread way of performing inpainting, ranging from attempts to fully automatic detection and removal of scratches in film [484, 485, 486], all the way to software tools that allow a sophisticated but mostly manual process.

This article is intended to be an overview of PDE based image inpainting algorithms, with emphasis in those models which were developed by the authors and that motivated a significant amount of effort in the area (some of the major contributions by other groups in image inpainting are briefly reviewed as well).

We should first note that classical image denoising algorithms do not apply to image inpainting. In common image enhancement applications, the pixels contain both information about the real data and the noise (e.g., image plus noise for additive noise), while in image inpainting, there is no significant information in the region to be inpainted. The information is mainly in the regions surrounding the areas to be inpainted. There is then a need to develop specific techniques to address these problems.

Mainly three groups of works can be found in the literature related to digital inpainting. The first one deals with the restoration of films, the second one is related to texture synthesis, and the third one is related to what we would call geometric inpainting.

Kokaram et al. [484, 485, 486] use motion estimation and autoregressive models to interpolate losses in films from adjacent frames. The basic idea is to copy into the gap the right pixels from neighboring frames. The technique can not be applied to still images or to films where the regions to be inpainted span many frames.

There are many works on texture synthesis, of which the most notable are based on Markov Random Fields after the pioneering work of Efros and Leung [297]. These techniques synthesize texture which is both stationary and local [869]. In [297] a new texture is incrementally synthesized by considering similar neighborhoods in the sample texture. Igehy and Pereira [416] replace image regions with synthesized texture [392, 745] according to a given mask. Ashikhmin [32] adds the constraint that the synthesized texture match a sample image. This yields the effect of rendering a given image with the texture appearance of a training texture. Efros and Freeman [296] introduce a simple and effective texture synthesis technique that synthesizes a new texture by stitching together blocks of existing sample texture. The results depend on the size of a block which is a parameter tuned by the user that varies according to the texture properties. Hirani and Totsuka [400] combine frequency and spatial domain information in order to fill a given region with a user-selected texture. We will later show how texture synthesis can be combined with PDE-based inpainting techniques to obtain state-of-the-art algorithms.

Finally, let us mention the geometric approaches used for filling-in the missing information in a region of the image. A pioneering contribution in the recovery of plane image geometry is due to D. Mumford, M. Nitzberg and T. Shiota [604]. They were not directly concerned with the problem of recovering the missing parts of the image, instead, they addressed the problem of segmenting the image into objects which should be ordered according to their depth in the scene. The segmentation functional should be able to find which are the occluding and the occluded objects while finding the occluded boundaries. For that they relied on a basic principle of Gestalt's psychology: our visual system is able to complete partially occluded boundaries and the completion tends to respect the *principle of good continuation* [453]. When an object occludes another the occluding and occluded boundaries form a particular configuration, called $T$-junction, which is the point where the visible part of the boundary of the occluded object terminates. Then our visual system smoothly continues the occluded boundary between $T$-junctions. In [604], the authors proposed an energy functional to segment a scene which took into account the depth of the objects in the scene and the energy of the occluded boundaries between $T$-junctions. They assumed that the completion curves should be as short as possible and should respect the principle of good (smooth) continuation. Thus, to define the energy of the missing curve they had to give a mathematical formulation of the above principles. Given two $T$-junction points $p$ and $q$ and the tangents $\tau_p$ and $\tau_q$ to the respective terminating edges, they proposed as smooth continuation curve Euler's elastica, i.e., the curve minimizing the energy

$$\int_C (\alpha + \beta\kappa^2) ds \tag{3.1}$$

where the minimum is taken among all curves $C$ joining $p$ and $q$ with tangents $\tau_p$ and $\tau_q$, respectively, $\kappa$ denotes the curvature of $C$, $ds$ its arc length, and $\alpha, \beta$ are positive constants. Let us mention that Euler's elastica has been frequently used in computer vision ([406, 511, 735, 795, 796, 821, 892, 893, 891]) and a beautiful account on it can be found in [589].

Inspired by the elastica, Masnou and Morel [551, 553, 552] proposed a variational formulation for the recovery of the missing parts of a grey level two-dimensional image and they referred to this interpolation process as *disocclusion*, since the missing parts can be considered as occlusions hiding the part of the image we want to recover. Their algorithm performs filling-in by joining with geodesic curves the points of the isophotes arriving at the boundary of the region to be inpainted.

Mumford's work on the Elastica Model and Masnou and Morel's contribution inspired Bertalmío, Sapiro, Caselles and Ballester [79] to propose an edge propagation PDE for the *Image Inpainting* formulation. Replicating basic art conservators techniques, a third order PDE propagates the level lines arriving at the missing region, and the completion tends to respect the *principle of good continuation*. Bertalmío , Bertozzi and Sapiro [77] showed the connection of this equation with Navier-Stokes equations, as well as a parallel among Image Processing and Fluid Dynamics quantities. On the other hand, Ballester, Bertalmío, Caselles, Sapiro and Verdera [46] introduce a relaxation of the Elastica functional which then can be minimized with a system of coupled PDE's: this is the first variational approach to the inpainting problem that complies with the *principle of good continuation* and is topologically independent.

The elastica has inspired most variational approaches to geometric image inpainting [46, 48, 47, 167, 544] and we shall discuss in detail some of them in Section 3.3. In particular, the approach in [47, 840] can be used for inpainting 3D images and surface hole reconstruction. Some other PDE methods for surface hole reconstruction will be discussed in Section 3.4.

This article is organized as follows. Section 3.2 discusses inpainting by propagation of information: PDE methods that propagate image quantities and do not explicitly minimize any functional. Section 3.3 discusses variational methods for inpainting: the inpainting problem is solved as the minimization of an energy functional. In Section 3.4 we show how can we use the Laplace and AMLE (Absolutely Minimizing Lipschitz Extension) interpolators in surface hole reconstruction. None of these purely-PDE-based methods can restore texture, so in Section 3.5 we discuss how to adapt those algorithms to deal with texture. Finally, in Section 3.6 we briefly mention some other recent works on the inpainting problem. We finish with Appendix 3.8 where we collect some notation and definitions used in the text.

## 3.2 Inpainting by Propagation of Information

### *3.2.1 Image Inpainting*

In [79], Bertalmío, Sapiro, Caselles and Ballester propose to translate into mathematical form the most basic techniques used by art conservators and restorators to inpaint, introducing also the art term 'inpainting' to the Image Processing and Graphics community.

Conservators at the Minneapolis Institute of Arts were consulted for this work and made it clear that inpainting is a very subjective procedure, different for each work of art and for each professional. There is no such thing as "the" way to solve the problem, but the underlying methodology is as follows: (1.) The global picture determines how to fill in the gap, the purpose of inpainting being to restore the unity of the work; (2.) The structure of the area surrounding the gap $\Omega$ is continued into it, contour lines are drawn via the prolongation of those arriving at the gap boundary $\partial\Omega$; (3.) The different regions inside $\Omega$, as defined by the contour lines, are filled with color, matching those of $\partial\Omega$; and (4.). The small details are painted (e.g. little white spots on an otherwise uniformly blue sky): in other words, "texture" is added.

The algorithm in [79] simultaneously, and iteratively, performs the steps (2.) and (3.) above. The gap $\Omega$ shrinks progressively by prolonging inward, in a smooth way, the lines arriving at the gap boundary $\partial\Omega$. The image beyond $\partial\Omega$ is not taken into account, and texture is not dealt with (yet) with this first technique.

The following exposition considers the grayscale case; for color images, the authors apply their method to each of the three channels separately, but using a color model like $CIE - Lab$ instead of $RGB$, to avoid color artifacts.

The digital inpainting procedure will construct a family of images $u(i,j,n)$ : $[0,M] \times [0,N] \times \mathbb{N} \rightarrow \mathbb{R}$ such that $u(i,j,0) = u_0(i,j)$ and $\lim_{n\to\infty} u(i,j,n) = u_R(i,j)$, where $u_0(i,j)$ is the image to inpaint and $u_R(i,j)$ is the output of the algorithm (inpainted image).

Any general algorithm of that form can be written as:

$$u^{n+1}(i,j) = u^n(i,j) + \Delta t u_t^n(i,j), \forall (i,j) \in \Omega \tag{3.2}$$

where the superindex $n$ denotes the inpainting "time" $n$, $(i,j)$ are the pixel coordinates, $\Delta t$ is the rate of improvement and $u_t^n(i,j)$ stands for the update of the image $u^n(i,j)$. Note that the evolution equation runs only inside $\Omega$, the region to be inpainted.

To design the update $u_t^n(i,j)$, the authors call $L^n(i,j)$ the information that needs to be propagated into the gap, and $\overrightarrow{N}^n(i,j)$ the propagation direction:

$$u_t^n(i,j) = \nabla L^n(i,j) \cdot \overrightarrow{N}^n(i,j), \tag{3.3}$$

With equation (3.3), they estimate the information $L^n(i,j)$ of the image and compute its change along the $\overrightarrow{N}^n$ direction. Note that when the algorithm converges, $u^{n+1}(i,j) = u^n(i,j)$ and from (3.2) and (3.3) we have that $\nabla L^n(i,j) \cdot \overrightarrow{N}^n(i,j) = 0$, meaning exactly that the information $L$ has been propagated in the direction $\overrightarrow{N}$.

Bearing in mind that the goal is to propagate contours and that the Laplacian has been frequently used as an edge detector, the authors choose for $L^n(i,j)$ a monotone increasing function of the Laplacian, the most simple one being the Laplacian itself. Thus, the proposed choice is $L^n(i,j) = \Delta u^n(i,j)$. Other edge detectors like Canny's edge detector which leads to the choice $L^n = \langle \nabla^2 u^n(\nabla u^n), \nabla u^n \rangle$ could be used.

For the field $\overrightarrow{N}$, the natural choice is the isophotes directions. This is a bootstrapping problem: having the isophotes directions inside $\Omega$ is equivalent to having the inpainted image itself, since we can easily recover the gray level image from its isophote direction field (see [460],[639]). They use then a time varying estimation of the isophotes direction field: $\overrightarrow{N}(i,j,n) = \nabla^{\perp} u^n(i,j)$

In terms of a continuous process, the inpainting procedure can be expressed as a third-order PDE:

$$\frac{\partial u(x,y,t)}{\partial t} = \nabla(\Delta u(x,y,t)) \cdot \nabla^{\perp} u(x,y,t)), \forall (x,y) \in \Omega \tag{3.4}$$

To ensure a correct evolution of the direction field, a diffusion process is interleaved with the image inpainting process described above. This diffusion cor-

Figure 3.1. Restoration of an old photograph.

Figure 3.2. Removal of superimposed text.

responds to the periodical curving of lines to avoid them from crossing each other, as art conservators do. The authors use anisotropic diffusion, [15, 642], in order to achieve this goal without losing sharpness in the reconstruction:

$$\frac{\partial u}{\partial t}(x,y,t) = \kappa(x,y,t)\,|\nabla u(x,y,t)|\,, \forall(x,y) \in \Omega \tag{3.5}$$

where $\kappa$ is the Euclidean curvature of the isophotes of $u$.

For the numerical implementation, a forward-time upwind scheme is used for (3.4) and a forward-time centered-space scheme for (3.5); see [618, 695] for details . To speed up the process, a non-linear scaling is applied to $u_t$ in (3.4): $u_t = \text{sign}(u_t)\,|u_t|^{\frac{1}{4}}$. With a time step $\Delta t$ of 0.1, one step of anisotropic diffusion is run every fifteen steps of inpainting. Convergence is typically achieved after a few thousands iterations, depending on the size of $\Omega$ and the initial condition inside it. The process may be sped-up by the use of multi-resolution for wide gaps, and by pre-processing by running a few steps of the Heat Equation inside $\Omega$ to get a good initial condition:

$$\frac{\partial u}{\partial t}(x,y,t) = \Delta u(x,y,t), \forall(x,y) \in \Omega \tag{3.6}$$

See examples in figures 3.1 and 3.2. In both cases, the algorithm is supplied only with the image to restore and a binary mask that specifies the region to restore. In figure 3.1, a deteriorated photograph is restored, the mask having been manually selected with a simple paintbrush-like program by a non-specialist. Observe that details in the nose and right eye of the middle girl could not be completely restored. This is in part due to the fact that the mask covers most of the relevant information, and there is not much to be done without the use of high level prior information (e.g., the fact that it is an eye). These minor errors can be corrected by the manual procedures mentioned in the introduction, and still the overall inpainting time would be reduced by orders of magnitude.

Figure 3.2 shows a color example: results are sharp and without color artifacts. This image is very ill-suited for texture synthesis algorithms, since the image gap $\Omega$ covers most of the image, which also has a very diverse background.

The technique presented above does not require any user intervention, once the region to be inpainted has been selected. The algorithm is able to simultaneously fill regions surrounded by different backgrounds, without the user specifying "what to put where." No assumptions on the topology of the region to be inpainted, or on the simplicity of the image, are made. The algorithm is devised for inpainting in structured regions (e.g., regions crossing through boundaries), though it is not devised to reproduce textured areas.

### 3.2.2 Navier-Stokes Inpainting

In [77], the authors propose an approach that uses ideas from classical fluid dynamics to propagate isophote lines continuously from the exterior into the region to be inpainted. The main idea is to think of the image intensity as a 'stream function' for a two-dimensional incompressible flow. The Laplacian of the image intensity plays the role of the vorticity of the fluid; it is transported into the region to be inpainted by a vector field defined by the stream function. The resulting algorithm is designed to continue isophotes while matching gradient vectors at the boundary of the inpainting region. The method is directly based on the Navier-Stokes equations for fluid dynamics, which has the immediate advantage of well-developed theoretical and numerical results. Existence and stability of the solution to the proposed algorithm follow from the Navier-Stokes theory, and the implementation is based on numerical methods used by the fluid dynamics community.

In [77], the authors start by re-introducing the inpainting method of [79]:

$$u_t = \nabla^{\perp} u \cdot \nabla \Delta u \tag{3.7}$$

and noting that its dynamics are those of a transport equation that convects the image intensity $u$ along level curves of the smoothness, $\Delta u$. This can be seen by noting that (3.7) is equivalent to $Du/Dt = 0$ where $D/Dt$ is the material derivative $\partial/\partial t + v \cdot \nabla$ for the velocity field $v = \nabla^{\perp} \Delta u$. In particular $u$ is convected by the velocity field $v$ which is in the direction of level curves of the smoothness $\Delta u$.

Next, the authors introduce an analogy to transport of vorticity in incompressible fluids. Incompressible Newtonian fluids are governed by the Navier-Stokes equations, which couple the velocity vector field $v$ to a scalar pressure $p$ [195]:

$$v_t + v \cdot \nabla v = -\nabla p + \nu \nabla^2 v, \quad \nabla \cdot v = 0. \tag{3.8}$$

In two space dimensions, the divergence free velocity field $v$ possesses a stream function $\Psi$ satisfying $\nabla^{\perp}\Psi = v$. In addition, in 2D the vorticity, $\omega = \nabla \times v$, satisfies a very simple advection diffusion equation, which can be computed by taking the curl of the first equation in (3.8) and using some basic facts about the geometry in 2D:

$$\omega_t + v \cdot \nabla \omega = \nu \nabla^2 \omega. \tag{3.9}$$

Note here that in 2D the vorticity is a scalar quantity that is related to the stream function through the smoothness or Laplacian operator, $\Delta\Psi = \omega$. In the absence of viscosity $\nu = 0$, we obtain the Euler equations for inviscid flow.

Both the inviscid and viscous problems, with appropriate boundary conditions, are globally well-posed in two space dimensions. Solutions exist for any smooth initial condition and they depend continuously on the initial and boundary data [500].

In terms of the stream function, equation (3.9) implies that steady state inviscid flows must satisfy

$$\nabla^{\perp}\Psi \cdot \nabla\Delta\Psi = 0 \tag{3.10}$$

which says that the Laplacian of the stream function, and hence the vorticity, must have the same level curves as the stream function. The analogy to image inpainting in the previous section is now clear: the stream function for inviscid fluids in 2D satisfies the same equation as the steady state image intensity equation (3.7).

The authors then procceed to present a 'Navier-Stokes' based method for image inpainting. In this method the fluid dynamic quantities have the following parallel to quantities in the inpainting method:

| Navier-Stokes | Image inpainting |
|---|---|
| stream function $\Psi$ | Image intensity $u$ |
| fluid velocity $v = \nabla^{\perp}\Psi$ | isophote direction $\nabla^{\perp}u$ |
| vorticity $\omega = -\Delta\Psi$ | smoothness $w = \Delta u$ |

where they denote by $w$ the smoothness $\Delta u$ of the image intensity. Instead of solving a transport equation for $u$ as in (3.7), they solve a vorticity transport equation for $w$:

$$\partial w/\partial t + v \cdot \nabla w = \nu\nabla \cdot (g(|\nabla w|)\nabla w), \tag{3.11}$$

where the function $g$ allows for anisotropic diffusion of the smoothness $w$.

The image intensity $u$ which defines the velocity field $v = \nabla^{\perp}u$ in (3.11) is recovered by solving simultaneously the Poisson problem

$$\Delta u = w, \quad u|_{\partial\Omega} = u_0. \tag{3.12}$$

For $g = 1$, the direct numerical solution of of (3.11-3.12) is a classical way to solve both the dynamic fluid equations and to evolve the dynamics towards a steady state solution [644].

When using any PDE-based method to do inpainting, the issue of boundary conditions becomes very important. In order to produce a result which, to the eye, does not distiguish where the inpainting has taken place, we must at the very least propagate both the image intensity and direction of the isophote lines continuously into the inpainting region.

This means that any PDE-based method involving the image intensity $u$ must enforce Dirichlet (fixed $u$) boundary conditions as well as a condition on the direction of $\nabla u$ on the boundary. Immediately we see that this poses a problem for lower-order PDE-based methods. Indeed, any first or second order PDE (including anisotropic diffusion) for the scalar $u$ could typically only enforce one of these boundary conditions, the result being an inpainting with discontinuities in the slope of the isophote lines, or a method with a jump in $u$ itself on the boundary [172]. From a mathematical point of view, to fix this, one can either go to a higher order equation for $u$, as in [79], that requires more boundary conditions, or consider a vector evolution for $\nabla u$, which is the idea of the Navier-Stokes method.

The Navier-Stokes analogy guarantees, in a very natural way, continuiuty of the image intensity function $u$ and its isophote directions across the boundary of the inpainting region. First, consider a solution of the Navier-Stokes equation (3.8) in primitive variables form satisfying the classical no-slip condition $v = 0$ on the boundary $\partial\Omega$. This condition guarantees two features: (a) that the stream function $\Psi$ must be constant on the boundary, since the boundary is trivially a streamline of the flow; (b) that the direction of the fluid velocity $v$ is always tangent to the boundary.

A general form of the no-slip boundary condition, for which well-posedness is known, is to prescribe the velocity vector $v = v_0$ on the boundary. This would be the natural choice for a moving boundary. Specifying the velocity on the boundary is equivalent to specifying both the normal and tangential derivatives of the stream function $\Psi$ on the boundary, since $v = \nabla^{\perp}\Psi$. However, specifying the tangential derivative of $\Psi$ determines $\Psi$ on the boundary up to a constant of integration, by simply integrating around the boundary with respect to its arc length. Similarly this information determines the direction of flow on the boundary. The result is that if we solve the Navier-Stokes equations with $v$ fixed on the boundary, we obtain a solution with a stream function $\Psi$ and velocity field $v$ both of which are continuous up to the boundary. For the Navier-Stokes inpainting method, we inherit the continuity across the boundary. For example, suppose we fix $\nabla^{\perp}u$ on the boundary. Then solving the Navier-Stokes inpainting equation with these boundary conditions will not only result in continuous isophotes, but also will produce an image intensity function that is continuous across $\partial\Omega$.

As for well-posedness and uniqueness of solutions, the authors note that without the presence of viscosity in the method there is not a unique *steady-state* solution. They expect that Navier-Stokes based inpainting may inherit some of the stability and uniqueness issues known for incompressible fluids, although the effect of anisotropic diffusion is not clear.

## 3.3 Variational Models for Filling-In

This section is a review of variational models for filling-in. We start with the elastica-based disocclusion model introduced by Masnou and Morel [551, 553]. Then we present the filling-in approach by joint interpolation of vector fields and gray levels proposed by Ballester et al. in [46, 48, 47]. The connections of this model with T. Chan and J. Shen approach [167] are then considered.

### 3.3.1 Elastica-based Reconstruction of Level Lines

We review the main assumptions of Masnou's approach to disocclusion [551, 553]. An image is usually modeled as a function defined in a bounded domain $D \subseteq \mathcal{R}^N$ (typically $N = 2$ for usual snapshots, $N = 3$ for medical images or movies) with values in $\mathcal{R}^k$ ($k = 1$ for grey level images, or $k = 3$ for color images). For simplicity, we shall consider only the case of grey level images. Any real image is determined in a unique way by its upper (or lower) level sets $X_\lambda u := \{x \in D : u(x) \geq \lambda\}$ ($X'_\lambda u := \{x \in D : u(x) \leq \lambda\}$). Indeed we have the reconstruction formula

$$u(x) = \sup\{\lambda \in R : x \in X_\lambda u\}. \tag{3.13}$$

The basic postulate of Mathematical Morphology prescribes that the geometric information of the image $u$ is contained in the family of its level sets [371, 723], or in a more local formulation, in the family of connected components of the level sets of $u$ [154, 723, 726]. We shall refer to the family of connected components of the upper level sets of $u$ as the topographic map of $u$.

In the case that $u$ is a function of bounded variation in $D \subseteq \mathcal{R}^2$, i.e., $u \in BV(D)$ (see Appendix and [19, 301, 926]), its topographic map has a description in terms of Jordan curves [18]. With an adequate definition of connected components, the essential boundary of a connected component of a rectifiable subset of $\mathcal{R}^2$ consists, modulo an $\mathcal{H}^1$ null set, of an exterior Jordan curve and an at most countable family of interior Jordan curves which may touch in a set of $\mathcal{H}^1$-null Hausdorff measure [18]. Since almost all level sets $X_\lambda u$ of a function $u$ of bounded variation are rectifiable sets, its essential boundary, $\partial^* X_\lambda u$, consists of a family of Jordan curves called the level lines of $u$. Thus, the topographic map of $u$ can be described in terms of Jordan curves. In this case, the monotone family of upper level sets $X_\lambda u$ suffices to have the reconstruction formula (3.13) which holds almost everywhere [371].

Let $D$ be a square in $\mathcal{R}^2$ and $\tilde{\Omega}$ be an open bounded subset of $D$ with Lipschitz continuous boundary. Suppose that we are given an image $u_0 : D \setminus \overline{\tilde{\Omega}} \longrightarrow [a, b]$, $0 \leq a < b$. Using the information of $u_0$ on $D \setminus \overline{\tilde{\Omega}}$ we want to reconstruct the image $u_0$ inside $\tilde{\Omega}$. We shall call $\tilde{\Omega}$ the *hole* or *gap*. We shall assume that the function $u_0$ is a function of bounded variation in $D \setminus \overline{\tilde{\Omega}}$. Then the topographic structure of the image $u_0$ outside $\tilde{\Omega}$ is given by a family of Jordan curves. Generically, by slightly increasing the hole, we may assume that, for almost all levels $\lambda$, the level lines of $X_\lambda u_0$ transversally intersect the boundary of the hole in a finite number of points [551]. Let us call $\Lambda \subseteq \mathcal{R}$ the family of such levels. As formulated by Masnou [551, 553, 552], the disocclusion problem consists in reconstructing the topographic map of $u_0$ inside $\tilde{\Omega}$. Given $\lambda \in \Lambda$ and two points $p, q \in X_\lambda u_0 \cap \partial\tilde{\Omega}$

whose tangent vector at the level line $X_\lambda u_0$ is $\tau_p$ and $\tau_q$, respectively, the optimal completion curve proposed in [551, 553] is a curve $\Gamma$ contained in $\tilde{\Omega}$ minimizing the criterion

$$\int_\Gamma (\alpha + \beta|\kappa|^p) d\mathcal{H}^1 + (\tau_p, \tau_\Gamma(p)) + (\tau_q, \tau_\Gamma(q)) \tag{3.14}$$

where $\kappa$ denotes the curvature of $\Gamma$, $\tau_\Gamma(p)$ and $\tau_\Gamma(q))$ denote the tangents to $\Gamma$ at the points $p$ and $q$, respectively, and $(\tau_p, \tau_\Gamma(p))$, $(\tau_q, \tau_\Gamma(q))$ denote the angle formed by the vectors $\tau_p$ and $\tau_\Gamma(p)$, and, respectively, for $q$. Here $\alpha, \beta$ are positive constants, and $p \geq 1$. The optimal disocclusion is obtained by minimizing the energy functional

$$\int_{-\infty}^{+\infty} \sum_{\Gamma \in F_\lambda} \left( \int_\Gamma (\alpha + \beta|\kappa|^p) d\mathcal{H}^1 + (\tau_p, \tau_\Gamma(p)) + (\tau_q, \tau_\Gamma(q)) \right) d\lambda \tag{3.15}$$

where $F_\lambda$ denotes the family of completion curves associated to the level set $X_\lambda u_0$. As we noted above, the family $F_\lambda$ is generically finite, thus the sum in (3.15) is generically finite. In [551, 553] the authors proved that for each $p \geq 1$ there is an optimal disocclusion in $\tilde{\Omega}$ and proposed an algorithm based on dynamic programming to find optimal pairings between compatible points in $\partial X_\lambda u_0 \cap \partial \tilde{\Omega}$ for $p = 1$, curves which are straight lines, thus finding in this case the minimum of (3.15) [551, 552]. In [20] the authors proposed a slight variation of the disocclusion energy functional (3.15). First, they observed that by computing the criterion $\int_\Gamma (\alpha + \beta|\kappa|^p) d\mathcal{H}^1$ not only on the completion curve but also in a small piece of the associated level line outside $\tilde{\Omega}$, the criterion (3.15) can be written as

$$\int_{-\infty}^{+\infty} \sum_{\Gamma \in F_\lambda} \left( \int_\Gamma (\alpha + \beta|\kappa|^p) d\mathcal{H}^1 \right) d\lambda \tag{3.16}$$

where now the curves in $F_\lambda$ are union of a completion curve and a piece of level line of $u_0$ in $\Omega \setminus \tilde{\Omega}$ for a domain $\Omega \supset \tilde{\Omega}$. This requires that the level lines of $u_0$ are essentially in $W^{2,p}$ in $\Omega \setminus \tilde{\Omega}$. Then, at least for $C^2$ functions $u$, (3.16) can be written as

$$\int_\Omega |\nabla u| (\alpha + \beta \left| \operatorname{div} \frac{\nabla u}{|\nabla u|} \right|^p) \, dx \tag{3.17}$$

with the convention that the integrand is 0 when $|\nabla u| = 0$. In [20], the authors considered this functional when the image domain $D$ and the hole $\tilde{\Omega}$ are subsets in $\mathcal{R}^N$ whit $N \geq 2$ and they studied the relaxed functional, proving that it coincides with

$$\int_R \int_{\partial[u \geq t]} (\alpha + \beta|H_{[u \geq t]}|^p) \, d\mathcal{H}^{N-1} \, dt \tag{3.18}$$

for functions $u \in C^2(\Omega)$, $N \geq 2$, $p > N - 1$, and $H_{[u \geq t]}$ denotes the mean curvature of $[u \geq t]$.

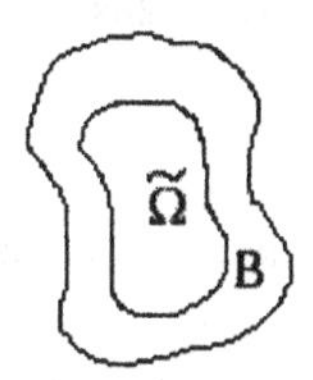


Figure 3.3. The hole and the band

### 3.3.2 *Joint Interpolation of Vector Fields and Gray Levels*

In [46, 48, 47], Ballester et al. proposed to fill-in the hole $\tilde{\Omega}$ using both the gray level and the vector field of tangents (or normals) to the level lines of the image outside the hole. Let $D$ be a hyperrectangle in $\mathcal{R}^N$, $N \geq 2$, which represents the image domain, and let $\Omega, \tilde{\Omega}$ be two open bounded domains in $\mathcal{R}^N$ with Lipschitz boundary. Suppose that $\overline{\tilde{\Omega}} \Subset \Omega \Subset D$ (for simplicity, we assume that $\Omega$ does not touch the boundary of the image domain $D$). Suppose that the image $u_0$ is given in $D \setminus \tilde{\Omega}$. Let $B := \Omega \setminus \overline{\tilde{\Omega}}$. The set $B$ will be called the band around $\tilde{\Omega}$ (see Figure 3.3).

To fill-in the hole $\tilde{\Omega}$ we shall use the information of $u_0$ contained in $B$, mainly the gray level and the vector field of normals (or tangents) to the level lines of $u_0$ in $B$. We attempt to continue the level sets of $u_0$ in $B$ inside $\tilde{\Omega}$ taking into account the principle of good continuation. Let $\theta_0$ be the vector field of directions of the gradient of $u_0$ on $D \setminus \tilde{\Omega}$, i.e., $\theta_0$ is a vector field with values in $\mathcal{R}^2$ satisfying $\theta_0(x) \cdot Du_0(x) = |Du_0(x)|$ and $|\theta_0(x)| \leq 1$. We shall assume that $\theta_0(x)$ has a trace on $\partial\Omega$.

We pose the image disocclusion problem in the following form: Can we extend (in a reasonable way) the pair of functions $(u_0, \theta_0)$ from the band $\Omega \setminus \overline{\tilde{\Omega}}$ to a pair of functions $(u, \theta)$ defined inside $\tilde{\Omega}$ ? Of course, we will have to precise what we mean by a reasonable way.

The data $u_0$ is given on the band $B$ and we should constrain the solution $u$ to be near the data on $B$. The vector field $\theta$ should satisfy $\theta \cdot \nu^{\Omega} = \theta_0 \cdot \nu^{\Omega}$, $|\theta| \leq 1$ on $\Omega$ and should be related to $u$ by the constraint $\theta \cdot Du = |Du|$, i.e., we should impose that $\theta$ is related to the vector field of directions of the gradient of $u$. The condition $|\theta(x)| \leq 1$ should be interpreted as a relaxation of this. Indeed, it may happen that $\theta(x) = 0$ (flat regions) and then we cannot normalize the vector field to a unit vector (the ideal case would be that $\theta = \frac{Du}{|Du|}$, $u$ being a smooth function with $Du(x) \neq 0$ for all $x \in \Omega$). Finally, we should impose that the vector field $\theta_0$ in $D \setminus \overline{\Omega}$ is smoothly continued by $\theta$ inside $\Omega$. Note that if $\theta$ represents the directions of the normals to the level lines of $u$, i.e., of the hypersurfaces $u(x) = \lambda$, $\lambda \in \mathcal{R}$, then $\mathrm{div}(\theta)$ represents its mean curvature. We shall impose the smooth continuation of the levels lines of $u_0$ inside $\Omega$ by requiring that $\mathrm{div}\,(\theta) \in L^p(\Omega)$, $p > 1$.

Interpreting the elastica functional in this framework, we propose to minimize the functional

$$
\begin{aligned}
&\text{Minimize} \int_{\Omega} |\mathrm{div}(\theta)|^p (\gamma + \beta |\nabla K * u|) dx \\
&|\theta| \leq 1,\ |Du| - \theta \cdot Du = 0 \text{ in } \Omega \\
&|u| \leq M \\
&u = u_0 \text{ in } B,\ \theta \cdot \nu^{\Omega}|_{\partial\Omega} = g_0,
\end{aligned}
\tag{3.19}
$$

where $p > 1, \gamma > 0, \beta \geq 0, g_0 = \theta_0 \cdot \nu^{\Omega}$, $K$ denotes a regularizing kernel of class $C^1$ such that $K(x) > 0$ a.e., $M = \sup_{x \in B} |u_0(x)|$, $\nu^{\Omega}$ and denotes the outer unit normal to $\Omega$. The convolution of $Du$ with the kernel $K$ in (3.19) is necessary to be able to prove the existence of a minimum of (3.19).

The functional can be interpreted as a formulation of the principle of good continuation and amodal completion as formulated in the Gestalt's theory of vision.

**Comments on model (3.19).**

A) Could we fill-in the hole without the band? To discuss this suppose that we are given the image of Figure 3.4.a, which is a gray band on a black background partially occluded by a square $\tilde{\Omega}$. We suppose that the sides of the square hole $\tilde{\Omega}$ are orthogonal to the level lines of the original image. In these conditions, the normal component of the vector field $\theta_0$ outside $\tilde{\Omega}$ is null at $\partial\tilde{\Omega}$. Thus if the boundary data is just $\theta_0 \cdot \nu^{\tilde{\Omega}}|_{\partial\tilde{\Omega}}$, we would have that $\theta_0 \cdot \nu^{\tilde{\Omega}}|_{\partial\tilde{\Omega}} = 0$. In particular, the vector field $\theta = 0$ satisfies this condition. If we are not able to propagate $\theta$ inside $\tilde{\Omega}$ this may become an unpleasant situation, since this would mean that we do no propagate the values of $u$ at the boundary. If we write the functional (3.23) with $\theta = 0$, $\alpha = 1$, it turns out to be the Total Variation [695]. The decision of extending the gray band or filling-in the hole with the black gray level would be taken as a function of the perimeter of the discontinuities of the function in the hole. Then the result of interpolating Figure 3.4.a, using Total Variation would be that of Figure 3.4.b, and not the one in Figure 3.4.c, because the interpolating lines in Figure 3.4.b, are shorter than the ones in Figure 3.4.c. To overcome this situation we introduce the band around the hole. The introduction of the band permits us to effectively incorporate in the functional the information given by the data $u_0$ and the vector field $\theta$ outside $\tilde{\Omega}$. In Figure 3.4.b, we display the result of the interpolation with $\theta = 0$ on $\tilde{\Omega}$. In Figure 3.4.c, we display the result of the interpolation using (3.23), which takes into account the band $B$ and computes the vector field $\theta$ in $\Omega$.

In practice, we suppose that only a narrow band around the hole influences what happens inside the hole, even if, in principle, it could be extended to all the known part of the image.

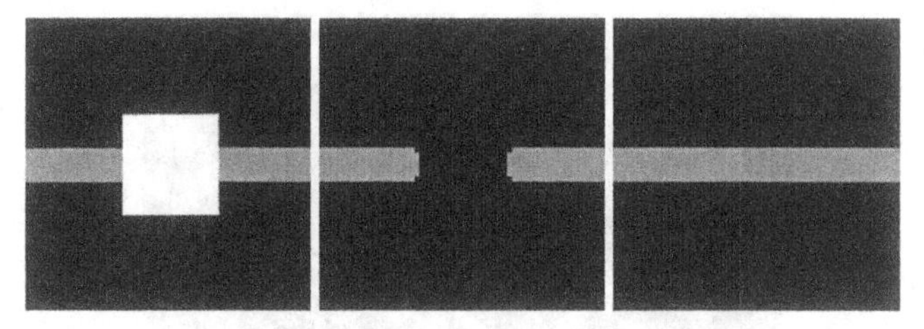

Figure 3.4. a) Left: a strip with a hole. b) Middle: image disocclusion obtained using Total Variation. c) Right: Image disocclusion obtained using functional (3.23).

B) If $N = 2$ and $u$ is the characteristic function of the region enclosed by a smooth ($C^2$) curve $C$ then the terms

$$\beta \int_\Omega |\mathrm{div}(\theta)|^p |Du| + \alpha \int_\Omega |Du| \tag{3.20}$$

can be written as $\int_C (\alpha + \beta|\kappa|^p) ds$, where $\kappa$ is the Euclidean curvature (of the level-sets). If $p = 2$, this coincides with Euler's elastica (3.1). Euler's elastica (3.1) was proposed in [604] as a technique for removing occlusions with the goal of image segmentation, since this criterion yields smooth, short, and not too curvy curves. In terms of characteristic functions, Euler's elastica can be written as

$$\int |\nabla u| \left( \alpha + \beta \left| div \left( \frac{\nabla u}{|\nabla u|} \right) \right|^2 \right). \tag{3.21}$$

In [70], it was shown that the elastica functional is not lower semicontinuous. As shown in [20], the functional proposed by Masnou and Morel [551, 552, 553] can be interpreted as a relaxation of it, since it integrates functionals like the elastica along the level lines of the function $u$. Our functional can be also considered as a relaxed formulation of the energy of the elastica. For that, we introduced $\theta$ as a independent variable, and we tried to couple it to $u$ by imposing that $\theta \cdot Du = |Du|$. Finally, let us say that to be able to prove the existence of a minimum for (3.23) we have convolved the $Du$ term of (3.20). This permits to avoid some of the mathematical difficulties involved in the study of (3.21).

C) Both coefficients $\gamma$ and $\beta$ are required to be $> 0$. The positivity of $\gamma$ gives us an $L^p$ bound on $\mathrm{div}(\theta)$ which implies the regularity of the level lines of $u$ ([554, 47]). If we do not take $\beta > 0$, $\theta = 0$ a.e. on $B$ (or on $\Omega$) in the image of Figure 3.4.a (since $\theta = 0$ except on some curves) and the term $\int_\Omega |\mathrm{div}(\theta)|^p dx$ would produce a null value since $\mathrm{div}(\theta) = 0$. If $\beta > 0$ we take into account the contribution of a power of the curvature on the level line corresponding to the boundary of the object.

D) In practice, functional (3.23) is used to interpolate shapes, i.e., to interpolate level sets. The image is decomposed into upper level sets $[u_0 \geq \lambda]$, which are interpolated using (3.23) to produce the level sets $X_\lambda u$ of a function $u$, which is reconstructed inside $\Omega$ by using the reconstruction formula (3.13). To guarantee

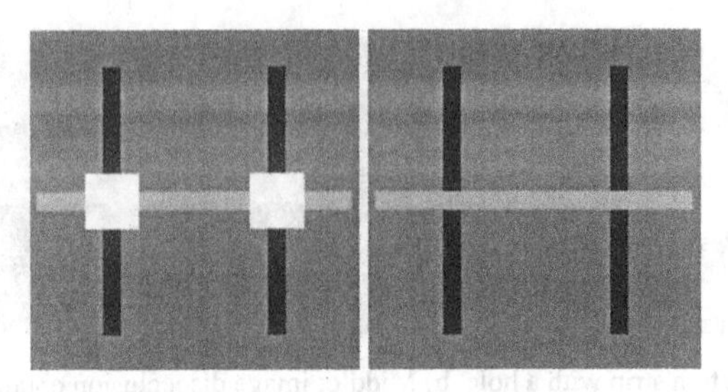

Figure 3.5. a) Left: a double cross with holes. b) Right: reconstructed image using functional (3.23). Observe that due to our choice of upper level sets to decompose and reconstruct the image, the white bar goes above the black ones.

that the reconstructed level sets correspond to the level sets of a function $u$, they should satisfy that $X_{\lambda+1}u \subseteq X_\lambda u$. In practice, we force our solution to satisfy this property.

Functional (3.23) could be used to interpolate functions. But, discontinuities of the image have a contribution to the energy which is proportional to the jump. This gives different weights to discontinuities of different sizes and, as a consequence, they are not treated in the same manner. When taking level sets, we treat all shapes equally, and the parameters of the functional weight geometric quantities (like length, total curvature) and decide which interpolation is taken as a function of them. This approach is less diffusive than directly interpolating the gray levels. A numerical implementation of (3.19) is possible using the scheme in [46].

E) The choice of decomposing the image $u_0$ into upper level sets, interpolating them and reconstructing the function $u$, introduces a lack of symmetry (of upper level sets versus lower level sets). This can be seen in Figure 3.5. Figure 3.5.a displays the image to be interpolated. The choice we made gives Figure 3.5.b as solution, favoring that the object whose level is 210 goes above the object whose level is 0. But, in that case, the "true" information is lacking and we selected one of the possible reasonable solutions.

### 3.3.3 *A Variant and Mathematical Results*

For the purposes of mathematical analysis and comparison with the implementation in [167], we write the boundary conditions in (3.19) in a relaxed way. In particular, the condition $u = u_0$ in $B$ will add the term $\int_B |u - u_0|^q \, dx$ in (3.19). To be able to handle noisy data in $B$ and to include the boundary condition $\theta \cdot \nu^{\Omega}|_{\partial\Omega} = g_0$ in a variational framework, we add the term $\int_{\widetilde{\Omega}} |Du| - \int_{\partial\Omega} g_0 u$.

Before continuing, let us make precise the functional analytic model for $u$ and $\theta$. We assume that $\Omega$ is a domain of class $C^1$. We assume that $u_0 \in BV(D \setminus \overline{\widetilde{\Omega}})$, and $\theta_0 : D \setminus \widetilde{\Omega} \longrightarrow \mathcal{R}^N$ is the vector field of directions of the gradient of $u_0$, i.e., a vector field $\theta_0 \in L^\infty(D \setminus \widetilde{\Omega}, \mathcal{R}^N)$, such that $|\theta_0| \leq 1$ and

$$\operatorname{div} \theta_0 \in L^p(B), \qquad \theta_0 \cdot Du_0 = |Du_0|, \tag{3.22}$$

where the last identity is understood in the sense of measures in $B$ (therefore, a.e.).

Let us denote by $\mathcal{E}_p(\Omega, B, \theta_0)$ the space of couples $(u, \theta)$ where $u \in BV(\Omega)$, $\theta$ is a bounded measurable vector field from $\Omega$ to $R^N$, $|\theta| \leq 1$, div $(\theta) \in L^p(\Omega)$, $\theta \cdot Du = |Du|$, $u|_B \in L^q(B)$, $\theta \cdot \nu^\Omega = g_0$ on $\partial\Omega$.

If $(u, \theta) \in \mathcal{E}_p(\Omega, B, \theta_0)$ we define

$$\begin{aligned} E_p(u,\theta) &= \int_{\tilde{\Omega}} |\mathrm{div}(\theta)|^p(\gamma + \beta|\nabla K * u|)dx \\ &\quad + \alpha \int_{\tilde{\Omega}} |Du| - \alpha \int_{\partial\Omega} g_0 u + \lambda \int_B |u - u_0|^q \, dx \end{aligned} \tag{3.23}$$

where $\gamma, \alpha, \lambda > 0$, $\beta \geq 0$, $p > 1$, $q \geq 1$.

We propose to interpolate the pair $(\theta, u)$ in $\Omega$ by solving the minimization problem

$$\text{Minimize } E_p(u,\theta), \qquad (u,\theta) \in \mathcal{E}_p(\Omega, B, \theta_0) \tag{3.24}$$

**Theorem.** *Assume that* $\sup_{x\in\partial\Omega} |g(x)| < 1$. *If* $p > 1$, $q \geq 1$, $\gamma, \alpha, \lambda > 0$, *and* $\beta \geq 0$, *then there is a minimum* $(u, \theta) \in \mathcal{E}_p(\Omega, B, \theta_0)$ *for the problem (3.24).*

The case $p = 1$ is is particularly interesting, in that case we should consider div $\theta$ to be a Radon measure and we do not know if an existence theorem holds in this case.

The assumption $\|g_0\|_\infty < 1$ does not permit the level lines of the topographic map of the image to be tangent to the boundary of the hole $\tilde{\Omega}$. To ensure it, we may slightly change the topographic map by replacing the level lines which are near to the tangent one by a constant gray level, and this gives us more freedom to choose the vector field $\theta_0$. On the other hand, the assumption $\|g_0\|_\infty < 1$ permits to prove the convergence (after subsequence extraction) of the minima of the functionals

$$\text{Minimize} \int_\Omega \left|\mathrm{div}\left(\frac{Du}{\sqrt{\epsilon^2 + |Du|^2}}\right)\right|^p (\gamma + \beta|\nabla K * u|)dx +$$

$$+ \alpha \int_\Omega |Du| - \alpha \int_{\partial\Omega} g_0 u + \lambda \int_B |u - u_0|^q dx \tag{3.25}$$

$$\frac{Du}{\sqrt{\epsilon^2 + |Du|^2}} \cdot \nu^\Omega = \frac{Du_0}{\sqrt{\epsilon^2 + |Du_0|^2}} \cdot \nu^\Omega.$$

That is, the minimizers of (3.25) converge (modulo a subsequence) to a minimum of (3.24) as $\epsilon \longrightarrow 0+$. For that, we proved in [47] the existence of minimizers for both problems and we studied the two operators div $\left(\frac{Du}{|Du|}\right)$ and div $\left(\frac{Du}{\sqrt{\epsilon^2+|Du|^2}}\right)$ which appear in (3.23) and (3.25), respectively. Notice that the convergence of minima of (3.25) to minima of (3.24) establishes a connection between the numerical approach of T. Chan and J Shen [167] which is based on the direct minimization of (3.25) and ours. Let us also mention that the authors of [167]

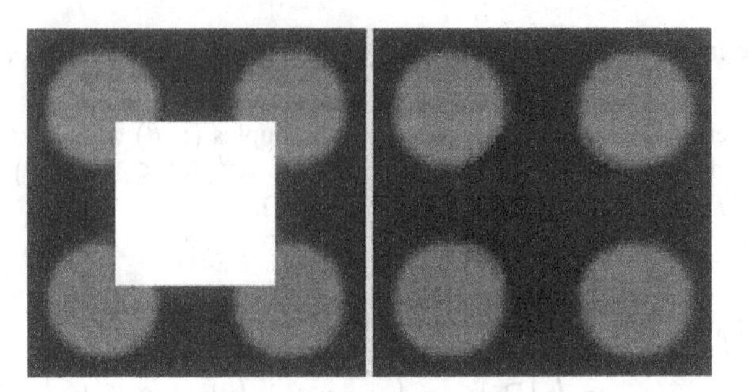

Figure 3.6. Left: Four circles. Right: Reconstructed image.

compared model (3.25) with previous curvature driven diffusion and Total Variation based inpaintings [172, 171]. Their analysis in [172] showed that a curvature term was necessary to have a connectivity principle.

Let us finally mention that a regularity result for the level lines of minimizers of (3.19) or (3.23) has been proved in [47].

### *3.3.4 Experimental Results*

**Examples in 2D.** In the following experiments we show the results of the joint interpolation of gray level and the vector field of directions using functional (3.19). The experiments have been done with $p = 1$ and/or $p = 2$. The results are quite similar and, unless explicitly stated, we display the results obtained with $p = 1$.

Figure 3.6 displays an image made of four circles covered by a square (left image) and the result of the interpolation (right image) obtained with $p = 2$. Figure 3.7.a is a detail of the mouth of Lena with a hole. Figures 3.7.b displays the result of the interpolation using (3.19). Figure 3.7.c shows the result of interpolating the hole of Figure 3.7.a by using a simple algorithm: the value of pixels at distance $k$ from the boundary is the average of its neighboring pixels at distance $k - 1$ from the boundary. In Figure 3.7.b we see the effect of continuing the level lines along the mouth, which is not the case in Figure 3.7.c. Figure 3.8.a is an image of a woman with a flower. In Figure 3.8.b we have represented a hole covering the region of the flower. In Figure 3.8.c we display the result of interpolating the hole of Figure 3.8.b using (3.19).

Figure 3.9.a displays an image with text to be removed. Figure 3.9.b displays the corresponding reconstructed result.

**Examples in 3D.** Let us describe how to use functional (3.19) to inpaint (fill-in) holes (or gaps) on surfaces $\mathcal{S}$, which we assume to be embedded in $\mathcal{R}^3$. To avoid any confusion with our previous use of the word hole, let us use the word gap of the surface. Assume, to fix ideas, that $\mathcal{S}$ is a smooth compact connected surface, and $\mathcal{M}$ is a part of $\mathcal{S}$ which is unknown or could not be obtained during scanning. Let us identify $\mathcal{S}$ with its known part. Let us choose a bounding box $Q$ in $\mathcal{R}^3$

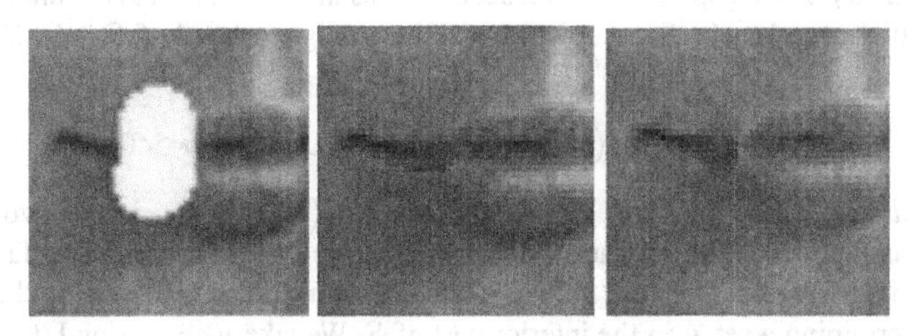

Figure 3.7. a) Left: Detail of the mouth of Lena. b) Middle: Reconstructed mouth using (3.19). c) Right: Result of interpolating the hole in a) by means of a propagation of neighbouring values.

Figure 3.8. a) Left: woman with flower. b) Middle: woman with a mask on the flower representing the hole. c) Right: Result of interpolation using (3.19)

Figure 3.9. Removing the text on an image. a) Left: original image, b) Right: reconstructed image.

strictly containing the gap $\mathcal{M}$ and part of $\mathcal{S}$ (see Figs. 3.10.a, 3.10.b). Let $\partial\mathcal{M}$ be the boundary of the gap (a curve or a set of curves in $\mathcal{R}^3$). Even if $\mathcal{M}$ is unknown, its relative boundary in $\mathcal{S}$ is known. Let $\mathcal{F}$ be a neighborhood of $\mathcal{S} \cap Q$ defined by

$$\mathcal{F} = \{x \in Q : d(x, \mathcal{S} \cap Q) < \alpha d(x, \partial\mathcal{M})\}, \quad \alpha > 0.$$

where $d$ denotes the distance. We assume that $\mathcal{F} \setminus (\mathcal{S} \cap Q)$ consists of two connected components, which can be identified as the two sides of the surface $\mathcal{S}$. With this information, we are able to complete an initial surface closing the gap and determining a set $A$ in the interior part of $\mathcal{S}$. We take $u_0 = \chi_A$ and $\theta_0$ as the outer unit normal vector field to the known part of $\mathcal{S}$ in $Q$ [840].

With the purpose of adapting them to our algorithm, the data, originally given as a triangulated surface, were converted to an implicit representation in a regularly spaced 3D grid. The result was visualized again as a triangulated surface. Figures 3.10.a, 3.10.b display some particular holes with a bounding box isolating them (taken from a scanned version of Michelangelo's David [516]). Figures 3.10.c, 3.10.e display the triangulated surface (the data) around the hole. The reconstructed surface is displayed in Figures 3.10.d, 3.10.f. These images have been rendered using the AMIRA Visualization and Modeling System [24].

The pioneering work [249] addressed the problem of hole filling via isotropic diffusion of volumetric data (that is, iterative Gaussian convolution of some distance function to the known data). The approach proposed by these authors addresses holes with complicated topology, a task very difficult with mesh representations. Most algorithms on reconstructing surfaces from range data are point-cloud reconstruction based and treat holes as regions with low sampling density, thereby interpolating across them [21, 42, 76, 294, 404]. Of course, these algorithms do not distinguish between a real hole in the data and one due to the lack of sampling, and equally fill or fail to fill both cases in the same fashion. Other point-cloud methods evolve a surface over time until it approximates the data [186, 888, 918], or fit a set of 3D radial basis functions to the data, compute a weighted sum of them and use a level set of this last function as reconstructed surface [270, 150]. Mesh based methods for surface reconstruction [819, 240, 886] can perform hole filling as a post-process or integrate hole filling into surface reconstruction [240].

## 3.4 Surface Reconstruction: The Laplace and the Absolute Minimizing Lipschitz Extension Interpolation

In [158] we studied and classified the interpolation algorithms which satisfy a reasonable set of axioms in terms of the solution of a partial differential equation. Two particular examples are: the Absolutely Minimizing Lipschitz Extension,

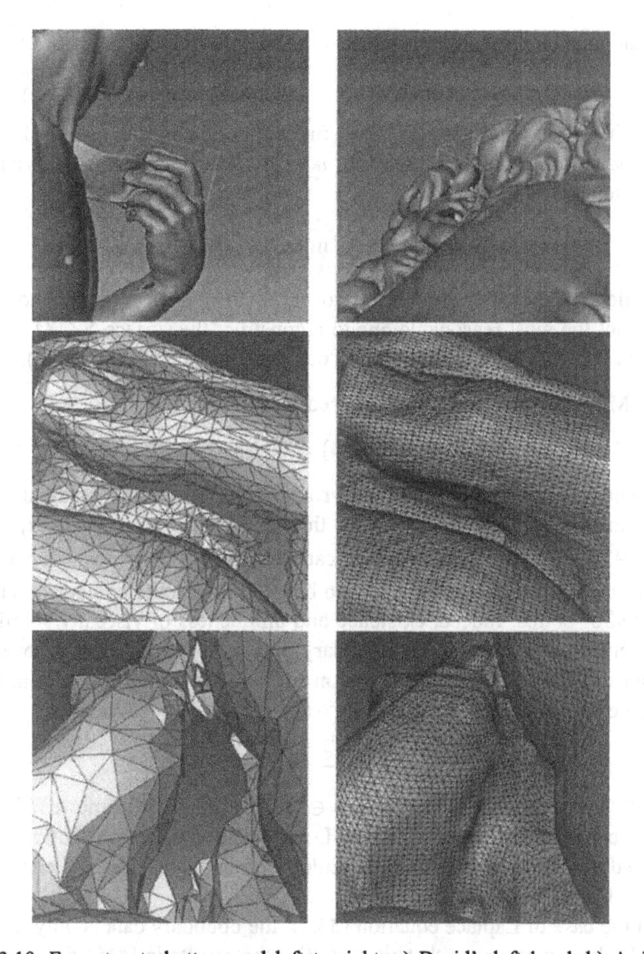

Figure 3.10. From top to bottom and left to right: a) David's left hand. b) A detail of its hair. c) A zoomed detail of a) showing the triangulated surface with the hole. d) The reconstruction of the hole in c) displayed as a triangulated surface. e) A zoomed detail of b) showing the triangulated surface with the hole. f) The reconstruction of the hole in e) displayed as a triangulated surface.

denoted as AMLE in the sequel, and the Laplacian interpolation. We study the applicability of both of them to the problem of surface reconstruction.

We use the notation introduced in section 3.3.4. As we said there, we assume that $\mathcal{F} \setminus (\mathcal{S} \cap Q)$ consists of two connected components, which can be identified as the two sides of the surface $\mathcal{S}$. By changing the sign of the distance function in one of them, we may define the signed distance function to $\mathcal{S} \cap Q$ which we denote by $d_s(x)$. Let us denote $Q_{\mathcal{F}} = Q \setminus \mathcal{F}$.

The Laplacian interpolation is based on solving the PDE

$$-\Delta u = 0 \qquad \text{in } Q_{\mathcal{F}}, \tag{3.26}$$

with specified boundary data on $\partial Q_{\mathcal{F}}$. Indeed, boundary data is only known in $\partial\mathcal{F} \cap Q$ where we should impose that $u = d_s$. Thus, a reasonable assumption would be to assume that

$$\frac{\partial u}{\partial \nu} = 0 \qquad \text{in } \partial Q_{\mathcal{F}} \setminus \partial\mathcal{F} \tag{3.27}$$

where $\nu$ denotes the outer unit normal to $\partial Q_{\mathcal{F}} \setminus \partial\mathcal{F}$. Even if this boundary condition is not the most reasonable one to reconstruct the surface $\mathcal{S} \cap Q$ (which is defined as $\partial[u > 0]$), we have used it in our experiments (see the result).

The AMLE interpolation ([31]) is based on solving the PDE

$$D^2u\,(Du, Du) = 0 \quad \text{in } Q_{\mathcal{F}}. \tag{3.28}$$

with boundary data on $\partial Q_{\mathcal{F}}$ (here $Du$ and $D^2u$ denote the gradient and the Hessian matrix of $u$, respectively, so that in coordinates, $D^2u\,(Du, Du) = \sum_{i,j=1}^{N} \frac{\partial^2 u}{\partial x_i \partial x_j} \frac{\partial u}{\partial x_i} \frac{\partial u}{\partial x_j}$). This equation can be solved with general domains and boundary data, in particular the data can be given in a finite number of surfaces, curves and/or points. Indeed, existence and uniqueness of viscosity solutions of (3.28) were proved in [434] for boundary data $\varphi \in C(\partial Q_{\mathcal{F}})$. Moreover, as it is proved in [434], the viscosity solution of (3.28) is an absolutely minimizing Lipschitz extension of $\varphi$, i.e., $u \in W^{1,\infty}(Q_{\mathcal{F}}) \cap C(\overline{Q_{\mathcal{F}}})$ and satisfies

$$\|Du\|_{L^\infty(Q';\mathcal{R}^N)} \leq \|Dw\|_{L^\infty(Q';\mathcal{R}^N)} \tag{3.29}$$

for all $Q' \subseteq Q_{\mathcal{F}}$ and $w$ such that $u - w \in W_0^{1,\infty}(Q')$. Finally, the $AMLE$ is locally Lipschitz continuous in $Q_{\mathcal{F}}$ [434]. Let us mention that the AMLE model was introduced by Aronsson in [31] as the Euler-Lagrange equation of the variational problem (3.29).

As in the case of Laplace equation (3.26), the boundary data is only known in $\partial\mathcal{F} \cap Q$ where we impose that $u = d_s$ (by the results in [445] there exist absolutely minimizing Lipschitz extensions of $d_s|_{\partial\mathcal{F}\cap Q}$ and satisfy (3.28) but there is no uniqueness result for them). In practice we impose the Neumann boundary condition (3.27) in $\partial Q_{\mathcal{F}} \setminus \partial\mathcal{F}$. We observe again that even if this boundary condition is not the most reasonable one to reconstruct the surface $\mathcal{S} \cap Q$ (which is defined as $\partial[u > 0]$), we have used it in our experiments (see the result).

### 3.4.1 Experimental Results

We display the results obtained using the 3D Laplace and AMLE interpolators on some holes of Michelangelo's David [516]. The result are visualized again as a triangulated surface (using the AMIRA Visualization and Modeling System [24]). Figures 3.11.a, 3.11.b display the original images with holes. Figures 3.11.c, 3.11.d display the result obtained using the Laplace interpolator. Figures

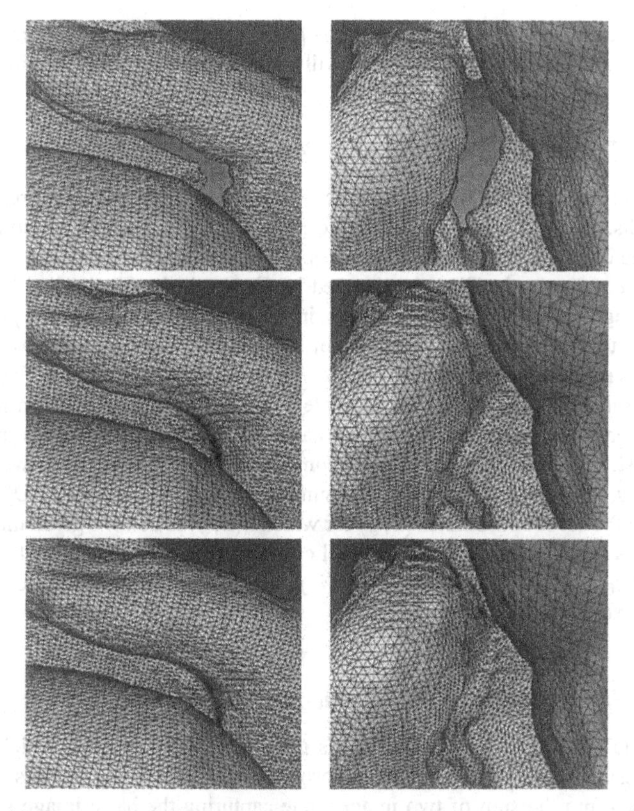

Figure 3.11. From top to bottom and left to right: a) David's left hand with a hole. b) A detail of its hair with a hole. c) and d) The results obtained with Laplace interpolator. e) and f) Results obtained using AMLE interpolator.

3.11.e, 3.11.f display the result obtained with the AMLE. Observe that the result obtained with AMLE interpolation is less regular.

## 3.5 Dealing with texture

All the PDE-based approaches to inpainting share the shortcoming that they cannot restore texture. The notion of texture implies a repetitive pattern, a missing portion of which may usually not be restored just by propagating the level lines into the gap in any clever way. On the other hand, there are a number of very good texture synthesis algorithms, which in turn do not give as good results when applied to gaps in 'structured' (as opposed to 'textured') regions. In this section we will comment on two methods to perform inpainting on images with textured and/or structured regions. Both methods use the remarkable algorithm introduced by Efros and Leung for texture synthesis [297], which gives excellent results for the inpainting problem as well, so we will start by discussing this algorithm.

### 3.5.1 *Texture Synthesis by Non-Parametric Sampling*

This algorithm [297] is fully automatic and produces very good texture synthesis results. It is also very well suited to natural images when the regions to be inpainted cover a large variety of textures.

Let the region to be filled be denoted by $\Omega$. $\Omega$ will be filled, pixel by pixel, proceeding from the border $\partial\Omega$ inwards, in an 'onion-peel' fashion. Let $p(i, j)$ be the pixel to fill-in next. We consider a $n \times n$ neighborhood of this pixel, call it $N_{ij}$. This neighborhood will typically contain several empty pixels. With only the filled pixels of $N_{ij}$, we build the template $T_{ij}$. Next we compare $T_{ij}$ with all the posible templates $T_{xy}$, centered at $(x, y)$ and shaped like $T_{ij}$, that are completely outside $\Omega$. This comparison is done by computing a distance $d(x, y)$ between both templates, which uses the normalized sum of squared differences (SSD) metric. We keep the set of coordinates $(x, y)$ for which $d(x, y)$ is below a given threshold. From this set, we randomly pick a pixel coordinate $(x_0, y_0)$, and copy the image value $I(x_0, y_0)$ to $I(i, j)$. Then, pixel $(i, j)$ is filled and we procceed to the next empty pixel at the boundary.

### 3.5.2 *Inpainting with Image Decomposition*

The basic idea of this algorithm [80] is presented in Figure 3.12, which shows a real result from this approach. The original image (first row, left) is first decomposed into the sum of two images, one capturing the basic image structure and one capturing the texture (and random noise), second row. This follows the work by Vese and Osher reported in [842]. The first image is inpainted following any of the PDE-based approaches described before, while the second one is filled-in with a texture synthesis algorithm, third row. The two reconstructed images are then added back together to obtain the reconstruction of the original data, first row, right. In other words, the general idea is to perform structure inpainting and texture synthesis not on the original image, but on a set of images with very different characteristics that are obtained from decomposing the given data. The decomposition is such that it produces images suited for these two reconstruction algorithms. This approach outperforms both image inpainting and texture synthesis when applied separately. Indeed, a separate reconstruction of missing blocks in wireless JPEG transmission was proposed in [669].

As for the decomposition step, the authors in [842], inspired by [567], propose a model to express any given image $I$ as the sum of two images $u$ and $v$, where $u$ will be a sketchy or cartoon image of $I$ (with sharp edges) and $v$ will be the the remainder (a term with noise, oscillations, texture.)Expressing then $I(x, y) = u(x, y) + v(x, y)$ and $v(x, y) = \nabla \cdot (v_1, v_2)$, the authors in [842] propose a minimization problem to find $u, v_1, v_2$, whose Euler-Lagrange equations

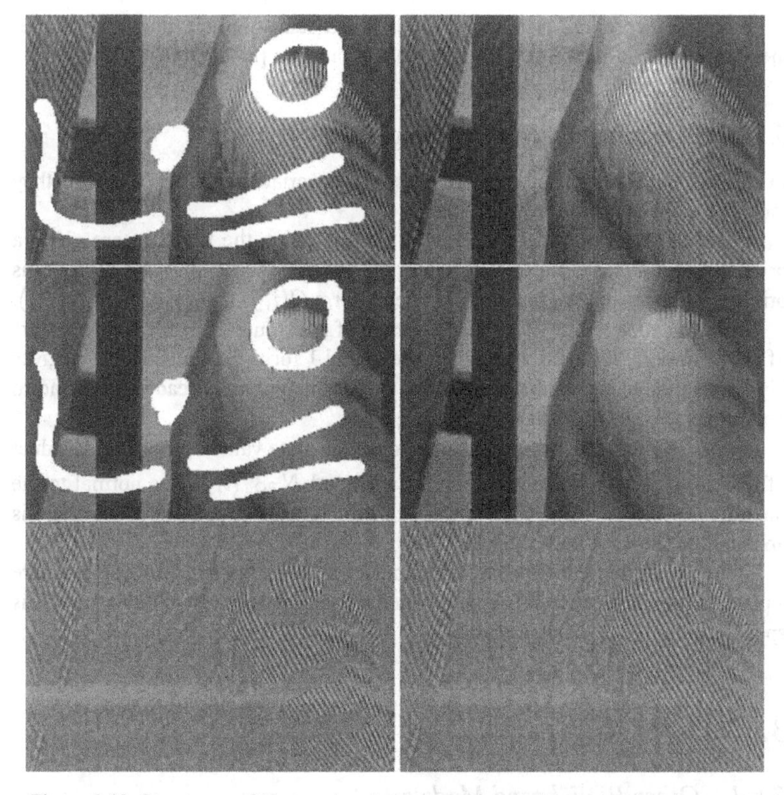

Figure 3.12. Structure and Texture inpainting using image decomposition (see text.)

are

$$
\begin{aligned}
u &= I - \partial_x g_1 - \partial_y g_2 + \frac{1}{2\lambda}\operatorname{div}\left(\frac{\nabla u}{|\nabla u|}\right), && (3.30)\\
\mu \frac{g_1}{\sqrt{g_1^2+g_2^2}} &= 2\lambda\left[\frac{\partial}{\partial x}(u-I) + \partial_{xx}^2 g_1 + \partial_{xy}^2 g_2\right], && (3.31)\\
\mu \frac{g_2}{\sqrt{g_1^2+g_2^2}} &= 2\lambda\left[\frac{\partial}{\partial y}(u-I) + \partial_{xy}^2 g_1 + \partial_{yy}^2 g_2\right]. && (3.32)
\end{aligned}
$$

For some theoretical results and the detailed semi-implicit numerical implementation of the above Euler-Lagrange equations, see [842].

### 3.5.3 *Exemplar-based Inpainting*

In this work [238], Criminisi et al. propose a variation on [297], where they modify the fill order of the algorithm.

Instead of the 'onion-peel' of [297], patches along the fill front are given a priority value $P(i,j)$, which determines the order in which they are filled. This priority $P(i,j)$ is the product of a confidence term $C(i,j)$ and a data term $D(i,j)$.

The confidence term $C(i,j)$ is an average of the values of $C$ for the neighbors of $(i,j)$; initially, $C$ is 0 for pixels inside $\Omega$ and 1 for pixels outside. So $C$ gives higher priority to pixels that have more of their neighbors already filled, and to pixels that are closer to $\partial\Omega$.

The data term $D(i,j)$ is proportional to the absolute value of the scalar product of $\nabla^{\perp}I(i,j)$, the isophote direction at $(i,j)$, and $\overrightarrow{N}_{\partial\Omega_n}(i,j)$, the normal to the boundary of the fill front. So $D$ gives higher priorities to patches where there is an isophote 'flowing into' the gap.

Finally, $\Omega$ is filled not one pixel at a time as in [297], but patch by patch, where a patch is the intersection of a $n \times n$ window (typically $n = 9$) with the gap. This speeds up the process considerably.

## 3.6 Other Approaches

### 3.6.1 *Other PDE-based Models*

Other PDE based models have been proposed by Chan and Shen [171, 172, 167]. In [172] the authors proposed an anisotropic diffusion model (called (CCD)) with curvature dependent diffusion coefficient. In [171] they compared several models, namely, TV based inpainting, segmentation-based inpaintings, and the (CCD) model. Finally, in [167], the authors proposed to minimize the Elastica model written as in (3.21) leading to a fourth order PDE gradient descent equation. The connection of this model with model (3.23) has been mentioned in Section 3.3.3. Esedoglu and Shen proposed in [300] an inpainting functional based on Mumford-Sha's functional plus some terms which approximate the Elastica.

Inspired by the real Ginzburg-Landau equation which develops homogeneous areas separated by phase transition regions (that are interfaces of minimal area), H. Grossauer and O. Scherzer proposed to use the complex Ginzburg-Landau equation for inpainting [369]. As we did above, we denote by $\Omega$ the hole to be inpainted and we suppose that the given image $u_0 : D \longrightarrow R$ has been extended in rough way to $\Omega$. Normalizing $u_0$ to take values in $[-1,1]$, the authors defined $v_0 = \sqrt{1-|u_0|^2}$, and $\mathbf{u_0} = (u_0, v_0)$. Then the authors solve the equation (which corresponds to the gradient descent method applied to the Ginzburg-Landau functional)

$$\frac{\partial \mathbf{u}}{\partial t} = \Delta \mathbf{u} + \frac{1}{\epsilon^2}(1 - |\mathbf{u}|^2)\mathbf{u} \quad \text{in } \Omega, \tag{3.33}$$

with initial condition $\mathbf{u}(0) = \mathbf{u_0}$ and boundary condition

$$\mathbf{u}|_{\partial\Omega} = \mathbf{u_0}|_{\partial\Omega}.$$

As an interesting feature of (3.33) let us mention that the solution corresponding to the image in Figure 3.5 would be the symmetric one: half gray and half black forming an X in the hole.

Inpainting models based on probability diffusion of orientations are proposed in [844]. Indeed, the authors define the function $P(x,\theta)$ as the probability that there is a level line passing through $x$ with direction $\theta$ and propose to compute $P(x,\theta)$ as the asymptotic state of the PDE

$$P_t + P(\cos\theta, \sin\theta) \cdot \nabla_x P = \alpha P_{\theta\theta} + \beta \Delta_x P \quad \text{in } \Omega,$$

where $P(x,\theta)(\cos\theta,\sin\theta)$ represents a probability distribution for the tangent direction. This equation also includes an spatial diffusion of the probability $P(x,\theta)$. Knowing $P(x,\theta)$, the authors define the orthogonal orientation of the level line through $x$ as the expectation of $P(x,\theta)$, i.e., as the vector $z(x) := \int_0^{2\pi}(-\sin\theta, \cos\theta)P(x,\theta)\,d\theta$. Then the authors reconstruct the image inside $\Omega$ using $z(x)$ and the value of the image on $\partial\Omega$ [844].

A related model has been used in [892, 893] for the completion of illusory contours. The connection between both models is given by the completion of level lines as if they were illusory contours. The model in [892, 893] was inspired by the work of [589] who interpreted the elastica as the *mode* of the probability distribution underlying the stochastic process given by the differential equations $\dot{x} = (\cos\theta, \sin\theta)$, $\theta$ being a normally distributed random variable with zero mean and given variance.

Let us finally mention that a finite element implementation of the Willmore functional $\int_{\mathcal{M}} \mathbf{H}^2\, dS$ has been used in [202] for surface restoration. As explained in Section 3.3.2, this functional (in a relaxed form) is a term in functional (3.19).

### 3.6.2 *Miscellaneous*

Finally, let us briefly mention some other approaches to the inpainting problem.

Jia and Tang [437] perform a texture-based segmentation of the image. Then they find the curves in $\Omega$ that connect texture boundaries arriving at $\partial\Omega$: these curves, boundaries between different texture regions, are found with a robust tensor voting algorithm that extrapolates curve shape. Then texture is synthesised inside each region, also with a tensor voting algorithm, where texture at pixel $(i, j)$ is encoded as a vector of length $N = n \times n + 1$ whose components are the image intensity values at the $n \times n$ neighborhood of $(i, j)$.

Levin et al. [515] use global information to guide the inpainting process. They choose features like the norm of the gradient, compute the histogram of these features over the whole image, define a probability taking these histograms into account, and find an integrable gradient field inside $\Omega$ that maximizes that probability and satisfies the boundary conditions at $\partial\Omega$.

Kim and Kim [466] use genetic algorithms to approximate the solution to the problem of minimizing the elastica inside $\Omega$, given the image and curvature values at $\partial\Omega$.

Tan et al. [786] perform highlight removal with a proposed variant of inpainting where the region to fill-in $\Omega$ is not empty but has some useful information, from which the highlights must be substracted.

Patwardhan and Sapiro [634] use wavelets in a Projection Onto Convex Sets (POCS) setting similar to Hirani and Totsuka's [400], but without the need for user-selection of similar neighborhoods. It is an iterative process where in each step the image is wavelet-transformed, its coefficients constrained, then wavelet-inverse-transformed, the resulting image values also constrained.

## 3.7 Concluding Remarks

In this chapter we have reviewed the area of image inpainting, which has received a significant amount of attention from the image processing, computer vision, computer graphics, and applied mathematics communities; following the early works of Masnou-Morel [553] and Bertalmío-Sapiro-Caselles-Ballester [79, 46]. We can not forget of course also one of the first works in the area, [605], where the famous Laplacian Pyramid is used to fill-in holes.

Although image inpainting still has many open problems, the main challenges are in the extension of this work to other visual sources, such as video [438, 885] and sensor arrays [905]. Preliminary and very promising results are starting to appear in this subject, and many important advances are expected in forthcoming years.

## 3.8 Appendix

Let $Q$ be an open subset of $R^N$. By $C_0^\infty(Q)$ (resp. $C_0^\infty(Q; R^N)$) we denote the space of functions (resp., vector fields with values in $R^N$) with are $C^\infty$ and have compact support in $Q$. By $L^p(Q), 1 \leq p < \infty$, we denote the space of measurable functions $f : Q \longrightarrow R$ whose $p$-power is integrable (in the sense of Lebesgue). $L^\infty(Q)$ denotes the space of measurable functions in $Q$ which are essentially bounded. By $W^{1,p}(Q)$, $1 \leq p \leq \infty$, we denote the space of functions $u \in L^p(Q)$ such that $\nabla u \in L^p(Q)$. By $W_0^{1,p}(Q)$ we denote the closure of $C_0^\infty(Q)$ in $W^{1,p}(Q)$. Saying that $u \in W_0^{1,p}(Q)$ is a way of saying that $W^{1,p}(Q)$ and $u = 0$ on the boundary of $Q$. By $C(\overline{Q})$ we denote the space of continuous functions in $\overline{Q}$.

A function $u \in L^1(Q)$ whose gradient $Du$ in the sense of distributions is a (vector valued) Radon measure with finite total variation in $Q$ is called a function of bounded variation. The class of such functions will be denoted by $BV(Q)$. The total variation of $Du$ on $Q$ turns out to be

$$\sup\left\{\int_Q u \operatorname{div} z \, dx : z \in C_0^\infty(Q; R^N), \sup_{x \in Q} |z(x)| \leq 1\right\}, \tag{3.34}$$

(where for a vector $v = (v_1, \ldots, v_N) \in R^N$ we set $|v|^2 := \sum_{i=1}^N v_i^2$) and will be denoted by $|Du|(Q)$ or by $\int_Q |Du|$. The total variation of $u$ on a Borel set $B \subseteq Q$ is defined as $\inf\{|Du|(A) : A \text{ open}, B \subseteq A \subseteq Q\}$.

A measurable set $E \subseteq R^N$ is said to be of finite perimeter in $Q$ if (3.34) is finite when $u$ is substituted with the characteristic function $\chi_E$ of $E$. The perimeter of $E$ in $Q$ is defined as $P(E, Q) := |D\chi_E|(Q)$. We shall use the notation $P(E) := P(E, R^N)$. For sets of finite perimeter $E$ one can define the essential boundary $\partial^* E$, which is countably $(N-1)$ rectifiable with finite $\mathcal{H}^{N-1}$ measure, and compute the outer unit normal $\nu^E(x)$ at $\mathcal{H}^{N-1}$ almost all points $x$ of $\partial^* E$, where $\mathcal{H}^{N-1}$ is the $(N-1)$ dimensional Hausdorff measure. Moreover, $|D\chi_E|$ coincides with the restriction of $\mathcal{H}^{N-1}$ to $\partial^* E$.

If $u \in BV(Q)$ almost all its level sets $[u \geq \lambda] = \{x \in Q : u(x) \geq \lambda\}$ are sets of finite perimeter. Thus at almost all points of almost all level sets of $u \in BV(Q)$ we may define a normal vector $\theta(x)$ which coincides $|Du|$-a.e. with the Radon-Nikodym derivative of the measure $Du$ with respect to $|Du|$, hence it formally satisfies $\theta \cdot Du = |Du|$, and also $|\theta| \leq 1$ a.e. (see [19], 3.9). For further information concerning functions of bounded variation we refer to [19, 301, 926].

## 3.9 Acknowledgments

We thank Marc Levoy and the Stanford Michelangelo Project for data provided for this work. We also thank our collaborators in the work on image inpainting. The first three authors acknowledge partial support by the Departament d'Universitats, Recerca i Societat de la Informació de la Generalitat de Catalunya and by PNPGC project, reference BFM2003-02125. The first author acknowledges partial support by Programa Ramón y Cajal. The fourth author is partially supported by the Office of Naval Research, the National Science Foundation, the National Geospatial-Intelligence Agency, and the McKnight Foundation.

# Part II

# Boundary Extraction, Segmentation and Grouping

# Chapter4

# Levelings: Theory and Practice

F. Meyer

Abstract

Connected operators enlarge the flat zones of an image and never create a contour where no contour was present. This definition is too vague to be useful in practice, except for binary images. For grey-tone images a more precise characterization has to be given in order to be operational. This leads to the introduction of floodings, razings, flattenings and levelings. Extending the notion of a flat zone and of a contour leads to extended connected operators. The chapter concludes by showing the versatility and power of these operators in practice.

## 4.1 Introduction

Filtering is ubiquitous in image processing before compression or segmentation, for suppressing noise or simplifying images. An ideal filter should suppress noise and unwanted details without degrading in any other respect the image. For instance it should not blur or displace the contours if one wishes to segment the filtered image. It should not create spurious structures such as minima or maxima if the aim is to describe the topography of a relief or to construct its watershed line. Each element in the filtered image should be traceable in the initial image.

It seems difficult to design a filter complying with all these constraints. Linear filters produce a blurring of the image. The problem is to find a good trade-off between smoothing and localization of the contours: a large smoothing simplifies the detection but creates poorly localized contours whereas a reduced smoothing does not suppress enough noise. Non linear smoothing techniques [642] avoid smoothing across object boundaries. However, depending on a number of parameters, they are difficult to tune. Alternate sequential filters based on openings and closings also displace the contours [724].

Connected operators do not suffer from this drawback, they enlarge the existing flat zones and produce new ones [726]. They are specially designed for simplifying images without blurring or displacing contours. The simplest ones suppress particles or holes in binary images [725]. Clipping peaks and filling valleys until a plateau of a given size is produced constitutes the area openings and closings introduced by Luc Vincent [845]. Particle reconstruction allows to suppress all connected particles not containing a marker. Applied on each threshold of a grey tone image, one obtains reconstruction openings and closings. [367, 846]. They are both members of a larger family, operating symmetrically on peaks and valleys, which comprises flattenings and razings [564],[555]. Their scale space properties and PDE formulation are studied in [566].

The present paper gives an insight in the nature and construction of these operators and illustrates their use. As we are concerned with practical applications we will restrict ourselves to a digital framework. Let $\mathcal{T}$ be some complete totally ordered lattice, and let $\mathcal{D}$,$\mathcal{E}$ be arbitrary sets in the discrete space. We call $O$ the smallest element and $\Omega$ the largest element of $\mathcal{T}$. $\text{Fun}(\mathcal{D},\mathcal{T})$ represents the image defined on the support $\mathcal{D}$ with value in $\mathcal{T}$. The value of function $f$ at pixel $p$ will be written $f_p$. A presentation of levelings and flattenings in the continuous space may be found in [555], [565].

## 4.2 Binary connected operators

The functions $f$, $g$, $h$ met in this section are binary and are the indicator functions of binary sets, being equal to 1 in the particles and to 0 in the holes. A binary connected operator suppresses particles and/or fills holes:

**Definition 1.** *A connected operator transforms an image $f$ into an image $g$ in such a way that the following relation is verified for all pairs of neighboring pixels : $\forall\,(p,q)$ neighbors: $f_p = f_q \Rightarrow g_p = g_q$ or equivalently $g_p \neq g_q \Rightarrow f_p \neq f_q$ (1).*

The relation (1) expresses that any contour between the pixels $p$ and $q$ in the destination image $g$ corresponds to a contour in the initial image $f$ at the same place. There is however no coupling between the directions of the transitions: between $p$ and $q$, the function $g$ may for instance be increasing and $f$ decreasing. Relation (1) may be rewritten as $g_p > g_q \Rightarrow f_p > f_q$ or $f_p < f_q$(1bis). As an example, the complementation of a binary image is a connected operator. This shows that a connected operator may turn a regional minimum into a maximum and vice-versa. If a function $g$ and a function $f$ verify relation (1) for all pairs of pixels, we say that by definition $g$ is a planing of $f$.

Planings may be specialized in 3 ways :

- A planing verifying $g \geq f$ only suppresses holes and is called flooding. It is characterized by $g \geq f$ and $\forall\,(p,q)$ neighbors: $g_p > g_q \Rightarrow f_p = g_p (= 1)$ (2)
- Planings which only suppress particles are called razings and verify $g \leq f$. They are characterized by: $g \leq f$ and $\forall\,(p,q)$ neighbors: $g_p > g_q \Rightarrow f_q = g_q (= 0)$ (3)

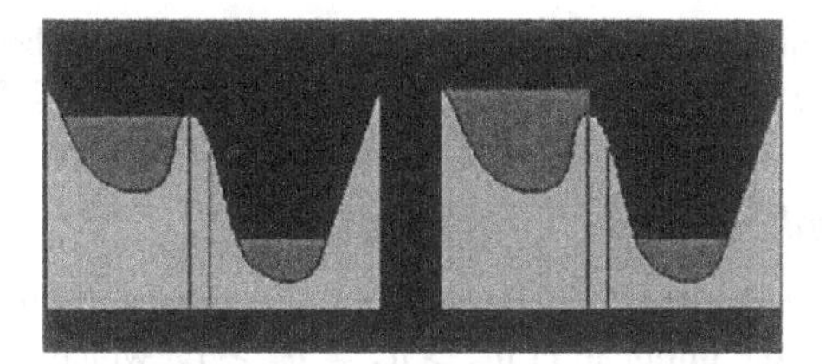

Figure 4.1. $g_q < g_p \Rightarrow g_p = f_p$

- Monotone planings are called levelings. They may suppress both particles and holes but if a hole and a particle are adjacent, the hole cannot become a particle and simultaneously the particle a hole. Levelings introduce a coupling between the directions of the transitions: between $p$ and $q$ : $g_p > g_q \Rightarrow f_p > f_q$ (4).

When applied to each threshold of a grey-tone function, these binary operators generate interesting grey-tone operators.

## 4.3 Flat grey-tone connected operators

### *4.3.1 Level by level construction*

The definitions of planings and monotone planings given in the preceding section still make sense if $g$ and $h$ are grey-tone functions. Relations (2) and (3) fully specify floodings and razings for grey-tone functions. Relation (2) has an obvious physical meaning. Fig.4.1A and Fig.4.1B represent respectively a possible and an impossible flooding $g$ of a relief $f$: if for two comparable pixels a lake verifies $g_q < g_p$, then the highest pixel is necessarily at ground level ($g_p = f_p$), otherwise the lake presents an unconstrained wall of water as in fig.4.1B.

On the contrary the relations (1) and (4) indicate that to any contour of $g$ corresponds a contour of $f$ at the same location, but do not establish a relation between the values of the functions themselves. However, applying the corresponding binary operators on each threshold of a grey-tone function produces a well constrained operator: a function $g$ is a flattening (resp. leveling) of a function $f$ if and only if for each $t$, $X^t\,(g)$ is a planing (resp. monotone planing) of $X^t\,(f)$ (where $X^t\,(f) = \{x \mid f(x) \leq t\}$). We derive the following criteria:

- An image $g$ is a flattening of the image $f$ iff $\forall\,(p,q)$ neighbors:

$$g_p > g_q \Rightarrow \begin{bmatrix} f_p \geq g_p \text{ and } g_q \geq f_q \\ \text{or} \\ f_q \geq g_p \text{ and } g_q \geq f_p \end{bmatrix} \quad (5)$$

* An image $g$ is a leveling of the image $f$ iff $\forall\,(p,q)$ neighbors:
$g_p > g_q \Rightarrow f_p \geq g_p$ and $g_q \geq f_q$ (6).

Basically relation (5) means that any transition in the destination image $g$ is bracketed by a larger variation in the source image. If furthermore the direction of the transitions is always the same as in relation (6), flattenings become levelings. Flattenings are floodings if they verify $g \geq f$ and razings if $g \leq f$.

### 4.3.2 A morphological characterization

Interesting characterizations may be derived from the relations (6) and (7). As an example consider the implication $[g_p > g_q \Rightarrow g_q \geq f_q]$ which is part of relation (6). Recalling that the logical meaning of $[A \Rightarrow B]$ is $[notA \text{ or } B]$ it may interpreted as $[g_p \leq g_q \text{ or } g_q \geq f_q] \Leftrightarrow [g_q \geq f_q \wedge g_p]$. As $p$ may be any element of the neighborhood $N_q$ of the central point $q$, we obtain $g_q \geq f_q \wedge \bigvee_{x \in N_q} g_x$ equivalent to $g_q \geq f_q \wedge \left( g_q \vee \bigvee_{x \in N_q} g_x \right) = f_q \wedge \delta g_q$, where $\delta$ represents the elementary morphological dilation with a flat structuring element containing the central point and all its neighbors. Taking into account the complete relation (6) yields the following criterion for levelings: $f \wedge \delta g \leq g \leq f \vee \varepsilon g$.

Since $g \leq \delta g$ and $\varepsilon g \leq g$, the preceding criterion is equivalent with $(f \wedge \delta g) \vee \varepsilon g \leq g \leq (f \vee \varepsilon g) \wedge \delta g$. But $(f \wedge \delta g) \vee \varepsilon g = (f \vee \varepsilon g) \wedge \delta g$, giving another criterion for levelings: $g = (f \wedge \delta g) \vee \varepsilon g = (f \vee \varepsilon g) \wedge \delta g$, known as the morphological centre [724] between $\delta g$ and $\varepsilon g$.

The criterion characterizing flattenings, floodings and razings may be established in a similar way:

* A function $g$ is a flattening of $f$ if and only if : $f \wedge \delta(f \wedge g) \leq g \leq f \vee \varepsilon(f \vee g)$ (9)
* A function $g$ is a flooding of $f$ if and only if : $g = f \vee \varepsilon g$
* A function $g$ is a razing of $f$ if and only if : $g = f \wedge \delta g$

In the next stage of generalization, the operators no longer commute with anamorphosis, as it is the case for operators constructed threshold by threshold.

## 4.4 Extended connected operators

Replacing $(\delta, \varepsilon)$ by a more general adjunction $(\alpha, \beta)$, where $\beta$ is an arbitrary erosion verifying $\beta \leq Id$ and $\alpha \geq Id$ its adjunct dilation, we get a generalized leveling $g = (f \vee \beta g) \wedge \alpha g = (f \wedge \alpha g) \vee \beta g$. For which type of flat zones is it a connected operator ?

We have the equivalence $g = (f \vee \beta g) \wedge \alpha g \Leftrightarrow f \wedge \alpha g \leq g \leq f \vee \beta g$.

A pixel $p$ verifying $g_p \leq (f \vee \beta g)_p$ also verifies the following equivalent expressions: $\left\{ g_p \leq (\beta g)_p \text{ or } g_p \leq f_p \right\} \Leftrightarrow \left\{ g_p > (\beta g)_p \Rightarrow g_p \leq f_p \right\}$ (10)

The relation $g_p > (\beta g)_p$ means that eroding the function $g$ with the erosion $\beta$ decreases the value of $g$ at pixel $p$, indicating that $p$ has a lower neighbor for the function $g$. In order to find this neighbor, we have to introduce the pulse functions $\uparrow_h^t = \left\{ \begin{array}{l} t \text{ if } x = h \\ O \text{ if } x \neq h \end{array} \right\}$ and $\downarrow_h^t (x) = \left\{ \begin{array}{l} t \text{ if } x = h \\ \Omega \text{ if } x \neq h \end{array} \right\}$. Every image $g$ of Fun($\mathcal{D}$,$\mathcal{T}$) can be written $g = \bigvee_{x\in\mathcal{D}} \uparrow_x^{g_x} = \bigwedge_{x\in\mathcal{D}} \downarrow_x^{g_x}$ and $(\beta g)_p = \left[\beta\left(\bigwedge_{x\in\mathcal{D}} \downarrow_x^{g_x}\right)\right]_p = \bigwedge_{x\in\mathcal{D}} \left[\beta\left(\downarrow_x^{g_x}\right)\right]_p$. The minimal value in this expression is attained at a pixel $x = q$. This pixel $q$ is the lower "neighbor" of $p$ we are looking for, and we write $g_q \sqsubset g_p \Leftrightarrow g_p > \beta_{qp}(g_q)$, where $\beta_{qp}(g_q) = \left[\beta\left(\downarrow_q^{g_q}\right)\right]_p$ is an erosion. Since $g_p < \Omega$, the relation $g_p > \beta_{qp}(g_q)$ also indicates that $\beta_{qp}(g_q) < \Omega$. When this is the case, we consider that $p$ and $q$ are $\alpha\beta-$neighbors for the adjunction $(\alpha, \beta)$. $\beta_{qp}$ has an adjunct dilation $\alpha_{p,q}(g_p) = \left[\alpha\left(\uparrow_p^{g_p}\right)\right]_q$, verifying : $g_p > \beta_{q,p}(g_q) \Leftrightarrow \alpha_{p,q}(g_p) > g_q$. Finally relation (10) may be rewritten for any $\alpha\beta-$neighbors $p$ and $q$: $g_q \sqsubset g_p \Rightarrow g_p \leq f_p$.

The inequality $f \wedge \alpha g \leq g$ may be treated in the same manner and putting everything together, we obtain the characterization of levelings: $g$ is a leveling of $f$ if $\forall(p,q)$ $\alpha\beta$-neighbors $g_q \sqsubset g_p \Rightarrow f_p \geq g_p$ and $g_q \geq f_q$, quite similar to relation (6).

As a summary we have found a general mechanism for defining transitions between pixels for a given function, based on an adjunction $(\alpha, \beta)$. Definitions and characterizations of extended flattenings, floodings, razings and levelings are obtained by simply replacing $(\delta, \varepsilon)$ by $(\alpha, \beta)$ and the relation $<$ by the relation $\sqsubset$ in all relations and definitions of the previous sections.

Negating the relation $\sqsubset$ yields the relation $\sqsupseteq$: for $(p,q)$ $\alpha\beta$-neighbors, $g_q \sqsupseteq g_p$ if only if $g_p \leq \beta_{q,p}(g_q) \Leftrightarrow \alpha_{p,q}(g_p) \leq g_q$. When the relations $\{f_y \sqsupseteq f_x\}$ and $\{f_x \sqsupseteq f_y\}$ are simultaneously true, we obtain a symmetrical relation written $f_x \diamond f_y$, expressing that there is a smooth transition between $f_y$ and $f_x$ or that $f_x$ and $f_y$ are at the same $\alpha\beta$-level : $\{f_x \diamond f_y\} \Leftrightarrow \{O < \alpha_{x,y}(f_x) \leq f_y \leq \beta_{x,y}(f_x) < \Omega\}$ $\Leftrightarrow \{O < \alpha_{y,x}(f_y) \leq f_x \leq \beta_{y,x}(f_y) < \Omega\}$.

We are now able to define smooth zones based on arcwise connectivity.

**Definition 2.** *We say that two values $f_x$ and $f_y$ are smoothly linked and we write $f_x \bowtie f_y$ if there exists a series of pixels $\{x_0 = x, x_1, x_2, ...x_n = y\}$ such that $f_{x_i} \diamond f_{x_{i+1}}$.*

**Definition 3.** *A set $X$ is a smooth zone of an image $f$ if and only if $f_x \bowtie f_y$ for any two pixels $x$ and $y$ in $X$.*

The relation $\bowtie$ is an equivalence relation. The associated equivalence classes are the maximal smooth zones. It is easy to verify that the smooth zones of $f$ form a connection of $\mathcal{D}$ [725]. For the pair of elementary dilation and erosion $(\delta, \varepsilon)$, one obtains ordinary flat zones.

**Definition 4.** *A set $X$ is uniformly smooth if $f_x \diamond f_y$ for any couple $(x,y)$ of $\alpha\beta$-neighbors in $X$.*

For the slope dilation $\delta_1(g) = g \vee (\delta g - 1)$, $X = \begin{matrix} 3 & 2 \\ 4 & 1 \end{matrix}$ is a smooth zone since there exists a path with a slope smaller or equal to 1 between any couple of pixels. However, there exists within $X$ a sharp transition between values 1 and 4, hence $X$ is not a uniformly smooth zone.

**Levelings enlarge smooth zones:** $g_q \sqsubset g_p \Rightarrow f_q \sqsubset f_p$ is equivalent with $f_q \sqsupset f_p \Rightarrow g_q \sqsupset g_p$ from which we derive $f_q \bowtie f_p \Rightarrow g_q \bowtie g_p$; this last relation shows that any smooth (resp. uniformly smooth) zone for $f$ is also a smooth zone (resp. uniformly smooth) for $g$. Levelings are indeed connected operators [700].

**Levelings create smooth zones:** Any zone where $\{g > f\}$ (resp. $\{g < f\}$) is uniformly smooth.

**Regional minima**

If $(\alpha, \beta)$ are flat operators, then the leveling based on $(\alpha, \beta)$ does not create regional minima or maxima. More precisely if $g$ is a leveling of $f$, and $X$ a regional minimum of $g$, then there exists a set $Z \subset X$, which is a regional minimum for $f$. However, this is not true if $(\alpha, \beta)$ are not flat operators.

### *4.4.1 Construction of floodings, razings, flattenings and levelings*

We call Inter $(g, f)$ the class of functions $h \in T^E$, verifying $g \wedge f \leq h \leq g \vee f$. We say that $g$ is farther away from $f$ than $h$, or that $g$ is bigger than $h$ *in the order* $f$ and we write $g >_f h$ if and only if $h \in$ Inter $(g, f)$ [555].

**Proposition 1.** *$>_f$ is an order relation on $T^E$. For $a, f \subset T^E$,* Inter $(a, f)$ *is a complete lattice for the order $f$. The function $a$ is then the highest element. For any family $h_i$ of* Inter $(a, f)$*:*

$$\bigwedge_f h_i = \left| \begin{array}{l} \vee h_i \text{ on } \{a \leq f\} \\ \wedge h_i \text{ on } \{a \geq f\} \end{array} \right| ; \bigvee_f h_i = \left| \begin{array}{l} \wedge h_i \text{ on } \{a \leq f\} \\ \vee h_i \text{ on } \{a \geq f\} \end{array} \right|$$

Considering a pair of functions $f$ and $h$ we will now study the family of floodings, razings and levelings of $f$ within Inter $(f, h)$.

#### 4.4.1.1 Construction of floodings, razings, flattenings and levelings

Each flooding of $f$ verifies $g \geq f$. For this reason the order relations $>_f$ and $>$ are identical. If $(g_i)$ is a family of floodings of $f$, then $\bigvee g_i$ also is a flooding of $f$. The family of floodings of $f$ belonging to Inter $(f, h)$ is not empty and its maximal element is written Fl$(f, h)$. It is obtained by finite iteration until stability of $h_n = f \vee \beta h_{n-1}$, with $h_0 = f \vee h$. We recognize the usual reconstruction closing if $\beta = \varepsilon$ [846].

Similarly the largest razing of $f$ for the order relation $>_f$ in Inter $(f, h)$, which is also the smallest razing for the order relation $>$ is equal to $\bigwedge h_n$, where $h_n = f \wedge \alpha h_{n-1}$, with $h_0 = f \wedge h$ ; we write Rz$(f, h)$. It is obtained by finite iteration until $h_{n+1} = h_n$.

The supremum for $\vee_f$ of a family of flattenings belonging to Inter $(f, h)$ is still a flattening. The largest flattening of Inter $(f, h)$ is also the supremum between the largest flooding and the largest razing within Inter $(f, h)$: $\Xi(f, h) =$ Fl$(f, h) \vee_f$ Rz$(f, h)$.

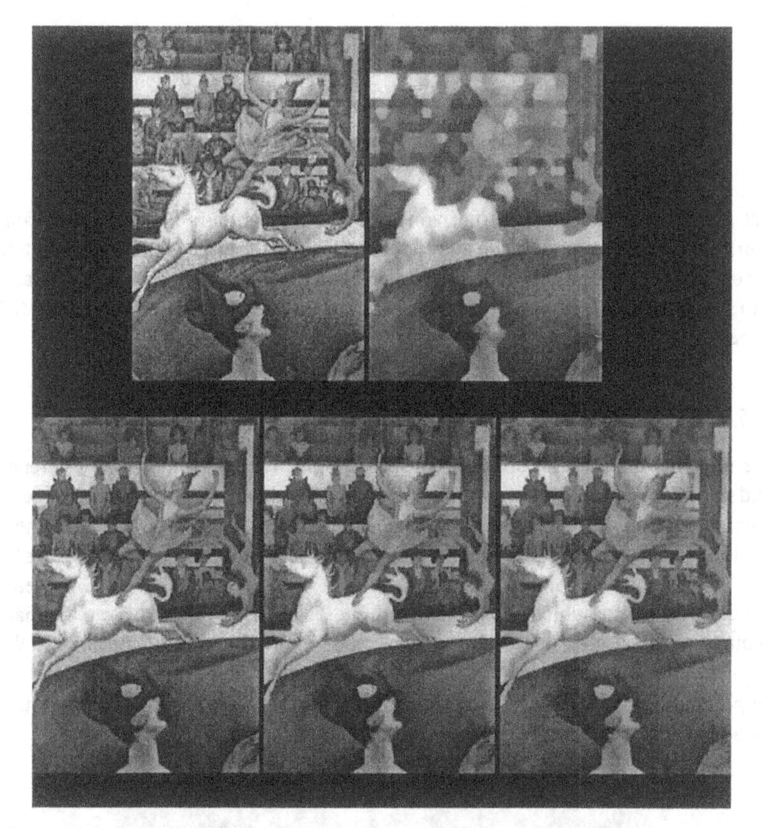

Figure 4.2. Levelings with increasing slopes of the same reference and marker functions.

The supremum $\vee_f$ of two levelings is not necessarily a leveling but a flattening. However if we replace $h$ by $k = \alpha h \wedge_f \beta h$, then all flattenings in Inter $(f, k)$ are levelings. Hence we will define the leveling of $f$ constrained by $h$ and write $\Lambda(f, h)$ as the largest flattening contained in Inter $(f, k)$: $\Lambda(f, h) = \Xi(f, k) = \mathrm{Fl}(f, k) \vee_f \mathrm{Rz}(f, k)$.

Fast algorithms, based for instance on hierarchical queues [563] exist for reconstruction closings and openings, producing respectively floodings and razings. Since flattenings and levelings rely on floodings and razings, their construction is fast also.

## 4.5 Levelings for image simplification

Floodings $\mathrm{Fl}(f, h)$, razings $\mathrm{Rz}(f, h)$, flattenings $\Xi(f, h)$ and levelings $\Lambda(f, h)$ are all functions of two arguments and depend on these two arguments. Their

Figure 4.3. Non connected structuring element

flat zones are larger than the flat zones of $f$, their contours correspond to contours of $f$ ; at the same time they are as close to $h$ as possible in the lattice Inter $(f, h)$. Furthermore, for each choice of an adjunction $(\alpha, \beta)$ a new operator can be constructed, to which is associated a particular type of contours and flat zones.

### 4.5.1 *Varying* $(\alpha, \beta)$

We will first explore the effect of various couples $(\alpha, \beta)$ on the same reference and marker images. Starting with the ordinary flat dilation $\delta$ (maximum value in a neighborhood of size 1), we define the slope dilation $\delta_\lambda = Id \vee (\delta - \lambda)$, where $Id$ is the identity) . The adjunct slope erosion is defined by $\varepsilon_\lambda = Id \wedge (\varepsilon + \lambda)$. Two neighboring pixels $p$ and $q$ are at level if $|f_p - f_q| \leq \lambda$. Fig.4.2 presents a picture by Seurat which is extremely grainy. The marker function is an alternate sequential filter of size 4, giving a very crude approximation of the image. We compare the results of 3 levelings ; the first being flat, the next being obtained for slopes 1 and 2. Increasing the slope produces much larger flat zones and a much smoother image. Nevertheless the contours remain sharp.

Figure 4.4. Left : $f$ =original image. The marker image $h$ is completely black with a white dot on the left hand of the girl.
Center : leveling associated to the dilation $\delta^{++}$ and erosion $\varepsilon^{--}$;
Right : leveling associated to the dilation $\delta$ and erosion $\varepsilon$ ; without jumps, the reconstruction is much less complete (see for instance the books)

In our second example we compare two levelings based respectively on a non connected and a connected structuring element. The first leveling is associated to the dilation $\delta^{++}$ and its adjunct erosion $\varepsilon^{--}$ and is based on a non connected structuring element consisting of a hexagon and two pixels at a distance of 4 pixels apart on each side (see fig. 4.3). The central part cares for the normal connectivity reconstructions whereas the couple of added pixels permits jumps from one zone

Figure 4.5. Levelings obtained after Gaussian blurring

to another. The second leveling is the basic flat leveling based on $(\delta, \varepsilon)$. Both levelings are applied on the same reference image $f$ (see fig.4.4left) and marker image $h$ (not illustrated here: it is completely black with a white dot on the hand holding the telephone). Indeed the ordinary leveling based on $(\varepsilon, \delta)$ illustrated by fig.4.4right is unable to reconstruct some parts of the image, although it uses the same marker ; it is unable to jump from one book to the next on the shelf in the background as is the case in fig.4.4center, where a leveling based on $(\delta^{++}, \varepsilon^{--})$ has been used. As expected, since $\delta^{++} > \delta$, the $(\varepsilon, \delta)$ leveling has larger flat zones than the $(\varepsilon^{--}, \delta^{++})$ leveling.

### *4.5.2 Varying the marker function h*

Gaussian blurring has a manifold of good properties from a theoretical point of view. It remarkably simplifies images. It has however one drawback: it blurs the contours. The greater the simplification, the larger the blurring. For this reason, levelings nicely finish off the work of Gaussian blurring by restoring all contours, while keeping the simplification. This effect is illustrated in fig.4.5 where blurrings with kernels of size 2 and 5 are restored by a slope leveling (slope 1).

In the two previous examples we have used a coarse simplification of the image as marker, either after an alternate sequential filter, or after a Gaussian blurring. We will now present ways to stress interesting features of the image. The first example stresses the contrast of the peaks. As marker we take a vertically shifted copy of the image $f$ itself, by subtracting a constant value ; a razing constructed with this marker function clips all peaks (fig.4.6left).

Some of them touch the marker functions, others do not. Let $X$ be the set where the razing and the marker function take the same value. In order to restore these peaks to their original height we construct a second razing of the initial image, but with a new marker function. equal to $f$ on $X$ and equal to 0 elsewhere (fig.4.6right). This process has been applied to the Seurat picture and illustrated in the first row of fig.4.7. First a razing has been applied clipping the peaks with the

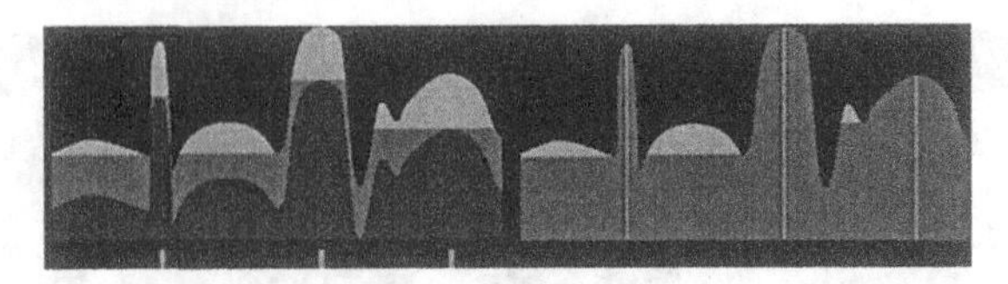

Figure 4.6. Two successive levelings permit to stress all peaks with the highest contrast.

lowest contrast (central image) but leaving the valleys unchanged. The resulting image is then submitted to the dual operator, filling the valleys.

The next example also stresses the contrast of the picture, from the point of view of the gradient. The gradient modulus is approximated as $\delta f - \varepsilon f$ and thresholded, yielding a binary set $X$ containing the sharp transitions in the image. The marker is equal to the original image within $X$ and black outside as illustrated in the bottom row of fig.4.7. A first razing produces the "black contour leveling" image. A second marker is the image constructed with again the original image within $X$, but with white outside. Applied on the result of the first razing, this flooding produces the final image, where the salient contours are completely restored and the rest of the image is smoothed out.

A last example shows the potential of levelings in the domain of selective image compression. When a video sequence has to be compressed and transmitted, it is worthwhile to compress the background more than the foreground, the face of a person for instance. Leveled images can be compressed economically, as they offer large smooth zones ; on the other hand, as the contours of the objects are precisely restored, they remain perceptually attractive even for high degrees of compression. In fig.4.8 we have constructed a composite marker image, made of an alternate sequential filter of varying size: a large size for the background, a low size for the foreground. The background is severely distorted. After leveling, contours of the background are restored and the face of the person appears undistorted.

## 4.5.3 *Multiscale filtering*

**Order relation between levelings, floodings and razings**

The relation {being a leveling of} is a preorder relation. The relations {being a flooding of} and {being a razing of} are order relations. Increasing floodings and increasing levelings are ideal tools for hierarchical segmentation, where for the same picture a series of segmentations with increasing coarseness is produced, each contour present at a coarse scale being also present in each finer scale.

### 4.5.3.1 Construction of a hierarchy based on increasing floodings

The watershed transform is the tool of choice for detecting contours ; generally it is used on a gradient image, associated to a set of markers. We flood the gradient image and as the flooding increases, adjacent basins progressively merge, pro-

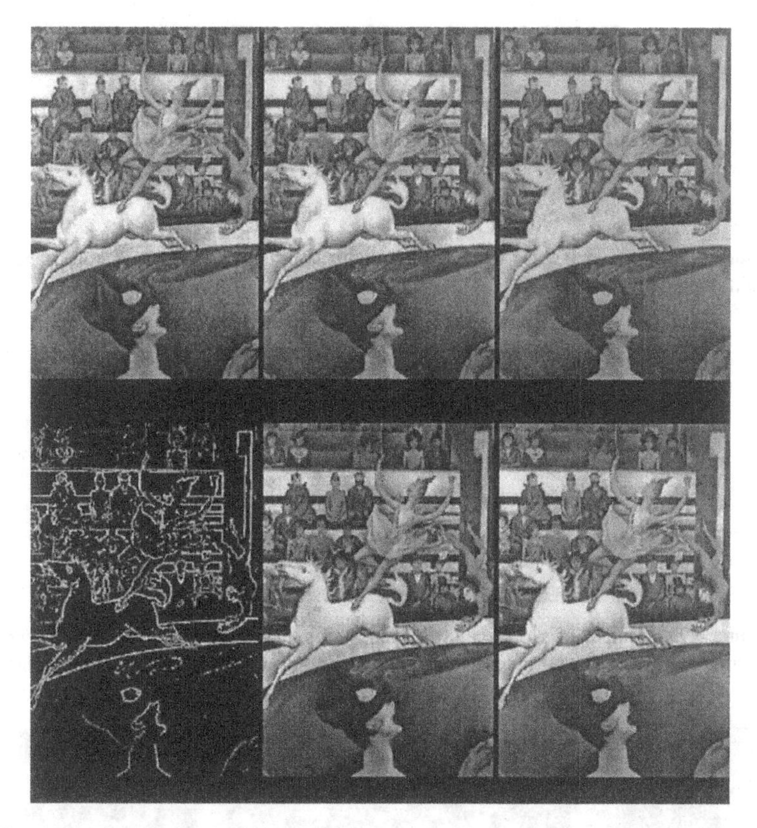

Figure 4.7. First row: successive levelings according to the contrast of peaks and valleys. Second row: the marker are the most contrasted contour zones.

ducing coarser and coarser segmentations. Depending on the law governing the progression of the flooding, one obtains different results. Size oriented flooding [368, 826] is produced by placing sources at each minimum and flooding the surface in such a way that all lakes share some common measure (height, volume or area of the surface). As the flooding proceeds, the level of some lakes cannot grow any further, as the level of the lowest path point has been reached. In the fig.4.9, a flooding starts from all minima in such a way that all lakes always have uniform depth. Size oriented flooding allows to produce hierarchical segmentation with good psychovisual properties. The depth criterion ranks the regions according to their contrast, the area according to their size and the volume offers a nice balance between size and contrast. The topographical surface to be flooded is a color gradient of the initial image (maximum of the morphological gradients computed in each of the R, G and B color channels). Synchronous volumic flooding has been

Figure 4.8. Marker image and leveling for a high simplification of the background and a faithful reproduction of the face.

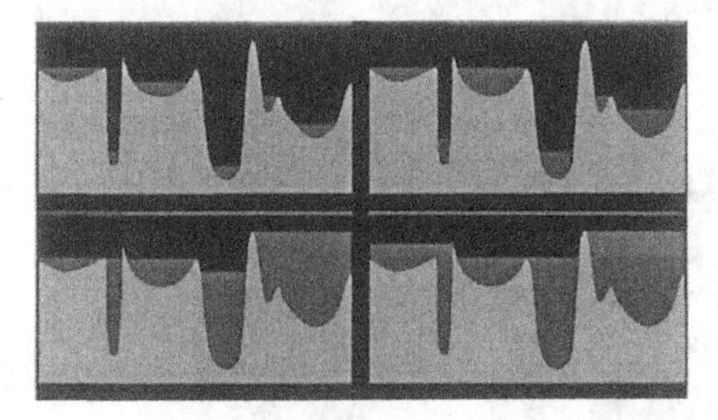

Figure 4.9. Example of a height synchronous flooding. Four levels of flooding are illustrated ; each of them is topped by a figuration of the corresponding catchment basins.

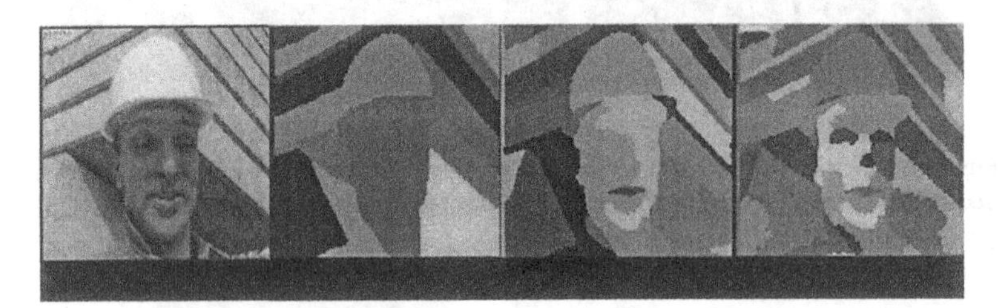

Figure 4.10. Initial image and 3 levels of a multiscale segmentation

used, and 3 levels of fusions have been represented, corresponding respectively to 15, 35 and 60 regions.

#### 4.5.3.2 Construction of a hierarchy based on quasi-flat zones

Since levelings enlarge quasi flat zones, the quasi-flat zones of a family of increasing levelings itself form a hierarchy. Fig.4.5.3.2 presents the construction. A slope leveling is produced associated to an alternated sequential filter. The quasi-flat zones are detected. However, as fig.4.5.3.2 shows, the quasi-flat zones have two different natures: on one hand large homogeneous zones, and in the transition

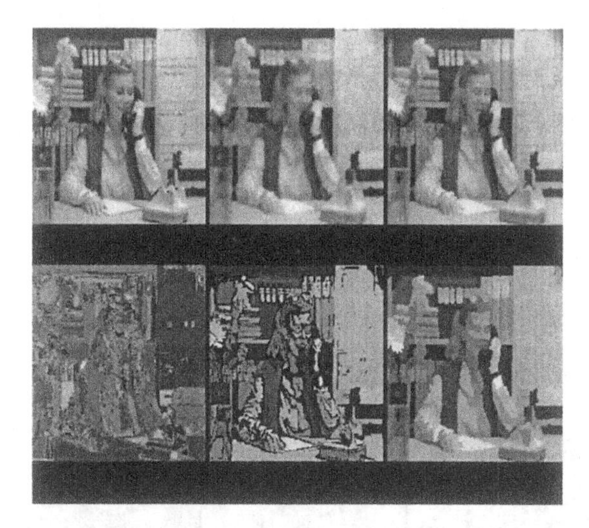

zones of high gradient tiny quasi-flat zones. For this reason, a more useful hierarchy is obtained if one gets rid of these transition zones. Only the largest of them are retained as markers of a watershed segmentation, yielding the final result.

The process may then be repeated for a cascade of levelings based on coarser and coarser alternate sequential filters. Fig.4.11 presents in the first row 3 increasing slope levelings associated to alternate sequential filters of sizes 3, 6 and 9 and in the second row the associated segmentations.

## 4.6 Conclusion

Floodings, razings and levelings have very interesting properties for image segmentation. They do not blur nor displace the contours, do not create spurious minima or maxima, may be cascaded in order to create a multiscale simplification of the image. The family is extremely large, since a leveling can be associated to each adjunction $(\alpha, \beta)$. Furthermore, a leveling also depends on the choice of a marker function, offering a unique possibility in the family of filters to inject in the filtering process a selection of the features one desires to stress.

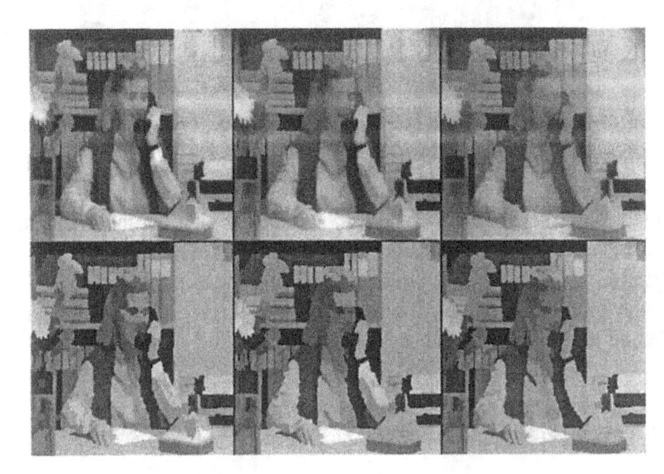

Figure 4.11. Hierarchy associated to increasing levelings. Line1 : 3 increasing levelings Line2 : Associated increasing partitions

# Chapter5

# Graph Cuts in Vision and Graphics: Theories and Applications

**Y. Boykov and O. Veksler**

Abstract

Combinatorial min-cut algorithms on graphs have emerged as an increasingly useful tool for problems in vision. Typically, the use of graph-cuts is motivated by one of the following two reasons. Firstly, graph-cuts allow geometric interpretation; under certain conditions a cut on a graph can be seen as a hypersurface in N-D space embedding the corresponding graph. Thus, many applications in vision and graphics use min-cut algorithms as a tool for computing optimal hypersurfaces. Secondly, graph-cuts also work as a powerful energy minimization tool for a fairly wide class of binary and non-binary energies that frequently occur in early vision. In some cases graph cuts produce globally optimal solutions. More generally, there are iterative techniques based on graph-cuts that produce provably good approximations which (were empirically shown to) correspond to high-quality solutions in practice. Thus, another large group of applications use graph-cuts as an optimization technique for low-level vision problems based on global energy formulations.

This chapter is intended as a tutorial illustrating these two aspects of graph-cuts in the context of problems in computer vision and graphics. We explain general theoretical properties that motivate the use of graph cuts, as well as show their limitations.

## 5.1 Introduction

Graph cuts remain an area of active research in the vision and graphics communities. Besides finding new applications, in the last years researchers have discovered and rediscovered interesting links connecting graph cuts with other combinatorial algorithms (dynamic programming, shortest paths [107, 477]), Markov random fields, statistical physics, simulated annealing and other regularization techniques [362, 113, 424], sub-modular functions [491], random walks and electric circuit theory [356, 357], Bayesian networks and belief propagation [790], integral/differential geometry, anisotropic diffusion, level sets and other variational methods [767, 109, 28, 477].

Graph cuts have proven to be a useful multidimensional optimization tool which can enforce piecewise smoothness while preserving relevant sharp discontinuities. This paper is mainly intended as a survey of existing literature and a tutorial on graph cuts in the context of vision and graphics. We present some basic background information on graph cuts and discuss major theoretical results, some fairly new and some quite old, that helped to reveal both strengths and limitations of these surprisingly versatile combinatorial algorithms. This chapter does not provide any new research results, however, some applications are presented from a point of view that may differ from the previous literature.

The organization of this chapter is as follows. Chapter 5.2 provides necessary background information and terminology. In their core, combinatorial min-cut/max-flow algorithms are binary optimization methods. Chapter 5.3 presents a simple binary problem that can help to build basic intuition on using graph cuts in computer vision. Then, graph cuts are discussed as a general tool for exact minimization of certain binary energies.

Most publications on graph cuts in vision and graphics show that, despite their binary nature, graph-cuts offer significantly more than "binary energy minimization". Chapter 5.4 shows that graph cuts provide a viable geometric framework for approximating continuous hypersurfaces on N-dimensional manifolds. This geometric interpretation of graph cuts is widely used in applications for computing globally optimal separating hypersurfaces. Finally, Chapter 5.5 presents generalized (non-binary) graph cuts techniques applicable to exact or approximate minimization of multi-label energies. In the last decade, such non-binary graph cut methods helped to significantly raise the bar for what is considered a good quality solution in many early vision problems.

## 5.2 Graph Cuts Basics

First, we introduce some basic terminology. Let $\mathcal{G} = \langle \mathcal{V}, \mathcal{E} \rangle$ be a graph which consists of a set of nodes $\mathcal{V}$ and a set of directed edges $\mathcal{E}$ that connect them. The nodes set $\mathcal{V} = \{s, t\} \cup \mathcal{P}$ contains two special *terminal* nodes, which are called the *source*, $s$, and the *sink*, $t$, and a set of non-terminal nodes $\mathcal{P}$. In Figure 5.1(a) we show a simple example of a graph with the terminals $s$ and $t$. Such N-D grids are typical for applications in vision and graphics.

Each graph edge is assigned some nonnegative weight or cost $w(p,q)$. A cost of a directed edge $(p,q)$ may differ from the cost of the reverse edge $(q,p)$. An edge is called a *t-link* if it connects a non-terminal node in $\mathcal{P}$ with a terminal. An edge is called a *n-link* if it connects two non-terminal nodes. A set of all (directed) n-links will be denoted by $\mathcal{N}$. The set of all graph edges $\mathcal{E}$ consists of n-links in

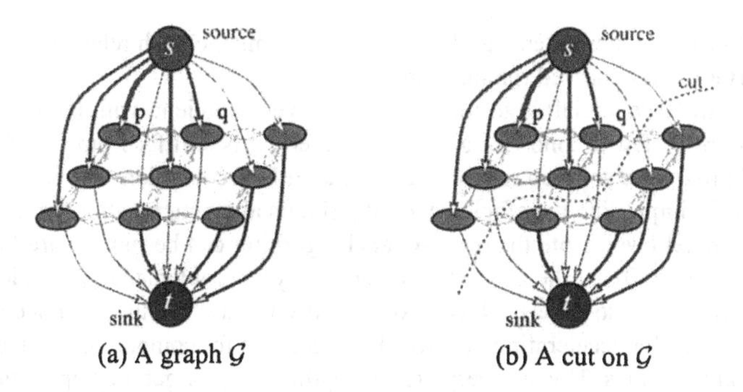


(a) A graph $\mathcal{G}$ (b) A cut on $\mathcal{G}$

Figure 5.1. Graph construction in Greig et. al. [362]. Edge costs are reflected by thickness.

$\mathcal{N}$ and t-links $\{(s,p),(p,t)\}$ for non-terminal nodes $p \in \mathcal{P}$. In Figure 5.1 t-links are shown in red and blue, while n-links are shown in yellow.

## 5.2.1 The Min-Cut and Max-Flow Problem

An $s/t$ cut $C$ (sometimes we just call it a *cut*) is a partitioning of the nodes in the graph into two disjoint subsets $\mathcal{S}$ and $\mathcal{T}$ such that the source $s$ is in $\mathcal{S}$ and the sink $t$ is in $\mathcal{T}$. Figure 5.1(b) shows one example of a cut. The cost of a cut $C = \{\mathcal{S}, \mathcal{T}\}$ is the sum of costs/weights of "boundary" edges $(p,q)$ such that $p \in \mathcal{S}$ and $q \in \mathcal{T}$. If $(p,q)$ is a boundary edge, then we sometimes say that cut $C$ severs edge $(p,q)$. The *minimum cut* problem is to find a cut that has the minimum cost among all cuts.

One of the fundamental results in combinatorial optimization is that the minimum $s/t$ cut problem can be solved by finding a *maximum flow* from the source $s$ to the sink $t$. Speaking informally, maximum flow is the maximum "amount of water" that can be sent from the source to the sink by interpreting graph edges as directed "pipes" with capacities equal to edge weights. The theorem of Ford and Fulkerson [324] states that a maximum flow from $s$ to $t$ saturates a set of edges in the graph dividing the nodes into two disjoint parts $\{\mathcal{S}, \mathcal{T}\}$ corresponding to a minimum cut. Thus, min-cut and max-flow problems are equivalent. In fact, the maximum flow value is equal to the cost of the minimum cut.

## 5.2.2 Algorithms for the Min-Cut and Max-Flow Problem

There are many standard polynomial time algorithms for min-cut/max-flow[217]. These algorithms can be divided into two main groups: "push-relabel" style methods [350] and algorithms based on augmenting paths. In practice the push-relabel algorithms perform better for general graphs. In vision applications, however, the most common type of a graph is a two or a higher dimensional grid. For the grid graphs, Boykov and Kolmogorov [110] developed a fast augmenting path algorithm which often significantly outperforms the push relabel algorithm. Furthermore, its observed running time is linear.

While the (sequential) algorithm in [110] is very efficient, with the execution time of only a few seconds for a typical problem, it is still far from real time. A possible real time solution may come from a GPU acceleration that has become popular for improving the efficiency of algorithms allowing parallel implementations on pixel level. Note that push-relabel algorithm can be run in parallel over graph nodes [350]. In the context of image analysis problems, graph nodes typically correspond to pixels. Thus, pixel based GPU architecture is a seemingly perfect match for accelerating push-relabel algorithm for computing graph cuts in vision and graphics. This is a very promising direction for getting applications of graph cuts up to real time.

## 5.3 Graph Cuts for Binary Optimization

In this section we concentrate on graph cuts as a binary optimization tool. In fact, min-cut/max-flow algorithms are inherently binary techniques, and so binary problems constitute the most basic case for graph cuts. In Section 5.3.1 we discuss the earliest known example where graph cuts were used in vision, which also happens to be a particularly clear binary problem. The example illustrates that graph cuts can effectively enforce spatial coherence on images. Section 5.3.2 presents the general case of binary energy minimization with graph cuts.

### 5.3.1 *Example: Binary Image Restoration*

The earliest use of graph cuts for energy minimization in vision is due to Greig et.al. [362]. They consider the problem of binary image restoration. Given a binary image corrupted by noise, the task is to restore the original image. This problem can be formulated as a simple optimization over binary variables corresponding to image pixels. In particular, [362] builds a graph shown in Figure 5.1(a) where non-terminal nodes $p \in \mathcal{P}$ represent pixels while terminals $s$ and $t$ represent two possible intensity values. To be specific, source $s$ will represent intensity 0 and sink $t$ will represent intensity 1. Assume that $I(p)$ is the observed intensity at pixel $p$. Let $D_p(l)$ be a fixed penalty for assigning to pixel $p$ some "restored intensity" label $l \in \{0, 1\}$. Naturally, if $I(p) = 0$ then $D_p(0)$ should be smaller than $D_p(1)$, and vice versa. To encode these "observed data" constraints, we create two t-links for each pixel node in Figure 5.1. The weight of t-link $(s, p)$ is set to $D_p(1)$ and the weight of $(p, t)$ is set to $D_p(0)$. Even though t-link weights should be non-negative, the restriction $D_p \geq 0$ for data penalties is not essential.

Now we should add regularizing constraints that help to remove image noise. Such constraints enforce spatial coherence between neighboring pixels by minimizing discontinuities between them. In particular, we create n-links between neighboring pixels using any (e.g. 4- or 8-) neighborhood system. The weight of these n-links is set to a smoothing parameter $\lambda > 0$ that encourages a minimum cut to sever as few n-links as possible.

Remember that a cut $C$ (Figure 5.1(b)) is a binary partitioning of the nodes into subsets $\mathcal{S}$ and $\mathcal{T}$. A cut can be interpreted as a binary labeling $f$ that assigns labels $f_p \in \{0, 1\}$ to image pixels: if $p \in \mathcal{S}$ then $f_p = 0$ and if $p \in \mathcal{T}$ then $f_p = 1$. Obviously, there is a one-to-one correspondence between cuts and binary labelings of pixels. Each labeling $f$ gives a possible image restoration result.

Consider the cost of an arbitrary cut $C = \{\mathcal{S}, \mathcal{T}\}$. This cost includes weights of two types of edges: severed t-links and severed n-links. Note that a cut severs exactly one t-link per pixel; it must sever t-link $(p, t)$ if pixel $p$ is in the source component $p \in \mathcal{S}$ or t-link $(s, p)$ if pixel $p$ is in the sink component $p \in \mathcal{T}$. Therefore, each pixel $p$ contributes either $D_p(0)$ or $D_p(1)$ towards the t-link part of the cut cost, depending on the label $f_p$ assigned to this pixel by the cut. The cut cost also includes weights of severed n-links $(p, q) \in \mathcal{N}$. Therefore,

$$|C| = \sum_{p \in \mathcal{P}} D_p(f_p) + \sum_{\substack{(p,q) \in \mathcal{N} \\ p \in \mathcal{S}, q \in \mathcal{T}}} w(p, q)$$

The cost of each $C$ defines the "energy" of the corresponding labeling $f$:

$$E(f) := |C| = \sum_{p \in \mathcal{P}} D_p(f_p) \quad + \quad \lambda \cdot \sum_{(p,q) \in \mathcal{N}} \mathcal{I}(f_p = 0, f_q = 1), \qquad (5.1)$$

where $\mathcal{I}(\cdot)$ is the identity function giving 1 if its argument is true and 0 otherwise. Stated simply, the first term says that pixel labels $f_p$ should agree with the observed data while the second term penalises discontinuities between neighboring pixels. Obviously, a minimum cut gives labeling $f$ that minimizes energy (5.1).

Note that parameter $\lambda$ controls the relative importance of the data constraints versus the regularizing constraints. Note that if $\lambda$ is very small, an optimal labeling assigns each pixel $p$ a label $f_p$ that minimizes its own data cost $D_p(f_p)$. In this case, each pixel chooses its own label independently from the other pixels. If $\lambda$ is big, then all pixels must choose one label that has a smaller average data cost. For intermediate values of $\lambda$, an optimal labeling $f$ should correspond to a balanced solution with compact spatially coherent clusters of pixels who generally like the same label. Noise pixels, or outliers, should conform to their neighbors.

Before [362], exact minimization of energies like (5.1) was not possible. Researches still used them, but had to approach them with iterative algorithms like simulated annealing [341]. In fact, Greig et.al. published their result mainly to show that in practice simulated annealing reaches solutions very far from the global minimum even in simple binary cases. Unfortunately, the result of Greig et.al. remained unnoticed in the vision community for almost 10 years probably because the binary image restoration looked too restrictive as an application.

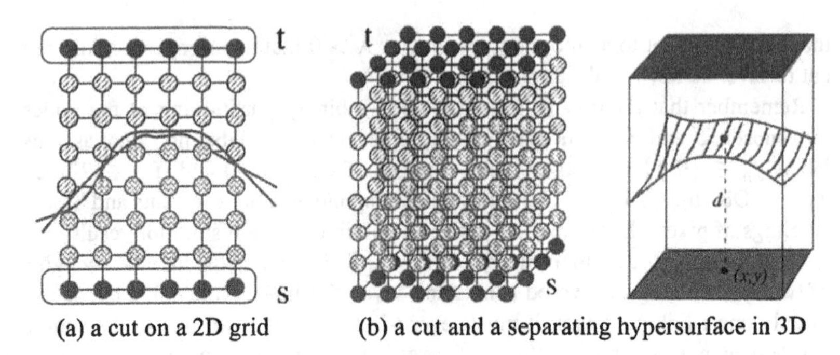


(a) a cut on a 2D grid (b) a cut and a separating hypersurface in 3D

Figure 5.2. s-t cut on a grid corresponds to binary partitioning of N-D space where the grid is embedded. Such space partitioning may be visualized via a separating hypersurface. As shown in (a), multiple hypersurfaces may correspond to the same cut. However, such hypersurfaces become indistinguishable as the grid gets finer.

### *5.3.2 General Case of Binary Energy Minimization*

In general, graph construction as in Figure 5.1 can be used for other binary "labeling" problems. Suppose we are given a penalty $D_p(l)$ that pixel $p$ incurs when assigned label $l \in \mathcal{L} = \{0, 1\}$ and we need to find a spatially coherent binary labeling of the whole image. We may wish to enforce spatial regularization via some global energy function that generalizes (5.1)

$$E(f) = \sum_{p \in \mathcal{P}} D_p(f_p) \; + \sum_{(p,q) \in \mathcal{N}} V_{pq}(f_p, f_q) \tag{5.2}$$

The question is: can we find a globally optimal labeling $f$ using some graph cut construction? There is a definitive answer to this question for the case of binary labelings. According to [491], a globally optimal binary labeling for (5.2) can be found via graph cuts if and only if the pairwise interaction potential $V_{pq}$ satisfies

$$V_{pq}(0,0) + V_{pq}(1,1) \leq V_{pq}(0,1) + V_{pq}(1,0)$$

which is called the regularity condition. The theoretical result in [491] is constructive and they show the corresponding graph. It has the same form as the graph of Greig et.al. in Figure 5.1, however, edge weights are derived differently.

## 5.4 Graph Cuts as Hypersurfaces

Solution of many problems in vision, image processing and graphics can be represented in terms of optimal hypersurfaces. This section describes a geometric interpretation of graph-cuts as hypersurfaces in N-D manifolds that makes them an attractive framework for problems like image segmentation, restoration, stereo, photo/video editing, texture synthesis, and others.

We show a basic idea allowing s-t cuts to be viewed as hypersurfaces, discuss interesting theories that make various connections between discrete graph cuts and hypersurfaces in continuous spaces, and we also provide a number of recently published examples where a hypersurface view of graph cuts has led to interesting applications in computer vision, medical imaging, or graphics.

### *5.4.1 Basic idea*

Consider two simple examples in Figure 5.2. Throughout Section 5.4 we assume that a graph has no "soft" t-links, that is the source and the sink terminals are directly connected only to some of the graph nodes via infinity cost t-links. In fact, all nodes hardwired to two terminals can be effectively treated as multiple sources and multiple sinks that have to be separated by a cut. Figure 5.2 shows these sources and sinks in dark red and dark blue colors. Such sources and sinks provide hard constraints or boundary conditions for graph cuts; any feasible cut must separate sources from sinks. Other nodes are connected to the sources and sinks via n-links.

Without loss of generality (see Section 5.4.2), we can concentrate on feasible cuts that partition the simple 4- and 6- nearest neighbor grid-graphs in Figure 5.2 into two connected subsets of nodes: source component and sink component. Continuous 2D and 3D manifolds where the grid nodes are embedded can be split into two disjoint contiguous regions, one containing the sinks, and the other containing the sources. A boundary between two such regions are separating hypersurfaces shown in green color. As illustrated in Figure 5.2(a), there are many separating hypersurfaces that correspond to the same cut. They should all correctly separate the grid nodes of the source and the sink components, but they can "freely move" in the space between the grid nodes. Without getting into mathematical details, we will identify a class of all hypersurfaces corresponding to a given cut with a single hypersurface. In particular, we can choose a hypersurface that follows boundaries of "grid cells", or we can choose "the smoothest" hypersurface. Note that the finer the grid, the harder it is to distinguish two separating hypersurfaces corresponding to the same cut.

Thus, any feasible cut on a grid in Figure 5.2 corresponds to a separating hypersurface in the embedding continuous manifold. Obviously, the opposite is also true; any separating hypersurface corresponds to a unique feasible cut. Generalization of examples in Figure 5.2 would establish correspondence between $s - t$ graph-cuts and separating hypersurfaces in case of "fine" locally connected grids embedded in N-D spaces. Following ideas in [109], one can set a cost (or area) of each continuous hypersurface based on the cost of the corresponding cut. This defines a *cut metric* introduced in [109] for continuous N-D manifold embedding a graph. By changing weights of n-links at graph nodes located in any particular point in space, one can tune local costs of all separating hypersurfaces that pass through such locations. In practical applications a cut metric can be easily tuned to attract (repel) hypersurfaces to (from) certain locations on N-D manifolds. A cut metric is a simple, yet sufficiently general tool. In particular, according to

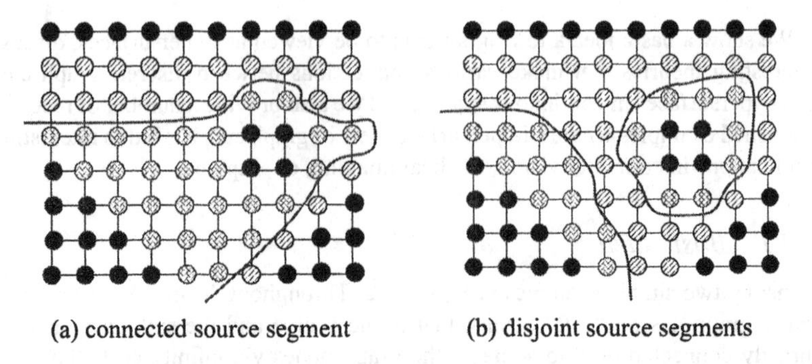


Figure 5.3. Separating hypersurfaces can have different topological properties for the same set of hard constraints. Separating hypersurfaces in (a) and (b) correspond to two distinct feasible $s-t$ cuts. Min-cut/max-flow algorithms compute a globally optimal hypersurface/cut without any restrictions on its topological properties as long as the sources and the sinks are separated.

[109] a cut metric on 2D and 3D manifolds can approximate any given continuous Riemannian metric. Finally, standard combinatorial algorithms for computing minimum cost $s-t$ cuts (see Section 5.2.2) become numerical tools for extracting globally optimal separating hypersurfaces.

## *5.4.2 Topological properties of graph cuts*

The adjective "separating" implies that a hypersurface should satisfy certain hard constraints or boundary conditions; it should separate source and sink grid cells (seeds). Note that there are many freedoms in setting boundary conditions for graph cuts. Depending on hard constraints, topological properties of separating hypersurfaces corresponding to $s-t$ cuts may vary.

For example, we can show that the boundary conditions in Figure 5.2 guarantee that any feasible cut corresponds to topologically connected separating hypersurface. For simplicity, we assume that our graphs are connected, that is, there are no "islands" of disconnected nodes. In Figure 5.2 all source and all sink nodes form two connected components. In such cases a minimum cost cut must partition the graph into exactly two connected subsets of nodes; one containing all sources and the other containing all sinks. Assuming that the minimum cost cut creates three or more connected components implies that some of these components contain neither sources, nor sinks. This contradicts minimality of the cut; linking any "no-source/no-sink" subset back to the graph corresponds to a smaller cost feasible cut.

Examples in Figure 5.3 illustrate different topological properties for separating hypersurfaces in more general cases where multiple disjoint components of sources and sinks (seeds) are present. Note that feasible $s-t$ cuts may produce topologically different separating hypersurfaces for the same set of boundary conditions.

In fact, controlling topological properties of separating hypersurfaces by setting up appropriate hard constraints is frequently a key technical aspect of applications using graph cuts. As discussed in Section 5.4.3, appropriate positioning of sources and sinks is not the only tool to achieve desired topology. As shown in Figure 5.4, certain topological properties of separating hypersurfaces can be enforced via infinity cost n-links.

### *5.4.3 Applications of graph cuts as hypersurfaces*

Below we consider several examples from recent publications where graph cuts are used as a method for extracting optimal hypersurfaces with desired topological properties.

Methods for object extraction [107, 96, 683, 903] take full advantage of topological freedom of graph-cut based hypersurfaces. In particular, they allow to segment objects of arbitrary topology. The basic idea is to set as sources (red seeds) some image pixels that are known (a priori) to belong to an object of interest and to set as sinks (blue seeds) some pixels that are known to be in the background. A separating hypersurface should coincide with a desirable object boundary separating object (red) seeds from background (blue) seeds, as demonstrated in Figure 5.3. A cut metric can be set to reflect image gradient. Pixels with a high image gradient would imply a low cost of local n-links and vice versa. Then, minimal separating hypersurfaces tend to adhere to object boundaries with high image gradients. Another practical strength of object extraction methods based on graph cuts is that they provide practical solutions for organ extraction problems in N-D medical image analysis [107]. One limitation of this approach to object extraction is that it may suffer from a bias to "small cuts", but this can often be resolved with proper constraining of the solution space.

Stereo was one of the first applications in computer vision where graph cuts were successfully applied as a method for optimal hypersurface extraction. Two teams, Roy&Cox [693, 692] and Ishikawa&Geiger [425], almost simultaneously proposed two different formulations of the stereo problem where disparity maps are interpreted as separating hypersurfaces on certain 3D manifolds. Their key technical contribution was to show that disparity maps (as optimal hypersurfaces) can be efficiently computed via graph cuts.

For example, Roy&Cox [693, 692] proposed a framework for stereo where disparity maps are separating hypersurfaces on 3D manifolds similar to one in Figure 5.2(b). Points of this bounded rectangular manifold are interpreted as points in 3D "disparity space" corresponding to a pair of rectified stereo images. This disparity space is normally chosen with respect to one of the images, so that each 3D point with coordinates $(x, y, d)$ represents correspondence between pixel $(x, y)$ in the first stereo image and pixel $(x + d, y)$ in the second image. Then, solution of stereo problem is a hypersurface $d = f(x, y)$ on 3D manifold in Figure 5.2(b) that represents a disparity map assigning certain disparity $d$ to each pixel $(x, y)$ in the first image. Note that hypersurface $d = f(x, y)$ separates the bottom and the top (facets) of 3D manifold in Figure 5.2(b). Then, an optimal

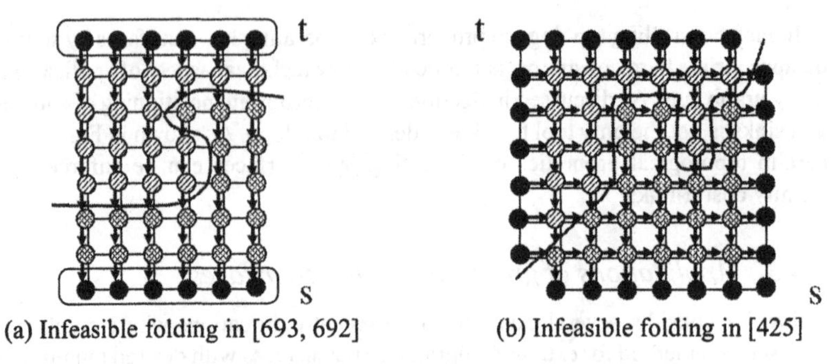


(a) Infeasible folding in [693, 692] (b) Infeasible folding in [425]

Figure 5.4. Graph-cuts approach allows to impose certain additional topological constraints on separating hypersurfaces, if necessary. For example, [426, 111] proposed infinity cost directed n-links, shown in brown color in (a), that forbid folds on separating hypersurfaces in Figure 5.2. In particular, a hypersurface in Figure 5.2(b) without such folds corresponds to a disparity map $d = f(x, y)$ according to [693, 692]. Also, [425] impose monotonicity/ordering constraint on their disparity maps by adding infinity cost directed n-links (in brown color) that make illegal topological folds shown in (b). For clarity, examples in (a) and (b) correspond to single slices of 3D manifolds in Figure 5.2(b) and 5.5(a).

disparity map can be computed using graph cuts as an efficient discrete model for extracting minimal separating hypersurfaces.

According to [693], cut metric on 3D "disparity space" manifold in Figure 5.2(b) is set based on color consistency constraint between two stereo cameras. Weights of n-links at node $(x, y, d)$ are set as follows: if intensities of pixels $(x, y)$ and $(x+d, y)$ in two cameras are similar then the likelihood that two pixels see the same 3D object point is high and the cost of n-links should be small. Later, [426, 692, 111] suggested anisotropic cut metric where vertical n-links are based on the same likelihoods as above but horizontal n-links are fixed to a constant encouraging smoother disparity maps that avoid unnecessary disparity level jumps.

In general, separating hypersurfaces in Figure 5.2(b) can have folds that would make them inappropriate as disparity maps $d = f(x, y)$. If a minimum hypersurface computed via graph cuts has a fold then we did not find a feasible disparity map. Therefore, [426, 111] propose a set of hard constraints that make topological folds (see Figure 5.4(a)) prohibitively expensive. Note that additional infinity cost vertical n-links (directed down) make folds infeasible. This topological hard constraint takes advantage of the "directed" nature of graph cuts; a cost of a cut includes only severed directed edges that go from the (red) nodes in the source component to the (blue) nodes in the sink component. A cut with an illegal fold in Figure 5.4(a) includes one infinity cost n-link.

Ishikawa&Geiger [425] also solve stereo by computing optimal separating hypersurfaces on a rectangular 3D manifold. However, their interpretation of the manifold and boundary conditions are different. As shown in Figure 5.5(a), they interpret a separating hypersurface $z = f(x, y)$ as a "correspondence mapping"

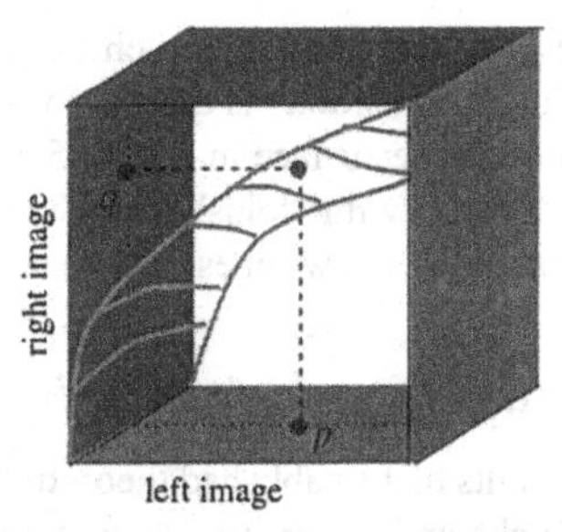


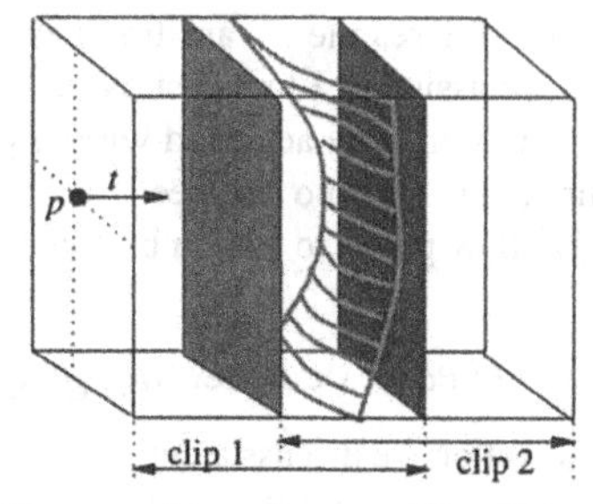


(a) Hypersurface as correspondence (b) Hypersurface separates two video clips

Figure 5.5. Two more examples of graph cuts as separating hypersurfaces. Formulation of stereo problem in [425] computes pixel correspondences represented by a separating hypersurface on a 3D manifold in (a). A smooth transition between two video clips is performed in [499] via graph cuts computing globally optimal separating hypersurface in a 3D region of overlap between two clips in (b).

between pixels $p = (x, y)$ in the left image and pixels $q = (f(x, y), y)$ in the right image (of a rectified stereo pair). Assignment of correspondences may be ambiguous if a hypersurface has folds like one in Figure 5.4(b). In order to avoid ambiguity, [425] introduce monotonicity (or ordering) constraint that is enforced by directed infinity cost n-links shown in brown color. Note that a cut in Figure 5.4(b) severs two brown n-links that go from a (red) node in a source component to a (blue) node in a sink component. Thus, the cost of the cut is infinity and the corresponding separating hypersurface with a fold becomes infeasible.

Similar to [693, 692], the cut metric on manifold in Figure 5.5(a) is based on color consistency constraint: a 3D points $(x, y, z)$ on the manifold has low n-link costs if intensity of pixel $(x, y)$ in the left image is close to intensity of pixel $(z, y)$ in the right image. Note that hyperplanes parallel to diagonal crossection (from bottom-left to top-right corners) of manifold in Figure 5.5(a) give correspondence mappings with constant stereo disparity/depth levels. Thus, spatial consistency of disparity/depth map can be enforced with anisotropic cut metric where diagonal n-links (from left-bottom to right-top corner) are set to a fixed constant representing penalty for jumps between disparity levels.

Another interesting example of graph-cuts/hypersurface framework is a method for video texture synthesis in [499]. The technique is based on computing a seamless transition between two video clips as illustrated in Figure 5.5(b). Two clips are overlapped in 3D (pixel-time) space creating a bounded rectangular manifold where transition takes place. A point in this manifold can be described by 3D coordinates $(x, y, t)$ where $p = (x, y)$ is a pixel and $t$ is time or video frame number. The transition is represented by a separating hypersurface $t = f(x, y)$ that specifies for each pixel when to switch from clip 1 to clip 2. During transition a frame may have a mix of pixels from each clip. The method in [499] suggest a specific cut metric that for each point $(x, y, t)$ in the overlap region depends on intensity difference between two clips. Small difference indicates a good moment (in space and time) for seamless transition between the clips and n-links at such $(x, y, t)$ points are assigned a low cost. Note that "seamless transition" is a purely visual effect and it may be achieved with any separating hypersurface in Figure 5.5(b). In this case there is no real need to avoid hypersurfaces with "folds" which would simply allow pixels to switch between clip 1 and clip 2 a few times.

### 5.4.4 *Theories connecting graph-cuts and hypersurfaces in $R^n$*

In this section we discuss a number of known results that established theoretically solid connections between cuts on discrete graphs and hypersurfaces in continuous spaces. It has been long argued in computer vision literature that discrete algorithms on graphs, including graph cuts, may suffer from metrication artifacts. Indeed, 4- and 6- nearest neighbor connections on 2D and 3D grids may produce "blocky" segments. Such geometric artifacts are due to "Manhattan distance" metrication errors. It turns out that such errors can be easily corrected, resolving the long-standing criticism of graph cuts methods. Boykov&Kolmogorov [109] showed that regular grids with local neighborhood systems of higher order can produce a cut metric that approximates any continuous Riemannian metric with arbitrarily small error. Using powerful results from integral geometry, [109] shows that weights of n-links from a graph node embedded at point $p$ of continuous N-D manifold are solely determined by a given $N \times N$ positive-definite matrix $D(p)$ that defines local metric/distance properties at point $p$ according to principles of Riemannian geometry. This result is quite intuitive as weights of n-links at this graph node define local measure for area/distance for hypersurfaces according to the corresponding cut metric. It is also interesting that results in [109] apply to arbitrary Riemannian metrics including anisotropic cases where local metric could be direction-sensitive.

So far in Section 5.4 we followed the general approach of [109] where hypersurfaces on N-D manifolds have implicit representation via cuts on embedded graphs. As illustrated in Figure 5.2, a cut only "implies" a separating hypersurface. A specific hypersurface can be obtained through additional conventions, as discussed in Section 5.4.1. More recently, [477] proposed an explicit approach to hypersurface representation by graph cuts that, in a way, is dual to [109]. The basic idea in [477] is to bisect a bounded N-D manifold with a large number of (random) hyperplanes. These hyperplanes divide the manifold into small cells (polyhedra) which can be thought of as irregular voxels. Then, [477] build an irregular "random-grid" graph where each cell is represented by a node. Two cells are connected by an n-link if and only if they touch through a common facet. Clearly, there is a one-to-one correspondence between a set of all n-links on the graph and a set of all facets between cells. A cut on this graph explicitly represents a unique hypersurface formed by facets corresponding to severed n-links. Obviously, a cost of any cut will be equal to the area of the corresponding hypersurface (in any metric) if weights of each n-link is equal to the area of the corresponding facet (in that metric). Thus, the model for representing hypersurfaces via graph-cuts in [477] can be applied to any metric. In their case, min-cut/max-flow algorithms will compute a minimum separating hypersurface among all explicitly represented hypersurfaces satisfying given boundary conditions.

Cuts on a graph in [477] represent only a subset of all possible hypersurfaces on an embedding manifold. If one keeps bisecting this bounded manifold into finer cells then the number of representable hypersurfaces increases. [477] proves that bisecting the manifold with a countably infinite number of random hyperplanes would generate small enough cells so that their facets can represent any continuous[1] hypersurface with an arbitrarily small error. This demonstrates that their approach to graph-cut/hypersurface representation is also theoretically solid.

Intuitively speaking, theoretical results in [109] and [477] imply that both approaches to representing continuous hypersurfaces via discrete graph cuts models have reasonable convergence properties and that minimum cost cuts on finer graphs "in the limit" produce a minimum separating hypersurfaces for any given metric. Results such as [109] and [477] also establish a link between graph cuts and variational methods such as level-sets [729, 616, 702, 617] that are also widely used for image segmentation.

There is (at least) one more interesting theoretical result linking graph cuts and hypersurfaces in continuous spaces that is due to G. Strang [767]. This result was established more than 20 years ago and it gives a view somewhat different from [109, 477]. Strang describes a continuous analogue of the min-cut/max-flow paradigm. He shows that maximum flow problem can be redefined on a bounded continuous domain $\Omega$ in the context of a vector field $\bar{f}(p)$ representing the speed of a continuous stream/flow. A constraint on discrete graph flow that comes from edge capacities is replaced by a "speed limit" constraint $|\bar{f}(p)| \leq c(p)$ where $c$ is a given non-negative scalar function[2]. Discrete flow conservation constraint for nodes on a graph has a clear continuous interpretation as well: a continuous stream/flow is "preserved" at points inside the domain if vector field $\bar{f}$ is divergence-free $div \bar{f} = 0$. Strang also gives appropriate definition for sources and sinks on the boundary of the domain[3]. Then, the continuous analogue of the maximum flow problem is straightforward: find a maximum amount of water that continuous stream $\bar{f}$ can take from sources to sinks across the domain while satisfying all the constraints.

The main topic of this sections connects to [767] as follows. Strang defines a "real" cut on $\Omega$ as a hypersurface $\gamma$ that divides the domain into two subsets. The minimum cut should separate sources and sinks and have the smallest possible cost $\int_\gamma c$ which can be interpreted as a length of hypersurface $\gamma$ in isotropic metric defined by a scalar function $c$. Strang also establishes duality between continuous versions of minimum cut and maximum flow problems that is analogous to the discrete version established by Ford and Fulkerson [324]. On a practical note,

[1]piece-wise twice differentiable, see [477] for more details.

[2]More generally, it is possible to set an anisotropic "speed limit" constraint $\bar{f}(p) \in c(p)$ where $c$ is some convex set defined at every point $p \in \Omega$.

[3]Sources and sinks can also be placed inside the domain. They would correspond to points in $\Omega$ where $div \bar{f}$ is non-null, t.e. where stream $\bar{f}$ has an in-flow or out-flow.

a recent work by Appleton&Talbot [28] proposed a finite differences approach that, in the limit, converges to a globally optimal solution of continuous min-cut/max-flow problem defined by Strang. Note, however, that they use graph cuts algorithms to "greatly increase the speed of convergence".

## 5.5 Generalizing Graph Cuts for Multi-Label Problems

In this section, we show that even though graph cuts provide an inherently binary optimization, they can be used for multi-label energy minimization. In some cases, minimization is exact, but in more interesting cases only approximate minimization is possible. There is a direct connection between the exact multi-label optimization and a graph cut as a hypersurface interpretation of Section 5.4. We begin by stating the general labeling problem, then in Section 5.5.1 we describe the case when optimization can be performed exactly. Finally, Section 5.5.2 describes the approximate minimization approaches and their quality guarantees.

Many problems in vision and graphics can be naturally formulated in terms of multi-label energy optimization. Given a set of sites $\mathcal{P}$ which represent pixels/voxels, and a set of labels $\mathcal{L}$ which may represent intensity, stereo disparity, a motion vector, etc., the task is to find a labeling $f$ which is a mapping from sites $\mathcal{P}$ to labels $\mathcal{L}$. Let $f_p$ be the label assigned to site $p$ and $f$ be the collection of such assignments for all sites in $\mathcal{P}$.

We can use the same general form of energy (5.2) that was earlier introduced in the context of binary labeling problems. The terms $D_p(l)$ are derived from the observed data and it expresses the label preferences for each site $p$. The smaller the value of $D_p(l)$, the more likely is the label $l$ for site $p$. Since adding a constant to $D_p(l)$ does not change the energy formulation, we assume, without loss of generality, that $D_p(l)$'s are nonnegative. The pairwise potential $V_{pq}(l_p, l_q)$ expresses prior knowledge about the optimal labeling $f$. In general, prior knowledge can be arbitrarily complex, but in graph cuts based optimization, we are essentially limited to different types of spatial smoothness priors. Typically $V_{pq}(l_p, l_q)$ is a nondecreasing function of $||l_p - l_q||$[4]. Different choices of $V_{pq}(l_p, l_q)$ imply different types of smoothness, see Sections 5.5.1 and 5.5.2 .

### 5.5.1 *Exact Multi-Label Optimization*

In this section, we describe the only known case of exact multi-label minimization of energy (5.2) via graph cuts. The corresponding graph construction is not covered by the general theoretical result in [491], which applies to binary labeling cases only. We have to make the assumption that labels are linearly ordered. This assumption limits the applicability of the method. For example, it cannot be directly used for motion estimation, since motion labels are 2 dimensional and

[4]Here we used the norm $|| \cdot ||$ notation because, in general, $l_p$ may be a vector

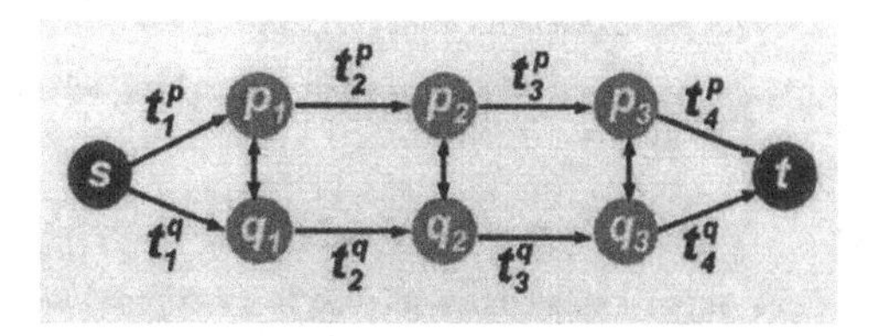


Figure 5.6. Part of the graph construction for energy minimization in 5.3 , $|\mathcal{L}| = 4$

cannot be linearly ordered[5]. Without loss of generality, assume that labels are integers in the range $\mathcal{L} = \{1, ..., k\}$. Let $V_{pq} = \lambda_{pq}|f_p - f_q|$. Then the energy is:

$$E(f) = \sum_{p \in \mathcal{P}} D_p(f_p) + \sum_{(p,q) \in \mathcal{N}} \lambda_{pq}|f_p - f_q|, \tag{5.3}$$

In vision, [425, 111] were the first to minimize energy (5.3) with a minimum cut on a certain graph $\mathcal{G}$. In fact, this graph is topologically similar to a graph of Roy&Cox [693] where separating hypersurface on 3D manifold gives a stereo disparity map, see Section 5.4.3.

The graph is constructed as follows. As usual, vertices $\mathcal{V}$ contain terminals $s$ and $t$. For each site $p$, create a set of nodes $p_1, ..., p_{k-1}$. Connect them with edges $\{t_1^p, ..., t_k^p\}$, where $t_1^p = (s, p_1)$, $t_j^p = (p_{j-1}, p_j)$, and $t_k^p = (p_{k-1}, t)$. Each edge $t_j^p$ has weight $K_p + D_p(j)$, where $K_p = 1 + (k-1)\sum_{q \in \mathcal{N}_p} \lambda_{pq}$. Here $\mathcal{N}_p$ is the set of neighbors of $p$ . For each pair of neighboring sites $p, q$ and for each $j \in \{1, ..., k-1\}$, create an edge $(p_j, q_j)$ with weight $\lambda_{pq}$. Figure 5.6 illustrates the part of $\mathcal{G}$ which corresponds to two neighbors $p$ and $q$. For each site $p$, a cut on $\mathcal{G}$ severs at least one edge $t_i^p$. The weights for $t_i^p$ are defined sufficiently large so that the minimum cut severs exactly one of them for each $p$. This establishes a natural correspondence between the minimum cut and an assignment of a label to $p$. If the minimum cut severs edge $t_i^p$, assign label $i$ to $p$. It is straightforward to show that the minimum cut corresponds to the optimum $f$ [111].

Ishikawa [424] generalized the above construction to minimize any energy function with convex $V_{pq}$'s. His construction is similar to the one in this section, except even more edges between $p_i$'s and $q_j$'s have to be added. Unfortunately, a convex $V_{pq}$ is not suitable for the majority of vision applications, especially if the number of labels is large. Typically, object properties tend to be smooth everywhere except the object boundaries, where discontinuities may be present. Thus in vision, a piecewise smooth model is more appropriate than the everywhere smooth model. However using a convex $V_{pq}$ essentially corresponds to the everywhere smooth model. The penalty that a convex $V_{pq}$ imposes on a sharp jumps in labels is so large, that in the optimal $f$ discontinuities are smoothed out with a "ramp". It is much cheaper to create a few small jumps in $f$ rather than one large jump.

[5]Iterative application of the algorithm described here was used for motion in [694]

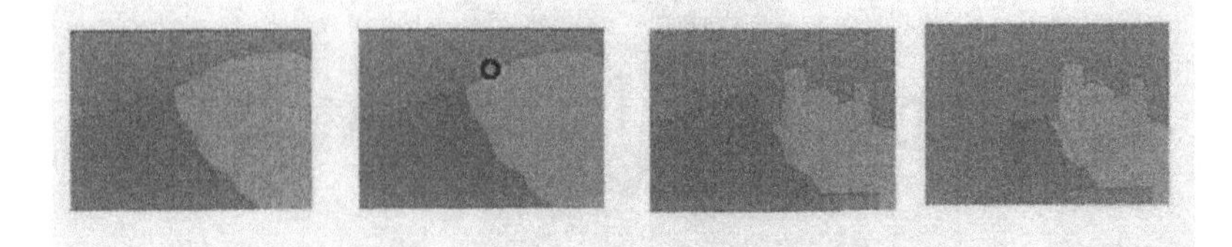

Figure 5.7. From left to right: a labeling $f$, a labeling within one standard move of $f$ (the changed site is highlighted by a black circle), labeling within one green-yellow swap of $f$, labeling within one green expansion of $f$.

Of all the convex $V_{pq}$, the one in (5.3) works best for preserving discontinuities. Nevertheless in practice, it oversmooths disparity boundaries [837].

### 5.5.2 *Approximate Optimization*

The potential $V_{pq}$ in the previous section is not discontinuity preserving because $V_{pq}$ is allowed to grow arbitrarily large. One way to construct a discontinuity preserving $V_{pq}$ is to cap its maximum value. Perhaps the simplest example is the Potts model $V_{pq} = \lambda_{pq} \cdot \mathcal{I}(f_p \neq f_q)$ [113]. We have already seen Potts $V_{pq}$ in Section 5.3.1[6], and it corresponds to the piecewise constant prior on $f$. Unfortunately, energy minimization with Potts $V_{pq}$ is NP-hard [113], however graph cuts can be used to find an answer within a factor of 2 from the optimum [113].

In this section, we describe two approximation methods, the expansion and the swap algorithms [113]. According to the results in [491], the swap algorithm may be used whenever $V_{pq}(\alpha,\alpha) + V_{pq}(\beta,\beta) \leq V_{pq}(\alpha,\beta) + V_{pq}(\beta,\alpha)$ for all $\alpha, \beta \in \mathcal{L}$, which we call the swap inequality. The expansion algorithm may be used whenever $V_{pq}(\alpha,\alpha)+V_{pq}(\beta,\gamma) \leq V_{pq}(\alpha,\gamma)+V_{pq}(\beta,\alpha)$ for all $\alpha, \beta, \gamma \in \mathcal{L}$, which we call the expansion inequality. Any $V_{pq}$ which satisfies the expansion inequality also satisfies the swap inequality, hence the expansion inequality is more restrictive.

Both swap and expansion inequalities admit discontinuity preserving $V_{pq}$'s. The truncated linear $V_{pq}(\alpha,\beta) = min(T, ||\alpha - \beta||)$ satisfies the expansion inequality. The truncated quadratic $V_{pq}(\alpha,\beta) = min(T, ||\alpha - \beta||^2)$ satisfies the swap inequality. Here $T$ is a positive constant, which is the maximum penalty for a discontinuity. The truncated linear and truncated quadratic $V_{pq}$ correspond to a piecewise smooth model. Small deviations in labels incur only a small penalty, thus the smoothness is encouraged. However sharp jumps in labels are occasionally permitted because the penalty $T$ is not too severe to prohibit them.

[6]In the binary case, it is typically called the Ising model.

#### 5.5.2.1 Local Minimum with Respect to Expansion and Swap Moves

Both the expansion and the swap algorithms find a local minimum of the energy function. However, in discrete optimization, the meaning of "a local minimum" has to be defined. For each $f$, we define a set of moves $M_f$. Intuitively, these are the moves to other labelings that are allowed from $f$. Then we say that $f$ is a local minimum with respect to the set of moves, if for any $f' \in M_f$, $E(f') \geq E(f)$. Most discrete optimization methods (e.g. [341, 81]) use *standard* moves, defined as follows. Let $H(f, f')$ be the number of sites for which $f$ and $f'$ differ. Then for each $f$, standard moves are $M_f = \{f' | H(f, f') \leq 1\}$. Thus a standard move allows to change a label of only one site in $f$, and hence $|M_f|$ is linear in the number of sites, making it is easy to find a local minimum with respect to the standard moves. The result, however is very dependent on the initial point since a high dimensional energy has a huge number of such local minima. In particular, the solution can be arbitrarily far from the global minimum.

We now define the swap moves. Given a labeling $f$ and a pair of labels $\alpha$ and $\beta$, a move $f^{\alpha\beta}$ is called an $\alpha$-$\beta$ swap if the only difference between $f$ and $f^{\alpha\beta}$ is that some sites that were labeled $\alpha$ in $f$ are now labeled $\beta$ in $f^{\alpha\beta}$, and some sites that were labeled $\beta$ in $f$ are now labeled $\alpha$ in $f^{\alpha\beta}$. $M_f$ is then defined as the collection of $\alpha$-$\beta$ swaps for all pairs of labels $\alpha, \beta \in \mathcal{L}$.

We now define the expansion moves. Given a labeling $f$ and a label $\alpha$, a move $f^\alpha$ is called an $\alpha$-expansion if the only difference between $f$ and $f^\alpha$ is that some sites that were not labeled $\alpha$ in $f$ are now labeled $\alpha$ in $f^\alpha$. $M_f$ is then defined as the collection of $\alpha$-expansions swaps for all labels $\alpha \in \mathcal{L}$. Figure 5.7 shows an example of standard move versus $\alpha$-expansion and $\alpha$-$\beta$ swap. Notice that a standard move is a special case of an $\alpha$-expansion and a $\alpha$-$\beta$ swap. However there are $\alpha$-expansion moves which are not $\alpha$-$\beta$ swaps and vice versa.

The expansion (swap) move algorithm finds a local minimum with respect to expansion (swap) moves. The number of expansion (swap) moves from each labeling is exponential in the number of sites. Thus direct search for an optimal expansion (swap) move is not feasible. This is where graph cuts are essential. It is possible to compute the optimal $\alpha$-expansion or the optimal $\alpha$-$\beta$ swap with the minimum cut on a certain graph. This is because computing an optimal $\alpha$-expansion (optimal $\alpha$-$\beta$ swap) is a binary minimization problem which happens to be regular [491] when the expansion (swap) inequality holds.

The expansion (swap) algorithms are iterative. We start with an initial labeling $f$. We then cycle in random order until convergence over all labels $\alpha \in \mathcal{L}$ (pairs of $\alpha, \beta \in \mathcal{L}$), find the optimal $f^\alpha$ ($f^{\alpha\beta}$) out of all $\alpha$-expansions ($\alpha$-$\beta$-swaps), and change current labeling to $f^\alpha$ ($f^{\alpha\beta}$). Obviously this cannot lead to an increase in energy, and at convergence we found the local minimum with respect to expansion (swap) moves. Thus the key step is how to find the optimal $\alpha$-expansion ($\alpha$-$\beta$ swap), which is performed by finding a minimum cut on a certain graph $\mathcal{G} = (\mathcal{V}, \mathcal{E})$. The actual graph constructions can be found in [113].

The criteria for a local minimum with respect to the expansions (swaps) are so strong that there are significantly fewer of such minima in high dimensional spaces compared to the standard moves. Thus the energy function at a local minimum is likely to be much lower. In fact, it can be shown that the local minimum with respect to expansion moves is within a constant factor of optimum. The best approximation is in case of the Potts model, where this factor is 2. It is not surprising then that most applications based on graph cuts use the expansion algorithm with the Potts model [111, 88, 489, 490, 895, 499, 521, 403, 10, 900].

# Chapter6

# Minimal Paths and Fast Marching Methods for Image Analysis

L. Cohen

Abstract

We present an overview of part of our work on minimal paths. Introduced first in order to find the global minimum of active contours' energy using Fast Marching [210], we have then used minimal paths for finding multiple contours for contour completion from points or curves in 2D or 3D images. Some variations allow to decrease computation time, make easier initialization and centering a path in a tubular structure. Fast Marching is also an efficient way to solve balloon model evolution using level sets. We show applications like for road and vessel segmentation and for virtual endoscopy.

## 6.1 Introduction

Deformable models have been the object of considerable studies and variations since their introduction in [456]. Most of the approaches that were introduced since then tried to overcome the main drawbacks of this model: initialization, minimization and topology changes. The model requires the user to input an initial curve close to the goal. Using the balloon model [204] allows a less demanding initialization. Level sets approaches have the same property [152, 538, 157]. A region-based approach (for example [207, 205]) also makes the solution less sensitive to local minima and initialization. Also, a priori knowledge included in a parametric deformable model (for example [51, 203]) allows to be more robust.

However, for images like the one in figure 6.4, a very precise initialization is needed to avoid the active contour being trapped by an insignificant local minimum of the energy [205, 204]. In order to find a global minimum for the energy, authors of [210] have introduced a minimal path approach. This is based on previous work by [472, 469] in a different framework. Curve initialization is replaced by just giving two endpoints. The numerical method has the advantages of being consistent (see [210]), fast and efficient, using the *Fast-Marching* algorithm introduced in [730].

This chapter contains various improvements of the original method, relevant in 2D or 3D. Some of the problems we dealt with for segmentation and contour extraction, finding trajectories and perceptual grouping are presented in this paper as follows:

- Minimal path between two points: The solution proposed in [209, 210] with Fast Marching is reviewed in Section 6.2.
- Minimal paths between an ordered list of points or a given set of pairs of points is a simple application of the previous case.
- Minimal paths for a given unstructured set of points: we propose a way to find pairs of linked neighbors and paths between them [206] (Section 6.3).
- Minimal paths between an unknown set of key points to be determined from a larger set of admissible points [206].
- Minimal paths for an unstructured set of connected components, by extending the previous approaches to determine pairs of regions to be linked. [266] (Section 6.4).
- Segmentation of 2D and 3D tubular and tree structures [264, 265] (sections 6.4 et 6.5).
- Finding a centered path inside a tubular structure and application to virtual endoscopy [264] (section 6.6).

## 6.2 Minimal Paths

### 6.2.1 *Geometrical optics*

In order to understand *Fermat* Principle which is the physical interpretation of minimal paths described afterwards, we illustrate light propagation in two simple cases.

According to *Fermat* Principle, the path followed by monochromatic light to go from a point $p_0$ to a point $p_1$ is the path which takes least time. In the case of an homogeneous medium, light speed is constant, and thus light follows a straight line, since shortest time is proportional to distance, as seen on figure 6.1-left. Sets of points that are reached at a given time are circles.

Let us now consider a non homogeneous medium composed of two homogeneous regions separated by a horizontal line in the middle, like in Figure 6.1-middle. Assuming that light speed is larger in the bottom rectangle, the trajectory will "prefer" to remain in this rectangle as much as possible. As a consequence, trajectories are submitted to a refraction effect, as seen on a few trajectories shown in the figure. Angles between the two lines and the normal to

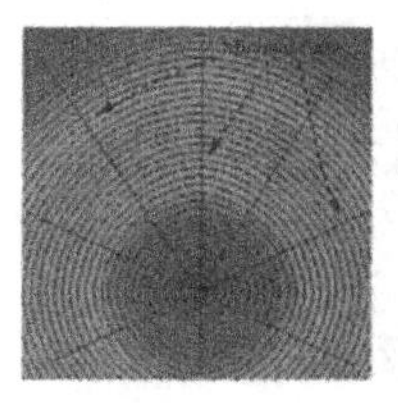 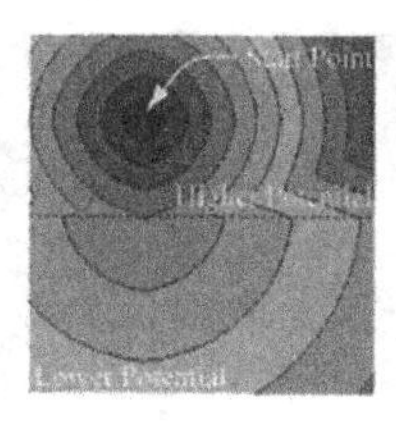 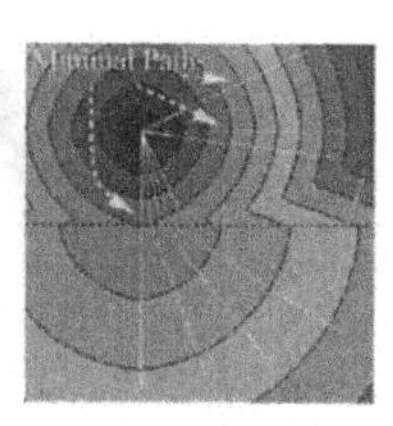 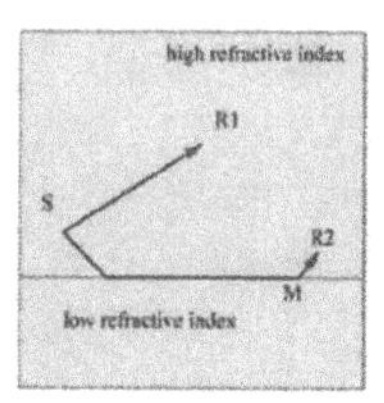


Figure 6.1. Cost function by front propagation and minimal paths for a potential with one or two values. See text.

the interface between the two media satisfy Snell-Descartes'law (ratio of their sines is equal to the ratio of refraction indices). The refraction index $n > 1$ is the ratio between light speed in emptiness $c$ and its speed in the considered medium $v$. From this definition, travel time T between two points is the integral along the followed path of the inverse of the speed $\frac{1}{v} = \frac{n}{c}$. The followed path is a minimum for $T = \frac{1}{c}\int_{p_0}^{p_1} n ds$. The Eikonal equation (see section 6.2.4) was obtained for this minimization by Hamilton, as a special case of Hamilton-Jacobi equations.

One of the trajectories shown again on figure 6.1-right illustrates the well known mirage effect. Light source $S$ is visible from points $R_1$ et $R_2$. But the path followed between $S$ and $R_2$ is not a straight line, since light "prefers" going through the smaller refraction index area to go faster. This is a common phenomenon when temperature variations are large enough between the ground and atmosphere, making believe an observer at $R_2$ there is an oasis in the desert. Similarity will be obvious in the following sections where active contours potential $P$ takes the same place as refraction index $n$.

### 6.2.2 *Global Minimum for active contours*

We present in this section the basic ideas of the method introduced in [210] to find the global minimum of the active contour energy using minimal paths. The energy to minimize is similar to classical deformable models (see [456]) where it combines smoothing terms and image features attraction term:

$$E(C)=\int_{\Omega}\left\{w_1\|C'(s)\|^2+ w_2\|C''(s)\|^2+P(C(s))\right\}ds \tag{6.1}$$

where $C(s)$ represents a curve drawn on a 2D image and $\Omega$ is its domain of definition. The method of [210] improves energy minimization since the problem is transformed in a way allowing to find the global minimum.

### 6.2.3 *Problem formulation*

As explained in [210], skipping second order term, we are lead to minimize

$$E(C) = \int_{\Omega=[0,L]} \{w + P(C(s))\}ds, \tag{6.2}$$

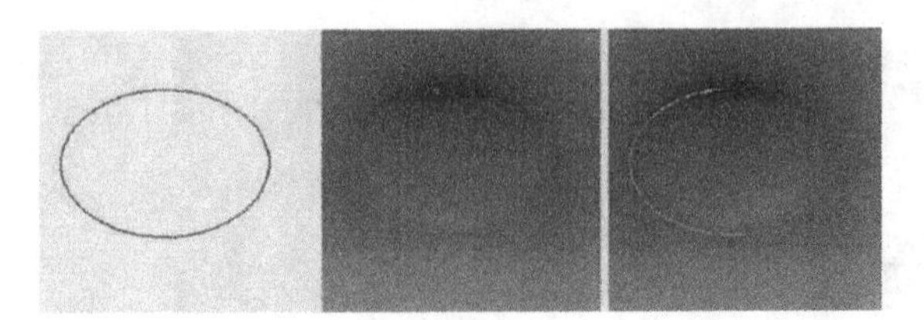

Figure 6.2. On the left, the potential is defined to be minimal on the ellipse. In the middle, the minimal action or weighted distance to the marked point. On the right, minimal path using backpropagation from the second point.

where $s$ is the arclength parameter ($\|C'(s)\| = 1$). The regularization of this model is now achieved by the constant $w > 0$ (see [210] for details). Given a potential $P \geq 0$, the energy is like a distance weighted by $\tilde{P} = P + w$. The minimal action $\mathcal{U}$ is defined as the minimal energy integrated along a path between starting point $p_0$ and any point $p$:

$$\mathcal{U}(p) = \inf_{\mathcal{A}_{p_0,p}} E(C) = \inf_{\mathcal{A}_{p_0,p}} \left\{ \int_{\Omega} \tilde{P}(C(s))ds \right\} \tag{6.3}$$

where $\mathcal{A}_{p_0,p}$ is the set of all paths between $p_0$ and $p$. The minimal path between $p_0$ and any point $p_1$ in the image can be easily deduced from this action map by a simple back-propagation (gradient descent on $\mathcal{U}$) starting from $p_1$ until $p_0$ is reached. This backpropagation step is made possible due to the fact that $\mathcal{U}$ has no local minimum except point $p_0$, therefore the descent converges to $p_0$ for any $p_1$. More accurate gradient descent methods like Runge-Kutta midpoint algorithm or Heun's method can be used.

### 6.2.4 Fast Marching Resolution

In order to compute $\mathcal{U}$, a front-propagation equation related to Eqn. (6.3) is solved: $\frac{\partial C}{\partial t} = \frac{1}{\tilde{P}} \vec{n}$. It evolves a front $C$ starting from an infinitesimal circle shape around $p_0$ until each point inside the image domain is assigned a value for $\mathcal{U}$. The value of $\mathcal{U}(p)$ is the time $t$ at which the front passes over $p$. The *Fast Marching* technique, introduced in [730], was used in [209, 210] noticing that the map $\mathcal{U}$ satisfies the *Eikonal* equation $\|\nabla \mathcal{U}\| = \tilde{P}$ and $\mathcal{U}(p_0) = 0$. The relation with this equation will be explained in section 6.5. Since classic finite difference schemes for this equation are unstable, an up-wind scheme was proposed by [730]:

$$\begin{aligned}(\max\{u - \mathcal{U}_{i-1,j}, u - \mathcal{U}_{i+1,j}, 0\})^2 + \\ (\max\{u - \mathcal{U}_{i,j-1}, u - \mathcal{U}_{i,j+1}, 0\})^2 = \tilde{P}^2_{i,j}.\end{aligned} \tag{6.4}$$

The improvement made by the *Fast Marching* is to introduce order in the selection of the grid points. This order is based on the fact that information is propagating *outward*, because the action can only grow due to the quadratic Eqn. (6.4).

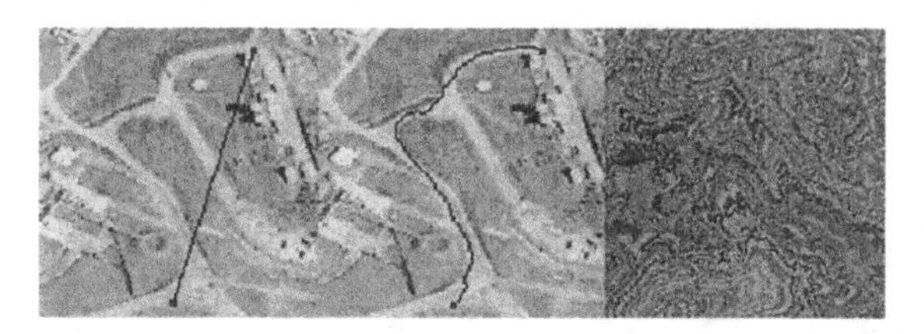

Figure 6.3. Global minimum of active contour model. After giving two points on the left, the minimal path between them is found in the middle image. On the right we show the cost function from the start point. Notice faster propagation along the roads. Potential is defined as a decreasing function of the gray level.

The main idea is similar to the construction of minimum length paths in a graph between two given nodes introduced in [269] (see discussion in [210]).

Complexity of *Fast Marching* on a grid with $N$ nodes is bounded by $O(N \log_2 N)$ for the *Fast Marching* on a grid with $N$ nodes. The algorithm is

---

**Algorithm for 2D Fast Marching for minimal action $\mathcal{U}$**

Definitions:

- *Alive* set: grid points at which values of $\mathcal{U}$ have been reached and will not be changed;
- *Trial* set: next grid points (4-connexity neighbors) to be examined. An estimate $U$ of $\mathcal{U}$ is computed using Eqn. (6.4) from alive neighbors only;
- *Far* set: all other grid points, there is not yet an estimate for $U$;

Initialization:

- *Alive* set: start point $p_0$, $U(p_0) = \mathcal{U}(p_0) = 0$;
- *Trial* set: four neighbors $p$ of $p_0$ with initial value $U(p) = \tilde{P}(p)$ ($\mathcal{U}(p) = \infty$);
- *Far* set: all other grid points, $\mathcal{U} = U = \infty$;

Loop:

- Let $p = (i_{min}, j_{min})$ be the *Trial* point with the smallest action $U$;
- Move it from the *Trial* to the *Alive* set;
- For each neighbor $(i, j)$ of $(i_{min}, j_{min})$:
  - If $(i, j)$ is *Far*, add it to the *Trial* set;
  - If $(i, j)$ is *Trial*, update $U_{i,j}$ with Eqn. (6.4).

---

Table 6.1. *Fast Marching* algorithm

detailed in Table 6.1. Examples are shown in Fig. 6.2 to 6.4. Solving Eqn. (6.4) is detailed next.

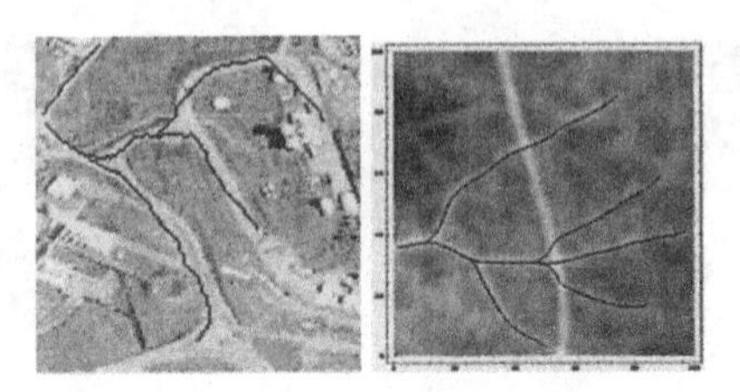

Figure 6.4. Many minimal paths are obtained from a same start point and many end points. This allows extracting the set of roads in the aerial image on the left and vessels in the eye fundus image on the right.

### 6.2.5 2D Up-Wind Scheme

Notice that for solving Eqn. (6.4), only values of alive neighbor points are considered (Table 6.1). Considering the neighbors of grid point $(i, j)$ in 4-connexity, we note $\{A_1, A_2\}$ and $\{B_1, B_2\}$ the two couples of opposite neighbors such that we get the ordering $\mathcal{U}(A_1) \leq \mathcal{U}(A_2), \mathcal{U}(B_1) \leq \mathcal{U}(B_2)$, and $\mathcal{U}(A_1) \leq \mathcal{U}(B_1)$. Considering that we have $u \geq \mathcal{U}(B_1) \geq \mathcal{U}(A_1)$, the equation derived is

$$(u - \mathcal{U}(A_1))^2 + (u - \mathcal{U}(B_1))^2 = \tilde{P}^2_{i,j} \tag{6.5}$$

Based on testing the discriminant $\Delta$ of Eqn. (6.5), one or two neighbors are used to solve it:

1. If $\tilde{P}_{i,j} > \mathcal{U}(B_1) - \mathcal{U}(A_1)$, solution of Eqn. (6.5) is $u = \frac{\mathcal{U}(B_1)+\mathcal{U}(A_1)+\sqrt{2\tilde{P}^2_{i,j}-(\mathcal{U}(B_1)-\mathcal{U}(A_1))^2}}{2}$.

2. else $u = \mathcal{U}(A_1) + \tilde{P}_{i,j}$.

### 6.2.6 Minimal Paths in 3D

A 3D extension of the *Fast Marching* algorithm was presented in [264]. Similarly to previous section, the minimal action $\mathcal{U}$ is defined as $\mathcal{U}(p) = \inf_{\mathcal{A}_{p_0,p}} \left\{ \int_\Omega \tilde{P}(C(s))ds \right\}$ where $\mathcal{A}_{p_0,p}$ is now the set of all 3D paths between $p_0$ and $p$. Given a start point $p_0$, in order to compute $\mathcal{U}$ we start from an initial infinitesimal spherical front around $p_0$. The 2D scheme of equation (6.4) is extended to 3D, leading to:

$$(\max\{u - \mathcal{U}_{i-1,j,k}, u - \mathcal{U}_{i+1,j,k}, 0\})^2 + (\max\{u - \mathcal{U}_{i,j-1,k}, u - \mathcal{U}_{i,j+1,k}, 0\})^2 + (\max\{u - \mathcal{U}_{i,j,k-1}, u - \mathcal{U}_{i,j,k+1}, 0\})^2 = \tilde{P}^2_{i,j,k} \tag{6.6}$$

giving the correct viscosity-solution $u$ for $\mathcal{U}_{i,j,k}$. An example is given in figure 6.13 of section 6.6.

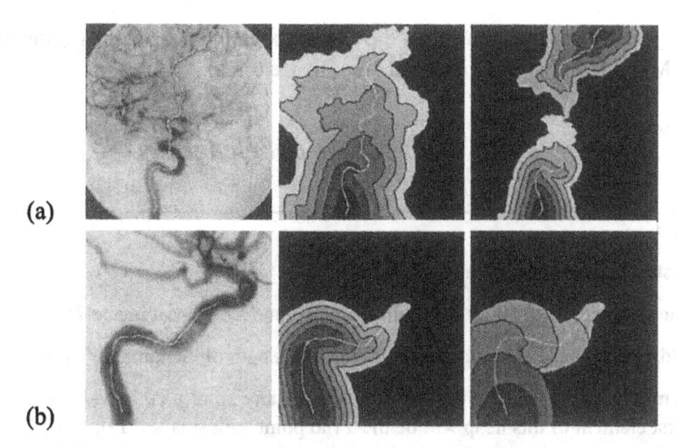


Figure 6.5. (a) Simultaneous propagation: The left image is the data set, used as potential to extract a path in a vessel. In the middle, the action map is obtained from the first point till second point is reached. The right image shows the action map resulting from a simultaneous propagation from both extremities points, and the two paths from the intersection point. (b) Simultaneous estimate of the path length. On the left, potential; In the middle, minimal action map; on the right, length of the minimal path. These maps are computed only until a given length is reached.

### 6.2.7 Simultaneous Front Propagation

The idea is to propagate simultaneously a front from each end point $p_0$ and $p_1$ [264]. Let us consider the first grid point $p$ where those fronts meet. This point has to be on the minimal path between $p_0$ and $p_1$. Since during propagation the action can only grow, propagation can be stopped at this step in order to make backpropagation. Adjoining the two paths, respectively between $p_0$ and $p$, and $p_1$ and $p$, gives an approximation of the exact minimal action path between $p_0$ and $p_1$. Since $p$ is a grid point, the exact minimal path might not go through it, but in its neighborhood. Precise location can be obtained through interpolation between grid points like in [643]. This algorithm is described in table 6.2. This approach

---

**Algorithm**

- Compute the minimal action maps $U_0$ and $U_1$ to respectively $p_0$ and $p_1$ until the two fronts have an *Alive* point $p_2$ in common;
- Compute the minimal path between $p_0$ and $p_2$ by back-propagation on $U_0$ from $p_2$;
- Compute the minimal path between $p_1$ and $p_2$ by back-propagation on $U_1$ from $p_2$;
- Join the two paths found.

---

Table 6.2. Minimal Path from two action maps

allows a parallel implementation of the two propagations. Also, the region covered by Fast Marching is greatly reduced (see Figure 6.5.a).

### 6.2.8 Simultaneous estimate of the path length

---

**Notations**

- a start point $p_0$ is manually set;
- the minimal energy map $U$, a min-heap $\mathcal{H}_U$ and a potential image $P$;
- a distance map $D$ to compute the Euclidean length of the minimal path ;
- a min-heap $\mathcal{H}_D$, where the ordering key for any point $p$ is the value of $D(p)$ (the first element of this heap will be the *Trial* point with smallest $D$);

**Initialization**

- initialize the front propagation method, by setting $U(p_0) = D(p_0) = 0$ and storing $p_0$ in both min-heaps $\mathcal{H}_U$ and $\mathcal{H}_D$;

**Loop:** at each iteration, consider $p_{min}$ the *Trial* point with smallest $U$

- Move it to *Alive* set, and remove it from both $\mathcal{H}_U$ and $\mathcal{H}_D$
- for each neighbor $p$ of $p_{min}$:
  - proceed according to the classical Fast Marching algorithm: update $U(p)$ and re-balance $\mathcal{H}_U$;
  - update $D(p)$ according to $||\nabla D|| = 1$ using the same neighbors of $p$ that were involved in updating $U(p)$ and re-balance $\mathcal{H}_D$

---

Table 6.3. Computing the Euclidean Distance traveled by the front.

In some cases, like for giving extremities in a 3D image, it is easier for the user to give only one start point and the second should be found automatically. We now describe an approach which builds a path given a starting point and a given path length to reach [264]. We are able to compute simultaneously at each point of the front energy $U$ of the minimal path and its length. The end point is then chosen as the first point that reach the expected length. Propagation is stopped when this point is reached and minimal path is computed from it. Since the front propagates faster along small values of the potential, the interesting paths are longer among all paths which have same minimal action $U$. When the front propagates in a tubular structure, all points who reach first the given length are in a same region of the image, far from the starting point and inside the tubular shape. This gives a justification for this choice of end point (see Figure 6.5.b).

Once the path is extracted by gradient descent, we can easily compute its length. But this is a very time consuming process to systematically do this at each point visited. Therefore we proposed to compute on-the-fly an approximation of the distance traveled by the front. We use the property that when propagating a front with a constant speed equal to one, the minimal energy obtained at each point

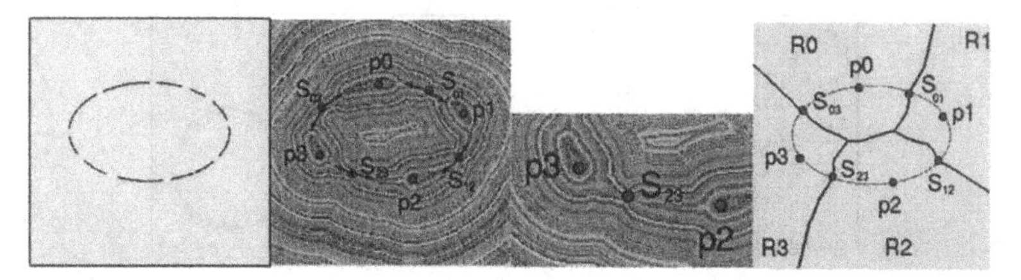


Figure 6.6. Ellipse. From left to right: potential is an incomplete ellipse and points $p_k$ are given; level sets of minimal action $U$ from $p_k$'s; zoom on a *saddle point*; backpropagation from selected *saddle points* to their two source points give the set of paths and voronoi diagram.

represents the *Euclidean* distance $D$ to the starting point. The Euclidean length of the path is found solving $\|\nabla D\| = 1$ using with the same neighbors involved for $P$ in Eqn. (6.5). The corresponding algorithm is described in table 6.3. This algorithm was used for reducing user-intervention in the $Virtual$ Endoscopy process presented in section 6.6 by giving only one point [264].

## 6.3 Minimal paths from a set of endpoints $p_k$

---

**Minimal paths between points $p_k$, minimal action $V = \mathcal{U}_{\{p_k, 0 \leq k \leq N\}}$**

- Initialization:
  - $p_k$'s are given; $\forall k, V(p_k) = 0; l(p_k) = k$ is the front index, $p_k$ *alive*.
  - $\forall p \notin \{p_k\}, V(p) = \infty; l(p) = -1;$ $p$ is *far* except 4-connexity neighbors of $p_k$'s that are *trial* with estimate $U$ using Eqn. (6.4).
- Loop for computing $V = \mathcal{U}_{\{p_k, 0 \leq k \leq N\}}$:
  - Let $p = (i_{min}, j_{min})$ be the *Trial* point with the smallest action $U$;
  - Move it from the *Trial* set to the *Alive* set with V(p) = U(p);
  - Update $l(p)$ with the same index as point $A_1$ in formula (6.5). If $l(A_1) \neq l(B_1)$ and we are in case 1 of section 6.2.5 where both points are used and if this is the first time regions of labels $l(A_1)$ and $l(B_1)$ meet, $S(p_{l(A_1)}, p_{l(B_1)}) = p$ is set as the *saddle point* between $p_{l(A_1)}$ and $p_{l(B_1)}$. If these points have not yet two *linked neighbors*, they are put as *linked neighbors* and $S(p_{l(A_1)}, p_{l(B_1)}) = p$ is selected,
    For each neighbor $(i, j)$ of $(i_{min}, j_{min})$:
    * If $(i, j)$ is *Far*, add it to the *Trial* set;
    * If $(i, j)$ is *Trial*, update action $U_{i,j}$.
- Obtain all paths between selected *linked neighbors* by backpropagation each way from their *saddle point*.

---

Table 6.4. Algorithm for unstructured set of points.

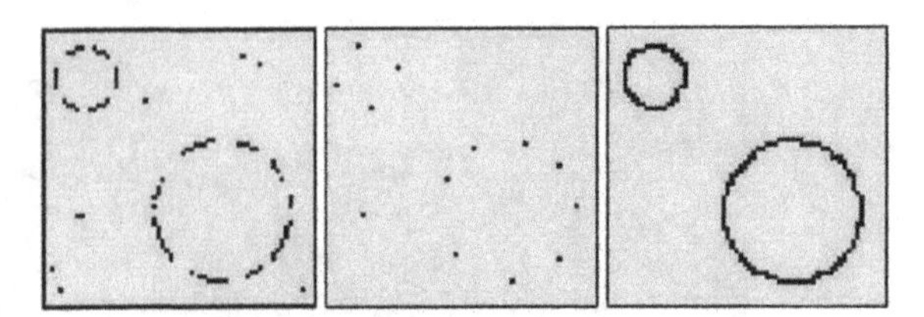

Figure 6.7. Two circles: from left to right: incomplete noisy data set; the set of found $p_k$'s; multiple minimal paths between $p_k$'s.

## *Multiple minimal paths*

We propose to use the minimal path approach to extract a set of contours from an unstructured set of points given on an image. In order to find the set of most representative contours on the image, we are looking for minimal paths between pairs of points. We describe briefly the method when points $p_k$ are already known. An approach to automatically find points $p_k$ that are most representative among a larger set of admissible points was introduced in [206], based on an iterative farthest point strategy relative to the weighted distance. Such a strategy was used later on to find adaptive or uniform remeshing of a surface using fast marching [643].

We assume here that points $p_k$ are known. If we knew as well which pairs of points have to be linked among $p_k$'s, finding all contours is a trivial application of section 6.2. The problem we are interested in here is also to find out which pairs of points have to be connected by a contour. Since the set of points $p_k$'s is assumed to be given unstructured, we do not know in advance how the points connect. This is the key problem that is solved here using a minimal action map.

The main goal of our method is to obtain all significant paths joining the given points. However, each point should not be connected to all other points, but only to those that are closer to them in the energy sense. There are many possibilities to decide which pairs of points have to be linked. It depends on data and on the application in view. In some cases, it is necessary to detect closed curves and avoid bifurcation, or T-junctions. The criterion is then to constrain a point $p_k$ to be linked to at most two other points among $p'_k s$, in order to generate a closed curve. In case we are looking for tree structures, the criterion is different, as in section 6.4.

For perceptual grouping, potential $P$ to be minimized along the paths is often a binary image of edge points, that form incomplete contours, as on figure 6.6-left. Attraction potential to the set of edge points can be defined (see [208]) in order to have lower values along edge points and higher values in the background.

## *Main ideas of the approach*

Our approach is similar to computing the distance map to a set of points and their Voronoi diagram. However, we use here a weighted distance defined through the

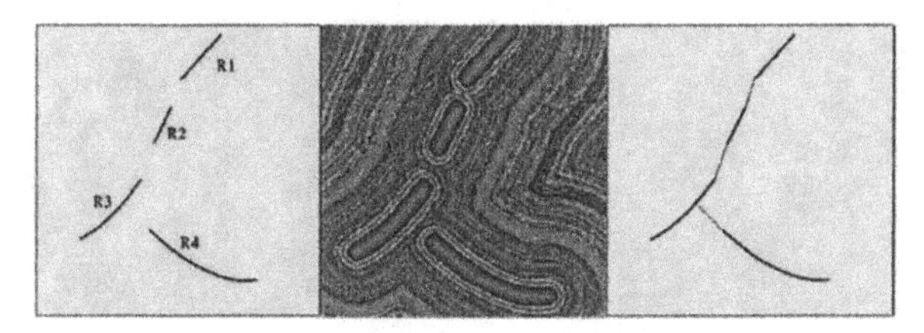


Figure 6.8. From left to right: examples of regions to link; level sets of the minimal action from the 4 regions; minimal paths obtained from the 3 selected *saddle points.*

potential $P$. This distance is obtained as the minimal action with respect to $P$ with zero value at all points $p_k$. Instead of computing a minimal action map for each pair of points, as in Section 6.2.3, we only need to compute one minimal action map in order to find all paths. At the same time the action map is computed we determine the pairs of points that have to be linked together by finding meeting points of the propagation fronts. These are *saddle points* of the minimal action $\mathcal{U}$.

Although the minimal action is computed using fast marching, the level sets of $\mathcal{U}$ give the evolution of the front. During the fast marching algorithm, the boundary of the set of alive points also gives the position of the front.

Figure 6.6 illustrates the steps of the algorithm. Figure 6.7 shows the result with points $p_k$ found automatically. More details can be found in [206].

## 6.4 Multiple minimal paths between regions $R_k$

We consider perceptual grouping and contour completion from an unstructured set of regions in a 2D or 3D image. As an extension of previous section 6.3, complete curves are obtained as minimal paths between pairs of regions [266]. This approach is extended to finding a set of minimal paths that connect a set of 3D regions in 3D images. This makes use of *Fast-Marching* in a 3D image, as in section 6.2.6 [264, 263].

### *Minimal path between 2 regions*

Defining a minimal path between two regions is an easy extension of [210]. Consider two connected regions, the start region $R_0$ and the set of end points $R_1$. The problem is finding a minimal path among all paths starting from a point in $R_0$ and ending on $R_1$. Minimal action is then defined as:

$$\mathcal{U}(p) = \inf_{\mathcal{A}_{R_0,p}} E(C) = \inf_{p_0 \in R_0} \inf_{\mathcal{A}_{p_0,p}} E(C) \tag{6.7}$$

where $\mathcal{A}_{R_0,p}$ is the set of paths starting from a point in $R_0$ and ending at $p$. This is computed using Fast Marching as in table 6.1, with initial set of *Alive* points being $R_0$, with $\mathcal{U} = 0$. In order to find a minimal path between $R_1$ and $R_0$, we

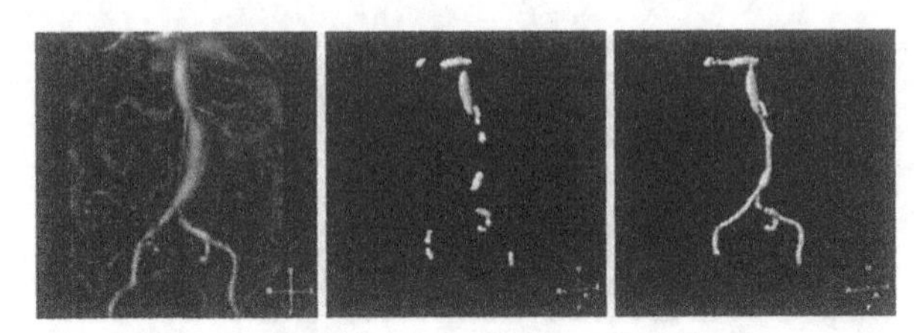

Figure 6.9. Perceptual grouping in the 3D aorta image: MIP view of vascularity potential; detection of regions in the aorta; vascular tree completion by minimal paths relatively to vascularity potential.

determine a point $p_1 \in R_1$ such that $\mathcal{U}(p_1) = \min_{p \in R_1} \mathcal{U}(p)$. We then make backpropagation from $p_1$ to $R_0$.

### *Tubular structures*

Linking regions can be useful when these regions are for example connected components obtained after edge detection. In the example of Fig. 6.8, which represent a tree structure, regions are selected in a way that they do not form together a closed curve. In medical imaging, finding vessels is a very important problem. Regions can then be defined from thresholding a vascularity criterion of [326] to detect tubular regions in a vessel image. In Figure 6.9, we show a MIP view of the vascularity potential [326] obtained from 3D MRI of the aorta with contrast product. We obtain a set of regions by thresholding the multiscale criterion. Our method helps completing these region and finding the structure of the vascular tree.

## 6.5 Segmentation by Fast Marching

Several approaches are possible to segment the boundary surface of an object starting from points inside. We can use for example a balloon model [204] or its level-sets implementation, as in [538]. In fact, this kind of region growing method can also be solved fast using the *Fast Marching* algorithm [535]. This allows to make a segmentation step in the same framework as minimal path finding. Having searched for the minimal action from one point, the algorithm provides the following regions:

- Inside : the points whose action is set, labeled *Alive*;
- Outside : the points not yet examined, labeled *Far*;
- the points at the interface between *Alive* and *Far* points, whose actions are not set, labeled *Trial*.

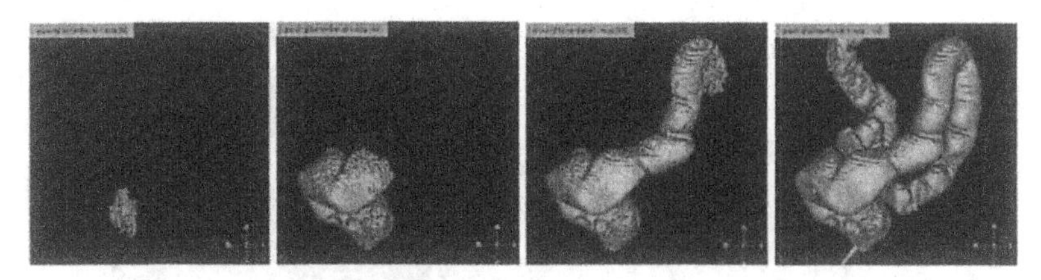

Figure 6.10. Propagation inside colon using Fast Marching.

This last region, on the boundary of the visited points, is a contour in 2D and a surface in 3D. If the potential is a lot higher along edges than it is inside the shape, the edges will act as an obstacle to the front propagation. In this case the *Trial* points define a surface which segments the object.

In order to see the precise relation between fast marching propagation and active contours, consider the usual evolution equation of an interface (2D curve or 3D surface) that appears in level sets methods $\frac{\partial C}{\partial t}(p) = F(\kappa)\mathbf{n}$ and $C(p, 0) = C_0(p)$. Assume the speed $F = \frac{1}{P} > 0$, and thus the front moves always outwards in the normal direction $\mathbf{n}$, like an inflating balloon [204], but with a speed which is not necessarily constant. A way to characterize the interface is to compute at each point $\mathbf{x}$ of the image the arrival time $T(\mathbf{x})$ of the interface $C(t)$ when it sweeps the domain. Using the classical properties of a level set of $T$ that its normal is in the direction of $\nabla T$, the following equation is obtained from the evolution of interface $C(t)$:

$$T(C(\mathbf{x}, t)) = t \Rightarrow \nabla T \cdot C_t = 1 \Rightarrow \nabla T \cdot \left( F \frac{\nabla T}{\|\nabla T\|} \right) = 1 \Rightarrow F \cdot \|\nabla T\| = 1$$

where we recognize the Eikonal equation seen above in section 6.2.4. This equation was thus solved by fast marching in [535] for surface segmentation since it has the same advantages as the level set formulation, but is much faster. This equation is solved using 3D Fast Marching (see section 6.2.6) in the example for segmentation of the colon shown in figure 6.10, [264].

When the front propagates in a long and thin structure for which the potential contrast between inside and outside is not sufficient, the front will likely flood out of the object during propagation. Indeed, when the front propagates in the tubular structure, there is only a small part of the front, which we could call the "head" of the front, that really moves. Most of the front is located close to the boundary of the structure and moves very slowly. For example voxels that are close to the starting point, the "tail" of the front, are moving very slowly. However, since the structure may be very long, in order for the "head" voxels to reach the end of the structure, the "tail" voxels may flow out of the boundary since their speed is always positive, and integrated over a long time. This is illustrated in the example of Figure 6.11.

We introduced in [265] an approach where points of the front are "frozen" when a distance criterion is satisfied. This makes use of the length of the minimal paths

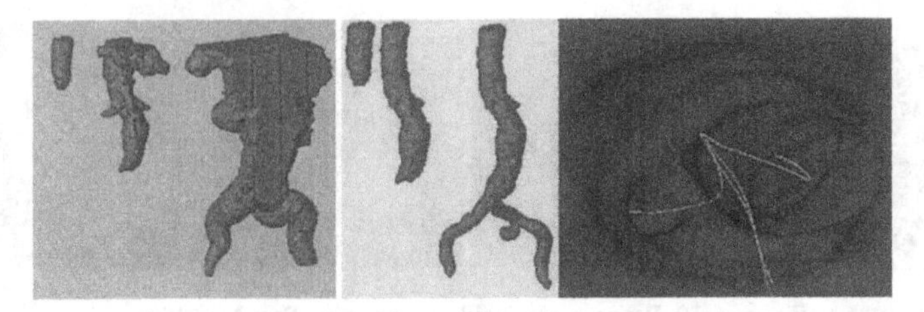

Figure 6.11. Front Propagation in a 3D MR image of the aorta. On the left it floods, in the middle, freezing prevents flooding. On the right, virtual endoscopy in the tree structure, with visible paths.

computed as in section 6.2.8. Figure 6.11 shows the result with freezing which gives a correct segmentation.

## 6.6 Centered Minimal Paths and virtual endoscopy

A minimal path minimizes the integral of the potential in equation (6.2). If the potential is constant in some areas, like inside a tubular object, it will lead to a shortest geodesic path. The same thing happens when the potential does not vary much inside a tubular shape. The minimal path extracted is often tangential to the edges, as shown on the left of figure 6.12, and this is a problem when looking for a trajectory for virtual endoscopy [264]. A centered path is more relevant. The method we proposed to obtain a centered path in a tubular shape first segments the tubular region and then looks for a path inside as far as possible from the walls, using a distance map. The complete method is detailed in [264], here are the main steps:

1. Segmentation: compute the weighted distance map by front propagation from the given start point till reaching the end point, which can be found automatically using a length criterion of section 6.2.8.
2. Segmentation: set of *trial* points, as described in section 6.5.
3. Centering Potential : compute inside the tubular object the distance map $\mathcal{D}$ to the surface previously obtained (fast marching with $P = 1$).
4. Centered path : this is the minimal path between start and end points relatively to a *decreasing* function of the distance $\mathcal{D}$. The path locates as far as possible from the walls, which means in the center where distance to the boundary is larger. The final step is to make back-propagation from the end point using the last action map.

Figure 6.12 compares the resulting path with classical potential and centering potential on brain vessels. Figure 6.13 shows an example of the centered minimal path obtained in a 3D colon image. This path is used as a trajectory

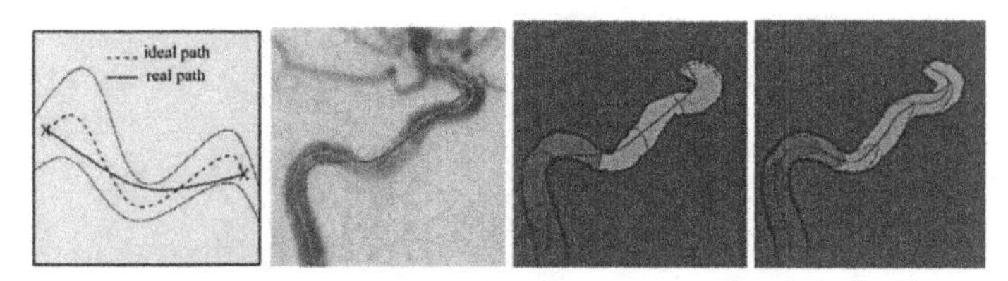


Figure 6.12. Centered path in a vessel: Two images on the left show both paths on a sketch and original image. Two images on the right show propagation and path for classical potential and centering distance potential obtained.

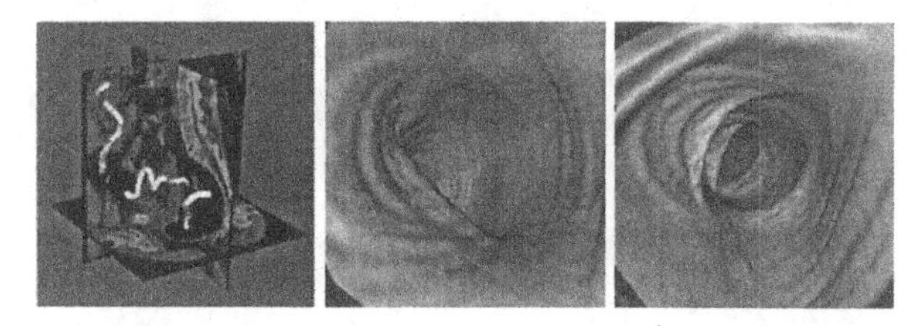

Figure 6.13. On the left, example of a minimal path on a 3D image of colon. On the right, virtual endoscopy through the colon (colonoscopy).

for a virtual camera by image rendering at each point of the path from the 3D image data giving a virtual endoscopy. Movies are available on the website http://www.ceremade.dauphine.fr/~cohen/MPEG This approach can be extended [265] to extraction of a set of paths in a tree structure and the possibility of virtual endoscopy where the user can choose at each bifurcation the path he wishes to follow (figure 6.11).

**Acknowledgements.** A large part of the presented work was done in collaboration with R. Kimmel or during PhD supervision of T.Deschamps and I thank both of them greatly.

## 6.7 Conclusion

We have presented various aspects of minimal paths methods and their applications, in particular for medical imaging. These approaches allow to extract a contour or a set of contours in a 2D image, as well as tubular structures, or tree structures in 2D and 3D images. The Fast marching algorithm makes the task much easier and also allows to segment curves or surfaces in an image very fast. Let us quote some of our more recent related work : surface segmentation defined as a set of minimal paths, [30], image segmentation from a set of source points using an extension of the definition of minimal action [29] and fast marching on a triangulated surface used for adaptive remeshing [643].

# Chapter7

# Integrating Shape and Texture in Deformable Models: from Hybrid Methods to Metamorphs

**D. Metaxas, X. Huang and T. Chen**

Abstract

In traditional shape-based deformable models, the external image forces come primarily from edge or gradient information. Such reliance on edge information, however, makes the models prone to get stuck in local minima due to image noise and various other image artifacts. Integrating region statistics constraints has been a centerpiece of the efforts toward more robust, well-behaved deformable models in boundary extraction and segmentation. In this chapter, we review previous work on the loose coupling of boundary and region information in two major classes of deformable models: the parametric models and the geometric models. Then, we propose a new class of deformable shape and texture models, which we term "Metamorphs". The novel formulation of the Metamorph models tightly couples shape and interior texture and the dynamics of the models are derived in a unified manner from both boundary and region information in a variational framework.

## 7.1 Introduction

Automated image segmentation is a fundamental problem in computer vision and medical image analysis applications. Object texture, image noise, intensity inhomogeneity and variations in lighting, to name a few, add to the problem complexity. To address these difficulties, deformable model-based segmentation methods have been extensively studied and widely used, with promising results.

Deformable models are curves or surfaces that move under the influence of internal smoothness and external image forces. In the literature, there are two major classes of deformable models. The first is the parametric (explicit) deformable models that explicitly represent deformable curves and surfaces in their parametric form during the segmentation process. Examples are Active Contour Models

[455] and their extensions in both 2D and 3D [562, 756, 208, 558, 255, 631, 887]. The evolution of these parametric models is derived either in a energy-minimization process [455, 897] or through a dynamic-force formulation [208]. The energy-minimization formulation has the advantage that its solution satisfies a minimum principle; while the dynamic force formulation provides the flexibility of applying different types of external forces onto the deformable model. The external forces can be potential forces such as image forces, non-potential forces such as balloon forces, and the combination of both. The other class of deformable models is the geometric (implicit) deformable models [155, 538, 908, 902, 174]. These models represent curves and surfaces implicitly as the level set of a higher-dimensional scalar function [728, 617], and the model evolution is based on the theory of curve evolution, with speed function specifically designed to incorporate image information. Comparing the two classes of deformable models, the parametric deformable models have a compact representation, and allow fast implementation, while the geometric deformable models can handle topological changes naturally.

Although the parametric and geometric deformable models differ both in their formulations and in their implementations, both classes use primarily edge (image gradient) information to derive external image forces to drive a shape-based model. Such reliance on edge information, however, makes the models sensitive to image noise and various other image artifacts. For instance, a model may leak through small or large gaps on the object boundary, or it may get stuck in local minima due to spurious edges inside the object or clutter around the true boundary.

To address these limitations, there have been significant efforts in the literature to integrate region information into both parametric [680, 922] and geometric deformable models [626, 841, 807]. The integration frameworks however, are still imperfect. In the case of parametric models, region information and boundary information are often treated separately in different energy minimization processes, thus parameters of region intensity statistics can not be updated simultaneously with the boundary shape parameters. In the case of geometric models, the integrations are mostly based on solving reduced cases of the minimal partition problem in the Mumford and Shah model for segmentation [591]. Variational frameworks are proposed to unifying boundary and region-based information sources, and level set approaches are used to implement the resulting PDE systems. However, these frameworks assume piecewise constant, or Gaussian intensity distributions within each partitioned region. This limits their applicability and robustness in finding objects whose interiors have high noise level, intensity inhomogeneity, and/or complex multi-modal intensity distributions.

In this chapter, we focus on presenting the work from our group on the integration of region statistics constraints into shape-based deformable models, which includes: (1) a hybrid framework that loosely couples a region-based module and a boundary deformable model-based module, and (2) Metamorphs, a recently developed new class of deformable models that possess both shape and interior texture and integrate boundary and region information in a unified manner within a variational framework.

In [181], we proposed a hybrid segmentation framework which integrates a region-based segmentation module driven by Gibbs prior models, a boundary-based module using deformable models and the marching cubes method which connects these two modules. The region-based and boundary based modules work recursively: The region segmentation results are used to initialize the deformable model and the deformable fitting results are used to update the parameters of the region segmentation. This way, the two modules can help each other out of local minima. The quality of the segmentation output also improves when we update the Gibbs model's parameters using more accurate region and boundary information at the end of each iteration. To accommodate 3D segmentation applications, we integrate the marching cubes method into our method, which can construct deformable meshes based on 3D binary masks.

One limitation in the hybrid framework, however, is that the region information and the boundary/shape information are still treated separately instead of being integrated in driving model deformations. To utilize information from both sources in a unified manner, we have developed, recently, a new class of deformable models called "Metamorphs" [412].

Metamorphs integrate dynamically shape and interior texture. The resulting lagrangian formulation is derived from both boundary and region information based on a novel variational framework. These new models bridge the gap between parametric and geometric deformable models by borrowing the best features of both worlds. The model shapes are embedded in a higher dimensional space of distance transforms, thus represented by distance map "images". (This is similar to the implicit shape representation in geometric level-set based models). The model deformations are efficiently parameterized using a space warping technique, the cubic B-spline based Free Form Deformations (FFD) [22, 51, 413]. [1] The interior intensity statistics of the models are captured using nonparametric kernel-based approximations, which can represent complex multi-modal distributions. When finding object boundaries in images, the dynamics of the Metamorph models are derived from an energy functional consisting of both edge (which encodes gradient information) and region intensity energy terms. In our formulation, both types of energy terms are differentiable with respect to the model deformation parameters. This allows for a unified gradient-descent based deformation parameter updating paradigm using both boundary and region information. Furthermore, our Metamorph model deformations are constrained in such way that the interior statistics of the model after deformation is consistent with the statistics learned from the past history of the model interiors. A Metamorph model can be initialized far-away from the object boundary and efficiently converge to an optimal solution. The proposed energy functional enables the model to pass small spurious edges and prevents it from leaking through large boundary gaps, hence makes the boundary finding robust to image noise and inhomogeneity.

[1] Note that we separate the shape representation, which is implicit in a higher dimension, and model deformation, which is explicitly parameterized by FFD.

In the remainder of the chapter, we will first present our hybrid segmentation framework and then the new form of deformable shape and texture models - Metamorphs.

## 7.2 Hybrid Segmentation Method

In the framework we proposed in [181], we segment an object as follows. First we use the Gibbs model to get a rough binary mask of the object. Then we use the marching cubes method to construct the deformable mesh and make the mesh deform to fit the object surface using the gradient information. The Gibbs parameters need to be updated from iteration to iteration to improve the segmentation results. By doing so, we integrated the region information into deformable models. In the following, we present the modules that comprise the hybrid segmentation approach.

### 7.2.1 *Gibbs Models*

Most medical images are Markov Random Field images, that is, the statistics of a pixel in the medical image are related to the statistics of pixels in its neighborhood. According to the Equivalence Theorem proved by Hammersley and Clifford [379], a Markov Random Field is equivalent to a Gibbs field under certain restrictions. Therefore the joint distribution of a medical image with MRF property can be written in the Gibbsian form as follows.

$$\Pi(X) = Z^{-1} \exp(-H(X)) \tag{7.1}$$

where $\mathbf{X}$ is the set of all possible configurations of the image $X$, $z$ is an image in the set of $\mathbf{X}$, $Z = \sum_{z \in \mathbf{X}} \exp(-H(z))$ is a normalizing factor, and $H(X)$ is the energy function of image $X$. The local and global properties of MRF images are incorporated into the model by designing an appropriate energy function $H(X)$ and minimizing it. The lower the value of the energy function, the better the image fits to the prior distribution. Therefore the segmentation procedure corresponds to the minimization of the energy function.

$$H_{prior}(X) = H_1(X) + H_2(X) \tag{7.2}$$

where $H_1(X)$ models the piecewise pixel homogeneity statistics and $H_2(X)$ models the object boundary continuity. In general, the homogeneity term $H_1(X)$ has a smoothing effect on pixels inside the object and will leave boundary features beyond the threshold unchanged. The boundary continuity term in the energy function $H_2(X)$ has the following form:

$$H_2(X) = \vartheta_2 \sum_{s \in X} \sum_{i=1}^{N} W_i(s) \tag{7.3}$$

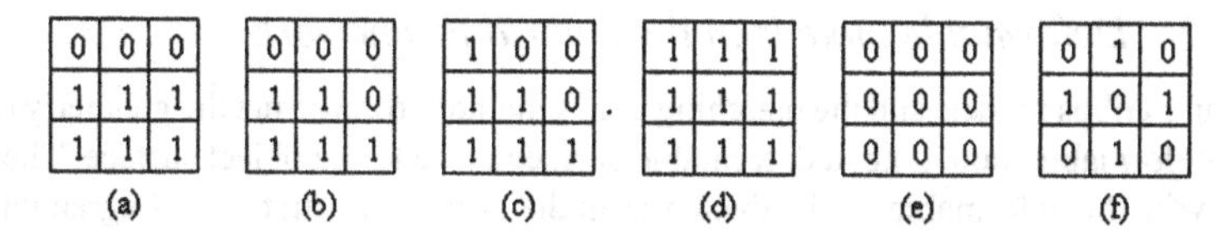


Figure 7.1. Clique definitions: cliques can be classified into clique types of a) smooth boundary, b) smooth boundary with angle, c) smooth boundary in diagonal direction, d) object interior, e) outside the object, f) irregular boundaries or noisy regions. Pixels labelled 1 are in the object, while pixels labelled 0 are out of the object.

where $s$ is a pixel, $\vartheta_2$ is the weight term for the boundary continuity, $N$ is the number of local configurations, and $W_i(s)$ are weight functions (also called the potential functions) of local configurations. In our model, the potential functions are defined on a neighborhood system based on cliques with the size of 3 by 3 pixels. There are altogether $2^9$ possible local configurations in a clique including 3 by 3 pixels. We can classify them into 6 clique types. Among these 6 types, three of them contain configurations at smooth boundaries (Fig.7.1.a, .b, .c), one type for the homogeneous region inside (Fig.7.1.d) and one for such region outside (Fig.7.1.e) the object respectively, and one clique type includes all local configurations that lead to noisy regions or irregular boundaries (Fig.7.1.f). Cliques that belong to the same clique type share the same potential value. We assign lower potential values to clique configurations that are located at smooth and continuous boundaries. Therefore, when we minimize $H_2(X)$, pixels in the image (especially those near the boundary) will alter their intensities to form clique configurations of lower potentials. These alternations make the currently estimated boundaries smoother, the weak boundaries stronger, and extend boundaries into image regions without strong gradient information. Hence the minimization of the energy function will lead to continuous and smooth object boundaries.

We use the Bayesian framework to get a MAP estimation of the object region. In a Bayesian framework, the segmentation problem can be formulated as the maximization of the posterior probability $P(X|Y)$, which can also be written as an energy functional:

$$H_{posterior}(X,Y) = H_{prior}(X) + H_{observation}(X,Y) \tag{7.4}$$

where $H_{observation}(X,Y)$ is the constraint from the observation of the original image. Using $H_{posterior}(X,Y)$ instead of $H_{prior}(X)$ in the energy minimization, we get a MAP estimation of the object region. The constraint of the observation will compete with the prior distribution during the minimization process. Hence the result of the minimization process will still be close enough to the original observation, while important image features, such as irregular edges, will be kept regardless of the prior distribution.

The output of the Gibbs prior model includes region information so that when its output is used to initialized the geometric form of the deformable, the region information will be passed to the deformable.

### 7.2.2 *Deformable models in the Hybrid Framework*

We use Gibbs models and the marching cubes method to construct the geometry of the deformable model, i.e, a deformable surface close to the object surface. Then we write the deformable model dynamics in the form of the first order Lagrangian equation:

$$\dot{\mathbf{d}} + \mathbf{Kd} = \mathbf{f}_{ext} \tag{7.5}$$

where $\dot{\mathbf{d}} = \frac{\partial \mathbf{X}}{\partial t}$. $\mathbf{K}$ is the stiffness matrix. $f_{ext}$ is the external force.

According to equation (5), the deformable model deforms under the effect of the internal force $\mathbf{Kd}$ and the external force. The internal force keeps the deformable model surface smooth and continuous during its deformation. If the object boundary in the image to be segmented is weak, the internal force will act as a surface constraint that prevents the model from being trapped into local minima or overflowing beyond the boundary. The external force will lead the model to the object surface using image information such as the gradient.

In our framework, we use the second order derivative gradient as the external force. It is defined as:

$$f_G(x,y,z) = -\nabla P(x,y,z) = -\nabla(w_e|\nabla[G_\sigma(x,y,z) * I(x,y,z)]|)^2 \tag{7.6}$$

where $I(x,y,z)$ is the original image, $w_e$ is a positive weighting parameter, $G_\sigma(x,y,z)$ is a three dimensional Gaussian function with standard deviation $\sigma$, $\nabla$ is the gradient operator, and $*$ is the convolution operator. We use the Gaussian filter to blur the original image in order to remove small noisy regions and expand the effective range of the gradient-derived force. In a second order gradient flow field, all gradient vectors point to the location of edge features so that they can lead the model to the object surface directly. During the fitting process, we calculate the dot product of the second order gradient vector and the normal vector at every node on the deformable surface. It yields a positive value if the model node locates inside the edge feature and a negative value if the model node locates outside. We can define the magnitude of the external force as the magnitude of the dot product and the direction of the force vector as the direction of the normal vector at the node.

We now can calculate the derivative of displacements of every node on the deformable model surface using Eqn. (7.5). The displacements will then be updated using the Euler equation:

$$\mathbf{d}_{new} = \dot{\mathbf{d}} \cdot \Delta t + \mathbf{d}_{old} \tag{7.7}$$

where $\Delta t$ is the time step. The deformation stops when the forces equilibrate or vanish.

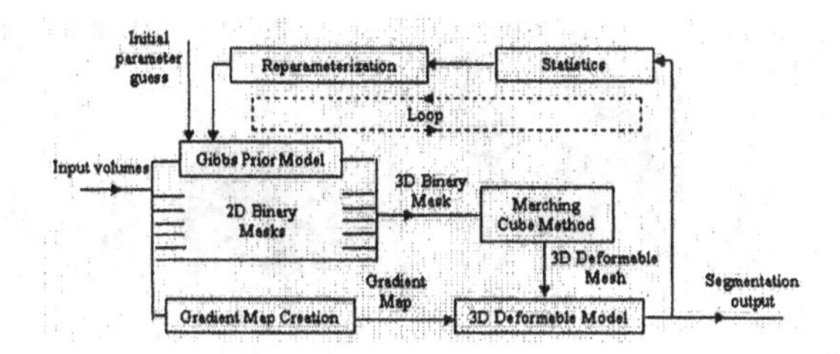


Figure 7.2. Flow-Chart for 3D-segmentation hybrid framework.

### 7.2.3 Integration of Deformable Models and Gibbs Models

Fig. 7.2 shows internal modules and the data flow of our 3D hybrid segmentation framework. In the first iteration of the recursive hybrid framework, the parameters of the Gibbs prior models are set to default values. Using the segmentation result of the deformable model in the current iteration, we update the Gibbs prior parameters before restarting the Gibbs models in the following iterations to improve their segmentation performance. Besides updating regional parameters such as the mean intensity and the standard deviation of the object, we also update potentials of local configurations in Eqn. (7.3). The clique potentials of the Gibbs Prior model are set to be proportional to the number of appearances of each type of cliques in the deformable model segmented binary image.

We illustrate our hybrid segmentation framework by applying it to segment the tumor region in a 3D MR image volume of human brain (See Fig. 7.3). Fig. 7.3(a) shows one slice of the volume. The image volume size is 256 by 256 by 32 pixels (preprocessing has been applied to remove slices that do not contain the structure of interest). We use 32 2D Gibbs Prior models to create a 3D binary mask for the tumor region. The initial edge threshold is set to 6, the potential weight for smooth boundaries are set to 0.0, and the potential for other local configurations are set to 5.0. We then use the marching cube method to create a surface mesh for the deformable model to begin with. During the deformable model fitting, the time step is set to 0.07, and the gradient magnitude $w_e$ is 1.0. The hybrid segmentation process stops after two iterations. Fig. 7.3(d) shows the final segmentation result of the hybrid framework. For quality evaluation purposes, we overlay the segmentation result onto the original image 7.3(a) as in Fig. 7.3(e). We show the initial deformable mesh surface constructed by the marching cube method in Fig. 7.3(f), and the 3D reconstruction of the tumor region based on final segmentation result in Fig. 7.3(g), (h). Notice that the segmentation result of the Gibbs model is improved by using updated parameters. The fact that in Fig. 7.3(g) and (h) the deformable model fits well at concavities and convexities proves that our hybrid framework has a good performance in segmenting complex object surfaces. The total segmentation time is about 6 minutes for 2 iterations, which is much shorter than the method described in [901].

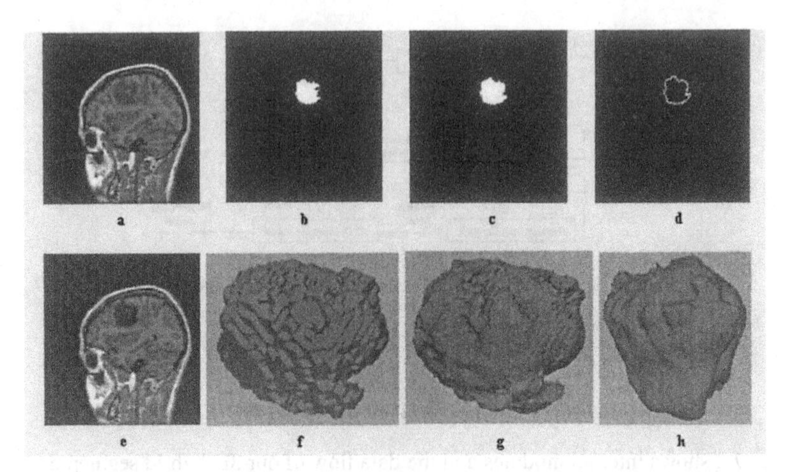


Figure 7.3. Segmentation of a tumor in the brain from MR image, (a) the original image; (b & c) the Gibbs model segmentation result in the first and second iterations; (d) the final segmentation result of the hybrid framework; (e) the segmentation result overlaid upon the original image; (f) the initial deformable surface; (g, h) 2 views of the final segmentation result in 3D. Data courtesy of Prof. Kikinis's group at Harvard University.

## 7.3 Metamorphs: Deformable Shape and Texture Models

A limitation in the hybrid segmentation framework introduced in section 7.2 is that, the region-based module and the boundary-based module are used separately, thus the information from both sources are not integrated during the evolution of a deformable model. Furthermore, the region-based module produces an initialization mesh to start a deformable model, which makes the final segmentation result highly dependent on this initialization. To address these limitations, we present our recent work [412] on a new class of deformable shape and texture models, which we call "Metamorphs". The formulation of Metamorphs naturally integrates both shape and interior texture, and the model dynamics are derived coherently from both boundary and region information during the whole course of model evolution in a common variational framework.

### *7.3.1 The Metamorphs Model representations*

#### 7.3.1.1 The Model's Shape Representation

The model's shape is embedded implicitly in a higher dimensional space of distance transforms. The Euclidean distance transform is used to embed an evolving model as the zero level set of a higher dimensional distance function. In order to facilitate notation, we consider the 2D case. Let $\Phi : \Omega \rightarrow R^{+}$ be a Lipschitz function that refers to the distance transform for the model shape $\mathcal{M}$. The shape defines a partition of the domain: the region that is enclosed by $\mathcal{M}$, $[\mathcal{R}_\mathcal{M}]$, the background $[\Omega - \mathcal{R}_\mathcal{M}]$, and on the model, $[\partial\mathcal{R}_\mathcal{M}]$ (In practice, we consider a narrow band around the model $\mathcal{M}$ in the image domain as $\partial\mathcal{R}_\mathcal{M}$). Given these definitions the following implicit shape representation is considered:

$$\Phi_\mathcal{M}(\mathbf{x}) = \begin{cases} 0, & \mathbf{x} \in \partial\mathcal{R}_\mathcal{M} \\ +ED(\mathbf{x}, \mathcal{M}) > 0, & \mathbf{x} \in \mathcal{R}_\mathcal{M} \\ -ED(\mathbf{x}, \mathcal{M}) < 0, & \mathbf{x} \in [\Omega - \mathcal{R}_\mathcal{M}] \end{cases}$$

where $ED(\mathbf{x}, \mathcal{M})$ refers to the min Euclidean distance between the image pixel location $\mathbf{x} = (x, y)$ and the model $\mathcal{M}$.

Such treatment makes the model shape representation a distance map "image", which greatly facilitates the integration of boundary and region information. This shape representation in 3D is similarly defined in a volumetric embedding space.

#### 7.3.1.2 The Model's Deformations

The deformations that Metamorph models can undergo are defined using a space warping technique, the Free Form Deformations (FFD) [719]. The essence of FFD is to deform an object by manipulating a regular control lattice $F$ overlaid on its volumetric embedding space. In Metamorphs, we consider an Incremental Free Form Deformations (IFFD) formulation using the cubic B-spline basis [413].

Let us consider a regular lattice of control points

$$F_{m,n} = (F^x_{m,n}, F^y_{m,n});\ m = 1, ..., M,\quad n = 1, ..., N$$

overlaid to a region $\Gamma_c = \{\mathbf{x}\} = \{(x, y) | 1 \leq x \leq X, 1 \leq y \leq Y\}$ in the embedding space that encloses the model in its object-centered coordinate system. Let us denote the initial configuration of the control lattice as $F^0$, and the deforming control lattice as $F = F^0 + \delta F$. Under these assumptions, the incremental FFD parameters, which are also the deformation parameters for the model, are the deformations of the control points in both directions $(x, y)$:

$$\mathbf{q} = \{(\delta F^x_{m,n}, \delta F^y_{m,n})\};\ (m, n) \in [1, M] \times [1, N]$$

The deformed position of a pixel $\mathbf{x} = (x, y)$ given the deformation of the control lattice from $F^0$ to $F$, is defined in terms of a tensor product of Cubic B-spline polynomials:

$$D(\mathbf{q}; \mathbf{x}) = \mathbf{x} + \delta D(\mathbf{q}; \mathbf{x}) = \sum_{k=0}^{3}\sum_{l=0}^{3} B_k(u)B_l(v)F^0_{i+k,j+l} + \delta F_{i+k,j+l}) \tag{7.8}$$

where $i = \lfloor \frac{x}{X} \cdot (M-1) \rfloor + 1$, $j = \lfloor \frac{y}{Y} \cdot (N-1) \rfloor + 1$. The terms of the deformation component refer to:

- $\delta F_{i+l,j+l}$, $(k, l) \in [0, 3] \times [0, 3]$ are the deformations of pixel $\mathbf{x}$'s (sixteen) adjacent control points,

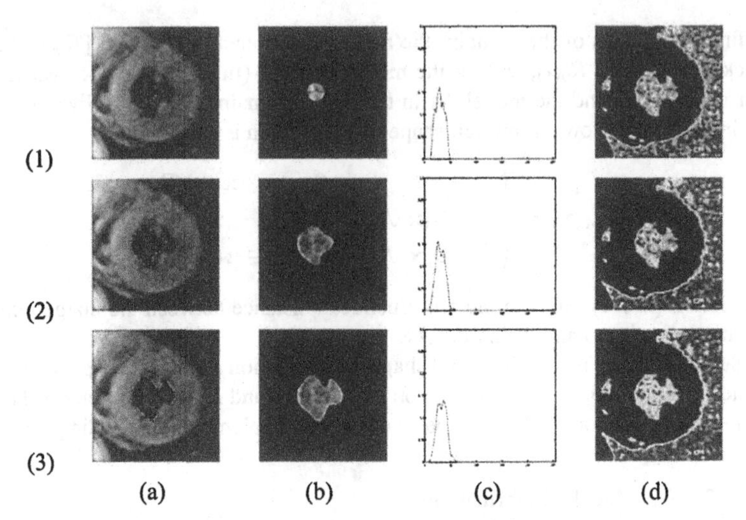


Figure 7.4. The Left Ventricle Endocardium segmentation. (1) Initial model. (2) Intermediate result. (3) Final converged result. (a) The evolving model drawn in colored lines (blue or red) on original image. (b) Interior of the evolving model. (c) The intensity p.d.f of the model interior. The X axis is the intensity value in the range of $[0, 255]$ and the Y axis is the probability value in the range of $[0, 1]$. (d) The image probability map based on the p.d.f of the model interior.

- $B_k(u)$ is the $k^{th}$ basis function of a Cubic B-spine, defined by:

$$B_0(u) = (1-u)^3/6,\ B_1(u) = (3u^3 - 6u^2 + 4)/6$$
$$B_2(u) = (-3u^3 + 3u^2 + 3u + 1)/6,\ B_3(u) = u^3/6$$

  with $u = \frac{x}{X} \cdot (M-1) - \lfloor \frac{x}{X} \cdot (M-1) \rfloor$. $B_l(v)$ is similarly defined.

- $\delta D(\mathbf{q}; \mathbf{x}) = \sum_{k=0}^{3} \sum_{l=0}^{3} B_k(u) B_l(v) \delta F_{i+k,j+l}$ is the incremental deformation for pixel $\mathbf{x}$.

The extension of the models to account for deformations in 3D is straightforward, by using control lattices in the 3D space and a 3D tensor product of B-spline polynomials.

#### 7.3.1.3 The Model's Texture

Rather than using traditional statistical parameters (such as mean and variance) to approximate the intensity distribution of the model interior, we model the distribution using a nonparametric kernel-based method. The nonparametric approximation is differentiable, more generic and can represent complex multi-modal intensity distributions.

Suppose the model is placed on an image $I$, the image region bounded by current model $\Phi_{\mathcal{M}}$ is $\mathcal{R}_{\mathcal{M}}$, then the probability of a pixel's intensity value $i$ being

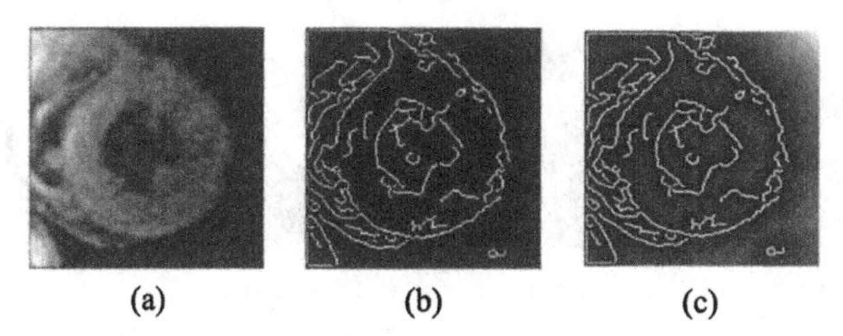

(a) (b) (c)

Figure 7.5. The effect of small spurious edges inside the object of interest (endocardium of the Left Ventricle) on the "shape image". (a) The original MR image. (b) The edge map of the image. (c) The derived "shape image", with edges points drawn in yellow. Note the effect of the small spurious edges on the "shape image" inside the object.

consistent with the model interior intensity can be derived using a Gaussian kernel as:

$$\mathbf{P}(i|\Phi_{\mathcal{M}}) = \frac{1}{V(\mathcal{R}_{\mathcal{M}})} \iint_{\mathcal{R}_{\mathcal{M}}} \frac{1}{\sqrt{2\pi}\sigma} e^{\frac{-(i-I(\mathbf{y}))^2}{2\sigma^2}} d\mathbf{y} \tag{7.9}$$

where $V(\mathcal{R}_{\mathcal{M}})$ denotes the volume of $\mathcal{R}_{\mathcal{M}}$, and $\sigma$ is a constant specifying the width of the gaussian kernel.

Using this nonparametric approximation, the intensity distribution of the model interior gets updated automatically while the model deforms. The initialization of the model texture is flexible. We can either start with a small model inside the texture region to be segmented, or use supervised learning to specify the desired texture a Priori. One example of the model interior texture representation can be seen in [Fig. (7.4)]. In the figure, we show the zero level set of the current model $\Phi_{\mathcal{M}}$ in colored lines [Fig. (7.4).a], the model interior region $\mathcal{R}_{\mathcal{M}}$ [Fig. (7.4).b], the probability density function (p.d.f.) for the intensity of current model interior $\mathbf{P}(i|\Phi_{\mathcal{M}})$ for $i = 0, ...255$ [Fig. (7.4).c], and the probability map of every pixel's intensity in the image according to the model interior distribution [Fig. (7.4).d].

## 7.3.2 The Metamorph Dynamics

The motion of the model is driven by both boundary (edge) and region (intensity) energy terms derived from the image. The overall energy functional $E$ consists of two parts – the shape data terms $E_S$, and the intensity data terms $E_I$:

$$E = E_S + kE_I \tag{7.10}$$

where $k$ is a constant balancing the contribution of the two parts. Next, we derive the shape and intensity data terms respectively.

### 7.3.2.1 The Shape Data Terms

We encode the gradient information of an image using a "shape image" $\Phi$, which is derived from the un-signed distance transform of the edge map of the image. In [Fig. (7.5).c], we can see the "shape image" of an example MR heart image.

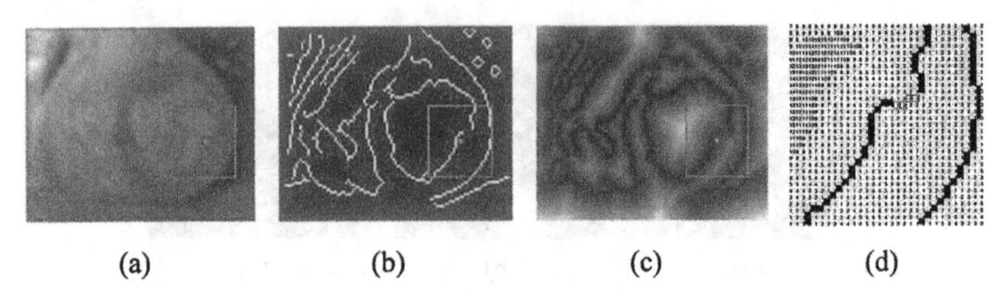


Figure 7.6. The boundary shape data term constraints at small gaps in the edge map. (a) Original Image. (b) The edge map, note the small gap inside the red square region. (c) The "shape image". (d) Zoom-in view of the region inside the red square. The numbers are the "shape image" values at each pixel location. The red dots are edge points, the blue squares indicate a path favored by the boundary term for a Metamorph model.

To evolve a Metamorph model toward image edges, we define two shape data terms – an interior term $E_{S_i}$ and a boundary term $E_{S_b}$:

$$E_S = E_{S_i} + aE_{S_b} \tag{7.11}$$

In the interior shape data term of the model, we aim to minimize the Sum-of-Squared-Differences between the implicit shape representation values in the model interior and the underlying "shape image" values at corresponding deformed positions. This can be written as:

$$E_{S_i} = \frac{1}{V(\mathcal{R}_\mathcal{M})} \iint_{\mathcal{R}_\mathcal{M}} \left(\Phi_\mathcal{M}(\mathbf{x}) - \Phi(D(\mathbf{q};\mathbf{x}))\right)^2 d\mathbf{x} \tag{7.12}$$

During optimization, this term will deform the model along the gradient direction of the underlying "shape image". Thus it will expand or shrink the model accordingly, serving as a two-way balloon force without explicitly introducing such forces, and making the attraction range of the model large.

To make the model deformation more robust to small spurious edges detected within an object due to texture, we consider a separated boundary shape data term, which allows higher weights for pixels in a narrow band around the model boundary $\partial\mathcal{R}_\mathcal{M}$.

$$E_{S_b} = \frac{1}{V(\partial\mathcal{R}_\mathcal{M})} \iint_{\partial\mathcal{R}_\mathcal{M}} \left(\Phi(D(\mathbf{q};\mathbf{x}))\right)^2 d\mathbf{x} \tag{7.13}$$

Intuitively, this term will encourage the deformation that maps the model boundary to the image edge locations where the underlying "shape image" distance values are as small (or as close to zero) as possible. One additional advantage of this term is that, at an edge with small gaps, this term will constrain the model to go along the "geodesic" path, which coincides with the smooth shortest path connecting the two open ends of a gap. This behavior can be seen from [Fig. (7.6)]. Note that at a small gap of the edge map, the boundary term will favor a path with the smallest accumulative distance values to the edge points.

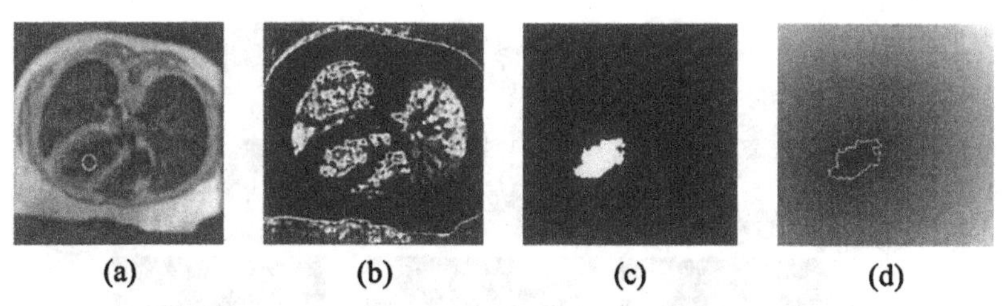

(a) (b) (c) (d)

Figure 7.7. Deriving the "region of interest" intensity data term. (a) The model shown (in yellow) on the original image. (b) The intensity probability map based on the model interior statistics. (c) The region of interest (ROI) derived from the thresholded probability map. The threshold is the mean probability over the entire image. (d) The "shape image" encoding boundary information of the ROI.

#### 7.3.2.2 The Intensity Data Terms

In our current framework, the intensity energy function $E_I$ consists of two intensity data terms – a "Region Of Interest" (ROI) term $E_{I_r}$, and a Maximum Likelihood term $E_{I_m}$:

$$E_I = E_{I_r} + bE_{I_m} \tag{7.14}$$

In the "Region Of Interest" (ROI) term $E_{I_r}$, we aim to evolve the model toward the boundary of current region of interest, which is determined based on current model interior intensity distribution. Given a model $\mathcal{M}$ on image $I$ [Fig. (7.7).a], we first compute the image intensity probability map $P_I$ [Fig. (7.7).b], based on the model interior intensity statistics (see section 7.3.1.3). Then a small threshold (typically the mean probability over the entire image domain) is applied on $P_I$ to produce a binary image $BP_I$, in which pixels with probabilities higher than the threshold have value 1. Morphological operations are used to fill in small holes in $BP_I$. We then take the connected component on this binary image overlapping the model as current region of interest (ROI). Suppose the binary mask of this ROI is $BI_r$ [Fig. (7.7).c], we encode its boundary information by computing the "shape image" of $BI_r$, which is the un-signed distance transform of the region boundary [Fig. (7.7).d]. Denote this "shape image" as $\Phi_r$, the ROI intensity data term is defined as follows:

$$E_{I_r} = \frac{1}{V(\mathcal{R_M})} \iint_{\mathcal{R_M}} \left(\Phi_{\mathcal{M}}(\mathbf{x}) - \Phi_r(D(\mathbf{q};\mathbf{x}))\right)^2 d\mathbf{x} \tag{7.15}$$

This ROI intensity data term is the most effective in countering the effect of small spurious edges inside the object of interest (e.g. in Figs. (7.5,7.9). It also provides implicit balloon forces to quickly deform the model toward object boundary.

To achieve better convergence when the model gets close to the object boundary, we design another Maximum Likelihood (ML) intensity data term that constrains the model to deform toward areas where the pixel probabilities of belonging to the model interior intensity distribution are high. This ML term is formalized by maximizing the log-likelihood of pixel intensities in a narrow band

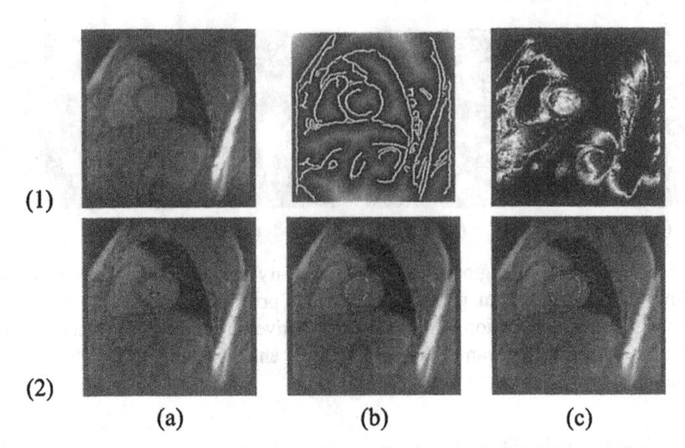


Figure 7.8. Segmentation of the Endocardium of the Left Ventricle in a MR image with a large portion of the object boundary edge missing. (1.a) The original image. (1.b) The "shape image" derived from edge map. (1.c) The intensity probability map based on the initial model. (2.a) Initial model (zero level set shown in blue). (2.b) Intermediate model (zero level set shown in red). (2.c) converged model.

around the model after deformation:

$$\begin{aligned} E_{I_m} &= -\frac{1}{V(\partial\mathcal{R}_\mathcal{M})}\iint_{\partial\mathcal{R}_\mathcal{M}} log\mathbf{P}(I(D(\mathbf{q};\mathbf{x}))|\Phi_\mathcal{M})d\mathbf{x} \\ &= -\frac{1}{V(\partial\mathcal{R}_\mathcal{M})}\iint_{\partial\mathcal{R}_\mathcal{M}}\Big[log\frac{1}{V(\mathcal{R}_\mathcal{M})} + log\frac{1}{\sqrt{2\pi}\sigma} \\ &\quad + log\iint_{\mathcal{R}_\mathcal{M}} e^{\frac{-(I(D(\mathbf{q};\mathbf{x}))-I(\mathbf{y}))^2}{2\sigma^2}}d\mathbf{y}\Big]d\mathbf{x} \end{aligned} \tag{7.16}$$

During model evolution, when the model is still far away from object boundary, this ML term generates very little forces to influence the model deformation. When the model gets close to object boundary, however, the ML term generates significant forces to prevent the model from leaking through large gaps (e.g. in Fig. 7.8), and help the model to converge to the true object boundary.

## 7.3.3 *Model Evolution*

In our formulations above, both shape data terms and intensity data terms are differentiable with respect to the model deformation parameters $\mathbf{q}$, thus a unified gradient-descent based parameter updating scheme can be derived using both boundary and region information. Based on the definitions of the energy functions, one can derive the following evolution equation for each element $\mathbf{q}_i$ in the model deformation parameters $\mathbf{q}$:

$$\frac{\partial E}{\partial \mathbf{q}_i} = \left(\frac{\partial E_{S_i}}{\partial \mathbf{q}_i} + a\frac{\partial E_{S_b}}{\partial \mathbf{q}_i}\right) + k\left(\frac{\partial E_{I_r}}{\partial \mathbf{q}_i} + b\frac{\partial E_{I_m}}{\partial \mathbf{q}_i}\right) \tag{7.17}$$

The detailed derivations for each term can be found in [412].

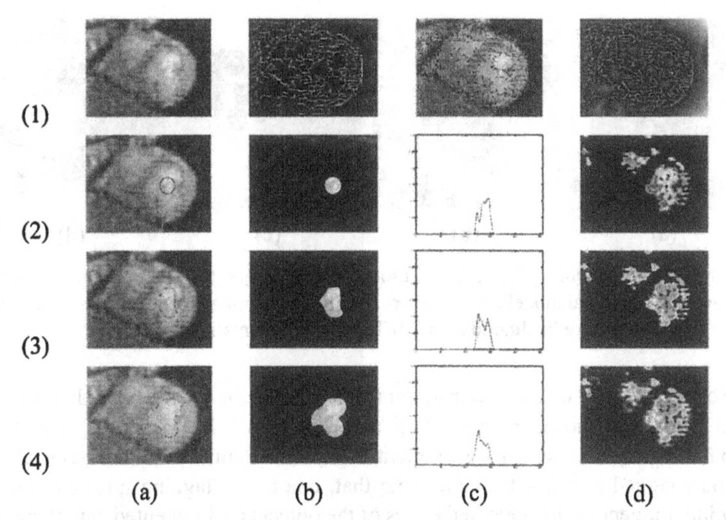


Figure 7.9. The tagged MR heart image. (1.a) The original image. (1.b) The edge map. (1.c) The edge points overlaid on original image. (1.d) The "shape image". (2) Initial model. (3) Intermediate result. (4) Final model (after 50 iterations). (2-4)(a) The evolving model. (2-4)(b) The model interior. (2-4)(c) The model interior intensity probability density. (2-4)(d) The intensity probability map of the image based on the p.d.f in (c).

### *7.3.4 The Model Fitting Algorithm and Experimental Results*

The overall model fitting algorithm consists of the following steps:

1. Initialize the deformation parameters $\mathbf{q}$ to be $\mathbf{q}^0$, which indicates no deformation.
2. Compute $\frac{\partial E}{\partial \mathbf{q}_i}$ for each element $\mathbf{q}_i$ in the deformation parameters $\mathbf{q}$.
3. Update the parameters $\mathbf{q}_i' = \mathbf{q}_i - \lambda \cdot \frac{\partial E}{\partial \mathbf{q}_i}$.
4. Using the new parameters, compute the new model $\mathcal{M}' = D(\mathbf{q}'; \mathcal{M})$.
5. Update the model. Let $\mathcal{M} = \mathcal{M}'$, re-compute the implicit representation of the model $\Phi_{\mathcal{M}}$, and the new partitions of the image domain by the new model: $[\mathcal{R}_{\mathcal{M}}]$, $[\Omega - \mathcal{R}_{\mathcal{M}}]$, and $[\partial \mathcal{R}_{\mathcal{M}}]$. Also re-initialize a regular FFD control lattice to cover the new model, and update the "region of interest" shape image $\phi_r$ based on the new model interior.
6. Repeat steps 1-5 until convergence.

In the algorithm, after each iteration, both shape and interior intensity statistics of the model get updated based on the model dynamics, and deformation parameters get re-initialized for the new model. This allows continuous, both large-scale and small-scale deformations for the model to converge to the energy minimum.

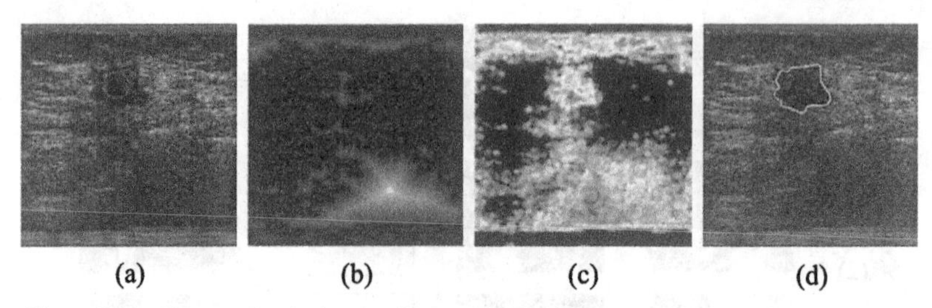


Figure 7.10. Segmenting lesion in ultrasound breast image. (a) The original ultrasound image, with the initial model drawn on top, (b) The shape image based on edge map of the image, (c) The texture likelihood map, (d) The final segmentation result.

Some examples of using our Metamorph models for boundary finding in images have been shown in [Fig. (7.4)] and [Fig. (7.8)]. In [Fig. (7.9)], we show another example in which we segment the Endocardium of the left ventricle in a noisy tagged MR heart image. Note that, due to the tagging lines and intensity inhomogeneity, the detected edges of the object are fragmented, and there are spurious small edges inside the region. In this case, the integration of both shape and texture information is critical in helping the model out of local minima. In [Fig. (7.10)], a metamorph model is used to extract the boundary of a lesion in an ultrasound image of the breast. On natural images, we show an example using the pepper image in [Fig. (7.11)]. Starting from a small model initialized inside the object, the model quickly deforms to the object boundary.

The Metamorph model evolution is computationally efficient, due to our use of the nonparametric texture representation and FFD parameterization of the model deformations. For all the examples shown, the segmentation process takes less than $200ms$ to converge on a 2Ghz PC station.

## 7.4 Conclusions

In this chapter, we have reviewed traditional shape-based deformable models, and introduced new frameworks that integrate region texture information into deformable models.

The new class of deformable models we proposed, Metamorphs, possess both boundary shape and interior intensity statistics. In Metamorphs the boundary and region information are intergated within a common variational framework to compute the deformations of the model towards the correct object boundaries. There is no need to learn statistical shape and appearance models *a priori*. In our formulation, the model deformations are constrained so that the interior model statistics as it deforms remain consistent with the statistics learned from the past evolution of the model's interior. This framework represents a generalization of previous *parametric* and *geometric* deformable models, by exploiting the best features of both worlds. Segmentation using Metamorph models can be straightforwardly applied

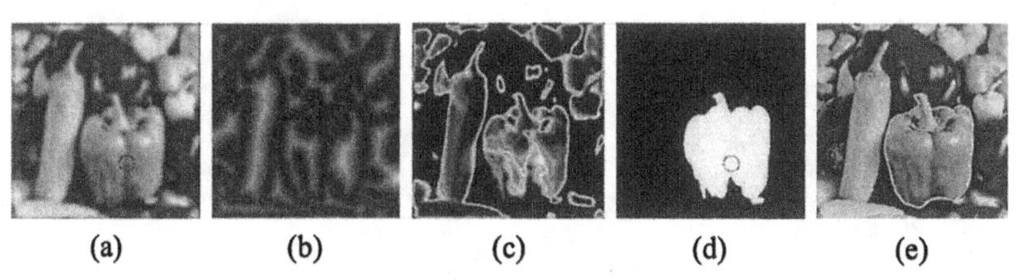

(a) (b) (c) (d) (e)

Figure 7.11. Boundary finding in the pepper image. (a) Original image, with initial model drawn in blue. (b) The shape image derived from edge map, with edges drawn in yellow. (c) The intensity probability map derived based on model interior statistics. (d) Region of Interest (ROI) extracted. (e) Final segmentation result.

in 3D, and can handle efficiently the merging of multiple models that are evolving simultaneously.

## Acknowledgement

This research has been funded by an NSF-ITR-0205671 grant to the first author. We also would like to acknowledge many stimulating discussions with Prof. Nikos Paragios and Dr. Chenyang Xu.

# Chapter8

# Variational Segmentation with Shape Priors


## M. Bergtholdt, D. Cremers and C. Schnörr



Abstract

We discuss the design of shape priors for variational region-based segmentation. By means of two different approaches, we elucidate the critical design issues involved: representation of shape, use of perceptually plausible dissimilarity measures, Euclidean embedding of shapes, learning of shape appearance from examples, combining shape priors and variational approaches to segmentation. The overall approach enables the appearance-based segmentation of views of 3D objects, without the use of 3D models.


## 8.1 Introduction

Variational models [456, 591] are the basis of established approaches to image segmentation in computer vision. The key idea is to generate a segmentation by locally optimizing appropriate cost functionals defined on the space of contours. The respective functionals are designed to maximize certain criteria regarding the low-level information such as edge consistency or (piecewise) homogeneity of intensity, color, texture, motion, or combinations thereof.

Yet, in practice the imposed models only roughly approximate the true intensity, texture or motion of specific objects in the image. Intensity measurements may be modulated by varying and complex lighting conditions. Moreover, the observed images may be noisy and objects may be partially occluded. In such cases, algorithms which are purely based on low-level properties will invariably fail to generate the desired segmentation.

An interpretation of these variational approaches in the framework of Bayesian inference shows that the above methods all impose a prior on the space of contours which favors boundaries of minimal length. While the resulting length constraint in the respective cost functionals has a strongly regularizing effect on the generated contour evolutions, this purely geometric prior lacks any experimental evidence. In practical applications, an algorithm which favors shorter boundaries may lead to the cutting of corners and the suppression of small-scale structures.

Given one or more silhouettes of an object of interest, one can construct shape priors which favor objects that are in some sense *familiar*. In recent years, it was suggested to enhance variational segmentation schemes by imposing such object-specific shape priors. This can be done either by adding appropriate shape terms to the contour evolution [513, 808] or in a probabilistic formulation which leads to an additional shape term in the resulting cost functional [237, 688, 573]. By extending segmentation functionals with a shape prior, knowledge about the appearance of objects can be directly combined with clues given by the image data in order to cope with typical difficulties of purely data-driven image processing caused by noise, occlusion, etc.

The design of shape priors strongly depends on ongoing work on statistical shape models [223, 284, 459]. In particular, advanced models of shape spaces, shape distances, and corresponding shape transformations have been proposed recently [912, 336, 785, 177, 478, 736]. Concerning variational segmentation, besides attempting to devise "intrinsic" mathematical representations of shape, further objectives which have to be taken into account include the gap between mathematically convenient representations and representations conforming to properties of human perception [820, 588, 61], the applicability of statistical learning of shape appearance from examples, and the overall variational approach from the viewpoint of optimization.

The objective of this paper is to discuss these issues involved in designing shape priors for region-based variational segmentation by means of two representative examples: (i) non-parametric statistics applied to the standard Euclidean embedding of curves in terms of shape vectors, and (ii) perceptually plausible matching functionals defined on the shape manifold of closed planar curves. Both approaches are powerful, yet quite different with respect to the representation of shape, and of shape appearance. Their properties will be explained in the following sections, in view of the overall goal – variational segmentation.

Section 8.2 discusses both the common representation of shapes by shape vectors, and the more general representation by dissimilarity structures. The latter is mathematically less convenient, but allows for using distance measures which conform to findings of psychophysics. Learning of shape appearance is described in Section 8.3. The first approach encodes shape manifolds globally, whereas the second approach employs structure-preserving Euclidean embedding and shape clustering, leading to a collection of locally-linear representations of shape manifolds. The incorporation of corresponding shape priors into region-based variational approaches to segmentation is discussed in Section 8.4.

We confine ourselves to parametric planar curves and do not consider the more involved topic of shape priors for implicitly defined and multiply connected curves – we refer the reader to [513, 808, 187, 688, 177, 236] for promising advances in this field. Nevertheless, the range of models addressed are highly relevant from both the scientific and the industrial viewpoint of computer vision.

## 8.2 Shape Representation

One generally distinguishes between *explicit* (parametric) and *implicit* contour representations. In the context of image segmentation, implicit boundary representations have gained popularity due to the introduction of the level set method, which allows to propagate implicitly represented interfaces by appropriate partial differential equations acting on the corresponding embedding surfaces. The main advantages of representing and propagating contours implicitly are that one does not need to deal with control/marker point regridding and can elegantly (without heuristics) handle topological changes of the evolving boundary.

On the other hand, explicit representations also have several advantages. In particular, they provide a compact (low-dimensional) representation of contours and concepts such as intrinsic alignment, group invariance and statistical learning are more easily defined. Moreover, as we shall see in this work, the notion of corresponding contour points (and contour parts) arises more naturally in an explicit representation. In this work, we will only consider explicit simply-connected closed contours.

### *8.2.1 Parametric Contour Representations, Geometric Distances, and Invariance*

Let

$$\mathbf{c} : [0, 1] \to \Omega \subset \mathcal{R}^2 \tag{8.1}$$

denote a parametric closed contour in the image domain $\Omega$. Throughout this paper, we use the finite-dimensional representation of 2D-shapes in terms of uniform periodic cubic B-splines [304]:

$$\mathbf{c}(s) = \sum_{m=1}^{M} \mathbf{p}_m B_m(s) = \mathbf{P}\mathbf{v}(s) , \tag{8.2}$$

with control points $\{\mathbf{p}_i\}$ and basis functions $\{B_i(s)\}$:

$$\mathbf{P} = \begin{bmatrix} \mathbf{p}_1 & \mathbf{p}_2 & \dots & \mathbf{p}_M \end{bmatrix} , \qquad \mathbf{v}(s) = \begin{pmatrix} B_1(s) & B_2(s) & \dots & B_M(s) \end{pmatrix}^{\mathsf{T}}$$

Well-known advantages of this representation include the compact support of the basis functions and continuous differentiability up to second order. Yet, most of our results also hold for alternative explicit contour representations.

Using the natural uniform sampling $\{s_1, \dots, s_M\}$ of the parameter interval, we stack together the corresponding collection of curve points, to form *shape vectors* representing the contour. For simplicity, and with slight abuse of notation, we

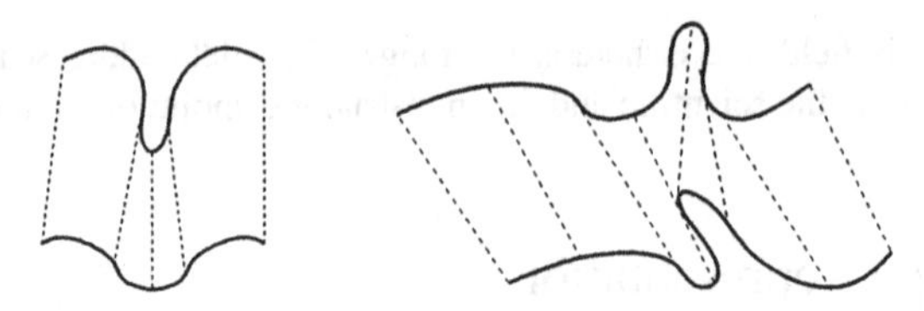

Figure 8.1. Stretching and bending of contours does not affect perceptually plausible matchings.

denote them again with[1]:

$$\mathbf{c} := \left(\mathbf{c}(s_1)^\top, \dots, \mathbf{c}(s_M)^\top\right)^\top \in \mathcal{R}^{2M} \tag{8.3}$$

Note, that there is a one-to-one correspondence between shape vectors $\mathbf{c}$ and corresponding control points $\{\mathbf{p}_i\}_{i=1,\dots,M}$ through the symmetric and sparse positive-definite matrix: $\mathbf{B} = \left(\mathbf{v}(s_1) \quad \dots \quad \mathbf{v}(s_M)\right)^\top$.

We consider a simple geometric distance measure between contours which is invariant under similarity transformations:

$$d^2(\mathbf{c}_1, \mathbf{c}_2) = \min_{s,\theta,\boldsymbol{t}} |\mathbf{c}_1 - s\mathbf{R}_\theta \mathbf{c}_2 - \boldsymbol{t}|^2 \tag{8.4}$$

Here, the planar rotation $\mathbf{R}_\theta$ and translation $\boldsymbol{t}$ are defined according to the definition (8.3) of shape vectors:

$$\mathbf{R}_\theta = \mathbf{I}_M \otimes \begin{pmatrix} \cos\theta & -\sin\theta \\ \sin\theta & \cos\theta \end{pmatrix}, \quad \boldsymbol{t} = \left(t_1, t_2, \dots, t_1, t_2\right)^\top ,$$

and $s$ is the scaling parameter. The solution to (8.4) can be computed in closed-form [284, 459]. Extensions of this alignment to larger transformation groups such as affine transformations are straight-forward. Furthermore, since the locations of the starting points $\mathbf{c}_1(0), \mathbf{c}_2(0)$ are unknown, we minimize (8.4) over all cyclic permutations of the contour points defining $\mathbf{c}_2$.

### *8.2.2 Matching Functionals and Psychophysical Distance Measures*

It is well-known that there is a gap between distance measures with mathematically convenient properties like (8.4), for example, and distance measures which conform with findings of psychophysics [820]. In particular, this observation is relevant in connection with shapes [588].

Given two arc-length parametrized curves $\mathbf{c}_1(t), \mathbf{c}_2(s)$, along with a diffeomorphism $t = g(s)$ smoothly mapping the curves onto each other, then corresponding studies [61] argued that matching functionals for evaluating the quality of the

[1]In the following, it will be clear from the context whether $\mathbf{c}$ denotes a contour (8.1) or a shape vector (8.3).

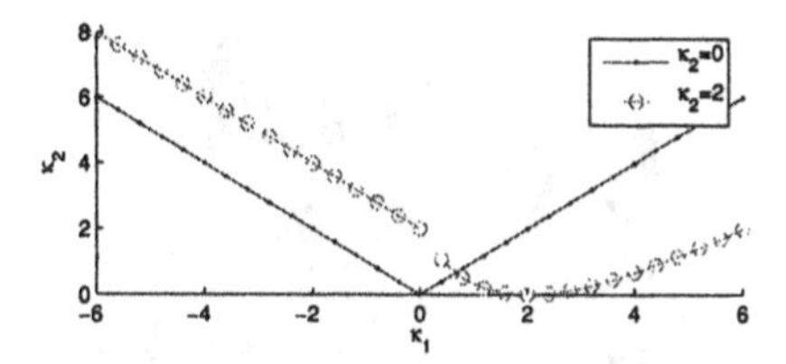

Figure 8.2. Local matching cost. The local cost for bending of the matching functional (8.5) as a function of the $\kappa_1$, for two values of $\kappa_2$. Note how in the case $\kappa_2 = 2$, relatively lower costs for $\kappa_1 \approx 2$ allow for significant bending, without affecting matching too much.

mapping $g$ based on low-order derivatives, should involve stretching $g'(s)$ and bending (change of curvature) of the curves (cf. Figure 8.1).

As a representative, we consider the matching functional [61]:

$$E(g;\, \mathbf{c}_1, \mathbf{c}_2) = \int_0^1 \frac{[\kappa_2(s) - \kappa_1(g(s))g'(s)]^2}{|\kappa_2(s)| + |\kappa_1(g(s))g'(s)|} ds + \lambda \int_0^1 \frac{|g'(s) - 1|^2}{|g'(s)| + 1} ds \quad (8.5)$$

where $\kappa_1(t), \kappa_2(s)$ denote the curvature functions of the contours $\mathbf{c}_1, \mathbf{c}_2$. The two terms in (8.5) take into account the bending and stretching of contours, respectively (see Figure 8.2).

Functional (8.5) favors perceptually plausible matchings because it accounts that often objects are structured into nearly convex-shaped parts separated by concave extrema. In particular, for non-rigid objects, parts are likely to articulate, and the matching functional produces articulation costs only at part boundaries.

From the mathematical viewpoint, functional (8.5) is invariant to rotation and translation of contours, and also to scaling provided both contours are normalized to length one. This is always assumed in what follows below. Furthermore, by taking the $q$-th root of the integral of local costs, where $q > 2.4$, (8.5) defines a *metric* between contours [61]:

$$d_E(\mathbf{c}_1, \mathbf{c}_2) := \min_g E(g;\, \mathbf{c}_1, \mathbf{c}_2)^{1/q} \quad (8.6)$$

Clearly, this distance measure is mathematically less convenient than (8.4). This seems to be the price for considering findings of psychophysics. However, regarding variational segmentation, we wish to work in this more general setting as well. For a discussion of further mathematical properties of matching functionals, we refer to [806].

The minimization in (8.6) is carried out by dynamic programming over all piecewise-linear and strictly monotonously increasing functions $g$. Figure 8.3 illustrates the result for two human shapes.

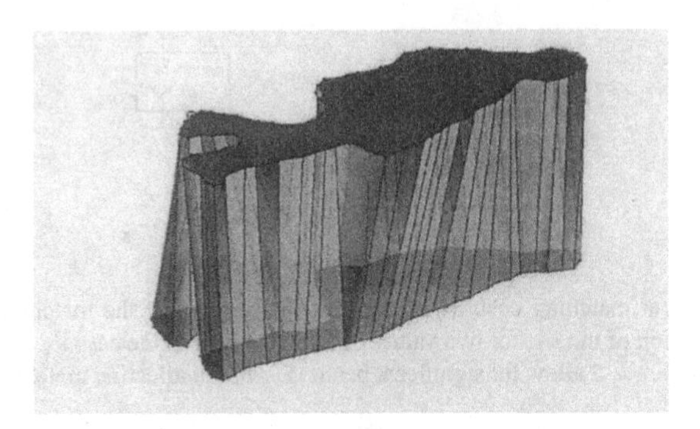

Figure 8.3. Matching by minimizing (8.5) leads to an accurate correspondence of parts of non-rigid objects, here illustrated for two human shapes.

## 8.3 Learning Shape Statistics

Based on the shape representations described in Section 8.2, we consider in this section two approaches to the statistical learning of shape appearance from examples. The common basis for both approaches are Euclidean embeddings of shapes.

The first approach uses the embedding of shape vectors into Reproducing Kernel Hilbert Spaces by means of kernel functions, leading to a non-parametric global representation of shape manifolds. The second approach uses embeddings of dissimilarity structures by multidimensional scaling, along with a cluster-preserving modification of the dissimilarity matrix. Subsequent clustering results in a collection of local encodings of shape manifolds, and in corresponding aspect graphs of 3D objects in terms of prototypical object views.

### *8.3.1 Shape Distances in Kernel Feature Space*

Let $\{\mathbf{c}_n\}_{n=1,\dots,N} \in \mathcal{R}^{2M}$ denote the shape vectors associated with a set of training shapes. In order to model statistical shape dissimilarity measures, it is commonly suggested to approximate the distribution of training shapes by a Gaussian distribution, either in a subspace formed by the first few eigenvectors [223], or in the full $2M$-dimensional space [237]. Yet, for more complex classes of shapes – such as the various silhouettes corresponding to different 2D views of a 3D object – the assumption of a Gaussian distribution fails to accurately represent the distribution underlying the training shapes.

In order to model more complex (non-Gaussian and multi-modal) statistical distributions, we propose to embed the training shapes into an appropriate Reproducing Kernel Hilbert Space (RKHS) [851], and estimate Gaussian densities there – see Figure 8.4 for a schematic illustration.

A key assumption in this context is that only scalar products of embedded shape vectors $\phi(\mathbf{c})$ have to be evaluated in the RKHS, which is done in terms of a kernel

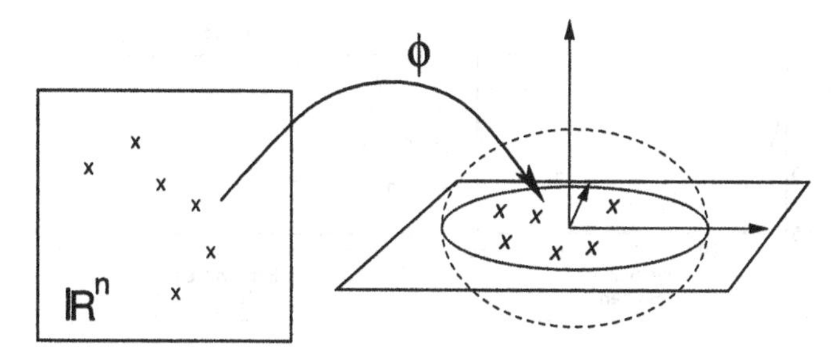


Figure 8.4. Gaussian density estimate upon nonlinear transformation to features space.

function:

$$K(\mathbf{c}_1, \mathbf{c}_2) = \langle \phi(\mathbf{c}_1), \phi(\mathbf{c}_2) \rangle \tag{8.7}$$

Knowledge of the embedding map $\phi(\mathbf{c})$ itself is not required. Admissible kernel functions, including the Gaussian kernel, guarantee that the Gramian matrix

$$\mathbf{K} = \left\{ K(\mathbf{c}_i, \mathbf{c}_j) \right\}_{i,j=1,\ldots,N} \tag{8.8}$$

is positive definite [851]. This "non-linearization strategy" has been successfully applied in machine learning and pattern recognition during the last decade, where the RKHS is called *feature space*.

Based on this embedding of given training shapes, we use the following Mahalanobis distance:

$$J_S(\mathbf{c}) = (\phi(\mathbf{c}) - \phi_0)^\top \, \Sigma_\phi^{-1} (\phi(\mathbf{c}) - \phi_0) \,, \tag{8.9}$$

where $\phi_0$ is the empirical mean, and $\Sigma_\phi$ is the corresponding covariance matrix. Note that all evaluations necessary to compute $J_S(\mathbf{c})$ in (8.9) can be traced back to evaluations of the kernel function according to (8.7). Furthermore, by exploiting the spectral decomposition of the kernel matrix $\mathbf{K}$ in (8.8), we regularize the covariance matrix $\Sigma_\phi$ with respect to its small and vanishing eigenvalues, thus defining two orthogonal subspaces as illustrated in Figure 8.4 on the right. For further details, we refer to [233].

## 8.3.2 Structure-Preserving Embedding and Clustering

Based on the matching functional (8.5) and the corresponding distance measure $d_E(\mathbf{c}_1, \mathbf{c}_2)$ defined in (8.6), we consider an arbitrary sample set $\{\mathbf{c}_n\}_{n=1,\ldots,N}$. To perform statistical analysis, we wish to compute an Euclidean embedding $\{\mathbf{x}_n\}_{n=1,\ldots,N}$ such that $\|\mathbf{x}_i - \mathbf{x}_j\| = d_E(\mathbf{c}_i, \mathbf{c}_j)\,,\ \forall i, j$. Such an embedding exists iff the matrix $\mathbf{K} = -\frac{1}{2}\mathbf{QDQ}$, with the dissimilarity matrix $\mathbf{D} = (d_E(\mathbf{c}_i, \mathbf{c}_j)^2)$ and the centering matrix $\mathbf{Q} = \mathbf{I} - \frac{1}{M}\mathbf{e}\mathbf{e}^\top$, is positive semidefinite [231]. The vectors $\mathbf{x}_n$ representing the objects (contours) $\mathbf{c}_n$ of our data structure can then be computed by a Cholesky factorization of $\mathbf{K}$.

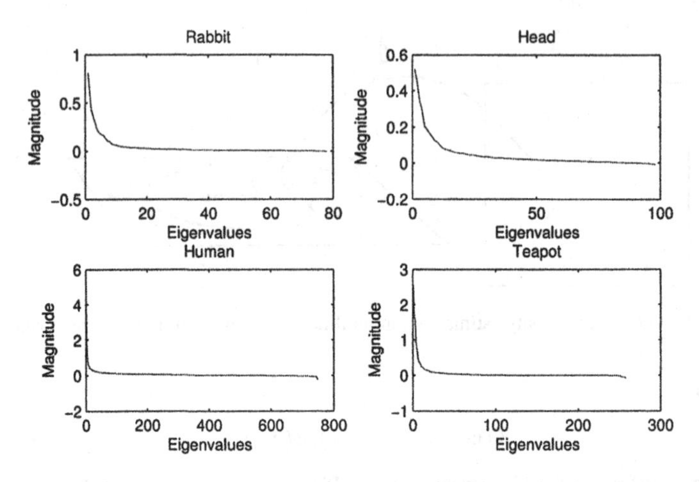


Figure 8.5. Eigenvalues of the matrices $\mathbf{K}$ corresponding to the shapes of four different objects.

Figure 8.5 shows the eigenvalues of $\mathbf{K}$ for four different objects. The graphs illustrate that the contours are "almost embeddable" since only few and small eigenvalues are negative. This fact is caused by the powerful matching which tightly groups given curves, and is performed by evaluating the distance measure $d_E$. The standard way then is to take the positive eigenvalues only, and to compute a *distorted* embedding.

In view of subsequent clustering, however, a better alternative is to regularize the data structure by shifting the off-diagonal elements of the dissimilarity matrix: $\tilde{\mathbf{D}} = \mathbf{D} - 2\lambda_N(\mathbf{e}\mathbf{e}^\top - \mathbf{I})$. For the resulting embedding, it has been shown [682] that the group structure with respect to subsequent k-means clustering is preserved.

Figure 8.6 shows a low-dimensional – and thus a heavily distorted – projection of the embedded shapes of the rabbit. For the purpose of illustration, only shapes corresponding to a single (hand-held) walk around the view-sphere are shown on the left, along with cluster centers as prototypical views of the object. In this way, we compute high-quality aspect graphs for *general* objects, without any restrictions discussed in the literature [106, 674].

On the right, Figure 8.6 also shows a clustering of 750 human shapes. In general, when using simple geometric distance measures, the many degrees of freedom of articulated shapes would require many templates for an accurate representation. The matching distance (8.6), however, accounts for part structure and, therefore, the principal components of the measure seem to be closer related to topological shape properties. For example, the clusters on the left are all "single-leg" prototypes, whereas on the right we find only clusters with two legs. The second principal component seems to account for the viewing direction of the

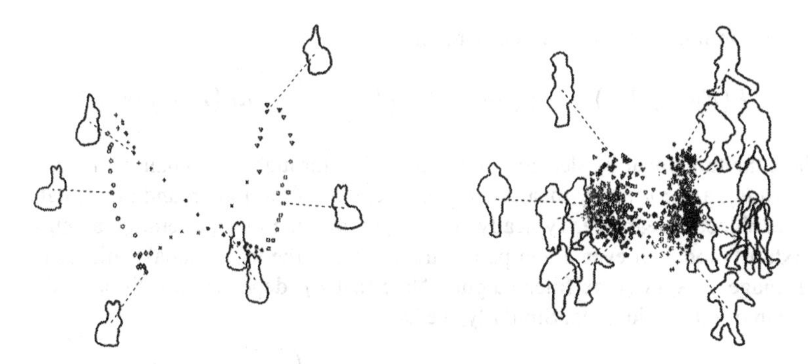

Figure 8.6. Clustering of the views of the rabbit sequence and the human shapes, projected to the first two principal components. The clusters are indicated by prototypical shapes (cluster centers) dominating a range of corresponding views.

human, which changes from left to right along a vertical direction through the plot.

## 8.4 Variational Segmentation and Shape Priors

### *8.4.1 Variational Approach*

We consider partitions $\Omega = \overline{\Omega(F)} \cup \Omega(B)$ of the image domain into foreground and background, respectively. Our objective is to compute an optimal partition in terms of a planar closed curve $\mathbf{c}(s) = \partial\Omega(F)$ based on the corresponding restrictions of the image function $F = I|_{\Omega(F)}, B = I|_{\Omega(B)}, G = I|_{\mathbf{c}(s)}$, and by using models $\mathcal{H} = (\mathcal{H}_F, \mathcal{H}_B, \mathcal{H}_G, \mathcal{H}_S)$ for these components, including a shape prior $\mathcal{H}_S$ for the separating curve $\mathbf{c}(s)$.

The variational approach is to compute the *Maximum A-Posteriori (MAP)* estimate of the contour $\mathbf{c}$, given the image data $I$, and using the models $\mathcal{H}$:

$$\hat{\mathbf{c}}(s) = \arg\max_{\mathbf{c}(s)} P(\mathbf{c}(s)|I, \mathcal{H}) \tag{8.10}$$

We use Bayes' rule to obtain:

$$\begin{aligned} P(\mathbf{c}(s)|I, \mathcal{H}) &= \frac{P(I|\mathbf{c}(s), \mathcal{H})P(\mathbf{c}(s)|\mathcal{H})}{P(I|\mathcal{H})} \\ &\propto P(F|\mathbf{c}(s), \mathcal{H}_F)P(B|\mathbf{c}(s), \mathcal{H}_B)P(G|\mathbf{c}(s), \mathcal{H}_G)P(\mathbf{c}(s)|\mathcal{H}_S)\,, \end{aligned}$$

where we have also split up the image likelihood $P(I|\mathbf{c}(s), \mathcal{H})$ into three parts, assuming independence of these parts, given the contour $\mathbf{c}(s)$. Moreover, we assume independence of the various models. This assumption is appropriate in the single object – single object class scenario considered here.

The common form of the foreground model is:

$$P(F|\mathbf{c}(s), \mathcal{H}_F) \propto \exp(-J_F) \,, \quad J_F(\mathbf{c}) = \int_{\Omega(F)} d_F\big(F(\mathbf{x})\big) d\mathbf{x} \,,$$

where the functional $J_F$ depends on the contour $\mathbf{c}$ through the domain of integration $\Omega(F)$, and $d_F$ is any measure of homogeneity of the foreground image data $F$, i.e. object appearance. Typically, $d_F$ is a parametric model, a semi-parametric (mixture) model, or even a non-parametric model of the local spatial statistics of the image data, or some filter outputs. Note that $d_F$ depends on $\mathbf{c}$ through the domain of integration, too. Similarly, we have:

$$P(B|\mathbf{c}(s), \mathcal{H}_B) \propto \exp(-J_B) \,, \quad J_B(\mathbf{c}) = \int_{\Omega(B)} d_B\big(B(\mathbf{x})\big) d\mathbf{x} \,,$$

$$P(G|\mathbf{c}(s), \mathcal{H}_G) \propto \exp(-J_G) \,, \quad J_G(\mathbf{c}) = \oint_{\mathbf{c}} d_G\big(G(\mathbf{x})\big) ds$$

In the following, we do not consider boundary models $P(G|\mathbf{c}(s), \mathcal{H}_G)$, but focus in the following two sections on shape models $P(\mathbf{c}(s)|\mathcal{H}_S)$, the main topic of this paper.

In order to solve (8.10), we minimize $-\log P(\mathbf{c}(s)|I, \mathcal{H})$, which entails to compute the derivatives of the above functionals with respect to $\mathbf{c}$, that is changes of the shape of the domain $\Omega(F)$. Let $\boldsymbol{v}(\mathbf{x})$ be a small and smooth vector field such that $(I+\boldsymbol{v})(\mathbf{x})$ is a diffeomorphism of the underlying domain. Then standard calculus [741, 256] yields:

$$\langle J_F'(\mathbf{c}), \boldsymbol{v}\rangle = \int_{\Omega(F)} d_F'\big(F(\mathbf{x})\big) d\mathbf{x} + \oint_{\mathbf{c}} d_F\big(F(\mathbf{x})\big)(\mathbf{n}\cdot\boldsymbol{v}) ds \,, \tag{8.11}$$

where $\mathbf{n}$ is the outer unit normal vector of $\Omega(F)$. Analogously, we compute the derivative of the background functional $J_B$.

If $d_F$ depends on *parameters* which are estimated within $\Omega(F)$, then computing $d_F'$ amounts to apply the chain rule until we have to differentiate (functions of) image data which do *not* depend on the domain (see, e.g., [432] for examples). As a result, the right hand side of (8.11) involves boundary integrals only. If, however, $d_F$ more generally depends on *functions* which, in turn, depend on the shape of $\Omega(F)$, e.g. through some PDE, then the domain integral in (8.11) involving the unknown domain derivative $d_F'$ can be evaluated in terms of a boundary integral by using an "adjoint state". See [715] for details and a representative application.

Finally, we set the normal vector field $v_n := \mathbf{n}\cdot\boldsymbol{v}$ equal to the *negative* integrand of the overall boundary integral resulting from the computation of $J_F'$, $J_B'$, and evolve the contour:

$$\dot{\mathbf{c}} = v_n \mathbf{n} \quad \text{on } \partial\Omega(F) \tag{8.12}$$

Inserting (8.2) yields a system of ODEs which are solved numerically.

Evolution (8.12) constitutes the data-driven part of the variational segmentation approach (8.10), conditioned on appearance models of both the foreground object and the background. In the following two sections, we describe how this approach is complemented in order to take into account statistical shape knowledge of object appearance.

### 8.4.2 Kernel-based Invariant Shape Priors

Based on the shape-energy (8.9), the shape-prior takes the form:

$$P(\mathbf{c}|\mathcal{H}_S) \propto \exp(-J_S)$$

Invariance with respect to similarity transforms is achieved by restricting the shape energy functional $J_S$ to *aligned shapes* $\hat{\mathbf{c}} = \hat{\mathbf{c}}(\mathbf{c})$ with respect to the mean shape, which result from given shapes $\mathbf{c}$ by applying to them the translation, rotation and scaling parameters defining the invariant distance measure (8.4):

$$J_S(\mathbf{c}) = J_S[\hat{\mathbf{c}}(\mathbf{c})]$$

To incorporate the statistical shape-knowledge into the variational segmentation approach, we perturb the evolution (8.12) by adding a small vector field directed towards the negative gradient of $J_S$:

$$v = -\varepsilon \frac{\mathrm{d}J_S}{\mathrm{d}\hat{\mathbf{c}}} \frac{\mathrm{d}\hat{\mathbf{c}}}{\mathrm{d}\mathbf{c}}$$

For further details, we refer to [233].

### 8.4.3 Shape Priors based on the Matching Distance

Related to the KPCA approach (Sections 8.3.1, 8.4.2), we use a non-parametric density estimate for the posterior of $\mathbf{c}$ given the training samples $\mathbf{c}_1, \ldots, \mathbf{c}_N$:

$$P(\mathbf{c}|\mathcal{H}_S) = p(\mathbf{c}|\mathbf{c}_1, \ldots, \mathbf{c}_N)$$

Given the Euclidean embedding $\mathbf{x}_1, \ldots, \mathbf{x}_N$ of the training samples (cf. Section 8.3.2), the kernel-estimate of the probability density evaluated at $\mathbf{x}$ reads:

$$p(\mathbf{x}) \approx p_N(\mathbf{x}) = \frac{1}{N} \sum_{n=1}^{N} \frac{1}{V} K\left(\frac{\mathbf{x} - \mathbf{x}_n}{h}\right), \tag{8.13}$$

where $K(\cdot)$ is a normalized non-negative smoothing kernel. A kernel with compact support, favored in practice, is the Epanechnikov kernel in $d$-dimensions:

$$K(\mathbf{x}) = \begin{cases} \frac{1}{2} V_d^{-1}(d+2)(1 - \mathbf{x}^\top \mathbf{x}) & \text{if } \mathbf{x}^\top \mathbf{x} < 1 \\ 0 & \text{otherwise} \end{cases}$$

where $V_d$ is the volume of the $d$-dimensional unit sphere. To increase the posterior probability of $\mathbf{c}$, we have to move in the gradient direction of the density estimate:

$$\nabla p(\mathbf{x}) = \frac{k}{N h^d V_d} \frac{d+2}{h^2} \left( \frac{1}{k} \sum_{\mathbf{x}_i \in \mathcal{B}_h(\mathbf{x})} \mathbf{x}_i - \mathbf{x} \right)$$

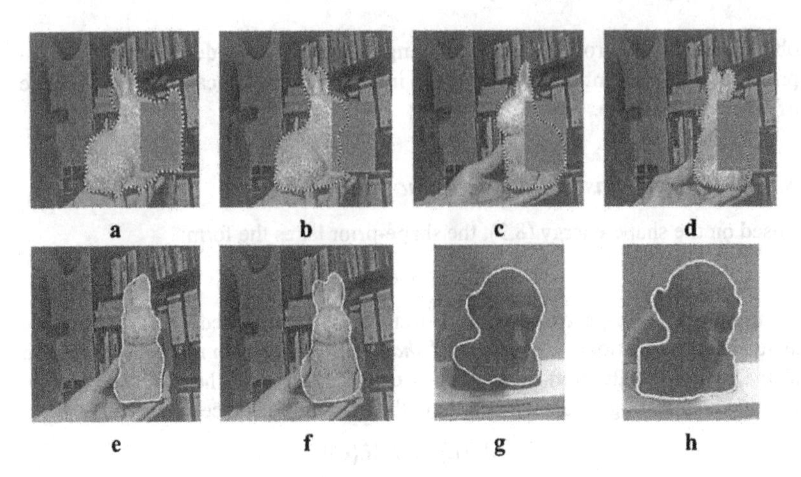


Figure 8.7. Top row: prior from Section 8.4.2, segmentation without the prior (**a**), with the prior (**b**), two more views with the prior (**c**), (**d**). Bottom row: prior from Section 8.4.3, segmentation without the prior (**e**), (**g**) and, with the prior (**f**), (**h**)

where $\mathcal{B}_h(\mathbf{x})$ is the ball with radius $h$ centered at $\mathbf{x}$, and $k$ is the number of samples $\mathbf{x}_k$ in $\mathcal{B}_h(\mathbf{x})$. This leads to the well-known *mean-shift* $\mathbf{x} \longrightarrow \frac{1}{k}\sum_{\mathcal{B}_h(\mathbf{x})} \mathbf{x}_i$ [333, 191].

By virtue of the embedding $\|\mathbf{x}_i - \mathbf{x}_j\| = d_E(\mathbf{c}_i, \mathbf{c}_j)$ (see Section 8.3.2), we may interpret this as computing the *Fréchet mean* [504]:

$$\hat{\mathbf{c}} = \arg\min_{\tilde{\mathbf{c}}} \int d_E(\tilde{\mathbf{c}}, \mathbf{c})^2 d\mu(\mathbf{c})$$

of the empirical probability measure $\mu$ on the space of contours $\mathbf{c}$, which is equipped with the metric (8.6). As a result, we perturb the evolution (8.12) by adding a small vector field $v = \varepsilon(\hat{\mathbf{c}} - \mathbf{c})\,,\ 0 < \varepsilon \in \mathcal{R}$, and thus incorporate statistical shape-knowledge into the variational segmentation approach.

## 8.4.4 Experimental Results

Both approaches to the design of shape priors allow to encode the appearance of objects. Applying the variational framework for segmentation, the models are automatically invoked by the observed data and, in turn, provide missing information due to noise, clutter, or occlusion. This bottom-up top-down behavior was verified in our segmentation experiments.

In Figure 8.7 we see segmentation results for two image sequences showing a rabbit and a head, computed with and without a shape prior. We can see that both shape priors can handle the varying point of view and stabilize the segmentation. Where data evidence is compromised by occlusion (**a**)-(**d**), shadows (**e**)-(**f**), or difficult illumination (**g**)-(**h**), the shape prior can provide the missing information. For the segmentation in Figure 8.8, we learned the shape prior model 8.4.3 using

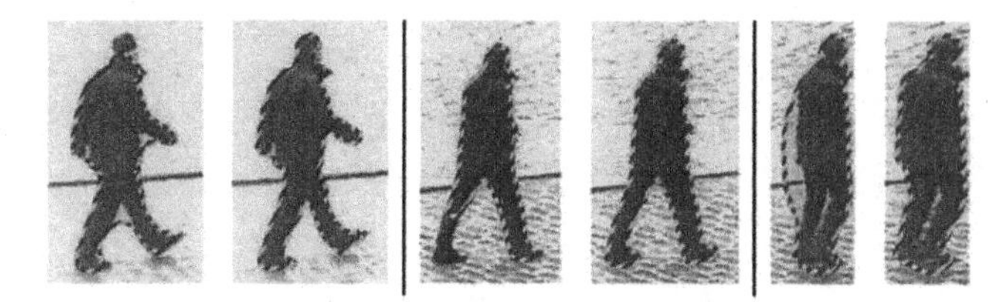

Figure 8.8. Sample screen shots of a human walking sequence. First image is without a shape prior, second image is result obtained with a shape prior, for each image pair respectively.

750 human shapes. The shapes in the sequence are not part of the training set. The obtained results encourage the use of shape-priors for the segmentation and tracking of articulated body motion as well.

## 8.5 Conclusion and Further Work

We investigated the design of shape priors as a central topic of variational segmentation. Two different approaches based on traditional shape-vectors, and on contours as elements of a metric space defined through a matching functional, respectively, illustrated the broad range of research issues involved. The use of shape priors allows for the variational segmentation of scenes where pure data-driven approaches fail.

Future work has mainly to address the categorization of shapes according to classes of objects, and the application of this knowledge for the interpretation of scenes with multiple different objects.

**Acknowledgment.** We thank Dr. Dariu Gavrila, DaimlerChrysler Research, for making available the database with human shapes to the CVGPR group.

# Chapter9

# Curve Propagation, Level Set Methods and Grouping

**N. Paragios**

Abstract

Image segmentation and object extraction are among the most well addressed topics in computational vision. In this chapter we present a comprehensive tutorial of level sets towards a flexible frame partition paradigm that could integrate edge-drive, regional-based and prior knowledge to object extraction. The central idea behind such an approach is to perform image partition through the propagation planar curves/surfaces. To this end, an objective function that aims to account for the expected visual properties of the object, impose certain smoothness constraints and encode prior knowledge on the geometric form of the object to be recovered is presented. Promising experimental results demonstrate the potential of such a method.

## 9.1 Introduction

Image segmentation has been a long term research initiative in computational vision. Extraction of prominent edges [381] and discontinuities between inhomogeneous image regions was the first attempt to address segmentation. Statistical methods that aim to separate regions according to their visual characteristics was an attempt to better address the problem [341], while the snake/active contour model [455] was a breakthrough in the the domain.

Objects are represented using parametric curves and segmentation is obtained through the deformation of such a curve towards the lowest potential of an objective function. Data-driven as well as internal smoothness terms were the components of such a function. Such a model refers to certain limitations like, the initial conditions, the parameterisation of the curve, the ability to cope with structures with multiple components, and the estimation of curve geometric properties.

Balloon models [204] where a first attempt to make the snake independent with respect to the initial conditions, while the use of regional terms forcing visual homogeneity [922] was a step further towards this direction. Prior knowledge was also introduced at some later point [756] through a learning stage of the snake coefficients. Geometric alternatives to snakes [152] like the geodesic active contour model [155] were an attempt to eliminate the parameterisation issue.

Curves are represented in an implicit manner through the level set method [618]. Such an approach can handle changes of topology and provide sufficient support to the estimation of the interface geometric properties. Furthermore, the use of such a space as an optimisation framework [917], and the integration of visual cues of different nature [622] made these approaches quite attractive to numerous domains [617]. One can also point recent successful attempts to introduce prior knowledge [513, 688] within the level set framework leading to efficient object extraction and tracking methods [689].

To conclude, curve propagation is an established technique to perform object extraction and image segmentation. Level set methods refer to a geometric alternative of curve propagation and have proven to be a quite efficient optimisation space to address numerous problems of computational vision. In this chapter, first we present the notion of curve optimisation in computer vision, then establishes a connection with the level set method and conclude with the introduction of ways to perform segmentation using edge-driven, statistical clustering and prior knowledge terms.

## 9.2 On the Propagation of Curves

Let us consider a planar curve $\Gamma : [0,1] \to \mathcal{R} \times \mathcal{R}$ defined at a plane $\Omega$. The most general form of the snake model consists of:

$$E(\Gamma) = \int_0^1 \left(\alpha E_{int}(\Gamma(p)) + \beta E_{img}(\mathcal{I}(\Gamma(p))) + \gamma E_{ext}(\Gamma(p))\right) dp \qquad (9.1)$$

where $\mathcal{I}$ is the input image, $E_{int}[= w_1|\Gamma'| + w_2|\Gamma''|]$ imposes smoothness constraints (smooth derivatives), $E_{img}[= -|\nabla \mathcal{I}|]$ makes the curve to be attracted from the image features (strong edges), $E_{ext}$ encodes either user interaction or prior knowledge and $\alpha, \beta, \gamma$ are coefficients that balance the importance of these terms.

The calculus of variations can be used to optimise such a cost function. To this end, a certain number of control points are selected along the curve, and the their positions are updated according to the partial differential equation that is recovered through the derivation of $E(\Gamma)$ at a given control point of $\Gamma$. In the most general case a flow of the following nature is recovered:

$$\Gamma(p;\tau) = \underbrace{\left(\alpha F_{gm}(\Gamma) + \beta F_{img}(\mathcal{I}) + \gamma F_{pr}(\Gamma)\right)}_{F} \mathcal{N} \qquad (9.2)$$

where $\mathcal{N}$ is the inward normal and $F_{gm}$ depends on the spatial derivatives of the curve, the curvature, etc. On the other hand, $F_{img}$ is the force that connects the propagation with the image domain and $F_{pr}(\Gamma)$ is a speed term that compares the evolving curve with a prior and enforces similarity with such a prior. The tangential component of this flow has been omitted since it affects the internal position of the control points and doesn't change the form of the curve itself.

Such an approach refers to numerous limitations. The number and the sampling rule used to determined the position of the control points can affect the final segmentation result. The estimation of the internal geometric properties of the curve is also problematic and depends on the sampling rule. Control points move according to different speed functions and therefore a frequent re-parameterisation of the contour is required. Last, but no least the evolving contour cannot change the topology and one cannot have objects that consist of multiple components that are not connected.

### *9.2.1 Level Set Method*

The level set method was first introduced in [261] and re-invented in [618] to track moving interfaces in the community of fluid dynamics and then emerged in computer vision [152, 537]. The central idea behind these methods is to represent the (closed) evolving curve $\Gamma$ with an implicit function $\phi$ that has been constructed as follows:

$$\phi(s) = \begin{cases} 0, s \in \Gamma \\ -\epsilon, s \in \Gamma_{in} \\ +\epsilon, s \in \Gamma_{out} \end{cases}$$

where $epsilon$ is a positive constant, $\Gamma_{in}$ the area inside the curve and $\Gamma_{out}$ the area outside the curve as shown in [Figure (9.1)]. Given the partial differential equation that dictates the deformation of $\Gamma$ one now can derive the one for $\phi$ using the chain rule according to the following manner:

$$\frac{\partial}{\partial \tau}\phi(\Gamma(p;\tau)) = \frac{\phi(\Gamma(p;\tau))}{\partial \Gamma} \underbrace{\frac{\partial \Gamma(p;\tau)}{\partial \tau}}_{F\mathcal{N}} + \frac{\partial \phi}{\partial \tau} = F\left(\nabla \phi \cdot \mathcal{N}\right) + \phi_\tau = 0 \quad (9.3)$$

Let us consider the arc-length parameterisation of the curve $\Gamma(c)$. The values of $\phi$ along the curve are 0 and therefore taking the derivative of $\phi$ along the curve $\Gamma$ will lead to the following conditions:

$$\frac{\partial \phi(\Gamma(c))}{\partial c} = 0 \rightarrow \frac{\partial \phi}{\partial \Gamma}(\Gamma(c)) \cdot \frac{\partial \Gamma}{\partial c} = 0 \rightarrow \nabla \phi(c) \cdot \mathcal{T}(c) = 0 \quad (9.4)$$

where $\mathcal{T}(c)$ is the tangential vector to the contour. Therefore one can conclude that $\nabla \phi$ is orthogonal to the contour and can be used (upon normalisation) to replace the inward normal $\left[\mathcal{N} = -\frac{\nabla \phi}{|\nabla \phi|}\right]$ leading to the following condition on

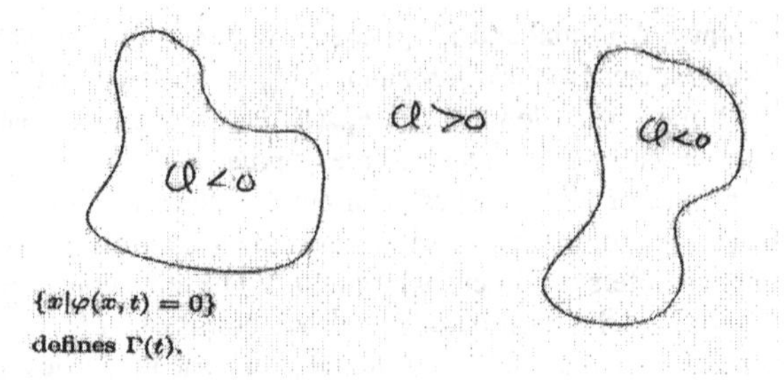


Figure 9.1. Level set method and tracking moving interfaces; the construction of the (implicit) $\phi$ function [figure is courtesy of S. Osher].

the deformation of $\phi$:

$$-F\ |\phi| + \phi_\tau = 0\ \rightarrow\ \phi_\tau = F\ |\phi| \tag{9.5}$$

Such a flow establishes a connection between the family of curves $\Gamma$ that have been propagated according to the original flow and the ones recovered through the propagation of the implicit function $\phi$. The resulting flow is parameter free, intrinsic, implicit and can change the topology of the evolving curve under certain smoothness assumptions on the speed function $F$. Last, but not least, the geometric properties of the curve like its normal and the curvature can also be determined from the level set function [618]. One can see a demonstration of such a flow in [Figure (9.2)].

In practice, given a flow and an initial curve the level set function is constructed and updated according to the corresponding motion equation in all pixels of the image domain. In order to recover the actual position of the curve, the marching cubes algorithm [526] can be used that is seeking for zero-crossings. One should pay attention on the numerical implementation of such a method, in particular on the estimation of the first and second order derivatives of $\phi$, where the ENO schema [618] is the one to be considered. One can refer to [728] for a comprehensive survey of the numerical approximation techniques.

In order to decrease computational complexity that is inherited through the deformation of the level set function in the image domain, the narrow band algorithm [194] was proposed. The central idea is update the level set function only within the evolving vicinity of the actual position of the curve. The fast marching algorithm [727, 815] is an alternative technique that can be used to evolve curves in one direction with known speed function. One can refer to earlier contribution in this book [Chapter 7] for a comprehensive presentation of this algorithm and its applications. Last, but not least semi-implicit formulations of the flow that guides the evolution of $\phi$ were proposed [351, 873] namely the additive operator splitting. Such an approach refers to a stable and fast evolution using a notable time step under certain conditions.

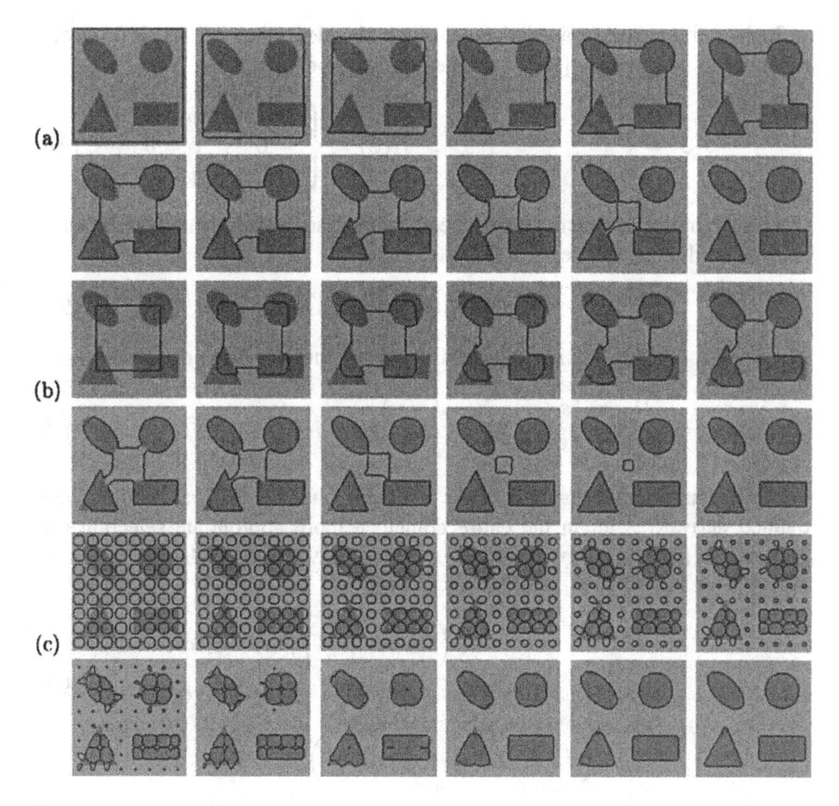


Figure 9.2. Demonstration of curve propagation with the level set method; handling of topological changes is clearly illustrated through various initialization configurations (a,b,c).

## 9.2.2 *Optimisation and Level Set Methods*

The implementation of curve propagation flows was the first attempt to use the level set method in computer vision. Geometric flows or flows recovered through the optimisation of snake-driven objective functions were considered in their implicit nature. Despite the numerous advantages of the level set variant of these flows, their added value can be seen as a better numerical implementation tool since the definition of the cost function or the original geometric flow is the core part of the solution. If such a flow or function does not address the desired properties of the problem to be solved, its level set variant will fail. Therefore, a natural step forward for these methods was their consideration in the form of an optimisation space.

Such a framework was derived through the definition of simple indicator functions as proposed in [917] with the following behaviour

$$\delta(\phi) = \begin{cases} 0 & , \phi \neq 0 \\ 1 & , \phi = 0 \end{cases}, \quad \mathcal{H}(\phi) = \begin{cases} 1 & , \phi > 0 \\ 0 & , \phi = 0 \\ 0 & , \phi < 0 \end{cases} \tag{9.6}$$

Once such indicator functions have been defined, an evolving interface $\Gamma$ can be considered directly on the level set space as

$$\Gamma = \{s \in \Omega : \delta(\phi) = 1\} \tag{9.7}$$

while one can define a dual image partition using the $\mathcal{H}$ indicator functions as:

$$\begin{array}{l} \Gamma_{in} = \{s \in \Omega : \mathcal{H}(-\phi) = 1\} \\ \Gamma_{out} = \{s \in \Omega : \mathcal{H}(-\phi) = 0\} \end{array}, \quad \Gamma_{in} \cup \Gamma_{out} = \Omega \tag{9.8}$$

Towards continuous behaviour of the indicator function $[\mathcal{H}]$ , as well as well-defined derivatives $[\delta]$ in the entire domain a more appropriate selection was proposed in [917], namely the DIRAC and the HEAVISIDE distribution:

$$\begin{array}{l} \delta_\alpha(\phi) = \begin{cases} 0 & , |\phi| > \alpha \\ \frac{1}{2\alpha}\left(1 + \cos\left(\frac{\pi\phi}{\alpha}\right)\right) & , |\phi| < \alpha \end{cases} \\ \mathcal{H}_\alpha(\phi) = \begin{cases} 1 & , \phi > \alpha \\ 0 & , \phi < -\alpha \\ \frac{1}{2}\left(1 + \frac{\phi}{\alpha} + \frac{1}{\pi}\sin\left(\frac{\pi\phi}{\alpha}\right)\right) & , |\phi| < \alpha \end{cases} \end{array} \tag{9.9}$$

Such an indicator function has smooth, continuous derivatives and the following nice property:

$$\frac{\partial}{\partial\phi}\mathcal{H}_\alpha(\phi) = \delta_\alpha(\phi)$$

Last, but not least one consider the implicit function $\phi$ to be a signed distance transform $D(s,\Gamma)$,

$$\phi(s) = \begin{cases} 0 & , \quad s \in \Gamma \\ D(s,\Gamma) & , \quad s \in \Gamma_{in} \\ -D(s,\Gamma) & , \quad s \in \Omega - \Gamma_{in} = \Gamma_{out} \end{cases} \tag{9.10}$$

Such a selection is continuous and supports gradient descent minimisation techniques. On the other hand it has to be maintained, and therefore frequent re-initialisations using either the fast marching method [727] or PDE-based approaches [774] were considered. In [353] the problem was studied from a different perspective. The central idea was to derive the same speed function for all level lines - the one of the zero level set - an approach that will preserve the distance function constraint.

## 9.3 Data-driven Segmentation

The first attempt to address such task was made in [537] where a geometric flow was proposed to image segmentation. Such a flow was implemented in the level set space and aimed to evolve an initial curve towards strong edges constrained by the curvature effect. Within the last decade numerous advanced techniques have taken advantage of the level set method for object extraction.

### *9.3.1 Boundary-based Segmentation*

The geodesic active contour model [155, 462] - a notable scientific contribution in the domain - consists of

$$E(\Gamma) = \int_0^1 g\left(|\nabla \mathcal{I}_\sigma\left(\Gamma(p)\right)|\right) |\Gamma'(p)|\, dp \tag{9.11}$$

where $\mathcal{I}_\sigma$ is the output of a convolution between the input image and a Gaussian kernel and $g$ is a decreasing function of monotonic nature. Such a cost function seeks a minimal length geodesic curve that is attracted to the desired image features, and is equivalent with the original snake model once the second order smoothness component was removed. In [155] a gradient descent method was used to evolve an initial curve towards the lowest potential of this cost function and then was implemented using the level set method.

A more elegant approach is to consider the level set variant objective function of the geodesic active contour;

$$E(\phi) = \int\int_\Omega \delta_\alpha(\phi(\omega)) g\left(|\nabla \mathcal{I}_\sigma(\omega)|\right) |\nabla \phi(\omega)| d\omega \tag{9.12}$$

where $\Gamma$ is now represented in an implicit fashion with the zero-level set of $\phi$. One can take take the derivative of such a cost function according to $\phi$:

$$\phi_\tau = \delta_\alpha(\phi) \mathrm{div}\left(g(;) \frac{\nabla \phi}{|\nabla \phi|}\right) \tag{9.13}$$

where $\omega$ and $|\nabla \mathcal{I}_\sigma(\omega)|$ were omitted from the notation. Such a flow aims to shrink an initial curve towards strong edges. While the strength of image gradient is a solid indicator of object boundaries, initial conditions on the position of the curve can be issue. Knowing the direction of the propagation is a first drawback (the curve has either to shrink or expand), while having the initial curve either interior to the objects or exterior is the second limitation. Numerous provisions were proposed to address these limitations, some of them aimed to modify the boundary attraction term [627], while most of them on introducing global regional terms [922].

### *9.3.2 Region-based Segmentation*

In [623] the first attempt to integrate edge-driven and region-based partition components in a level set approach was reported, namely the geodesic active region model. Within such an approach, the assumption of knowing the expected intensity properties (supervised segmentation) of the image classes was considered. Without loss of generality, let us assume an image partition in two classes, and let $r_{in}(\mathcal{I})$, $r_{out}(\mathcal{I})$ be regional descriptors that measure the fit between an observed intensity $\mathcal{I}$ and the class interior $[r_{in}(\mathcal{I})]$ and exterior to $[r_{out}(\mathcal{I})]$ the curve. Under such an assumption one can derive a cost function that separates the image domain into two regions:

- according to a minimal length geodesic curve attracted by the regions boundaries,
- according to an optimal fit between the observed image and the expected properties of each class,

$$\begin{aligned} E(\phi) &= w \iint_{\Omega} \delta_{\alpha}(\phi(\omega)) g\left(|\nabla \mathcal{I}_{\sigma}(\omega)|\right) |\nabla \phi(\omega)| d\omega \\ &+ \iint_{\Omega} \mathcal{H}_{\alpha}(-\phi(\omega)) r_{in}(\mathcal{I}) d\omega + \iint_{\Omega} (1 - \mathcal{H}_{\alpha}(-\phi(\omega))) r_{out}(\mathcal{I}) d\omega \end{aligned} \tag{9.14}$$

where $w$ is a constant balancing the contributions of the two terms. One can see this framework as an integration of the geodesic active contour model [155] and the region-based growing segmentation approach proposed in [922]. The objective is to recover a minimal length geodesic curve positioned at the object boundaries that creates an image partition that is optimal according to some image descriptors. Taking the partial derivatives with respect to $\phi$, one can recover the flow that is to be used towards such an optimal partition:

$$\phi_{\tau} = \delta_{\alpha}(\phi)(r_{in}(\mathcal{I}) - r_{out}(\mathcal{I})) + w\delta_{\alpha}(\phi)\text{div}\left(g(;)\frac{\nabla \phi}{|\nabla \phi|}\right) \tag{9.15}$$

where the term $\delta_{\alpha}(-\phi)$ was replaced with $\delta_{\alpha}(\phi)$ since it has a symmetric behaviour. In [623] such descriptor function was considered to be the -log of the intensity conditional density $[p_{in}(\mathcal{I}), p_{in}(\mathcal{I})]$ for each class

$$r_{in}(\mathcal{I}) = -\log\left(p_{in}(\mathcal{I})\right), \quad r_{out}(\mathcal{I}) = -\log\left(p_{out}(\mathcal{I})\right)$$

In [701] the case of supervised image segmentation for more than two classes was considered using the frame partition concept introduced in [917]. One can also refer to other similar techniques [16]. Promising results were reported from such an approach for the case of image in [624] [Figure (9.3)] and for supervised texture segmentation in [625].

However, segmentation often refers to unconstrained domains of computational vision and therefore the assumption of known appearance properties for the objects to be recovered can be unrealistic. Several attempts were made to address

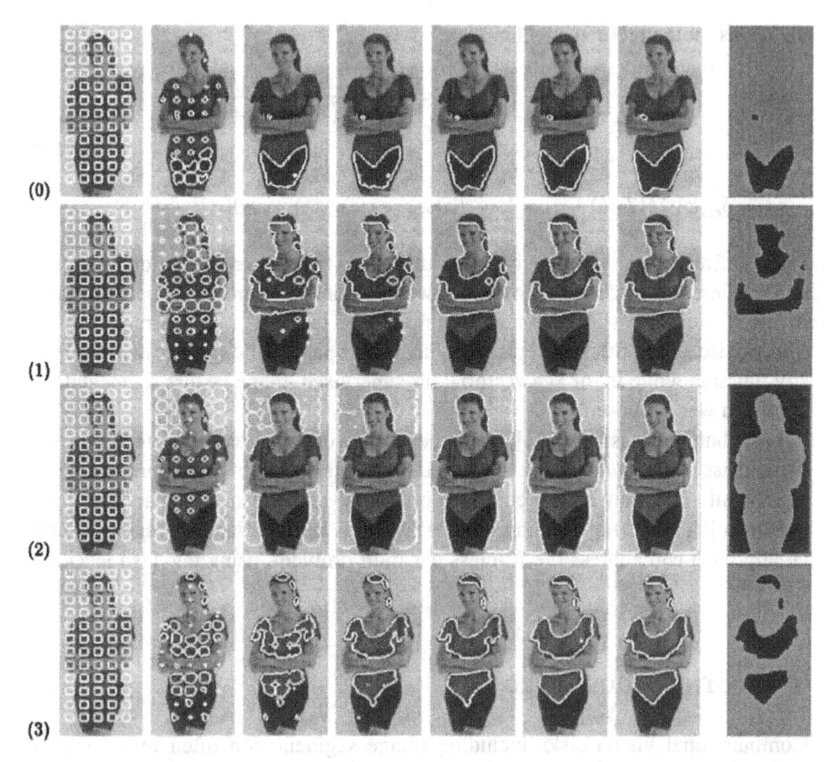


Figure 9.3. Multi-class image segmentation [624] through integration of edge-driven and region-based image metrics; The propagation with respect to the four different image classes as well as the final presentation result is presented.

this limitation. To this end, in [173, 909] an un-supervised region based segmentation approach based on the Mumford-Shah [590] was proposed. The central idea behind these approaches of bi-modal [173] and tri-modal [909] segmentation was that image regions are piece-wise constant intensity-wise.

The level set variant of the Mumford-Shah [590] framework consists of minimising

$$\begin{aligned} E(\phi, \mu_{in}, \mu_{out}) = \\ w \int\int_{\Omega} \delta_\alpha(\phi(\omega))|\nabla\phi(\omega)|d\omega + \int\int_{\Omega} \mathcal{H}_\alpha(-\phi(\omega))(\mathcal{I}(\omega) - \mu_{in})^2 d\omega \\ + \int\int_{\Omega} (1 - \mathcal{H}_\alpha(-\phi(\omega)))(\mathcal{I}(\omega) - \mu_{out})^2 d\omega \end{aligned} \quad (9.16)$$

where both the image partition $[\phi]$ and the region descriptors $[\mu_{in}, \mu_{out}]$ for the inner and the outer region are to be recovered. The calculus of variations with respect to the curve position and the piece-wise constants can be consider to recover the lowest potential of such a function,

$$\mu_{in} = \frac{\int\int_{\Omega} \mathcal{H}(-\phi)\mathcal{I}(\omega)d\omega}{\int\int_{\Omega} \mathcal{H}(-\phi)d\omega}, \quad \mu_{out} = \frac{\int\int_{\Omega}(1-\mathcal{H}(-\phi))\mathcal{I}(\omega)d\omega}{\int\int_{\Omega}(1-\mathcal{H}(-\phi))d\omega}$$

$$\phi_{\tau} = \delta_{\alpha}(\phi)\left[((\mathcal{I}(\omega)-\mu_{in}))^2 - (\mathcal{I}(\omega)-\mu_{out})^2)) + w\text{div}\left(\frac{\nabla\phi}{|\nabla\phi|}\right)\right] \tag{9.17}$$

Such a framework was the basis to numerous image segmentation level set approaches, while certain provisions were made to improve its performance. In [465] the simplistic Gaussian assumption of the image reconstruction term (piece-wise constant) was replaced with a non-parametric approximation density function while in [685] a vectorial un-supervised image/texture segmentation approach was proposed.

Last, but not least in [841] the same framework was extended to deal with multi-class segmentation. The most notable contribution of this approach is the significant reduction of the computational cost and the natural handling (opposite to [917]) of not forming neither vacuums nor overlapping regions. Such an approach can address the $N$-class partition problem, using $\log_2(N)$ level set functions.

## 9.4 Prior Knowledge

Computational vision tasks including image segmentation often refer to constrained environments. Medical imaging is an example where prior knowledge exists on the structure and the form of the objects to be recovered. One can claim that the level set method is among the most promising framework to model-free segmentation. Introducing prior knowledge within such a framework is a natural extension that could make such level sets an adequate selection to numerous applications like object extraction, recognition, medical image segmentation, tracking, etc. In [513] a first attempt to perform knowledge-based segmentation was reported, while later numerous authors have proposed various alternatives [188, 808, 688, 236].

### *9.4.1 Average Models*

Statistical representation of shapes is the first step of such an approach. Given a set of training examples, one would like to recover a representation of minimal length that can be used to reproduce the training set. To this end, all shapes of the training set should be registered to the same pose. Numerous methods can be found in the literature for shape registration, an adequate selection for building shape models in the space of implicit functions is the approach proposed in [413] where registration is addressed on this space. Without loss of generality we can assume that registration problem has been solved.

Let $\mathcal{S}_{\mathcal{A}} = \{\phi_1, \phi_2, ..., \phi_n\}$ be the implicit representations of $n$ training samples according to a signed Euclidean distance transform. Simple averaging of the shape belonging to the training set can be used to determine a mean model

$$\phi_{\mathcal{M}} = \frac{1}{n} \sum_{i=1}^{n} \phi_i \tag{9.18}$$

that was considered in [513, 808, 236]. Such a model is a not an signed Euclidean implicit function, an important limitation. However, one can recover a mean model in the form of a planar curve $\Gamma_{\mathcal{M}}$ through the marching cubes algorithm [526]. Once such a model has been determined, one can impose shape prior knowledge through the constraint that the object to be recovered at the image plane $\Gamma$ that is a clone of the average shape $\Gamma_{\mathcal{M}}$ according to some transformation:

$$\Gamma = \mathcal{A}(\Gamma_{\mathcal{M}}) \tag{9.19}$$

where $\mathcal{A}$ can be a linear or non-linear transformation. In [188] prior knowledge has been considered in the form of a mean represented with a signed distance function. Once such a model was recovered, it was used [188] within the geodesic active contour model [155] to impose prior knowledge in the level set space:

$$E(\phi, \mathcal{A}) = \int\int_{\Omega} \delta_{\alpha}(\phi) \left( g(|\nabla \mathcal{I}|)|\nabla \phi| + \lambda \phi_{\mathcal{M}}^2(\mathcal{A}(\omega)) \right) d\omega \tag{9.20}$$

where $\mathcal{A} = (\mathrm{s}, \theta, (\mathcal{T}_x, \mathcal{T}_y))$ is a similarity transformation that consists of a scale factor $[\mathrm{s}]$, a rotation component $[\theta]$ and a translation vector $(\mathcal{T}_x, \mathcal{T}_y)$. $\phi_{\mathcal{M}}$ is an implicit representation of the mean model according to a distance function and $\lambda$ is a constant that determines the importance of the prior term. Such an objective function aims at finding a minimal length geodesic curve that is attracted to the object boundaries and is not far from being a similarity transformation of the prior model:

$$\phi_{\mathcal{M}}(\mathcal{A}(\Gamma_{\mathcal{M}})) \to 0$$

Such an approach can be very efficient when modelling shapes of limited variation. On the other hand, one can claim that for shapes with important deviation from the mean model the method could fail. Furthermore, given the small number of constraints when determining the transformation between the image and the model space the estimation $[\mathcal{A}]$ could become a quite unstable task.

Towards a more stable approach to determine the optimal transformation between the evolving contour and the average model, in [688] a direct comparison between the contour implicit function and the model distance transform was used to enforce prior knowledge:

$$\phi(\omega) = \phi_{\mathcal{M}}(\mathcal{A}(\omega))$$

Despite the fact that distance transforms are robust to local deformations, invariant to translation and rotation, they are not invariant to scale variations. Slight modification of the above condition [629] could also lead to scale invariant term:

$$\mathrm{s}\phi(\omega) = \phi_{\mathcal{M}}(\mathcal{A}(\omega))$$

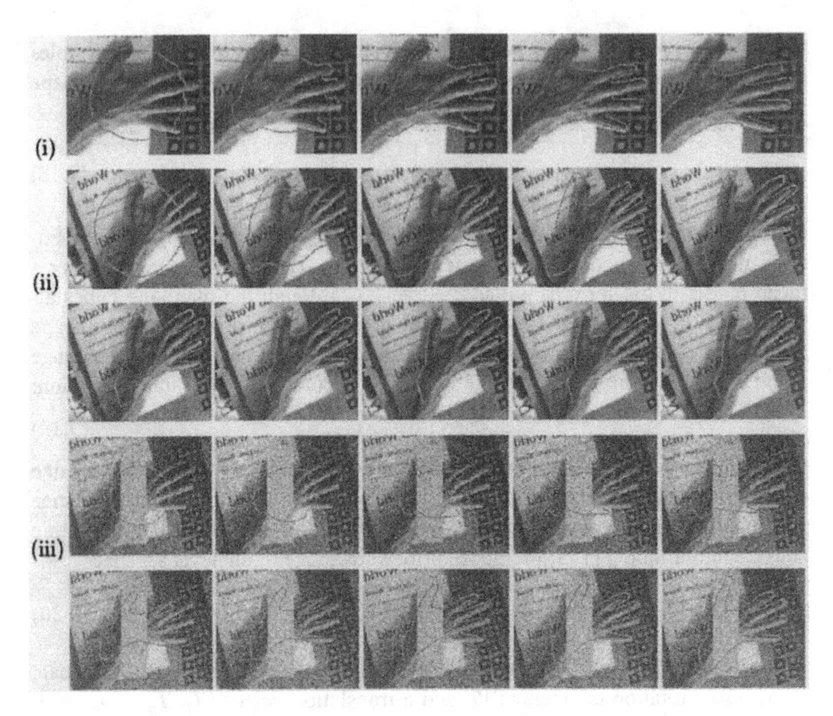


Figure 9.4. Level set methods, prior knowledge, average models and similarity invariant object extraction [688] in various pose conditions (i,ii, iii).

The minimisation of the SSD between the implicit representations of the evolving contour and the distance transform of the average prior model can be considered to impose prior knowledge, or

$$E(\phi, \mathcal{A}) = \int\int_{\Omega} \delta_\alpha(\phi)\left(s\phi(\omega) - \phi_{\mathcal{M}}\left(\mathcal{A}(\omega)\right)\right)^2 d\omega \tag{9.21}$$

a term that is evaluated within the vicinity of the zero level-set contour (modulo the selection of $\alpha$). The calculus of variations within a gradient descent method can provide the lowest potential of the cost function. Two unknown variables are to be recovered, the object position (form of function $\phi$),

$$\frac{d}{d\tau}\phi = -\underbrace{\left[\frac{\partial}{\partial\phi}\delta_\alpha(\phi)\right]((s\phi - \phi_{\mathcal{M}}(\mathcal{A})^2}_{area\ force} - 2\underbrace{\delta_\alpha(\phi)s(s\phi - \phi_{\mathcal{M}}(\mathcal{A}))}_{shape\ consistency\ force} \tag{9.22}$$

This flow consists of two terms: (i) a shape consistency force that updates the interface towards a better local much with the prior and (ii) a force that aims at updating the level set values such that the region on which the objective functions is evaluated $(-\alpha, \alpha)$ becomes smaller and smaller in the image plane. In order to better understand the influence of this force, one can consider a negative $\phi$ value, within the range of $(-\alpha, \alpha)$; Such a term does not change the position of the interface and therefore it could be omitted:

$$\frac{d}{d\mathcal{T}}\phi = -2\delta_{\alpha}(\phi)\mathrm{s}(\mathrm{s}\phi - \phi_{\mathcal{M}}(\mathcal{A})) \tag{9.23}$$

Towards recovering the transformation parameters $[\mathcal{A}]$ between the evolving contour and the average model, a gradient descent approach could be considered in parallel: $\mathcal{A}$

$$\begin{cases} \frac{d}{dt}\theta = 2\int_{\Omega} \delta_{\epsilon}(\phi)(\mathrm{s}\phi - \phi_{\mathcal{M}}(\mathcal{A}))(\nabla\phi_{\mathcal{M}}(\mathcal{A}) \cdot \frac{\partial}{\partial\theta}\mathcal{A})d\Omega \\ \frac{d}{dt}T_x = 2\int_{\Omega} \delta_{\epsilon}(\phi)(\mathrm{s}\phi - \phi_{\mathcal{M}}(\mathcal{A}))(\nabla\phi_{\mathcal{M}}(\mathcal{A}) \cdot \frac{\partial}{\partial T_x}\mathcal{A})d\Omega \\ \frac{d}{dt}T_y = 2\int_{\Omega} \delta_{\epsilon}(\phi)(\mathrm{s}\phi - \phi_{\mathcal{M}}(\mathcal{A}))(\nabla\phi_{\mathcal{M}}(\mathcal{A}) \cdot \frac{\partial}{\partial T_y}\mathcal{A})d\Omega \\ \frac{d}{dt}\mathrm{s} = 2\int_{\Omega} \delta_{\epsilon}(\phi)(\mathrm{s}\phi - \phi_{\mathcal{M}}(\mathcal{A}))(-\phi + \nabla\phi_{\mathcal{M}}(\mathcal{A}) \cdot \frac{\partial}{\partial \mathrm{s}}\mathcal{A})d\Omega \end{cases} \tag{9.24}$$

One can refer to very promising results - as shown in [Figure (9.4)] - on objects that refer to limited shape variability using such a method [688]. However, often the object under consideration presents important shape variations that cannot be accounted for with simple average models. Decomposition and representation of the training set through linear shape spaces is the most common method to address such a limitation.

### 9.4.2 Prior Knowledge through Linear Shape Spaces

In [513] a principal component analysis on the registered set of the space of distance functions (training examples) was considered to recover a model that can account for important shape variations. Similar approach was consider in [808, 116, 689]. Principal component analysis refers to a linear transformation of variables that retains - for a given number $n$ of operators - the largest amount of variation within the training data.

Let $\phi_{i=1...n}$ be a column vector representation of the training set of $n$ implicit function elements registered to the same pose. We assume that the dimensionality of this vector is $d$. Using the technique introduced in [688] one can estimate a mean vector $\phi_{\mathcal{M}}$ that is part of the space of implicit functions and subtract it from the input to obtain zero mean vectors $\{\tilde{\phi}_i = \phi_i - \phi_{\mathcal{M}}\}$.

Given the set of training examples and the mean vector, one can define the $d \times d$ covariance matrix:

$$\Sigma_{\tilde{\phi}} = E\{\tilde{\phi}_i \tilde{\phi}_i^T\} \tag{9.25}$$

It is well known that the principal orthogonal directions of maximum variation are the eigenvectors of $\Sigma_{\tilde{\phi}}$.

One can approximate $\Sigma_{\tilde{\phi}}$ with the sample covariance matrix that is given by $[\tilde{\phi}_N \tilde{\phi}_N^T]$, where $\tilde{\phi}_N$ is the matrix formed by concatenating the set of implicit

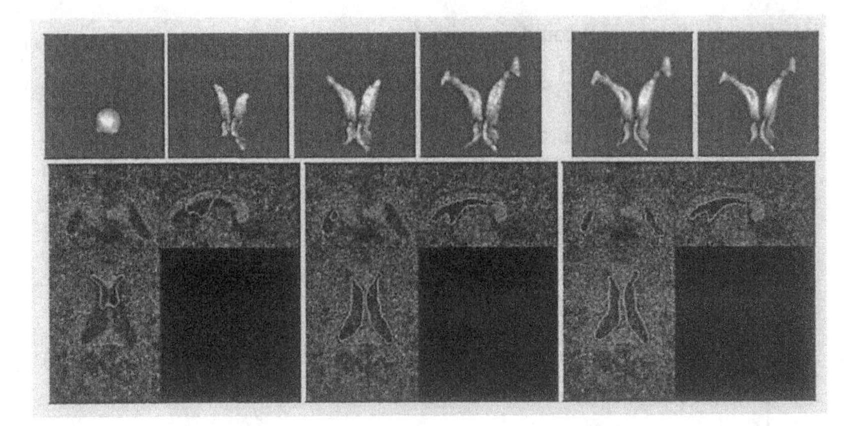

Figure 9.5. Level set methods, prior knowledge, linear shape spaces and Object Extraction [689]; segmentation of lateral brain ventricles (Top Left) surface evolution, (Top Right) projected surface in the learning space and ground-truth surface (from the training set), (Bottom) surface cut and its projection in the learning space during surface evolution.

functions $\{\tilde{\phi}_i\}_{i=1\ldots n}$. Then, the eigenvectors of $\Sigma_{\tilde{\phi}}$ can be computed through the singular value decomposition (SVD) of $\tilde{\phi}_N$ :

$$\tilde{\phi}_N = \mathbf{U}\mathbf{D}\mathbf{U}^T \tag{9.26}$$

The eigenvectors of the covariance matrix $\Sigma_{\tilde{\phi}}$ are the columns of the matrix $\mathbf{U}$ (referred to as the basis vectors henceforth) while the elements of the diagonal matrix $D$ are the square root of the corresponding eigenvalues and refer to the variance of the data in the direction of the basis vectors. Such information can be used to determine the number of basis vectors $(m)$ required to retain a certain percentage of the variance in the data.

Then, one can consider a linear shape space that consists of the $(m)$ basis vectors required to retain a certain percentage of the training set:

$$\phi = \phi_{\mathcal{M}} + \sum_{j=1}^{m} \lambda_j \, \mathbf{U}_j \tag{9.27}$$

Such linear space can now be used as prior model that refers to a global transformation $\mathcal{A}$ of the average model $\phi_{\mathcal{M}}$ and its local deformation $\lambda = (\lambda_1, \ldots, \lambda_m)$ through a linear combination of the the basis vectors $\mathbf{U}_j$. Then, object extraction is equivalent with finding a shape for which there exists such a transformation that will map each value of current representation to the "best" level set representation belonging to the class of the training shapes:

$$E(\phi, \mathcal{A}, \lambda) = \int_{\Omega} \delta_\epsilon(\phi) \left( \mathbf{s}\phi - \left( \phi_{\mathcal{M}}(\mathcal{A}) + \sum_{j=1}^{m} \lambda_j \, \mathbf{U}_j(\mathcal{A}) \right) \right)^2 d\Omega \tag{9.28}$$

where the rotation factor $\mathbf{U}_j(\mathcal{A})$ has to be accounted for when applying the principal modes of variations to deform the average shape.

In order to minimise the above functional with respect to the evolving level set representation, the global linear transformation $\mathcal{A}$ and the modes weights $\lambda_j$, we use the calculus of variations. The deformation of $\phi$ is guided by a flow similar to (9.22) that is also the case with respect to the pose parameters $\mathcal{A}$ as shown in (). Last, but not least he differentiation with respect to the coefficients $\lambda = (\lambda_1, \ldots, \lambda_m)$ leads to a linear system that has a closed form solution $\bar{V}\lambda = b$ with:

$$\begin{cases} \bar{V}(i,j) = \int_\Omega \delta_\epsilon(\phi)\mathbf{U}_i(\mathcal{A})\mathbf{U}_j(\mathcal{A}) \\ b(i) = \int_\Omega \delta_\epsilon(\phi)(s\phi - \phi_{\mathcal{M}}(\mathcal{A}))\mathbf{U}_i(\mathcal{A}) \end{cases} \tag{9.29}$$

where $\bar{V}$ is a $m \times m$ positive definite matrix. Such an approach as shown in [Figure (9.5)] - can cope with important shape variations under the assumption that the distribution of the training set is Gaussian and therefore its PCA is valid.

## 9.5 Discussion

In this chapter, we have presented an approach to object extraction through the level set method that is implicit, intrinsic, parameter free and can account for topological changes. First, we have introduced a connection between the active contours, propagation of curves and their level set implementation. Then, we have considered the notion of implicit functions to represent shapes and define objective functions in such spaces to perform object extraction and segmentation. Edge-driven as well as global statistical-based region-defined segmentation criteria were presented. In the last part of the chapter we have presented prominent techniques to account for prior knowledge on the object to be recovered. To this end, we have introduced constraints of increasing complexity proportional to the spectrum of expected shape deformations that constraints the evolving interface according to the prior knowledge. Therefore one can conclude that the level set method is an efficient technique to address object extraction, is able to deal with important shape deformations, topological changes, can integrate visual cues of different nature and can account for corrupted, incomplete and occluded data.

# Chapter10

## On a Stochastic Model of Geometric Snakes

**A. Yezzi, D. Nain, G. Unal, O. Zeitouni and A. Tannenbaum**

Abstract

It this note, we give a formulation of a stochastic snake model based the theory of interacting particle systems and hydrodynamic limits. Curvature flows have been extensively considered from a deterministic point of view. They have been shown to be useful for a number of applications including crystal growth, flame propagation, and computer vision. In some previous work [71], we have described a random particle system, evolving on the discretized unit circle, whose profile converges toward the Gauss-Minkowsky transformation of solutions of curve shortening flows initiated by convex curves. The present note shows that this theory may be implemented as a new way of evolving curves as a possible alternative to level set methods.

## 10.1 Introduction

In this paper, we describe a model of stochastic snakes based on the theory of interacting particle systems. In some previous work Ben-Arous, Tannenbaum, and Zeitouni [71], described a stochastic interpretation of curve shortening flows. This brought together the theories of curve evolution and hydrodynamical limits, and as such impacted on the growing use of joint methods from probability and pde's in the image processing and computer vision. In this present note we will indicate how this theory may be implemented to forge a novel stochastic curve evolution algorithm.

We should note that there have been other models of stochastic active contours and geometric flows; see [443] and the references therein. These approaches are very different than ours. In [443], the authors consider stochastic perturbations of mean curvature flows and applications to computer vision. Their model is continuous (macroscopic). Our model is inherently microscopic as we will elucidate below.

Following [71], we will now set the background for our results, to which we refer the reader for all the technical details. Let $\mathcal{C}(p,t) : S^1 \times [0,T) \mapsto \mathbb{R}^2$ be a family of embedded curves where $t$ parameterizes the family and $p$ parameterizes each curve. We consider stochastic interpretations of certain *curvature driven flows*, i.e., starting from an initial embedded curve $\mathcal{C}_0(p)$ we consider the solution (when it exists) of an equation of the form

$$\frac{\partial \mathcal{C}(p,t)}{\partial t} = \hat{V}(\kappa(p,t))\mathcal{N}, \quad \mathcal{C}(\cdot,t) = \mathcal{C}_t(\cdot), \tag{10.1}$$

where $\kappa(p,t)$ denotes the curvature and $\mathcal{N}$ denotes the inner unit normal of the curve $\mathcal{C}(\cdot,t)$ at $p$. Of particular interest is the case in which $\hat{V}(x) = \pm x^\alpha$. Note that the case $\hat{V}(x) = x$ corresponds to the *Euclidean* curve shortening flow [334] while $\hat{V}(x) = x^{1/3}$ corresponds to the *affine* curve shortening, which is of strong relevance in computer vision and image processing [706]. Since in both cases we get gradient flows and resulting heat equations, a stochastic interpretation seems quite natural.

We will be dealing with convex curves here and so we employ the standard parameterization via the Gauss map, that is fixing $p = \theta$, the angle between the exterior normal to the curve and a fixed axis. It is well known that the Gauss map can be used to map smooth convex curves $\mathcal{C}(\cdot)$ into positive functions $m(\cdot)$ on $S^1$ such that $\int_{S^1} e^{2\pi i\theta} m(\theta)d\theta = 0$, and that this map can be extended to the *Gauss-Minkowsky* bijection between convex curves with $\mathcal{C}(0) = 0$ and positive measures on $S^1$ with zero barycenter; see [140, Section 8] for details. We denote by $\mathcal{M}_+^0$ the latter set of measures. Under this parameterization, a convex curve $\mathcal{C}(\theta)$ can be reconstructed from a $\mu \in \mathcal{M}_+^0$ by the formula $\mathcal{C}(\theta) = \int_0^\theta e^{2\pi i\Theta}\mu(d\Theta)$, using linear interpolation over jumps of the function $\mathcal{C}(\theta)$. Further, whenever $\mu$ possesses a strictly positive density $\rho(\theta)d\theta$ then the curvature of the curve at $\theta$ is $\kappa(\theta) = 1/\rho(\theta)$.

Our interest is in constructing stochastic approximations to the solutions of curvature driven flows and from this to derive a new stochastic snake model. Approximations corresponding to polygonal curves have been discussed in the literature under the name "crystalline motion"; see [824] for a description of recent results and references. The approach in [71] is different and can be thought of as a *stochastic* crystalline algorithm: we construct a stochastic particle system whose profile defines an *atomic* measure on $S^1$, such that the corresponding curve is a convex polygon. Applying standard tools from hydrodynamic limits, it is proven in [71] that the (random) evolution of this polygonal curve converges, in the limit of a large number of particles, to curve evolution under the curve shortening flow.

## 10.2 Overview of Geodesic Snake Models

The snake model we develop here is based on so-called *geodesic* or *conformal* snakes developed by [157, 463, 733]. The underlying flow for these models is given by

$$\mathcal{C}_t = (\phi\kappa - \nabla\phi \cdot \mathcal{N})\mathcal{N},$$

where $\phi$ is a stopping term, $\kappa$ is curvature and $\mathcal{N}$ is the unit normal. (See more details about this below.) The curvature based term is used as regularization term as well as directing the flow inward to capture the object of interest. The term involving $\nabla\phi \cdot \mathcal{N}$ acts to pull the contour into the potential well defined by the object via the flow and to push it out when it passes the object of interest.

Our stochastic snake model will be based completely on an outward flow whose underlying density evolution $\rho$ is a **linear** heat equation (see equation (10.3) below). The corresponding curve evolution equation is certainly nonlinear and expanding, and would be difficult to implement in a stable manner using a deterministic scheme. The linear heat equation of course is very easy to model stochastically, and so leads to a straightforward implementation of our expanding flow. All this will be explained in Section 10.7 below.

## 10.3 Birth and Death Zero Range Particle Systems

We first set-up some notation. As above let

$$\mathcal{C}(p,t) : S^1 \times [0,T] \rightarrow \mathbb{R}^2$$

be a family of embedded curves where $p$ parametrizes the curve and $t$ the family. Then as above we consider curvature-driven flows of the form

$$\frac{\partial \mathcal{C}}{\partial t} = \hat{V}(\kappa(p,t))\mathcal{N}, \tag{10.2}$$

where $\kappa$ denotes the curvature and $\mathcal{N}$ the inward unit normal.

Since we are interested in a stochastic interpretation, we consider the evolution of a "density" corresponding to Equation (10.2). Accordingly, using the standard angle parametrization $\theta$, we interpret $\rho(\theta,t) := 1/\kappa(\theta,t)$ as a density, and compute its evolution to be:

$$\begin{aligned} \frac{\partial \rho(\theta,t)}{\partial t} &= -\frac{\partial^2 V(\rho(\theta,t))}{\partial t^2} - V(\rho(\theta,t)), \\ V(x) &:= \hat{V}(1/x). \end{aligned} \tag{10.3}$$

In Equation (10.3), the first term on the right hand side is called the ***diffusion term*** and the second term the ***reaction term.***

The approximations we use are based on so-called **birth and death zero range particle systems.** To get a flavor of the simplicity of the algorithm, we write down this system down in some detail. Full details may be found in [71].

Let $T_K = \mathbb{Z}/K\mathbb{Z}$ denote the discrete circle. Let $g : \mathrm{N} \longrightarrow \mathbb{R}_+$ (the *jump rate*, with $g(0) = 0$), $b : \mathrm{N} \longrightarrow \mathbb{R}_+$ (the *birth rate*), $d : \mathrm{N} \longrightarrow \mathbb{R}_+$ (the *death rate*, with $d(0) = 0$) be given, and define the Markov generator on the particle configuration $E_K = \mathrm{N}^{T_K}$ by

$$(\mathcal{L}^K f)(\eta) = K^2(\mathcal{L}_0 f)(\eta) + (\mathcal{L}_1 f)(\eta)\,, \quad f \in C_b(E_K)\,,$$

where

$$(\mathcal{L}_0 f)(\eta) = \frac{1}{2}\sum_{i\in T_K} g(\eta(i))\left[f(\eta^{i,i+1}) + f(\eta^{i,i-1}) - 2f(\eta)\right]$$

$$(\mathcal{L}_1 f)(\eta) =$$

$$\sum_{i\in T_K}\left[b(\eta(i))\left[f(\eta^{i,+}) - f(\eta)\right] + d(\eta(i))\left[f(\eta^{i,-}) - f(\eta)\right]\right],$$

and

$$\eta^{i,i\pm 1}(j) = \begin{cases} \eta(j)+1, & j = i\pm 1, \eta(i) \neq 0, \\ \eta(j)-1, & j = i, \eta(i) \neq 0, \\ \eta(j), & \text{else}, \end{cases}$$

$$\eta^{i,+}(j) = \begin{cases} \eta(j)+1, j = i, \\ \eta(j), \text{else}, \end{cases}$$

$$\eta^{i,-}(j) = \begin{cases} \eta(j)-1, j = i, \eta(i) > 0, \\ \eta(j), \text{else}. \end{cases}$$

Note that the **zero-range part** $\mathcal{L}_0$ approximates diffusion term of equation (10.3) while the **birth-death part** $\mathcal{L}_1$ approximates the reaction term of (10.3).

## 10.4 Poisson System Simulation

We assume that we have a system in state $\eta \in \mathrm{N}^{T_K}$ at time $t_o$. We will suppress the dependence on $\eta$ (unless absolutely necessary). We are given 4 rates at site $i$: $b_i$=birth, $d_i$=death, $g_i{}^+$=jump to right, $g_i{}^-$=jump to left. We let $E_i := \{b_i, d_i, g_i{}^+, g_i{}^-\}$ be the set of possible transition rates for system in state $\eta$ at site $i$. We let $e_i \in E_i$.

A bit more notation:

$$E := \cup_{i\in T_K} E_i,$$

$$\lambda(\eta, e_i) := e_i(\eta(i)),$$

$$U(\eta, i) := \sum_{e_i\in E_i} \lambda(\eta, e_i).$$

Then there are three standard ways of getting the Poisson system for simulating the Markov process described above.

**Algorithm 1: Per Site Transition**

1. Get values for $T_i \sim$ exponential$(U(\eta, i))$. (By this of course we mean that the $T_i$'s are exponential random variables with parameter $U(\eta, i)$.)

2. Set
$$T := \min_{i \in T_K} \{T_i\} =: T_{i^*}.$$
$i^*$ is the site where the transition occurs at time $t_o + T$.

3. To find the event in $E_i$, we then take $e_i \in E_i$ with probability
$$\frac{\lambda(\eta, e_i)}{U(\eta, e_i)},$$
the (conditional) transition probability.

**Algorithm 2: Per Event Transition**

1. Get $T(e_i) \sim$ exponential$(\lambda(\eta, e_i))$ for all $i, e_i$.

2. Set
$$T := \min_{i, e_i} \{T(e_i))\}.$$

3. Then the next event time is $t_o + T$ and the next event is
$$\arg\min\{T(e_i)\}.$$

**Algorithm 3: Summing all the Rates**
This is the method we use, so we only briefly describe it here. See our discussion below in Section 10.5. The basic idea is as follows:

1. We sum all the rates
$$U(\eta) := \sum_i U(\eta, i).$$

2. Choose an event $e \in E$ with probability $\lambda(\eta, e)/U(\eta)$. The time for this event would be $T$ $exp(U(\eta))$. Note that this way you need only one exponential random variable per transition, while the choice of $e$ only requires one uniform random variable.

## 10.5 Choosing a Random Event

We now outline in detail an $O(\log K)$ implementation of Algorithm 3 (summing all the rates) in the previous section to choose which random event $E_k$ to carry out at each step in the simulation of the stochastic particle system. Note that an event $E_k$ denotes a particular event type (birth, death, jump left, jump right) at a particular site location.

### *10.5.1 Using a List of Event Tokens*

A conceptually simple method to simulate a random event utilizes a list of "event tokens" together with a uniform random number generator such as rand() in the standard C library. The method proceeds as follows. We first generate a list of tokens corresponding to events $E_k$. Given that there are four different event types, a particle system with $K$ sites will admit a total of $4K$ distinct event tokens. Note however that to ensure the proper likelihood ratios between different events, the list will in general *not contain exactly one occurrence of each event token*. Instead, $U_k$ tokens will be included in the list for each event $E_k$, where $U_k$ is chosen to be proportional to the event's transition rate. Next a random element of the list is selected with uniform probability, and the event corresponding to the selected token is performed.

While conceptually simple, there are some practical difficulties in implementing this token list method. First, it is only possible to choose token counts $U_k$ which are all exactly proportional to the transition rates of their respective events $E_k$ if the the total set of $4K$ transition rates has a common divisor. Calculating a common divisor, assuming one even exists, can be expensive. Second, once an event occurs, the transition rates change for events at the corresponding lattice site (as well as a neighboring site in the case of a jump event). Thus, a new common divisor must be computed and the number of tokens $U_k$ must be redetermined for every event $E_k$.

If we opt to use a constant small $\epsilon > 0$ as an approximate common divisor, then we may calculate token counts $U_k$ which are approximately proportional to the transition rates of their associated events $E_k$ by integer truncation of the quotients between each transition rate and this constant divisor. In this way, we avoid having to change the number of tokens in the list for events whose rates have not changed. Updating the list for events whose rates have increased is easy and efficient since this amounts to adding new tokens (note that we do not have to keep the list sorted, so new tokens may simply be appended to the end of the list). However, updating the list for events whose rates have decreased is much more expensive since this amounts to removing tokens and therefore requires searching the list for the tokens we want to remove.

A final difficulty in the implementation difficulty stems from the fact that the size of the list changes dynamically as the particle system evolves. However, if we can estimate a reasonable upper-bound in advance, then we may avoid having to perform multiple memory allocations to maintain the event token list.

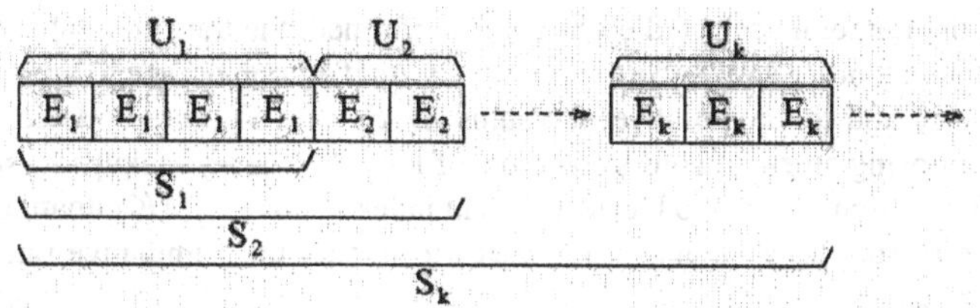


Figure 10.1. Sorted virtual list of "event tokens" used when simulating a random event.

### *10.5.2 Virtual Token List Method*

We now outline a more efficient algorithm, closely modelled after the token list method described above, which avoids the need to physically instantiate and maintain the token list. The method will be based upon a "virtual token list" which has the additional property that its event tokens are sorted in increasing order according to the index $k$ of the associated events $E_k$. The fact that the list is sorted means that tokens belonging to the same event must occur consecutively within the list (see Fig. 10.1).

The algorithm will utilize an array (of size $4K$) of nondecreasing accumulator variables $S_k$ defined by

$$S_k = \sum_{i=1}^{k} U_i$$

recalling that $U_k$ denotes the number of tokens stored in the list for event $E_k$. Notice that the size of the virtual token list is equal to value of $S_{4K}$ (recalling that $K$ denotes that number of lattice sites) and that the first token for event $E_k$, assuming $U_k \neq 0$, occurs at site $S_{k-1} + 1$. We may now choose a random token from this list by generating a random integer $n$ between 1 and $S_{4K}$ and selecting the $n$'th list element. It is possible to determine the event $E_k$ associated with the $n$'th token in the list using only the set of accumulator variables $S_1, S_2, \ldots, S_{4K}$ by noting that

$$k = \min\{i \mid n \leq S_i\}. \tag{10.4}$$

We may easily locate this event index $k$ by traversing the array of accumulator variables until the first $S_k$ is encountered such that $S_k \geq n$. We may further capitalize on the monotonicity of the $S_k$ values and use a bisection technique to locate the index $k$.

We therefore see that the only data structure we need to maintain is the array of accumulators. Furthermore, since we don't actually instantiate the list of event tokens, we are free to use non-integer values of $U_k$ (thereby circumventing the problem of finding a common divisor) and can directly equate each $U_k$ to the transition rate for event $E_k$ and accordingly set each $S_k$ to the cumulative sum of the first $k$ transition rates. $n$ is then chosen as a random positive floating point number between 0 and $S_{3K}$ which is the sum of all the event rates, and the event $E_k$ is still chosen according to the criterion (10.4).

Once the randomly selected event $E_k$ is performed, the transition rates for the corresponding lattice site (and its neighboring site if a jump occurred) are updated if necessary and the cumulative rate sums $S_1, S_2, \ldots, S_{4K}$ are updated. Thus, cost of choosing an event consists at most of $\log_2(4K)$ array lookups (assuming a bisection search is used) to locate the event index $k$ and $\log_2(4K)$ floating point additions (assuming a binary tree is used) to update the cumulative rate sums $S_1, \ldots, S_{4K}$.

## 10.6 Similarity Invariant Flows

For the stochastic snake model, we will use a flow which is scale-invariant. Accordingly, in this section, we describe a flow which are invariant relative to the scale-invariant versions of the Euclidean group, namely the similarity flow.

We begin with the heat flow for the similarity group (rotations, translations, and isotropic scalings). This flow was first presented and analyzed in [705]. We assume for the remainder of this section that our curves are strictly convex ($\kappa > 0$). Accordingly, let $\mathcal{C}$ be a smooth strictly convex plane curve, $p$ the curve parameter, and as above, let $\mathcal{N}, \mathcal{T}$, and $ds$ denote the Euclidean unit normal, unit tangent, and Euclidean arc-length, respectively. Let

$$\sigma := \frac{\partial s}{\partial p}$$

be the speed of parametrization, so that

$$\mathcal{C}_p = \sigma \mathcal{T}, \quad \mathcal{C}_{pp} = \sigma_p \mathcal{T} + \sigma^2 \kappa \mathcal{N}.$$

Then clearly,

$$\mathcal{C}_p \cdot \mathcal{C}_p = \sigma^2,$$

$$[\mathcal{C}_p, \mathcal{C}_{pp}] = \sigma^3 \kappa.$$

For the similarity group (in order to make the Euclidean evolution scale-invariant), we take a parametrization $p$ such that

$$\mathcal{C}_p \cdot \mathcal{C}_p = [\mathcal{C}_p, \mathcal{C}_{pp}], \tag{10.5}$$

which implies that

$$\sigma = 1/\kappa.$$

Therefore the similarity group invariant arc-length is the standard angle parameter $\theta$, since

$$\frac{d\theta}{ds} = \kappa,$$

where $ds$ is the Euclidean arc-length. (Note that $\mathcal{T} = [\cos\theta, \sin\theta]^T$.) Thus the *similarity normal* is $\mathcal{C}_{\theta\theta}$, and the *similarity invariant flow* is

$$\mathcal{C}_t = \mathcal{C}_{\theta\theta}. \tag{10.6}$$

Projecting the similarity normal into the Euclidean normal direction, the following flow is obtained

$$C_t = \frac{1}{\kappa}\mathcal{N}, \tag{10.7}$$

and both (10.6) and (10.7) are geometrically equivalent flows.

Instead of looking at the flow given by (10.7) (which may develop singularities), we reverse the direction of the flow, and look at the expanding flow given by

$$\frac{\partial C}{\partial t} = -\frac{1}{\kappa}\mathcal{N}. \tag{10.8}$$

We should also note that $-\mathcal{N}/\kappa$ is the normal to the curve $C$ where the derivatives are computed with respect to $\theta$.

For completeness, we state the following results for the flow (10.8) (the proofs are given in [705]):

**Theorem 10.6.1.** *1. A simple convex curve remains simple and convex when evolving according to the similarity invariant flow (10.8).*

*2. The solution to (10.8) exists (and is smooth) for all* $0 \leq t < \infty$.

**Lemma 1.** *Changing the curve parameter from p to $\theta$, we obtain that the radius of curvature $\rho$, $\rho := 1/\kappa$, evolves according to*

$$\rho_t = \rho_{\theta\theta} + \rho. \tag{10.9}$$

**Theorem 10.6.2.** *A simple (smooth) convex curve converges to a circle when evolving according to (10.8).*

**Sketch of Proof:**
So this result is so important to our construction of stochastic snakes, we briefly sketch the proof. The idea is that since the equation (10.9) is a linear heat equation, we can separate variables and see that in the standard manner $\rho(\theta, t)$ converges to a constant as $t \to \infty$. This means that the curvature converges to a constant, i.e., we get convergence to a circle. □

## 10.6.1 Heat Equation and Similarity Flows

The equation (10.9) will be the basis for our stochastic model of snakes. In the equation $\rho$ will be interpreted as a density. It is important to note that it is a linear heat equation (even though the underlying curvature flow (10.8) is nonlinear.

The stochastic model of (10.9) also gives a simple way of implementing the flow (10.8). Indeed, one can easily show that the stochastic rates for (10.9) are

$g(n) = b(n) = n$ and that $d(n) = 0$. This means that the interacting particle system is based on a classical random walk with a birth rate equal to the number of particles at a given site.

### *10.6.2 Gradient Flow*

We now state the fundamental flow underpinning the segmentation method. We state it in general even though we will only apply it to planar curves. See [740] for another derivation.

Let $R$ be an open connected bounded subset of $\mathbb{R}^n$ with smooth boundary $\partial R$. Let $\psi^t : R \to \mathbb{R}^n$ be a family of embeddings, such that $\psi^0$ is the identity. Let $\lambda : \mathbb{R}^n \to \mathbb{R}$ be a $C^1$ function. We set $R(t) := \psi^t(R)$ and $S(t) := \psi^t(\partial R)$. We consider the family of $\phi$-weighted volumes

$$\begin{aligned}\mathcal{H}(t) :=& \int_R \lambda(\psi^t(x))d\psi^t(x)\\ =& \int_{R(t)} \lambda(y)dy.\end{aligned}$$

Set $X = \frac{\partial \psi^t}{\partial t}|_{t=0}$ then using the area formula [742] and then by the divergence theorem, the first variation is

$$\begin{aligned}\frac{d\mathcal{H}}{dt}\Big|_{t=0} =& \int_R div(\lambda X)dx\\ =& -\int_{\partial R}(\lambda X)\cdot \mathcal{N}dy,\end{aligned}$$

where $\mathcal{N}$ is the inward unit normal to $\partial R$.

We now specialize the discussion to planar curves. In this case we have that if we define the functional

$$A_\lambda(t) := -\int_0^{L(t)} \langle C, \mathcal{N}\rangle \lambda dv,$$

the first variation is

$$A'_\lambda(t) = \int_0^{L(t)} [C_t, C_\theta]\lambda d\theta.$$

Then notice if we take

$$C_t = -\lambda C_{\theta\theta}$$

and using the relation (10.5), we get that

$$A'_\lambda(t) = \int [C_\theta, C_{\theta\theta}]\lambda^2 d\theta = \int \|C_\theta\|^2\lambda^2 d\theta,$$

which implies that the flow

$$C_t = -\lambda\mathcal{N}/\kappa, \tag{10.10}$$

is a gradient flow for increasing $\lambda$-weighted area.

Following the discussion about geodesic snakes in Section 10.7, we will choose

$$\lambda := \phi + \nabla\phi\cdot\mathcal{N}. \tag{10.11}$$

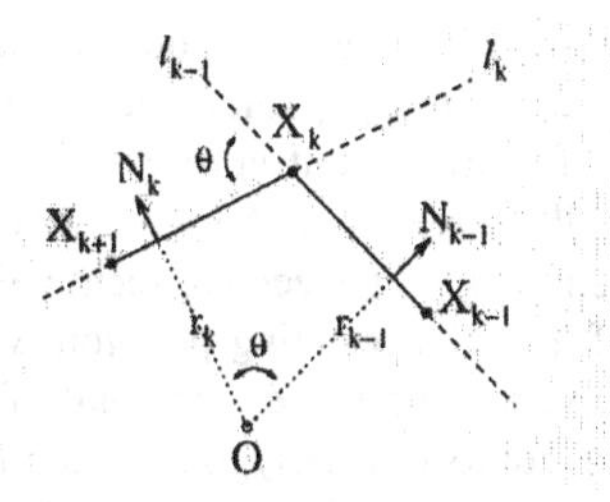


Figure 10.2. Envelope representation of a convex polygon

Here $\phi$ is the conformal stopping term. Notice that for a $\lambda$ evaluated inside a object (expanding snakes), it will be positive.

This will be used in the formulation of stochastic snakes.

## 10.7 Stochastic Snakes

In this section, we describe our formulation of a stochastic geometric active contour model. For the geometric active contour model case which we considered in this study, the density function evolves according to

$$\rho_t = (\lambda\rho)_{\theta\theta} + \lambda, \tag{10.12}$$

where $\lambda$ is as in (10.11), and subscripts indicate partial derivatives. This corresponds to the curvature driven flow (10.11).

The rates of the interacting particle system corresponding to the equation (10.12) are given by $\lambda n K^2 \times$ (mass of particle) for the diffusion, and by $\lambda n \times$ mass of particle for the birth/death, where $n$ is the number of particles at the given site. With these rates, we use the method outlined in section 10.5 to choosing an event (site and type) to simulate in each single iteration of the Markov process (i.e. one evolution step for the stochastic snake). We now turn our attention to the remaining implementation details. In particular, how do we construct an evolving snake from the evolving particle system?

### *10.7.1 Polygon representation and construction*

Here we describe a representation of polygons that connects in a particularly convenient way with our particle system model. They key point is that each site in the particle system corresponds to a polygonal edge with a fixed angle. As the particle system evolves, only the lengths of the polygonal edges change. Since the angle is always a fixed property associated with each site, we wish to exploit this in our mathematical representation of the evolving polygon.

**Envelope Representation**

One way to represent a $K$-sided convex polygon is as the inner envelope of a set of $K$ oriented lines $l_0,\ldots,l_{K-1}$ in the plane, where the orientation of each line $l_k$ is given by a choice of outward unit normal $N_k$. We assume that the lines are ordered according to the angle made between their unit normals and the $x$ axis and that the changes in angle between consecutive unit normals are all positive with a total sum of $2\pi$. The resulting polygon will consist of $K$ vertices $X_0,\ldots,X_{K-1}$ where each vertex is given by the point of intersection $X_k$ between the lines $l_{k-1}$ and $l_k$. Each edge of the polygon will in turn correspond to the segment of the line $l_k$ between the points $X_k$ and $X_{k+1}$. A minimal set of parameters to describe a particular polygon in this representation would be the unit normals $N_0,\ldots,N_{K-1}$ of each line (or equivalently their angles with the $x$-axis) and the distances $r_0,\ldots,r_{K-1}$, between each line $l_0,\ldots,l_{K-1}$ and the origin $O$. Note that these distances are *signed* to indicate whether the origin lies on the inner or outer side of each line according to the orientation of its unit outward normal.

Notice that this representation is particularly convenient in conjunction with or particle system since we may associate each line $l_k$ to a lattice site $k$ and that the unit normal $N_k$ is a function of the lattice site only, not the number of particles $\eta(k)$ at that site. Assuming an equally spaced lattice, the angle $\theta$ between consecutive unit normals will be fixed and given by $\theta = 2\pi/K$. This prescribes all of the unit normals once the first one is chosen. Thus, as the particle system evolves, the only parameters that need to be determined in this representation are the signed distances $r_0,\ldots,r_{K-1}$. Next we will show how to compute these distances based upon the particle configuration. First, however, we refer the reader to Fig. 10.2 which illustrates the representation and notation discussed in this section.

**Least-squares construction**

Let us denote by $L_k$ the length of the polygon segment on the line $l_k$ between the vertices $X_k$ and $X_{k+1}$. Note that to relate the polygon ideally to the particle system, each edge length $L_k$ should be proportional to the number of particles $\eta(k)$ at the site $k$. If we let $\Delta L$ denote the proportionality factor (i.e. the per-particle-length), then the ideal relationship between the polygon and the particle system is:

$$L_k = \hat{L}_k := \eta(k)\,\Delta L \qquad \text{(ideal case)} \tag{10.13}$$

However, this is not always realizable in the form of a closed polygon with the prescribed unit normals $N_k$. As such it is not always possible to choose the parameters $r_k$ to satisfy the constraint (10.13) for all $0 \le k \le K{-}1$. We will instead try to satisfy the constraints in a least squares sense by choosing

$$\{r_0^*,\ldots,r_{K-1}^*\} = \arg\min E(r_0,\ldots,r_{K-1})$$

where

$$E(r_0,\ldots,r_{K-1}) = \frac{1}{2}\sum_{i=0}^{K-1}\left(L_i - \eta(i)\,\Delta L\right)^2 \tag{10.14}$$

We may generate an expression for each vertex point $X_k$ in terms of the unit normals $N_0,\ldots,N_{K-1}$ and the distances $r_0,\ldots,r_{K-1}$ by noting that $X_k$ is given by the intersection of lines $l_{k-1}$ and $l_k$ (see Fig. 10.2) and therefore satisfies both line equations ($X_k \cdot N_k = r_k$ and $X_k \cdot N_{k-1} = r_{k-1}$). Hence

$$X_k^T = \begin{bmatrix} r_k & r_{k-1} \end{bmatrix} \begin{bmatrix} N_k & N_{k-1} \end{bmatrix}^{-1} = \frac{1}{\sin\theta}\left(\begin{bmatrix} T_{k-1} & T_k \end{bmatrix} \begin{bmatrix} r_k \\ -r_{k-1} \end{bmatrix}\right)^T$$

where $T_k$ denotes the unit tangent vector of the line $l_k$ (by clockwise rotation of its outward unit normal $N_k$). This of course yields

$$X_k = \frac{r_k T_{k-1} - r_{k-1} T_k}{\sin\theta} \tag{10.15}$$

and

$$L_k = (X_{k+1} - X_k) \cdot T_k = \frac{r_{k-1} - 2 r_k \cos\theta + r_{k+1}}{\sin\theta} \tag{10.16}$$

from which we can now see that the partial derivatives of $E$ are given by

$$\begin{aligned} \frac{\partial E}{\partial r_k} &= \frac{r_{k-2} - 4 r_{k-1}\cos\theta + (2+4\cos^2\theta) r_k - 4 r_{k+1}\cos\theta + r_{k+2}}{\sin^2\theta} \\ &- \frac{\hat{L}_{k-1} - 2\cos\theta \hat{L}_k + \hat{L}_{k+1}}{\sin\theta}. \end{aligned}$$

Setting $\frac{\partial E}{\partial r_k} = 0$ yields the following optimality criteria for the distances $r_0^*,\ldots,r_{K-1}^*$:

$$\begin{aligned} &(r_{k-2}^* + r_{k+2}^*) - 4\cos\theta (r_{k-1}^* + r_{k+1}^*) \\ +(2+4\cos^2\theta) r_k^* &= \Delta L \sin\theta \Big(\eta(k-1) - 2\cos\theta \eta(k) + \eta(k+1)\Big) \end{aligned} \tag{10.17}$$

## 10.8 Experimental Results

In this section, we describe illustrate our algorithm on a real data set. We used the stochastic implementation of equation 10.12 as described in Section 10.7 above. Specifically, we considered the problem of segmenting the left ventricle (short-axis view) of a heart from an MRI data set gotten from the the Department of Radiology of the Emory Medical School.

Our results are shown in Figure 10.3. We start from a polygonal initial curve and let the contour grow according to the stochastic snake model given above. "Green" indicates birth and "red" indicates death. Notice that one gets a death process when the contour leaks over the boundary which pushes it back to a steady state position.

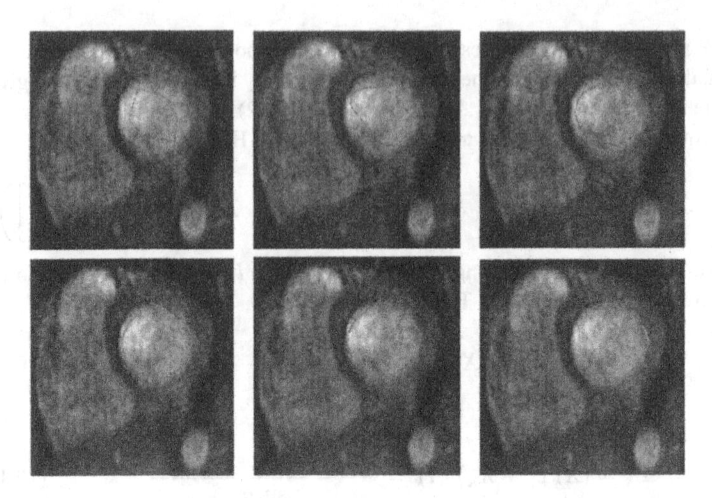

Figure 10.3. *Stochastic Snake Capturing Left Ventricle of Heart from MRI*

## 10.9 Conclusions and Future Research

In this paper, we proposed a novel approach to active contours based on a stochastic interpretation of curvature-driven flows. There are a number of extensions which we would like to consider in some future work.

First of all, the theory is now restricted to convex objects. Using the gradient term, we were able to overcome this difficulty, however we are considering other approaches based on first principles. One way would be to use negatively weighted particles (particles with "negative" mass) for concavities in the given curve.

Secondly, we are interested in extending our work to active surfaces. There is a theory of stochastic flows for surfaces; see [476]. However, the extension would certainly be nontrivial. We would need to consider the theory on the 2D discrete torus.

## Acknowledgement

This work was supported in part by grants from the National Science Foundation, AFOSR, ARO, NIH, NSF, MURI, MRI-HEL.

# Part III

# Shape Modeling & Registration

# Chapter11

## Invariant Processing and Occlusion Resistant Recognition of Planar Shapes

**A. Bruckstein**

Abstract

This short paper surveys methods for planar shape smoothing and processing and planar shape recognition invariant under viewing distortions and even partial occlusions. It is argued that all the results available in the literature on these problems implicitly follow from successfully addressing two basic problems: invariant location of points with respect to a given shape (a given set of points in the plane) and invariant displacement of points with regard to the given shape.

## 11.1 Introduction

Vision is an extremely complex process aimed at extracting useful information from images: recognizing three-dimensional shapes from their two-dimensional projections, evaluating distances and depths and spatial relationships between objects are tantamount to what we commonly mean by seeing. In spite of some irresponsible promises, made by computer scientists in the early 60's, that within a decade computers will be able "to see", we are not even close today to having machines that can recognize objects in images the way even the youngest of children are capable to do. As a scientific and technological challenge, the process of vision has taught us a lesson in modesty: we are indeed quite limited in what we can accomplish in this domain, even if we call to arms deep mathematical results and deploy amazingly fast and powerful electronic computing devices. In order to address some practical technological image analysis questions and in order to appreciate the complexity of the issues involved in "seeing" it helps to consider simplified vision problems such as "character recognition" and other "model-based planar shape recognition" problems and see how far our theories (i.e. our "brain-power") and experiments (or our "number-crunching power") can take us toward working systems that accomplish useful image analysis tasks. As a result of such scientific and commercial efforts we do have a few vision systems that work and there is a vast literature in the "hot" field of computer vision dealing with representation, approximation, completion, enhancement, smoothing, exaggeration, characterization and recognition of planar shapes. This paper surveys methods for planar shape recognition and processing (smoothing, enhancement, exaggeration etc.) invariant under distortions that occur when looking at planar shapes from various points of view. These distortions are the Euclidean, Similarity, Affine and Projective maps of the plane to itself and model the possible viewing projections of the plane where a shape is assumed to reside into the image plane of a pinhole camera, that captures the shape from arbitrary locations. A further problem one must often deal with when looking at shapes is occlusion. If several planar shapes are superimposed in the plane or are floating in 3D-space they can and will (fully, or partially) occlude each other. Under full occlusion there is of course no hope for recognition, but how about partial occlusion? Can we recognize a planar shape from a partial glimpse of its contour? Is there enough information in a portion of the projection of a planar shape to enable its recognition? We shall here address such questions too. The main goal of this paper will be to point out that all methods proposed to address the above mentioned topics implicitly require the solution of two fundamental problems: distortion-invariant location of points with respect to given planar shape (which for our purposes can be a planar region with curved or polygonal boundaries or in fact an arbitrary set of points) and invariant displacement, motion or relocation of points with respect to the given shape.

## 11.2 Invariant Point Locations and Displacements

A planar shape $S$, for our purpose, will be a set of points in $R^2$ points that most often specify a connected a planar region with a boundary that is either smooth or polygonal. The viewing distortions are classes of transformations $V_\phi : R^2 \longrightarrow R^2$ parameterized by a set of values $\phi$, and, while the class of transformations is assumed to be known to us, the exact values of the parameters are not. The classes of transformations considered are continuous groups of transformations modeling various imaging modalities, the important examples being:

- The Euclidean motions (parameterized by a rotation angle $\theta$ and a two-dimensional translation vector $(t_x, t_y)$, i.e. $\phi$ has 3 parameters).

$$\mathrm{V}_\phi^E : \ (x,y) \rightarrow (x,y) \begin{bmatrix} \cos\theta & sin\theta \\ -sin\theta & \cos\theta \end{bmatrix} + (t_x, t_y)$$

- Similarity transformations (Euclidean motions complemented by uniform scaling transformations, i.e. $|\phi| = 4$ parameters).

$$\mathrm{V}_{\phi}^{S} : \ (x, y) \rightarrow (x, y) \begin{bmatrix} \cos\theta & sin\theta \\ -sin\theta & \cos\theta \end{bmatrix} \alpha + (t_x, t_y)$$

- Equi-Affine and Affine Mappings (parameterized by $2 \times 2$ matrix - 4 parameters - or 3, if the matrix has determinant 1 - and a translation vector, i.e. $|\phi| = 6$ or 5 parameters).

$$\mathrm{V}_{\phi}^{A} : \ (x, y) \rightarrow (x, y) \begin{bmatrix} a_{11} & a_{12} \\ a_{21} & a_{22} \end{bmatrix} + (t_x, t_y)$$

- Projective Transformations (modeling the perspective projection with $|\phi| =$ 8 parameters).

$$\mathrm{V}_{\phi}^{P} : \ (x, y) \rightarrow \frac{1}{a_{31}x + a_{32}y + 1}(x, y, 1) \begin{bmatrix} a_{11} & a_{21} \\ a_{12} & a_{22} \\ a_{13} & a_{23} \end{bmatrix}$$

Given a planar shape $S \subset R^2$ and a class of viewing distortions $V_\phi : R^2 \longrightarrow R^2$ we consider the following problem:

> Two observers $A$ and $B$ look at $S_A = V_{\phi_A}(S)$ and at $S_B = V_{\phi_B}(S)$ respectively without knowing $\phi_A$ and $\phi_B$. In other words $A$ and $B$ look at $S$ from different points of view and the details of their camera location orientation and settings are unknown to them (See Figure 11.1). Observer $A$ chooses a point $P_A$ in its image plane $R^2$, and wants to describe its location w.r.t. $V_{\phi_A}(S)$ to observer $B$, in order to enable him to locate the corresponding point $P_B = V_{\phi_B}(V_{\phi_A}^{-1}(P_A))$. $A$ knows that $B$ looks at $S_B = V_{\phi_B}(S) = V_{\phi_B}(V_{\phi_A}^{-1}(S_A))$, but this is all the information available to $A$ and $B$. How should $A$ describe the location of $P_A$ w.r.t. $S_A$ to $B$?
> Solving this problem raises the issue of characterizing a position $(P_A)$ in the plane of $S_A$ in a way that is invariant to the class of transformations $V_\Phi$.

Let us consider a very simple example: take $S$ be a set of indistinguishable points in the plane $\{P_1, P_2, \ldots, P_N\}$ and $V_\phi$ be the class of Euclidean motions. A new point $\tilde{P}$ should be described to observers of this point constellation, under arbitrary viewing distortions $V_\phi$, i.e. observers of

$$V_\phi\{P_1, \ldots, P_N\} = \{V_\phi(P_1), V_\phi(P_2) \ldots V_\phi(P_N)\}$$

so that they will be able to locate $V_\phi(\tilde{P})$ in their respective "images". How should we do this? Well, we shall have to describe $\tilde{P}$'s location w.r.t. $\{P_1, P_2, \ldots, P_N\}$ in an Euclidean-invariant way. We know from elementary geometry that Euclidean motions preserve lengths and angles between line segments so there are several ways to provide invariant coordinates in the plane w.r.t. the shape $S$. The origin of an invariant coordinate system could be the Euclidean-invariant (in fact even

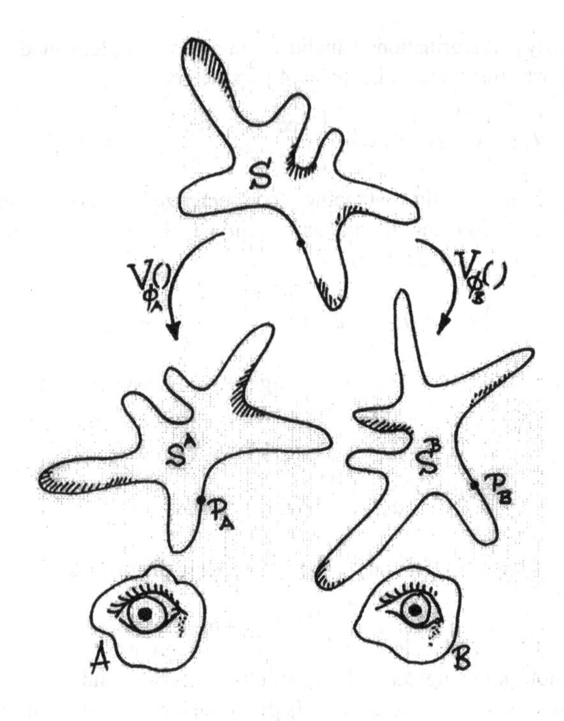


Figure 11.1.

Affine-invariant) centroid of the points $S$, i.e. $O_S = \left(\sum_{i=i}^{N} P_i\right)/N$. As one of axes (say the x-axis) of a "shape-adapted invariant" coordinate system, one may choose the longest or shortest (or closest in length to the "average" length) vector among $\{O\overline{P}_i\}$ for $i = 1, 2, \ldots, N$. This being settled, the y-axis can be defined as a 90°- rotation counter-clockwise and all one has to do is to specify $\tilde{P}$ in this adapted and Euclidean-invariant coordinate system with origin at $O_s$ and orthogonal axes chosen as described above. Note that many other solutions are possible. We here assumed that the points of $S$ are indistinguishable, otherwise the problem would be even simpler. Note also that ambiguous situations can and do arise. In case all the points of $S$ form a regular $N$-gon, there are $N$ equal length vectors $\{OP_i\}$ $i = 1, 2, \ldots, N$ and we can not specify uniquely an x-axis. But, a moment of thought will reveal that in this case the location of any point in the plane is inherently ambiguous up to rotations of $2\pi/N$.

Contemplating the above-presented simple example one realizes that solving the problem of invariant point location is heavily based on the invariants of the continuous group of transformations $V_\phi$. The centroid of the point constellation $(S)$, $O_S$, an invariant under $V_\phi$, enabled the description of $\tilde{P}$ using a distance

$d(O_S, \tilde{P})$, the length of vector $O_S\tilde{P}$ (again a $V_\phi$-invariant), up to a further parameter that locates $\tilde{P}$ on the circle centered at $O_S$ with radius $d(O_S P)$, and then the "variability" or inherent "richness" of the geometry of $S$ enabled the reduction of the remaining ambiguity.

Suppose next that we want not only to locate points in ways that are invariant under $V_\phi$ but we also want to perform invariant motions. This problem is already completely addressed in the above presented example, once an "$S$-shape-adapted" coordinate system became available. Any motion can be defined with respect to this coordinate system and hence invariantly reproduced by all viewers of $S$. In fact, when we establish an adapted frame of references we implicitly determine the transformation parameters, $\phi$, and can effectively undo the action of $V_\phi$.

To complicate the matters further consider the possibility that the shape $S$ will be partially occluded in some of its views. Can we, in this case, establish the location of $\tilde{P}$ invariantly and perform some invariant motions as before? Clearly, in the example when $S$ is a point constellation made of $N$ indistinguishable points, if we assume that occlusion can remove arbitrarily some of the points, the situation may become rather hopeless. However, if the occlusion is restricted to wiping out only points covered by a disk of radius limited to some $R_{max}$, or alternatively, we can assume that we shall always see all the points within a certain radius around an (unknown) center point in the plane, the prospects of being able to solve the problem, at least in certain lucky instances, are much better. Indeed, returning to our simple example, assume that we have many indistinguishable landmark points (forming a "reference" shape $S$ in the plane), and that a mobile robot navigates in the plane, and has a radius of sensing or visibility of $R_{max}$. At each location of the robot in the plane, $\tilde{P}$, it will see all points of $S$ whose distance from $\tilde{P}$ is less than $R_{max}$, up to an arbitrary rotation. Hence, the question of being able to specify $\tilde{P}$ from this data becomes the problem of robotic self location w.r.t the landmarks. So given a reference map (showing the "landmark"points of $S$ in some "absolute" coordinate system), we want the robot to be able to determine its location on this map from what it sees (i.e. a portion of the points of $S$ translated by $\tilde{P}$ and seen in an arbitrary rotated coordinate system). Clearly, to locate itself the robot can do the following:

> Using the arbitrarily rotated constellation of points of $S$ within its radius of sensing, i.e.
>
> $$\tilde{S}(\tilde{P}, R) = \{\Omega_\theta(P_i - \tilde{P})/P_i \in S,\ d((P_i\tilde{P}) \leq R\}$$
>
> when $\Omega_\theta$ is a rotation matrix $2 \times 2$ about $\tilde{P}$, "search" in $S$ for a similar constellation by checking various center points (2 parameters: $x_{\tilde{p}}, y_{\tilde{p}}$) and rotations (1 parameter: $\theta$). As stated this solution involves a horrendous 3-dimensional search and it must be avoided by using various available tricks like invariant geometric signatures and (geometric) hashing based on "distances" from $\tilde{P}$ to $\Omega_\theta(P_i - \tilde{P})$

and distances and angles between the $P_i$'s seen from $\tilde{P}$. This leads to much more efficient, Hough-Transform like solutions, for the self location problem. By the way, this is exactly how satelites determine their orientation in space with respect to the constellations of distant stars acting as landmark points!

In the above discussed problem it would help if the points of $S$ would be ordered on a curve, forming, say, a polygonal boundary of a planar region, or would be discrete landmarks on a continuous but clearly visible and definable boundary curve in the plane. Fortunately for those addressing planar shape analysis problems this is most often the case.

## 11.3 Invariant Boundary Signatures for Recognition under Partial Occlusions

If the shape $S$ is a region of $R^2$ with a boundary curve $\partial S = C$ that is either smooth or polygonal, we shall have to address the problem of recognizing the shape $S$ from $V_\phi$-distorted portions of its boundary. Portions of the boundary, and not the entire boundary because, we must remember, we are dealing with a scenario of possible occlusions. Our claim is that if we can effectively solve the problem of locating a point $\tilde{P}$ on the curve $C$ in a $V_\phi$-invariant way based on the "local behavior" of $C$ in a neighborhood of $\tilde{P}$, then we also have a way to detect the possible presence of the shape $S$ from a portion of its boundary. How can we locate $\tilde{P}$ based on the local behavior of $C$ in $V_\phi$-invariant ways? We shall have to associate to $\tilde{P}$ a set of numbers ("co-ordinates" or "signature" values) that are invariant under the class of $V_\phi$-transformations. To do so, one again has to rely on known geometric invariants of the group of viewing transformation assumed to act on $S$ to produce its image. The fact that we live on a curve $C$ makes our life quite a bit easier.

As an example, consider first the case where $C$ is a polygonal curve and $V_\phi$ is the group of Affine-transformations. Since all the viewing transformations map lines into lines and hence the vertices of the poly-line $C$ into vertices of a transformed poly-line $V_\phi(C)$ we can define the local neighborhood of each vertex $C(i)$ of $C$, as the "ordered" constellation of $2n+1$ points $\{C(i-n), \ldots, C(i-1), C(i), C(i+1), \ldots, C(i+n)\}$ and associate to $C(i)$ invariants of $V_\phi$ based on this constellation of points. Affine transformations are known to scale areas by the determinant of their associated $2 \times 2$ matrix, $\mathbf{A}$, of "shear and scale" parameters, hence we know that ratios of corresponding areas will be affine invariant. Therefore we could consider the areas of the triangles $\Delta_1 = [(C(i-1)C(i)C(i+1)], \Delta_2 = [C(i-2)C(i)C(i+2)] \cdots \Delta_n = [C(i-n), C(i), C(i+n)]$ and associate to $C(i)$ a vector of ratios of the type $\{\Delta_k/\Delta_l | k, l =$

$\{1, 2, \ldots, n\}, k \neq l\}$ (See Figure 11.2). This vector will be invariant under the affine group of viewing transformation and will (hopefully) uniquely characterize the point $C(i)$ in an affine-invariant way.

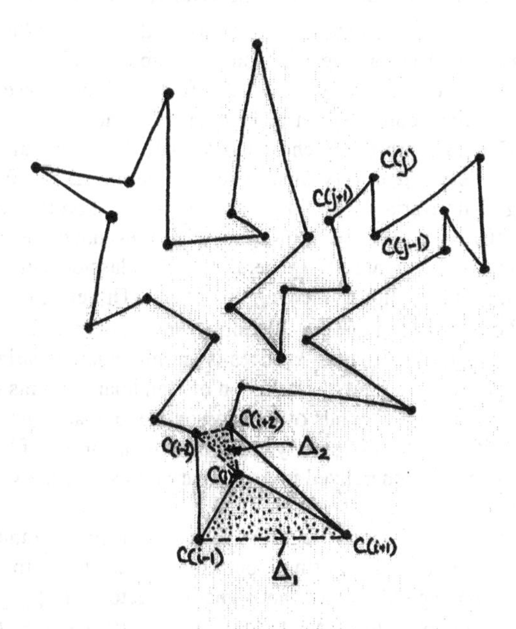


Figure 11.2.

The ideas outlined above provide us a procedure for invariantly characterizing the vertices of a poly-line, however, we can use similar ideas to also locate intermediate points situated on the line segments connecting them. Note that the number $n$ in the example above is a locality-parameter : smaller $n$'s imply more local characterization in terms of the size of neighborhoods on the curve $C$. Contemplating the foregoing example we may ask how to adapt this method to smooth curves where there are no vertices to enable us to count "landmark" points to the left and to the right of the chosen vertex in view-invariant ways. There is a beautiful body of mathematical work on invariant differential geometry providing differential invariants associated to smooth curves and surfaces, work that essentially carried out Klein's famous Erlangen program for differential geometry, and is reported on in books and papers that appeared many years ago, see ([97], [890], [378], [501] and [136]). Differential invariants enable us to determine a $V_\phi$-invariant metric, i.e. a way to measure "length" on the curve $C$ invariantly with respect to the viewing distortion, similar to the way one has, in a straightforward manner, the Euclidean-invariant arclength on smooth curves. If we have an invariant metric, we claim that our problem of invariant point characterizations on $C$ can be readily put in the same framework as in the example of a poly-line. Indeed we can now use the invariant metric to locate to the left and right of $\tilde{P}$ on $C$ (if we define $\tilde{P} \triangleq C(0)$, and describe $C$ as $C(\mu)$ where $\mu$ is the invariant metric parameterization of $C$ about $C(0) = \tilde{P}$) the points $\{C(0-n\Delta), \ldots, C(0-\Delta), C(0+\Delta), \ldots, C(0+n\Delta\}$, and these $2n+1$ points now form an invariant constellation of landmarks anchored at $\tilde{P} = C(0)$ (See Figure 11.3). Here $\Delta$ is arbitrarily chosen as a small "invariant" distance in terms of the invariant metric. It is very nice to see that letting $\Delta \searrow 0$ one often recovers, from global invariant quantities that were defined on the constellation of points about $C(0) = \tilde{P}$, differential invariant quantities that correspond to known "generalized invariant curvatures" (generalizing the classical curvature obtained if $V_\phi$ is the simplest, Euclidean viewing distortion). Therefore to invariantly locate a point $\tilde{P}$ on $C$, we can use the existing $V_\phi$ invariant metrics on $C$ (note that if $C$ is a polygon - the ordering of vertices is an immediate invariant metric!) to determine about $\tilde{P}$ an invariant constellation of "landmark" points on the boundary curve and use global invariants of $V_\phi$ to associate to $\tilde{P}$ an "invariant signature vector" $I_{\tilde{P}}(\Delta)$. If $\Delta \searrow 0$ this vector yields, for quite a variety of "good" choices of invariant quantities "generalized invariant curvatures" for the various viewing groups of transformations $V_\phi$.

We however do not propose to let $\Delta \searrow 0$. $\Delta$ is a locality parameter (as was $n$ before) and we could use several small, but finite, values for $\Delta$ to produce (what we can call) a "scale-space" of invariant signature vectors $\{I_{\tilde{P}}\}_{\Delta_i \in Range(of \Delta's)}$.

This freedom allows us to associate to a curve $C(\mu)$, parameterized in terms of its "invariant metric or arclength", a vector valued scale space of signature functions $\{I_P(\mu)\}_{\Delta_i \in Range}$, that will characterize it in both a localized and view-invariant ways. This characterization being local (its locality being in fact under our control via $\Delta$ and $n$) is useful to recognize portions of boundaries in scenes where planar shapes appear both distorted and partially occluded. The recognition process becomes, in terms of the vector-valued signature function, a partial matching algorithm, see [129].

## 11.4 Invariant Processing of Planar Shapes

Smoothing and other processes of modifying and enhancing planar shapes involves moving their points to new locations. Here we are naturally led to define planar shape deformations or evolutions, by motions of points on the shape boundaries that are small and based on the local geometry, i.e. the geometry of the constellation of other boundary points in the neighborhood. In the spirit of the discussion above, we want to do this in "viewing-distortion-invariant" ways. To do so we have to locate the points of a shape $S$ (or of its boundary $C = \delta S$) and then invariantly move them to new locations in the plane. The discussions of the

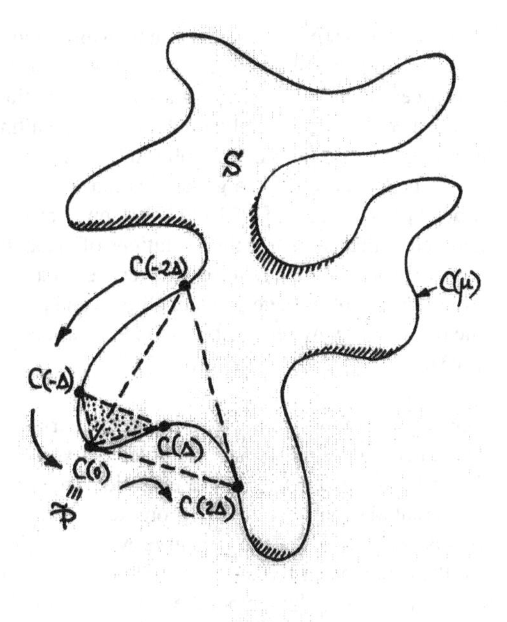


Figure 11.3.

previous sections showed us various ways to invariantly locate points on $S$ or in the plane of $S$. Moving points around is not much more difficult. We shall have to associate to each point (of $S$, or in the plane of $S$) a vector $M$ whose direction and length have been defined so as to take us to another point, in a way that is $V_\phi$-invariant. In the example of $S$ being a constellation of points, with a robot using the points of $S$ to locate itself at $\tilde{P}$, we may also want it to determine a new place to go, i.e. to determine a point $\tilde{P}_{new} = \tilde{P}+M$, so as to have the property that from $V_\phi(\tilde{P})$ a robot using the points $\{V_\phi(P_1)\ldots V_\phi(P_N)\}$ will be able to both locate itself and move to $V_\phi(\tilde{P}_{new})$. Of course, on shapes we shall have to do motions that achieve certain goals like smoothing the shape or enhancing it in desirable ways as discussed in [761] . To design view-distortion invariant motions, we can (and indeed must) rely on invariant point characterizations. Suppose we are at a point $\tilde{P}$ on the boundary $C = \delta S$ of a shape $S$, and we have established a constellation of landmark points about $\tilde{P}$. We can use the invariant point constellation about $\tilde{P}$ to define a $V_\phi$-invariant motion from $\tilde{P}$ to $\tilde{P}_{new}$ (See Figure 11.4).

Let us first consider again a very simple example: if $V_\phi$ is the Affine group of viewing transformations, the centroid of the point constellations about $\tilde{P}$ is an invariantly defined candidate for $\tilde{P}_{new}$. Indeed it is an average of points around $\tilde{P}$ and the process of moving $\tilde{P}$ to such a $\tilde{P}_{new}$ or, differentially, toward such a new position can (relatively easily) be proved to provide an affine invariant shape smoothing operation. If $S$ is a polygonal shape, i.e. $\partial S = C$ is a poly-line, then moving the vertices according to such a smoothing operation can be shown to shrink any shape into a polygonal ellipse, "the affine image of a regular polygon" with the same number of vertices as the original shape. This beautiful result is a generalization of the very early work of Darboux [244], see also [130] and [717], on "a problem in elementary geometry" that addresses the evolutions of planar polygons under an iterative process which replaces the vertices of a polygon by the (ordered) midpoints of its edges. In fact ellipses and polygonal ellipses are the results of many reasonably defined invariant averaging processes [703].

If we are dealing with a smooth continuous boundary curve $C$ and we move the points infinitesimally according to a local "velocity" vector invariantly defined we are in the realm of "geometric" curve evolution processes described by nonlinear partial differential equations. A very prominent recent example of such an evolution process is the Euclidean invariant curve evolution moving the smooth boundary points in the direction of the local normal vector $N_c$ proportionally to the local Euclidean curvature $k_c$. The temporal evolution of simple closed curves (boundaries of planar shapes) under this rule, described by

$$\frac{\partial}{\partial t} C_t = k_{c_t} N_{c_t} \qquad C_0 = original\ boundary.$$

was thoroughly analyzed and it was proved to smoothly deform and shrink any original curve into an infinitesimal circle, see e.g. [334] and [361]. This nice mathematical result, together with the fact that other Euclidean invariant motions like Blum's prairie fire evolution model [100], [142] which postulates constant velocity motion in the direction of the local normal and leads to "shocks" or "wavefront" collisions that were found useful for shape descriptions since they produce the so-called "shape skeletons", generated a lot of interest and activity in the computer vision community. This activity culminated with the realization that a variety of geometric and viewpoint invariant shape evolutions exist and may be useful in invariant shape analysis and classification. Note that $k_c N_c = \frac{\partial^2 C}{\partial s^2}$, i.e. that in the Euclidean case the invariant vector $k_c N_c$ associated to a point $C$ on a smooth curve is the second derivative of the curve with respect to the Euclidean invariant arclength. This observation yields a very nice interpretation of this invariant evolution: a point on the curve is simply replaced by weighted average of points of the curve of $C(s)$ in the neighborhood with averaging weight dependent on their distance measured in the view invariant metric, from the anchor part $C$. If the averaging kernel is Gaussian then if is readily seen that a "geometric" diffusion process results, but recall that a variety of local processing and averaging operations are readily available and implementable and should be considered as viable alternatives in generating useful shape evolutions. The process of deriving the invariant evolutions is, of course, readily generalized to more complex viewing transformations and distortions [131], [703].

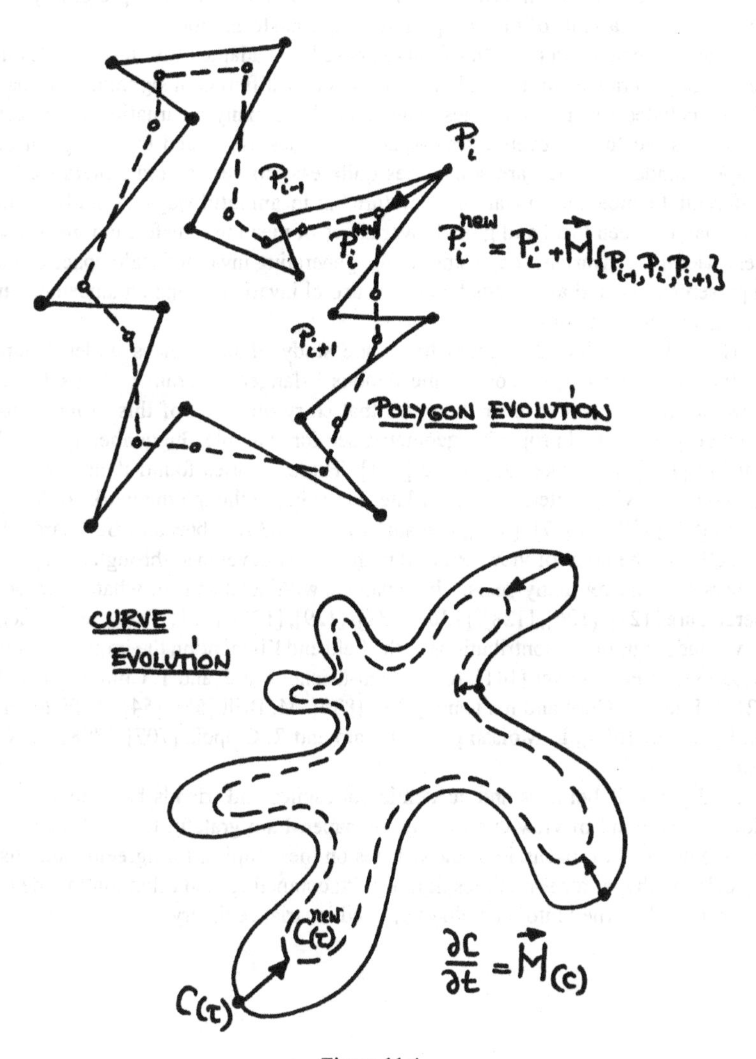


Figure 11.4.

## 11.5 Concluding Remarks

The main point of this paper is the thesis that in doing "practical" view-point invariant shape recognition or shape processing for smoothing or enhancement, one has to rely on the interplay between global and local (and preferably not differential) invariants of the group of viewing transformations.

Invariant reparameterization of curves based on "adapted metrics" enables us to design generalized and local but not necessarily differential signatures for partially occluded recognition. These signatures have many incarnations, they can be scalars, vectors or even a scale-space of values associated to each point on shape boundaries. They are sometimes quite easy to derive, and generalize the differential concept of "invariant curvature" in meaningful ways. A study of the interplay between local and global invariances of viewing transformations is also very useful for invariant shape smoothing, generating invariant scale-space shape representations, and also leads to various useful invariant shape enhancement an exaggeration operations.

The point of view that geometry is the study of invariances under groups of transformations is, of course, the famous Erlangen program of Felix Klein. Several books appeared over the years that carry out parts of this program for Euclidean affine and projective geometry, see for example Guggenheimer [370], Buchin [136], Blaschke [97], Lane [501]. These theories found their way into the computer vision literature rather late, for example though the works of Weiss [875],[875] [876], [877], [677], Cygansky [242], [833], Abter and Burkhardt [1] and others. The point of view exposed in this paper developed through a series of papers written over many years. These papers, with the details of what is exposed herein, are [127], [126], [128], [130], [131], [129], [132], [761]. Other researchers have made significant contributions to the field and I'll mention the important contributions of Peter Olver [141], [610], Jean-Michel Morel and T. Cohignac [211], [212], Luc Van Gool and his team [829], [830], M. Brill [55], [54], Z. Pizlo and A. Rosenfeld [650], L. Moisan [582], J. Sato and R. Cippola [707], [708] and O. Faugeras [307].

Students, collaborators and academic colleagues and friends have helped me develop the point of view exposed in this paper. I am grateful to all of them for the many hours of discussions and debates on these topics, for agreeing and disagreeing with me, for sometimes fighting and competing, and often joining me on my personal journey into the field of applied invariance theory.

# Chapter12

## Planar Shape Analysis and Its Applications in Image-Based Inferences

**A. Srivastava, S. Joshi, D. Kaziska and D. Wilson**

Abstract

Shapes of boundaries can play an important role in characterizing objects in images. We describe an approach for statistical analysis of shapes of closed curves using ideas from differential geometry. A fundamental tool in this shape analysis is the construction and implementation of geodesic paths between shapes. We use geodesic paths to accomplish a variety of tasks, including the definition of a metric to compare shapes, the computation of intrinsic statistics for a set of shapes, and the definition of probability models on shape spaces. We demonstrate this approach using three applications: (i) automated clustering of objects in an image database according to their shapes, (ii) interpolation of heart-wall boundaries in echocardiographic image sequences, and (iii) a study of shapes of human silhouettes in infrared surveillance images.

## 12.1 Introduction

Detection, extraction and recognition of objects in an image is an important area of research. Objects can be characterized using a variety of features: textures, edges, boundaries, colors, motion, shapes, locations, etc. These features are often used in a statistical framework to perform image analysis. In particular, one defines a feature space, trains probability models on these spaces using past data, and uses them to conduct statistical inferences on future data. Shape often provides an important clue for determining how an object appears in an image. For example, we have displayed the images of four animals in the top panels of Figure 12.1. The lower panels show the silhouettes of these animals in the corresponding images. It is easy to see that the shapes of these silhouettes can help shortlist, or even identify, the animals present in these images. Tools for shape analysis can

Figure 12.1. Analysis of shapes of objects' boundaries in images can help in computer vision tasks such as object recognition.

prove important in several applications including medical image analysis, human surveillance, military target recognition, finger-print analysis, space exploration, and underwater search. One reason for pursuing shape analysis is the possibility that an efficient representation and analysis of shapes can help even in situations where the observations are corrupted, e.g. when objects are partially obscured or corrupted by excess clutter. Shape is a global feature that can help overcome loss of some local data. This possibility, along with the development of statistical methods, has led to the idea of *Bayesian shape analysis*. In this approach a contextual knowledge is used to impose prior probabilities on shape spaces, followed by the use of posterior probabilities to perform inferences from images.

In order to perform statistical analysis of shapes, one needs tools to address the following questions:

1. How can an object be represented by the shape of its boundary?
2. How can dissimilarities between the shapes of two closed curves be quantified?
3. How to compute summary statistics, such as mean, covariance, etc, for a given collection of observed shapes?
4. What family of probability models can be used to describe variability in a collection of shapes?
5. How to solve an optimization problem, e.g. estimation of maximum a-posteriori (MAP) shape, on a shape space?
6. Given an observed shape, how to decide which family of shapes does it belong to?

In summary, one needs tools for representation, comparison, clustering, learning, estimating, and testing of shapes. Solutions to several of these questions exist as shapes have been an important topic of research over the past decade. However, a comprehensive approach for analysis of shapes in $\mathbb{R}^2$ has emerged only recently. A significant part of the past efforts has been restricted to "landmark-based" analysis, where shapes are represented by a coarse, discrete sampling of the object contours[284, 747]. Since automatic detection of landmarks is not straightforward and the ensuing shape analysis depends heavily on the landmarks chosen, this approach is limited. In addition, shape interpolation with geodesics in this framework lacks a physical interpretation. A similar approach, called *active shape models*, uses principal component analysis (PCA) of landmarks to model shape variability [223]. Despite its simplicity and efficiency, its scope is rather limited because it ignores the nonlinear geometry of shape space. Grenander's formulation [364] considers shapes as points on infinite-dimensional manifolds, where the variations between the shapes are modeled by the action of Lie groups (diffeomorphisms) on these manifolds [366]. In summary, the majority of previous work on analyzing shapes of planar *curves* involves either a discrete collection of points or diffeomorphisms on $\mathbb{R}^2$. Seldom have shapes been studied as *closed curves*!

In contrast, a recent approach [478, 755] considers the shapes of **continuous**, closed curves in $\mathbb{R}^2$, without any need for landmarks, diffeomorphisms, or level sets to model shape variations. We summarize this approach in Section 12.2, and present three applications of this approach in later sections. First, in the area of computer vision, one is interested in automated partitioning of an observed set of shapes into clusters of similar shapes, which is useful in applications such as image retrieval, organization of large databases of images, and learning of probability models on the shape space. We describe a method for clustering shapes where dissimilarities between shapes are quantified using geodesic lengths on the shape space. Second, we look at a problem in ecocardiographic image analysis where shapes of epicardial and endocardial boundaries are studied to determine the extent and progression of disease in a patient's heart. We focus on the specific problem of interpolating these boundaries in image sequences when an expert provides contours for the first and last frames in the sequence. Lastly, we will present an application involving human surveillance with a goal of detecting humans in low-quality night-vision (infrared) images. Our approach is to use a statistical analysis of shapes of human silhouettes in detection, and we present a statistical model to capture human shapes.

The rest of this chapter is organized as follows. In Section 2 we present a differential-geometric representation of shapes that leads to natural and efficient statistical analysis. In Sections 12.3-12.5, we describe the three applications and present a summary in Section 6.

## 12.2 A Framework for Planar Shape Analysis

We start with a basic question of how to represent shapes of closed curves. Our approach is to identify a space of closed curves, remove shape-preserving transformations from it, impose a Riemannian structure on it, and treat the resulting quotient space as the shape space. Using the Riemannian structure of this space, we have developed algorithms for computing geodesic paths on these shape

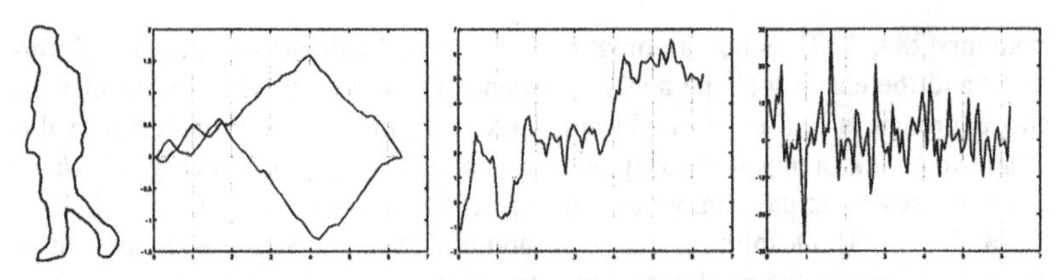

Figure 12.2. Alternate representations of a closed curve (left panel) via $x$ and $y$ coordinate functions $\alpha$ (second panel), angle function $\theta$ (third panel), or curvature function $\kappa$ (last panel).

spaces. We summarize the main ideas here and refer to the recent paper by Klassen et al [478] for details.

1. **Geometric Representation of Shapes**: Consider the boundaries or silhouettes of the imaged objects as closed, planar curves in $\mathbb{R}^2$ (or equivalently in $\mathbb{C}$) parameterized by the arc length. Define the angle function as follows: note the angle, made by the velocity vector with the positive $x$-axis, as a function of arc length. Coordinate function $\alpha(s)$ relates to the angle function $\theta(s)$ according to $\dot{\alpha}(s) = e^{j\,\theta(s)}$, $j = \sqrt{-1}$. The curvature function of this curve is given by $\kappa(s) = \dot{\theta}(s)$. A curve can be represented by its coordinate function $\alpha$, the angle function $\theta$, or the curvature function $\kappa$, as demonstrated in Figure 12.2.

In this approach, we choose angle functions to represent and analyze shapes. The direction function of a unit circle is given by $\theta_0(s) = s$. For any other closed curve of rotation index 1, the direction function takes the form $\theta = \theta_0 + h$, where $h \in \mathbb{L}^2$, and $\mathbb{L}^2$ denotes the space of all real-valued functions with period $2\pi$ and square integrable on $[0, 2\pi]$. The next issue is to account for equivalence of shapes. As shown in Figure 12.3, shape is a characteristic that is invariant to rigid motions (translation and rotation) and uniform scaling. Additionally, for closed curves, shape is also invariant to the placement of origin (or starting point) on the curves. To build representations that allow such invariances, we proceed as follows. We remove the scale variations by forcing all curves to be of length $2\pi$. The translation is already removed since the angle function $\theta$ is invariant to the translation of the curve in $\mathbb{R}^2$. To make shapes invariant to rotation, restrict to $\theta \in \{\theta_0 + \mathbb{L}^2\}$ such that, $\frac{1}{2\pi}\int_0^{2\pi} \theta(s)ds = \pi$. Also, for a closed curve, $\theta$ must satisfy the *closure condition*: $\int_0^{2\pi} \exp(j\,\theta(s))ds = 0$. Summarizing, one restricts to the set $\mathcal{C} = \{\theta \in \theta_0 + \mathbb{L}^2 | \ \frac{1}{2\pi}\int_0^{2\pi} \theta(s)ds = \pi, \ \int_0^{2\pi} e^{j\theta(s)}ds = 0\}$. Furthermore, to remove the re-parametrization group (relating to different placements of the origin), define the quotient space $\mathcal{S} \equiv \mathcal{C}/\mathbb{S}^1$ as the space of continuous, planar shapes, where $\mathbb{S}^1$ denotes the unit circle in $\mathbb{R}^2$. $\mathcal{C}$ is called the *pre-shape space* and $\mathcal{S}$ is called the *shape space*.

For the purpose of shape analysis, the incidental variables such as scale, location, orientation, etc, are termed as *nuisance variables*, and are removed from the analysis as described above. In contrast, detection and recognition of objects in

Figure 12.3. Shape is a characteristic that is invariant to rigid rotation and translation, and uniform scaling. Shape spaces are always quotient spaces.

images requires estimation of both their shapes and their nuisance variables. In this case, the shape and the nuisance variables may have independent probability models. Let $\mathcal{Z} = (SO(2) \times \mathbb{R}^2 \times \mathbb{R}_+)$ be the space of nuisance variables, and let $(\theta, z)$ be a representation of a closed curve $\alpha$ such that $\theta \in \mathcal{S}$ is its shape and $z \in \mathcal{Z}$ are its nuisance variables.

2. **Geodesic Paths Between Shapes**: An important tool in a Riemannian analysis of shapes is to construct geodesic paths between arbitrary shapes. Klassen et al. [478] approximate geodesics on $\mathcal{S}$ by successively drawing infinitesimal line segments in $\mathbb{L}^2$ and projecting them onto $\mathcal{S}$, as depicted in the top panel of Figure 12.4. For any two shapes $\theta_1, \theta_2 \in \mathcal{S}$, one uses a *shooting method* to construct a geodesic between them. The basic idea is to search for a tangent direction $g$ at the first shape $\theta_1$, such that a geodesic in that direction reaches the second shape $\theta_2$ (called the target shape) in unit time. This search is performed by minimizing a "miss function", defined as the chord length or the $\mathbb{L}^2$ distance between the shape reached and $\theta_2$, using a gradient process. The geodesic metric is $\langle g_1, g_2 \rangle = \int_0^{2\pi} g_1(s) g_2(s) ds$ on the tangent space of $\mathcal{S}$. This choice implies that a geodesic between two shapes is the path that uses **minimum energy to bend one shape into the other**. Shown in the bottom two rows are examples of geodesic paths connecting the two end shapes. We will use the notation $\Psi_t(\theta, g)$ for a geodesic path starting from $\theta \in \mathcal{S}$, in the direction $g \in T_\theta(\mathcal{S})$, as a function of time $t$. Here $T_\theta(\mathcal{S})$ denotes the space of functions tangents to $\mathcal{S}$ at the point $\theta$. If $g \in T_{\theta_1}(\mathcal{S})$ is the shooting direction to reach $\theta_2$ in unit time from $\theta_1$, then the following holds: $\Psi_0(\theta_1, g) = \theta_1$, $\Psi_1(\theta_1, g) = \theta_2$, and $\dot{\Psi}_0(\theta_1, g) = g$. The length of this geodesic is given by $d(\theta_1, \theta_2) = \sqrt{\langle g, g \rangle}$.

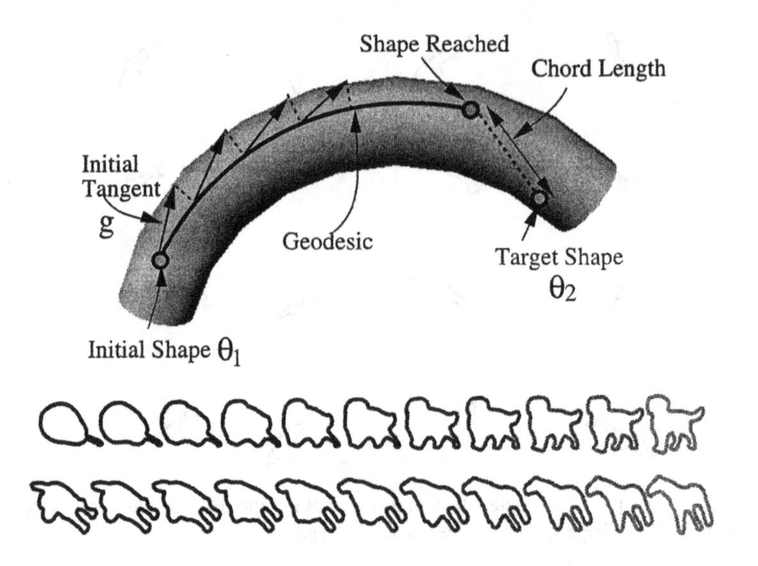


Figure 12.4. Top: A cartoon diagram of a shooting method to find geodesics in shape space. Bottom two rows: Examples of a geodsic path in $\mathcal{S}$.

3. **Mean Shape in** $\mathcal{S}$: For a collection $\theta_1, \ldots, \theta_n$ in $\mathcal{S}$, and $d(\theta_i, \theta_j)$ the geodesic length between $\theta_i$ and $\theta_j$, the Karcher mean is defined as the element $\mu \in \mathcal{S}$ that minimizes the quantity $\sum_{i=1}^{n} d(\theta, \theta_i)^2$. A gradient-based, iterative algorithm for computing the Karcher mean is presented in [503, 454] and is particularized to $\mathcal{S}$ in [478].

This approach provides a comprehensive framework for a statistical analysis of planar shapes. In the next three sections, we present some applications of this framework to problems of practical interest.

## 12.3 Clustering of Shapes

In order to facilitate training of probability models for shape families, one needs to organize the observed shapes into clusters of similar shapes. One of the popular techniques for clustering points in Euclidean spaces is $k$-mean clustering [429]. In this method, $n$ given points are clustered into $k$ groups, for a given $k$, in such a way that the sum of within cluster-variances is minimized. Since computing means of shapes is expensive, we modify this procedure so that it avoids computing cluster means at every iteration.

Our approach is to divide $n$ given shapes into $k$ clusters in such a way that a cumulative dispersion within the clusters is minimized. Let a configuration $C$ consist of clusters denoted by $C_1, C_2, \ldots, C_k$. If $n_i$ is the size of $C_i$, then the cost

$Q$ associated with a cluster configuration $C$ is given by [401]:

$$Q(C) = \sum_{i=1}^{k} \frac{2}{n_i} \left( \sum_{\theta_a \in C_i} \sum_{b<a, \theta_b \in C_i} d(\theta_a, \theta_b)^2 \right) . \tag{12.1}$$

We seek configurations that minimize $Q$, i.e., $C^* = \text{argmin}\, Q(C)$. This cost function differs from the usual variance function and avoids the need for updating means of clusters at every iteration.

In [755], we utilize a stochastic search process to find an optimal configuration. The basic idea is to start with a random configuration of $n$ shapes into $k$ clusters, and use a sequence of moves, performed probabilistically, to re-arrange that configuration into an optimal one. The moves are restricted to be of two different kinds: move a shape from one cluster to another, or swap two shapes from two different clusters. The probabilities of performing these moves are set to the negative exponential of the resulting $Q$ function. Additionally, a temperature variable $T$ is decreased slowly in each iteration to simulate annealing so that this process converges to an optimal configuration in due time. Next we present the algorithm for clustering of $n$ planar shapes into $k$ clusters.

**Algorithm 1.** *1. Compute pairwise geodesic distances between all $n$ shapes. This requires $n(n-1)/2$ geodesic computations.*

*2. With equal probabilities pick one of two moves:*

*(a)* **Move a shape***:*

*i. Pick a shape $\theta_j$ randomly. If it is not a singleton in its cluster, then compute $Q_j^{(i)}$, the cost obtained after moving $\theta_j$ to $C_i$, for all $i = 1, 2, \ldots, k$.*

*ii. Compute the probability $P_M(j, i; T)$ according to*

$$P_M(j, i; T) = \frac{\exp(-Q_j^{(i)}/T)}{\sum_{l=1}^{k} \exp(-Q_j^{(l)}/T)} , \quad i = 1, 2, \ldots, k ,$$

*and re-assign $\theta_j$ to a cluster chosen according to the probability $P_M$,*

*(b)* **Swap two shapes***:*

*i. Select two clusters randomly, and select a shape from each of them. Let $Q^{(1)}$ and $Q^{(2)}$ be the configuration costs before and after the swap, respectively.*

*ii. Compute the probability $P_S(T)$, where*

$$P_S(T) = \frac{\exp(-Q^{(2)}/T)}{\sum_{i=1}^{2} \exp(-Q^{(i)}/T)} ,$$

*and swap the two shapes according to that probability.*

*3. Update temperature using $T = T/\beta$ and return to Step 2. We have used $\beta = 1.0001$ in our experiments.*

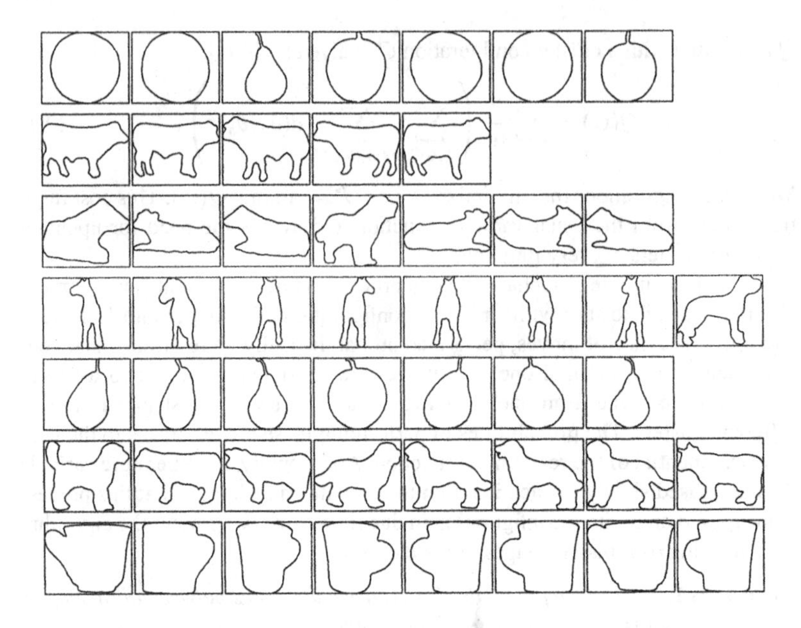

Figure 12.5. Clustering of 50 shapes from ETH-80 dataset using Algorithm 1. Each row represents a cluster.

Displayed in Figure 12.5 are the results of Algorithm 1, where a collection of 50 shapes have been sorted into into seven clusters. All the shapes in a cluster have been placed in the same row. (These shapes are taken from the ETH database.) With only a few exceptions (e.g. the pear in row one or the dog in row four) similar shapes have been clustered together. Shown in Figure 12.6 is an evolution of algorithm (left panel) and a histogram of $Q(C^*)$ values resulting from 200 runs of the algorithm, each starting at a different random initial condition. Additional examples of clustering databases, consisting of thousands of shapes, are presented in [755]. Once the shapes are clustered, the next goal is to develop probability models that efficiently capture variability within clusters. Another extension is to form a hierarchy, where one organizes shapes into a tree structure. The mean is computed for each cluster at each each level of the tree. The clusters of these means are used to form the next level of the tree [755].

## 12.4 Interpolation of Shapes in Echocardiographic Image-Sequences

Shape analysis continues to play a major role in medical diagnostics using non-invasive imaging. Shapes and shape variations of anatomical parts are often important factors in deciding normality/abnormality of imaged patients. For ex-

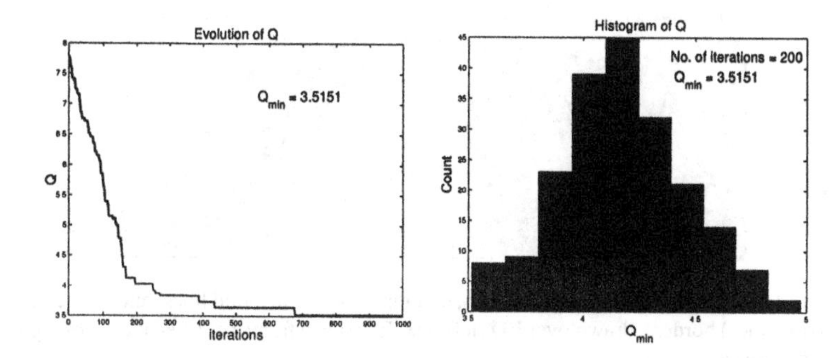


Figure 12.6. Left panel shows the evolution of Q under Algorithm 1. Right panel shows the histogram of minimum Q values obtained in 200 runs.

ample, the two images displayed in Figure 12.7 were acquired as the end diastolic (ED) and end systolic (ES) frames from a sequence of echocardiographic images during systole, taken from the apical four chamber view. Note that systole is the squeezing portion of the cardiac cycle and that the typical acquisition rate in echocardiography is 30 image frames/second. Superimposed on both images are expert tracings of the epicardial (solid lines) and endocardial borders (broken lines) of the left ventricle of the heart. From these four borders, indices of cardiac health, including chamber area, fractional area change, and wall thickness, can be easily computed. Since a manual tracing of these borders is too time consuming to be practical in a clinical setting, these borders are currently generated for research purposes only. The current clinical practice is to estimate these indices subjectively or (at best) make a few one-dimensional measurements of wall thickness and chamber diameter.

A major goal in echocardiographic image analysis has been to develop and implement automated methods for computing these two sets of borders as well as the sets of borders for the 10-12 image frames that are typically acquired between ED and ES. Different aspects of past efforts [896, 188, 187] include both the construction of geometric figures to model the shape of the heart as well as validation. While it is difficult for cardiologists to generate borders for all the frames, it is possible for them to provide borders for the first and the last frames in a cardiac cycle. Since it is not uncommon for the heart walls to exhibit diskinetic (i.e. irregular) motion patterns, the boundary variations in the intermediate frames can be important in a diagnosis. Our goal is to estimate epicardial and endocardial boundaries in the intermediate frames given the boundaries at the ED and ES frames.

As stated earlier, a closed contour $\alpha$ has two sets of descriptors associated with it: a shape descriptor denoted by $\theta \in \mathcal{S}$ and a vector $z \in \mathcal{Z}$ of nuisance variables. In our approach, interpolation between two closed curves is performed via interpolations between their shapes and nuisance components, respectively. The interpolation of shape is obtained using geodesic paths, while that

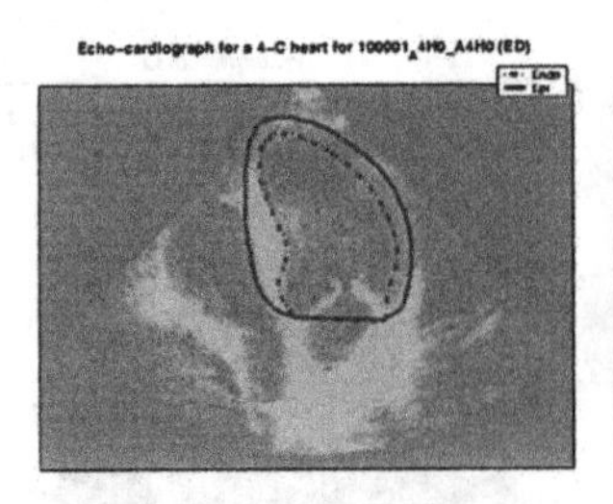 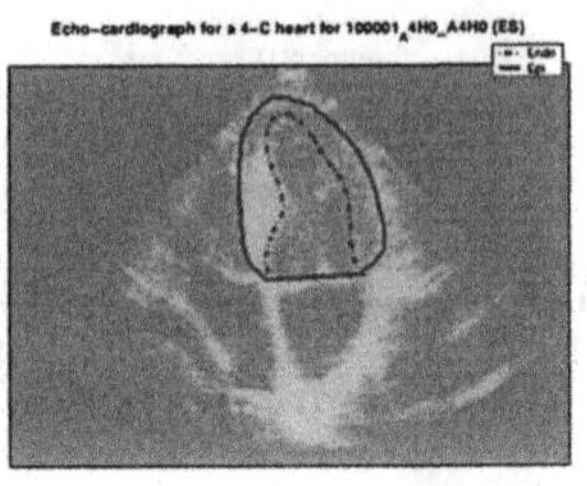

Figure 12.7. Expert generated boundaries, denoting epicardial (solid lines) and endocardial (broken lines) borders, drawn over ED (left) and ES (right) frames of an echocardiographic image sequence.

of the nuisance components is obtained using linear methods. Let $\alpha_1 = (\theta_1, z_1)$ and $\alpha_2 = (\theta_2, z_2)$ be the two closed curves, and our goal is to find a path $\Phi : [0,1] \mapsto \mathcal{S} \times \mathcal{Z}$ such that $\Phi_0 = (\theta_1, z_1)$ and $\Phi_1 = (\theta_2, z_2)$. For example, in Figure 12.7, the endocardial boundary (broken curves) of the ED and ES frames can form $\alpha_1$ and $\alpha_2$, respectively. Alternatively, one can treat the epicardial boundaries (solid curves) of ED and ES frames as $\alpha_1$ and $\alpha_2$ as well. The different components are interpolated as follows:

1. **Shape Component**: Given the two shapes $\theta_1$ and $\theta_2$ in $\mathcal{S}$, we use the shooting method to find the geodesic that starts from the first and reaches the other in unit time. This results in the flow $\Psi_t(\theta_1, g)$ such that $\Psi_0(\theta_1, g) = \theta_1$ and $\Psi_1(\theta_1, g) = \theta_2$. This also results in a re-parametrization of $\theta_2$ such that the origins (points where $s = 0$) on the two curves are now registered. With a slight abuse of notation we will also call the new curve $\theta_2$. Let a shape along this path be given by $\theta_t = \Psi_t(\theta_1, g)$. Since the path $\theta_t$ lies in $\mathcal{S}$, the average value of $\theta_t$ for all $t$ is $\pi$.

2. **Translation**: If $p_1, p_2$ represent the locations of the initial points on the two curves, i.e. $p_i = \alpha_i(0)$, $i = 1, 2$, then the linear interpolation between them is given by $p(t) = (1-t)p_1 + tp_2$.

3. **Orientation**: For a closed curve $\alpha_i$, the average orientation is defined by $\phi_i = \frac{1}{2\pi}\int_0^{2\pi} \frac{1}{j}\log(\dot{\alpha}_i(s))ds$, $i = 1, 2$, $j = \sqrt{-1}$. Given $\phi_1$ and $\phi_2$, a linear interpolation between them is $\phi(t) = (1-t)\phi_2 + t\tilde{\phi}_2$, where $\tilde{\phi}_2 = \text{argmin}_{\phi \in \{\phi_2 - 2\pi, \phi_2, \phi_2 + 2\pi\}} |\phi - \phi_1|$.

4. **Scale**: If $\rho_1$ and $\rho_2$ are the lengths of the curves $\alpha_1$ and $\alpha_2$, then a linear interpolation on the lengths is simply $\rho(t) = (1-t)\rho_1 + t\rho_2$.

Using these different components, the resulting geodesic on the space of closed curves is given by $\{\Phi_t : t \in [0,1]\}$ where:

$$\Phi_t(s) = p(t) + \rho(t)\int_0^s \exp(j(\theta_t(\tau) - \pi + \phi(t)))d\tau \, .$$

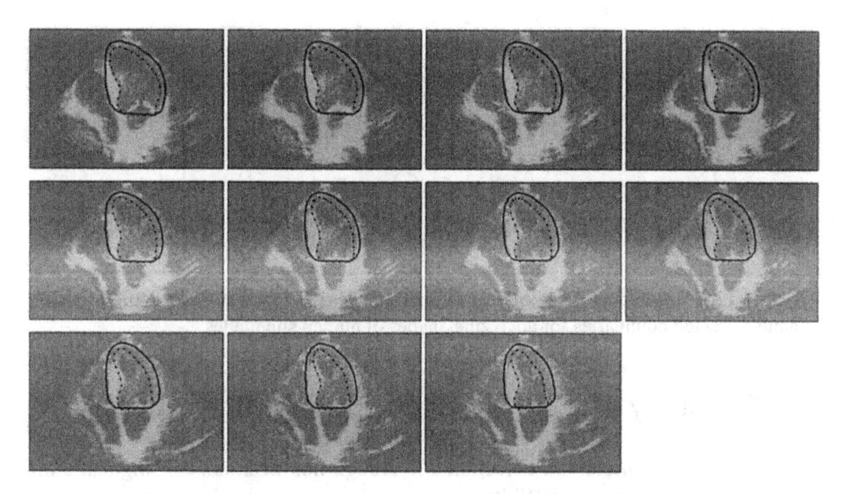

Figure 12.8. Interpolated shapes using geodesic paths in shape space.

Shown in Figure 12.8 is a sequence of 11 image frames for the same patient as displayed in Figure 12.7. Again, each image frame has a set of epicardial and endocardial borders overlaid on the image. In Figure 12.8, borders in the first and last frames have been traced by an expert, while the borders on the intermediate frames have been generated using the path $\Phi_t$, one each for epicardial and endocardial boundaries. Note that the endocardial border is more distorted than the epicardial border in the transition. In view of the geodesic paths in $\mathcal{S}$ relating to the minimum bending energy, this method provides a smoother interpolation for the endocardial borders, as compared to a direct linear interpolation of coordinates.

We foresee a number of uses for this idea. First, this method could be included in an acquisition system so that if an expert traces sets of borders at ED and ES, then the borders for the intermediate frames can be generated automatically. Since the technique for generating the intermediate borders uses no image information, they may not always be acceptable. However, one can implement software that allows the expert to adjust the intermediate contours manually to reflect a better match with the images. In this way, models will be available for both computer-based automated methods as well as validation and testing. As a future extension, one might modify the proposed interpolation to include image information. That is, formulate a boundary-value problem in $\mathcal{S}$ that seeks an optimal path under an image-based energy function, while fixing the expert generated boundaries as the end points.

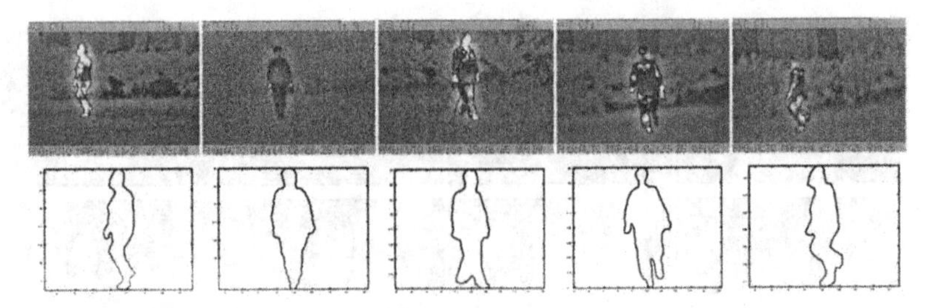

Figure 12.9. Top panels: Examples of infrared images of human subjects. Bottom panels: hand extracted boundaries for analyzing shapes of human silhouettes.

## 12.5 Study of Human Silhouettes in Infrared Images

There is a great interest in detection and recognition of humans using static images and video sequences. While most applications use visible-spectrum cameras for imaging humans, certain limitations, such as large illumination variability, has shifted interest towards cameras that operate in bandwidths beyond the visual spectrum. In particular, night vision cameras, or infrared cameras, have been found important in human detection and tracking, especially in surveillance and security environments. These cameras capture emissivity, or thermal states, of the imaged objects, and are largely invariant to ambient illumination. In this section, we investigate the use of infrared images in detection of human silhouettes. Although, we are generally interested in the full problem of detection, tracking, and recognition, here we restrict ourselves to two specific subproblems: (i) building statistical shape models for human silhouettes, and (ii) their use in improving silhouette detection.

Using a hand-held Raytheon Pro250 IR camera, we have hand-generated a database of human silhouettes. Shown in Figure 12.9 are some examples: the top panels show five IR images and the bottom panels show the corresponding hand-extracted human silhouettes. Furthermore, the database has been partitioned into clusters of similar shapes. These clusters correspond to front views with legs appearing together, side views with legs apart, side views with leg together, etc, and an example cluster is shown in Figure 12.10.

### *12.5.1 TPCA Shape Model*

Our first goal is to "train" probability models by assuming that elements in the same cluster are samples from the same probability model. These models can then be used for future Bayesian discoveries of shapes or for classification of new shapes. To train a probability model amounts to estimating a probability density function on the shape space $\mathcal{S}$, a task that is rather difficult to perform precisely. The two main difficulties are: nonlinearity and infinite-dimensionality of $\mathcal{S}$, and they are handled here as follows. $\mathcal{S}$ is a nonlinear manifold, so we impose a prob-

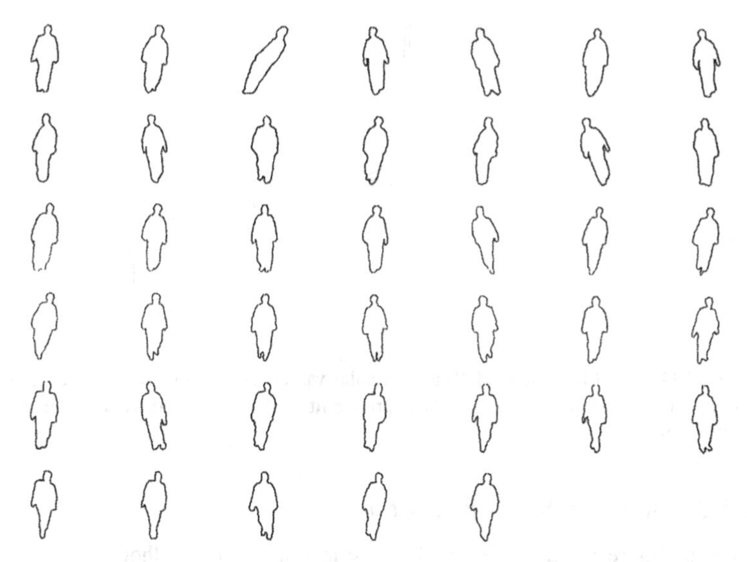

Figure 12.10. An example of a cluster of human silhouettes.

ability density on a tangent space instead. For a mean shape $\mu \in \mathcal{S}$, $T_\mu(\mathcal{S}) \subset \mathbb{L}^2$, is a **vector space** and more conventional statistics applies. Next, we approximate a tangent function $g$ by a finite-dimensional vector, e.g. a vector of Fourier coefficients, and thus characterize a probability distribution on $T_\mu(\mathcal{S})$ as that on a finite-dimensional vector space. Let a tangent element $g \in T_\mu(\mathcal{S})$ be represented by its approximation: $g(s) = \sum_{i=1}^{m} x_i e_i(s)$, where $\{e_i\}$ is a complete orthonormal basis of $T_\mu(\mathcal{S})$ and $m$ is a large positive integer. Using the identification $g \equiv \mathbf{x} = \{x_i\} \in \mathbb{R}^m$, one can define a probability distribution on elements of $T_\mu(\mathcal{S})$ via one on $\mathbb{R}^m$. The simplest model is a multivariate normal probability imposed as follows. Using principal component analysis (PCA) of the elements of $\mathbf{x}$, determine variances of the principal coefficients, and impose independent Gaussian models on the these coefficients with zero means and estimated variances. This imposes a probability model on $T_\mu(\mathcal{S})$, and through the exponential map ($\exp_\mu : T_\mu(\mathcal{S}) \mapsto \mathcal{S}$ defined by $\exp_\mu(g) = \psi_1(\mu, g)$) leads to a probability model on $\mathcal{S}$. We term this model "Tangent PCA" or TPCA.

Consider the set of 40 human silhouettes displayed in Figure 12.10. Their Karcher mean $\mu$ is shown in the top-left panel of Figure 12.11. For each observed shape $\theta_i$, we compute a tangent vector $g_i$, such that $\Psi_1(\mu, g_i) = \theta_i$. Using TPCA model we obtain a normal probability model on the tangent space $T_\mu(\mathcal{S})$. Shown in the bottom row of Figure 12.11 are 12 examples of random shapes generated by this probability model.

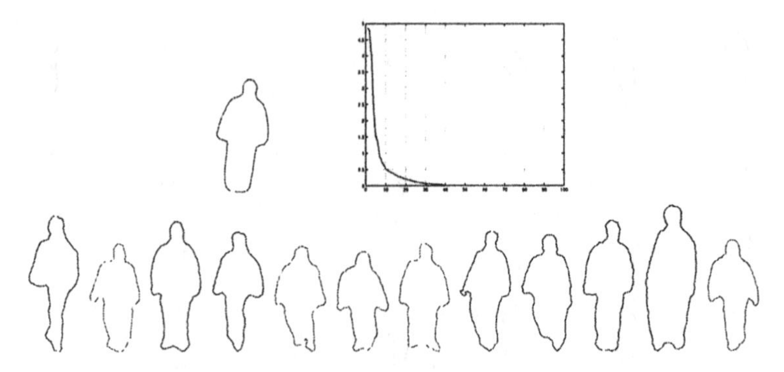

Figure 12.11. Top: Mean shape (left) and singular values (right) of covariance in $T_\mu(\mathcal{S})$. Bottom: Random samples from a Gaussian probability model on the principal coefficients of $g \in T_\mu(\mathcal{S})$.

### *12.5.2 Bayesian Shape Estimation*

Shown in Figure 12.12 is an example of estimating a human silhouette in an infrared image. The left panel shows the observed image $I$, and we seek a closed curve $\alpha^* = \text{argmax}_\alpha P(\alpha|I) = \text{argmax}_\alpha P(\alpha)P(I|\alpha)$. The prior $P(\alpha)$ comes from the TPCA model described previously. For the likelihood function $P(I|\alpha)$ there are a variety of choices: Kullback-Leibler divergence between interior and exterior pixel histograms, absolute difference between entropies of interior and exterior pixel densities, Gaussian models for pixels, etc. In this paper, we use a simple function that measures the proportion of saturated pixels, i.e. pixels with highest possible value, inside the contour. The remaining three panels in Figure 12.12 show the evolution of $\alpha$ as $P(\alpha|I)$ is maximized.

## 12.6 Summary & Discussion

We have described a geometric approach for statistical analysis of planar shapes, and its use in image-based inferences. Shapes of closed curves are represented by their angle functions, restricted appropriately to remove shape-preserving transformations. Geodesic paths on the resulting shape space, under the classical $\mathbb{L}^2$ Riemannian metric, are used to impose a metric on the shape space. The use of geodesic paths also leads to a framework for statistical modeling of shape variability, including an intrinsic technique to compute sample statistics (means, covariances, etc) of a given set of shapes. We have demonstrated this framework using three applications of shape analysis in clustering, medical image analysis, and human surveillance.

One limitation of the proposed model is its assumption of arc-length parametrization for all shapes, which does not allow local stretching or compressing of shapes. In some situations, it is preferable to match shapes via local stretching/shrinking,

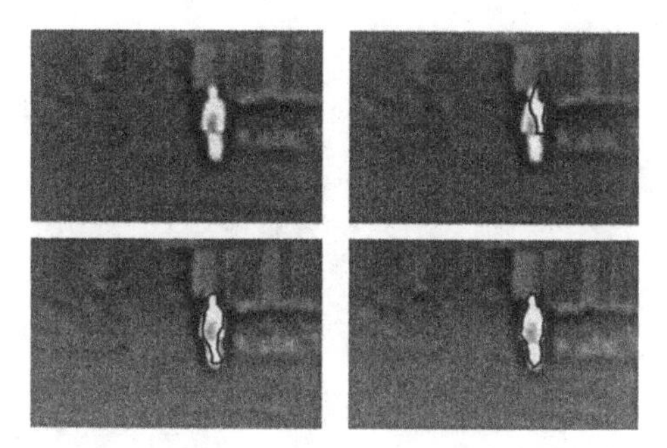

Figure 12.12. Bayesian shape extraction: Left panel shows an IR image $I$ and the remaining three panels show the evolution of a search that maximizes the posterior $P(\alpha|I)$. Estimated curves are drawn over the image in black.

and not be limited to bending only. A recent paper [572] describes an extension that uses a different Riemannian metric on shape spaces to allow for both bending and local stretching/shrinking.

# Acknowledgement

This research was supported in part by the grants NSF (FRG) DMS-0101429, NSF (ACT) DMS-0345242, and ARO W911NF-04-01-0268. We gratefully acknowledge this support. We thank the producers of different datasets used in this paper for making them available to us.

# Chapter13

# Diffeomorphic Point Matching

**H. Guo, A. Rangarajan and S. Joshi**

Abstract

In medical imaging and computer vision, the problem of registering point-sets that differ by an unknown non-rigid transformation frequently arises. We discuss the matching problem of shapes parameterized by point sets. Mathematical models of diffeomorphic landmark matching and diffeomorphic point shape matching are formulated. After formulating an objective function for diffeomorphic point matching, we give numerical algorithms to solve the objective. Results are shown for 2D corpus callosum shapes.

## 13.1 Introduction

Point matching and correspondence problems arise in various application areas such as computer vision, pattern recognition, machine learning and especially in computational anatomy and biomedical imaging. Point representation of image data is widely used in all areas and there is a huge amount of point feature data acquired in various modalities, including MRI, CT and Diffusion Tensor Images (DTI) [223, 315, 199]. The advantage of point set representations of shapes over other forms like curves and surfaces is that the point set representation is a universal representation of shapes regardless of the topologies of the shapes. This is especially useful in biomedical imaging because it has the ability to fuse different types of anatomical features in a single uniform representation.

Point matching in general is a difficult problem because, as with many other problems in computer vision, like image registration and segmentation, it is often ill-posed. In this chapter, we attempt to formulate a precise mathematical model for point matching. There are two important cases that need to be distinguished. When the two point-sets are of equal cardinality and when the correspondences are known, we have the *landmark matching problem*. This problem is not as difficult as the case when the correspondences are unknown. When we have two point-sets of unequal cardinality and when the correspondences are unknown, we have the *point shape matching problem*. The presence of outliers in either/both sets makes the correspondence problem even more difficult. In the following, we will first discuss the landmark matching problem and then the point shape matching problem.

## 13.2 Diffeomorphic Landmark Matching

We assume the image domain is the $d$-dimensional Euclidean space $\mathbf{R}^d$. Usually $d = 2$ or $d = 3$. In landmark based registration, we assume that we have two corresponding sets of feature points, or landmarks, $\{p_i \in \Omega_1 | \, i = 1, 2, ..., n\}$ and $\{q_i \in \Omega_2 | \, i = 1, 2, ..., n\}$ where $\Omega_1 \subseteq \mathbf{R}^d$ and $\Omega_2 \subseteq \mathbf{R}^d$. We need to find a transformation $f : \Omega_1 \rightarrow \Omega_2$ such that $\forall i = 1, 2, ..., n$, $f(p_i) = q_i$.

In many applications, we are required to find the transformation within some restricted groups, like rigid transformations, similarity transformations, affine transformations, projective transformations, polynomial transformations, B-spline transformations and "non-rigid" transformations. Different transformation groups have different degrees of freedom, namely, the number of parameters needed to describe a transformation in the group. This also determines the number of landmark pairs that the transformation can exactly interpolate. Let us look at some examples. In two dimensional space, where $d = 2$, a rigid transformation, which preserves Euclidean distance, has 3 degrees of freedom and cannot interpolate arbitrary landmark pairs. The landmark pairs to be matched must be subject to some constraints. That is, they have to have the same Euclidean distance. A similarity transformation has 4 degrees of freedom and can map any 2 points to any 2 points. An affine transformation has 6 degrees of freedom and can map any 3 non-degenerate points to any 3 non-degenerate points. A projective transformation has 8 degrees of freedom and can map any 4 non-degenerate points to any 4 non-degenerate points. In three dimensional space, where $d = 3$, a rigid transformation has 6 degrees of freedom. A similarity transformation has 7 degrees of freedom. An affine transformation has 12 degrees of freedom and can map any 4 non-degenerate points to any 4 non-degenerate points. A projective transformation has 15 degrees of freedom and can map any 5 non-degenerate points to any 5 non-degenerate points.

The term "non-rigid" transformation is often used in a narrower sense. Although similarity, affine and projective transformations do not preserve Euclidean distance, they all have finite degrees of freedom. In the literature, "non-rigid" transformations usually refer to a transformation with infinite degrees of freedom, which can potentially map any finite number of points to the same number of points. So we immediately see a big difference between finite degree of freedom transformations and non-rigid transformations. Given a fixed number of landmark pairs to be interpolated, the former is easily over constrained but the latter is always under constrained. This is one of the reasons why the non-rigid point matching problem is much more difficult. To find a unique non-rigid transformation, we need further constraints. This is termed regularization in the computer vision and medical image analysis literature [347].

Two desirable properties of non-rigid transformations are smoothness and topology preservation. Again, let $\Omega_1 \subseteq \mathbf{R}^d$ and $\Omega_2 \subseteq \mathbf{R}^d$. A transformation $f : \Omega_1 \to \Omega_2$ is said to be smooth if all partial derivatives of $f$, up to certain orders, exist and are continuous. A transformation $f : \Omega_1 \to \Omega_2$ is said to preserve the topology if $\Omega_1$ and $\text{Img}(f) = \{p_2 \in \Omega_2 | \exists p_1 \in \Omega_1, p_2 = f(p_1)\}$ have the same topology. A transformation that preserves topology is called a *homeomorphism* and its definition is: A transformation $f : \Omega_1 \to \Omega_2$ is a homeomorphism if $f$ is a bijection and if it is continuous and if its inverse is also continuous. A smooth transformation $f : \Omega_1 \to \Omega_2$ may not preserve the topology. There are several cases when this is true. First, the smooth map $f$ is a bijection but the inverse is not continuous. Second, the smooth map $f$ may fail to be a bijection. That is, multiple points may be mapped to the same point and we call this the folding of space. There are two sub-cases here, one sub-case is that at some point, the tangent map of $f$ is not an isomorphism. The other sub-case is that the tangent map of $f$ is an isomorphism at every point but globally it is not a bijection. On the other hand, a homeomorphism may not be smooth because in the definition, we only require continuity in both $f$ and its inverse but we do not require differentiability. A transformation $f$ that is both smooth and topology preserving is called a *diffeomorphism*. The diffeomorphism $f : \Omega_1 \to \Omega_2$ is defined as a bijection that is smooth and its inverse is also smooth. Now let us look at an example of a smooth transformation, namely, the Thin-Plate Spline (TPS) interpolation [851].

For simplicity, we discuss the problem in 2-D space. Everything in the 2-D formulation easily applies to 3-D except we have a different kernel in 3-D. The original thin-plate spline interpolation problem is formulated as: find a smooth function $f : \Omega \to \mathbf{R}$, such that the thin-plate energy $\int\int_\Omega \left[ (\frac{\partial^2 f}{\partial x^2})^2 + 2(\frac{\partial^2 f}{\partial x \partial y})^2 + (\frac{\partial^2 f}{\partial y^2})^2 \right] dxdy$ is minimized, subject to constraints at $n$ control points $\{p_i \in \Omega | \, i = 1, 2, ..., n\}$

$$f(p_i) = v_i \, , \, p_i \in \Omega, \, v_i \in \mathbf{R}, \, i = 1, 2, ..., n \, . \tag{13.1}$$

The reproducing kernel Hilbert space (RKHS) method is used to solve this problem. We assume $f$ is in the Sobolev space $W^{k,2}(\Omega)$. Let $||f||^2 = E = \int\int_\Omega \left[ (\frac{\partial^2 f}{\partial x^2})^2 + 2(\frac{\partial^2 f}{\partial x \partial y})^2 + (\frac{\partial^2 f}{\partial y^2})^2 \right] dxdy$, where $||f||$ is the norm of $f$ in $W^{k,2}(\Omega)$. Since $W^{k,2}(\Omega)$ is a Hilbert space, from the Riesz representation theorem, for any $p \in \Omega$, the evaluation linear functional

$$\delta_p : W^{k,2}(\Omega) \to \mathbf{R}, \, \delta_p(f) = f(p) \tag{13.2}$$

has a representer [467] $u_p \in W^{k,2}(\Omega)$ such that

$$\delta_p(f) = f(p) = < u_p, f > . \tag{13.3}$$

Now the original problem is transformed to the problem: find a function $f \in W^{k,2}(\Omega)$ with minimal norm $||f||$, subject to constraints

$$< u_{p_i}, f >= v_i \, , i = 1, 2, ..., n \, . \tag{13.4}$$

For $p_a, p_b \in \Omega$, $u(p_a, p_b) = u_{p_a}(p_b)$ is the kernel of the reproducing kernel Hilbert space.

Let $T$ be the linear subspace spanned by $u_{p_i}$, $i = 1, 2, ..., n$. Any function $f \in W^{k,2}(\Omega)$ can be decomposed into $f = f_T + f_\perp$ where $f_T \in T$ and $f_\perp$ is in the orthogonal complement of $T$ and hence $< u_{p_i}, f_\perp >= 0$. We know if $f_T$ satisfies (13.4), then $f$ also satisfies (13.4) only with $||f|| > f_T$ if $f_\perp \neq 0$. So we only need to search for the solution in $T$. The general solution can thus be written as

$$f(p) = a_0 + a_1 x + a_2 y + \sum_{i=1}^{n} w_i u(p_i, p) \, , \tag{13.5}$$

where $a_0, a_1, a_2, w_i \in \mathbf{R}$ and functions of the form $a_0 + a_1 x + a_2 y$ span the null space.

With this form, $E$ can be rewritten as

$$E = \sum_{i=1,j=1}^{n} w_i U_{ij} w_j = W U W^+, \tag{13.6}$$

where $W = (w_1, ..., w_n)$ and $U$ is the matrix with elements $U_{ij} = u(p_i, p_j)$.

Bookstein [101, 102] applied thin-plate splines to the landmark interpolation problem. The goal is to find a smooth transformation $f : \Omega \to \Omega$ that interpolates $n$ pairs of landmarks $\{p_i \in \Omega | \, i = 1, 2, ..., n\}$ and $\{q_i \in \Omega | \, i = 1, 2, ..., n\}$ and also minimize the thin-plate bending energy

$$E = \sum_{h=1}^{2} \int \int_{R^2} \left[ (\frac{\partial^2 f_h}{\partial x^2})^2 + 2(\frac{\partial^2 f_h}{\partial x \partial y})^2 + (\frac{\partial^2 f_h}{\partial y^2})^2 \right] dx dy, \tag{13.7}$$

where $f_1$ and $f_2$ are the $x$ and $y$ components of the mapping. If we interpret each of $f_1$ and $f_2$ as the bending in the $z$ direction of a metal sheet, or thin plate, extending in the $x$-$y$ plane, the energy in (13.7) is the analog of the thin plate bending energy. The kernel in this case is

$$U(r) = r^2 \log r^2 \, , \tag{13.8}$$

where $r$ is the distance $\sqrt{x^2 + y^2}$. We also denote

$$P = \begin{bmatrix} 1 & x_1 & y_1 \\ 1 & x_2 & y_2 \\ ... & ... & ... \\ 1 & x_n & y_n \end{bmatrix} \text{, which is } 3 \times n, \tag{13.9}$$

Duchon [286] proved that if $P$ has maximum column rank, then the solution exists and is unique and the general solution is of the form

$$f(x, y) = a_1 + a_x x + a_y y + \sum_{i=1}^{n} w_i U(|p_i - (x, y)|). \tag{13.10}$$

Because an affine transformation has no contribution to the bending energy, the transformation allows for a free affine transformation. Define the matrices

$$K = \begin{bmatrix} 0 & U(r_{12}) & \dots & U(r_{1n}) \\ U(r_{21}) & 0 & \dots & U(r_{2n}) \\ \dots & \dots & \dots & \dots \\ U(r_{n1}) & U(r_{n2}) & \dots & 0 \end{bmatrix}, which\, is\, n \times n, \tag{13.11}$$

and

$$L = \begin{bmatrix} K & P \\ P^T & O \end{bmatrix}, \text{which is } (n+3) \times (n+3), \tag{13.12}$$

where the symbol $^T$ is the matrix transpose operator and $O$ is a $3 \times 3$ matrix of zeros.

Let $V = (v_1, ..., v_n)$ be any $n$-vector and write $Y = (V \,|\, 0\,0\,0)^T$. The coefficients $W = (w_1, ..., w_n)$ and $(a_1, a_x, a_y)$ can be found by

$$L^{-1}Y = (W \,| a_1\, a_x\, a_y)^T. \tag{13.13}$$

A numerically stable solution in a different form is given by Wahba [851] using a QR decomposition.

While the preceding development is somewhat appealing, there is no mechanism to guarantee a diffeomorphic transformation. Intuitively this problem is known as the folding of space.

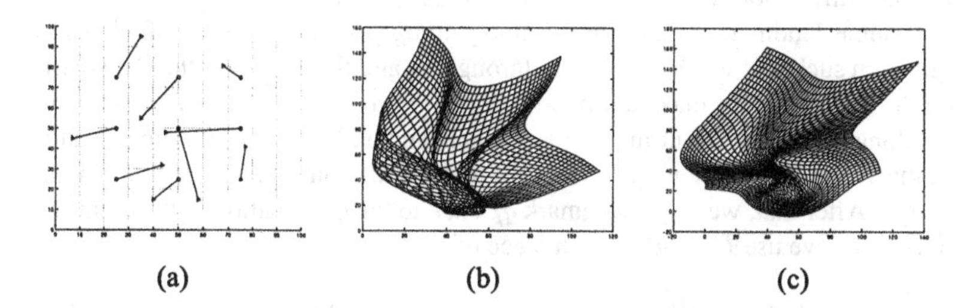

(a) (b) (c)

Figure 13.1. The folding problem in TPS and the desirable diffeomorphism.

Figure 13.1a shows the displacement of landmarks. Figure 13.1b is the thin-plate spline interpolation. We can see the folding of space. This is the drawback of thin-plate spline interpolation. Due to the folding of space, features in the template may be smeared in the overlapping regions. And furthermore, the transformation is not invertible. A diffeomorphic transformation is strongly desirable, which preserves the features, the topology and which is smooth as shown in Figure 13.1c. Next we show that such a diffeomorphism always exists.

**Theorem.** *A diffeomorphic transformation that interpolates arbitrary numbers of $n$ pairs of landmarks always exists.*

*Proof:*

We show the existence by construction. We construct a simple, although most likely undesirable in most of the applications, diffeomorphism. The intuitive idea is to dig canals connecting the landmark pairs. We first choose the first pair of landmarks $p_1$ and $q_1$. For simplicity, we assume the dimension $d$ of space is 2. The proof is similar for $d > 2$. First assume no other landmarks lie on the line connecting $p_1$ and $q_1$. Establish a coordinate system such that $p_1$ and $q_1$ are on the $x$ axis, shown in Figure 13.2, where dots are source landmarks and squares are target landmarks. Let the signed distance from $p_1$ to $q_1$ be $a$. Construct the transformation $f_1 : \Omega_1 \to \Omega_2$ such that $f_1(x, y) = (x', y')$,

$$\begin{aligned} x' &= x + ae^{-v^2} \\ y' &= y \end{aligned} \tag{13.14}$$

where $v = \tan(\frac{\pi}{2\epsilon}y)$, for any arbitrarily small $\epsilon$. We choose $\epsilon$ to be sufficiently small so that any other landmarks do not lie in the belt
$\{(x, y) \in \mathbf{R}^2 | \, |y| < \epsilon\}$.

It is easy to show that $f_1$ is a diffeomorphism and that it maps $p_1$ to $q_1$ and keeps all other landmarks $q_2,..., q_n$ fixed. This is very much like the flow of viscous fluid in a tube. Similarly we can construct a diffeomorphism $f_i$ that maps $p_i$ to $q_i$ and keeps all other landmarks fixed, for $i = 1, 2, ..., n$. The composition of this series of diffeomorphisms

$$f = f_n \circ \cdots f_2 \circ f_1 \tag{13.15}$$

is also a diffeomorphism and obviously $f$ maps $p_i$ to $q_i$, for $i = 1, 2, ..., n$.

If some landmark $q_k$ lies on the line joining $p_i$ and $q_i$, we can find such a direction such that we draw a line $l_k$ through $q_k$ and there are no other landmarks on the line. Then we make a diffeomorphism $h$ transporting $q_k$ to a nearby point $q_k'$ along the line without moving any other landmarks, using the same canal as in the viscous fluid technique. Then we make a diffeomorphism $f_i$ as described before. After that, we move landmark $q_k'$ back to the old position with the inverse of $h^{-1}$. So we use $F_i = h^{-1} f_i h$ in place of $f_i$.

One straightforward approach to find a diffeomorphism for practical use is to remedy the thin-plate spline so that it does not fold. We can restrict our search space to the set of diffeomorphisms and the ideal one should minimize the thin-plate energy. We make the observation that if the Jacobian of the transformation $f$ changes sign at a point, then there is folding. We can place a constraint requiring the Jacobian to always be positive. There is some literature on this approach but most of these approaches do not guarantee that the transformation is smooth [440, 197].

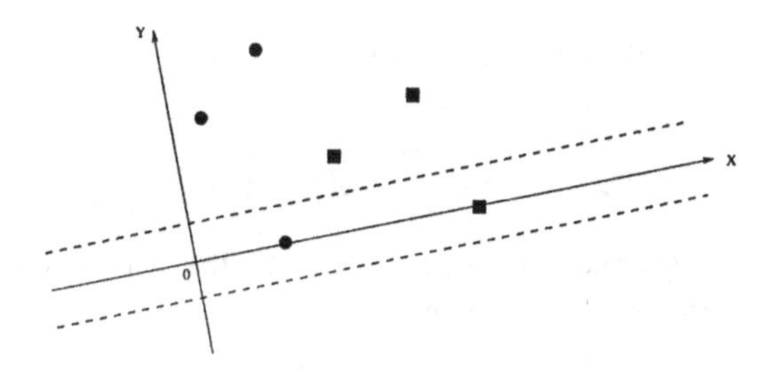


Figure 13.2. Diffeomorphism construction.

Another approach is to utilize the flow field [289, 442, 571]. We introduce one parameter, the time $t$ into the diffeomorphism. Let $\phi_t : \Omega \to \Omega$ be the diffeomorphism from $\Omega$ to $\Omega$ at time $t$. A point $x$ is mapped to the point $\phi_t(x)$. Sometimes we also denote this as $\phi(x,t)$. It is easy to verify that for all the values of $t$, $\phi_t$ forms a one parameter diffeomorphism group. If $x$ is fixed, then $\phi(x,t)$ traces a smooth trajectory in $\Omega$. The interpolation problem becomes: find the one parameter diffeomorphic group $\phi(\cdot,t) : \Omega \to \Omega$ such that given $p_i \in \Omega$ and $q_i \in \Omega\ \forall i = 1,2,...,n$, $\phi(x,0) = x$ and $\phi(p_i,1) = q_i$. We introduce the velocity field $v(x,t)$ and construct a dynamical system by the transport equation

$$\frac{\partial \phi(x,t)}{\partial t} = v(\phi(x,t),\, t)\,. \tag{13.16}$$

The integral form of the relation between $\phi(x,t)$ and $v(x,t)$ is

$$\phi(x,1) = x + \int_0^1 v(\phi(x,t),t)dt\,. \tag{13.17}$$

Obviously, such a $\phi(x,t)$ is not unique and there are infinitely many such solutions. With the analogy to the TPS, it is natural that we require the desirable diffeomorphism results in minimal space deformation. Namely we require the deformation energy

$$\int_0^1 \int_\Omega ||Lv(x,t)||^2 dxdt \tag{13.18}$$

to be minimized, where $L$ is a given linear differential operator.

The following theorem [442] states the existence of such a velocity field and shows a way to solve for it.

**Theorem (Joshi and Miller).** *Let $p_i \in \Omega$ and $q_i \in \Omega\ \forall i = 1,2,...,n$. The solution to the energy minimization problem*

$$\hat{v}(\cdot) = \arg\min \int_0^1 \int_\Omega ||Lv(x,t)||^2 dxdt \tag{13.19}$$

*subject to*

$$\phi(p_i, 1) = q_i, \qquad \forall i = 1, 2, ..., n \tag{13.20}$$

*where*

$$\phi(x, 1) = x + \int_0^1 v(\phi(x, t), t)dt \tag{13.21}$$

*exists and defines a diffeomorphism* $\phi(\cdot, 1) : \Omega \to \Omega$. *The optimum velocity field* $\hat{v}$ *and the diffeomorphism* $\hat{\phi}$ *are given by*

$$\hat{v}(x, t) = \sum_{i=1}^{n} K(\phi(x_i t), x) \sum_{j=1}^{n} (K(\phi(t))^{-1})_{ij} \dot{\hat{\phi}}(x_j, t) \tag{13.22}$$

*where*

$$K(\phi(t)) = \begin{pmatrix} K(\phi(p_1, t), \phi(p_1, t)) & \cdot \ \cdot \ \cdot & K(\phi(p_1, t), \phi(p_n, t)) \\ \cdot & \cdot \ \cdot \ \cdot & \cdot \\ \cdot & \cdot \ \cdot \ \cdot & \cdot \\ \cdot & \cdot \ \cdot \ \cdot & \cdot \\ K(\phi(p_n, t), \phi(p_1, t)) & & K(\phi(p_n, t), \phi(p_n, t)) \end{pmatrix} \tag{13.23}$$

*with* $(K((\phi(t))_{ij}$ *denoting the* $ij$, $3 \times 3$ *block entry* $(K(\phi(t))_{ij} = K(\phi(p_i, t), \phi(p_j, t))$, *and*

$$\hat{\phi}(p_i, \cdot) \quad = \quad \arg \min_{\phi(p_i, \cdot)} \int_0^1 \sum_{ij} \dot{\phi}(p_i, t)^T (K(\phi(t))^{-1})_{ij} \dot{\phi}(p_j, t) dt \tag{13.24}$$

*subject to* $\phi(p_i, 1) = q_i$, $i = 1, 2, ..., N$ *with the optimal diffeomorphism given by*

$$\hat{\phi}(x, 1) = x + \int_0^1 \hat{v}(\hat{\phi}(x, t), t)dt\,. \tag{13.25}$$

The proof [442] is omitted here. With this theorem, we can convert the original optimization problem on the vector field $\hat{v}(x, t)$ to a problem of finite dimensional optimal control with end point conditions.

This problem is called the *exact matching problem* because we required the given set of points $p_i, i = 1, 2, ..., n$ map exactly to the other given set of points $q_i, i = 1, 2, ..., n$. The exact matching problem is symmetric with respect to two sets of landmarks or two point shapes. When the two point sets $\{p_i \in \Omega_1 | i = 1, 2, ..., n\}$ and $\{q_i \in \Omega_2 | i = 1, 2, ..., n\}$ are swapped, the new optimal diffeomorphism is the inverse of the old diffeomorphism. This is stated more formally in the following theorem.

**Theorem.** *If* $\phi(x_k, 1) = y_k$ *and* $\phi(x, t)$ *and* $v(x, t)$ *minimize the energy* $E = \int_0^1 \int_\Omega ||Lv(x, t)||^2 dxdt$, *then the inverse mapping maps the landmarks backward* $\phi^{-1}(y_k, 1) = x_k$ *and* $\phi^{-1}(x, t)$ *and* $-v(x, -t)$ *also minimize the energy* $E$.

*Proof:* First, from the known property of the diffeomorphism group of such a dynamical system, $\phi(x, t_1 + t_2) = \phi(\phi(x, t_1), t_2)$, it is easy to show that $\phi^{-1}(x, t) = \phi(x, -t)$. This is because

$$\begin{aligned} & \phi(., -t) \circ \phi(., t)(x) \\ = \; & \phi(., t) \circ \phi(., -t)(x) \\ = \; & \phi(\phi(x, t), -t) \\ = \; & \phi(x, t + (-t)) \\ = \; & \phi(x, 0) \\ = \; & x \, . \end{aligned}$$

Furthermore, $\phi(x, -t)$and $-v(x, -t)$ also satisfy the transport equation

$$\frac{\partial \phi(x, -t)}{\partial t} = -v(\phi(x, -t), -t)$$

Suppose $\phi(x, t)$ and $v(x, t)$ minimize the energy

$$E = \int_0^1 \int_\Omega ||Lv(x, t)||^2 dx dt$$

but $\phi^{-1}(x, t) = \phi(x, -t)$ and $-v(x, -t)$ *do not* minimize the energy

$$E = \int_0^1 \int_\Omega ||Lv(x, t)||^2 dx dt$$

Let the minimizer be $\psi(x, t)$ and $u(x, t)$ such that $\forall k$, $\psi(y_k) = x_k$ and $\int_0^1 \int_\Omega ||Lu(x, t)||^2 dx dt < \int_0^1 \int_\Omega ||Lv(x, t)||^2 dx dt$. Then, we can construct

$$\psi^{-1}(x, t) = \psi(x, -t)$$

such that $\psi^{-1}(x, t)$ and $-u(x, -t)$ satisfy the transport equation and $\psi^{-1}(x_k, 1) = y_k$. However $\int_0^1 \int_\Omega ||Lu(x, t)||^2 dx dt < \int_0^1 \int_\Omega ||Lv(x, t)||^2 dx dt$ contradicts the assumption that $v(x, t)$ is the minimizer of the energy $E$.

The exact matching problem can be generalized to the *inexact matching problem.* In the inexact matching problem, we do not require that the points exactly match. Instead, we seek a compromise between the closeness of the matching points and the deformation of space. We minimize

$$\int_0^1 \int_\Omega ||Lv(x, t)||^2 dx dt + \lambda \sum_{i=1}^{n} ||q_i - \phi(p_i, 1)||^2 , \tag{13.26}$$

which can be similarly solved.

## 13.3 Diffeomorphic Point Shape Matching

In the diffeomorphic point matching problem, the points are samples from the shape and we have a point representation of the shape. When we have two such shapes represented by points, usually the cardinality of the points in the two shape point-sets are different and there is no point-wise correspondence. We want to find the correspondence between the two shapes. The approach we take is clustering. The two point shapes are clustered simultaneously and we assume there is a one-to-one correspondence between the clusters. The correspondences between the two sets of clusters are, unfortunately, also unknown. We put the correspondence and the diffeomorphism together and by minimizing an objective function which has both the clustering energy and the diffeomorphic deformation energy, we are able to find the clustering, the correspondence between cluster centers and the diffeomorphism in space simultaneously. The objective function is

$$\begin{aligned} & E(M^x, M^y, r, s, v, \phi) \\ = & \sum_{i=1}^{N_1}\sum_{k=1}^{N} M^x_{ik}||x_i - r_k||^2 + \sum_{j=1}^{N_2}\sum_{k=1}^{N} M^y_{jk}||y_j - s_k||^2 \\ & + \sum_{k=1}^{N} ||s_k - \phi(r_k, 1)||^2 + \lambda \int_0^1 \int_\Omega ||Lv(x,t)||^2 dx dt. \end{aligned} \tag{13.27}$$

In the above objective function, the $M^x$ and $M^y$ are the cluster membership matrices, which satisfy $M^x_{ik} \in [0,1], \forall ik$ and $M^y_{jk} \in [0,1], \forall jk$ and $\sum_{k=1}^{N} M^x_{ik} = 1$, $\sum_{k=1}^{N} M^y_{jk} = 1$. The matrix entry $M^x_{ik}$ is the membership of data point $x_i$ in cluster $k$ whose center is at location $r_k$. The matrix entry $M^y_{jk}$ is the membership of data point $y_j$ in cluster $k$ whose center is at position $s_k$. Point-set $X$ has $N_1$ points, $Y$ has $N_2$ points and the number of shared cluster centers is $N$.

The diffeomorphic deformation energy in $\Omega$ is induced by the landmark displacements from $r$ to $s$, where $x \in \Omega$ and $\phi(x,t)$ is the one parameter diffeomorphism: $\Omega \to \Omega$. Since the original point-sets differ in point count and are unlabeled, we cannot immediately use the diffeomorphism objective functions as in [442] or [145] respectively. Instead, the two point-sets are clustered and the landmark diffeomorphism objective is used between two sets of cluster centers $r$ and $s$ whose indices are always in correspondence. The diffeomorphism $\phi(x,t)$ is generated by the velocity field $v(x,t)$. $\phi(x,t)$ and $v(x,t)$ together satisfy the transport equation $\frac{\partial \phi(x,t)}{\partial t} = v(\phi(x,t),t)$ and the initial condition $\forall x$, $\phi(x,0) = x$ holds. This is in the inexact matching form and the displacement term $\sum_{k=1}^{N} ||s_k - \phi(r_k,1)||^2$ plays an important role here as the bridge between the two systems. This is also the reason why we prefer the deformation energy in this form because the coupling of the two sets of clusters appear naturally through the inexact matching term and we don't have to introduce external coupling terms as in [372]. Another advantage of this approach is that in this dynamic system described by the diffeomorphic group $\phi(x,t)$, the landmarks trace a trajectory exactly on the flow lines dictated by the field $v(x,t)$. Also, the feedback coupling is no longer needed as in the previous approach because with this deformation energy described above, due to the above theorem, if $\phi(x,t)$ is the minimizer of this energy, then $\phi^{-1}(x,t)$ is the inverse mapping which also minimizes the same energy.

We are now ready to give an algorithm that simultaneously finds the cluster centers, the correspondence and the diffeomorphism.

The joint clustering and diffeomorphism estimation algorithm has two components: i) diffeomorphism estimation and ii) clustering. For the diffeomorphism estimation, we expand the velocity field in term of the kernel $K$ of the $L$ operator

$$v(x,t) = \sum_{k=1}^{N} \alpha_k(t) K(x, \phi_k(t)) \tag{13.28}$$

where $\phi_k(t)$ is notational shorthand for $\phi(r_k, t)$ and we also take into consideration the affine part of the mapping when we use thin-plate spline kernel with matrix entry $K_{ij} = r_{ij}^2 \log r_{ij}$ and $r_{ij} = \| x_i - x_j \|$. After discretizing in time $t$, the objective in 13.27 is expressed as

$$\begin{aligned} E = & \sum_{i=1}^{N_1} \sum_{k=1}^{N} M_{ik}^x \|x_i - r_k\|^2 + \sum_{j=1}^{N_2} \sum_{k=1}^{N} M_{jk}^y \|y_j - s_k\|^2 \\ & + \sum_{k=1}^{N} \|s_k - r_k - \sum_{l=1}^{N} \sum_{t=0}^{S} [P(t) d_l(t) + \alpha_l(t) K(\phi_k(t), \phi_l(t))] \|^2 \\ & + \lambda \sum_{k=1}^{N} \sum_{l=1}^{N} \sum_{t=0}^{S} < \alpha_k(t), \alpha_l(t) > K(\phi_k(t), \phi_l(t)) \end{aligned} \tag{13.29}$$

where

$$P(t) = \begin{pmatrix} 1 & \phi_1^1(t) & \phi_1^2(t) \\ \cdot & \cdot & \cdot \\ \cdot & \cdot & \cdot \\ \cdot & \cdot & \cdot \\ 1 & \phi_N^1(t) & \phi_N^2(t) \end{pmatrix} \tag{13.30}$$

and $d$ is the affine parameter matrix. We then perform a QR decomposition on $P$,

$$P(t) = (Q_1(t) : Q_2(t)) \begin{pmatrix} R(t) \\ 0 \end{pmatrix}. \tag{13.31}$$

We iteratively solve for $\alpha_k(t)$ and $\phi_k(t)$ using an alternating algorithm. When $\phi_k(t)$ is held fixed, we use the following approximation to solve for $\alpha_k(t)$. The solutions are

$$d(t) = R^{-1}(t) [Q_1(t)\phi(t+1) - Q_1(t) K(\phi(t)) Q_2(t) \gamma(t)] \tag{13.32}$$

$$\alpha(t) = Q_2(t) \gamma(t) \tag{13.33}$$

where $K(\phi(t))$ denotes the thin-plate spline kernel *matrix* evaluated at $\phi(t) \stackrel{\text{def}}{=} \{\phi(r_k, t) | k = 1, \ldots, N\}$ and

$$\gamma(t) = (Q_2^T(t)K(\phi(t))Q_2(t) + \lambda)^{-1}Q_2^T(t)\phi(t+1). \tag{13.34}$$

When $\alpha_k(t)$ is held fixed, we use gradient descent to solve for $\phi_k(t)$:

$$\frac{\partial E}{\partial \phi_k(t)} = 2\sum_{l=1}^{N} < \alpha_k(t), \alpha_l(t) - 2W_l > \nabla_1 K(\phi_k(t), \phi_l(t)) \tag{13.35}$$

where $W_l = s_l - r_l - \sum_{m=1}^{N} \int_0^1 \alpha_m(t)K(\phi_m(t), \phi_l(t))dt$.

The clustering of the two point-sets is handled by a deterministic annealing EM algorithm which iteratively estimates the cluster memberships $M^x$ and $M^y$ and the cluster centers $\mathbf{r}$ and $\mathbf{s}$. The update of the memberships is the very standard E-step of the EM algorithm [199] and is performed as shown below.

$$M_{ik}^x = \frac{\exp(-\beta\|x_i - r_k\|^2)}{\sum_{l=1}^{N} \exp(-\beta\|x_i - r_l\|^2)}, \forall ik \text{ and} \tag{13.36}$$

$$M_{jk}^y = \frac{\exp(-\beta\|y_j - s_k\|^2)}{\sum_{l=1}^{N} \exp(-\beta\|y_j - s_l\|^2)}, \forall jk \tag{13.37}$$

where $\beta = \frac{1}{T}$ is the inverse temperature. The cluster center update is the M-step of the EM algorithm. This step is not the typical M-step. We use a closed-form solution for the cluster centers which is an approximation. From the clustering standpoint, we assume that the change in the diffeomorphism at each iteration is *sufficiently small so that it can be neglected.* After making this approximation, we get

$$r_k = \frac{\sum_{i=1}^{N_1} M_{ik}^x x_k + s_k - \sum_{l=1}^{N} \int_0^1 \alpha_l(t)K(\phi_l(t), \phi_k(t))dt}{1 + \sum_{i=1}^{N_1} M_{ik}^x}, \tag{13.38}$$

$$s_k = \frac{\sum_{j=1}^{N_2} M_{jk}^y y_j + \phi(r_k, 1)}{1 + \sum_{j=1}^{N_2} M_{jk}^y}, \forall k. \tag{13.39}$$

In the clustering and diffeomorphic estimation steps, we let $\lambda$ vary proportionately with the temperature. This controls the rigidity of the mapping, starting from an almost rigid mapping while we obtain good correspondence and gradually softens so that good clustering is achieved. In this way both clustering and diffeomorphism are obtained simultaneously at convergence.

The overall algorithm is described below.

- **Initialization:** Initial temperature $T = 0.5(\max_i \|x_i - x_c\|^2 + \max_j \|y_j - y_c\|^2)$ where $x_c$ and $y_c$ are the centroids of $X$ and $Y$ respectively.
- **Begin A:** While $T > T_{\text{final}}$
  - **Step 1:** Clustering

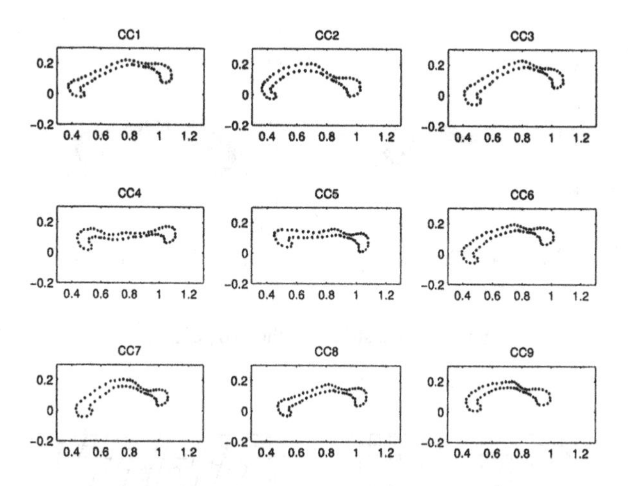


Figure 13.3. Point sets of nine corpus callosum images.

Update memberships according to (13.36), (13.37).
Update cluster centers according to (13.38), (13.39).

- **Step 2:** Diffeomorphism
  Update $(\phi, v)$ by minimizing

$$E_{\text{diff}}(\phi, v) = \sum_{k=1}^{N} ||s_k - \phi(r_k, 1)||^2 + \lambda T \int_0^1 \int_\Omega ||Lv(x,t)||^2 dx dt$$

  according to (13.32)(13.33) and (13.35).
- **Step 3:** Annealing. $T \leftarrow \gamma T$ where $\gamma < 1$.

- **End**

Next we show the experimental results applying the algorithm to nine sets of 2D corpus callosum slices. The feature points were extracted with the help of a neuroanatomical expert. Figure 13.3 shows the nine corpus callosum 2D images, labeled CC1 through CC9. In our experiments, we first did the simultaneous clustering and matching with the corpus callosum point sets CC5 and CC9. The clustering of the two point sets is shown in Figure 13.4. There are 68 cluster centers. The circles represent the centers and the dots are the data points. The two sets of cluster centers induce the diffeomorphic mapping of the 2D space. The warping of the 2D grid under this diffeomorphism is shown in Figure 13.5. Using this diffeomorphism, we calculated the after-image of original data points and compared them with the target data points. Due to the large number of cluster centers,

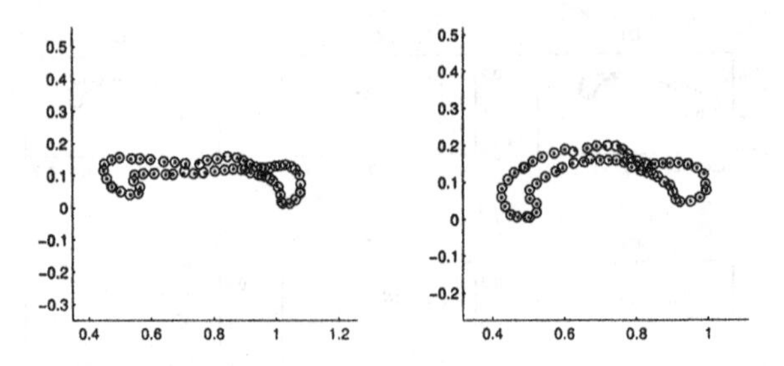


Figure 13.4. Clustering of the two point sets.

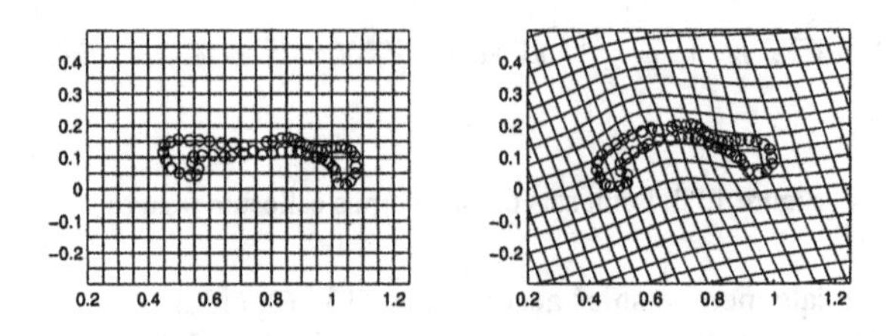


Figure 13.5. Diffeomorphic mapping of the space.

the cluster centers nearly coincide with the original data points and the warping of the original data points is not shown in the figure. The correspondences (at the cluster level) are shown in Figure 13.6. The algorithm allows us to simultaneously obtain the diffeomorphism and the correspondence.

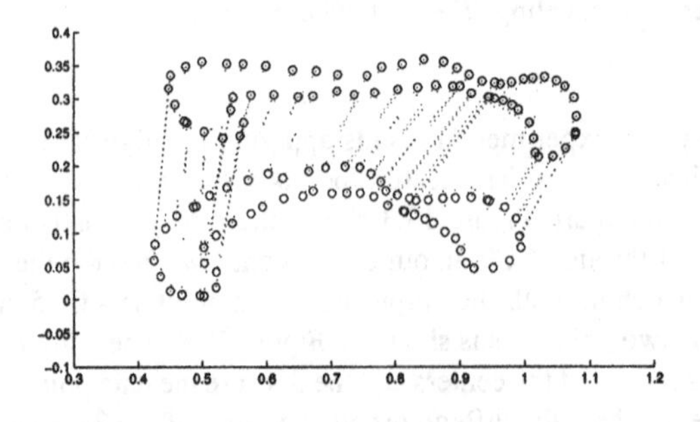


**Figure 13.6. Matching between the two point sets.**

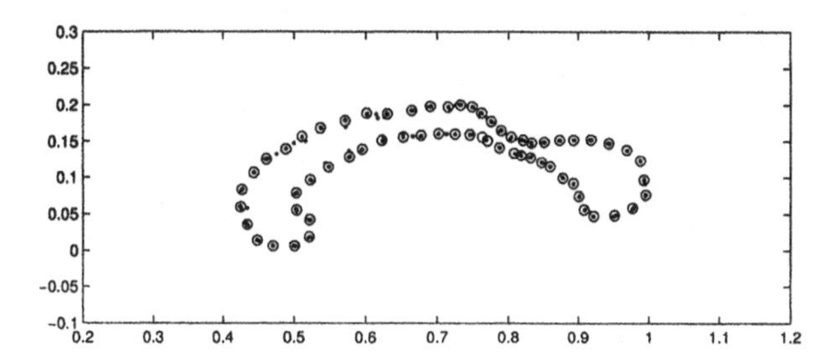


Figure 13.7. Overlay of the after-images of eight point sets with the ninth set.

## 13.4 Discussion

There are other approaches to the diffeomorphic point matching problem which we have not considered here. One indirect approach is to use distance transforms to convert the point matching problem into an image matching problem. There are as yet no theoretical and/or experimental comparisons between distance transforms-based diffeomorphisms and our approach. Also, there are other approaches to diffeomorphic landmark matching [145, 372]. While we have only provided results for 2D diffeomorphic point matching, the theoretical formulation presented here extends to 3D. Finally, the joint clustering and matching formulation is not the only approach that in principle can marry diffeomorphisms and correspondence [198]. However, it appears to be the simplest formulation that does not require us to establish point correspondences via estimation of permutations.

## Acknowledgements

We acknowledge support from the National Science Foundation (NSF IIS 0307712).

# Chapter 14

# Uncertainty-Driven, Point-Based Image Registration

C. Stewart

Abstract

Point-based registration is the problem of computing the transformation that best aligns two point sets, such as might be obtained using range scanners or produced by feature extraction algorithms. The Iterative Closest Points (ICP) algorithm and its variants are the most commonly used techniques for point-based registration. The ICP algorithm may be derived as the solution to a global optimization problem. A commonly-used linearization of the distance function in this optimization problem produces a useful approximation to the covariance matrix of the ICP-estimated transformation parameters. Two recent algorithms exploit this covariance matrix to improve ICP registration. One uses the covariance matrix to sample the correspondences so that the estimate is well-constrained in all directions in parameter space. A second uses the covariance matrix to guide a region-growing and model-selection technique that "grows" accurate estimates from low-order initial estimates that are only accurate in small image regions. Both show substantial improvements over standard ICP on challenging alignment problems.

## 14.1 Introduction

Point-based registration techniques have been used in many applications, ranging from 3d modeling and industrial inspection to medical imaging. In point-based registration, the data are geometric point sets, $\mathcal{P}$ and $\mathcal{Q}$, such as image feature locations or 3d range measurements. The points are treated as samples from curves or surfaces in $\mathbb{R}^n$, and they may have associated attributes such as intensity values or normal vectors. The goal of point-based registration is to compute the transformation, $\mathbf{M} : \mathbb{R}^n \rightarrow \mathbb{R}^n$, that best aligns the point sets. Of particular interest here are parametric transformation models of the form $\mathbf{M}(\mathbf{p}; \boldsymbol{\theta})$, where $\mathbf{p} \in \mathbb{R}^n$ is point location, and $\boldsymbol{\theta}$ is the vector of transformation mapping parameters to be

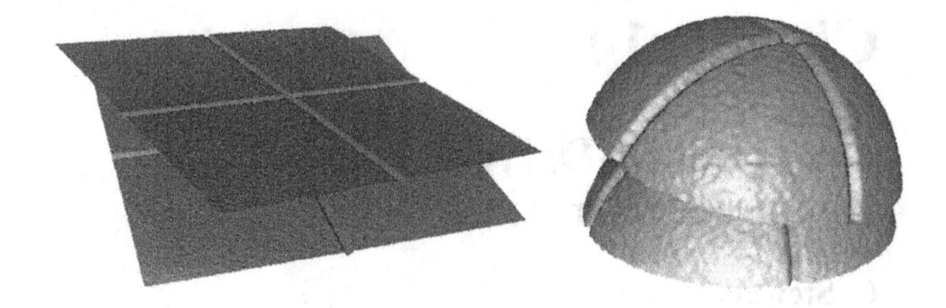

Figure 14.1. Synthetic range data sets illustrating the challenges that arise when the set of surfaces being aligned differ significantly in size. In the example on the left two planar surfaces have 1 mm deep groves cut into them. When a small amount noise is added to the data, constraints from matching points on the much larger planar surface prevent matches along the surface of the grooves from rotating the ICP alignment into place. A similar effect occurs with the alignment of two data sets from a spherical shell, shown on the right.

estimated. Similarity, affine, projective and quadratic transformations all fit into this category of parametric models.

Most approaches to point-based registration require establishing correspondence between points from $\mathcal{P}$ and $\mathcal{Q}$. If reliable correspondences are known, estimating the optimal set of transformation parameters is well-understood. On the other hand, given an accurate estimate of $\boldsymbol{\theta}$, establishing correspondence is straightforward. This poses a classic "chicken-and-egg" problem. This problem is widely addressed using the Iterative Closest Points (ICP) algorithm, discovered almost simultaneously in the early 1990's by several groups [82, 164, 185, 560, 916]. The idea of ICP is straightforward: (1) given a transformation parameter estimate, $\hat{\boldsymbol{\theta}}$, apply the transformation to a subset of $\mathcal{P}$, and for each transformed point find the closest point from $\mathcal{Q}$; (2) from these (temporary) correspondences, compute a new transformation parameter estimate $\hat{\boldsymbol{\theta}}$. These two steps are repeated until an appropriate convergence criteria is met. Important variations on ICP are discussed and analyzed in [697].

While initialization of ICP is clearly an important issue, the primary focus of this chapter is convergence. Ensuring proper convergence of ICP is challenging. Two reasons for this are illustrated in Figures 14.1 and 14.2. First, when there are significant variations in the sizes and the orientations of the surfaces to be registered, correspondence constraints from large surfaces can impede the alignment of smaller surfaces, mostly due to the effects of noise. Second, when the point sets represent complicated curve or surface patterns, such as in the vascular structure of the retina (Figure 14.2), misalignments early in the ICP process can cause mismatches that drive the algorithm to an incorrect local minimum. These mismatches often have relatively small alignment errors and therefore are not eliminated easily using robust estimation.

These two problems — one caused by a lack of balance in the constraints and one caused by incorrect correspondences — have been addressed recently in papers from the 3d modeling literature [339], and from the medical imaging literature [765]. Underlying both is the use of uncertainty in the transformation estimate that is computed by ICP. Unlike earlier work, which studied the influence of uncertainty in point locations [276] and evaluated the uncertainty of the final ICP result [766], these two new techniques use uncertainty to guide the ICP estimation process itself. This new theme in registration could have important implications for developing more reliable and more general-purpose algorithms.

The goal of this chapter is to present this uncertainty-driven approach to registration. Section 14.2 formulates the point-based registration problem and derives both the ICP algorithm and the commonly-used normal distance form of ICP. Section 14.3 derives the transformation estimation equations and resulting approximate covariance matrix. This is used as a measure of uncertainty in the two algorithms described in Sections 14.4 and 14.5. The chapter concludes with a summary of the techniques and an outline of important questions suggested by the uncertainty-driven approach.

## 14.2 Objective Function, ICP and Normal Distances

Given are two point sets, $\mathcal{P}$ and $\mathcal{Q}$. These points sets are generally discrete, but they may be formed into a mesh. For expository purposes, however, they may be modeled in the continuous domain using an implicit function, e.g. $f : \mathbb{R}^n \to \mathbb{R}$, such that $\mathcal{Q} = \{\mathbf{q} \mid f(\mathbf{q}) = 0\}$. The point set registration objective function may be defined based on the proximity between transformed points from $\mathcal{P}$ and the set $\mathcal{Q}$:

$$F(\theta; \mathcal{P}, \mathcal{Q}) = \sum_{\mathbf{p}_i \in \mathcal{P}} \min_{\mathbf{q} \in \mathcal{Q}} \|\mathbf{M}(\mathbf{p}_i; \theta) - \mathbf{q}\|^2. \tag{14.1}$$

The goal of registration, now stated more formally, is to find the parameter estimate $\hat{\theta}$ minimizing this objective function.

Several approaches to minimizing $F(\theta; \mathcal{P}, \mathcal{Q})$ are possible. Here are two:

- The approach taken in the ICP algorithm alternates steps of solving the two minimization problems. The inner minimization (the matching step) in (14.1) is solved for fixed $\theta$ to produce a correspondence set $C = \{\mathbf{p}_i, \mathbf{q}_i\}$, and then the outer minimization is solved in slightly altered form by replacing the inner minimization with just the distance $\|\mathbf{M}(\mathbf{p}_i; \theta) - \mathbf{q}_i\|^2$. If infinitesimal steps are taken in $\mathbf{q}_i$ and in $\theta$, this converges to a local minimum of the objective function.

- $\mathcal{Q}$ is represented implicitly using a distance function in $\mathbb{R}^n$ that is 0 at locations $\mathbf{q}$ where $f(\mathbf{q}) = 0$. Example representations include Chamfer distance measures [103] and octree splines [164]. Derivatives of the objective function (14.1) may be computed based on computing derivatives of the distance function without explicitly identifying the closest point in $\mathcal{Q}$.

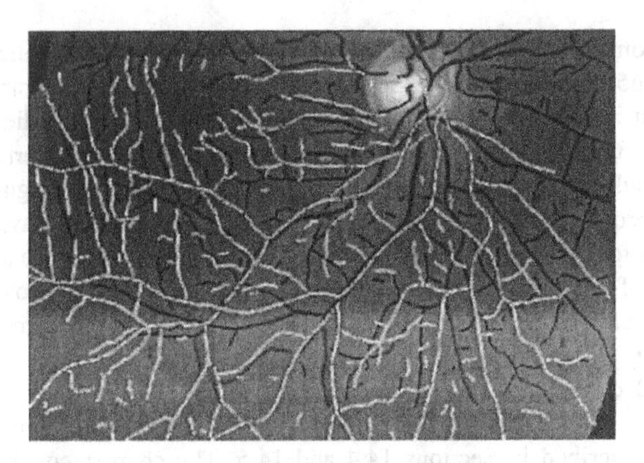

Figure 14.2. Example of misregistration of retinal images. Contours in black are blood vessel centerlines detected in one retinal image and contours in white are blood vessel centerlines detected in a second retinal image (of the same eye). The complexity of the structure of the vessels, together with a small initial misalignment, causes ICP to mismatch a significant fraction of the contours and converge to an incorrect estimate.

The focus of this chapter is on the ICP approach, which has been used widely, especially in the range image literature [697].

With the focus on ICP, the matching step must be examined in more detail. Using the implicit function definition of $\mathcal{Q}$, the minimization

$$\min_{\mathbf{q}\in\mathcal{Q}} \|\mathbf{M}(\mathbf{p}_i;\boldsymbol{\theta}) - \mathbf{q}\|^2 \tag{14.2}$$

becomes

$$\min \|\mathbf{M}(\mathbf{p}_i;\boldsymbol{\theta}) - \mathbf{q}\|^2 \text{ subject to } f(\mathbf{q}) = 0.$$

Writing this using Lagrange multipliers and introducing the simplifying notation $\mathbf{p}'_i \cong \mathbf{M}(\mathbf{p}_i;\boldsymbol{\theta})$ creates the function

$$h(\mathbf{q},\lambda) = \|\mathbf{p}'_i - \mathbf{q}\|^2 - 2\lambda f(\mathbf{q}),$$

which must be minimized simultaneously over $\mathbf{q}$ and $\lambda$. Computing partial derivatives $\partial h/\partial \mathbf{q}$ and $\partial h/\partial \lambda$ and setting the results equal to 0 yields

$$\begin{aligned}(\mathbf{p}'_i - \mathbf{q}) - \lambda \nabla f(\mathbf{q}) &= 0\\ f(\mathbf{q}) &= 0\end{aligned} \tag{14.3}$$

Solving this, in turn, requires an iterative technique. Let $\mathbf{q}_i$ be the current best estimate of the closest point. After the iterations converge it will be the corresponding point for $\mathbf{p}_i$ in ICP. Linearizing $f$ around $\mathbf{q}_i$ produces

$$f(\mathbf{q}) = (\mathbf{q} - \mathbf{q}_i)^T \boldsymbol{\eta}_i = 0 \qquad \text{and} \qquad \nabla f(\mathbf{q}_i) = \boldsymbol{\eta}_i,$$

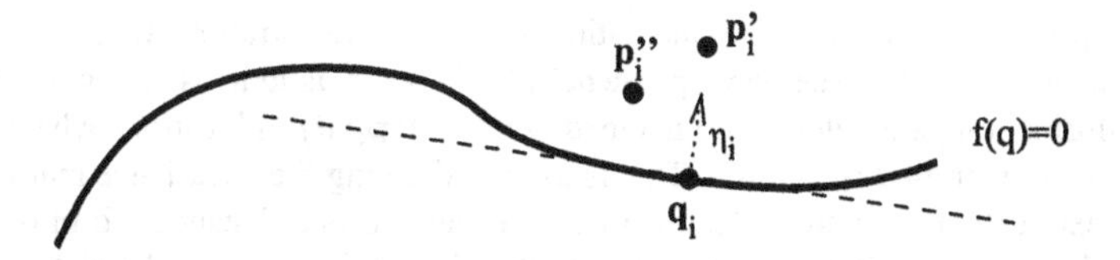


Figure 14.3. Illustrating the linearization of implicit function $f$ that defines point set $\mathcal{Q}$. Let $\mathbf{p}_i'$ be a transformed point from $\mathcal{P}$, let $\mathbf{q}_i$ be the closest point from $\mathcal{Q}$, and let $\eta_i$ be the local surface normal. (The linearization is pictured as the dashed line segment.) A small change in the transformation that moves $\mathbf{p}_i'$ to $\mathbf{p}_i''$ does not require recomputation of the closest point from $\mathcal{Q}$ in order to compute the (approximate) distance from $\mathbf{p}_i''$ to $\mathcal{Q}$.

where $\eta_i$ is the normal to $f$ at $\mathbf{q}_i$. Substituting these into (14.3) produces the system of equations

$$\begin{pmatrix} \mathbf{I} & \eta_i \\ \eta_i^T & 0 \end{pmatrix} \begin{pmatrix} \mathbf{q} \\ \lambda \end{pmatrix} = \begin{pmatrix} \mathbf{p}_i' \\ \eta_i^T \mathbf{q}_i \end{pmatrix}.$$

Solving yields

$$\mathbf{q} = \mathbf{p}_i' - \eta_i \eta_i^T \mathbf{p}_i' + \eta_i \eta_i^T \mathbf{q}_i. \tag{14.4}$$

This produces an update $\mathbf{q}_i \leftarrow \mathbf{q}$. This point, however, does not satisfy $f(\mathbf{q}) = 0$, a problem that must be solved by moving along the constraint surface in direction $\mathbf{q} - \mathbf{q}_i$ rather than directly making the substitution $\mathbf{q}_i \leftarrow \mathbf{q}$. This important detail is not a concern here, however, because the current focus is on approximating the objective function.

The approximate closest point in (14.4) may be substituted back into the distance calculation equation (14.2) to yield a simplified but approximate calculation of distance. After some manipulation this yields,

$$\min_{\mathbf{q} \in \mathcal{Q}} \|\mathbf{M}(\mathbf{p}_i; \boldsymbol{\theta}) - \mathbf{q}\|^2 = a[(\mathbf{M}(\mathbf{p}_i; \boldsymbol{\theta}) - \mathbf{q}_i)^T \eta_i]^2, \tag{14.5}$$

where $a = \eta_i^T \eta_i$. When $f(\mathbf{q})$ is a distance function, $a \approx 1$ because a unit step normal to the surface produces a unit change in distance. This is equivalent to assuming $\eta_i$ is a unit vector, an assumption made throughout the remainder of this chapter. As illustrated in Figure 14.3, equation (14.5) simply reflects the fact that computing the minimum distance between a point and a linear structure does not require knowing the closest point on the linear structure; all that is needed is any point from the structure and the normal vector.

Turning back to the original problem of estimating the transformation parameters, (14.5) may be substituted into the original objective function (14.1) to obtain the approximation

$$F(\boldsymbol{\theta}; \mathcal{P}, \mathcal{Q}) = \sum_{\mathbf{p}_i \in \mathcal{P}} [(\mathbf{M}(\mathbf{p}_i; \boldsymbol{\theta}) - \mathbf{q}_i)^T \eta_i]^2. \tag{14.6}$$

This approximation allows the calculation of the point-registration objective function without updating the correspondences.[1] It is valid as long as changes in the transformation parameters keep mapped points $\mathbf{M}(\mathbf{p}_i;\boldsymbol{\theta})$ in locations where the linearization around $\mathbf{q}_i$ is valid. This is used in deriving the covariance matrix in the next section. Equation 14.6 also leads to the "normal-distance" form of the ICP algorithm, originally proposed in [185]. The summation on the right-hand side of (14.6) is minimized for a fixed set of correspondences to estimate the next set of transformation parameters. The fact that this is a closer approximation to the true underlying objective function shows why use of normal distance constraints causes much faster and more reliable convergence of ICP [697].

## 14.3 Parameter Estimates and Covariance Matrices

The next step is to derive equations for estimating the transformation parameters given a fixed set of correspondences, $C = \{(\mathbf{p}_i, \mathbf{q}_i)\}$. This leads directly to an approximation for the covariance matrix of the resulting estimate.

The derivation starts with a simplified form of the transformation model:

$$\mathbf{M}(\mathbf{p};\boldsymbol{\theta}) = \mathbf{p} + \mathbf{X}(\mathbf{p})\boldsymbol{\theta}. \tag{14.7}$$

A few examples will clarify this revised form. For a 3D rigid transformation using a small angle approximation (see [339], e.g.),

$$\mathbf{M}(\mathbf{p};\boldsymbol{\theta}) = \mathbf{R}\mathbf{p} + \mathbf{t} \approx \mathbf{p} + \mathbf{r} \times \mathbf{p} + \mathbf{t} = \mathbf{p} + \begin{pmatrix} \mathbf{S} & \mathbf{I} \end{pmatrix} \begin{pmatrix} \mathbf{r} \\ \mathbf{t} \end{pmatrix}.$$

Here, $\mathbf{r}$ is the vector of small angle approximations, $\mathbf{t}$ is the translation, and $\mathbf{S}$ is the skew-symmetric matrix such that $\mathbf{S}\mathbf{r} = \mathbf{r} \times \mathbf{p}$. This form is used for estimating incremental estimates of a rigid transformation. Writing an affine transformation in the form (14.7) is straightforward.[2] A 2D quadratic transformation is written

$$\mathbf{M}(\mathbf{p};\boldsymbol{\theta}) = \mathbf{p} + \begin{pmatrix} \mathbf{x}(\mathbf{p})^T & \mathbf{0}^T \\ \mathbf{0}^T & \mathbf{x}(\mathbf{p})^T \end{pmatrix} \boldsymbol{\theta}.$$

Here $\boldsymbol{\theta}$ is a 12x1 vector and if $\mathbf{p} = (u, v)^T$ then $\mathbf{x}(\mathbf{p}) = (1, u, v, u^2, uv, v^2)^T$.

Using the form of (14.7), the normal-distance ICP equation (14.6) for a fixed set of correspondences becomes

$$F(\boldsymbol{\theta}; C) = \sum_{(\mathbf{p}_i, \mathbf{q}_i) \in C} [(\mathbf{p}_i + \mathbf{X}(\mathbf{p}_i)\boldsymbol{\theta} - \mathbf{q}_i)^T \boldsymbol{\eta}_i]^2. \tag{14.8}$$

---

[1] See [576] for a recent generalization to second-order approximations.

[2] Planar homographies may not be written in this form because side constraints must be imposed on the parameter vector. Different derivations of the estimation equations and covariance matrices are needed, combining the normal-distance form of (14.6) with the covariance derivations in [389, Ch. 4].

Rewriting,

$$F(\boldsymbol{\theta}; \mathcal{C}) = \sum_{(\mathbf{p}_i, \mathbf{q}_i) \in \mathcal{C}} [\eta_i^T \mathbf{X}(\mathbf{p}_i)\boldsymbol{\theta} - \eta_i^T(\mathbf{q}_i - \mathbf{p}_i)]^2$$
$$= (\mathbf{X}\boldsymbol{\theta} - \mathbf{y})^T(\mathbf{X}\boldsymbol{\theta} - \mathbf{y}) \quad (14.9)$$

where

$$\mathbf{X} = \begin{pmatrix} \eta_1^T \mathbf{X}(\mathbf{p}_1) \\ \vdots \\ \eta_k^T \mathbf{X}(\mathbf{p}_k) \end{pmatrix} \quad \text{and} \quad \mathbf{y} = \begin{pmatrix} \eta_1^T(\mathbf{q}_1 - \mathbf{p}_1) \\ \vdots \\ \eta_k^T(\mathbf{q}_k - \mathbf{p}_k) \end{pmatrix}.$$

Taking the derivative with respect to $\boldsymbol{\theta}$, setting the result to $\mathbf{0}$, and solving yields the estimate,

$$\hat{\boldsymbol{\theta}} = (\mathbf{X}^T\mathbf{X})^{-1}\mathbf{X}^T\mathbf{y}. \quad (14.10)$$

This has the structure of a linear regression problem. Making the simplifying assumption (discussed below) that $\mathbf{y}$ is the only random variable, the expected value of the estimate is

$$\overline{\boldsymbol{\theta}} = E[\hat{\boldsymbol{\theta}}] = (\mathbf{X}^T\mathbf{X})^{-1}\mathbf{X}^T E[\mathbf{y}].$$

Moreover, if $\mathbf{y}$ is independent and identically distributed (i.i.d.), with covariance matrix $\sigma^2\mathbf{I}$, then the covariance matrix of the parameter estimate is

$$\Sigma_{\boldsymbol{\theta}} = E[(\hat{\boldsymbol{\theta}} - \overline{\boldsymbol{\theta}})(\hat{\boldsymbol{\theta}} - \overline{\boldsymbol{\theta}})^T] = \sigma^2(\mathbf{X}^T\mathbf{X})^{-1} \quad (14.11)$$

When robust weighting of the correspondences is added (see, e.g. [764]), the estimate becomes

$$\hat{\boldsymbol{\theta}} = (\mathbf{X}^T\mathbf{W}\mathbf{X})^{-1}\mathbf{X}^T\mathbf{W}\mathbf{y}. \quad (14.12)$$

where $\mathbf{W}$ is a diagonal matrix of the weights of the individual constraints. The parameter estimate covariance matrix is then approximately

$$\Sigma_{\boldsymbol{\theta}} = E[(\hat{\boldsymbol{\theta}} - \overline{\boldsymbol{\theta}})(\hat{\boldsymbol{\theta}} - \overline{\boldsymbol{\theta}})^T] = \sigma^2(\mathbf{X}^T\mathbf{W}\mathbf{X})^{-1} \quad (14.13)$$

The approximate covariance matrix has been used in a number of algorithms, including the ones described here. Before proceeding to these, it is important to examine the assumptions and approximations underlying the foregoing derivation.

- The derivation of the covariance matrix that started from (14.6) is based on a fixed correspondence set. The prior derivation leading to (14.6) showed that (14.6) is a good approximation to the original objective function (which involves changing correspondences) when changes in the transformation are not large enough to invalidate the linearization around the points $\mathbf{q}_i$. This is true in particular as the overall algorithm — not just the estimate for a fixed set of correspondences — nears convergence.

- In deriving (14.11) from the estimate equation (14.10), the matrix $\mathbf{X}$ is assumed to depend only on deterministic quantities. For this to hold, point locations $\mathbf{p}_i$ are treated as deterministic. While this clearly underestimates the uncertainty, the effects of this should be small since the $\mathbf{p}_i$ values themselves will be much larger than errors in $\mathbf{p}_i$.

- Errors in the normal directions are assumed to be small enough that any resultant errors in projections onto the normal vectors — as in $\eta_i^T \mathbf{X}(\mathbf{p}_i)$ and $\eta_i^T(\mathbf{p}_i - \mathbf{q}_i)$ — are relatively insignificant. Since the errors in these projections will be proportional to the error in the orientation and since for small error angles, $\phi$, $\cos\phi \approx 1$, this is reasonable, especially as the algorithm converges.

- Weight matrix $\mathbf{W}$ is also assumed to be non-random. Since each $w_i$ depends on the error in the correspondence and therefore in the transformation itself, this is again an oversimplification.

- Finally, $\eta_i^T(\mathbf{p}_i - \mathbf{q}_i)$ is assumed to be i.i.d. In part this says that all errors in the point positions are along the normal direction. On the negative side, this ignores errors that depend on the sensor direction [276]. On the positive side, since the point sets are treated as sets of samples from continuous manifolds, the errors in the point positions $\mathbf{q}$ tangent to the manifold keep the points (almost) on the manifold and do not change the distance measurement significantly.

Overall, it should be clear that the derived covariance matrix (a) is only a rough approximation of the true covariance matrix, (b) the approximation becomes more accuracte as the ICP estimation process nears the minimum, and (c) the primary effect of the approximation is that the magnitude of the covariance matrix is under-estimated.

## 14.4 Stable Sampling of ICP Constraints

This section and the next present applications of the covariance matrix estimate in ICP algorithms that address the two problems described in the introduction. This section considers the situation (Figure 14.1) where the ICP correspondences match points from the same surface in the two different data sets and are therefore in a sense "correct", but they still do not pull the estimate in the direction needed to correctly align the surfaces.

This problem is addressed in [339] by using the covariance matrix to select a subset of the correspondences that will constrain the transformation estimate as uniformly as possible in all directions. This sampling strategy is governed by a spectral decomposition of the parameter estimate covariance matrix and its inverse:

$$\Sigma_\theta^{-1} = \frac{1}{\sigma^2}(\mathbf{X}^T\mathbf{W}\mathbf{X}) = \sum_{j=1}^{m} \lambda_j \gamma_j \gamma_j^T, \qquad \Sigma_\theta = \sum_{j=1}^{m} (1/\lambda_j)\gamma_j\gamma_j^T. \quad (14.14)$$

The $\lambda_j$'s and $\gamma_j$'s are the eigenvalues and eigenvectors, respectively, of the inverse covariance matrix, ordered so that $\lambda_1 \geq \lambda_2 \geq \cdots \geq \lambda_m \geq 0$. The $\lambda_j$ values represents the "stability" — the inverse of the variance — in direction $\gamma_j$ in parameter space. Ideally, the stability values for each direction should be approximately equivalent. Stated another way, the condition number $\lambda_1/\lambda_m$ should be as small as possible.

Consider the constraints from Equation (14.9) and in particular consider the projection of the constraint for correspondence $i$ onto eigenvector $j$:

$$\eta_i^T \mathbf{X}(\mathbf{p}_i)\gamma_j. \quad (14.15)$$

The magnitude of this projection tells how much the $i$th point correspondence constrains the transformation in the $j$th direction in parameter space. Given a subset $C'$ of the correspondence set $C$, the value

$$s_j^2 = \sum_{(\mathbf{p}_i,\mathbf{q}_i)\in C'} [\eta_i^T \mathbf{X}(\mathbf{p}_i)\gamma_j]^2/|C'| \quad (14.16)$$

is roughly proportional to the inverse variance of the estimate in the $j$th direction based on the subset. The goal of the stable sampling algorithm is to find a subset that makes these $s_j^2$ values as close to equal as possible, thereby constraining the estimate equally-well in all directions.

The steps involved are:

1. Compute the inverse covariance matrix and its eigenvector decomposition from a small initial set $C'$ of correspondences in the region where the data sets overlap. These correspondences and the overlap region must be computed using an earlier ICP parameter estimate.

2. Compute $s_j^2$ for each eigenvector based on the initial set.

3. For eigenvector $j$ with the smallest $s_j^2$, choose the correspondence from the overlap region that has the greatest magnitude of (14.15), add it to the correspondence set $C'$, and update $s_j^2$ (14.16) for all eigenvectors. Note that the chosen correspondence is taken from $C - C'$.

4. Repeat until a sufficient number of correspondences have been selected or until the addition of a new correspondence starts to increase the approximate condition number — the ratio between the largest $s_j$ value and the smallest. The second condition tests if the constraints available to increase the stability of the smallest eigenvalue have been exhausted.

For details of the data structures and search algorithms that make this computation efficient, see [339]. Two other important details should be mentioned here, however.

- The parameter vector $\boldsymbol{\theta}$ involves parameters of different units, including rotation angles and terms of differing orders. Numerically, the individual components of $\boldsymbol{\theta}$ are not comparable; they can differ by several orders of magnitude. To solve this, the subsets of $\mathcal{P}$ and $\mathcal{Q}$ that form the correspondences should each be centered and then normalized so that the average magnitude of corresponding points $\mathbf{p}_i$ and $\mathbf{q}_i$ are each 1 [339, 389]. All computations of the sample selection technique should be done in the centered and normalized system.
- The constraint $\eta_i^T \mathbf{X}(\mathbf{p}_i)$ depends on a point location from $\mathcal{P}$ and a normal from $\mathcal{Q}$. This means sampling must be applied after correspondences are formed, even though many correspondences will not be used. This wasted computation may be avoided easily. Observe that after the ICP algorithm has removed the worst of the misalignments, the surface normals of the transformed points $\mathbf{p}_i$ should be roughly parallel to the normals from $\mathbf{q}_i$. Therefore, the transformed normals from $\mathbf{p}_i$ can be used in place of the normals from $\eta_i$ in the above calculations. This means the sampling can be computed prior to establishing correspondence.

The overall computation places a third step in each iteration of ICP: (1) apply stable sampling to select a subset of the points in the overlap region, (2) establish matches (correspondences) for these points, and (3) compute the new transformation estimate using the correspondences.

Using this technique, the two problem examples shown in Figure 14.1 are each correctly aligned. For the iteration starting from the positions shown in the figure, the condition numbers dropped from 66.1 to 3.7 for the planes and from 26.9 to 4.1 for the spheres using stable sampling. The RMS alignment errors after ICP converged using stable sampling were in each case a factor of 3 lower than when a spatially-uniform sampling of point set $\mathcal{P}$ was used. See [339] for more examples.

## 14.5 Dual-Bootstrap ICP

The second algorithm that exploits the covariance matrix during the registration process is designed to avoid the problem of mismatches due to poor initialization. The problem occurs in particular in the registration of retinal images because of the complexity of the vascular structure and the effects of disease.

The Dual-Bootstrap algorithm described here uses points detected along the centers of blood vessel curves [146, 331] as the registration point sets $\mathcal{P}$ and $\mathcal{Q}$. Registration is initialized using matches between landmarks — branching and cross-over points of the vessels — detected in the two images. Unfortunately, images with significant pathologies sometimes have very few landmarks and even fewer that match correctly for initialization. Therefore, the approach taken is a hypothesize-and-test method, where single correspondences are generated to form initial transformation estimates that are only accurate in small image regions. The Dual-Bootstrap algorithm tests each small region and initial estimate separately

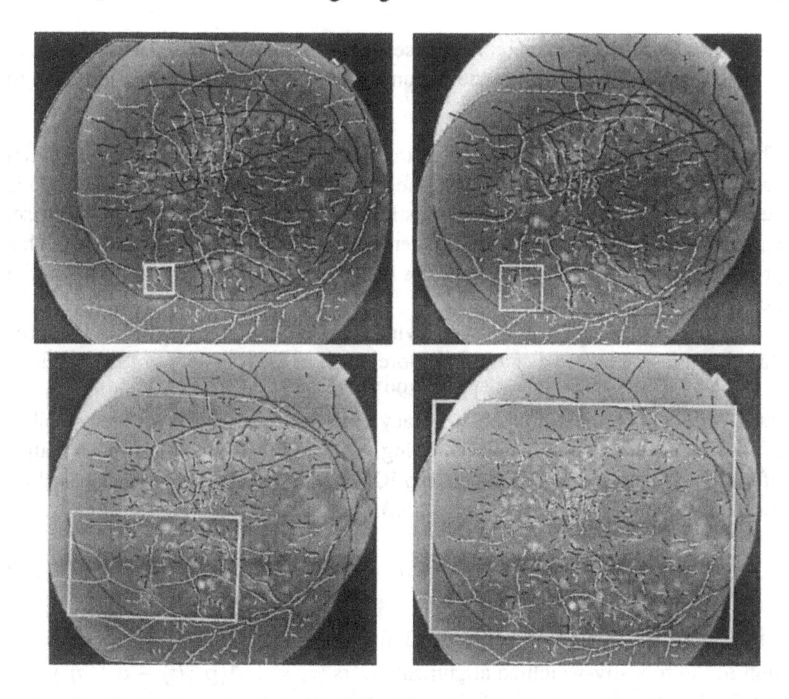

Figure 14.4. Initial (upper left), intermediate (upper right, lower left), and final (lower right) results of the Dual-Bootstrap ICP algorithm on a pair of retinal images. Vessel centerlines forming the point sets $\mathcal{P}$ and $\mathcal{Q}$ are shown using white and black contours. The rectangle drawn on top of the images shows the current region, $R$. The images are well-aligned within $R$ in each iteration, and as $R$ is expanded to cover the entire overlap region, the overall estimate converges to an accurate alignment.

by "growing" an image-wide transformation estimate. If the initial transformation is moderately accurate the Dual-Bootstrap algorithm rarely fails to produce an accurate result.

Dual-Bootstrap ICP works by iterating three steps, illustrated in Figure 14.4:

1. It applies one iteration of ICP using only points from the current region, $R$ (the highlighted rectangle in the panels of Figure 14.4).

2. Based on the correspondences and the covariance matrix, the best transformation model is selected from among a set of possible models. Initially, when the region is small, there are only sufficient constraints for a similarity transformation. The eventual image-wide transformation is a quadratic model [147]. In between, the algorithm can select an affine transformation or a simplified version of the quadratic transformation.

3. The Dual-Bootstrap algorithm uses the uncertainty in the transformation to expand the boundary of the region, $R$. More stable transformations lead to faster region growth.

These steps are repeated until the entire process converges for the given initial estimate. If the final estimate covers the apparent overlap between images and is sufficiently accurate and stable, the estimate is accepted as correct. Otherwise, another starting landmark correspondence and associated region is tried. This greedy process terminates and indicates that no alignment is possible if the initial possibilities are exhausted.

The model selection and region growing steps are most relevant to the theme of this chapter, so they are discussed in more detail in the remainder of this section.

Model selection techniques [135, 799] choose the model that optimizes the trade-off between the alignment accuracy of high-order models and the stability of low-order models, with stability being measured using the covariance matrix of the parameters. The Dual-Bootstrap ICP model selection criteria is based on the expression (see [135] for a derivation):

$$\frac{d}{2} \log 2\pi - \sum_i w_i r_i^2 + \frac{1}{2} \log \det(\Sigma_{\hat{\theta}}), \tag{14.17}$$

where $d$ is the number of degrees of freedom in the model, $\sum_i w_i r_i^2$ is the sum of the robustly-weighted alignment errors ($r_i = (\mathbf{M}(\mathbf{p}_i; \hat{\theta}) - \mathbf{q}_i)^T \eta_i$), and $\det(\Sigma_{\hat{\theta}})$ is the determinant of the parameter estimate covariance matrix. Intuitively, for higher-order models $d$ increases, $-\sum_i w_i r_i^2$ increases (because the residuals decrease), and $\det(\Sigma_{\hat{\theta}})$ decreases because the models are less stable. In choosing the best model, (14.17) is evaluated for a set of models using a fixed correspondence set. The model with the greatest value of (14.17) is chosen.

The growth of the region in step 3 of the Dual-Bootstrap algorithm is based on the uncertainty in the mapping of point locations on the boundary of the regions. This uncertainty is computed from the covariance of the transformation parameter estimate using fairly standard covariance propagation techniques, often called the "transfer error" [389, Ch. 4] in the computer vision literature. As before, let $\mathbf{p}' = \mathbf{M}(\mathbf{p}, \hat{\theta})$ be the mapping of point location $\mathbf{p}$. The covariance of this mapping is approximately

$$\Sigma_{\mathbf{p}'} = \mathbf{J} \Sigma_{\theta} \mathbf{J}^T$$

where

$$\mathbf{J} = \frac{\partial \mathbf{M}}{\partial \theta}(\hat{\theta}) = \mathbf{X}(\mathbf{p}),$$

using the definition of $\mathbf{M}$ from (14.7). No uncertainty in $\mathbf{p}$ is considered because $\mathbf{p}$ is treated simply as a position in the coordinate system of set $\mathcal{P}$, not an estimated point location.

The transfer error is used to expand each of the four sides of region rectangle $R$ (Figure 14.5). Let $\mathbf{p}_s$ be one of these points, described in a coordinate system

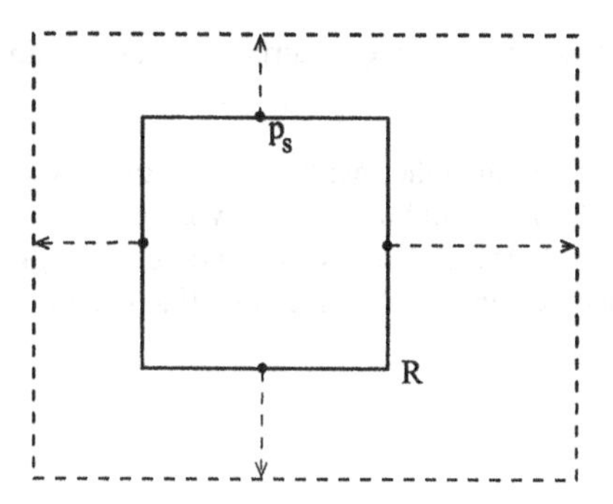


Figure 14.5. Expansion of the region $R$ in the Dual-Bootstrap ICP algorithm. The center of each side of the region rectangle is pushed outward in inverse proportion to the transfer error variance. This means that more certainty in the transformation leads to faster growth in $R$. The new region is the axis-aligned rectangle formed by the four outwardly-moved points.

centered on the rectangle, and let $\mathbf{p}'_s$ be its mapping into the coordinate system of $\mathcal{Q}$. Let $\eta_s$ be the outward normal of the side of the rectangle and let $\eta'_s$ be the mapping of this normal into the coordinate system of $\mathcal{Q}$. The variance of $\mathbf{p}'_s$ in the outward direction is $\sigma_s^2 = {\eta'_s}^T \Sigma_{\mathbf{p}'_s} \eta'_s$. Using this, the outward movement of $\mathbf{p}_s$ is:

$$\Delta \mathbf{p}_s = \beta \frac{(\mathbf{p}_s^T \eta_s)}{\max(1, \sigma_s^2)} \tag{14.18}$$

This growth is proportional to the current distance $(\mathbf{p}_s^T \eta_s)$ of $\mathbf{p}_s$ from the center of $R$, and is inversely proportional to the transfer error in the normal direction. The lower bound of 1 in the denominator prevents growth from becoming too fast. The center of each side of $R$ is expanded outward independently using Equation 14.18, and the new region is the axis-aligned rectangle formed by the resulting four points (Figure 14.5). Parameter $\beta$ controls the growth rate; the setting used in practice, $\beta = \sqrt{2} - 1$, ensures that the area of $R$ at most doubles in each iteration.

The Dual-Bootstrap ICP algorithm has been tested on thousands of retinal image pairs, including images of unhealthy eyes in various stages of disease progression [765, 810]. Overall, when there is at least 30% overlap between images, at least one starting correspondence, and enough extracted vessels to form a stable covariance matrix, the algorithm never fails. Together, the region growth and model selection techniques work to keep the algorithm near the optimal estimate within region $R$. More detail about the behavior of these techniques is as follows:

- Model selection is imperfect. The algorithm tends to switch to higher-order models too early. Estimation errors in these higher-order models may lead to more mismatches, especially on the region periphery. Empirically, the implementation uses the heuristic that the quadratic model may not be used until the region has grown to 20% of the image size. A likely cause of this problem is that the covariance matrix used underestimates the amount of uncertainty.

- Region growth, on the other hand, works extremely well. One measure of this is that halving or doubling the growth rate does not change the effectiveness of the algorithm. Removing region growth altogether, on the other hand, reduces the number of image pairs that the algorithm is able to align by 16%.

## 14.6 Discussion and Conclusion

This chapter has addressed the problem of point-based registration, focusing on the use of covariance-based techniques to improve the performance of the iterative closest point (ICP) algorithm in both range image and retinal image registration. The chapter started by formulating the objective function and then deriving the normal distance version of ICP. This provides a locally-accurate approximation to the overall objective function without the need for rematching. This approximation was then used to derive the equations for estimating the transformation parameters and the covariance matrix of this estimate. Several simplifying assumptions were used in deriving this matrix. These assumptions lead to an underestimate in the overall amount of uncertainty, but are a reasonable approximation as the overall ICP process nears convergence.

The chapter then summarized two algorithms in which the covariance matrix is used to modify the behavior of ICP. In the stable sampling algorithm of [339], the covariance matrix is used to guide the selection of correspondences, ensuring that all directions in parameter space are well-constrained. Geometrically, this allows the ICP algorithm to accurately align small-scale surfaces. In the Dual-Bootstrap ICP algorithm of [765], the covariance matrix is used to grow a transformation estimate and its associated region, starting from a small region surrounding a single correspondence. The covariance matrix helps avoid mismatches between vascular structures by controling the growth of the region and the selection of transformation models. Empirical results show that in both algorithms the use of the covariance matrix substantially improves the registration results.

The algorithms work well despite the approximations needed to compute the covariance matrix. The main reason for this effectiveness is that the covariance matrix plays its most important role as the algorithms near convergence. Stable sampling only has a significant effect when the dominant structures of the data are well-aligned — the small surface misalignments then appear in the eigenvectors of the smaller eigenvalues and may therefore be corrected through the sampling procedure. In the Dual-Bootstrap algorithm, the alignment is always close to convergence in region $R$, even when the alignment appears to be poor throughout the image. This means that the parameter estimate covariance matrix may be used to guide the growth and model selection based only on points from $R$.

The work described here offers a new approach to improving the performance of registration algorithms — using the uncertainty in the estimates being computed to guide further steps in the overall algorithm. This is reminiscent of recursive estimation techniques such as the Kalman filter [637], but in the new algorithms uncertainty is used more broadly, beyond the estimation equations themselves. This could point toward the development of a variety of new algorithms. Moving in this direction requires that a number of issues be addressed. On the theoretical side, a new and more accurate approximation of the covariance matrix is needed that depends on fewer assumptions. One approach might be the use of resampling methods such as the bootstrap technique from statistics [295]. On the more applied side, a second advance would be integrating uncertainty-driven methods with approaches to initialization based on keypoint matching [121]. A third advance would be incorporating uncertainty information into deformable registration, one of the most important problems in medical image analysis.

## Acknowledgements

Portions of this research were supported by National Science Foundation Experimental Partnerships grant EIA-0000417, by the Center for Subsurface Sensing and Imaging Systems under the Engineering Research Centers Program of the National Science Foundation (Award Number EEC-9986821), and by the US Army INSCOM. Thanks go to Rich Radke and Charlene Tsai for comments that significantly enhanced the presentation and to Natasha Gelfand for the images in Figure 14.1.

# Part IV

# Motion Analysis, Optical Flow & Tracking

# Chapter15

# Optical Flow Estimation

**D. Fleet and Y. Weiss**

Abstract

This chapter provides a tutorial introduction to gradient-based optical flow estimation. We discuss least-squares and robust estimators, iterative coarse-to-fine refinement, different forms of parametric motion models, different conservation assumptions, probabilistic formulations, and robust mixture models.

## 15.1 Introduction

Motion is an intrinsic property of the world and an integral part of our visual experience. It is a rich source of information that supports a wide variety of visual tasks, including 3D shape acquisition and oculomotor control, perceptual organization, object recognition and scene understanding [319, 346, 393, 525, 542, 596, 754, 822, 865]. In this chapter we are concerned with general image sequences of 3D scenes in which objects and the camera may be moving. In camera-centered coordinates each point on a 3D surface moves along a 3D path $\boldsymbol{X}(t)$. When projected onto the image plane each point produces a 2D path $\boldsymbol{x}(t) \equiv (x(t), y(t))^T$, the instantaneous direction of which is the velocity $d\boldsymbol{x}(t)/dt$. The 2D velocities for all visible surface points is often referred to the *2D motion field* [407]. The goal of *optical flow* estimation is to compute an approximation to the motion field from time-varying image intensity. While several different approaches to motion estimation have been proposed, including correlation or block-matching (e.g, [25]), feature tracking, and energy-based methods (e.g., [5]), this chapter concentrates on gradient-based approaches; see [59] for an overview and comparison of the other common techniques.

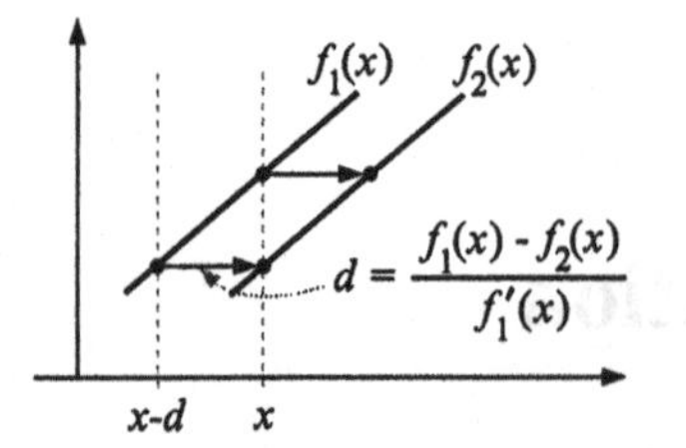


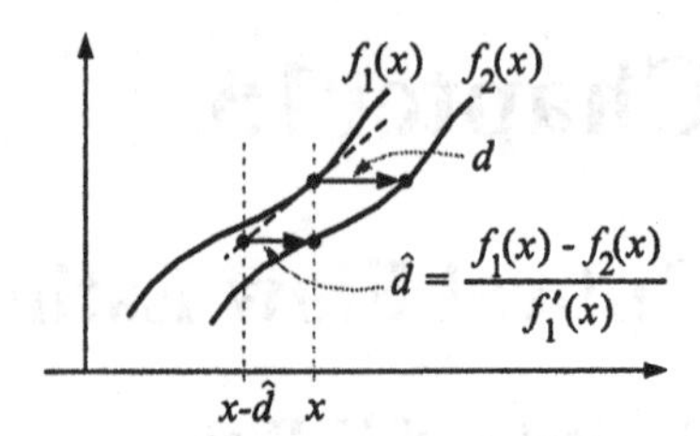


Figure 15.1. The gradient constraint relates the displacement of the signal to its temporal difference and spatial derivatives (slope). For a displacement of a linear signal (left), the difference in signal values at a point divided by the slope gives the displacement. For nonlinear signals (right), the difference divided by the slope gives an approximation to the displacement.

## 15.2 Basic Gradient-Based Estimation

A common starting point for optical flow estimation is to assume that pixel intensities are translated from one frame to the next,

$$I(\boldsymbol{x}, t) = I(\boldsymbol{x} + \boldsymbol{u},\, t + 1)\,, \tag{15.1}$$

where $I(\boldsymbol{x}, t)$ is image intensity as a function of space $\boldsymbol{x} = (x, y)^T$ and time $t$, and $\boldsymbol{u} = (u_1, u_2)^T$ is the 2D velocity. Of course, *brightness constancy* rarely holds exactly. The underlying assumption is that surface radiance remains fixed from one frame to the next. One can fabricate scenes for which this holds; e.g., the scene might be constrained to contain only Lambertian surfaces (no specularities), with a distant point source (so that changing the distance to the light source has no effect), no object rotations, and no secondary illumination (shadows or inter-surface reflection). Although unrealistic, it is remarkable that the brightness constancy assumption (15.1) works so well in practice.

To derive an estimator for 2D velocity $\boldsymbol{u}$, we first consider the 1D case. Let $f_1(x)$ and $f_2(x)$ be 1D signals (images) at two time instants. As depicted in Fig. 15.1, suppose further that $f_2(x)$ is a translated version of $f_1(x)$; i.e., let $f_2(x) = f_1(x - d)$ where $d$ denotes the translation. A Taylor series expansion of $f_1(x - d)$ about $x$ is given by

$$f_1(x - d) = f_1(x) - d\, f_1'(x) + O(d^2 f_1'')\,, \tag{15.2}$$

where $f' \equiv d\,f(x)/dx$. With this expansion we can rewrite the difference between the two signals at location $x$ as

$$f_1(x) - f_2(x) = d\, f_1'(x) + O(d^2 f_1'')\,.$$

Ignoring second- and higher-order terms, we obtain an approximation to $d$:

$$\hat{d} = \frac{f_1(x) - f_2(x)}{f_1'(x)}\,. \tag{15.3}$$

The 1D case generalizes straightforwardly to 2D. As above, assume that the displaced image is well approximated by a first-order Taylor series:

$$I(\boldsymbol{x}+\boldsymbol{u},t+1) \approx I(\boldsymbol{x},t)+\boldsymbol{u}\cdot\nabla I(\boldsymbol{x},t)+I_t(\boldsymbol{x},t)\,, \tag{15.4}$$

where $\nabla I \equiv (I_x, I_y)$ and $I_t$ denote spatial and temporal partial derivatives of the image $I$, and $\boldsymbol{u} = (u_1, u_2)^T$ denotes the 2D velocity. Ignoring higher-order terms in the Taylor series. and then substituting the linear approximation into (15.1), we obtain [409]

$$\nabla I(\boldsymbol{x},t)\cdot\boldsymbol{u}+I_t(\boldsymbol{x},t) = 0\,. \tag{15.5}$$

Equation (15.5) relates the velocity to the space-time image derivatives at one image location, and is often called the *gradient constraint equation*. If one has access to only two frames, or cannot estimate $I_t$, it is straightforward to derive a closely related gradient constraint, in which $I_t(\boldsymbol{x},t)$ in (15.5) is replaced by $\delta I(\boldsymbol{x},t) \equiv I(\boldsymbol{x},t+1) - I(\boldsymbol{x},t)$ [533].

### *Intensity Conservation*

Tracking points of constant brightness can also be viewed as the estimation of 2D paths $\boldsymbol{x}(t)$ along which intensity is conserved:

$$I(\boldsymbol{x}(t),t) = c\,, \tag{15.6}$$

the temporal derivative of which yields

$$\frac{d}{dt}\,I(\boldsymbol{x}(t),t) = 0\,. \tag{15.7}$$

Expanding the left-hand-side of (15.7) using the chain rule gives us

$$\frac{d}{dt}\,I(\boldsymbol{x}(t),t) = \frac{\partial I}{\partial x}\frac{dx}{dt}+\frac{\partial I}{\partial y}\frac{dy}{dt}+\frac{\partial I}{\partial t}\frac{dt}{dt} = \nabla I\cdot\boldsymbol{u}+I_t\,, \tag{15.8}$$

where the path derivative is just the optical flow $\boldsymbol{u} \equiv (dx/dt, dy/dt)^T$. If we combine (15.7) and (15.8) we obtain the gradient constraint equation (15.5).

### *Least-Squares Estimation*

Of course, one cannot recover $\boldsymbol{u}$ from one gradient constraint since (15.5) is one equation with two unknowns, $u_1$ and $u_2$. The intensity gradient constrains the flow to a one parameter family of velocities along a line in *velocity space*. One can see from (15.5) that this line is perpendicular to $\nabla I$, and its perpendicular distance from the origin is $|I_t|/||\nabla I||$.

One common way to further constrain $\boldsymbol{u}$ is to use gradient constraints from nearby pixels, assuming they share the same 2D velocity. With many constraints there may be no velocity that simultaneously satisfies them all, so instead we find the velocity that minimizes the constraint errors. The least-squares (LS) estimator minimizes the squared errors [533]:

$$E(\boldsymbol{u}) = \sum_{\boldsymbol{x}} g(\boldsymbol{x}) \left[\boldsymbol{u} \cdot \nabla I(\boldsymbol{x},t) + I_t(\boldsymbol{x},t)\right]^2 , \tag{15.9}$$

where $g(\boldsymbol{x})$ is a weighting function that determines the *support* of the estimator (the region within which we combine constraints). It is common to let $g(\boldsymbol{x})$ be Gaussian in order to weight constraints in the center of the neighborhood more highly, giving them more influence. The 2D velocity $\hat{\boldsymbol{u}}$ that minimizes $E(\boldsymbol{u})$ is the least squares flow estimate.

The minimum of $E(\boldsymbol{u})$ can be found from its critical points, where its derivatives with respect to $\boldsymbol{u}$ are zero; i.e.,

$$\frac{\partial E(u_1,u_2)}{\partial u_1} = \sum_{\boldsymbol{x}} g(\boldsymbol{x}) \left[u_1 {I_x}^2 + u_2 I_x I_y + I_x I_t\right] = 0$$

$$\frac{\partial E(u_1,u_2)}{\partial u_2} = \sum_{\boldsymbol{x}} g(\boldsymbol{x}) \left[u_2 {I_y}^2 + u_1 I_x I_y + I_y I_t\right] = 0 .$$

These equations may be rewritten in matrix form:

$$\mathrm{M}\,\boldsymbol{u} = \boldsymbol{b} , \tag{15.10}$$

where the elements of M and $\boldsymbol{b}$ are:

$$\mathrm{M} = \begin{bmatrix} \sum g\,{I_x}^2 & \sum g\,I_x I_y \\ \sum g\,I_x I_y & \sum g\,{I_y}^2 \end{bmatrix} , \quad \boldsymbol{b} = -\begin{pmatrix} \sum g\,I_x I_t \\ \sum g\,I_y I_t \end{pmatrix} .$$

When M has rank 2, then the LS estimate is $\hat{\boldsymbol{u}} = \mathrm{M}^{-1}\boldsymbol{b}$.

### *Implementation Issues*

Usually we wish to estimate optical flow at every pixel, so we should express M and $\boldsymbol{b}$ as functions of position $\boldsymbol{x}$, i.e., $\mathrm{M}(\boldsymbol{x})\,\boldsymbol{u}(\boldsymbol{x}) = \boldsymbol{b}(\boldsymbol{x})$. Note that the elements of M and $\boldsymbol{b}$ are local sums of products of image derivatives. An effective way to estimate the flow field is to first compute derivative images through convolution with suitable filters. Then, compute their products (${I_x}^2$, $I_x I_y$, ${I_y}^2$, $I_x I_t$ and $I_y I_t$), as required by (15.10). These quadratic images are then convolved with $g(\boldsymbol{x},)$ to obtain the elements of $\mathrm{M}(\boldsymbol{x})$ and $\boldsymbol{b}(\boldsymbol{x})$.

In practice, the image derivatives will be approximated using numerical differentiation. It is important to use a consistent approximation scheme for all three directions [303]. For example, using simple forward differencing (i.e., $\hat{I}_x = I(x,y) - I(x+1,y)$) will not give a consistent approximation as the $x$, $y$ and $t$ derivatives will be centered at different locations in the $xyt$-cube [407]. Another practicality worth mentioning is that some image smoothing is generally useful prior to numerical differentiation (and can be incorporated into the derivative filters). This can be justified from the first-order Taylor series approximation used to derive (15.5). By smoothing the signal, one hopes to reduce the amplitudes of higher-order terms in the image and to avoid some related problems with temporal aliasing.

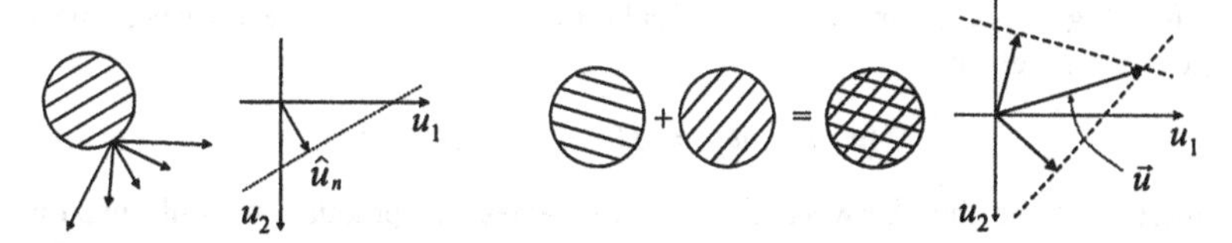


Figure 15.2. (left) A single moving grating viewed through a circular aperture is consistent with all 2D velocities along a line in velocity space. (right) With two drifting gratings there are multiple constraint lines that intersect to uniquely constrain the 2D velocity. (After [6])

### *Aperture Problem*

When M in (15.10) is rank deficient one cannot solve for $\boldsymbol{u}$. This is often called the aperture problem as it invariably occurs when the support $g(x)$ is sufficiently local. However, the important issue is not the width of support, but rather the dimensionality of the image structure. Even for large regions, if the image is one-dimensional then M will be singular. As depicted in Fig. 15.2 (left); when each image gradient within a region has the same spatial direction, it is easy to see that $rank[\mathrm{M}] = 1$. Moreover, note that a single gradient constraint only provides the normal component of $\boldsymbol{u}$,

$$\boldsymbol{u}_n = \frac{-I_t}{||\nabla I||} \frac{\nabla I}{||\nabla I||} .$$

When there exist constraints with two or more gradient directions, as depicted in Fig. 15.2 (right), then the different constraint lines intersect to uniquely constrain the 2D velocity.

## 15.3 Iterative Optical Flow Estimation

Equation (15.9) provides an optimal solution, but not to our original problem. Remember that we ignored high-order terms in the derivation of (15.3) and (15.5). As depicted in Fig. 15.1, if $f_1$ is linear then $d = \hat{d}$. Otherwise, to leading order, the accuracy of the estimate is bounded by the magnitude of the displacement and the second derivative of $f_1$:

$$|\hat{d} - d| \leq \frac{d^2 \, |f_1''(x)|}{2 \, |f_1'(x)|} + O(d^3) . \tag{15.11}$$

For a sufficiently small displacement, and bounded $|f_1''/f_1'|$, we expect reasonably accurate estimates. This suggests a form of Gauss-Newton optimization in which we use the current estimate to *undo* the motion, and then we reapply the estimator to the *warped* signals to find the residual motion. This continues until the residual motion is sufficiently small.

In 2D, given an estimate of the optical flow field $\boldsymbol{u}^0$, we create a *warped* image sequence $I^0(\boldsymbol{x}, t)$:

$$I^0(\boldsymbol{x},\, t+\delta t) \;=\; I(\boldsymbol{x} + \boldsymbol{u}^0 \delta t, t + \delta t)\,, \tag{15.12}$$

where $\delta t$ is the time between consecutive frames. (In practice, we only need to warp enough frames for temporal differentiation.) Assuming that $\boldsymbol{u} = \boldsymbol{u}^0 + \delta\boldsymbol{u}$, it is straightforward to see from (15.1) and (15.12) that

$$I^0(\boldsymbol{x}, t) \;=\; I^0(\boldsymbol{x} + \delta\boldsymbol{u}, t+1)\,. \tag{15.13}$$

If $\delta\boldsymbol{u} = 0$, then clearly $I^0$ would be constant through time (assuming brightness constancy). Otherwise, we can estimate the residual flow using

$$\delta\hat{u} = \mathrm{M}^{-1}\boldsymbol{b} \tag{15.14}$$

where M and $\boldsymbol{b}$ are computed by taking spatial and temporal derivatives (differences) of $I^0$. The refined optical flow estimate then becomes

$$\boldsymbol{u}^1 \;=\; \boldsymbol{u}^0 + \delta\hat{u}\,.$$

In an iterative manner, this new flow estimate is then used to rewarp the original sequence (as in (15.12)), and another residual flow can be estimated.

This iteration yields a sequence of approximate objective functions that converge to the desired objective function [91]. At iteration $j$, given the estimate $\boldsymbol{u}^j$ and the warped sequence $I^j$, our desired objective function is

$$E(\delta\boldsymbol{u}) \;=\; \sum_{\boldsymbol{x}} g(\boldsymbol{x})\,\left[I(\boldsymbol{x},t) - I(\boldsymbol{x} + \boldsymbol{u}^j + \delta\boldsymbol{u},\, t+1)\right]^2 \tag{15.15}$$

$$\;=\; \sum_{\boldsymbol{x}} g(\boldsymbol{x})\,\left[I^j(\boldsymbol{x},t) - I^j(\boldsymbol{x} + \delta\boldsymbol{u},\, t+1)\right]^2$$

$$\;\approx\; \sum_{\boldsymbol{x}} g(\boldsymbol{x})\,\left[\nabla I^j(\boldsymbol{x},t)\cdot\delta\boldsymbol{u} + I_t^j(\boldsymbol{x},t)\right]^2 \;\equiv\; \tilde{E}(\delta\boldsymbol{u})\,. \tag{15.16}$$

The gradient approximation to the difference in (15.15) gives an approximate objective function $\tilde{E}$. From (15.11) one can show that $\tilde{E}$ approximates $E$ to second-order in the magnitude of the residual flow, $\delta\boldsymbol{u}$. The approximation error vanishes as $\delta\boldsymbol{u}$ is reduced to zero. The iterative refinement with rewarping reduces the residual motion at each iteration so that the approximate objective function converges to the desired objective function, and hence the flow estimate converges to the optimal LS estimate (15.15).

The most expensive step at each iteration is the computation of image gradients and the matrix inverse in (15.14). One can, however, formulate the problem so that the spatial image derivatives used to form $\mathbf{M}$ are taken at time $t$, and as such, do not depend on the current flow estimate $\boldsymbol{u}^j$ [375]. To see this, note that the spatial deriatives are computed at time $t$ and it is straightforward to see that $I(\boldsymbol{x}, t) = I^j(\boldsymbol{x}, t)$. Of course $\boldsymbol{b}$ in (15.14) will always depend on the warped image sequence and must be recomputed at each iteration. In practice, when $\mathbf{M}$ is

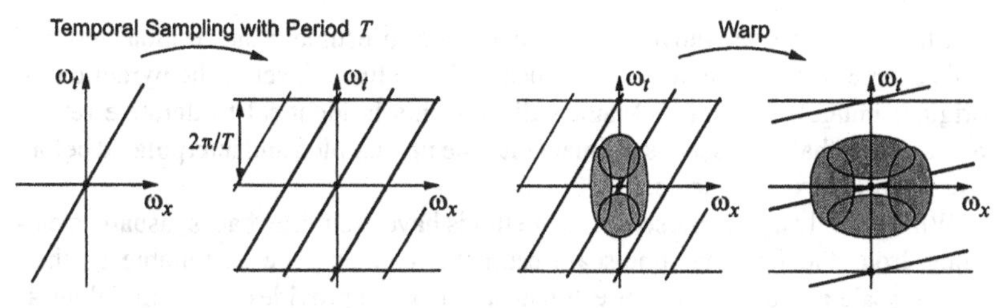


Figure 15.3. (Left) The spectrum of a translating signal is nonzero on a line in the frequency domain. Temporal sampling introduces spectral replicas, causing aliasing for higher speeds (steeper slopes). (Right) The problem may be avoided by blurring the images before computing derivatives. The spectra of such *coarse-scale* filters will be insensitive to the replicas. Velocity estimates from the coarse scale are used to warp the images, thereby undoing much of the motion. Finer-scale derivative filters can now be used to estimate the residual motion. (After [743])

not recomputed from the warped sequence then the spatial and temporal derivatives will not centered at the same location in $(x, y, t)$ and hence more iterations may be needed.

## *Temporal Aliasing and Coarse-To-Fine Refinement*

In practice, our images have temporal sampling rates lower than required by the sampling theorem to uniquely reconstruct the continuous signal. As a consequence, temporal aliasing is a common problem in motion estimation.

The spectrum of a translating signal is confined to a plane through the origin in the frequency domain [322, 866]. That is, if we construct a space-time signal $f(\boldsymbol{x}, t)$ by translating a 2D signal $f_0(\boldsymbol{x})$ with velocity $\boldsymbol{u}$, i.e., $f(\boldsymbol{x}, t) = f_0(\boldsymbol{x} - \boldsymbol{u}t)$, one can show that the space-time Fourier transform of $f(\boldsymbol{x}, t)$ is given by

$$F(\omega_x, \omega_y, \omega_t) = F_0(\omega_x, \omega_y)\, \delta(u_1\omega_x + u_2\omega_y + \omega_t)\,, \qquad (15.17)$$

where $F_0$ is the 2D Fourier transform of $f_0$ and $\delta()$ is a Dirac delta. Equation (15.17) shows that the spectrum is nonzero only on a plane, the orientation of which gives the velocity. When the continuous signal is sampled in time, replicas of the spectrum are introduced at intervals of $2\pi/T$ radians, where $T$ is the time between frames (see Fig. 15.3 (left)). It is easy to see how this causes problems; i.e., the derivative filters may be more sensitive to the spectral replicas at high spatial frequencies than to the original spectrum on the plane through the origin.

This suggests a simple approach to aliasing problems [25, 75]. Optical flow can be estimated at the coarsest scale of a Gaussian pyramid, where the image is significantly blurred, and the velocity is much slower (due to subsampling). The coarse-scale estimate can be used to warp the next (finer) pyramid level to *stabilize* its motion. Since the velocities after warping are slower, as shown in Fig. 15.3 (right)), a wider low-pass frequency band will be free of aliasing. One can therefore use derivatives at the finer scale to estimate the residual motion. This coarse-to-fine estimation continues until the finest level of the pyramid (the original image) is reached. Mathematically, this is identical to iterative refinement except that each scale's estimate must be up-sampled and interpolated before warping the next finer scale.

While widely used, coarse-to-fine methods have their drawbacks, usually stemming from the fact that fine-scale estimates can only be as reliable as their coarse-scale precursors; a poor estimate at one scale provides a poor initial guess at the next finer scale, and so on. That said, when aliasing does occur, one must use some mechanism such as coarse-to-fine estimation to avoid local minima in the optimization.

## 15.4 Robust Motion Estimation

The LS estimator is optimal when the gradient constraint errors, i.e.,

$$e(\boldsymbol{x}) \equiv \boldsymbol{u} \cdot \nabla I(\boldsymbol{x}, t) + I_t(\boldsymbol{x}, t) , \tag{15.18}$$

are mean-zero Gaussian, and the errors in different constraints are independent and identically distributed (IID). Not surprisingly, this is a fragile assumption. For example, brightness constancy is often violated due to changing surface orientation, specular reflections, or time-varying shadows. When there is significant depth variation in the scene, the constant motion model will be extremely poor, especially at occlusion boundaries.

LS estimators are not suitable when the distribution of gradient constraint errors is heavy-tailed, as they are sensitive to small numbers of measurement outliers [380, 518]. It is therefore often crucial that the quadratic estimator in (15.9) be replaced by a robust estimator, $\rho(\cdot)$, which limits the influence of constraints with larger errors (e.g., see [40, 89, 612]):

$$E(\boldsymbol{u}) = \sum_{x,y} g(\boldsymbol{x}) \, \rho(e(\boldsymbol{x}), \sigma) . \tag{15.19}$$

For example, Black and Anandan [89] used the redescending Geman-McClure estimator [342], $\rho(e, \sigma) = e^2/(e^2 + \sigma^2)$, where $\sigma^2$ determines the range of constraint errors for which influence is reduced.

Among the various ways one might minimize (15.19), one very useful approach takes the form of iteratively reweighted least-squares [518]. In short, this is an iterative solution in which the weights $g(\boldsymbol{x})$ in (15.9) are scaled by a weight function that downweights those constraints that are inconsistent (i.e., have large errors) with the current motion estimate. Often it is also useful to anneal the optimization, wherein $\sigma^2$ starts large, and is then slowly decreased to achieve greater robustness.

## 15.5 Motion Models

Thus far we have assumed that the 2D velocity is constant in local neighbourhoods. Unfortunately, even for small regions this is often a poor assumption. We now consider generalizations to more interesting motion models.

### *Affine Model*

General first-order affine motion is usually a better model of local motion than a translational model (e.g., [75, 89, 320]). An affine velocity field centered at location $\boldsymbol{x}_0$ can be expressed in matrix form as

$$\boldsymbol{u}(\boldsymbol{x};\boldsymbol{x}_0) \;=\; \mathrm{A}(\boldsymbol{x};\boldsymbol{x}_0)\,\boldsymbol{c}\,, \tag{15.20}$$

where $\mathrm{c} = (c_1, c_2, c_3, c_4, c_5, c_6)^T$ are the motion model parameters, and

$$\mathrm{A}(\boldsymbol{x};\,\boldsymbol{x}_0) \;=\; \begin{bmatrix} 1 & 0 & x-x_0 & y-y_0 & 0 & 0 \\ 0 & 1 & 0 & 0 & x-x_0 & y-y_0 \end{bmatrix}.$$

Combining (15.20) and (15.5) yields the gradient constraint equation

$$\nabla I(\boldsymbol{x},t)\,\mathbf{A}(\boldsymbol{x};\boldsymbol{x}_0)\,\boldsymbol{c} + I_t(\boldsymbol{x},t) \;=\; 0\,,$$

for which the LS estimate for the neighbourhood has the form

$$\hat{c} \;=\; \mathrm{M}^{-1}\,b\,, \tag{15.21}$$

where now M and $b$ are given by

$$\mathrm{M} = \sum_{\boldsymbol{x}} g\,\mathrm{A}^T \nabla I^T\, \nabla I\,\mathrm{A}\quad,\quad b = -\sum_{\boldsymbol{x}} g\,\mathrm{A}^T \nabla I^T\, I_t\,.$$

When **M** is rank deficient there is insufficient image structure to estimate the six unknowns. Affine models often require larger support than constant models, and one may need a robust estimator instead of the LS estimator.

Iterative refinement is also straightforward with affine motion models. Let the optimal affine motion be $\boldsymbol{u} = \mathbf{A}\,\mathbf{c}$, and let the affine estimate at iteration $j$ be $\boldsymbol{u}^j = \mathbf{A}\,\boldsymbol{c}^j$. Because the flow is linear in the motion parameters, it follows that $\delta\boldsymbol{u} \equiv \boldsymbol{u} - \boldsymbol{u}^j$ and $\delta\boldsymbol{c} \equiv \mathbf{c} - \boldsymbol{c}^j$ satisfy

$$\delta\boldsymbol{u} \;=\; \mathbf{A}\delta\boldsymbol{c}\,. \tag{15.22}$$

Accordingly, defining $I^j(\boldsymbol{x},t)$ to be the original sequence $I(\boldsymbol{x},t)$ warped by $\boldsymbol{u}^j$ as in (15.12) we use the same LS estimator as in (15.21), but with $I$ and $\hat{c}$ replaced by $I^j$ and $\delta\hat{c}$. The updated LS estimate is then $\boldsymbol{c}^{j+1} = \boldsymbol{c}^j + \delta\hat{c}$.

### *Low-Order Parametric Deformations*

There are many other polynomial and rational deformations that make useful motion models. *Similarity deformations*, comprising translation $(d_1, d_2)$, 2D rotation

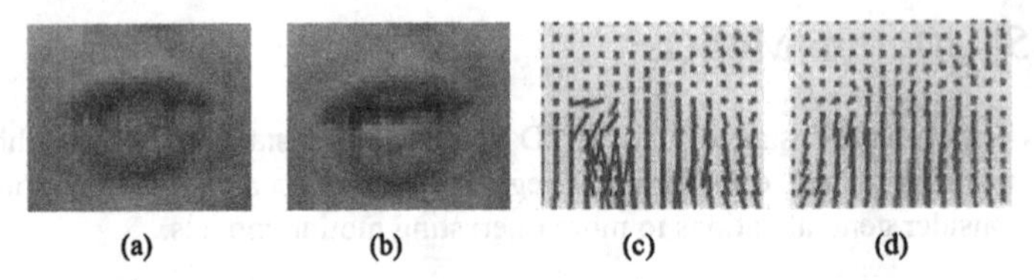

Figure 15.4. (a,b) Mouth regions of two consecutive images of a person speaking. (c) Flow field estimated using dense optical flow method. (d) Flow field estimated using the learned model with 6 basis flow fields. (After [319])

$\theta$, and uniform scaling by $s$ are a special case of the affine model, but still very useful in practice. In a neighbourhood centred at $\boldsymbol{x}_0$ it has the same form as (15.20), but with $\boldsymbol{c} = (d_1,\, d_2,\, s\,\cos\theta,\, s\,\sin\theta)^T$ and

$$\mathbf{A}(\boldsymbol{x};\, \boldsymbol{x}_0) \;=\; \begin{bmatrix} 1 & 0 & x - x_0 & -y + y_0 \\ 0 & 1 & y - y_0 & x - x_0 \end{bmatrix} .$$

With this linear form, one can solve directly for $\boldsymbol{c}$ using linear least-squares, and then compute the similarity parameters $d_1, d_2, s$, and $\theta$.

Another useful motion model is the *projective deformation* (or homography) [75], which captures image deformations of a 3D plane under camera rotation and translation. See in Chapter 17 for a discussion of homographies and related motion models.

## *Learned Subspace Models*

Many objects exhibit complex motions that are not well modeled by low-order polynomials. For example Fig. 15.4(a,b) shows two frames of a mouth during speech, for which non-rigidity, occlusion, and fast speeds make flow estimation difficult. Interestingly, the regression framework above extends to diverse types of complex 2D motions with the use of basis flow fields, $\{\boldsymbol{b}_j(\boldsymbol{x})\}_{j=1}^{J}$, such that the local optical flow field is expressed as

$$\boldsymbol{u}(\boldsymbol{x}) \;=\; \sum_{j=1}^{J} c_j\, \boldsymbol{b}_j(\boldsymbol{x}) \,. \tag{15.23}$$

In this context, optical flow estimation reduces to the estimation of the linear coefficients $\boldsymbol{c}$, analogous to the affine model discussed above.

In [319] a motion basis was learned for human mouths. This was accomplished by applying a robust estimator with a generic smoothness model [89] to mouths to obtain training data (e.g., see Fig. 15.4(c)). The principal components of the ensemble of training flow fields were then extracted and used as the basis. Figure 15.4(d) shows the optical flow obtained with the subspace model and a robust estimator. The model was found to greatly increase the quality of the optical flow estimates, and the temporal variation in the subspace coefficients were then used to recognize linguistic events [319].

### *General Differentiable Warps*

In general, one can formulate area-based regression in terms of warp functions $w(\boldsymbol{x}; \boldsymbol{p})$ that are not necessarily smooth in space, nor linear in the warp parameters $\boldsymbol{p}$. One can parametrize the warp as a function of time, or assume the two-frame case:

$$I(\boldsymbol{x}, t) \;=\; I(w(\boldsymbol{x}; \boldsymbol{p}),\, t+1)\,. \tag{15.24}$$

The warp functions must be differentiable with respect to $\boldsymbol{p}$. To develop an efficient estimation algorithm, one may need to further constrain $w$ to be invertible (e.g., see [375]).

## 15.6 Global Smoothing

While area-based regression is commonly used, some of the earliest formulations of optical flow estimation assumed smoothness through non-parametric motion models, rather than an explicit parametric model in each local neighbourhood (e.g., see [407, 593, 714]). Horn and Schunck [409] proposed an energy functional of the form:

$$E(\boldsymbol{u}) \;=\; \int (\nabla I \cdot \boldsymbol{u} + I_t)^2 \;+\; \lambda \left( ||\nabla u_1||^2 + ||\nabla u_2||^2 \right)\; dx\, dy\,. \tag{15.25}$$

A key advantage of global smoothing is that it enables propagation of information over large distances in the image. In image regions of nearly uniform intensity, such as a blank wall or tabletop, local methods will often yield singular (or poorly conditioned) systems of equations. Global methods can *fill in* the optical flow from nearby gradient constraints.

Equation (15.25) can be minimized directly with discrete approximations to the integral and the derivatives in (15.25). Thie yields a large system of linear equations that may be solved through iterative methods such as *Gauss-Seidel* or *SOR overrelaxation* [352]. Alternatively one can solve the corresponding Euler-Lagrange (PDE) equations under reflecting boundary conditions (e.g., [133, 714]). Recent extensions to global methods include robust penalty functions (for data and smoothness terms), the use of coarse-to-fine search for optimization, and the incorporation of stronger local constraints on the motion, resulting in impressive optical flow estimates [133].

The main disadvantage of global methods is computational efficiency. Even with more efficient optimization algorithms (e.g. [779, 878]) the computational cost is far higher than with local methods. Whether this is justified may depend on the image domain and the need for dense optical flow. Another problem is in the setting of the *regularization parameter* $\lambda$ that determines the amount of desired smoothing (similar problems arise in choosing the support width for area-based regression). Prior knowledge on the smoothness of flow can be useful here, and more sophisticated methods might be used to estimate (or marginalize) the regularization parameter.

## 15.7 Conservation Assumptions

All of the above formulations assumed intensity conservation. Nevertheless, gradient constraints may be used to track any differentiable image property.

### *Higher-Order Derivative Constraints*

Some techniques assume that image gradients are conserved (e.g., [593, 743, 823]). This provides two further constraints at each pixel, i.e.,

$$\begin{aligned} u_1 I_{xx} + u_2 I_{xy} + I_{xt} &= 0 \\ u_1 I_{xy} + u_2 I_{yy} + I_{yt} &= 0 \,. \end{aligned} \tag{15.26}$$

These are useful insofar as they provide more constraints with which to estimate motion parameters. Conversely, higher-order derivatives are often extremely noisy, and the conservation of $\nabla I$ implies that the motion field has no first-order deformation (e.g., rotation). Intensity conservation (15.7), by comparison, assumes only that the image motion is smooth.

### *Phase-Based Methods*

Phase-based methods [320, 321] are based on an initial decomposition of the image into band-pass channels, like those produced by quadrature-pair filters in steerable pyramids [330]. While multi-scale representations are commonly used for flow estimation, a further decomposition into orientation bands yields more local constraints, often with better signal-to-noise ratios. Complex-valued band-pass images can be represented as real and imaginary images, or in terms of amplitude and phase images. Figure 15.5 shows the real-part of a 1D band-pass signal, along with its amplitude and phase. Amplitude encodes the magnitude of local signal modulation, while phase encodes the local structure of the signal (e.g., zero-crossings, peaks, etc).

Phase-based methods assume conservation of phase in each band-pass channel. The phase-based gradient constraint, given a complex-valued band-pass channel, $r(\boldsymbol{x}, t)$, with phase $\phi(\boldsymbol{x}, t) \equiv \arg[r(\boldsymbol{x}, t)]$, is simply

$$\nabla\phi(\boldsymbol{x}, t) \cdot \boldsymbol{u} + \phi_t(\boldsymbol{x}, t) = 0 \,. \tag{15.27}$$

These may be combined to estimate optical flow using any of the estimators above. In practice, because phase is a multi-function, only uniquely defined on intervals of width $2\pi$, explicit differentiation is difficult. Instead, it is convenient to exploit the following identities for computing spatial derivatives and temporal differences,

$$\frac{\partial\phi(\boldsymbol{x}, t)}{\partial x} = \frac{\mathrm{Im}[r_x(\boldsymbol{x}, t)\, r^*(\boldsymbol{x})]}{|r(\boldsymbol{x})|^2} \,, \quad \delta\phi(\boldsymbol{x}, t) = \arg[r(\boldsymbol{x}, t+1)\, r^*(\boldsymbol{x}, t)] \,.$$

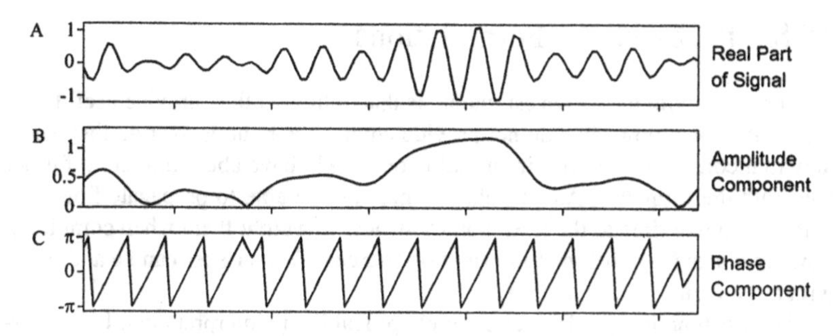


Figure 15.5. A band-pass filtered 1D signal can be expressed using its amplitude and phase signals. Note the linearity of phase over large spatial extents.

where $\mathrm{Im}[r]$ denotes the imaginary part of $r$, $r^*$ is the complex-conjugate of $r$, and $r_x \equiv \partial r / \partial x$. Compared to phase, $r(x, t)$ is relatively easy to differentiate and interpolate [322, 320].

Phase has a number of appealing properties for optical flow estimation. First, phase is amplitude invariant, and therefore quite stable when significant changes in contrast and mean intensity occur between frames. Second, phase is approximately linear over relatively large spatial extents, and has very few critical points where the gradient is zero. This is important as it implies that more gradient constraints may be available, and that the range of velocities that can be estimated is significantly larger than with image derivatives. This also improves the accuracy of gradient-based estimates, reducing the number of iterations required for refinement. Phase has also been shown to be stable with respect to first-order deformations of the image from one time to the next [321]. Both the expected spatial extent of phase linearity and the stability of phase are determined, in part, by filter bandwidth. The main disadvantages of phase concern the computational expense of the band-pass filters, and the spatial support of the filters near occlusion boundaries and fine-scale objects.

## *Brightness Variations*

While contrast normalization, or the use of phase, provides some degree of invariance with respect to deviations from brightness constancy, more significant variations in brightness must be modeled explicitly. The models may be object specific, to model objects under different lighting conditions [375], poses or configurations [91]. Alternatively, the models may be physics-based [390], or they may be generic models for smooth mean and contrast variations [595]. Despite the wide-spread use of brightness constancy these models may be extremely useful for certain domains.

## 15.8 Probabilistic Formulations

One problem with the above estimators is that, although they provide useful estimates of optical flow, they do not provide confidence bounds. Nor do they show how to incorporate any prior information one might have about motion to further constrain the estimates. As a result, one may not be able to propagate flow estimates from one time to the next, nor know how to weight them when combining flow estimates from different information sources. These issues can be addressed with a probabilistic formulation.

The cost function (15.16) has a simple probabilistic interpretation. Up to normalization constants, it corresponds to the log likelihood of a velocity under the assumption that intensity is conserved up to Gaussian noise.

$$I(\boldsymbol{x},t) = I(\boldsymbol{x}+\boldsymbol{u},t+1)+\eta\,. \tag{15.28}$$

If we assume that the same velocity $\boldsymbol{u}$ is shared by all pixels within a neighbourhood, that $\eta$ is white Gaussian noise with standard deviation $\sigma$, and uncorrelated at different pixels, we obtain the conditional density

$$p(I \mid \boldsymbol{u}) \propto e^{-\frac{1}{2\sigma^2}E(\boldsymbol{u})}\,, \tag{15.29}$$

where $E(\boldsymbol{u})$ is the LS objective function (15.16). To obtain further insight into this likelihood function, we again approximate $E$ to second order using $\tilde{E}$ as in (15.15). Under this approximation the likelihood function is Gaussian with mean $\mathrm{M}^{-1}\,\boldsymbol{b}$ and covariance matrix $\mathrm{M}^{-1}$.

The approximate covariance matrix $\mathrm{M}^{-1}$ defines an uncertainty ellipse around the estimated optical flow. These uncertainties can be propagated to subsequent frames, or to other spatial scales [744]. They can also be used directly in algorithms for 3D reconstruction [418]. (See [880] for a more detailed discussion of likelihood functions for probabilistic optical flow estimation.)

The probabilistic formulation also allows one to introduce *prior information*. Equation (15.29) can be combined with a prior probability distribution over local velocities. For example, a very useful prior model is that the local flow tends to be *slow* (e.g. [744]). This is convenient to model with a zero-mean Gaussian distribution,

$$p(\boldsymbol{u}) \propto e^{\frac{1}{2\sigma_p^2}-\|\boldsymbol{u}\|^2}\,. \tag{15.30}$$

Combining this prior probability with the approximate likelihood function (15.29) gives us a Gaussian posterior probability whose mean (and mode) is

$$\boldsymbol{u} = (\mathrm{M}+\lambda I)^{-1}\,\boldsymbol{b}\,, \tag{15.31}$$

where $\lambda$ is the ratio of the noise and prior variances, $\lambda = \sigma^2/\sigma_p^2$. Note that this Bayesian estimate will actually be biased, and will not correctly estimate the speed or direction of patterns where the local uncertainty is large. This has the benefit that it dampens the estimates to help avoid divergence in iterative refinement and tracking. Interestingly, many "illusions" in human motion perception can actually be explained with a prior favoring slow motions and a Bayesian model of inference [881].

### *Total Least-Squares*

When one assumes significant image noise that contaminates spatial as well as temporal derivatives, then the maximum likelihood motion estimate given a collection of space-time image gradients is given by *total-least-squares* (TLS) [598, 867]. If we view velocity as a unit direction in space-time, or in 3D homogeneous coordinates $\boldsymbol{v} \equiv \alpha(u_1, u_2, 1)$, $\alpha \in \mathcal{R}$, and denote the space-time image gradient $\boldsymbol{o}_k \equiv (\nabla I(\boldsymbol{x}_k, t), I_t(\boldsymbol{x}_k, t))^T$, then the gradient constraint becomes $\boldsymbol{o}_k{}^T \boldsymbol{v} = 0$. The sum or squared constraint errors is then

$$E(\boldsymbol{v}) = \boldsymbol{v}^T \mathrm{S}\, \boldsymbol{v} \ , \quad \text{where} \ \ \mathrm{S} = \sum_k \boldsymbol{o}_k \, \boldsymbol{o}_k{}^T . \tag{15.32}$$

The TLS solution is obtained by minimizing $E(\boldsymbol{v})$ in (15.32), subject to the constraint $||\boldsymbol{v}|| = 1$ to avoid the trivial solution. The solution is given by the eigenvector corresponding to the minimum eigenvalue of S. This approach has been called tensor-based, with S called the structure tensor [86, 390, 428], These methods have produced excellent optical flow results [305].

Different noise models yield different estimators. TLS is a ML estimator when the noise in $\boldsymbol{o}_k$ is additive, isotropic and IID. When the noise is anisotropic and not identically distributed the formulation becomes much more complex [597]. More complex noise models, especially those with correlated noise in local regions, remain a topic for future research.

## 15.9 Layered Motion

One common problem with area-based regression methods concerns the size of spatial support. With larger support there are more constraints for parameter estimation, but there is a greater risk that simple parametric motion models will be unsuitable. This is particularly serious near occlusion boundaries where multiple motions exist. For example, in the scene depicted in Fig. 15.6 the camera was translating, and therefore both the soda can and the background move with respect to the camera, but with different image velocities. To demonstrate this, Fig. 15.6 (right) shows a subset of the gradient constraints in the small region (marked in white) at the left side of the can. There are two points with a high density of constraint-line intersections, corresponding to the velocities of the can and the background.

One way to cope with regions with multiple motions is to explicitly model the *layers* in the scene. The layered model is like a cardboard cutout representation of a scene in which different cardboard surfaces correspond to different layers, and they are assumed to be able to move independently [435, 853]. Layered motion estimation can be formulated using probabilistic mixture models,

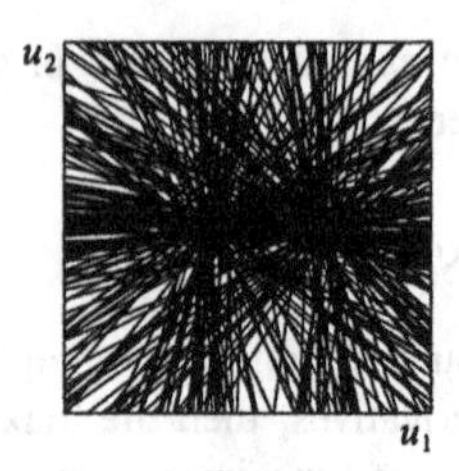


Figure 15.6. (left) The depth discontinuity at the left side of the can creates a motion discontinuity as the camera translates right. (right) Motion constraint lines in velocity space are shown from pixels within the white square. (After [435])

with the Expectation-Maximization (EM) algorithm for parameter estimation [38, 435, 878, 879].

## *Mixture Models*

Let there be a region of pixels $\{\boldsymbol{x}_k\}_{k=1}^{K}$ in which we suspect there are multiple velocities; e.g., the region might contain an occlusion boundary. By way of notation, let $\boldsymbol{u}(\boldsymbol{x};\, \boldsymbol{c})$ denote a parameterized flow field with parameters $\boldsymbol{c}$. Within a single region of the image we will assume that there are $N$ motions, parameterized by $\boldsymbol{c}_n$, for $1 \leq n \leq N$. Furthermore, according to the our *mixture model*, the individual motions occur with probability $m_n$. These *mixing probabilities* tell us what fraction of the $K$ pixels within the region we expect to be consistent with (i.e., *owned* by) each motion. Of course the mixing probabilities sum to 1.

Let us further assume that we have one gradient constraint per pixel within the region. Let $\boldsymbol{o}_k \equiv (\nabla I(\boldsymbol{x}_k, t), I_t(\boldsymbol{x}_k, t))^T$ denote the spatial and temporal image derivatives at pixel $\boldsymbol{x}_k$. As above, given the correct motion, we assume that the gradient constraint is satisfied up to Gaussian noise:

$$e(\boldsymbol{x}_k;\, \boldsymbol{c}_n) \;\equiv\; \nabla I(\boldsymbol{x}_k, t) \cdot \boldsymbol{u}_n(\boldsymbol{x};\, \boldsymbol{c}_n) + I_t(\boldsymbol{x}_k, t) \;=\; \eta\,,$$

where $\eta$ is a mean-zero Gaussian random variable with a standard deviation of $\sigma_v$. Thus, the likelihood of observing a constraint $\boldsymbol{o}_k$ given the $n^{th}$ flow model, is simply $p_n(\boldsymbol{o}_k \,|\, \boldsymbol{c}_n) = G(e(\boldsymbol{x}_k;\, \boldsymbol{c}_n);\, \sigma_v)$ where $G(e;\, \sigma)$ denotes a mean-zero Gaussian with standard deviation $\sigma$ evaluated at $e$.

Finally, given the mixing probabilities and likelihood functions, the mixture model expresses the probability of a gradient measurement $\boldsymbol{o}_k$, as

$$p(\boldsymbol{o}_k \,|\, \boldsymbol{m},\, \boldsymbol{c}_1,\, \ldots,\, \boldsymbol{c}_N) \;=\; \sum_{n=1}^{N} m_n\, p_n(\boldsymbol{o}_k \,|\, \boldsymbol{c}_n)\,.$$

The probability of observing $\boldsymbol{o}_k$ is a weighted sum of the probabilities of observing $\boldsymbol{o}_k$ from each of the individual motions. The joint likelihood of a collection of $K$ independent observations $\{\boldsymbol{o}_k\}_{k=1}^{K}$ is the product of the individual probabilities:

$$L(\boldsymbol{m}, \boldsymbol{c}_1, ..., \boldsymbol{c}_N) = \prod_{k=1}^{K} p(\boldsymbol{o}_k \,|\, \boldsymbol{m}, \boldsymbol{c}_1, ..., \boldsymbol{c}_N)\,. \qquad (15.33)$$

Our goal is to find the mixture model parameters (the mixture proportions and the motion model parameters) that maximize the likelihood (15.33). Alternatively, it is often convenient to maximize the log likelihood:

$$\log L(\boldsymbol{m}, \boldsymbol{c}_1, ..., \boldsymbol{c}_N) = \sum_{k=1}^{K} \log\left(\sum_{n=1}^{N} m_n\, p_n(\boldsymbol{o}_k \,|\, \boldsymbol{c}_n)\right)\,.$$

## *EM and Ownerships*

The EM algorithm is a general technique for maximum likelihood or MAP parameter estimation [257]. The approach is often explained in terms of a parametric model, some observed data, and some unobserved data. Our observed data are the gradient constraints. The model parameters are the motion parameters and mixing probabilities, and the unobserved data are the assignments of gradient measurements to motion models. Note that if we knew which measurements were associated with which motion, then we could solve for each motion independently from their respective constraints.

Roughly speaking, the EM algorithm is an iterative algorithm that iterates two steps that compute 1) the expected values of the unobserved data given the most recent estimate of the model parameters (the E Step), and then 2) the ML/MAP estimate for the model parameters given the observed data, and the expected values for the unobserved data.

A key quantity in this algorithm is called the *ownership probability*. An ownership probability, denoted $q_n(\boldsymbol{x}_k)$, is the probability that the $n^{th}$ motion model is responsible for the constraint (i.e., generated the observed data) at pixel $\boldsymbol{x}_k$. This is an important quantity as it effectively segments the region, telling us which pixels belong to which motions. Using Bayes' rule, the probability that $\boldsymbol{o}_k$ is owned by model $\mathcal{M}_n$ can be expressed as

$$p(\mathcal{M}_n \,|\, \boldsymbol{o}_k) = \frac{p(\boldsymbol{o}_k \,|\, \mathcal{M}_n)\, p(\mathcal{M}_n)}{p(\boldsymbol{o}_k)}\,.$$

In terms of the mixture model notation here, this becomes

$$q_n(\boldsymbol{x}_k) = \frac{m_n\, p_n(\boldsymbol{o}_k \,|\, \boldsymbol{c}_n)}{\sum_{n=1}^{N} m_n\, p_n(\boldsymbol{o}_k \,|\, \boldsymbol{c}_n)}\,. \qquad (15.34)$$

That is, the likelihood of the observation given the $n^{th}$ model is simply $p_n(\boldsymbol{o}_k \,|\, \boldsymbol{c}_n)$, and the probability of the $n^{th}$ model is just $m_n$. The denominator is the marginalization of the joint distribution $p(\boldsymbol{o}_k, \boldsymbol{c}_n)$ over the space of models. And of course it is easy to show that $\sum_n q_n(\boldsymbol{x}_k) = 1$. In the context of the EM algorithm these ownership probabilities can be viewed as soft assignments of data to models. Once these assignments are made we can perform a weighted regression to find the motion parameters of each model, using the same tools developed above for a single motion.

Given ownership probabilities, the updated mixing probability for model $\mathcal{M}_n$ is just the fraction of the total available ownership probability assigned to the $n^{th}$ model, $m_n = \frac{1}{K}\sum_{k=1}^{K} q_n(\boldsymbol{x}_k)$. The estimation of the motion model parameters is similarly straightforward. That is, given the ownership probabilities, we estimate the motion parameters for each model independently as a weighted area-based regression problem. For the case of a translational motion model, where the motion parameters are just $\boldsymbol{c}_n \equiv \boldsymbol{u}_n$, this is just the minimization of the weighted least-squares error

$$E(\boldsymbol{u}_n) = \sum_{k=1}^{K} q_n(\boldsymbol{x}_k)\,[\nabla I(\boldsymbol{x}_k,t)\cdot \boldsymbol{u}_n + I_t(\boldsymbol{x}_k,t)]^2 \,. \tag{15.35}$$

Because the mixture model likelihood function (15.33) will have multiple local minima, a starting point for the EM iterations is required. That is, to begin the iterative procedure one needs an initial guess of either the ownership probabilities, or of the model parameters (motion and mixture parameters). Often one starts by choosing random values for the initial ownership probabilities and then begin with the estimation of the mixing probabilities and the motion model parameters.

### *Outliers*

As above, we must expect outliers among the gradient constraint observations. Gradient measurements near an occlusion boundary, for example, may not be consistent with either of the two motions. As a result, it is often extremely useful to introduce an outlier model, $\mathcal{M}_0$, in addition to the motion models; the likelihood for this outlier layer may be modeled with a uniform density [435]. Figure 15.7 shows results for the region near the can with two motion models and an outlier model like that described here. For the region shown in Fig. 15.7, the measurement constraints owned by the outlier model are shown in the bottom-right plot.

## 15.10 Conclusions

This chapter surveys several approaches to optical flow estimation. It is therefore natural to ask what works best? While historically some techniques have been shown to outperform others [59], in recent years several different approaches have produced excellent results on benchmark data sets, provided one pays attention to detail. Some of the important details include (1) multiple scales to help avoid local minima, (2) iterative warping and estimate refinement, and (3) robust cost functions to handle outliers. Accordingly, many techniques work well up to the limits of the key assumptions, namely, brightness constancy and smoothness.

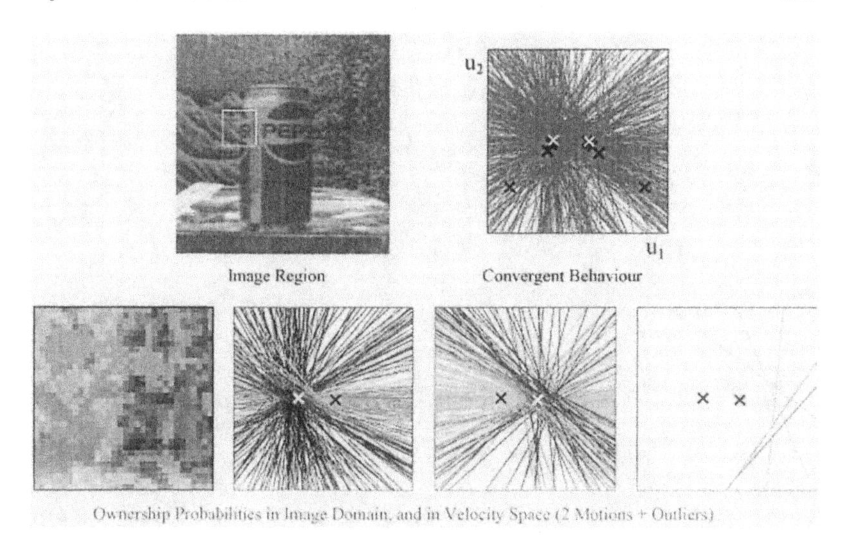


Figure 15.7. The top figures show a region at a depth discontinuity, and some of the constraint lines from pixels within that region. The black crosses in the upper-right show a sequence of estimates at EM iterations. White crosses depict the final the estimates. The bottom figures showing ownership probabilities. The bottom-left shows ownership probabilities at each pixel (based on the motion constraint at that pixel). The next two plots shown the velocity constraints where intensity depicts ownership (black denotes high ownership probability). The bottom-right plot shows constraint lines owned by the outlier model. (After [435])

---

Future research is needed to move beyond brightness constancy and smoothness. Detecting and tracking occlusion boundaries should greatly improve optical flow estimation. Similarly, prior knowledge concerning the expected form of brightness variations (e.g., given knowledge of scene geometry, lighting, or reflectance) can dramatically improve optical flow estimation. Brightness constancy is especially problematic over long image sequences where one must expect the appearance of image patches to change significantly. One promising area for future research is the joint estimation appearance and motion, with suitable dynamics for both quantities.

# Chapter16

## From Bayes to PDEs in Image Warping

**M. Nielsen and B. Markussen**

Abstract

In many disciplines of computer vision, such as stereo vision, flow computation, medical image registration, the essential computational problem is the geometrical alignment of images. In this chapter we describe how such an alignment may be obtained as statistical optimal through solving a partial differential equation (PDE) in the matching function. We treat different choices of matching criteria such as minimal square difference, maximal correlation, maximal mutual information, and several smoothness criteria. All are treated from a Bayes point of view leading to a functional minimization problem solved through an Euler-Lagrange formulation as the solution to a PDE. We try in this chapter to collect the most used methodologies and draw conclusions on their properties and similarities.

### 16.1 Motivation and problem statement

In many disciples in computer vision, the essential ill-posed problem is that of matching two images. The same problem has been given different names dependent on the context: matching, correspondence, flow, registration, warping, etc. No matter the name, the problem is to establish pairs $(x_1, x_2)$ of points so that $I_1(x_1) \approx I_2(x_2)$. Here the notation $\approx$ has been used for "corresponds to". At this point, we will denote the correspondence with a function $W$, disregarding problems in relation to whether it has unique values or whether it is defined in every point:

$$x_2 = W(x_1) : \Omega_1 \mapsto \Omega_2.$$

Using projection images $I_1, I_2$ as in stereo vision [653], optic flow computation [409], and x-ray imaging, a unique correspondence will not in general exist.

This is a serious problem of also practical importance since occlusion often appears in real world examples. However, this is due to the projection and not the fundamental problem of establishing a geometrical alignment. In this chapter, we will not deal with the problems originating from the projection (such as occlusions) leading to multi-valued or locally undefined functions $W$. We will limit ourselves to mappings originating from physical actions such as deformations, articulations, or viscous flows of objects. That is, taking the domains $\Omega_1$ and $\Omega_2$ as the $N$-dimensional Euclidean space $\mathbb{R}^N$,

$$x_2 = W(x_1), \quad x_1 \in \mathbb{R}^N, x_2 \in \mathbb{R}^N, W \in \mathcal{H},$$

where $\mathcal{H}$ is a set of admissible warps. We will later define the set of admissible warps as the set of *diffeomorphisms* $\mathcal{D}$: differentiable mappings where the inverse exists and also is differentiable. This setup is valid for all physical, non-projected, imaging situations of evolving objects. It also applies to the situation of mapping images of different objects of identical topology.

## 16.2 Admissible warps

The set $\mathcal{H}$ of admissible warps must from a physics/engineering point of view fulfill some basic requirements. Nevertheless, often for mathematical convenience and guaranteing unique solutions, these properties may be violated. In the following we list the required properties, the most popular choices of sets of admissible warps, and shortly list their pros and cons.

The required properties of admissible warps are:

**Realizability** A smooth development over time must be able to produce the warp. Otherwise an underlying physical process would not be realizable.

**Preservation** An admissible warp must be defined for all points in the source image. Otherwise points would disappear and have no image.

**Smoothness** The admissible warp must be continuous and differentiable. Otherwise non-physical situations like breaking objects under zero-viscosity would arise.

**Composition** The warp of a warp must be an admissible warp. Otherwise, warped images would not be images themselves, since they could not necessarily be warped again.

**Invertibility** The inverse of an admissible warp must also be an admissible warp. Otherwise, we could match $A$ to $B$, but not necessarily $B$ to $A$. Hence, all the above properties must also be fulfilled for the inverse warp.

Examples of these are shown in Figure 16.1.

The most general class of warps fulfilling all the above criteria is the set of positive *diffeomorphisms* $\mathcal{D}_+$: non-reflective differentiable mappings where the inverse exists and also is differentiable. They form a group so that the difference

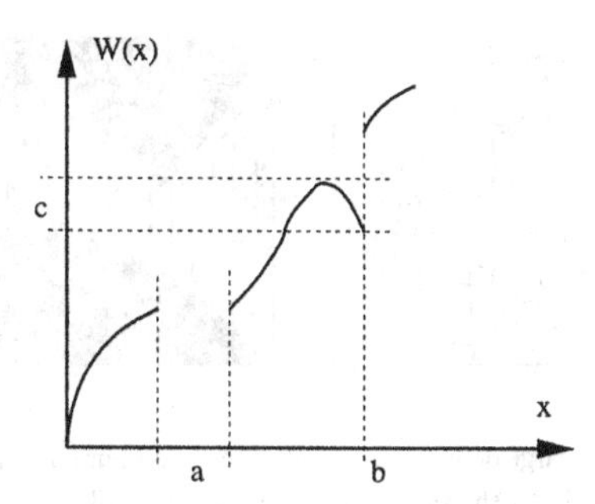


Figure 16.1. Example of a one dimensional matching function $W(x) : \mathbb{R} \mapsto \mathbb{R}$. At $a$ the preservation is violated, a $b$ the smoothness, and at $c$ the invertibility.

of any two warps exists: $W = W_2 \circ W_1^{-1}$ takes $W_1$ into $W_2$. A non-reflective diffeomorphism $W = \{w_n\}_{n \leq N}$ has positive Jacoby determinant with the Jacobean given by $J(W) = \left\{\partial w_n / \partial x_m\right\}_{n,m \leq N}$. In 2D this reads

$$W = \begin{pmatrix} u(x,y) \\ v(x,y) \end{pmatrix}, \quad \text{where } \forall x, y \in \mathbb{R}^2: \ \det(J(W)) = u_x v_y - u_y v_x > 0.$$

The group of diffeomorphisms is not always very easy to handle. Hence, other sets of admissible warps often are employed in practical algorithms. From a mathematical point of view, the two most "nasty" properties of diffeomorphisms are:
– It is not possible to decide whether a mapping is diffeomorphic from the individual coordinate functions.
– One cannot span all diffeomorphisms in a neighborhood by exponentiating the local affine connection. That is, one cannot view the group as a manifold.
The first nasty property makes it algorithmically necessary to treat all coordinate functions at the same time. One cannot decouple their computation and prove properties in simple seperable form. The second nasty property makes it complicated (though not impossible) to define sensible norms on diffeomorphisms since "geodesics will not be straight", to put it simply. Norms are often a very necessary ingredient in identifying warps, as one wishes to find the "smallest" warp fulling some data constraints.

The two nasty properties of diffeomorphism have lead to alternative choices of admissible sets of warps. The most popular choice has been (here stated in 2D):

$$W = \begin{pmatrix} u(x,y) \\ v(x,y) \end{pmatrix}, \quad \text{where } u, v \in S_n^2.$$

Here, $S_n^2$ is a Sobolev space of order $n$ (all derivatives up to order $n$ exist and are square integrable). We will denote this space of admissible warps $\mathcal{S}_n$. The popularity of $\mathcal{S}_n$ mainly comes from the fact that the Sobolev spaces are normed regular spaces, and the norm on $\mathcal{S}_n$ may simply be computed as the sum of the norm of the coordinate functions $(u, v)$. Using $\mathcal{S}_n^2$ (employing the $L_2$-norm) fur-

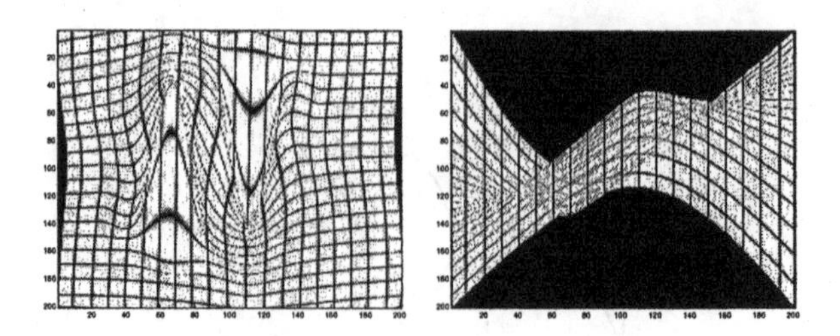

Figure 16.2. Two images of large deformations: Left is the maximum likelihood diffeomorphic warp, right is a Sobolev warp. The latter shows a folding and is not invertible.

thermore makes it possible to prove uniqueness and existence of solutions: the theory from Tikhonov regularization [797] of ill-posed problems carries over.

The price to pay for this mathematical and algorithmic convenience is that some of the above properties are violated. The definition on the full domain, continuity and differentiability (up to order $n$) of the warp, and the smooth development are fulfilled. While, we cannot any longer guarantee an inverse warp and composition of warps.

For small deformations, there may not be huge difference, from a practical point of view, of the warps minimizing any Sobolev norm or one minimizing a norm on diffeomorphisms. Nevertheless, it is evident from theory, and also practical experiments, that for larger deformations, the Sobolev norm minimizing warps suffer also from practical deficiencies such as foldings making several points map to the same point.

Lately, also the space of functions of bounded variation $BV$-space has been employed as warp functions. These also (in a zero-measure set) violate the continuity and differentiability criterion. They have mainly been employed to overcome problems due to a projective transformation (such as occlusion), and they will not be treated further here, but show nice practical properties in e.g. the optic flow setting [122].

## 16.3 Bayesian formulation of warp estimation

The problem of identifying a warp given a pair of images is ill-posed since a unique solution in general does not exist. Hence, an inference is necessary and a mere deduction is not viable. We employ statistical inference. The optimal warp may be identified on basis of the posterior:

$$p(W|I_1, I_2) = \frac{p(I_1, I_2|W)\, p(W)}{p(I_1, I_2)}.$$

Many different solution may be singled out from the posterior: the mean, generalized median, local modes, or global modes. We employ the MAP (Maximum A Posteriori) scheme being the minimum risk estimator when all wrong solutions are considered equally bad. That is, we wish to find the warp that maximizes the posterior.

We then ignore the *evidence* $p(I_1, I_2)$ in our quest for $W$, since it does not depend on $W$. Notice, if you wish to probe the posterior as a normalized probability density function computing e.g. mean and/or moments, normalization may easiest be obtained through explicit normalization by $\int p(W|I_1, I_2)\mathrm{d}W$. Our optimal estimate reads:

$$W^* = \arg\max_{W \in \mathcal{H}} p(W|I_1, I_2) = \arg\min_{W \in \mathcal{H}} \big( -\log p(I_1, I_2|W) - \log p(W)\big).$$

In which the first term is denoted the *log-likelihood* or matching or data term and the second term is denoted the *log-prior* or smoothness or regularization term.

Normally, we will assume a Markov property of both the likelihood and the prior so that the marginal distribution in a point only depends on a spatial neighborhood $\mathcal{N}$ of this point, and so that the total distribution may be written as independent marginal distributions. Hence,

$$W^* = \arg\min_{W \in \mathcal{H}} \int \Big( P_d\big(W(x' \in \mathcal{N}(x))\big) + P_s\big(W(x' \in \mathcal{N}(x))\big)\Big)\, \mathrm{d}x.$$

In the discrete setting, this translation using the Markov property is straightforward. In the continuous setting however, this is not the case since the distribution is a product integral and the space of warps on the finite neighborhood is still infinite dimensional. Without treating potential problems arising from this, we will simply assume that in the continuous domain we may write

$$P\big(W(x' \in \mathcal{N}(x))\big) \approx F\big(W(x), DW(x)\big),$$

where $D$ is an appropriate finite list of differential operators. In the discrete setting this is straightforward using appropriate finite difference operators.

In this way the functional $E[W] = E_d[W] + E_s[W] = -\log p(W|I_1, I_2)$ reads

$$E[W] = \int F_d\big(W(x), DW(x)\big)\mathrm{d}x + \int F_s\big(W(x), DW(x)\big)\, \mathrm{d}x.$$

As a conclusion, we have now cast the MAP estimation problem into a functional minimization problem that may be solved by a gradient descend algorithm:

$$\begin{aligned} W(x, 0) &= W_0(x), \\ \frac{\partial W(x, t)}{\partial t} &= -\frac{\delta E}{\delta W}, \end{aligned} \tag{16.1}$$

where $W_0$ is some initial guess, and typically the identity $W_0(x) = x$.

In the following we will examine different likelihood and prior terms, and to the degree the result is known to us, comment on problems regarding existence and uniqueness of the solution. In some special simple cases direct solution of the Euler Lagrange equation $\delta E/\delta W = 0$ may even be possible.

## 16.4 Likelihood: Matching criteria

In this section we go through some of the most commonly adopted matching criteria, their statistical justifications, and when appropriate their variational formulation to plug into the gradient descend PDE (Eq.16.1).

**Landmarks.** We will start with the simplest possible scenario where a number of landmarks $x_i$ and their match $y_i$ are known a priori:

$$p(I_1, I_2|W) = \prod_i \delta(y_i - W(x_i)).$$

This will normally be solved as a constraint minimization problem:

$$W^* = \underset{W \in \mathcal{H}: y_i = W(x_i)}{\arg\min} E_s[W]$$

either using Lagrange multipliers or a gradient projection algorithm [681]. For simple smoothness terms this may even be solved analytically [102]. This is the case for $\mathcal{H} = \mathcal{S}$ (see below).

Given imprecise landmark matches the likelihood reads:

$$p(I_1, I_2|W) = \prod_i G_{\sigma_i}(y_i - W(x_i)),$$

here using additive Gaussian noise of standard deviations $\sigma_i$. In this case the minus log-likelihood is quadratic in $W$:

$$E_d[W] = \sum_i \frac{(y_i - W(x_i))^2}{2\sigma_i}, \qquad \frac{\delta E_d}{\delta W} = -\sum_i \frac{y_i - W(x_i)}{\sigma_i}.$$

Like for the precise landmark matches this may be solved analytically for simple smoothness terms.

The situation where the landmarks are not matched up in pairs so that $y_{P(i)} - W(x_i)$ is the quantity for minimization and the minimization must also be performed over the permutation function $j = P(i)$ is not tractable to solve by gradient descend approaches, but must be combined with algorithms for solving the "optimal marriage problem" [349].

**Image noise models.** A more interesting class of likelihood terms does not require an identification of landmarks, but operate directly on the image functions. The simplest here is to assume a model of i.i.d. Gaussian additive noise: $I_2(x) = I_1(W(x)) + \eta(x)$. This leads to a data term

$$\begin{aligned} E_d[W] &= \frac{1}{2\sigma^2} \int \left(I_2(x) - I_1(W(x))\right)^2 \mathrm{d}x \\ &\Rightarrow \\ \frac{\delta E_d}{\delta W} &= -\int \frac{I_2(x) - I_1(W(x))}{\sigma^2} \nabla I_1(W(x)) \, \mathrm{d}x. \end{aligned}$$

Of course other i.i.d. noise models than Gaussian can be assumed leading to the more general form

$$\begin{aligned} E_d[W] &= \int \phi(I_2(x) - I_1(W(x)))\,\mathrm{d}x \\ &\Rightarrow \\ \frac{\delta E_d}{\delta W} &= \int \phi'(I_2(x) - I_1(W(x)))\nabla I_1(W(x))\,\mathrm{d}x. \end{aligned}$$

Furthermore the noise model may be generalized to stationary noise process, e.g. the Brownian noise model, where the local increments are i.i.d. Gaussian:

$$\begin{aligned} E_d[W] &= \frac{1}{2\sigma^2}\int |\nabla(I_2(x) - I_1(W(x)))|^2\,\mathrm{d}x \\ &\Rightarrow \\ \frac{\delta E_d}{\delta W} &= \int \frac{(\nabla\nabla^T)(I_2(x) - I_1(W(x)))}{\sigma^2}\nabla I_1(W(x))\,\mathrm{d}x. \end{aligned}$$

This term has, to our best knowledge, never been employed in practice, but would potentially allow for drift in intensities like in MR images with a smooth bias fields uncorrelated between the two images.

**Intensity transforms.** Another model directly taking into account that image intensities are not the result of the same formation process but may arise from different modalities is the correlation ratio:

$$E_d[W] = \int \frac{I_1(W(x))I_2(x)}{\sqrt{I_1(W(x))^2}\sqrt{I_2(x)^2}}\,\mathrm{d}x.$$

The probabilistic interpretation on this is very similar to the additive Gaussian noise process above, but with the additional degree of freedom that images may be multiplied with a free parameter so that $(aI_1 - bI_2)$ is the term for minimization. Here appropriate constraint on $a, b$ must be added, so that $ab = 1$, to avoid the trivial solution $a = b = 0$. Formal derivation of this leads to the correlation ratio being proportional to the log-likelihood. The variation of this with respect to the warp is computed by Hermosillo et al. [397].

The final data term we will present is the *mutual information* criterion [652]. This is not simplest derived from the Bayes point of view, but from the *Minimum Description Length* principle [676]. In general, Bayes and MDL inference has been shown to be identical and merely a reformulation of each other [430]. Some problems may easier be formulated in the Bayes language, while others are most easily formulated using the MDL language. Mutual information belongs to the latter category. The basic idea here is to find the warp so that $I_2$ may be communicated in the shortest message possible to a person already knowing $I_1$. Using Shannon's formalism of $\mathcal{I}(x) = -\log p(x)$, where $\mathcal{I}$ is the information shows that this does not change anything with respect to handling the prior term. The total object of minimization is the sum of the code length of the likelihood and the prior: $\mathcal{I} = \mathcal{I}_{\text{data}} + \mathcal{I}_{\text{prior}}$. The code length of $I_2$ knowing $I_1(W(x))$ may be written as a (neglectable) constant minus the *mutual information* of $I_1$ and $I_2$ so that

$$E_d[W] = -\int p_W(i_1, i_2) \log \frac{p_W(i_1, i_2)}{p_W(i_1)p(i_2)} \, \mathrm{d}i_1 \, \mathrm{d}i_2,$$

where $i_1, i_2$ are the intensities in image $I_1$ and $I_2$ respectively and $p(i)$ is the distribution of intensities in the image. $p(i_1, i_2)$ is the joint distribution of intensities in *corresponding positions*. Hence, it depends on the warping of $I_1$. In this case we find [397]:

$$\frac{\delta E_d}{\delta W} = \int \left( \frac{1}{p_W(i_1, i_2)} \frac{\partial p_W(i_1, i_2)}{\partial I_1} - \frac{1}{p(i_1)} \frac{\partial p_W(i_1)}{\partial I_1} \right) \nabla I_1(W(x)) \, \mathrm{d}x.$$

This concludes our journey through likelihood terms. A final comment is that all these are of course constructed *models* relying on the image formation process. In any situation, the choice of model comprises a compromise between the modeling capabilities and the model complexity. The models presented here are, to our opinion, gradually more complex allowing for more loose definitions of "correspondence".

## 16.5 Prior: Smoothness criteria

All the likelihood terms above are not sufficient to ensure a regular warp. The prior term ensures this, when regular warps are more probable than irregular warps. In this section we look at a number of priors and the properties they induce on the warps. In order to make a prior on warps ensuring the above mentioned properties, rather complex constructions must be used. Before going into such constructions, we handle the simplest and most popular cases. Some of these also have the nice property that they ensure the existence of solutions.

**Sobolev norms.** The most simple construction is a Brownian motion model for the coordinate functions so that

$$\begin{aligned} E_s(W) &= \int \frac{\|J(W(x))\|_2^2}{2\sigma^2} \, \mathrm{d}x \\ &\Rightarrow \\ \frac{\delta E_s}{\delta W} &= -\int \frac{\Delta W(x)}{\sigma^2} \, \mathrm{d}x, \end{aligned}$$

where $J$ is the Jacobean and $\Delta$ denotes the coordinate-wise Laplacean. This Brownian motion assumptions hence leads to a first order Sobolev norm $S_1^2$ inducing a gradient descend which is the heat equation in the coordinate functions, and thereby fulfill the realizability, preservation, and smoothness, but not necessarily the composition and invertibility criteria. We may denote this *warp diffusion*. It has been employed for simplifying 3D shape correspondences in the *geometry-constrained diffusion* [26], and in the optic flow setting in the original work by Horn and Schunck [409].

Next step of complication is to use the local second order structure for a Gaussian model to construct the prior. In this case

$$\begin{aligned} E_s(W) &= \int \frac{\|\mathrm{Tr}(H^2(W(x)))\|^2}{2\sigma^2}\, \mathrm{d}x \\ &\Rightarrow \\ \frac{\delta E_s}{\delta W} &= \int \frac{\Delta\Delta W(x)}{\sigma^2}\, \mathrm{d}x, \end{aligned}$$

where $H$ is the component-wise Hessian and Tr is the component-wise trace. In 2D with $W = (u, v)$, this reads

$$F_s(W) = u_{xx}^2 + 2u_{xx}u_{yy} + u_{yy}^2 + v_{xx}^2 + 2v_{xx}v_{yy} + v_{yy}^2.$$

This is also denoted the Thin-Plate energy as $E_s$ compares to the bending energy of a thin plate [102]. This is simply the second order Sobolev norm $S_2^2$. Since it does not depend on the first order structure, it is invariant to affine transformations of any of the source or destination images [348]. Furthermore, it remains the nice existence properties of the Sobolev norm regularizers, but still does not in general lead to invertible warps.

For the Sobolev norms it is in general possible to solve the Euler-Lagrange equation analytically by eigenfunction expansions. Especially in the exact landmark matching scenario this leads to very simple schemes just involving the inversion of an $NM \times NM$ matrix where $M$ is the number of landmarks and $N$ the dimensionality of the warp domain [102].

It has also been proved for more complex likelihood terms such as image difference, correlation ratio, and mutual information that the solution exists [397].

**Diffeomorphic warps.** We could now continue to Sobolev norms of even higher order. However, they have never proved their value in practice, and may mainly be seen as an exercise in symbol manipulation. Instead we turn to the construction of a sensible prior on the space of diffeomorphisms. This is a little more complex than the above quadratic separable energies that lead to simple separable linear PDEs in the warp. On the other hand, the motivation is to obtain provable invertible warps, that may even be constructed source-destination symmetric [600].

Above, we used Brownian motion models in the coordinate function to construct a prior. In the warp setting we may make another Brownian motion construction more natural for warps [601]. Assume a warp $W$ is constructed as the composition of $H$ small warps $W_h = W_h^{(H)}$:

$$W = W_H \circ \cdots \circ W_2 \circ W_1 = \prod_{h \leq H} \circ W_h.$$

Furthermore, assume that these warps are statistically independent. In this case such a sequence of warps may be considered a Brownian motion in the space of diffeomorphisms. If we only look at the first order structure $J(W(x_h))$ in a point $x_h$ following along the warps so that $x_h = W_h(x_{h-1})$ we find by chain rule

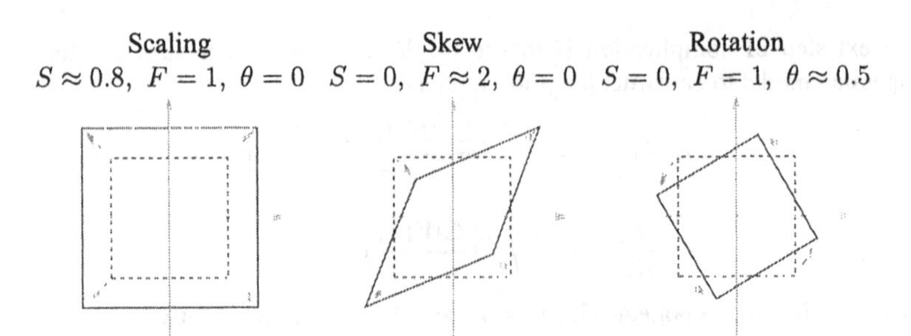


Figure 16.3. The independent action of the parameters on a unit square.

differentiation:

$$J(W(x_0)) = \prod_{h \le H} J(W_h(x_{h-1}))$$

Assuming statistical independence and finite variance of the individual entries in the Jacobians lead to a unique solution for the distribution $p_{A,b}(J)$ only dependent on the infinitesimal mean and variance

$$b = \lim_{H \to \infty} H \operatorname{mean}\big(J(W_h^{(H)}) - I\big), \qquad A = \lim_{H \to \infty} H \operatorname{var}\big(J(W_h^{(H)})\big).$$

The exact probability density function is a rather complicated entity and is to our knowledge only known explicitly in dimension $N = 2$, see [427]. In this case it for $b \equiv 0$ and $A = \sigma^2 I$ can be written as

$$p_{\sigma^2 I,0}(J) = G_\sigma(S)\, g(F, \theta),$$

where $G_\sigma$ is the Gaussian distribution with standard deviation $\sigma$ and $S, F, \theta$ are the local scaling, skew, and rotation respectively:

$$\text{Scaling} \qquad S = \log(\det(J(W)))$$

$$\text{Skewness} \qquad F = \frac{1}{2\det(J(W))} \|J(W)\|_2^2$$

$$\text{Rotation} \qquad \theta = \arctan\left(\frac{j_{12} - j_{21}}{j_{11} + j_{22}}\right)$$

The distribution may be approximated very well by the following expression independent in the parameters [601]:

$$p_{\sigma^2 I,0}(J) \approx G_\sigma(S)\, G_{\sigma/\sqrt{2}}(\theta)\, e^{-F/\sigma}.$$

Furthermore assuming spatial independence of the Jacobeans of the warps leads to the following energy and its variation:

$$\begin{aligned} F_s(W) &= \frac{S^2 + 2\theta^2 + 2\sigma F}{2\sigma^2} \\ &\Rightarrow \\ \frac{\delta E_s}{\delta W} &= \frac{1}{\sigma^2}\frac{2\log D - 2\sigma F}{D}\frac{\delta D}{\delta W} + \frac{1}{D}\frac{\delta \|J\|_2^2}{\delta W} + 2\theta\frac{\delta\theta}{\delta W}, \end{aligned}$$

where $J$ is the Jacoby matrix of $W$ and $D = \det(J)$.

This smoothness term provably leads to source-destination symmetric (and also thereby also invertible) solutions [600]. It fulfills all the above mentioned properties, but is more complicated to handle than the regularizers based on Sobolev norms as they lead to linear terms in the PDEs. Furthermore, existence of the solution still remains to be proved on the continuous domain.

The solution above is not the only one fulfilling all the properties. Any smoothness term on the following form will ensure this:

$$F_s(W) = \lambda_1\, f(S^2) + \lambda_2\, g(F) + \lambda_3\, h(\theta^2)$$

where $f, g, h$ are differentiable monotonically increasing functions. Especially the cases $(\lambda_1, \lambda_2, \lambda_3) = (1, 0, 0)$ leads to nearly incompressible (area preserving) warps in the landmark matching cases, whereas $(\lambda_1, \lambda_2, \lambda_3) = (0, 1, 0)$ leads to nearly conformal (angle preserving) mappings.

## 16.6 Warp time and computing time

The realizability requirement on admissible warps implicitly assumes the existence of a physical mechanism producing the warp. Suppose this physical process evolves over the time interval $[0, T]$. At time zero we have the initial image $I_1$ and at time $T$ we have the final image $I_2$. Similarly, the warp at time zero is simply the identity mapping $W_0(x) = x$ and the warp $W_T$ at time $T$ is the warp between the initial and final image. At other points of time $t$ intermediate warps $W_t$ exist. We refer to the time variable $t$ as the warping time, which is to be interpreted as a physical time. For any two points of time $s, t$ the physical mechanism produces the warp $W_{s,t} = W_t \circ W_s^{-1}$ from time $s$ to time $t$. The collection of warps $W_{s,t}$ must satisfy the following flow properties

$$W_{s,s}(x) = x, \qquad\qquad W_{s,t} = W_{u,t} \circ W_{s,u}.$$

The first property states that there is no physical action at a single time point. The second property states that the evolution from time $s$ to time $t$ can be realized by composing the evolution from time $s$ to time $u$ with the evolution from time $u$ to time $t$. The Brownian motion in the space of diffeomorphisms discussed in Section 16.5 uses warping time. Here the warp $W_h^{(H)}$ can be considered as the warp from time $(h-1)T/H$ to time $hT/H$. The statistical assumptions leading to the Brownian prior then state that the flow of warps is temporal homogeneous and stochastically independent over disjoint time intervals.

The introduction of the warping time can be impracticable for implementations since it requires the estimation of the entire flow of warps $W_t$ instead of the final warp $W = W_T$. If only the final warp is needed, then it can be beneficial to neglect the warping time. In this formulation an energy functional $E_s(W)$ of the warp is defined, e.g. implicitly using the warping time. The corresponding MAP estimation is often performed using gradient descend, see (Eq.16.1), over some artificial time variable $t$. We refer to such a time variable as the computing time.

## 16.7 From fluid registration to diffeomorphic minimizers

A classical PDE approach to image warping still left undiscussed is fluid registration. Let $\mathscr{L}$ be some partial differential operator, e.g. the linear elasticity operator with Lamé constants $\mu$ and $\lambda$ from the Navier-Stokes equation

$$\mathscr{L}v = \mu\nabla^2 v + (\lambda + \mu)\nabla(\nabla \cdot v).$$

The velocity field $v(x)$ is given by the PDE $\mathscr{L}v = b$ with appropriate boundary conditions, see [196]. Here the driving force $b$ is given by the direction minimizing the matching criteria. The fluid registration is given by the evolution

$$\frac{\partial W_t}{\partial t} = v_t, \qquad \mathscr{L}v_t = -\frac{\delta E_d(W)}{\delta W}\Big|_{W=W_t}.$$

Over the infinitesimal time increment $\mathrm{d}t$ the warp is evolved by $W_{t+\mathrm{d}t} = (I + v_t \mathrm{d}t \circ W_t^{-1}) \circ W_t$. This composition of warps ensures the admissibility at every point of time. As time increases the matching criteria $E_d(W_t)$ is minimized. If we define the energy of the velocity field by $\int \|\mathscr{L}v\|^2 \,\mathrm{d}x$, then the total warping energy is given by the squared length of the gradient descend path

$$E = \int_0^\infty \int_{\mathbb{R}^N} \|\mathscr{L}v_t\|^2 \,\mathrm{d}x\,\mathrm{d}t = \int_0^\infty \int_{\mathbb{R}^N} \left\| \frac{\delta E_d(W)}{\delta W}\Big|_{W=W_t} \right\|^2 \mathrm{d}x\,\mathrm{d}t.$$

Albeit linked to physics via the Navier-Stokes equation the temporal variable $t$ in the fluid registration method is a computing time. This is due to the fact that the driving force at time $t$ depends on the image $I_2$ at the infinite future and hence is non-physical. To reformulate the fluid registration method in a Bayesian framework and interpret the temporal variable as warping time the energy should be decoupled from the matching criteria. For this we define the velocity field on its own right, now over a finite time horizon $v(t,x)\colon [0,T] \times \mathbb{R}^N \longrightarrow \mathbb{R}^N$. The warp $W(x) = W_T(x)$ is connected to the velocity field via the transport equation

$$W_t(x) = x + \int_0^t v(s, W_s(x))\,\mathrm{d}s$$

and the corresponding smoothness criteria is defined via the energy

$$E_s(W) = \int_0^T \int_{\mathbb{R}^N} \|\mathscr{L} v(t,x)\|^2 \, \mathrm{d}x \, \mathrm{d}t,$$

see [442]. This functional may be interpreted as an iterated Sobolev norm. If $v^*$ denotes the MAP solution to the velocity field, then the corresponding warp $W^* = W_T^*$ is given by

$$W_t^*(x) = x + \int_0^t v^*\left(s, W_s^*(x)\right) \mathrm{d}s.$$

The Bayesian formulation of the fluid registration method is closely linked to the Brownian prior described at the end of Section 16.5. The statistical assumptions leading to the Brownian prior imply, see [495], that there exist coefficient functions $b, f_k \colon \mathbb{R}^N \longrightarrow \mathbb{R}^N$ such that the flow of warps is the solution to the stochastic differential equation

$$W_t(x) = x + \int_0^t b\big(W_s(x)\big) \, \mathrm{d}s + \sum_{k=1}^{\infty} \int_0^t f_k\big(W_s(x)\big) \, \mathrm{d}B_k(s).$$

Here the $B_k(s)$'s are stochastically independent Brownian motions and the integral is the so-called Itô stochastic integral. Let the function $a(x,y)$ be defined by $a(x,y) = \sum_{k=1}^{\infty} f_k(x) f_k(y)^*$. If the functions $b(x) \equiv b$ and $a(x,x) \equiv A$ are constant, then the distribution of the Jacobean $J(W_T)$ exactly equals the distribution $p_{A,b}(J)$ discussed above. Instead of using the Jacobeans it is possible to do the Bayesian analysis directly on the Brownian motions $B_k$, see [545]. Moreover, if the covariance function $a(x,y)$ is the Greens function for the square of the partial differential operator $\mathscr{L}$, i.e. $\mathscr{L}^* \mathscr{L} a(x,y) = \delta_{x=y}$, then the MAP warp in the fluid registration formulation equals the MAP warp in the Brownian prior formulation. This was proved explicitly for the landmark matching problem in [545].

## 16.8 Discussion and open problems

Above we have presented the warp-estimation process in the Bayes framework leading to PDEs as gradient descend algorithms in the minus log posterior. For all likelihood or data terms existence of a solution to the PDE has been proved using a Sobolev norm regularizer on the warp.

We have presented such Sobolev norm regularizers. Among which the second order "thin-plate bending energy" is one of the most popular choices in biomedical image registration. We have argued that the Sobolev norm regularizers has a deficiency in fulfilling the required property of invertibility. As a remedy we have made the connection of iterated Sobolev norms to more complex regularizers based on norms on diffeomorphisms. Basically, this indicates that all existence results may carry over to these more theoretically satisfactory albeit more complex norms. Nevertheless, from a formal point of view, many open problems still exists:

- Are iterated Sobolev norms formally identical to the Brownian warps?
- Does the more complex diffeomorphic norms guarantee existence of solution to the Euler-Lagrange equation?
- Can efficient algorithms for the diffeomorphic norm minimizers be constructed?
- Can a more formal connection between fluid flows and Brownian warps be established?
- How does scale arise in the representations? Can we carry over knowledge from weakly turbulent flows?

These are open problems of more theoretical nature. The final and maybe most important challenge is to characterize the solution implied by the different PDEs in order to make it possible to make qualified decision of which methodology to apply in which practical setting.

# Chapter17

# Image Alignment and Stitching

**R. Szeliski**

Abstract

Stitching multiple images together to create beautiful high-resolution panoramas is one of the most popular consumer applications of image registration and blending. In this chapter, I review the motion models (geometric transformations) that underlie panoramic image stitching, discuss direct intensity-based and feature-based registration algorithms, and present global and local alignment techniques needed to establish high-accuracy correspondences between overlapping images. I then discuss various compositing options, including multi-band and gradient-domain blending, as well as techniques for removing blur and ghosted images. The resulting techniques can be used to create high-quality panoramas for static or interactive viewing.

## 17.1 Introduction

Algorithms for aligning images and stitching them into seamless photo-mosaics are among the oldest and most widely used in computer vision. Image stitching algorithms have been used for decades to create the high-resolution photo-mosaics used to produce digital maps and satellite photos [570]. Frame-rate image alignment is used in every camcorder that has an image stabilization feature. Image stitching algorithms come "out of the box" with today's digital cameras and can be used to create beautiful high-resolution panoramas.

In film photography, special cameras were developed at the turn of the century to take wide-angle panoramas, often by exposing the film through a vertical slit as the camera rotated on its axis [559]. In the mid-1990s, image alignment techniques started being applied to the construction of wide-angle seamless panoramas from regular hand-held cameras [543, 180, 776]. More recent work in this area has addressed the need to compute globally consistent alignments [781, 709, 739], the removal of "ghosts" due to parallax and object movement [570, 248, 739, 825, 7], and dealing with varying exposures [543, 825, 7]. (A collection of some of these papers can be found in [74].) These techniques have spawned a large number of commercial stitching products [180, 710].

While most of the above techniques work by directly minimizing pixel-to-pixel dissimilarities, a different class of algorithms works by extracting a sparse set of *features* and then matching these to each other [148, 121]. Feature-based approaches have the advantage of being more robust against scene movement, and are potentially faster. Their biggest advantage, however, is the ability to "recognize panoramas", i.e., to automatically discover the adjacency (overlap) relationships among an unordered set of images, which makes them ideally suited for fully automated stitching of panoramas taken by casual users [121].

What, then, are the fundamental algorithms needed for image stitching? First, we must determine the appropriate *motion model* relating pixel coordinates in one image to pixel coordinates in another (Section 17.2). Next, we must somehow estimate the correct alignments relating various pairs of images, using either *direct* pixel-to-pixel comparisons combined with gradient descent or *feature-based* alignment techniques (Section 17.3). We must also develop algorithm to compute globally consistent alignments from large collections of overlapping photos (Section 17.4). Once the alignments have been estimated, we must choose a final compositing surface onto which to warp and place all of the aligned images (Section 17.5). We also need to seamlessly blend overlapping images, even in the presence of parallax, lens distortion, scene motion, and exposure differences (Section 17.6). In the last section of this chapter, I discuss additional applications of image stitching and open research problems. For a more detailed tutorial on all of these components, please consult [778].

## 17.2 Motion models

Before we can stitch images to create panoramas, we need to establish the mathematical relationships that map pixel coordinates from one image to another. A variety of such *parametric motion models* are possible, from simple 2D transforms, to planar perspective models, 3D camera rotations, and non-planar (e.g., cylindrical) surfaces [776, 781].

Figure 17.1 shows a number of commonly used 2D planar transformations, while Table 17.1 lists their mathematical form along with their intrinsic dimensionality. The easiest way to think of these is as a set of (potentially restricted) $3 \times 3$ matrices operating on 2D homogeneous coordinate vectors, $\boldsymbol{x}' = (x', y', 1)$ and $\boldsymbol{x} = (x, y, 1)$, s.t.

$$\boldsymbol{x}' \sim \boldsymbol{H}\boldsymbol{x}, \tag{17.1}$$

where $\sim$ denotes equality up to scale and $\boldsymbol{H}$ is one of the $3 \times 3$ matrices given in Table 17.1.

*2D translations* are useful for tracking small patches in videos and for compensating for instantaneous camera jitter. This simple two-parameter model is the

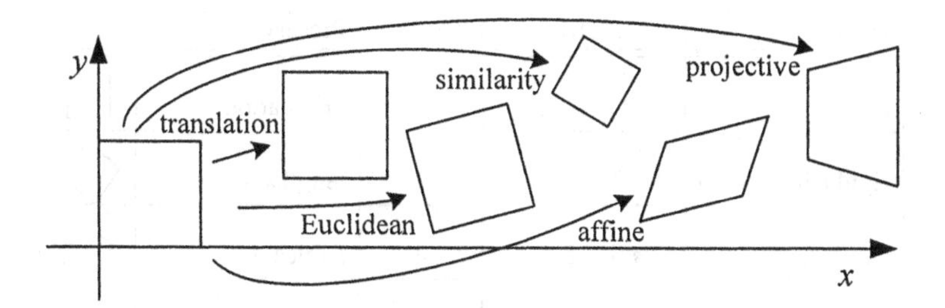


Figure 17.1. Basic set of 2D planar transformations

one most commonly associated with Lucas and Kanade's patch tracker [533], although, in fact, their paper also describes how to use an affine motion model.

The three-parameter rotation+translation (also known as *2D rigid body motion* or the *2D Euclidean transformation*) is useful for modeling in-plane rotations, for example when different portions of a larger image are scanned on a flatbed scanner.

Scaled rotation, also known as the *similarity transform*, adds a fourth isotropic scale parameter $s$. This is a good model for a slowly panning and zooming camera, especially when the camera has a long focal length. The similarity transform preserves angles between lines.

The six parameter affine transform uses a general $2 \times 3$ matrix (or equivalently, a $3 \times 3$ matrix where the bottom row is $\begin{bmatrix} 0 & 0 & 1 \end{bmatrix}$). It is a good model of local deformations induced by more complex transforms, and also models the 3D surface foreshortening observed by an orthographic camera. Affine transforms preserve parallelism between lines.

The most general planar 2D transform is the eight-parameter *perspective transform* or *homography* denoted by a general $3 \times 3$ matrix $\boldsymbol{H}$. The result of multiplying $\boldsymbol{H}\boldsymbol{x}$ must be normalized in order to obtain an inhomogeneous result, i.e.,

$$x' = \frac{h_{00}x + h_{01}y + h_{02}}{h_{20}x + h_{21}y + h_{22}} \quad \text{and} \quad y' = \frac{h_{10}x + h_{11}y + h_{12}}{h_{20}x + h_{21}y + h_{22}}. \tag{17.2}$$

Perspective transformations preserve straight lines, and, as we will see shortly, are an appropriate model for planes observed under general 3D motion and 3D scenes observed under pure camera rotation.

In 3D, the process of *central projection* maps 3D coordinates $\boldsymbol{x} = (x, y, z)$ to 2D coordinates $\boldsymbol{x}' = (x', y', 1)$ through a *pinhole* at the camera origin onto a 2D projection plane a distance $f$ along the $z$ axis,

$$x' = f\frac{x}{z}, \quad y' = f\frac{y}{z}. \tag{17.3}$$

Perspective projection can also be denoted using an upper-triangular $3 \times 3$ *intrinsic calibration matrix* $\boldsymbol{K}$ that can account for non-square pixels, skew, and a variable optic center location. However, in practice, the simple focal length scaling used above provides high-quality results when stitching images from regular cameras.

| Name | Matrix | # D.O.F. | Preserves: | Icon |
|---|---|---|---|---|
| translation | $\begin{bmatrix} I \mid t \end{bmatrix}_{2\times 3}$ | 2 | orientation $+\cdots$ | |
| rigid (Euclidean) | $\begin{bmatrix} R \mid t \end{bmatrix}_{2\times 3}$ | 3 | lengths $+\cdots$ | |
| similarity | $\begin{bmatrix} sR \mid t \end{bmatrix}_{2\times 3}$ | 4 | angles $+\cdots$ | |
| affine | $\begin{bmatrix} A \end{bmatrix}_{2\times 3}$ | 6 | parallelism $+\cdots$ | |
| projective | $\begin{bmatrix} H \end{bmatrix}_{3\times 3}$ | 8 | straight lines | |

Table 17.1. Hierarchy of 2D coordinate transformations. The $2 \times 3$ matrices are extended with a third $[\mathbf{0}^T\ 1]$ row to form a full $3 \times 3$ matrix for homogeneous coordinate transformations.

What happens when we take two images of a 3D scene from different camera positions and/or orientations? A 3D point $\boldsymbol{p} = (X, Y, Z, 1)$ gets mapped to an image coordinate $\boldsymbol{x}_0'$ through the combination of a 3D rigid-body (Euclidean) motion $\boldsymbol{E}_0$ and a perspective projection $\boldsymbol{K}_0$,

$$\boldsymbol{x}_0 \sim \boldsymbol{K}_0 \boldsymbol{E}_0 \boldsymbol{p} = \boldsymbol{P}_0 \boldsymbol{p}, \tag{17.4}$$

where the $3 \times 4$ matrix $\boldsymbol{P}_0$ is often called the *camera matrix*. If we have a 2D point $\boldsymbol{x}_0$, we can only project it back into a 3D *ray* in space. However, for a *planar scene*, we have one additional *plane equation*, $\hat{\boldsymbol{n}}_0 \cdot \boldsymbol{p} + d_0 = 0$, which we can use to augment $\boldsymbol{P}_0$ to obtain $\tilde{\boldsymbol{P}}_0$, which then allows us to invert the 3D→2D projection. If we then project this point into another image, we obtain

$$\boldsymbol{x}_1 \sim \boldsymbol{P}_1 \tilde{\boldsymbol{P}}_0^{-1} \boldsymbol{x}_0 = \boldsymbol{H}_{10} \boldsymbol{x}_0, \tag{17.5}$$

where $\boldsymbol{H}_{10}$ is a general $3 \times 3$ homography matrix and $\boldsymbol{x}_1$ and $\boldsymbol{x}_0$ are 2D homogeneous coordinates. This justifies the use of the 8-parameter homography as a general alignment model for mosaics of planar scenes [543, 776].

The more interesting case is when the camera undergoes pure rotation (which is equivalent to assuming all points are far from the camera). In this case, we get the more restricted $3 \times 3$ homography

$$\boldsymbol{H}_{10} = \boldsymbol{K}_1 \boldsymbol{R}_1 \boldsymbol{R}_0^{-1} \boldsymbol{K}_0^{-1} = \boldsymbol{K}_1 \boldsymbol{R}_{10} \boldsymbol{K}_0^{-1}. \tag{17.6}$$

In practice, we usually set $\boldsymbol{K}_k = \mathrm{diag}(f_k, f_k, 1)$. Thus, instead of the general 8-parameter homography relating a pair of images, we get the 3-, 4-, or 5-parameter *3D rotation* motion models corresponding to the cases where the focal length $f$ is known, fixed, or variable [781]. Estimating the 3D rotation matrix (and optionally, the focal length) associated with each image is intrinsically much more stable than estimating a full 8-d.o.f. homography, which makes this the method of choice for large-scale image stitching algorithms [781, 739, 121].

An alternative to using homographies or 3D rotations is to first warp the images into *cylindrical* coordinates and to then use a pure translational model to align them [180]. Unfortunately, this only works if the images are all taken with a level camera or with a known tilt angle. The equations for mapping between planar and cylindrical/spherical coordinates can be found in [781, 778].

## 17.3 Direct and feature-based alignment

Once we have chosen a suitable motion model to describe the alignment between a pair of images, we need to devise some method to estimate its parameters. One approach is to shift or warp the images relative to each other and to look at how much the pixels agree. Approaches such as these are often called *direct methods*, as opposed to the *feature-based methods* described a little later.

### *17.3.1 Direct methods*

To use a direct method, a suitable *error metric* must first be chosen to compare the images. Once this has been established, a suitable *search* technique must be devised. The simplest search technique is to exhaustively try all possible alignments, i.e., to do a *full search*. In practice, this may be too slow, so *hierarchical* coarse-to-fine techniques based on image pyramids have been developed [75]. Alternatively, Fourier transforms can be used to speed up the computation [778]. To get sub-pixel precision in the alignment, *incremental* methods based on a Taylor series expansion of the image function are often used [533]; these can also be applied to *parametric motion models* [533, 75]. Each of these techniques is described in more detail in [778] and summarized below.

The simplest way to establish an alignment between two images is to shift one image relative to the other. Given a *template* image $I_0(\boldsymbol{x})$ sampled at discrete pixel locations $\{\boldsymbol{x}_i = (x_i, y_i)\}$, we wish to find where it is located in image $I_1(\boldsymbol{x})$. A least-squares solution to this problem is to find the minimum of the *sum of squared differences* (SSD) function

$$E_{\text{SSD}}(\boldsymbol{u}) = \sum_i [I_1(\boldsymbol{x}_i + \boldsymbol{u}) - I_0(\boldsymbol{x}_i)]^2 = \sum_i e_i^2, \qquad (17.7)$$

where $\boldsymbol{u} = (u, v)$ is the *displacement vector* and $e_i = I_1(\boldsymbol{x}_i + \boldsymbol{u}) - I_0(\boldsymbol{x}_i)$ is called the *residual error*.

In general, the displacement $\boldsymbol{u}$ can be fractional, so a suitable interpolation function must be applied to image $I_1(\boldsymbol{x})$. In practice, a bilinear interpolant is often used, but bi-cubic interpolation may yield slightly better results.

We can make the above error metric more robust to outliers by replacing the squared error terms with a robust function $\rho(e_i)$ [764]. We can also model potential *bias and gain* variations between the images being compared, and to associate *spatially varying weights* with different pixels, which is a principled way to deal with partial overlap and regions that have been "cut out" from one of the images [43, 778]. The extended version of this chapter [778] also discusses correlation (and *phase correlation*) as an alternative to robust pixel difference matching. It also discusses how coarse-to-fine (*hierarchical*) techniques [75] and *Fourier transforms* can be used to speed up the search for optimal alignment. (Fourier transforms unfortunately only work for pure translation and for a very limited set of (small-motion) similarity transforms.)

*Incremental refinement*

To obtain better *sub-pixel* estimates, we can use one of several techniques. One possibility is to evaluate several discrete (integer or fractional) values of $(u, v)$ around the best value found so far and to *interpolate* the matching score to find an analytic minimum. A more commonly used approach, first proposed by Lucas and Kanade [533], is to do *gradient descent* on the SSD energy function (17.7), using a Taylor Series expansion of the image function,

$$E_{\text{LK-SSD}}(\boldsymbol{u} + \Delta\boldsymbol{u}) \approx \sum_i [\boldsymbol{J}_1(\boldsymbol{x}_i + \boldsymbol{u})\Delta\boldsymbol{u} + e_i]^2, \tag{17.8}$$

where

$$\boldsymbol{J}_1(\boldsymbol{x}_i + \boldsymbol{u}) = \nabla I_1(\boldsymbol{x}_i + \boldsymbol{u}) = (\frac{\partial I_1}{\partial x}, \frac{\partial I_1}{\partial y})(\boldsymbol{x}_i + \boldsymbol{u}) \tag{17.9}$$

is the *image gradient* at $\boldsymbol{x}_i + \boldsymbol{u}$.

The above least squares problem can be minimizing by solving the associated *normal equations*.

$$\boldsymbol{A}\Delta\boldsymbol{u} = \boldsymbol{b} \tag{17.10}$$

where

$$\boldsymbol{A} = \sum_i \boldsymbol{J}_1^T(\boldsymbol{x}_i + \boldsymbol{u})\boldsymbol{J}_1(\boldsymbol{x}_i + \boldsymbol{u}) \quad \text{and} \quad \boldsymbol{b} = -\sum_i e_i \boldsymbol{J}_1^T(\boldsymbol{x}_i + \boldsymbol{u}) \tag{17.11}$$

are called the *Hessian* and *gradient-weighted residual vector*, respectively.

The gradients required for $\boldsymbol{J}_1(\boldsymbol{x}_i + \boldsymbol{u})$ can be evaluated at the same time as the image warps required to estimate $I_1(\boldsymbol{x}_i + \boldsymbol{u})$, and in fact are often computed as a side-product of image interpolation. If efficiency is a concern, these gradients can be replaced by the gradients in the *template* image,

$$\boldsymbol{J}_1(\boldsymbol{x}_i + \boldsymbol{u}) \approx \boldsymbol{J}_0(\boldsymbol{x}), \tag{17.12}$$

since near the correct alignment, the template and displaced target images should look similar. This has the advantage of allowing the pre-computation of the Hessian and Jacobian images, which can result in significant computational savings [43].

*Parametric motion*

Many image alignment tasks, for example image stitching with handheld cameras, require the use of more sophisticated motion models. Since these models typically have more parameters than pure translation, a full search over the possible range of values is impractical. Instead, the incremental Lucas-Kanade algorithm can be generalized to parametric motion models and used in conjunction with a hierarchical search algorithm [533, 75, 43].

For parametric motion, instead of using a single constant translation vector $\boldsymbol{u}$, we use a spatially varying *motion field* or *correspondence map*, $\boldsymbol{x}'(\boldsymbol{x};\boldsymbol{p})$, parameterized by a low-dimensional vector $\boldsymbol{p}$, where $\boldsymbol{x}'$ can be any of the motion models presented in Section 17.2. The parametric incremental motion update rule now becomes

$$\begin{aligned} E_{\text{LK-PM}}(\boldsymbol{p}+\Delta\boldsymbol{p}) &= \sum_i [I_1(\boldsymbol{x}'(\boldsymbol{x}_i;\boldsymbol{p}+\Delta\boldsymbol{p})) - I_0(\boldsymbol{x}_i)]^2 \\ &\approx \sum_i [\boldsymbol{J}_1(\boldsymbol{x}'_i)\Delta\boldsymbol{p} + e_i]^2, \end{aligned} \tag{17.13}$$

where the Jacobian is now

$$\boldsymbol{J}_1(\boldsymbol{x}'_i) = \frac{\partial I_1}{\partial \boldsymbol{p}} = \nabla I_1(\boldsymbol{x}'_i)\frac{\partial \boldsymbol{x}'}{\partial \boldsymbol{p}}(\boldsymbol{x}_i), \tag{17.14}$$

i.e., the product of the image gradient $\nabla I_1$ with the Jacobian of correspondence field, $\boldsymbol{J}_{\boldsymbol{x}'} = \partial \boldsymbol{x}'/\partial \boldsymbol{p}$.

The derivatives required to compute the Jacobian can be derived directly from Table 17.1 and are given in [778].

The computation of the Hessian and residual vectors for parametric motion can be significantly more expensive than for the translational case. For parametric motion with $n$ parameters and $N$ pixels, the accumulation of $\boldsymbol{A}$ and $\boldsymbol{b}$ takes $O(n^2 N)$ operations [43]. One way to reduce this by a significant amount is to divide the image up into smaller sub-blocks (patches) $P_j$ and to only accumulate the simpler $2 \times 2$ quantities (17.11) at the pixel level [739, 43, 778].

For a complex parametric motion such as a homography, the computation of the motion Jacobian becomes complicated, and may involve a per-pixel division. Szeliski and Shum [781] observed that this can be simplified by first warping the target image $I_1$ according to the current motion estimate $\boldsymbol{x}'(\boldsymbol{x};\boldsymbol{p})$ and then comparing this *warped* image against the template $I_0(\boldsymbol{x})$. Baker and Matthews [43] call this the *forward compositional* algorithm, since the target image is being re-warped, and the final motion estimates are being composed, and also present an *inverse compositional* algorithm that is even more efficient.

### 17.3.2 Feature-based registration

As mentioned earlier, directly matching pixel intensities is just one possible approach to image registration. The other major approach is to first extract distinctive *features* from each image, to match individual features to establish a global correspondence, and to then estimate the geometric transformation between the images. This kind of approach has been used since the early days of stereo matching and has more recently gained popularity for image stitching applications [148, 121].

Schmid *et al.* [712] survey the vast literature on interest point detection and perform some experimental comparisons to determine the *repeatability* of feature detectors. They also measure the *information content* available at each detected feature point. Among the techniques they survey, they find that an improved version of the Harris operator works best.

More recently, feature detectors that are more invariant to scale [532] and affine transformations have been proposed. These can be very useful when matching images that have different scales or different aspects (e.g., for 3D object recognition).

After detecting the features (interest points), we must *match* them, i.e., determine which features come from corresponding locations in different images. In some situations, e.g., for video sequences or for stereo pairs that have been *rectified*, the local motion around each feature point may be mostly translational. In this case, the error metrics introduced previously can be used to directly compare the intensities in small patches around each feature point. (The comparative study by Mikolajczyk and Schmid [569] discussed below uses cross-correlation.)

If features are being tracked over longer image sequences, their appearance can undergo larger changes. In this case, it makes sense to compare appearances using an *affine* motion model. Because the features can appear at different orientations or scales, a more *view invariant* kind of representation must be used. Mikolajczyk and Schmid [569] review some recently developed view-invariant local image descriptors and experimentally compare their performance.

The simplest method to compensate for in-plane rotations is to find a *dominant orientation* at each feature point location before sampling the patch or otherwise computing the descriptor. Mikolajczyk and Schmid use the direction of the average gradient orientation, computed within a small neighborhood of each feature point. The descriptor can be made invariant to scale by only selecting feature points that are local maxima in scale space. Among the local descriptors that Mikolajczyk and Schmid compared, David Lowe's Scale Invariant Feature Transform (SIFT) [532] performed the best.

The simplest way to find all corresponding feature points in an image pair is to compare all features in one image against all features in the other, using one the local descriptors described above. Unfortunately, this is quadratic in the expected number of features, which makes it impractical for some applications. More efficient matching algorithms can be devised using different kinds of *indexing schemes*, many of which are based on the idea of finding nearest neighbors in high-dimensional spaces.

Once an initial set of feature correspondences has been computed, we need to find a set that is will produce a high-accuracy alignment. One possible approach is to simply compute a least squares estimate, or to use a robustified version of least squares. However, in many cases, it is better to first find a good starting set of *inlier* correspondences, i.e., points that are all consistent with some particular motion estimate. Two widely used solution to this problem are RANdom SAmple

Consensus (RANSAC) and *least median of squares* (LMS) [764]. Both techniques start by selecting a random subset of $k$ correspondences, which is then used to compute a motion estimate $\boldsymbol{p}$. The RANSAC technique then counts the number of *inliers* that are within $\epsilon$ of their predicted location. Least median of squares finds the median value of the $\|\boldsymbol{r}_i\|$ values. The random selection process is repeated $S$ times, and the sample set with largest number of inliers (or with the smallest median residual) is kept as the final solution.

*Geometric registration*

Once we have computed a set of matched feature point correspondences, we still need to estimate the motion parameters $\boldsymbol{p}$ that best register the two images. The usual way to do this is to use least squares, i.e., to minimize the sum of squared residuals given by

$$E_{\mathrm{LS}} = \sum_i \|\boldsymbol{r}_i\|^2 = \|\tilde{\boldsymbol{x}}_i'(\boldsymbol{x}_i;\boldsymbol{p}) - \hat{\boldsymbol{x}}_i'\|^2, \tag{17.15}$$

where $\tilde{\boldsymbol{x}}_i'$ are the *estimated* (mapped) locations, and $\hat{\boldsymbol{x}}_i'$ are the sensed (detected) feature point locations corresponding to point $\boldsymbol{x}_i$ in the other image.

Many of the motion models presented in Section 17.2, i.e., translation, similarity, and affine, have a *linear* relationship between the motion and the unknown parameters $\boldsymbol{p}$. In this case, a simple linear regression (least squares) using normal equations works well.

The above least squares formulation assumes that all feature points are matched with the same accuracy. This is often not the case, since certain points may fall in more textured regions than others. If we associate a variance estimate $\sigma_i^2$ with each correspondence, we can minimize *weighted least squares* instead,

$$E_{\mathrm{WLS}} = \sum_i \sigma_i^{-2}\|\boldsymbol{r}_i\|^2. \tag{17.16}$$

As discussed in [778], a covariance estimate for patch-based matching can be obtained by multiplying the inverse of the Hessian with the per-pixel noise estimate. Weighting each squared residual by the inverse covariance $\Sigma_i^{-1} = \sigma_n^{-2}\boldsymbol{A}_i$ (which is called the *information matrix*), we obtain

$$E_{\mathrm{CWLS}} = \sum_i \|\boldsymbol{r}_i\|^2_{\Sigma_i^{-1}} = \sum_i \boldsymbol{r}_i^T\Sigma_i^{-1}\boldsymbol{r}_i = \sum_i \sigma_n^{-2}\boldsymbol{r}_i^T\boldsymbol{A}_i\boldsymbol{r}_i, \tag{17.17}$$

where $\boldsymbol{A}_i$ is the *patch Hessian.*

If there are outliers among the feature-based correspondences, it is better to use a robust version of least squares, even if an initial RANSAC or MLS stage has been used to select plausible inliers. The robust least squares cost metric is then

$$E_{\mathrm{RLS}}(\boldsymbol{u}) = \sum_i \rho(\|\boldsymbol{r}_i\|_{\Sigma_i^{-1}}). \tag{17.18}$$

For motion models that are not linear in the motion parameters, *non-linear least squares* must be used instead. Deriving the Jacobian of each residual equation with respect to the motion parameters is relatively straightforward, once a suitable parameterization has been chosen [778].

### *17.3.3 Direct vs. feature-based*

Given that there are these two alternative approaches to aligning images, which is preferable?

My original work in image stitching was firmly in the direct (image-based) camp [776, 781, 739]. Early feature-based methods seemed to get confused in regions that were either too textured or not textured enough. The features would often be distributed unevenly over the images, thereby failing to match image pairs that should have been aligned. Furthermore, establishing correspondences relied on simple cross-correlation between patches surrounding the feature points, which did not work well when the images were rotated or had foreshortening due to homographies.

Today, feature detection and matching schemes are remarkably robust and can even be used for known object recognition from widely separated views [532]. Because they operate in scale-space and use a dominant orientation (or orientation invariant descriptors), they can match images that differ in scale, orientation, and even foreshortening. My own recent experience is that if the features are well distributed over the image and the descriptors reasonably designed for repeatability, enough correspondences to permit image stitching can usually be found.

The other major reason I used to prefer direct methods was that they make optimal use of the information available in image alignment, since they measure the contribution of *every* pixel in the image. Furthermore, assuming a Gaussian noise model (or a robustified version of it), they properly weight the contribution of different pixels, e.g., by emphasizing the contribution of high-gradient pixels. (See Baker *et al.* [43], who suggest that adding even more weight at strong gradients is preferable because of noise in the gradient estimates.)

The biggest disadvantage of direct techniques is that they have a limited range of convergence. Even though hierarchical (coarse-to-fine) techniques can help, it is hard to use more than two or three levels of a pyramid before important details start get blurred. For matching sequential frames in a video, the direct approach can usually be made to work. However, for matching partially overlapping images in photo-based panoramas, they fail too often to be useful.

Is there no rôle then for direct registration? I believe there is. Once a pair of images has been aligned with a feature-based approach, we can warp the two images to a common reference frame and re-compute a more accurate correspondence using patch-based alignment. Notice how there is a close correspondence between the patch-based approximation to direct alignment and the inverse covariance weighted feature-based least squares error metric (17.17).

In fact, if we divide the template images up into patches and place an imaginary "feature point" at the center of each patch, the two approaches return exactly the same answer (assuming that the correct correspondences are found in each case). However, for this approach to succeed, we still have to deal with "outliers", i.e., regions that do not fit the selected motion model due to either parallax or moving objects. While a feature-based approach may make it somewhat easier to reason about outliers (features can be classified as inliers or outliers), the patch-based approach, since it establishes correspondences more densely, is potentially more useful for removing local mis-registration (parallax).

## 17.4 Global registration

So far, I have discussed how to register pairs of images using both direct and feature-based methods. In most applications, we are given more than a single pair of images to register. The goal is to find a *globally consistent* set of alignment parameters that minimize the mis-registration between all pairs of images [781, 739, 709]. In order to do this, we need to extend the pairwise matching criteria to a global energy function that involves all of the per-image pose parameters. Once we have computed the global alignment, we need to perform *local adjustments* such as *parallax removal* to reduce double images and blurring due to local mis-registration. Finally, if we are given an unordered set of images to register, we need to discover which images go together to form one or more panoramas.

### *17.4.1 Bundle adjustment*

One way to register a large number of images is to add new images to the panorama one at a time, aligning the most recent image with the previous ones already in the collection [781], and discovering, if necessary, which images it overlaps [709]. In the case of 360° panoramas, accumulated error may lead to the presence of a *gap* (or excessive overlap) between the two ends of the panorama, which can be fixed by stretching the alignment of all the images using a process called *gap closing* [781]. However, a better alternative is to simultaneously align all the images together using a least squares framework to evenly distribute any mis-registration errors.

The process of simultaneously adjusting pose parameters for a large collection of overlapping images is called *bundle adjustment* in the photogrammetry community [805]. In computer vision, it was first applied to the general structure from motion problem [780], and then later specialized for panoramic image stitching [739, 709].

In this section, I formulate the problem of global alignment using a feature-based approach, since this results in a simpler system. An equivalent direct approach can be obtained by dividing images into patches and creating a virtual feature correspondence for each one [739].

Consider the feature-based alignment problem given in (17.15). For multi-image alignment, instead of having a single collection of pairwise feature correspondences, $\{(\boldsymbol{x}_i, \hat{\boldsymbol{x}}_i')\}$, we have a collection of $n$ features, with the location of the $i$th feature point in the $j$th image denoted by $\boldsymbol{x}_{ij}$ and its scalar confidence (inverse variance) denoted by $c_{ij}$. Each image also has some associated *pose* parameters.

In this section, I assume that this pose consists of a rotation matrix $\boldsymbol{R}_j$ and a focal length $f_j$, although formulations in terms of homographies are also possible [781, 709]. The equation mapping a 3D point $\boldsymbol{x}_i$ into a point $\boldsymbol{x}_{ij}$ in frame $j$ can be re-written from (17.4–17.6) as

$$\boldsymbol{x}_{ij} \sim \boldsymbol{K}_j \boldsymbol{R}_j \boldsymbol{x}_i \quad \text{and} \quad \boldsymbol{x}_i \sim \boldsymbol{R}_j^{-1} \boldsymbol{K}_j^{-1} \boldsymbol{x}_{ij}, \tag{17.19}$$

where $\boldsymbol{K}_j = \text{diag}(f_j, f_j, 1)$ is the simplified form of the calibration matrix. The motion mapping a point $\boldsymbol{x}_{ij}$ from frame $j$ into a point $\boldsymbol{x}_{ik}$ in frame $k$ is similarly given by

$$\boldsymbol{x}_{ik} \sim \boldsymbol{H}_{kj} \boldsymbol{x}_{ij} = \boldsymbol{K}_k \boldsymbol{R}_k \boldsymbol{R}_j^{-1} \boldsymbol{K}_j^{-1} \boldsymbol{x}_{ij}. \tag{17.20}$$

Given an initial set of $\{(\boldsymbol{R}_j, f_j)\}$ estimates obtained from chaining pairwise alignments, how do we refine these estimates?

One approach is to directly extend the pairwise energy to a multiview formulation,

$$E_{\text{all-pairs-2D}} = \sum_i \sum_{jk} c_{ij} c_{ik} \|\tilde{\boldsymbol{x}}_{ik}(\hat{\boldsymbol{x}}_{ij}; \boldsymbol{R}_j, f_j, \boldsymbol{R}_k, f_k) - \hat{\boldsymbol{x}}_{ik}\|^2, \tag{17.21}$$

where the $\tilde{\boldsymbol{x}}_{ik}$ function is the *predicted* location of feature $i$ in frame $k$ given by (17.20), $\hat{\boldsymbol{x}}_{ij}$ is the *observed* location, and the "2D" in the subscript indicates than an image-plane error is being minimized.

While this approach works well in practice, it suffers from two potential disadvantages. First, since a summation is taken over all pairs with corresponding features, features that are observed many times get overweighted in the final solution. Second, the derivatives of $\tilde{\boldsymbol{x}}_{ik}$ w.r.t. the $\{(\boldsymbol{R}_j, f_j)\}$ are a little cumbersome.

An alternative way to formulate the optimization is to use true bundle adjustment, i.e., to solve not only for the pose parameters $\{(\boldsymbol{R}_j, f_j)\}$ but also for the 3D point positions $\{\boldsymbol{x}_i\}$,

$$E_{\text{BA-2D}} = \sum_i \sum_j c_{ij} \|\tilde{\boldsymbol{x}}_{ij}(\boldsymbol{x}_i; \boldsymbol{R}_j, f_j) - \hat{\boldsymbol{x}}_{ij}\|^2, \tag{17.22}$$

where $\tilde{\boldsymbol{x}}_{ij}(\boldsymbol{x}_i; \boldsymbol{R}_j, f_j)$ is given by (17.19). The disadvantage of full bundle adjustment is that there are more variables to solve for, so both each iteration and the overall convergence may be slower. However, the computational complexity of each linearized Gauss-Newton step can be reduced using sparse matrix techniques [780, 739, 805].

An alternative formulation is to minimize the error in 3D projected ray directions [739], i.e.,

$$E_{\text{BA-3D}} = \sum_i \sum_j c_{ij} \|\tilde{\boldsymbol{x}}_i(\hat{\boldsymbol{x}}_{ij}; \boldsymbol{R}_j, f_j) - \boldsymbol{x}_i\|^2, \tag{17.23}$$

where $\tilde{\boldsymbol{x}}_i(\boldsymbol{x}_{ij}; \boldsymbol{R}_j, f_j)$ is given by the second half of (17.19).

However, if we eliminate the 3D rays $x_i$, we can derive a pairwise energy formulated in 3D ray space [739],

$$E_{\text{all-pairs-3D}} = \sum_i \sum_{jk} c_{ij} c_{ik} \|\tilde{x}_i(\hat{x}_{ij}; R_j, f_j) - \tilde{x}_i(\hat{x}_{ik}; R_k, f_k)\|^2. \quad (17.24)$$

This results in the simplest set of update equations [739], since the $f_k$ can be folded into the creation of the homogeneous coordinate vector. Thus, even though this formula over-weights features that occur more frequently, it is the method used both by Shum and Szeliski [739] and in my current feature-based aligner. In order to reduce the bias towards longer focal lengths, I multiply each residual (3D error) by $\sqrt{f_j f_k}$, which is similar to projecting the 3D rays into a "virtual camera" of intermediate focal length.

### *17.4.2 Parallax removal*

Once we have estimated the global orientations and focal lengths of our cameras, we may find that the images are still not perfectly aligned, i.e., the resulting stitched image looks blurry or ghosted in some places. This may be caused by a variety of factors, including unmodeled radial distortion, 3D parallax (failure to rotate the camera around its optical center), small scene motions such as waving tree branches, and large-scale scene motions such as people moving in and out of pictures.

Each of these problems can be treated with a different approach. Radial distortion can be estimated using one of several classic calibration techniques. 3D parallax can be attacked by doing a full 3D bundle adjustment. The 3D positions of the matched features points and cameras can then be simultaneously recovered, although this can be significantly more expensive that parallax-free image registration.

When the motion in the scene is very large, i.e., when objects appear and disappear completely, a sensible solution is to simply *select* pixels from only one image at a time as the source for the final composite [248, 7], as discussed in Section 17.6. However, when the motion is reasonably small (on the order of a few pixels), general 2-D motion estimation (optic flow) can be used to perform an appropriate correction before blending using a process called *local alignment* [739]. This same process can also be used to compensate for radial distortion and 3D parallax, although it uses a weaker motion model than explicitly modeling the source of error, and may therefore fail more often.

### *17.4.3 Recognizing panoramas*

The final piece needed to perform fully automated image stitching is a technique to determine which images actually go together, which Brown and Lowe call *recognizing panoramas* [121]. If the user takes images in sequence so that each image overlaps its predecessor, bundle adjustment combined with the process of *topology inference* can be used to automatically assemble a panorama [709]. However,

Figure 17.2. A set of images and the panorama discovered in them

users often jump around when taking panoramas, e.g., they may start a new row on top of a previous one, or jump back to take a repeated shot, or create 360° panoramas where end-to-end overlaps need to be discovered. Furthermore, the ability to automatically discover multiple panoramas taken by a user can be a big convenience.

To recognize panoramas, Brown and Lowe [121] first find all pairwise image overlaps using a feature-based method and then find connected components in the overlap graph to "recognize" individual panoramas (Figure 17.2). First, they use Lowe's Scale Invariant Feature Transform (SIFT features) [532] followed by nearest neighbor matching. RANSAC is then used to find a set of *inliers*, using pairs of matches to hypothesize similarity motion estimates. Once pairwise alignments have been computed, a global registration (bundle adjustment) stage is used to compute a globally consistent alignment for all of the images. Finally, a two-level Laplacian pyramid is used to seamlessly blend the images [121].

## 17.5 Choosing a compositing surface

Once we have registered all of the input images with respect to each other, we need to decide how to produce the final stitched (mosaic) image. This involves selecting a final compositing surface, e.g., flat, cylindrical, or spherical. It may also involve computing an optimal *reference view* to ensure that the scene appears to be *upright*, as described in [778].

If only a few images are stitched together, a natural approach is to select one of the images as the reference and to then warp all of the other images into the reference coordinate system. The resulting composite is called a *flat* panorama, since the projection onto the final surface is still a perspective projection, and hence straight lines remain straight.

For larger fields of view, however, we cannot maintain a flat representation without excessively stretching pixels near the border of the image. (In practice, flat panoramas start to look severely distorted once the field of view exceeds 90° or so.) The usual choice for compositing larger panoramas is to use a cylindrical [180] or spherical [781] projection. In fact, any surface used for *environment mapping* in computer graphics can be used, including a *cube map* that represents the full viewing sphere with the six square faces of a box [781].

The choice of parameterization is somewhat application dependent and involves a tradeoff between keeping the local appearance undistorted (e.g., keeping straight lines straight) and providing a reasonably uniform sampling of the environment. Automatically making this selection and smoothly transitioning between representations based on the extent of the panorama is an interesting topic for future research.

## 17.6 Seam selection and pixel blending

Once the source pixels have been mapped onto the final composite surface, we must decide how to blend them in order to create an attractive looking panorama. If all of the images are in perfect registration and identically exposed, this is an easy problem (any pixel combination will do). However, for real images, visible seams (due to exposure differences), blurring (due to mis-registration), or ghosting (due to moving objects) can occur.

Creating clean, pleasing looking panoramas involves both deciding which pixels to use and how to weight or blend them. The distinction between these two stages is a little fluid, since per-pixel weighting can be though of as a combination of selection and blending. In this section, I discuss spatially varying weighting, pixel selection (seam placement), and then more sophisticated blending.

*Feathering and center-weighting*

The simplest way to create a final composite is to simply take an *average* value at each pixel., However, this usually does not work very well, since exposure differences, mis-registrations, and scene movement are all very visible (Figure 17.3a). If rapidly moving objects are the only problem, taking a *median* filter (which is a kind of pixel selection operator) can often be used to remove them [417].

A better approach is to weight pixels near the center of the image more heavily and to down-weight pixels near the edges. When an image has some cutout regions, down-weighting pixels near the edges of both cutouts and edges is preferable. This can be done by computing a *distance map* or *grassfire transform*, where each valid pixel is tagged with its Euclidean distance to the nearest invalid pixel. Weighted averaging with a distance map is often called *feathering* [781, 825], and does a reasonable job of blending over exposure differences. However, blurring and ghosting can still be problems (Figure 17.3b).

One way to improve feathering is to raise the distance map values to some power. The weighted averages then become dominated by the larger values, i.e., they act like a *p-norm*. The resulting composite can often provide a reasonable tradeoff between visible exposure differences and blur.

Figure 17.3. Final composites computed by a variety of algorithms: (a) average, (b) feathered average, (c) weighted ROD vertex cover with feathering, (d) graph cut seams with Poisson blending. Notice how the regular average cuts off moving people near the edges of images, while the feathered average slowly blends them in. The vertex cover and graph cut algorithms produce similar results.

In the limit as $p \rightarrow \infty$, only the pixel with the maximum distance value gets selected, which is equivalent to computing the *Vornoi diagram*. The resulting composite, while useful for artistic guidance and in high-overlap panoramas (*manifold mosaics*) tends to have very hard edges with noticeable seams when the exposures vary.

*Optimal seam selection*

Computing the Vornoi diagram is one way to select the *seams* between regions where different images contribute to the final composite. However, Vornoi images totally ignore the local image structure underlying the seam.

A better approach is to place the seams in regions where the images agree, so that transitions from one source to another are not visible. In this way, the algorithm avoids "cutting through" moving objects, where a seam would look unnatural [248]. For a pair of images, this process can be formulated as a simple dynamic program starting from one edge of the overlap region and ending at the other [570, 248]. Unfortunately, when multiple images are being composited, the dynamic program idea does not readily generalize.

To overcome this problem, Uyttendaele *et al.* [825] observed that for well-registered images, moving objects produce the most visible artifacts, namely translucent looking *ghosts*. Their system therefore decides which objects to keep, and which ones to erase. First, the algorithm compares all overlapping input image pairs to determine *regions of difference* (RODs) where the images disagree. Next, a graph is constructed with the RODs as vertices and edges representing ROD pairs that overlap in the final composite. Since the presence of an edge indicates an area of disagreement, vertices (regions) must be removed from the final composite until no edge spans a pair of unremoved vertices. The smallest such set can be computed using a *vertex cover* algorithm. Since several such covers may exist, a *weighted vertex cover* is used instead, where the vertex weights are computed by summing the feather weights in the ROD. The algorithm therefore prefers removing regions that are near the edge of the image, which reduces the likelihood that partially visible objects will appear in the final composite. Once the required regions of difference have been removed, the final composite is created using a feathered blend (Figure 17.3c).

A different approach to pixel selection and seam placement was recently proposed by Agarwala *et al.* [7]. Their system computes the label assignment that optimizes the sum of two objective functions. The first is a per-pixel *image objective* $C_D$ that determines which pixels are likely to produce good composites. In their system, users can select which pixels to use by "painting" over an image with the desired object or appearance. Alternatively, automated selection criteria can be used, such as *maximum likelihood* that prefers pixels which occur repeatedly (for object removal), or *minimum likelihood* for objects that occur infrequently (for greatest object retention).

The second term is a *seam objective* $C_S$ that penalizes differences in labelings between adjacent images. For example, the simple color-based seam penalty used in [7] measures the color difference between corresponding pixels on both sides of the seam. The global energy function that is the sum of the data and seam costs can be minimized using a variety of techniques [778]. Agarwala *et al.* [7] use graph cuts, which involves cycling through a set of simpler *$\alpha$-expansion* relabelings, each of which can be solved with a graph cut (max-flow) polynomial-time algorithm [113].

For the result shown in Figure 17.3d, Agarwala *et al.* [7] use a large data penalty for invalid pixels and 0 for valid pixels. Notice how the seam placement algorithm avoids regions of differences, including those that border the image and which might result in cut off objects. Graph cuts [7] and vertex cover [825] often produce similar looking results, although the former is significantly slower since it optimizes over all pixels, while the latter is more sensitive to the thresholds used to determine regions of difference.

### *Laplacian pyramid blending*

Once the seams have been placed and unwanted objects removed, we still need to blend the images to compensate for exposure differences and other misalignments. An attractive solution to this problem was developed by Burt and Adelson [139]. Instead of using a single transition width, a frequency-adaptive width is used by creating a band-pass (Laplacian) pyramid and making the transition widths a function of the pyramid level. First, each warped image is converted into a band-pass (Laplacian) pyramid. Next, the *masks* associated with each source image are converted into a low-pass (Gaussian) pyramid and used to perform a per-level feathered blend of the band-pass images. Finally, the composite image is reconstructed by interpolating and summing all of the pyramid levels (band-pass images).

*Gradient domain blending*

An alternative approach to multi-band image blending is to perform the operations in the *gradient domain.* Here, instead of working with the initial color values, the image gradients from each source image are copied; in a second pass, an image that best matches these gradients is reconstructed [7]. Copying gradients directly from the source images after seam placement is just one approach to gradient domain blending. Levin *et al.* [514] examine several different variants on this approach, which they call *Gradient-domain Image STitching* (GIST). The techniques they examine include feathering (blending) the gradients from the source images, as well as using an L1 norm in performing the reconstruction of the image from the gradient field, rather than using an L2 norm. Their preferred technique is the L1 optimization of a feathered (blended) cost function on the original image gradients (which they call GIST1-$l_1$). While L1 optimization using linear programming can be slow, a faster iterative median-based algorithm in a multigrid framework works well in practice. Visual comparisons between their preferred approach and what they call *optimal seam on the gradients* (which is equivalent to Agarwala *et al.*'s approach [7]) show similar results, while significantly improving on pyramid blending and feathering algorithms.

*Exposure compensation*

Pyramid and gradient domain blending can do a good job of compensating for moderate amounts of exposure differences between images. However, when the exposure differences become large, alternative approaches may be necessary.

Uyttendaele *et al.* [825] iteratively estimate a local correction between each source image and a blended composite. First, a block-based quadratic transfer function is fit between each source image and an initial feathered composite. Next, transfer functions are averaged with their neighbors to get a smoother mapping, and per-pixel transfer functions are computed by *splining* between neighboring block values. Once each source image has been smoothly adjusted, a new feathered composite is computed, and the process is be repeated (typically 3 times). The results in [825] demonstrate that this does a better job of exposure compensation than simple feathering and can handle local variations in exposure due to effects like lens vignetting.

*High dynamic range imaging*

A more principled approach is to estimate a single *high dynamic range* (HDR) radiance map from of the differently exposed images [252, 577]. This approach assumes that the input images were taken with a fixed camera whose pixel values are the result of applying a parameterized *radiometric transfer function* $f(R, \mathbf{p})$

to scaled radiance values $c_k R(\boldsymbol{x})$. The exposure values $c_k$ are either known (by experimental setup, or from a camera's EXIF tags), or are computed as part of the parameter estimation process. After the transfer function has been estimated, radiance values from different exposures can be combined to emphasize reliable pixels.

Once a radiance map has been computed, it is usually necessary to display it on a lower gamut (i.e., 8-bit) screen or printer. A variety of *tone mapping* techniques have been developed for this purpose, which involve either computing spatially varying transfer functions or reducing image gradients to fit the the available dynamic range.

Unfortunately, most casually acquired images may not be perfectly registered and may contain moving objects. Kang *et al.* [452] present an algorithm that combines global registration with local motion estimation (optic flow) to accurately align the images before blending their radiance estimates. Since the images may have widely different exposures, care must be taken when producing the motion estimates, which must themselves be checked for consistency to avoid the creation of ghosts and object fragments.

## 17.7 Extensions and open issues

While image stitching has now reached a point where it is commonly used in consumer photo editing products, there are still a lot of open research problems that need to be addressed.

The first of these is improving the reliability of fully automated stitching. Whenever images contain small amounts of overlap, repeated textures, or large regions of difference because of moving objects, it becomes increasingly difficult to disambiguate between accidental and correct alignments. Global reasoning about a compatible set of correspondences might be the solution, as might be improvements in robust (partial) feature matching.

Dealing with motion and parallax is another important area, since pictures are often taken with handheld cameras in highly dynamic situations. At some point, full 3D reconstruction with moving object detection and layer extraction may be required, which also raises interesting issues in designing quick and easy user interfaces to specify the desired final output.

Dealing with images at different resolutions and zoom factors is another interesting area, especially since variable resolution image representations and viewers are not common. A related issue is super-resolution, i.e., enhancing image resolution through the combination of jittered photographs of the same region [543, 148]. Unfortunately, because of limitations in optics and motion estimation, there seems to be a very limited ($< 2\times$) improvement that can be achieved in practice.

Stitching videos is another area that is likely to grow as more digital cameras start to include the ability to take videos. Examples of stitching videos to obtain summary panoramas have been around for a while [417, 710]. In the future, we are likely to see the construction of "live" panoramas that include moving elements along with still portions [776].

Ultimately, image alignment and stitching will become part of a repertoire of computer vision algorithms used to merge multiple images (with different orientations, exposures, and other attributes) to create enhanced and innovative composite pictures and photographic experiences.

# Chapter18

# Visual Tracking: A Short Research Roadmap

**A. Blake**

Abstract

A research roadmap to many of the best known, and most used, contributions to visual tracking is set out. The scope includes simple appearance models, active contours, spatiotemporal filtering and briefly points to important further topics in tracking.

## 18.1 Introduction

Visual tracking is the repeated localisation of instances of a particular object, or class of objects, in successive frames of a video sequence. Video analysis may be causal or non-causal, but tracking is usually taken to be an online process, and therefore causal with some emphasis on efficient algorithms. The question of automatic initialisation, though sometimes important, is not addressed here. This is sensible in that there are plentiful applications where initialisation is not an issue, such as tracking vehicles on a highway, or indoor surveillance, in which initialisation can be effected by a simple motion trigger. The aim is to achieve location estimates at least as good as independent, exhaustive examinations of each frame [402]. Exploitation of object dynamics offers improved computational efficiency and more refined motion estimates. Perhaps most important of all, it offers extended capability to resolve ambiguity, as with a person in a crowd or a leaf on a bush (figure 18.1).

Figure 18.1. **Tracking in camouflage.** *The trail of tracked positions of a moving leaf, in heavy camouflage, at two different times in a sequence. FOr details of the method see section 18.4. Images reprinted from [94]. For related movies see* robots.ox.ac.uk/~vdg/dynamics.html.

# 18.2 Simple appearance models

## *18.2.1 Simple patches*

The most basic tracker consists of matching a template patch $T(\mathbf{r}), \mathbf{r} \in \mathcal{T}$ onto an image $I(\mathbf{r}$ under translation [533] by cross correlation. The aim is to minimise the misregistration error

$$\rho = \sum_{\mathbf{r} \in \mathcal{T}} [I(\mathbf{r}) - T(\mathbf{r} + \mathbf{u})]^2 \tag{18.1}$$

and this can be done to subpixel resolution using an estimate of the gradient $\mathbf{g}(\mathbf{r}) = \nabla I(\mathbf{r})$, computed using a suitable filter (such as a gradient of Gaussian filter). Then the iterative registration algorithm alternates two steps, to convergence:

1. Newton step on $\rho$

$$\hat{\mathbf{v}} = \sum_i \left(\mathbf{g}_i \cdot \mathbf{g}_i^{\top}\right)^{-1} \sum_n \mathbf{g}_i b_i$$

2. Recompute template offset

$$\mathbf{u} \rightarrow \mathbf{u} - \hat{\mathbf{v}}$$

More generally, the class of transformations can be generalised from translation $\mathbf{x} \rightarrow \mathbf{x} + \mathbf{u}$ to a larger class $\mathbf{x} \rightarrow W_{\mu}(\mathbf{x})$ in which $\gamma$ are the parameters of, for example, an affine transformation or a non-rigid spline mapping [101] — see later for more details of these transformations. Taking $\mu = \mu_0 + \delta\mu$ and linearising gives

$$I(\mathbf{r}) \approx T(W(\mathbf{r}, \mu_0)) + \delta\mu \cdot \frac{\partial W}{\partial \mu} \nabla T, \tag{18.2}$$

which can be solved iteratively for $\mu$, to perform generalised registration [533, 60, 376].

### 18.2.2 Blobs

An alternative approach to localising regions is to model only the gross properties of a region, modelling it as a "blob" [898], a Gaussian mixture model (GMM) in a joint $(\mathbf{r}, I)$ position and colour space. Thus a pixel $I(\mathbf{r})$ is modelled probabilistically as belonging to a model M with probability $p(\mathbf{r}, I(\mathbf{r}) \mid \mathrm{M})$ and in a new test image, each pixel is evaluated against each of a number of models $\mathrm{M} \in \mathcal{M}$. The model with the greatest likelihood is assigned to the pixel. The cluster of pixels with label M is deemed to be the new position of object M, whose moments (mean etc.) can be computed to represent the location of object M, and the GMM for M can also be updated periodically.

Recently a variation on the blob idea, "mean-shift" tracking [216] has been very influential because it allows progressive updating of object position without the obligation to visit all pixels of each and every frame. Successive approximations to the estimated locations of an object are obtained iteratively as:

$$\hat{\mathbf{r}}_t = \frac{1}{C} \sum_{\mathbf{r} \in \mathcal{T}} \mathbf{r}\, w(\mathbf{r}) g(\|\mathbf{r} - \hat{\mathbf{r}}_{t-1}\|^2) \tag{18.3}$$

where $C = \sum_{\mathbf{r} \in \mathcal{T}} \sqrt{w(\mathbf{r})} g(\|\mathbf{r} - \hat{\mathbf{r}}_{t-1}\|^2)$, $g$ is the derivative of a particular kernel function used to build spatial density functions, and $w(\mathbf{r})$ is a weight measuring the degree of prevalence of the color of pixel $\mathbf{r}$ in the template relative to its prevalence in the test object. The result, used over an image sequence, is a remarkably tenacious tracker (figure 18.2), despite its simplicity.

Figure 18.2. **Mean shift tracking** *A mean-shift tracker, (here in a particle filter form — see later) is used here to track player no 75 in a primitive form of sport. Image reprinted from [640].*

### 18.2.3 Background maintenance

Blobs represent foreground objects as distributions over colour (and space) but modelling a background, assuming it is largely static, is also useful as a guide to what is *not* part of an object. [592]. Just as blobs model the foreground as a mixture, so also modelling background pixels as mixture distributions is useful

[691, 757]. If $M_0$ is the background model, then pixels could be tested for their likelihood of belonging to the background in general by evaluating $p(I \mid M_0)$, and high scoring pixels removed from consideration as possible parts of any foreground object. What is more powerful still, when the background is static, is to model each background pixel individually by collecting statistics of colour over time from that pixel, and building a mixture model for $p(I \mid \mathbf{r}, M_0)$. These form typically narrow distributions which make powerful tests for background membership.

Having introduced some simple, though nonetheless very effective forms of tracker, the next section looks at some elaborations on the basic theme of matching shapes.

## 18.3 Active contours

An active contour is a parameterised curve $\mathbf{r}(s), 0 \le s \le 1$ in the plane that is set up to be attracted to features in an image $I(\mathbf{r})$. A detailed account of the development and mechanisms of active contours is given elsewhere [94], but here we summarise the main types. In section 18.4, explicitly dynamical forms of active contour $\mathbf{r}(s,t), t \ge 0$, attracted to an image sequence $I(t)$, are outlined. It focuses on the temporal filtering required to extract information most effectively over a sequence, exploiting fully the temporal coherence of the moving scene. This section is restricted to the static case and follows the development of active contours from snakes to parametric structures and affine contour models.

### *18.3.1 Snakes*

"Snakes" [455] have been one of the most influential ideas in computer vision. They were revolutionary in their time because they directed attention away from bottom up edge detection, an enterprise which had become stuck in a rut, towards top down, hypothesis driven search for object structures. The main idea is that the active contour $\mathbf{r}(s)$ is dropped into a potential energy field $F(\mathbf{r})$ which is itself a function of the image intensity landscape. For example $F(\mathbf{r}) = -|\nabla I|$ would generate an attraction of the snake towards high image contrast. An equilibrium configuration of the snake satisfies an (Euler-Lagrange) equation

$$\underbrace{\left(\frac{\partial(w_1\mathbf{r})}{\partial s} - \frac{\partial^2(w_2\mathbf{r})}{\partial s^2}\right)}_{\text{internal forces}} + \underbrace{\nabla E_{\text{ext}}}_{\text{external force}} = 0. \tag{18.4}$$

in which internal force parameters can be adjusted to give the curve a tendency towards smooth shapes. Such a system can be converted to a numerical scheme, for example using finite differences along a fine polygonal approximation to the curve $\mathbf{r}(s)$, with typically hundreds of variables corresponding to the polygon vertices $\mathbf{q}_i$, $i = 1, \ldots, M$. Equilibria are then sought by iterative solving. Alter-

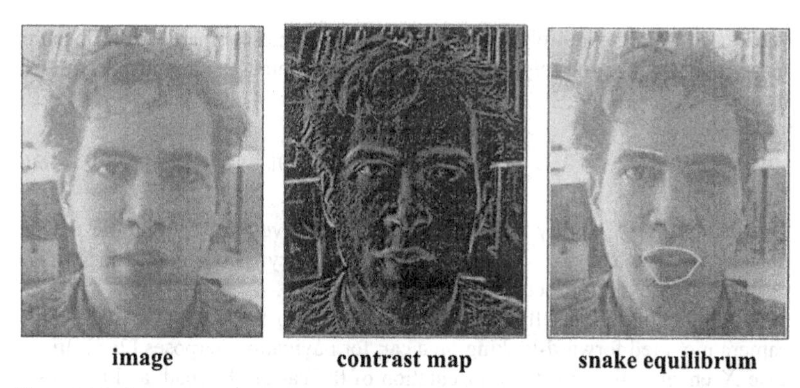


Figure 18.3. **Snakes** *An input image and a filter to extract a contrast map $F(\mathbf{r})$, serving as a potential field under which snakes can reach equilibrium. Images reprinted from [94].*

natively direct solution by dynamic programming [23] is also possible, with the added attraction that hard constraints can be incorporated easily.

So far the snake is defined with respect to a single image $I(\mathbf{r})$ but for shape tracking, its behaviour over an image sequence $I(\mathbf{r}, t)$ must be defined. This can be expressed as a Lagrangian dynamical system [793, 241] with distributed mass and viscosity, whose equations of motion could typically take the following form

$$\underbrace{\rho \mathbf{r}_{tt}}_{\text{inertial force}} = -\underbrace{\left(\gamma \mathbf{r}_t - \frac{\partial(w_1 \mathbf{r})}{\partial s} + \frac{\partial^2(w_2 \mathbf{r})}{\partial s^2}\right)}_{\text{internal forces}} + \underbrace{\nabla F}_{\text{external force}} \tag{18.5}$$

in which the additional parameters $\gamma$ and $\rho$ respectively govern viscosity of the medium and distributed mass along the contour.

Of course this leaves questions about how to choose parameters $w_1, w_2, \gamma, \rho$, which may be spatial functions, not just constants, unanswered. This is a problem that can be addressed effectively in a rather different framework, that of probabilistic temporal filtering (see section 18.4). This idea was first cast [793] in a space of state vectors consisting of vertices of the snake polygon $\{\mathbf{q}_i\}$. Practical implementation however, demands a much lower dimensional state space, not just for computational economy but for stability [93], and this is elaborated in section 18.4.

### *18.3.2 Parametric structures*

If a lower dimensional state space is essential for stable tracking, one way to construct such a state space is in terms of a state vector $X = (\lambda_1, \ldots, \lambda_K)$ whose components are physical degrees of freedom in the underlying object, representing a contour (or set of contours) $\mathbf{r}(s; X)$, $s \in [0, 1]$. For example $X$ could encode the position and orientation of a rigid object. Then the image locations $\mathbf{r}(s_i, X)$, $i = 1, \ldots, M$ of $M$ distinguished features on the curve (for example vertices of a polyhedral object) can be predicted, and compared with observed locations $\mathbf{r}_f(s_i)$. In principle $X$ can then be estimated by minimising an error measure such as

$$E = \sum_{i=1}^{M} \|\mathbf{r}(s_i, X) - \mathbf{r}_f(s_i)\|^2. \tag{18.6}$$

To include the possibility that the model contains vertices or multiple disconnected segments, $\mathbf{r}(s; X)$, $s \in [0, 1]$ need not be everywhere smooth, and may be discontinuous at a finite set of points along $s \in [0, 1]$.

A simple and highly effective example applies to the view of a road from a camera mounted forward-looking on a car, for navigation purposes [268]. In that case $X$ encodes the offset and orientation of the car on the road, and the observations are the road edges. Such a system resulted in the first autonomous, vision guided automobile to travel at realistic speeds on the open road. Other prominent examples of the parametric approach include real-time tracking of complex 3D wire-frame structures [384] and a hinged box [530], in which the prediction function $\mathbf{r}(s; X)$ applies perspective projection to map a canonical structure, in state $X$, onto the image plane. The state vector $X$ can also incorporate further parameters which allow adjustment of the underlying canonical structure, in addition to position and orientation, allowing tracking of any object from a given family of objects. This was successful for example with tracking automobiles in overhead views of the highway [487], in which the pose of the vehicle and also variations in automobile shape were encoded together in the state vector $X$.

### *18.3.3 Affine contours*

Another natural way to construct a low-dimensional state space for tracking is to specify parameters relating directly to image-based shape of the active contour. This is especially appealing because because, as we will see, the contour $\mathbf{r}(s; X)$ can then often be expressed as a linear function of $X$ and this considerably simplifies the task of curve fitting and (later) of temporal filtering [93]. One natural choice is the planar affine space in which $\mathbf{r}(s; X)$ sweeps out the space of 2D affine transformations of a base shape $\overline{\mathbf{r}}(s)$:

$$\mathbf{r}(s; X) = A\overline{\mathbf{r}}(s) + \mathbf{u} \tag{18.7}$$

where $A$ is a $2 \times 2$ matrix and $\mathbf{u}$ is a $2 \times 1$ vector. It is natural because it is known to span the space of outlines of a planar shape, in an arbitrary 3D pose, and viewed under affine projection (the approximation to image projection that holds when perspective effects are not too strong). It is linear because we can choose $X = (A, \mathbf{u})$ so that $\mathbf{r}(s; X)$ is linear in $X$, and this linear relation is denoted

$$\mathbf{r}(s; X) = H(s)X, \tag{18.8}$$

where $H(s)$ is a simple (linear) function of $\overline{\mathbf{r}}(s)$. For nonplanar 3D outlines, still under affine projection, there is a linear parameterisation of the form $X =$

$(A, \mathbf{u}, \mathbf{v})$ (see [94] for details) where $\mathbf{v}$ is another vector, so the dimensionality of $X$ increases from 6 to 8. Of course the underlying dimensionality of the space is still 6 — three parameters for 3D translation and 3 for rotation — and the additional 2 are the price of insisting on a linear parameterisation.

Having defined the linear parameterisation $\mathbf{r}(s; X)$ of image curves, a curve can now be fitted to a particular set of image data. Suppose the data itself is a curve $\mathbf{r}_f(s)$, then the least squares fit, the curve $\mathbf{r}(s; \hat{X})$ minimising

$$\int |\mathbf{r}(s; X) - \mathbf{r}_f(s)|^2 \, \mathrm{d}s, \tag{18.9}$$

is given simply by

$$\hat{X} = \mathcal{H}^{-1} \int H^\top(s) \mathbf{r}_f(s) \, \mathrm{d}s \text{ where } \mathcal{H} = \int H^\top(s) H(s) \, \mathrm{d}s, \tag{18.10}$$

provided the solution is unique. For better stability, regularisation on $\mathbf{r}(s; X)$ can also be introduced. The integrals in (18.10) have to be computed finitely in practice, and this can be achieved by a using finite parameterisation of the base curve $\bar{\mathbf{r}}(s)$ (and therefore also of $H(s)$): for example $\bar{\mathbf{r}}(s)$ can be modelled as a B-spline [93, 94] or simply as a polygon [223].

There remains one important issue. The fitting scheme above is correct only if correspondence between the curves is known — that is, for any given value of $s$, the point $\mathbf{r}(s; X)$ in the plane is supposed to correspond to the point $\mathbf{r}_f(s)$ on the data curve. In practice, of course, this is not the case: $\mathbf{r}_f(s)$ may be parameterised quite differently from $\mathbf{r}(s; X)$ so that in principle one should fit $\mathbf{r}(s; X)$ to $\mathbf{r}_f(g(s))$, for some unknown reparameterisation function $g$. In the case that the reparameterisation is not too severe, this is dealt with approximately by replacing total displacement in (18.9) by normal displacement [94, Ch. 6] , as in figure 18.4. Normal displacement is commonly used, for this reason, in tracking systems

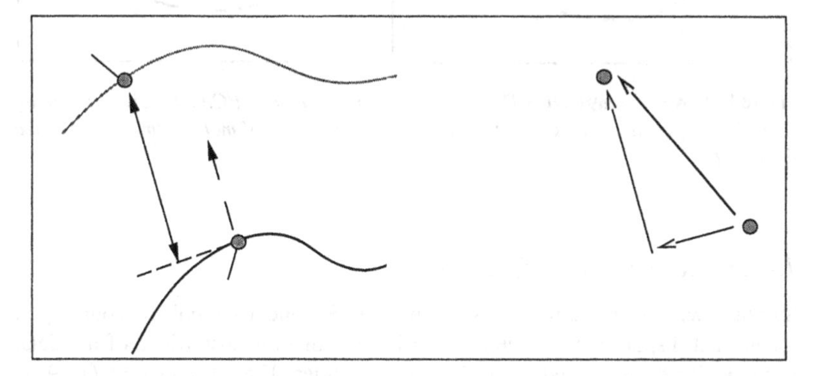

Figure 18.4. **Normal displacement** *a) Displacement along the normal from one curve to another, as shown, forms the basis for a measure of difference between curves that is approximately invariant to reparametrisation. b) Total displacement can be factored vectorially into two components, tangential and normal. Image reprinted from [94].*

[384, 223].

For full details on curve fitting, regularisation, recursive fitting and normal displacement see [94, ch. 6].

### 18.3.4 Nonrigidity

Nonrigid motions fall outside the affine families described above, but may still be captured by a suitable space of shapes. The widely used "Active Shape Model" (ASM) [223] does this by analysing a training set of contours, and constructing an eigen-space of shape by Principal Components Analysis (PCA). Initially the high-dimensional parameterisation $X = (\mathbf{q}_i,\ i = 1, \ldots, V)$ of polygon vertices is chosen. Then the training set $\{\mathbf{r}_1(s), \ldots, \mathbf{r}_{N_T}(s)\}$ of curves is encoded in terms of its polygon-vertex representation $X_1, \ldots, X_{N_T}$. Now the sample covariance matrix $\Sigma$ of the $X_1, \ldots, X_{N_T}$ is computed and, as usual in PCA, its dominant eigenvectors are retained, and form a compact basis for curve shape. Components in this basis form a new, low-dimensional curve parameter $X$ which captures nonrigidity. Finally it is possible to combine the rigid and the nonrigid approach by explicitly projecting out the affine variations in the training set $\{\mathbf{r}_1(s), \ldots, \mathbf{r}_{N_T}(s)\}$ of cirves, and using PCA to account only for the remaining nonrigid variability. In this way the curve parameter $X$ contains both affine components and, separately, components for nonrigid deformation as in figure 18.5.

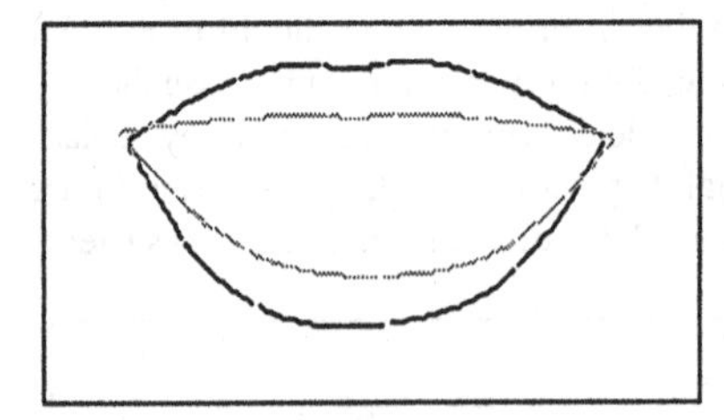
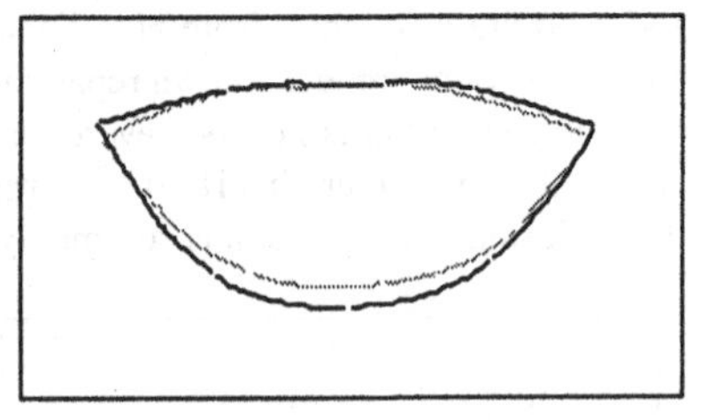

Figure 18.5. **ASM components** *The dominant eigenvectors from PCA analysis of a training set of lip shapes, describing the main non-rigid components of motion. Images reprinted from [94].*

### 18.3.5 Robust curve distances

Simple least squares error measures like (18.9), and its modified counterpart for normal displacement, have no built in robustness to distortions of the data, in particular those caused by occlusion and clutter. The advantage of (18.9) is its tractability, in that it is quadratic and so can be minimised in closed form. "Chamfer matching", which has been used with notable success in pedestrian detection [338], exchanges some tractability for robustness. In place of summing squared-distance (18.9), summing a truncated distance $\int d_\epsilon(\mathbf{r}(s; X) - \mathbf{r}_f(s))\, \mathrm{d}s$, where $d_\epsilon(x) = \min(|x|, \epsilon)$, is more tolerant to outliers. Furthermore, the ideal of minimising over possible parameterisations, previously approximated by normal displacements, can be fully restored to give an asymmetric distance

$$\rho = \int \min_{s'} d_\epsilon(\mathbf{r}(s; X) - \mathbf{r}_f(s'))\, \mathrm{d}s, \tag{18.11}$$

which can be expressed as

$$\rho = \int D(\mathbf{r}(s; X))\, \mathrm{d}s, \quad \text{where} \quad D(\mathbf{r}) = \min_{s'} d_\epsilon(\mathbf{r} - \mathbf{r}_f(s')). \tag{18.12}$$

The image $D(\mathbf{r})$ is the "chamfer image" which can be precomputed for a given observed data curve $\mathbf{r}_f(.)$. In this way, much of the computational load of computing $\rho$ is compiled, once for all, into the computation of $D(\mathbf{r})$. Then the marginal cost of multiple evaluations of $\rho$ for numerous different values of $X$ is very low, consisting simply of a summation along the curve $\mathbf{r}(s; X)$. This low marginal cost makes up considerably for the lack of closed form minimisation, and can be used to search efficiently over both pose and shape. Further organisation of shapes into a tree structure based on similarlity makes matching even more efficient by reducing the number of evaluations of $\rho$ required, and this has been very successful in matching even articulated shapes [338, 763].

A related distance measure [415], mentioned briefly here as a relative of the chamfer distance, is the Hausdorff distance $\min_s \min_{s'} |\mathbf{r}(s; X) - \mathbf{r}_f(s')|$ which is also asymmetric and, in its pure form, not robust. Robustness is dealt with in practice by replacing $\min_s$, which is frail in that it makes the Hausdorff distance dependent on the distance between two particular points on each of the curves, by a quantile over $s$.

## 18.4 Spatio-temporal filtering

The difference between tracking and localisation is that tracking exploits object dynamics, both for efficiency and for effectiveness.

### *18.4.1 Dynamical models*

Dynamical models can be more or less elaborate, according to the nature of the motion being modelled. Some motions, for example of vehicles, talking lips or human gait are often quite predictable and it makes sense to model them in some detail [66, 95]. In any case it is natural to think of a classes of motions, and a probability distributions over that class, which is very naturally represented as an AutoRegressive process (ARP) on the state vector $X$ at time $t$ (denoted $X_t$). A simple ARP on $X_t$, expressed in terms of a "driving" vector $\mathbf{w}_t$ of independent Gaussian noise variables, and constant square matrix $B$, takes the form (first order AR process)

$$X_t = F(X_{t-1}, \mathbf{w}_t), \tag{18.13}$$

with $F$ linear, and some examples follow.

**Tethered**: $X_t = B\mathbf{w}_t$

**Brownian**: $X_t = X_{t-1} + B\mathbf{w}_t$

**Constant velocity**: $X_t = X_{t-1} + B\mathbf{w}_t + \mathbf{v}$

**Constrained Brownian**: $X_t = aX_{t-1} + B\mathbf{w}_t$ with $|a| < 1$

**Damped oscillation**: $X_t = a_1 X_{t-1} + a_2 X_{t-2} + B\mathbf{w}_t$ with appropriate $a_1, a_2$.

The last is, of course, not a first-order AR process, but is 2nd order, of the form $X_t = F(X_{t-1}, X_{t-2}) + \mathbf{w}_t$. Details of the expressive power of various AR models, the roles of the various constants, and algorithms for learning them from training data are detailed in [94, Ch. 9]. Of course these are just a few of the possible linear dynamical models. More elaborate models may also be appropriate, and nonlinearity is also powerful for allowing switching between different kinds of motions [422] — effectively *mixtures* of AR models.

### *18.4.2 Kalman filter for point features*

Classically, the Kalman filter is the exact computational mechanism for incorporating predictions from an AR model of dynamics into a stream of observations, and in due course this important idea was introduced into machine vision [377, 343, 312]. The most straightforward setting is the tracking of point features, such as polyhedral vertices, used with an affinely deforming image structure [673] (recall section 18.3.3) or a 3D rigid body structure [383] (as section 18.3.2). In either case, it is essential to represent explicitly the *uncertainty* in the observation $\mathbf{r}_f(s_i)$ of each point, in terms of independent, two-dimensional standard Gaussian noise vectors $\nu_i$:

$$\mathbf{r}_f(s_i) = \mathbf{r}(s_i, X) + \sigma_i \nu_i \;\; i = 1, \ldots, M \tag{18.14}$$

where $\sigma_i$ is the magnitude of the positional uncertainty associated with the measured the image location $\mathbf{r}_f(s_i)$ of the $i^{\text{th}}$ feature. Measurement uncertainty can then be traded off with uncertainty in the (noise driven) AR predictions to achieve a natural and automatic balance between the influence of observations and of prediction. The result is that an estimate $\hat{X}_t$ of state $X_t$ is propagated in the following manner.

*At each clock tick, predict:*

$$\hat{X}_t = F(\hat{X}_{t-1}, \mathbf{0}). \tag{18.15}$$

— the ARP prediction equation (18.13) with zero noise.

*Each measurement* $\mathbf{r}_f(s_1, t), \ldots, \mathbf{r}_f(s_M, t)$ *is assimilated as:*

$$\hat{X}_t = \leftarrow \hat{X}_t + K_{i,t}(\mathbf{r}_f(s_i, t) - \mathbf{r}(s_i, \hat{X}_t)). \tag{18.16}$$

The "Kalman gains" $K_{i,t}$ are computed by an associated recursion whose details are omitted here, but see e.g. [268].

### 18.4.3 *Kalman filter for contours*

Kalman filtering for contour tracking [93] proceeds in a similar fashion as for point-features, but using the idea of normal displacement, introduced in section 18.3.3 and illustrated here in fig 18.6. Only the normal component of feature

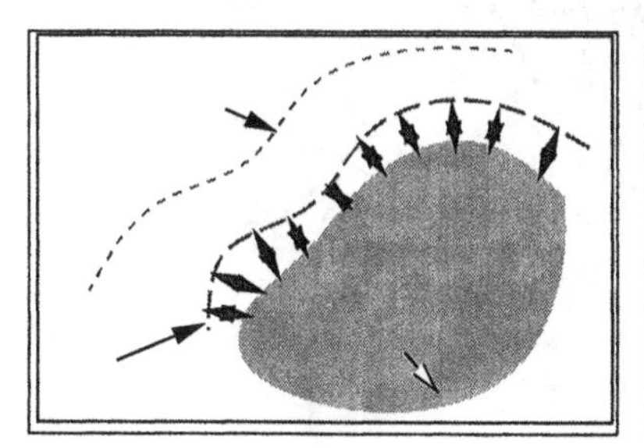
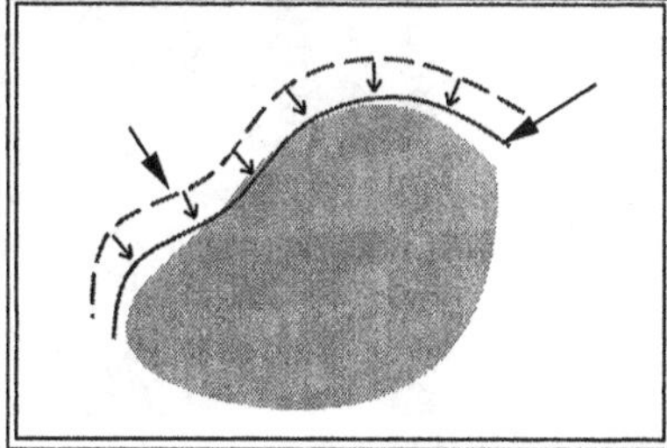

Figure 18.6. **Kalman filter for contours** *Prediction and measurement phases for contours, with observations (double arrows) of normal displacement. Images reprinted from [94].*

displacement is assimilated, so that step (18.16) above takes instead the form:

$$\hat{X}_t = \leftarrow \hat{X}_t + K'_{i,t}[\mathbf{n}(s_i, t) \cdot (\mathbf{r}_f(s_i, t) - \mathbf{r}(s_i, \hat{X}_t))], \tag{18.17}$$

where $\mathbf{n}(s_i, t)$ is the normal to the curve $\mathbf{r}(s, \hat{X}_t)$ at the $i^{\text{th}}$ sample point $s = s_i$. Unlike the case of point features, where the locations $s = s_i$ are locations on the contour of distinguished point features, here the $s = s_i$ are simply a convenient sampling pattern along the length of the contour, implementing a numerical approximation of the mean-square normal displacement.

### 18.4.4 *Particle filter*

The Kalman filter has two limitations that can prove very restrictive in relatively unconstrained tracking problems.

1. **Clutter:** it is limited to one observation $\mathbf{r}_f(s_i, t)$ for each contour location $\mathbf{r}(s_i, t)$. *Clutter* in the image tends to generate multiple observations at each location, as figure 18.7 shows.

2. **Dynamics:** the Kalman filter is limited to ARP models of dynamics. Mild non-linearities can be dealt with, in practice, by local linearisation. Hybrid dynamical models that switch between ARPs (e.g. flight/bouncing/rolling) demand a more powerful mechanism for temporal filtering.

Particle filters are a class of Monte-Carlo temporal filters that are more powerful than the Kalman filter in that they escape both from the restrictions of clutter

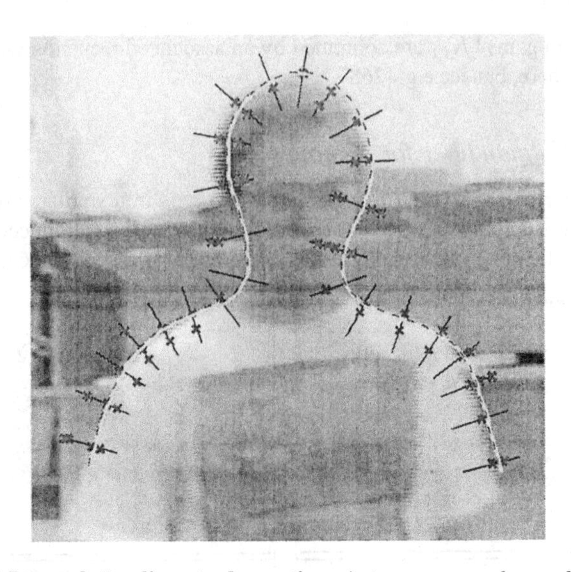

Figure 18.7. **Image clutter disrupts observations** *Active contour and normals are shown. Crosses mark observations of high contrast features, some of which are triggered by the true object outline while others are responding to clutter, both inside and outside the object. Image reprinted from [419].*

[419] and dynamics [422], but at the cost of being only approximate. The idea of sampling shapes in cluttered observations derives originally from static studies [365]. The earliest form of the particle filter was the "bootstrap filter" [355]. The more powerful form described here is based [421, 523] on importance sampling .

The essence of the particle filter is summarised in figure 18.8. In place of the single estimate $\hat{X}_t$ in the Kalman filter, particle filters maintain an entire set $\{X_{t-1}^n, n = 1, \ldots, N_S\}$ of possible estimated values of the state $X_t$. This is a robust approach that allows the explicit representation of ambiguity in a way that a Kalman filter simply cannot. For example in clutter, the ambiguity is generated by uncertainty as to which of many visible features is actually generated by the true object. With hybrid dynamics, the ambiguity reflects uncertainty as to which ARP model currently explains the observed motion; typically ambiguity is heightened around the time that the model switches. The particle set for time $t$ consists of the set of possible values $\{X_{t-1}^n\}$ along with a set of positive weights $\{\pi_{t-1}^n\}$.

The algorithm description explains how the particle set evolves from one timestep to the next. First new values ${X_t}^n$ are generated by sampling from a proposal distribution $q_t$. In the simplest CONDENSATION [419] or bootstrap [355] forms of the filter,

$$q_t(X_t \mid X_{t-1}^n) = p(X_t \mid X_{t-1} = X_{t-1}^n)$$

— the proposal is simply a simulation of the dynamical model itself. In other words, particles are generated by predicting the change of state from time-step

**Temporal update** for time step $t-1 \rightarrow t$

From the sample-set $\{X_{t-1}^n, \pi_{t-1}^n, n = 1, \ldots, N_S\}$ at time $t-1$, construct a new sample-set $\{X_t^n, \pi_t^n\}, n = 1, \ldots, N_S$ for time $t$, as follows.

1. **Select** samples $X_t{}^n$ by sampling from the "proposal distribution" $q_t(X \mid X_{t-1}^n)$.
2. **Weight** the new particles in terms of the vector of measured features $\mathbf{z}_t = \{\mathbf{r}_f(s_1, t), \ldots, \mathbf{r}_f(s_M, t)\}$:
$$\pi_t^n = \pi_{t-1}^n \frac{p(\mathbf{z}_t|X_t = X_t^n)\, p(X_t = X_t^n|X_{t-1} = X_{t-1}^n)}{q_t(X_t = X_t^n \mid X_{t-1} = X_{t-1}^n)}$$
3. **Resample**, at occasional time-steps, to avoid the distribution of weights becoming too uneven:
   (a) Sample, with replacement, from $\{X_t^n, n = 1, \ldots, N_S\}$, selecting $X_t^n$ with probability proportional to $\pi_t^n$, to form a new, resampled set $\{X_t^n, n = 1, \ldots, N_S\}$.
   (b) Reset all weights to $\pi_t^n = 1$.

Figure 18.8. **A Particle filter.** *Standard form of particle filter, following [641].*

$t-1$ to timestep $t$. In the case of ARP dynamics (18.13) this gives

$$X_t^n = F(X_{t-1}^n, \mathbf{w}_t^n), \tag{18.18}$$

where the $\mathbf{w}_t^n$, $n = 1, \ldots N$ are independent draws of a standard normal variable, thus using the ARP to make noisy predictions of object position. In this way, particles $X_t^n$ sweep out a set of *a priori* probably values for $X_t$. A more adventurous form of proposal distribution uses hints from the image — "importance sampling" — at time $t$ to generate probable values for $X_t$. For example, tracking hands or faces, a "pinkness" measure $q_t^{\text{pink}}(X)$ can be used to generate states likely to coincide with skin colouration in the image.

The second step of the algorithm generates the weights $\pi_t^n$ and in doing so achieves two things: i) it takes account of the new measurements $\mathbf{r}_f(s_i, t)$; and ii) it compensates for any bias in the proposal distribution $q_t(.)$. Again, the simplest case is the CONDENSATION filter, in which $q_t(.)$ is unbiased, and the formula for weights simplifies to

$$\pi_t^n = \pi_{t-1}^n p(\mathbf{z}_t|X_t = X_t^n). \tag{18.19}$$

A simple example of a measurement process was given earlier (18.14), and in that case the observation likelihood is the Gaussian

$$p(\mathbf{z}|X) \propto \exp - \sum_{i=1}^{M} \frac{1}{2\sigma^2} \|\mathbf{r}_f(s_i) - \mathbf{r}(s_i, X)\|^2. \tag{18.20}$$

Of course, part of the point of the particle filter is to be able to track in clutter, and then the simple likelihood (18.20) is replaced by something non-Gaussian with multiple modes [419].

The third step of the algorithm controls the efficacy of the particle set in representing the posterior distribution over $X_t$ via occasional reweightings. Details of how exactly reweighting is triggered are omitted here, but see [641].

Results of particle filtering for an active contour was given in figure 18.1. This example uses simple CONDENSATION [419] to track a blowing leaf in severe clutter. The figure shows a trail of estimated mean states $\overline{X}_t = [\sum_n \pi_t^n X_t^n]/[\sum_n \pi_t^n]$ over time.

## 18.5 Further topics

There are a number of further topics in tracking that build on the ideas already outlined, and go beyond them in various intriguing ways. There is no space here to explore them in the depth they deserve, so pointers and brief summaries will have to suffice.

**Fusing contour and appearance** Much of this roadmap has addressed contour tracking, and in section 18.2 we briefly outlined approaches to appearance tracking. More recently there have been breakthroughs in joint modelling and localisation of contour and appearance [221] and the related approach [718], without dynamics however. An alternative fusion of appearance and contour combines particle filtering of contours [640] with an observation model like the one used in mean-shift tracking.

**Filter Banks** Observations based around contours have drawbacks both from the point of view of the principles of good Bayesian inference and, as above, the need to fuse both contour and appearance information. A complementary approach is to model the observations as the joint output of a set of filter banks [340, 773], which harnesses both appearance from filters within the object contour, and contrast from those that straddle the contour. The approach becomes even more powerful when combined with background modelling [423]. Another impressively powerful variation models filter outputs as a hybrid [436], with each filter switching independently between models for stasis, steady motion, or random walk.

**Articulated and deformable structures** Modelling deformation has been discussed above, and there are numerous variations on the theme, for example "deformable templates" [317, 914]. Outright articulation — jointed assemblies of rigid bodies — can be dealt with effectively using greedy strategies [402, 672], though at considerable computational cost, which can be mitigated using observation-cost gradient information [115]. Alternatively, the ASM approach of section 18.3.4 can be used for articulation also [90]. Issues arising in image-based models when image topology changes as the body articulates have been addressed using several shape space models connected via "wormholes" [391], in a Markov network. Alternatively, cartoon-like catalogues of outline-exemplars with differing topologies [338, 801], also connected in a Markov network, and matched using chamfers, are a very effective memory-intensive approach.

**Persistence** Finally, there have been striking advances in trained recognisers for localising faces and walking figures, in a single frame [847, 848]. These are so powerful and efficient that, without any recourse to dynamical models, real-time performance can be achieved on a modern workstation. However, these too can benefit from a dynamical approach [37, 894], promising real-time tracking in the background of a desktop machine's process load, and on portable devices, in the future.

All of these issues and others will be treated in more detail in a forthcoming, long version of this roadmap article [92].

# Chapter19

# Shape Gradient for Image and Video Segmentation

**S. Jehan-Besson, A. Herbulot, M. Barlaud, G. Aubert**

Abstract

In this chapter, we propose to concentrate on the research of an optimal domain with regards to a global criterion including region and boundary functionals. A local shape minimizer is obtained through the evolution of a deformable domain in the direction of the shape gradient. Shape derivation tools, coming from shape optimization theory, allow us to easily differentiate region and boundary functionals. We more particularly focus on region functionals involving region-dependent features that are globally attached to the region. A general framework is proposed and illustrated by many examples involving functions of parametric or non parametric probability density functions (pdfs) of image features. Among these functions, we notably study the minimization of information measures such as the entropy for the segmentation of homogeneous regions or the minimization of the distance between pdfs for tracking or matching regions of interest.

## 19.1 Introduction

Active contours are powerful tools for image and video segmentation or tracking. They can be formulated in the framework of variational methods. The basic principle is to construct a PDE (Partial Differential Equation) from an energy criterion, including usually both region and boundary functionals. This PDE changes the shape of the current curve according to some velocity field which can be thought of as a descent direction of the energy criterion. Given a closed curve enclosing an initial region, one then computes the solution of this PDE for this initial condition. The corresponding family of curves decreases the energy criterion and converges toward a (local) minimum of the criterion hopefully corresponding to the objects to be segmented.

Originally, snakes [456], balloons [204] or geodesic active contours [157] are driven towards the edges of an image through the minimization of a boundary integral of features depending on edges. Active contours driven by the minimization of region functionals in addition to boundary functionals have appeared later. Introduced by [207] and [680], they have been further developed in [922, 174, 192, 626, 625, 254, 910]. Actually, the use of active contours for the optimization of a criterion including both region and boundary functionals appears to be powerful.

However, the PDE computation is not trivial when the energy criterion involves region functionals. This is mostly due to the fact that the set of image regions does not have a structure of vector space, preventing us to use in a straightforward fashion gradient descent methods. To circumvent this problem, we propose to take benefit of shape derivation principles developed by [751, 256]. This computation becomes more involved when global information about regions is present in the energy criterion, the so-called region-dependent case. It happens when statistical features of a region such as, for example, the mean or the variance of the intensity, are involved in the minimization. In this chapter, we propose a general framework based on shape derivation tools for the computation of the related evolution equation. Inside this theoretical framework, many descriptors based on parametric or non parametric pdfs of image features may be studied. We propose to give some results for both of them and some examples of applications.

Region and boundary functionals are presented in section 19.2 while shape derivation tools are presented in section 19.3. Statistical region-dependent descriptors based on parametric and non parametric probability density functions (pdfs) are studied in section 19.4.

## 19.2 Problem Statement

In many image processing problems, the issue is to find a set of image regions that minimize a given error criterion. The basic idea of active contours is to compute a Partial Differential Equation (PDE) that will drive the boundary of an initial region towards a local minimum of the error criterion. The key point is to compute the velocity vector at each point of the boundary at each time instant.

To fix ideas, in the two-dimensional case, the evolving boundary, or active contour, is modeled by a parametric curve $\Gamma(s,\tau) = (x_1(s,\tau), x_2(s,\tau))$, where $s$ may be its arc-length and $\tau$ is an evolution parameter. The active contour is then driven by the following PDE:

$$\Gamma_\tau \stackrel{def}{=} \frac{\partial \Gamma}{\partial \tau} = \mathbf{v} \quad \text{with } \Gamma(\tau = 0) = \Gamma_0,$$

where $\Gamma_0$ is an initial curve defined by the user and $\mathbf{v}$ the velocity vector of $\Gamma(s,\tau)$. This velocity is the unknown that must be differentiated from an error criterion so that the solution $\Gamma(.,\tau)$ converges towards a curve achieving a local minimum and thus, hopefully, towards the boundary of the object to be segmented, as $\tau \to \infty$.

Following the pioneer work of Mumford Shah [591], a segmentation problem may be formulated through the minimization of a criterion including both region and boundary functionals. Let $\mathcal{U}$ be a class of domains of $\mathcal{R}^n$, and $\Omega$ an element of $\mathcal{U}$ of boundary $\partial\Omega$. A boundary functional, $J_b$, may be expressed as a boundary integral of some scalar function $k_b$ of image features:

$$J_b(\partial\Omega) = \int_{\partial\Omega} k_b(\mathbf{x}, \partial\Omega)\, d\mathbf{a}(\mathbf{x}) \tag{19.1}$$

where $\partial\Omega$ is the boundary of the region and $d\mathbf{a}$ its area element.
The most classical example of boundary functional comes from the work of Caselles et al [157], where the authors minimize for an image in 2D:

$$J(\partial\Omega) = \int_{\partial\Omega} g(|\nabla I(\partial\Omega(s))|)ds$$

where $s$ represents the arc length of the curve $\partial\Omega$ and $g(r) = \frac{1}{1+r^m}$, $m = 1$ or 2. The function $g$ drives the curve towards the image edges characterized by high values of the image gradient.

A region functional, $J$, may be expressed as an integral, in a domain $\Omega$ of $\mathcal{U}$, of some function $k$ of some region features:

$$J(\Omega) = \int_{\Omega} k(\mathbf{x}, \Omega)d\mathbf{x} \tag{19.2}$$

Let us note that the scalar function $k$ in (19.2) is generally region-dependent. A classical example of region-dependent descriptor is the following one proposed by [174, 254]:

$$k(\mathbf{x}, \Omega) = (I(\mathbf{x}) - \mu(\Omega))^2$$

where $\mu(\Omega)$ represents the mean of the intensity values within the region $\Omega$. This dependency on the region must be taken into account when searching for a local minimum of the functional.

Generally one uses a linear combination of region-based and contour-based terms in order to perform a segmentation task. A simple example is the segmentation into two regions $\Omega_{in}$ and $\Omega_{out}$, which basically correspond to objects and background. An appropriate energy functional for this task would be:

$$J(\Omega_{in}, \Omega_{out}) = \int_{\Omega_{in}} k_{in}(\mathbf{x}, \Omega_{in})\, \mathrm{d}\mathbf{x} + \int_{\Omega_{out}} k_{out}(\mathbf{x}, \Omega_{out})\, \mathrm{d}\mathbf{x} + \int_{\partial\Omega_{in}} k_b(\mathbf{x})\, \mathrm{d}s$$

where $k_{in}$ is the descriptor for the object region, $k_{out}$ for the background region and $k_b$ the descriptor for the contour.

The choice of the descriptors is dependent on the application. In this article we propose to focus on statistical descriptors based on parametric or non parametric pdfs. Once this choice is made the terms have to be derived in order to calculate a velocity function that drives an initial contour towards a minimum. A detailed state of the art on region-based active contours can be found in [432]. Let us briefly note that some authors do not compute the theoretical expression of the velocity field but choose a deformation of the curve that will make the criterion decrease [159, 192]. Other authors [922, 625] compute the theoretical expression of the velocity vector from the Euler-Lagrange equations. The computation is performed in two main steps. First, region integrals representing region functionals are transformed into boundary integrals using the Green-Riemann theorem. Secondly, the corresponding Euler-Lagrange equations are derived, and used to define a dynamic scheme in order to make evolve the initial region. Another alternative is to keep the region formulation to compute the gradient of the energy criterion with respect to the region instead of reducing region integrals to boundary integrals. In [254], the authors propose to compute the derivative of the criterion while taking into account the discontinuities across the contour. In [431, 432] the computation of the evolution equation is achieved through shape derivation principles.

This computation becomes more difficult for region-dependent descriptors. It happens when statistical features of a region such as, for example, the mean or the variance of the intensity, are involved in the minimization. This case has been studied in [174, 254, 910, 465, 234]. In [431, 432] the authors show that the minimization of functionals involving region-dependent features can induce additional terms in the evolution equation of the active contour that are important in practice. These additional terms are easily computed thanks to shape derivation tools.

In the following, we present shape derivation tools for the computation of the evolution equation.

## 19.3 From shape derivation tools towards region-based active contours models

As far as the derivation is concerned, two main difficulties must be solved. First, the set of image regions, i.e. the set of regular open domains in $\mathcal{R}^n$, denoted by $\mathcal{U}$, does not have a structure of vector space, preventing us from using in a straightforward fashion gradient descent methods. To circumvent this problem, shape derivation methods [751, 256] can be brought to bear on the problem as detailed in this section. Secondly, the descriptors $k_r$ or $k_b$ may be region or boundary-dependent. Such a dependence must be taken into account in the derivation of the functionals as pointed out in [431, 432, 34, 335]. We here recall a theorem giving relation between derivatives that will be helpful for derivation of region functionals for both region-independent and region-dependent descriptors. We also give some details and references for the derivation of boundary-based terms using shape derivation tools.

### 19.3.1 Shape derivation tools

#### 19.3.1.1 Introduction of transformations

As it has already been pointed out, the optimization of the region functional $J(\Omega)$ is difficult since $\mathcal{U}$ does not have the structure of a vector space. Variations of a domain must then be defined in some way. Let us consider a reference domain $\Omega \in \mathcal{U}$ and the set $\mathbf{A}$ of applications $T : \Omega \to \mathcal{R}^n$, which are at least as regular as homeomorphisms (i.e. one to one with $T$ and $T^{-1}$ continuous). We define

$$\hat{\mathcal{A}} = \{T \text{ one to one}, T, T^{-1} \in W^{1,\infty}(\Omega, \mathcal{R}^n)\}$$

where:

$$\begin{aligned} W^{n,\infty}(\Omega, \mathcal{R}^n) &= \{T : \Omega \to \mathcal{R}^n \text{ such that} \\ &\quad T \in L^\infty(\Omega, \mathcal{R}^n) \text{ and } \partial_i T \in L^\infty(\Omega, \mathcal{R}^n),\ i = 1, \cdots, n\} \end{aligned}$$

Given a shape function $F : \mathcal{U} \to \mathcal{R}^+$, for $T \in \hat{\mathcal{A}}$, let us define $\hat{F}(T) = F(T(\Omega))$. The key point is that $W^{1,\infty}(\Omega, \mathcal{R}^n)$ is a Banach space. This allows us to define the notion of derivative with respect to the domain $\Omega$ as follows:

**Definition 19.3.1.** *$F$ is Gâteaux differentiable with respect to $\Omega$ if and only if $\hat{F}$ is Gâteaux differentiable with respect to $T$.*

In order to compute Gâteaux derivatives with respect to $T$ we introduce a family of deformation $(T(\tau))_{\tau \geq 0}$ such that $T(\tau) \in \mathbf{A}$ for $\tau \geq 0$, $T(0) = Id$, and $T(.) \in C^1([0, A]; W^{1,\infty}(\Omega, \mathcal{R}^n)$, $A > 0$.

For a point $\mathbf{x} \in \Omega$, we denote:

$$\begin{aligned} \mathbf{x}(\tau) = T(\tau, \mathbf{x}) &\quad \text{with} \quad T(0, \mathbf{x}) = \mathbf{x} \\ \Omega(\tau) = T(\tau, \Omega) &\quad \text{with} \quad T(0, \Omega) = \Omega \end{aligned}$$

Let us now define the velocity vector field $\mathbf{V}$ corresponding to $T(\tau)$ as

$$\mathbf{V}(\tau, \mathbf{x}) = \frac{\partial T}{\partial \tau}(\tau, \mathbf{x}) \quad \forall \mathbf{x} \in \Omega \ \forall \tau \geq 0$$

#### 19.3.1.2 Relations between the derivatives

We now introduce two main definitions:

**Definition 19.3.2.** *The Gâteaux derivative of $J(\Omega) = \int_\Omega f(\mathbf{x}, \Omega) d\mathbf{x}$ in the direction of $\mathbf{V}$, noted $dJ_r(\Omega, \mathbf{V})$, is equal to:*

$$dJ_r(\Omega, \mathbf{V}) = \lim_{\tau \to 0} \frac{J(\Omega(\tau)) - J(\Omega)}{\tau}$$

*This derivative is called the Eulerian derivative.*

**Definition 19.3.3.** *The shape derivative of $k(\mathbf{x}, \Omega)$, noted $k_s(\mathbf{x}, \Omega, V)$, is equal to:*

$$k_s(\mathbf{x}, \Omega, \mathbf{V}) = \lim_{\tau \to 0} \frac{k(\mathbf{x}, \Omega(\tau)) - k(\mathbf{x}, \Omega)}{\tau}$$

The following theorem gives a relation between the Eulerian derivative and the shape derivative for the region functional (19.2). The proof can be found in [751, 256], an elementary one is provided in [432] for completeness.

**Theorem 19.3.1.** *The Eulerian derivative of the functional $J(\Omega) = \int_\Omega k(\mathbf{x}, \Omega)\, d\mathbf{x}$ in the direction of* $\mathbf{V}$ *is the following:*

$$dJ_r(\Omega, \mathbf{V}) = \int_\Omega k_s(\mathbf{x}, \Omega, \mathbf{V}) d\mathbf{x} - \int_{\partial\Omega} k(\mathbf{x}, \Omega)(\mathbf{V}(\mathbf{x}) \cdot \boldsymbol{N}(\mathbf{x})) d\mathbf{a}(\mathbf{x})$$

*where* $\boldsymbol{N}$ *is the unit inward normal to* $\partial\Omega$ *and* $d\mathbf{a}$ *its area element.*

Note that Theorem 19.3.1 provides a necessary condition for a domain $\hat{\Omega}$ to be an extremum of $J(\Omega)$:

$$\int_{\hat{\Omega}} k_s(\mathbf{x}, \hat{\Omega}, \mathbf{V}) d\mathbf{x} - \int_{\partial\hat{\Omega}} k(\mathbf{x}, \hat{\Omega})(\mathbf{V}(\mathbf{x}) \cdot \mathbf{N}(\mathbf{x}))\, d\mathbf{a}(\mathbf{x}) = 0 \quad \forall \mathbf{V}.$$

## 19.3.2 *Derivation of boundary-based terms*

In the case of boundary-independent descriptors, the Eulerian derivative of $J_b = \int_{\partial\Omega} k_b(\mathbf{x}) d\mathbf{a}(\mathbf{x})$ in the direction $v_n = (\mathbf{V} \cdot \mathbf{N})$ is the following:

$$dJ_b(\partial\Omega, v_n) = \int_{\partial\Omega} (\nabla k_b(\mathbf{x}) \cdot \mathbf{N} - k_b(\mathbf{x})\, \kappa)(\mathbf{V} \cdot \mathbf{N}) d\mathbf{a} \tag{19.3}$$

where $\kappa$ is the mean curvature of $\partial\Omega$.

>From this Eulerian derivative, we can deduce the following evolution equation for the active contour:

$$\Gamma_\tau = (k_b(\mathbf{x})\, \kappa - \nabla k_b(\mathbf{x}) \cdot \mathbf{N})\mathbf{N} \quad \text{with} \quad \Gamma(\tau = 0) = \Gamma_0. \tag{19.4}$$

This evolution equation has been computed by Caselles et al [157] by using techniques of calculus of variations.

As far as boundary-dependent descriptors are concerned, the dependence on the boundary must be taken into account for the computation of the Eulerian derivative. In [335], the authors studied the following descriptor which represents the distance between the current boundary $\partial\Omega$ and a reference one $\partial\Omega_{ref}$:

$$k_b = d(\partial\Omega, \partial\Omega_{ref}).$$

The authors compute the evolution equation and they show that some terms appear coming from the dependency of the descriptor with the boundary. This descriptor has been used for the introduction of shape prior for segmentation. Let us note that the introduction of shape priors for segmentation using active contours has also been studied by [628, 233, 235]. Let us also note that in [399], the authors remind some theorems for the computation of the Eulerian derivative of boundary-dependent descriptors and in [177], the authors deal with shape metrics following considerations developed in [256].

### 19.3.3 *Derivation of region-based terms*

Let us now apply the previous results to differentiate the velocity vector of the active contour.

#### 19.3.3.1 Region-independent descriptors

We first consider the simple case where the function $k$ does not depend on $\Omega$, i.e. $k = k(\mathbf{x})$. In that case, the shape derivative $k_s$ is equal to zero and the Eulerian derivative of $J$ is simply (Theorem 19.3.1):

$$dJ_r(\Omega, \mathbf{V}) = -\int_{\partial\Omega} k(\mathbf{x})(\mathbf{V}(\mathbf{x}) \cdot N(\mathbf{x}))d\mathbf{a}(\mathbf{x})$$

This leads to the following evolution equation for region-independent descriptors:

$$\Gamma_\tau = k\mathbf{N} \quad \text{with} \quad \Gamma(\tau = 0) = \Gamma_0.$$

This is the classical result [922, 625] when $k$ has no region dependency. Let us now consider the more general case where the function $k$ has some region dependency.

#### 19.3.3.2 Region-dependent descriptors

Region-dependent descriptors of the form $J_r(\Omega) = \int_\Omega k(\mathbf{x}, \Omega)d\mathbf{x}$ are more complicated to differentiate. Using Theorem 19.3.1 one can obtains a derivative of the following form [432, 34] for some of them (see section 19.4):

$$dJ_r(\Omega, \mathbf{V}) = -\int_{\partial\Omega} (k(\mathbf{x}, \Omega) + A(\mathbf{x}, \Omega))(\mathbf{V} \cdot \mathbf{N})\, \mathrm{da} \tag{19.5}$$

This leads to the following evolution equation for these region-dependent descriptors:

$$\Gamma_\tau = (k + A)\mathbf{N} \quad \text{with} \quad \Gamma(\tau = 0) = \Gamma_0.$$

The term $A(\mathbf{x}, \Omega)$ is a term that comes from the region-dependence and so from the evaluation of the shape derivative $k_s$. We here propose a general framework for deriving some region-dependent descriptors based on parametric or non parametric statistics. The principle is to model region-dependent descriptors as follows:

$$J(\Omega) = \int_\Omega k(\mathbf{x}, G(\Omega))\, d\mathbf{x}, \quad \text{where} \quad G(\Omega) = \int_\Omega H(\mathbf{x}, \Omega)\, d\mathbf{x} \tag{19.6}$$

As shown in this equation, the function $H$ is itself region-dependent, more precisely:

$$H(\mathbf{x}, \Omega) \stackrel{def}{=} H(\mathbf{x}, K(\Omega))\,, \quad \text{and} \quad K(\Omega) = \int_\Omega L(\mathbf{x})\, d\mathbf{x} \tag{19.7}$$

Note that we have stopped the process at the second level but it could conceivably continue. We have chosen this special case of dependency because it often arises in applications, as shown in sections 19.4.2 and 19.4.1.

**Theorem 19.3.2.** *The Eulerian derivative in the direction of* $\mathbf{V}$ *of the functional* $J$ *defined in (19.9) is:*

$$d_r J(\Omega, \mathbf{V}) = -\int_{\partial\Omega} (A(\mathbf{x}, \Omega) + k(\mathbf{x}, \Omega))\ (\mathbf{V}(\mathbf{x}) \cdot \mathbf{N}(\mathbf{x}))da(\mathbf{x})$$

*where :*

$$A(\mathbf{x}, \Omega) = \left(\int_{\Omega} k_G(\mathbf{x}, G(\Omega))\, d\mathbf{x}\right) \left(L(\mathbf{x}) \int_{\Omega} H_K(\mathbf{x}, K(\Omega))\, d\mathbf{x} + H(\mathbf{x}, K(\Omega))\right)$$

*The terms* $k_G$ *and* $H_K$ *denote respectively the partial derivative of the function* $k$ *and* $H$ *with respect to their second argument.*

*Proof:* According to Theorem 19.3.1, we have:

$$d_r J(\Omega, \mathbf{V}) = \int_{\Omega} k_s\, d\mathbf{x} - \int_{\partial\Omega} k\ (\mathbf{V} \cdot \mathbf{N})da(\mathbf{x})$$

Let us first compute the shape derivative of $k$. From the chain rule we get:

$$k_s(\mathbf{x}, \Omega, \mathbf{V}) = k_G(\mathbf{x}, G) d_r G(\Omega, \mathbf{V}), \tag{19.8}$$

where $k_G$ denotes the partial derivative of the function $k$ with respect to its second argument.

Next we compute the Eulerian derivative of $G$ in the direction of $\mathbf{V}$. We apply again Theorem 19.3.1, and we get:

$$d_r G(\Omega, \mathbf{V}) = \int_{\Omega} H_s\, d\mathbf{x} - \int_{\partial\Omega} H\ (\mathbf{V} \cdot \mathbf{N})da(\mathbf{x}).$$

Plugging this into (19.8), we obtain:

$$\int_{\Omega} k_s\, d\mathbf{x} = \left(\int_{\Omega} k_G(\mathbf{x}, G(\Omega))\, d\mathbf{x}\right) \left(\int_{\Omega} H_s\, d\mathbf{x} - \int_{\partial\Omega} H(\mathbf{V} \cdot \mathbf{N})da(\mathbf{x})\right),$$

We also compute the shape derivative of $H$ thanks to Theorem 19.3.1:

$$H_s(\mathbf{x}, \Omega, \mathbf{V}) = H_K(\mathbf{x}, K) d_r K(\Omega, \mathbf{V})$$

The Eulerian derivative of $K$ in the direction of $\mathbf{V}$ is given by:

$$d_r K(\Omega, \mathbf{V}) = \int_{\Omega} L_s\, d\mathbf{x} - \int_{\partial\Omega} L(\mathbf{x})(V(\mathbf{x}) \cdot \mathbf{N}(\mathbf{x}))da(\mathbf{x})$$

Since $L$ does not depend on $\Omega$, we obtain $L_s = 0$ and we get the result.

We can now state the result for the general case where $k$ is described as a linear combination or region functionals as follows:

$$J(\Omega) = \int_{\Omega} k(\mathbf{x}, G_1(\Omega), G_2(\Omega), .., G_m(\Omega))\, d\mathbf{x}, \tag{19.9}$$

where the functionals $G_i$ are given by $G_i(\Omega) = \int_\Omega H_i(\mathbf{x},\Omega)\,d\mathbf{x} \quad i = 1..m$. As shown in this equation, the function $H_i$ is itself region-dependent, more precisely:

$$H_i(\mathbf{x},\Omega) \stackrel{def}{=} H_i(\mathbf{x}, K_{i1}(\Omega), K_{i2}(\Omega), .., K_{il_i}(\Omega)) \tag{19.10}$$

$$\text{where} \quad K_{ij}(\Omega) = \int_\Omega L_{ij}(\mathbf{x})\,d\mathbf{x} \quad j = 1..l_i \quad i = 1..m. \tag{19.11}$$

We have chosen this special case of dependency because it often arises in applications, as shown in sections 19.4.2 and 19.4.1.

**Theorem 19.3.3.** *The Eulerian derivative in the direction of* $\mathbf{V}$ *of the functional* $J$ *defined in (19.9) is:*

$$d_r J(\Omega,\mathbf{V}) = -\int_{\partial\Omega} (A(\mathbf{x},\Omega) + k(\mathbf{x},\Omega))\ (\mathbf{V}\cdot\mathbf{N})da.$$

*where* $A(\mathbf{x},\Omega) = \sum_{i=1}^m D_i \sum_{j=1}^{l_i}(B_{ij}\, L_{ij}(\mathbf{x})) + \sum_{i=1}^m (D_i\, H_i)$,

$$\begin{aligned} \textit{and} \quad D_i &= \int_\Omega k_{G_i}(x, G_1(\Omega), .., G_m(\Omega))\,d\mathbf{x} \quad i = 1..m \\ B_{ij} &= \int_\Omega H_{iK_{ij}}(x, K_{i1}(\Omega), .., K_{il_i}(\Omega))\,d\mathbf{x} \quad i = 1..m \quad j = 1..l_i \end{aligned}$$

## 19.4 Segmentation using Statistical Region-dependent descriptors

In this section, we are interested in the minimization of the region functional (19.2) for region-dependent descriptors. The general framework introduced in section 19.3.3.2 allows us to compute the derivative and the evolution equation for many descriptors based on parametric or non parametric statistics. Some examples of computation are given for descriptors based on parametric statistics in section 19.4.1, while a general computation of the derivative is proposed for non parametric statistics in section 19.4.2.

Let us first introduce some notations and some examples of region-dependent descriptors. We note $\mathbf{f}(\mathbf{x})$ the feature of interest of the image at location $\mathbf{x}$. This feature may be the intensity of the image, the motion vector, a shape descriptor and is a function $\mathbf{f} : \Omega_f \longrightarrow \mathbb{R}^m$ where $\Omega_f \subset \mathbb{R}^2$ is the image domain and $m$ is the dimension of the feature. If $\mathbf{f}$ is the image intensity, $m = 1$ for grayscale images and $m = 3$ for color images. If $\mathbf{f}$ is a motion vector, $m = 2$.

When considering the pdf of $\mathbf{f}$ within the region, denoted by $q(\mathbf{f}(\mathbf{x}),\Omega)$, we can choose the following general descriptor for segmentation:

$$k(\mathbf{x},\Omega) = \varphi(q(\mathbf{f}(\mathbf{x}),\Omega)) \tag{19.12}$$

When minimizing the -log-likelihood function for independent and identically distributed observations (iid) $\mathbf{f}(\mathbf{x})$, we have:

$$\varphi(q(\mathbf{f}(\mathbf{x}),\Omega)) = -\ln(q(\mathbf{f}(\mathbf{x}),\Omega) \quad (19.13)$$

When minimizing the entropy function, we get:

$$\varphi(q(\mathbf{f}(\mathbf{x}),\Omega)) = -q(\mathbf{f}(\mathbf{x}),\Omega)\ln(q(\mathbf{f}(\mathbf{x}),\Omega) \quad (19.14)$$

The concept entropy designates the average quantity of information carried out by a feature [229]. Intuitively the entropy represents some kind of diversity of a given feature.

These descriptors may be chosen to characterize the homogeneity of a region according to the feature. In both cases, the pdf may be parametric, i.e. it follows a prespecified law (Gaussian, Rayleigh ...) or non parametric. In the last case, no assumption is made on the underlying distribution.

As far as parametric pdfs are concerned, the descriptor (19.13) has first been introduced by [922] for the segmentation of homogeneous regions using region-based active contours and further developed by [625, 547]. In the case of parametric pdfs, the probability density function $q$ is indexed by one or more parameters, denoted by a vector $\theta$, describing the distribution model. For example, when using a one dimensional Gaussian distribution, we get:

$$q_\theta(f(x),\Omega)) = \frac{1}{\sqrt{2\pi}\sigma}\exp\frac{-(f(x)-\mu)^2}{2\sigma^2}$$

where $\theta = [\mu\ \sigma]^T$. The terms $\mu$ and $\sigma$ represent respectively the mean and the variance of the scalar feature f within the region $\Omega$. Note that the parameters of the distribution depend on $\Omega$ and that such a dependence must be taken into account during the derivation process. Some other descriptors for segmentation are derived from the development of the expression (19.13) for Gaussian distributions. For example, the descriptor $k(\mathbf{x},\Omega) = (\mathbf{I}(\mathbf{x}) - \mu)^2$ has been proposed by [174] for the segmentation of homogeneous regions, and the descriptor $k(\mathbf{x},\Omega) = \varrho(\sigma^2)$ by [432].

As far as non parametric pdfs are concerned, the expression of the pdf $q$ is given by the Parzen method [287]:

$$q(\mathbf{f}(\mathbf{x}),\Omega) = \frac{1}{|\Omega|}\int_\Omega K(\mathbf{f}(\mathbf{x}) - \mathbf{f}(\hat{\mathbf{x}}))\,d\hat{\mathbf{x}} \quad (19.15)$$

where K is the Gaussian kernel of the estimation with 0-mean and $\sigma$-variance and $|\Omega|$ the shape area. Non parametric pdfs have been introduced in region-based active contours in [34] for the minimization of the distance between two pdfs and in [465] for the minimization of information measures. The general descriptor (19.12) has been studied in [395, 396] and the descriptor (19.13) has been studied by [465, 464, 123].

## 19.4.1 *Examples of Descriptors based on parametric statistics*

In the case of parametric pdfs, the probability density function $q$ is indexed by one or more parameters, denoted by a vector $\theta$, describing the distribution model. The parameters $\theta$ depend on the domain $\Omega$ and such a dependence must be taken into account in the derivation process through the evaluation of the domain derivative. We propose here to give some results for the derivation of functions depending on simple statistical parameters such as the mean or the variance. This study can be extended to the derivation of the covariance matrix determinant.

### 19.4.1.1 Region-dependent descriptors using the mean

For a one-dimensional image feature $f$, let us choose:

$$k(\mathbf{x},\Omega) = \varrho(f(\mathbf{x}) - \mu) = \varrho(f(\mathbf{x}) - \frac{1}{|\Omega|}\int_{\Omega} f(\mathbf{x})\,d\mathbf{x}) \qquad (19.16)$$

where $\varrho : \mathbf{R} \to \mathbf{R}^+$ is a positive function of class $C^1$. The region functional can be expressed as in equation (19.9):

$$J(\Omega) = \int_{\Omega} k(\mathbf{x},\Omega)\,d\mathbf{x} = \int_{\Omega} \varrho(f(\mathbf{x}) - \mu)\,d\mathbf{x} = \int_{\Omega} \varrho(f(\mathbf{x}) - \frac{G_1(\Omega)}{G_2(\Omega)})\,d\mathbf{x},$$

where

$$G_1(\Omega) = \int_{\Omega} H_1(\mathbf{x},\Omega)\,d\mathbf{x} = \int_{\Omega} f(\mathbf{x})\,d\mathbf{x} \text{ and } G_2(\Omega) = \int_{\Omega} H_2(\mathbf{x},\Omega)\,d\mathbf{x} = \int_{\Omega} 1 d\mathbf{x}$$

In this case, the functions $H_i$, $i = 1,2$ do not depend on the region $\Omega$, $l_1 = l_2 = 0$ and $K_{ij}(\mathbf{x}) = 0 \quad \forall i, j$. The terms $D_j, j = 1,2$ can then be computed:

$$\begin{array}{rcl} D_1 & = & -\int_{\Omega} \frac{1}{G_2}\,\varrho'\left(f(\mathbf{x}) - \frac{G_1}{G_2}\right) d\mathbf{x} = \frac{-1}{|\Omega|}\int_{\Omega} \varrho'(f-\mu)d\mathbf{x} \\ D_2 & = & \int_{\Omega} \frac{G_1}{(G_2)^2}\,\varrho'\left(f(\mathbf{x}) - \frac{G_1}{G_2}\right) d\mathbf{x} = \frac{\mu}{|\Omega|}\int_{\Omega} \varrho'(f-\mu)d\mathbf{x} \end{array}$$

The terms $B_{ij}$ are equal to zero and the velocity vector of the active contour is then:

$$\Gamma_\tau = \left[k - \frac{(f-\mu)}{|\Omega|}\int_{\Omega} \varrho'(f-\mu)d\mathbf{x}\right] \mathbf{N}$$

In this example, the term coming from the region dependency of $f$ is equal to $\frac{(f-\mu)}{|\Omega|}\int_{\Omega} \varrho'(f-\mu)d\mathbf{x}$. Note that in the particular case of $\varrho(r) = r^2$, this term is equal to zero [174, 254].

### 19.4.1.2 Region-dependent descriptors based on the variance

Let us take another example of descriptor for one dimensional image feature. Consider the case where the function $k$ is a function of the variance given by:

$$k(\mathbf{x},\Omega) = \varrho(\sigma^2) = \varrho\left(\frac{1}{|\Omega|}\int_{\Omega}(f(\mathbf{x}) - \mu)^2\right) = \varrho\left(\frac{G_1(\Omega)}{G_2(\Omega)}\right)$$

where $\varrho : \mathbf{R}^+ \to \mathbf{R}^+$ is of class $C^1$.

We can then compute the velocity vector of the active contour from Theorem 19.3.3 using:

$$\begin{aligned} G_1(\Omega) &= \int_\Omega H_1(\mathbf{x},\Omega)\,d\mathbf{x}\,, \quad H_1(\mathbf{x},\Omega) = \left(f(\mathbf{x}) - \frac{K_{11}}{K_{12}}\right)^2, \quad l_1 = 2, \\ G_2(\Omega) &= \int_\Omega H_2(\mathbf{x},\Omega)\,d\mathbf{x}\,, \quad H_2(\mathbf{x},\Omega) = 1, \quad l_2 = 0\,, \end{aligned}$$

and we find:

$$\Gamma_\tau = \left[k + \varrho'(\sigma^2)\left((f-\mu)^2 - \sigma^2\right)\right] \mathbf{N}.$$

In this simple example, we notice that the dependency of the function on the region induces the term $A(\mathbf{x},\Omega) = \varrho'(\sigma^2)\left((f(\mathbf{x}) - \mu)^2 - \sigma^2\right)$ in the evolution equation, see [432] for details.

This result can be extended to a descriptor based on the covariance matrix determinant for multidimensional image features $\mathbf{f} = [f^1, f^2, ..., f^n]^T$. It can be a useful tool for the segmentation of homogeneous regions since minimizing the entropy is equivalent to minimize the determinant of the covariance matrix in the case of Gaussian distributions [360, 359]. The evolution equation can be computed using Theorem 19.3.3. Details of the computation as well as experimental results for the segmentation of the face in color video sequences may be found in [432].

## *19.4.2 Descriptors based on non parametric statistics*

### 19.4.2.1 Region-dependent descriptors based on non parametric pdfs of image features

We consider the following descriptor, where $\varphi$ is a function: $\mathbb{R}^+ \to \mathbb{R}^+$ and $q$ is given by (19.15):

$$k(\mathbf{x},\Omega) = \varphi\big(q(\mathbf{f}(\mathbf{x}),\Omega)\big) \tag{19.17}$$

**Theorem 19.4.1.** *The Eulerian derivative in the direction* $\mathbf{V}$ *of the functional* $J(\Omega) = \int_\Omega k(\mathbf{x},\Omega)d\mathbf{x}$ *where* $k$ *is defined in (19.17) is:*

$$dJ_r(\Omega, V) = -\int_{\partial\Omega} (k(\mathbf{x},\Omega) + A(\mathbf{x},\Omega))(\mathbf{V}\cdot\mathbf{N})\,\mathrm{da}(\mathbf{x})$$

*where* $$A(\mathbf{x},\Omega) = -\frac{1}{|\Omega|}\left[\int_\Omega \varphi'(q(\mathbf{f}(\hat{\mathbf{x}}),\Omega))[q(\mathbf{f}(\hat{\mathbf{x}}),\Omega) - K(\mathbf{f}(\hat{\mathbf{x}}) - \mathbf{f}(\mathbf{x}))]d\hat{\mathbf{x}}\right]$$

*Proof:* The criterion is differentiated using the methodology developed in section 19.3.3.2. We have:

$$J(\Omega) = \int_\Omega \varphi(\frac{G_1(\mathbf{x},\Omega)}{G_2(\Omega)})\,\mathrm{dx} = \int_\Omega f(G_1(\mathbf{x},\Omega), G_2(\Omega))dx$$

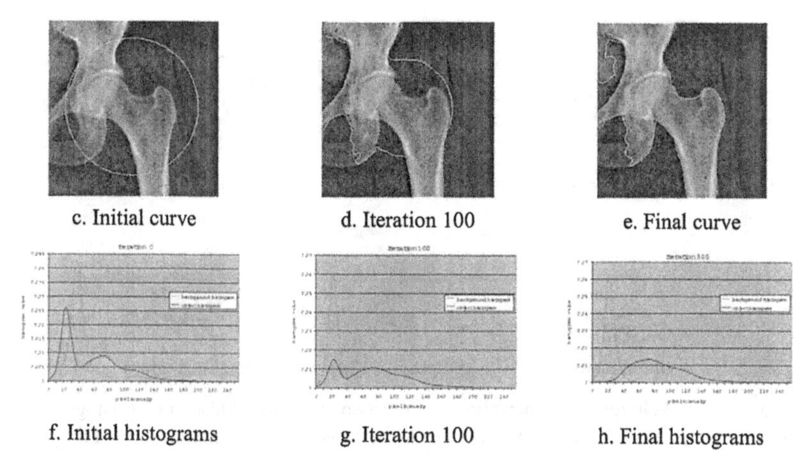

c. Initial curve d. Iteration 100 e. Final curve

f. Initial histograms g. Iteration 100 h. Final histograms

Figure 19.1. Evolution of segmentation and histograms with the minimization of the entropy for a grayscale image ($f = I$)

$$\begin{aligned} \text{with} \quad G_1(\mathbf{x},\Omega) &= \int_\Omega H_1(\mathbf{x},\hat{\mathbf{x}},\Omega)\,d\hat{\mathbf{x}}\,, \quad H_1(\mathbf{x},\hat{\mathbf{x}},\Omega) = K(\mathbf{f}(\mathbf{x}) - \mathbf{f}(\hat{\mathbf{x}})), \\ G_2(\Omega) &= \int_\Omega H_2(\hat{\mathbf{x}},\Omega)\,d\hat{\mathbf{x}}\,, \quad H_2(\hat{\mathbf{x}},\Omega) = 1\,, \end{aligned}$$

In comparison with the general results presented in section 19.3.3.2, we must pay attention to the fact that $H_1$ depends on $\mathbf{x}$ and $\hat{\mathbf{x}}$ during the derivation process.

These results can then be used for segmentation using information measures such as the entropy or the mutual information [395, 396]. If we choose to minimize the entropy as in [395], $\varphi(q) = -q\ln(q)$. In Figure 19.1, an example of segmentation of an osteoporosis image is given by minimizing $J(\Omega_{in},\Omega_{out}) = E(\Omega_{in}) + E(\Omega_{out}) + \lambda\int_\Gamma ds$ where $E(\Omega_{in})$ and $E(\Omega_{out})$ represent respectively the entropy of the one-dimensional feature $f(\mathbf{x}) = I(\mathbf{x})$ inside and outside the curve and $\lambda\int_\Gamma ds$ is the classical regularization term that minimizes the curve length balanced with a positive parameter $\lambda$. The Figure 19.1 shows the evolution of the segmentation and the evolution of the associated histograms (of the region $\Omega_{in}$ and $\Omega_{out}$) during iterations. Figure 19.2 shows an example of segmentation of color video by minimizing the joint entropy of a two dimensional feature $\mathbf{f}(\mathbf{x}) = [Y(\mathbf{x}), U(\mathbf{x})]^T$, where $Y$ is the luminance and $U$ is the chrominance. The joint entropy is computed by using the joint probabilities between each color channel. In Figure 19.2, we can see the evolution of the object histogram (histogram inside the region $\Omega_{in}$).

#### 19.4.2.2 Minimization of the distance between pdfs for tracking

We next assume that we have a function $\varphi : \mathcal{R}^+ \times \mathcal{R}^+ \to \mathcal{R}^+$ which allows us to compare two pdfs. This function is small if the pdfs are similar and large

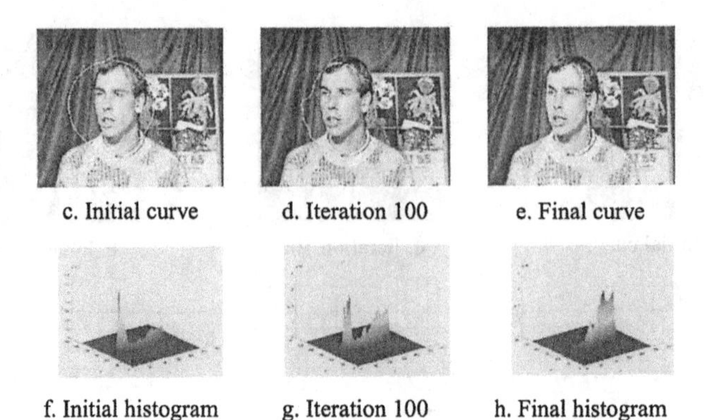


Figure 19.2. Evolution of segmentation and the associated object histogram (histogram of the two components color of the region inside the curve) with the minimization of the joint entropy

otherwise. It allows us to introduce the following functional which represents the "distance" between the two histograms:

$$D(\Omega) = \int_{\mathcal{R}^m} \varphi(\hat{q}(\mathbf{f}, \Omega), q(\mathbf{f}, \Omega_{ref}))\, d\mathbf{f} \tag{19.18}$$

The distance can be for example the Hellinger distance when $\varphi(\hat{q}, q) = \left(\sqrt{\hat{q}} - \sqrt{q}\right)^2$. Using the tools developed in section 19.3.3.2, we can compute the Eulerian derivative of the functional $D$. We have the

**Theorem 19.4.2.** *The Eulerian derivative in the direction* $\mathbf{V}$ *of the functional* $D$ *defined in (19.18) is:*

$$d_r D(\Omega, \mathbf{V}) = -\frac{1}{|\,\Omega\,|} \int_{\partial\Omega} \Big(\partial_1 \varphi(\hat{q}(.,\Omega), q(.,\Omega_{ref})) * K(\mathbf{f}(\mathbf{x})) - C(\Omega)\Big)(\mathbf{V}\cdot\mathbf{N}) d\mathbf{a}(\mathbf{x}),$$

*where* $\partial_1 \varphi(.,.)$ *is the derivative of* $\varphi$ *according to its first variable and* $C(\Omega) = \int_{\mathcal{R}^m} \partial_1 \varphi(\hat{q}(\mathbf{f}, \Omega), q(\mathbf{f}, \Omega_{ref}))\hat{q}(\mathbf{f}, \Omega)\, d\mathbf{f}$. *The first term under the integral,* $\partial_1 \varphi(\hat{q}(.,\Omega), q(.,\Omega_{ref})) * K$, *is the convolution of the function* $\partial_1 \varphi(\hat{q}(.,\Omega), q(.,\Omega_{ref})) : \mathcal{R}^m \to \mathcal{R}$ *with the kernel* $K$.

A proof of this theorem can be found in [34, 433]. An example of tracking is given in Figure 19.3 for a two-dimensional image feature $\mathbf{f}(\mathbf{x}) = [H(\mathbf{x}), V(\mathbf{x})]^T$, where $H$ is the hue and $V$ is the value of the color system $HSV$.

## 19.5 Discussion

In this article, we focus on the problem of finding local minima of a large class of region and boundary functionals by applying methods of shape derivation [256, 751]. We more particularly turn our attention to region-based functionals involving region-dependent descriptors. We propose a general methodology

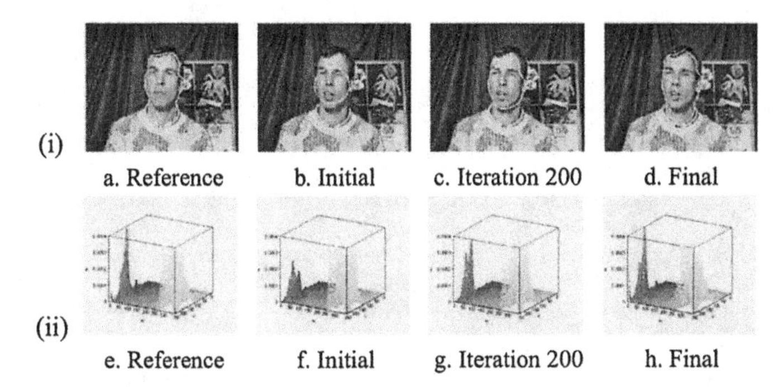


Figure 19.3. Example of tracking using the minimization between the current histogram and a reference one; (i) segmentation, (ii) histogram. Figure $a$ represents the reference segmentation and Figure $e$ the associated reference object histogram. Figures $b$, $c$ and $d$ show the evolution of the curve and Figures $f$, $g$, $h$ the evolution of the object histogram.

to derive region-based functionals based on parametric or non parametric pdfs. To illustrate our framework, some examples of derivation and computation of the evolution equation are given for parametric and non parametric statistical descriptors.

# Chapter20

# Model-Based Human Motion Capture

**I. Kakadiaris and C. Barrón**

Abstract

Human motion analysis is a challenging research area aimed at automating the study of human behavior. An important part of any such system is the component that performs the Human Motion Capture (HMC); in order for human motion to be processed and semantically analyzed, a mathematical representation of the observed motion needs to be extracted. There are two separate aspects to a HMC system; sensing (hardware) and processing (software). The processing itself comprises of an initialization (anthropometry and pose estimation) and a tracking phase. In this chapter, we present methods for three-dimensional model-based human motion capture from uncalibrated passive optical sensors with semi-automatic initialization and tracking. Such methods allow for non-intrusive capture of natural human behavior from video cameras or from archival recordings. We demonstrate the accuracy, advantages, and limitations of our methods for various classes of data.

## 20.1 Introduction

In computer vision, human motion analysis (HMA) is a term that describes a broad field with diverse applications. At its core, the goal of HMA is to provide automated systems that can recognize humans and their behavior. More specifically, its aim is to develop algorithms that can process image sequences in order to detect, track, and provide semantic context for the people recorded and the activities they are involved in. Activities is a very loose term in this context, and several diverse areas of applications have been studied that can be classified under the HMA heading, as indicated by the early seminal surveys of Aggarwal and Cai [8] or Gavrila [337], the later work of Moeslund and Granum [580] or the recent review by Wang *et al.* [854]. They include tele-presence (teleconferencing, interactive virtual worlds, avatar animation), perceptual user interfaces for control and command (gestures, sign-language, signaling), kinesiology (diagnosis, training, rehabilitation), ergonomics (robotics, product design, testing), content-based video storage and retrieval (from sports to choreography), and last but not least, smart visual surveillance. The last area alone is rapidly becoming a driving force behind advancements in the field, as there is an increasing awareness of its importance, ranging from parking lot safety to maintenance of strict access control environments, and beyond.

An important part of any successful HMA system is the component that performs the Human Motion Capture (HMC); in order for human motion to be semantically analyzed, it first needs to be captured. In other words, a mathematical representation of the motion observed needs to be extracted. This is a challenging task in its own right, and garners large interest as it includes fundamental and inherently difficult problems such as image segmentation, and shape and motion estimation. This is all compounded by the fact that the objects being studied are non-rigid bodies that are frequently occluded.

There are two separate aspects to a human motion capture system; sensing (hardware) and processing (software). Subsequently, the systems used for HMC can be classified into several different categories according to the methods used to carry out each of these tasks. Although electromechanic or electromagnetic sensing devices can be used, we restrict our discussion to optical sensors.

Optical sensors can be active or passive. The key difference is whether or not special equipment such as measuring devices or markers need to be worn by the subject. Active optical sensing operates by placing visible markers on the subject in the form of a body suit and employs arrays of calibrated infrared cameras in a predefined, restricted space. It allows for simpler processing and it is used successfully in highly controlled environments (e.g., movie production). On the other hand, passive sensing does not require special equipment suits; it captures motion from regular video sequences. Passive sensors operate in the visible or infrared spectrum. It is important here to differentiate, between single and multiple sensor systems, whether they are moving or stationary, and whether or not the sensors need to be calibrated before motion capture. Passive sensing systems, and in particular single sensors, are the preferred and often compelling alternative in terms of cost, reliability, ease of use, and adaptability. Another important aspect is that they are non-intrusive, allowing for natural human behavior capture from video cameras or from archival recordings.

Once the observed human motion has been recorded by the appropriate sensing devices, its mathematical representation can be extracted. This processing step entails an initialization phase (anthropometry and pose estimation) before the actual tracking can occur. In our context, the problem of anthropometry pose estimation from a single image can be formulated as follows:

> *Given a set of points in an image that correspond to the projection of landmark points of a human subject, estimate both the anthropometric measurements (up to a scale) of the subject and his/her pose that best match the observed image.*

Tracking can then be stated as follows:

> *Given an image sequence of a moving human, estimate his or her motion by estimating the corresponding pose of the human at each frame of the image sequence.*

As mentioned earlier, motion capture of non-rigid objects such as moving humans presents several challenging steps including segmentation of the human body from the background and into meaningful body parts, handling of occlusions, and tracking body parts along the image sequence. The approaches that have been proposed for HMC can be classified into two groups: model-based approaches and view-based approaches. Model-based approaches use *a priori* models explicitly defined in terms of kinematics and dynamics. They differ based on the types of models used (stick figures, surface or volume), the ways of modeling motion dynamics (kinetics, kinematics), and on whether the model is general or customized for the person under observation. In general, model-based approaches are preferred, as the use of predefined or acquired models introduces robustness that overcomes obstructions related to lighting conditions, clothes, rapid motion, occlusion, image quality, and problems with camera calibration.

The rest of the chapter is organized as follows: Section 20.2.1 summarizes our methods for human body model acquisition, while Section 20.2.2 summarizes our methods for human body tracking. Section 20.3 presents selected results, and Section 20.4 offers a brief reflection into the future.

## 20.2 Methods

The problem of human body model acquisition entails shape and motion estimation for the moving parts of a complex multi-part object. In earlier work, we have developed a *Part Segmentation Algorithm* (PSA) that recovers all the moving parts of a multi-part object by monitoring and reasoning over the deformation of its apparent contour [449]. This algorithm allows partial overlap between the parts and determination of their joint location (if any). We have employed this algorithm to build a 3D model of the person under observation. First, the apparent body contour of a moving subject is segmented into its constituent parts and then the 3D shape of a subject's body parts is estimated by fusing information from images taken from three cameras placed orthogonally [447]. Having obtained a geometric model of the person to be tracked the next step is motion estimation (i.e., tracking the human in the image sequence [448]).

Note that this technique applies to humans of any anthropometric dimension. However, it requires multiple cameras and the subject has to perform a set of movements according to a protocol that allows the integration of information from multiple views in order to estimate the 3D shape of all the major parts of the human body. Recently, we have developed methods for estimating shape (both an-

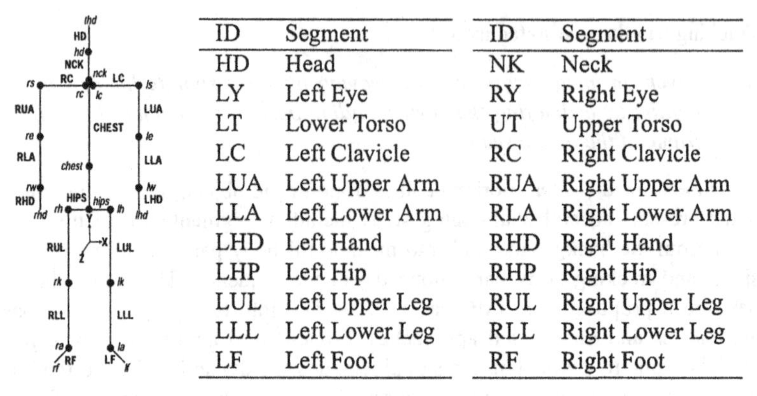


| ID | Segment | ID | Segment |
|---|---|---|---|
| HD | Head | NK | Neck |
| LY | Left Eye | RY | Right Eye |
| LT | Lower Torso | UT | Upper Torso |
| LC | Left Clavicle | RC | Right Clavicle |
| LUA | Left Upper Arm | RUA | Right Upper Arm |
| LLA | Left Lower Arm | RLA | Right Lower Arm |
| LHD | Left Hand | RHD | Right Hand |
| LHP | Left Hip | RHP | Right Hip |
| LUL | Left Upper Leg | RUL | Right Upper Leg |
| LLL | Left Lower Leg | RLL | Right Lower Leg |
| LF | Left Foot | RF | Right Foot |

Figure 20.1. Names of the VHM's segments

thropometry and pose) [57] and tracking [58] from a single uncalibrated camera. In the following, we present results for each of these methods separately.

### *20.2.1 Human body model acquisition*

We have developed a four-step technique for simultaneously estimating a human's anthropometric measurements (up to a scale parameter) and pose from a single uncalibrated image. The user initially selects a set of image points that constitute the projection of selected landmarks. Using this information, along with *a priori* statistical information about the human body, a set of plausible segment length estimates are produced. In the third step, a set of plausible poses are inferred using a geometric method based on joint limit constraints. In the fourth step, pose and anthropometric measurements are obtained by minimizing an appropriate cost function subject to the associated constraints. The novelty of our approach is the use of anthropometric statistics to constrain the estimation process which allows the simultaneous estimation of both anthropometry and pose.

For the purposes of our research, we have developed a generic virtual human model (VHM) and a statistical model for the distributions of various model measurements. This allows our algorithm to employ a hierarchical solver to estimate the parameters of a VHM whose projection most closely matches the image. Our VHM (Fig. 20.1) was inspired by the human body model employed at the Human Modeling and Simulation Center at the University of Pennsylvania [41]. Its skeleton consists of a set of sites/landmarks (Table 20.1) and a collection of segments (Fig. 20.1). Using the anthropometric measurements in [594], we build a cadre family for our statistical model, also known as a boundary family [39]. The cadre family is a multivariate representation of the extremes of the population distribution. It has the ability to span the multivariate space in a systematic fashion and to capture a significant amount of the variance in the space using a small number of sample human models. Our particular cadre family has 2187

Table 20.1. Information related to the joints of the Stick Model

| ID | Joint | From | To | DOF | PR |
|---|---|---|---|---|---|
| at | atlanto occipital | NK | HD | Tz*Rz*Ry*Rx | 3 |
| sp | solar plexus | UT | NK | Tz*Ry*Rz*x | 2 |
| la | left ankle | LLL | LF | Tx*Rz*Rx*Ry | 4 |
| lc | left clavicle | UT | LC | Tz*Rx*Ry | 3 |
| le | left elbow | LUA | LLA | Tz*Ry | 5 |
| lh | left hip | LT | LUL | Tz*Rz*Rx*Ry | 2 |
| lk | left knee | LUL | LLL | Tz*R-y | 3 |
| ls | left shoulder | LC | LUA | Tz*Rz*Rx*Ry | 4 |
| lw | left wrist | LLA | LHD | Tz*Ry*Rx*Rz | 6 |
| ra | right ankle | RLL | RF | Tx*R-z*R-x*Ry | 4 |
| rc | right clavicle | UT | RC | Tz*R-x*Ry | 3 |
| re | right elbow | RUA | RLA | Tz*Ry | 5 |
| rh | right hip | LT | RUL | Tz*R-z*R-x*Ry | 2 |
| rk | right knee | RUL | RLL | Tz*R-y | 3 |
| rs | right shoulder | RC | RUA | Tz*R-z*R-x*Ry | 4 |
| rw | right wrist | RLA | RHD | Tz*Ry*R-x*R-z | 6 |
| wt | waist | LT | UT | Tz*Ry*Rz*Rx | 1 |

Table 20.2. The segments used for computing the *covering set*

| $l_1$ | $l_2$ | $l_3$ | $l_4$ | $l_5$ | $l_6$ | $l_7$ | $l_8$ |
|---|---|---|---|---|---|---|---|
| UT+LT | LC | LUA | LLA | LHP | LUL | LLL | LF |

VHMs. Specifically, our algorithm has the following steps: 1) Selection of projected landmarks; 2) Initial anthropometric estimates; 3) Initial pose estimates; and 4) Iterative minimization over lengths and angles.

Step 1 is accomplished via a simple user interface that allows a user to select the projection of visible landmarks of the subject's body. In order to conduct anthropometric measurements, the user is also prompted to select pairs of segments from the covering set given in Table 20.2. These pairs need to be oriented either parallel to the image plane, or similarly with respect to the camera. Let $\mathcal{I}$ be the set of indices these segments have. Let $h_n$ ($n \in \mathcal{I}$) be the length of segment $n$ measured on the image, and let $l_n(q)$ ($q = 1, \ldots, 2187$) be the length of the same segment $n$ on the VHM indexed by $q$ in our cadre family. According to projective geometry, ratios of these measurements carry over to ratios of the corresponding measurements on the VHM. We fix an indexing set $\mathcal{K}$ for the possible ratios $s_k$, and always choose as denominator the segment with the smaller average length:

$$s_k = \begin{cases} h_m/h_n & \text{if } \mu(l_n) > \mu(l_m) \\ h_n/h_m & \text{otherwise} \end{cases}$$

where $\mu(l_n)$ is the average of the lengths $l_n(q)$ over all VHMs (indexed by $q$) in the cadre family.

Having established the set $\mathcal{K}$ of ratios that we are going to use, we next compute the corresponding length ratios $r_k(q)$, (where $k \in \mathcal{K}$ and $q = 1, \ldots, 2187$) on each VHM (indexed by $q$) in our cadre family, and let $\boldsymbol{O}$ be the covariance matrix of their ratios.

Step 2 identifies the VHM $q^\star$ from the cadre family whose length ratios $r_k(q^\star)$ are closest to the ratios $s_k$ using the Mahalanobis distance. It requires solving the following discrete selection problem:

$$q^\star = \arg\min_q \sum_{k\in\mathcal{K}} (r_k(q) - s_k)(\sum_{j\in\mathcal{K}} v_{kj}(r_j(q) - s_j)),$$

where $[v_{kj}] = \boldsymbol{O}^{-1}$.

We have now identified a VHM with the right proportions (correct up to scale). In the next two steps of the algorithm the variables we want to estimate are the lengths of the body segments (the scale factor) and their pose. Therefore, we will solve a system of equations where prior information about the human body (e.g., relations between lengths of segments) will provide constraints to an optimization that minimizes the discrepancy between the synthesized appearance of the VHM (for that pose) and the image data of the subject in the given image.

As mentioned earlier, the user selects a set of points on the image that correspond to the projection of sites of the VHM. For each of these points, we set up a point-to-line constraint, since the site will lie on a line that goes through the center of the camera and the projection of a landmark. Let $\boldsymbol{c}$ be the camera's center of projection, $\boldsymbol{m}_i$ be the position of a VHM's site, and $\boldsymbol{m}_i^p$ be the corresponding projection point selected by the user. The point-to-line constraint is $\boldsymbol{c}_i = \boldsymbol{c} + \lambda \boldsymbol{d}_i$, where $\boldsymbol{d}_i = \frac{(\boldsymbol{m}_i^p - \boldsymbol{c})}{\|\boldsymbol{m}_i^p - \boldsymbol{c}\|}$.

Gathering all these constraints together, the optimization problem becomes: minimize $\|(\boldsymbol{m}_i, \boldsymbol{c}_i)\|$ subject to $C_j(\boldsymbol{m}_i)(j = 1, 2, 3)$, where $C_1(\boldsymbol{m}_i)$ is a constraint derived from the range of motion of the VHM's joint; $C_2(\boldsymbol{m}_i)$ is a constraint that enforces symmetry between the left and right sides of the VHM (e.g., RC$\approx$LC, RF$\approx$LF, RUL$\approx$LUL), and $C_3(\boldsymbol{m}_i)$ proportional constraints (i.e., $r_k(q^\star) \approx s_k$).

We seek to minimize the value of this function using a BFGS nonlinear solver [919]. Due to the large number of degrees of freedom, we apply the solver in a hierarchical manner. Our method schedules an optimization process starting with the joints closer to the waist and moving outwards using the priorities given in Table 20.1 (PR column).

In order for the nonlinear solver not to get trapped into a local minimum, we use a geometric method to provide an initial estimate for the pose of the segments whose endpoints were selected by the user. We compute two initial estimates as follows. Let $\boldsymbol{m}_i^p$ be the projection of site $m_i$ in the image, $l_i > 0$ be the length of the segment of which this landmark is the end-effector, $\boldsymbol{j}$ be the position of the parent joint of that landmark on the VHM's skeleton, and $\boldsymbol{d}_i$ be the unit direction between the camera and $\boldsymbol{m}_i^p$. Then, the two possible initial guesses for $m_i$ are: $\boldsymbol{m}_{i1} = \boldsymbol{c} + \lambda_1 \boldsymbol{d}_i$ and $\boldsymbol{m}_{i2} = \boldsymbol{c} + \lambda_2 \boldsymbol{d}_i$, where $A = \sqrt{[\boldsymbol{d}_i \bullet (\boldsymbol{c} - \boldsymbol{j})]^2 - \|\boldsymbol{c} - \boldsymbol{j}\|^2 + l_i^2}$, $\lambda_1 = \boldsymbol{d}_i \bullet (\boldsymbol{j} - \boldsymbol{c}) + A$ and

$\lambda_2 = d_i \bullet (j - c) - A$. Information about the joint limits is used to prune the solutions that are not feasible.

### 20.2.2 *Model-based tracking*

Having obtained a geometric model of the person to be tracked the next step is to track the human in the image sequence. We achieve this by continuously comparing the difference between the actual image frame and the synthetic image computed by projecting the estimated VHM to the image plane. We assume that the VHM is described by a set of parameters $\Theta$. The proposed algorithm assumes that the similarity of appearance of the subject over the time of acquisition leads to the minimum of a convex function on $\Theta$. Specifically, the method searches for the best pose in each image by minimizing the discrepancies between the image under consideration and a synthetic image of an appropriate VHM. By including in the objective function penalty factors from the image segmentation step, the search focuses on the regions that belong to the subject. These penalty factors convert the objective function to a convex function, which guarantees that the minimization converges to the global minimum. In addition, we follow a hierarchical decomposition approach from the hip towards the limbs and the head using ensembles of no more than three segments and restrict the search on the meridian directions (as per Algorithm 6). By constraining movement to one meridian direction at a time, the minimization procedure reduces to a one-dimensional problem.

We now describe our method in more detail. In each frame, let $p$ be the coordinates of a pixel, $v_p$ be the intensity in location $p$ of the current image, $V_p$ be the intensity in location $p$ of the synthetic image of the projection of the VHM, $\lambda(V_p)$ be penalty factors for the projected values of the VHM's segments, and $\lambda(v_p)$ be penalty factors for the image frame. The values of the penalty factors $\lambda$ are close to 1 for pixels that belong to a region in the image that corresponds to the subject being tracked or to a region in the synthetic image that corresponds to the projection of the VHM's segment, and assume a large positive value otherwise. The tracking problem can then be described as determining the set of parameters $\Theta$ that minimize the value of the function $f(\Theta)$:

$$\text{minimize } f(\Theta) = \sum_p \lambda(V_p)\lambda(v_p)\left(V_p - v_p\right)^2 . \qquad (20.1)$$

Our hierarchical method of solving this is presented in Algorithm 3. We use an iterative forward and backward prediction algorithm, where the output of processing one frame is used as input for the next. We present this algorithm first. Let $t_f$ be the number of frames in the image sequence, $s_f$ denote the selected initial frame, $v_f$ a frame counter, $c_f$ denote the current frame, and $f_d$ denote the order of processing the frames (it can take only two values: 1 for forward or $-1$ for backward).

**Algorithm 2.** HUMAN MOTION TRACKING
**1:** $v_f = 1$, $f_d = 1$, *and* $c_f = s_f$.

**2: while** *($v_f \neq t_f$)* **do**
3: *Find* $\Theta$ *that minimizes* $f(\Theta)$ *(Algorithm 3).*
4: **if** $c_f = t_f$ **then** $f_d = -1$, $c_f = s_f + 1$.
5: **else** $v_f = v_f + 1$.
6: $c_f = c_f + f_d$.

The objective function $f(\Theta)$ described in Eq. 20.1 is nonlinear and non-convex, and the search space for $\Theta$ is high dimensional. The key to solving Eq. 20.1 is to restrict the search to subspaces, and proceed hierarchically to cover the whole search space. The subspaces in the search space correspond to the parameters that describe each area of the VHM: hips (HPS), chest-neck-head (CNH), left arm (LAR), right arm (RAR), left leg (LLG) and right leg (RLG). Thus, $\Theta = (\Theta_{HPS}, \Theta_{CNH}, \Theta_{LAR}, \Theta_{RAR}, \Theta_{LLG}, \Theta_{RLG})$. Each subspace S describes an ensemble of at most three articulated segments $(L_1^S, L_2^S, L_3^S)$. For each segment in an ensemble, all that is required is the estimation of the segment's rotation $\Theta_L = (\alpha, \beta, \gamma)$, since the position of its distal end has already been established in previous steps of the algorithm. The only exception is the initial subspace, HPS, which requires the determination of both positional and rotational information.

**Algorithm 3.** HIERARCHICAL DECOMPOSITION
1: *Update the VHM's appearance.*
2: *Segment the next image.*
3: *Find* $\Theta$ *restricted to HPS that minimizes* $f(\Theta)$
4: **for** *S in* $\{CND, LAR, RAR, LLG, RLG\}$.
5: **for** *L in* $\{L_1^S, L_2^S, L_3^S\}$.
6: *Predict projected angle (Algorithm 4).*
7: *Compute line-sphere intersection.*
8: *Perform convexity test (Algorithm 5).*
9: *Find* $\Theta$ *restricted to L that minimizes* $f(\Theta)$ *in two steps:*
*Compute* $\alpha, \beta$ *(meridian directions), and*
*Compute* $\gamma$ *(the segment's rotation with respect to its axis).*

We now describe each of the mentioned steps in more detail. For each ensemble S, the prediction algorithm searches a small sector of a circular region. The estimation of the projected angle is based on the continuity of a line that connects an active joint and its next joint or site over a segment. Let $\epsilon_1$ and $\epsilon_2$ be scaling parameters, $l_s$ be the projected length of a segment, $l_m$ be the length's lower bound, $l_M$ be the length's upper bound, $B$ be a set of points, and $\emptyset$ be the empty set. The steps predicting the projected angle are the following:

**Algorithm 4.** PREDICTION
1: $l_m = \epsilon_1 l_s$, $l_M = l_s$, $\delta_l = \epsilon_2 l_s$, and $B$=$\emptyset$.
2: **while** *($B$=$\emptyset$ and $l_m > 0$)* **do**
3: **for** *all the points on the sector of the circular region between* $l_m$ and $l_M$ *on the parent segment's orientation*
4: *Compute a line from the active joint to the point.*
5: **if** *the line lies inside the segment* **then** *add this point to* $B$.

6: **if** $B=\emptyset$ **then** $l_m = l_m - \delta_l$ and $l_M = l_M - \delta_l$.
7: **If** $B=\emptyset$ **then return** *no projected angle.*
8: *Estimate the center of mass of B.*
9: **return** *projected angle of the center of mass.*

Using a projected angle as input, the convexity test is performed on the meridian direction perpendicular to the segment's orientation. The objective is to verify that changes in the angle of the active joint result in a basin on $f(\Theta)$. Let $\epsilon_3$ be a parameter, $\theta$ be the projected angle, $\theta_l$ be the left limit, $\theta_r$ be the right limit, and $f_{R_1}(\theta) = f(\Theta)$ be the objective function restricted to angles on the meridian direction $R_1$ (see Algorithm 6). The steps to perform a convexity test are the following:

**Algorithm 5.** CONVEXITY TEST
1: $\theta_l = \theta - \epsilon_3$ and $\theta_r = \theta + \epsilon_3$.
2: $S_l = f_{R_1}(\theta_l)$, $S_r = f_{R_1}(\theta_r)$, and $S = f_{R_1}(\theta)$.
3: **if** $S_l > S \wedge S_r > S$ **then return convexity on** $[\theta_l, \theta_r]$.
4: **if** $S_l > S$ **then**
5: **repeat until** $|\theta_r - \theta_l| > 180°$
6: $\theta_r = \theta + \epsilon_3$, $S_r = f_{R_1}(\theta_r)$.
7: **if** $S_r > S$ **then return** *convexity on* $[\theta_l, \theta_r]$.
8: **else**
9: **repeat until** $|\theta_r - \theta_l| > 180°$
10: $\theta_l = \theta - \epsilon_3$, $S_l = f_{R_1}(\theta_l)$.
11: **if** $S_l > S$ **then return** *convexity on* $[\theta_l, \theta_r]$.
12: **return** *no convexity on* $[\theta_l, \theta_r]$.

Finally, we compute a rotation on the meridian directions as follows. Let us consider a joint-segment ensemble for which the coordinate system of the joint $\text{ZXY}(\alpha, \beta, \gamma)$ is rotated by a rotation matrix $M$. Let $\hat{\imath}$, $\hat{\jmath}$, and $\hat{k}$ be the unit vectors of the global coordinate systems, $\mathbf{c}$ be the camera's center of projection, $s$ be the current site position, and $j$ be the current joint position with respect to the global coordinate system. The spherical coordinates are denoted by $r, \psi, \omega$ where $r > 0$, $\psi \in [0, 2\pi]$ and $\omega \in [-\pi/2, \pi/2]$ centered at the origin of the global coordinate system. Note that when the position of a site moves along the direction $\psi$ this is equivalent to almost moving parallel over the image plane, while moving along the direction $\omega$ is equivalent to moving towards or away from the image plane. Let $\Delta\psi$, $\Delta\omega$ be the amount to rotate a site around the image plane or towards and backwards from the camera respectively. Then, the angles $\alpha, \beta$, and $\gamma$ to locally rotate the joint-segment ensemble are obtained by the following algorithm:

**Algorithm 6.** MERIDIAN MOTION
1: *Compute* $r_s$, $\psi_s$ and $\omega_s$, *the spherical coordinates of* $s$.
2: *Compute the desirable position* $s_1 = (r_s, \psi_s + \Delta\psi, \omega + \Delta\omega)$.
3: *Compute* $\mathbf{p}$ *the Cartesian coordinates of* $s_1$.
4: *Compute* $\mathbf{p}_1 = \mathbf{p} - j$, and $\hat{p}_1 = \frac{\mathbf{p}_1}{\|\mathbf{p}_1\|}$.
5: *Compute the rotated orthogonal unit vectors:*

Table 20.3. Accuracy of the length estimates for the subject *Vanessa*

| | $\frac{LC}{UT+LT}$ | $\frac{LLA}{LUA}$ | $\frac{LHP}{LUA}$ | $\frac{LF}{LUL}$ | $\frac{LF}{LLL}$ |
|---|---|---|---|---|---|
| Actual | 0.6279 | 0.8625 | 0.6949 | 0.5517 | 0.4778 |
| Estimated | 0.6402 | 0.8516 | 0.6728 | 0.5594 | 0.4888 |
| PE % | 1.9589 | 1.2638 | 3.1803 | 1.3957 | 2.3022 |

$\hat{i}_1 = M\hat{i}, \hat{j}_1 = M\hat{j}$, and $\hat{k}_1 = M\hat{k}$.

**6:** *If the segment's direction is negative set:*
$\hat{i}_1 = -\hat{i}_1, \hat{j}_1 = -\hat{j}_1$, and $\hat{k}_1 = -\hat{k}_1$.

**7:** *Compute the orthogonal projection of* $\hat{p}_1$ *on the rotated local coordinate system:* $x_1 = \hat{i}_1 \cdot \hat{p}_1$, $y_1 = \hat{j}_1 \cdot \hat{p}_1$, and $z_1 = \hat{k}_1 \cdot \hat{p}_1$.

**8:** *Compute the angles* $\alpha, \beta, \gamma$ *taking into consideration the orientation of the segment at its initial pose, i.e., for its angles equal to (0,0,0).*
**case ± X:** $\alpha = \arctan(\frac{y_1}{x_1})$ and $\gamma = -\arcsin(z_1)$.
**case ± Y:** $\alpha = -\arctan(\frac{x_1}{y_1})$ and $\beta = \arcsin(z_1)$.
**case ± Z:** $\beta = -\arctan(\frac{y_1}{z_1})$ and $\gamma = \arcsin(x_1)$

## 20.3 Results

We have performed numerous experiments to assess the accuracy, limitations, and advantages of our methods. Due to space considerations, we present selected results only. The real image sequences were recorded using a PULNiX TMC-9700 2/3" Color Progressive Scan, and a SONY Handycam Corder Hi8 at 30 frames per second. For our experiments, we selected $\epsilon_1 = 0.9$, $\epsilon_2 = 0.1$, and $\epsilon_3 = 5°$.

For the first experiment, we applied our technique to an image from the subject *Vanessa* whose anthropometric dimensions were manually measured. Fig. 20.2(a) depicts the selected points, Fig. 20.2(b) depicts the reconstructed model overlaid to the image, and Figs. 20.2(c,d) depict the reconstructed model from novel views. Table 20.3 captures the percentage errors (PE) in estimating the length ratios. We observe that the estimation of anthropometric information is within 3.2% of the anthropometric dimensions of the subject. Figure 20.3 summarizes results from a variety of application domains. Specifically, Figs. 20.3(a,d,g,i) depict a geologist, a basketball player, a tennis player and a golfer, respectively. Figs. 20.3(b,e) depict the reconstructed models overlaid to the images, and Figs. 20.3(c,f) depict novel views of the reconstructed VHM. The second experiment assessed the robustness of our method in the presence of occlusion. To that end, we have recorded an image sequence depicting a human drawing on a board (Fig. 20.4(a)). Fig. 20.4(b) depicts the results of our algorithm overlayed onto the original image sequence, while Fig. 20.4(d) depicts the estimated trajectories for lsh, le, lw, rsh, re, rw, and rhd. To validate our algorithm we compared the reconstructed coordinates of the markers' tips on the plane of the board using MatchMover (REALVIZ Products)

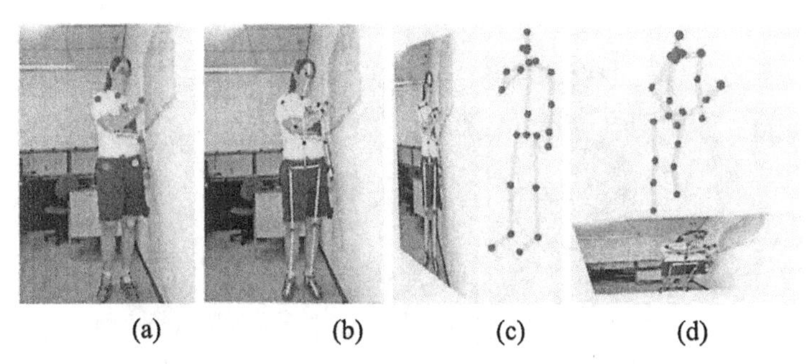

(a) (b) (c) (d)

Figure 20.2. Anthropometry and pose estimation for the subject Vanessa: (a) selected points, (b) reconstructed model overlaid to the image, and (c,d) novel views of the reconstructed 3D skeleton.

and our algorithm. Fig. 20.4(c) depicts the estimates of our algorithm for the right hand as compared with the estimates obtained with MatchMover. For the right hand there is occlusion at frames 44-46, 83-85 and 123-125. For these frames MatchMover requires manual intervention. However, our algorithm can cope with partial self-occlusions without any need for intervention.

In addition, we have tested our algorithm using a variety of video clips to assess the robustness of our method with respect to differences in lighting conditions, differences in motion cadence, and degraded image quality. As an example, a video clip obtained from David Carradine's Kung Fu Workout video, which presents difficulties due to variation of lighting, presence of shadows, rapid motion, and self-occlusion, is analyzed in Fig. 20.5. Figure 20.6 presents tracking results for a video clip (from http://www.fencing.net/) depicting two fencers in an action called parry-riposte. Both fencers were tracked independently. In all the clips tested, our algorithm successfully estimated the movement of the subjects. Currently, we are developing a vision-based interface that will allow an astronaut to remotely control ROBONAUT by tracking the astronaut's upper body movements. The ROBONAUT (ROBOtic astroNAUT) is an anthropomorphic robot with two arms, two hands, a head, a torso and a stabilizing leg, that is currently being developed at NASA Johnson Space Center to provide an astronaut substitute for EVA operations. We have developed a technique for estimating upper body motion from monocular images [550], estimating the motion parameters of the links by maximizing the conditional probability of the frame to frame intensity differences at observation points. Our contribution is that technique relates the frame to frame intensity difference to the motion parameters, we have considered also: a) the camera noise, b) the shape errors of the model, and c) the position errors due to the motion estimation errors resulting from the motion analysis of previous frames. Preliminary results in that domain indicate that for a camera noise level of PSNR=40 dB our algorithm achieves a reduction of the error variance of up to 40% for the estimated translation parameters and up to 35% for the rotation parameters.

Figure 20.3. Input images depicting (a) a geologist, (d) a basketball player, (g) a tennis player, and (j) a golfer along with the user selected input landmarks; (b,e) Reconstructed models overlaid to the images; (c,f) Novel views of the reconstructed VHM; and (i,k) Reconstructed 3D models.

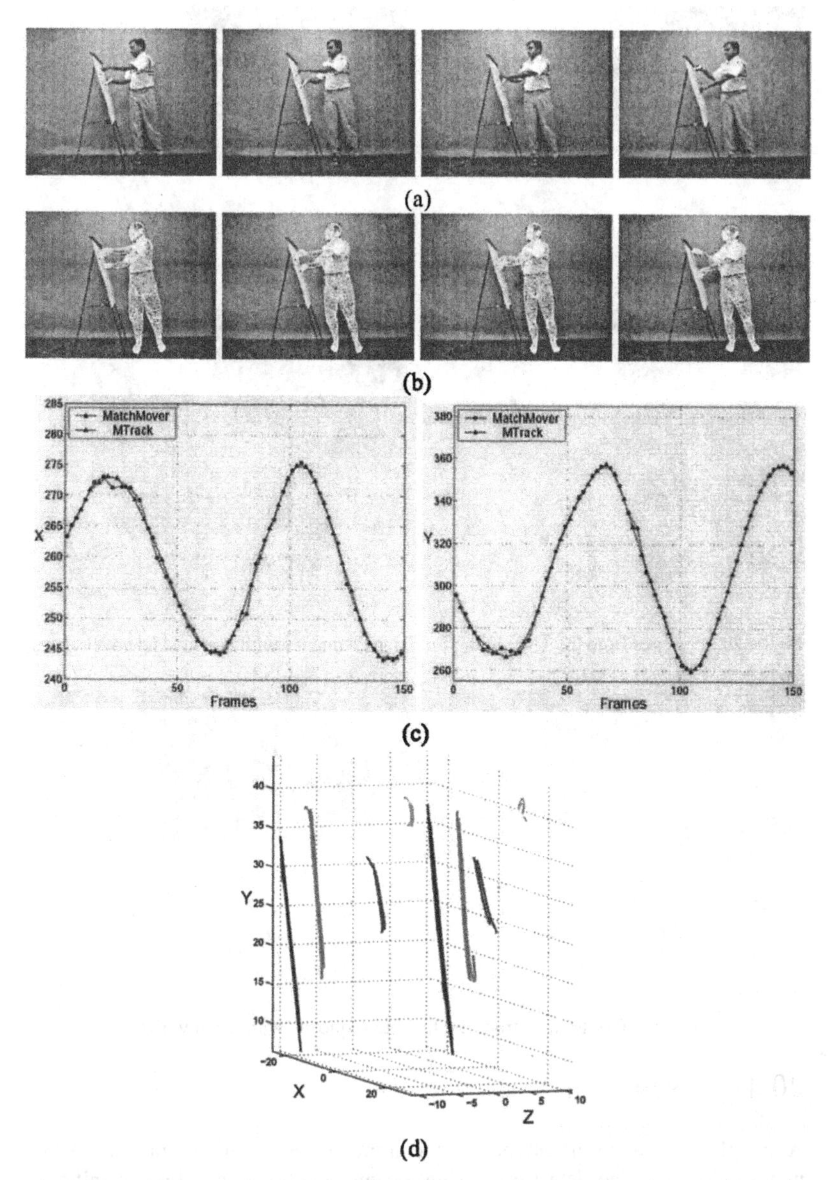


Figure 20.4. (a) Frames 0, 90, 120, and 149 from a video depicting a human writing on a board. (b) Overlay of our results onto the image sequence. (c) Estimated coordinates of the markers' tips on the plane of the board using MatchMover (REALVIZ Products) and our algorithm. (d) Estimated 3D trajectories for lsh, le, lw, lhd, rsh, re, rw, and rhd for the image sequence depicting a subject writing on the board.

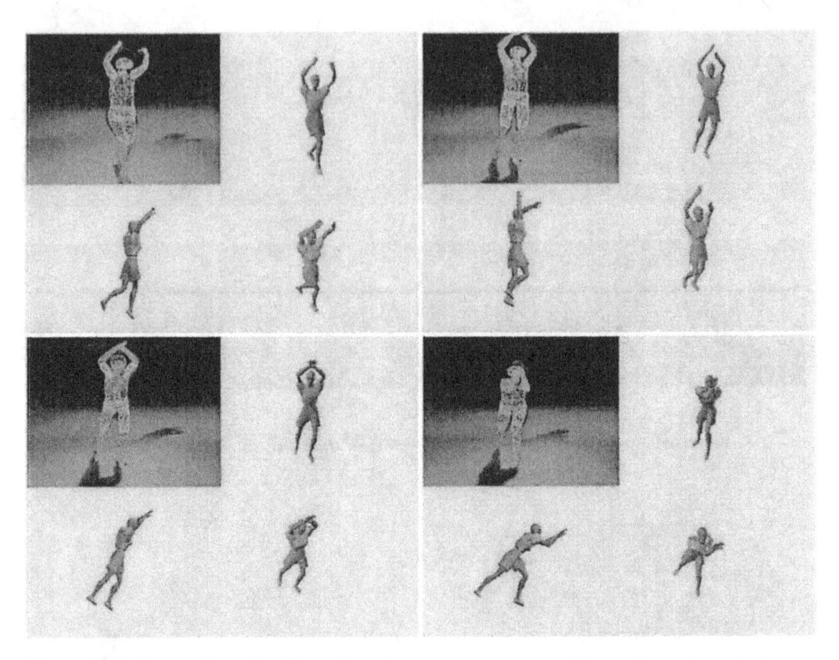

Figure 20.5. Frames from the Tiger Kung Fu moves sequence with overlay and novel views.

Figure 20.6. Frames from the fencing sequence with novel views.

## 20.4 Discussion

Although the problem of autonomous, continuous detection and tracking of human motion in video data from a single camera is far from solved, significant progress in this direction has been made in recent years. It is an area of active research that is receiving significant attention (as evidenced by the large number of papers in the area), especially after the recent increased awareness of its importance in relation to homeland security. As a result, several systems have been designed to tackle this problem, with varying degrees of success. Commer-

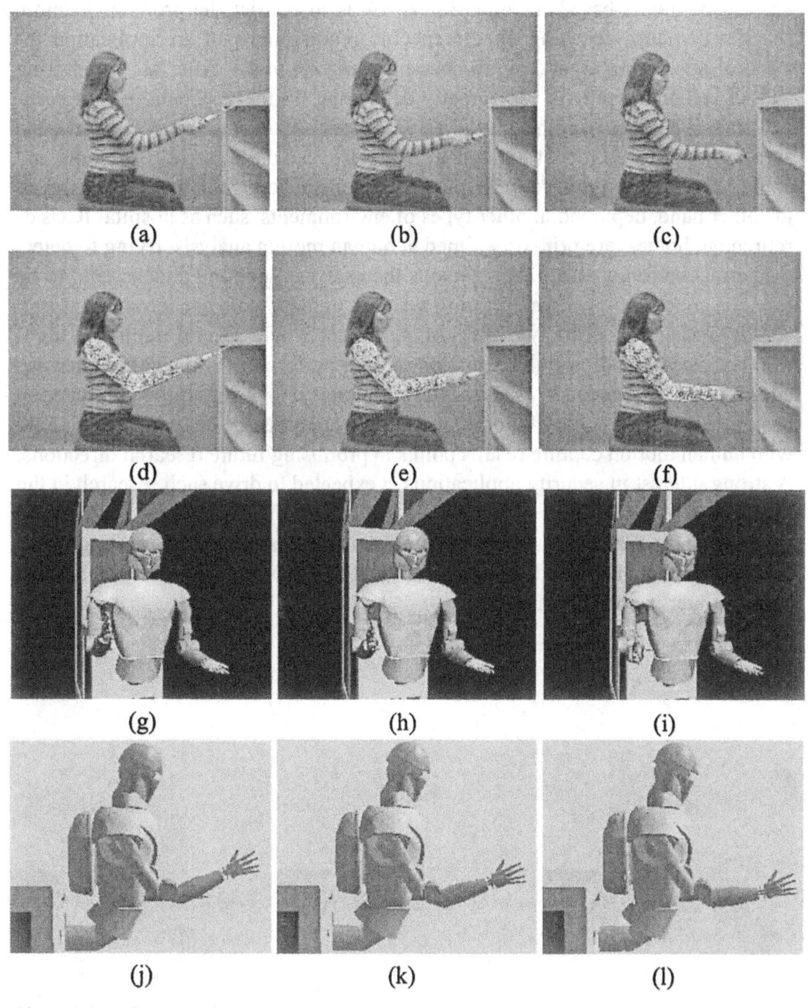


Figure 20.7. Commanding a ROBONAUT simulation developed at NASA–JSC with the estimated motion parameters of the HAZEL-B sequence. (a-c) Frames 1, 45, 90 from the sequence HAZEL-B. (d-f) Original frames with the model overlayed at the estimated position and orientation. (g-i) Coronal and (j-l) sagittal view of the postures corresponding to the frames 1, 45, 90 of the sequence.

cial systems have entered the market. Model-based approaches can effectively overcome some of the issues that have been hindering progress in the past, such as handling occlusions. As the computational efficiency and the availability of processing power increase, such algorithms become more attractive. The key to success of model-based methods for HMA is to restrict the problem in more specific domains, focusing on the specific requirements of an application instead of attempting a general, all encompassing approach. After all, monitoring and surveillance systems vary greatly depending on the application. In a security surveillance environment such as an access control area (e.g., airport), the system needs to be robust enough to be able to handle thwarting attempts and could be primarily focused on human motion detection. Monitoring systems on the other hand, deployed in other types of environments, such as hospital ICUs or retirement homes, are primarily aimed at human motion analysis, trying to detect dangerous behavior rather than persons themselves. Spoofing attacks tend to be less of a concern for such monitoring systems, and the main concern is accuracy.

Concerning validation, we are now at a point where several published algorithms exist, and a thorough evaluation of their performance on a comprehensive corpus of data is needed to assess issues of accuracy, robustness, and computation cost. In addition, the role of biometrics and how they can be integrated [645, 921] with human motion capture results points in promising future research directions. A strong interest in security applications is expected to drive such research in the near future.

In summary, research in human motion analysis has entered an exciting phase that will continue to provide challenging problems as well as inspired solutions to the computer vision community.

# Chapter21

# Modeling Dynamic Scenes: An Overview of Dynamic Textures

**G. Doretto and S. Soatto**

Abstract

Dynamic scenes with arbitrary radiometry and geometry present a challenge in that a physical model of their motion, shape, and reflectance cannot be inferred. Therefore, the issue of representation becomes crucial, and while there is no right or wrong representation, the task at hand should guide the modeling process. For instance, if the task is three-dimensional reconstruction, one can make assumptions on reflectance and illumination in order to recover shape and motion. If the task is synthesis, or reprojection, the correct shape is unimportant, as long as the model supports the generation of a valid view of the scene. If the task is detection or recognition, a physical model is not necessary as long as one can infer a statistical model that can be used to perform classification. We concentrate our attention on the two latter cases, and describe a modeling framework for dynamic scenes for the purpose of synthesis, detection and recognition. In particular, we restrict our attention to sequences of images of moving scenes that exhibit certain statistical stationarity properties, which have been called Dynamic Textures. They include sea-waves, smoke, foliage, whirlwind etc. In this chapter we describe a characterization of dynamic textures and pose the problems of modeling, learning, recognition and segmentation of dynamic textures using tools from time series analysis, and system identification theory.

## 21.1 Introduction

Consider a sequence of images of a moving scene. Each image is an array of positive numbers that depend upon the shape, pose, viewpoint (*geometry*), material reflectance properties, and light distribution (*radiometry*) of the scene, as well as upon the changes of all of these factors over time, i.e. upon the *dynamics* of the scene. In principle, to fully analyze and understand the properties of a video sequence, one would want to recover the physical model of the scene that could have generated the images. Unfortunately, it is well known that the joint reconstruction of radiometry, geometry, and dynamics of the scene (*visual reconstruction problem*) is an intrinsically ill-posed problem: From any number of images it is not possible to uniquely recover all the unknowns (shape, pose, reflectance, light distribution, and viewpoint). This means that it is always possible to construct scenes with different radiometry, geometry, and dynamics that give rise to the same images. For example, a video clip of the sea at sunset could have been originated by a very complex dynamic shape (the surface of the sea) with constant reflectance properties (homogeneous material, water), but also by a very simple shape (e.g. the plane of the television monitor) with a dynamic non-homogeneous radiance (the televised spatio-temporal signal). The ill-posedness of the visual reconstruction problem can be turned into a well-posed inference problem within the context of a specific task, and one can also use the extra degrees of freedom to the benefit of the application at hand by satisfying some additional optimality criterion (e.g. the minimum description length (MDL) principle [675] for compression). This way, even though one cannot infer "the" (physically correct) model of a scene, one can infer a representation of the scene that can be sufficient to support, for instance, control, or recognition tasks.

In this chapter we survey a series of recent papers that describe very simple statistical models that can explain the measured video signal, predict new measurements, and extrapolate new image data. These models are not models of the scene, but statistical models of the video signal. In general, they fail to capture the correct radiometry, geometry, and dynamics of the scene. Instead, they capture a mixture of the three that is equivalent to the underlying physical model of the scene, once the statistical model is "visualized" as a sequence of images. Hopefully, these models will provide a representation of geometry, radiometry and dynamics that is sufficient to support recognition and segmentation tasks.

We put the emphasis on sequences of images that exhibit some form of temporal regularity[1], such as sequences of fire, smoke, water, foliage or flowers in wind, clouds, crowds of waving people, etc., and we refer to them as *dynamic textures* [278]. In statistical terms, we assume that a dynamic texture is a sequence of images, that is a realization from a stationary stochastic process[2]. In Section 21.2 we describe a representation of dynamic textures introduced in [748] that is general (it accounts for every possible decomposition of images, and every possible dynamics of sequences), and precise (it allows making analytical statements and drawing from the rich literature on system identification). In Section 21.7 we describe a technique to learn model parameters using maximum likelihood or prediction error methods. Under the hypothesis of second-order stationarity, there is a closed-form sub-optimal solution of the learning problem. In Section 21.4 the model is tested on simulation and prediction, showing that even the simplest

[1]The case of sequences that exhibit temporal and spatial regularity is treated in [280].

[2]A stochastic process is stationary (of order $k$) if the joint statistics (up to order $k$) are time-invariant. For instance a process $\{I(t)\}$ is second-order stationary if its mean $\bar{I} \doteq E[I(t)]$ is constant and its covariance $E[(I(t_1) - \bar{I})(I(t_2) - \bar{I})]$ only depends upon $t_2 - t_1$.

instance of the model captures a wide range of dynamic textures. The algorithm is simple to implement, efficient to learn and fast to simulate; it allows generating infinitely long sequences from short input sequences, and to control the parameters in the simulation [281]. In Section 21.7 we investigate the discriminative power of the models and describe a classification scheme based on the $k$-nearest neighbor rule, as it has been proposed in [698]. Section 21.7 addresses the problem of segmenting the image plane of a video sequence into homogeneous regions characterized by constant spatio-temporal signatures, as introduced in [279]. We illustrate a region-based segmentation framework where we model the signatures with dynamic texture models and compare them by means of the distances proposed in Section 21.5.1.

### *21.1.1 Related work*

Statistical inference for analyzing and understanding general images has been extensively used for the last two decades. There has been a considerable amount of work in the area of 2D texture analysis, starting with the pioneering results of Julesz [444], until the more recent statistical models (see [658] and references therein).

There has been comparatively little work in the specific area of dynamic (or time-varying) textures. The problem has been first addressed by Nelson and Polana [596], who classify regional activities of a scene characterized by complex, non-rigid motion. Szummer and Picard's work [783] on temporal texture modeling uses the spatio-temporal auto-regressive model, which imposes a neighborhood causality constraint for both spatial and temporal domain. This restricts the range of processes that can be modeled, and does not allow to capture rotation, acceleration and other simple non translational motions. Bar-Joseph et al. [50] uses multi-resolution analysis and tree-merging for the synthesis of 2D textures and extends the idea to dynamic textures by constructing trees using a 3D wavelet transform.

Other related work [318] is used to register nowhere-static sequences of images, and synthesize new sequences. Parallel to these approaches there is the work of Wang and Zhu [855, 856] where images are decomposed by computing their primal sketch, or by using a dictionary of Gabor or Fourier bases to represent image elements called "movetons." Such models capture the temporal variability of either the graph describing the sketches, or the movetons. Finally, in [913] feedback control is used to improve the rendering performance of the dynamic texture model we describe in this chapter.

The problem of modeling dynamic textures for the purposes of synthesis has been tackled also by computer graphics researchers. The typical approach is to synthesize new video sequences using procedural techniques that essentially entail clever concatenation or repetition of training image data. The reader is referred to [716, 869, 499, 83] and references therein.

## 21.2 Representation of dynamic textures

The intuitive notion of a dynamic texture is that of a sequence of images that exhibits temporal regularity. Individual images are clearly not independent realizations from a stationary distribution, for there is a temporal coherence intrinsic in the process that needs to be captured. Therefore, the underlying assumption is that the temporal correlation of sequences can be modeled by the integration of independent and identically distributed (IID) samples from a stationary distribution. In other words, a sequence of images can be modeled as the output of a dynamical system. We follow [278] and now make this concept precise.

Let $\{I(t)\}_{t=1\dots\tau}$, $I(t) \in \mathbb{R}^m$, be a sequence of $\tau$ images. Suppose that at each instant of time $t$ we can measure a noisy version of the image, $y(t) = I(t) + w(t)$, where $w(t) \in \mathbb{R}^m$ is an IID sequence drawn from a known distribution $p_w(\cdot)$ (that can be inferred from the physics of the imaging device), resulting in a positive measured sequence $\{y(t)\}_{t=1\dots\tau}$. We say that *the sequence* $\{I(t)\}$ *is a (linear) dynamic texture* if there exists a set of $n$ spatial filters $\phi_\alpha : \mathbb{R} \to \mathbb{R}^m$, $\alpha = 1 \dots n$ and a stationary distribution $q(\cdot)$ such that, defining $x(t) \in \mathbb{R}^n$ such that $I(t) = \phi(x(t))$ (where $\phi(\cdot)$ indicates the combination of the output of the $n$ filters $\{\phi_\alpha\}$ respectively applied to each of the $n$ state components) we have $x(t) = \sum_{i=1}^{k} A_i x(t-i) + v(t)$, with $v(t) \in \mathbb{R}^n$ an IID realization from the density $q(\cdot)$, for some choice of matrices, $A_i \in \mathbb{R}^{n\times n}$, $i = 1, \dots, k$, and initial condition $x(0) = x_0$. Without loss of generality, we can assume $k = 1$ since we can augment the state of the above model to be $\bar{x}(t) \doteq [x(t)^T\, x(t-1)^T \dots x(t-k)^T]^T$. Therefore, a linear dynamic texture is associated to the dynamical system

$$\begin{cases} x(t+1) = Ax(t) + v(t) \\ y(t) = \phi(x(t)) + w(t) \end{cases} \tag{21.1}$$

with $x(0) = x_0$, $v(t) \overset{IID}{\sim} q(\cdot)$ unknown, $w(t) \overset{IID}{\sim} p_w(\cdot)$ given, such that $I(t) = \phi(x(t))$. One can easily generalize the definition to an arbitrary non-linear model of the form $x(t+1) = f(x(t), v(t))$, leading to the concept of a *non-linear dynamic texture*.

## 21.3 Learning dynamic textures

Given a sequence of noisy images $\{y(t)\}_{t=1\dots\tau}$, learning the dynamic texture model (21.1) amounts to identifying the model parameter $A$, the filters $\phi(\cdot)$, and the distribution of the input $q(\cdot)$. This is a form of *system identification problem* [524], where one has to infer a dynamical model from a time series. The maximum-likelihood formulation of the dynamic texture learning problem can be posed as follows:

$$\begin{aligned} &\text{given } y(1), \dots, y(\tau), \text{ find} \\ &\hat{A}, \hat{\phi}(\cdot), \hat{q}(\cdot) = \arg\max_{A,\phi,q} \log p(y(1), \dots, y(\tau)) \\ &\text{subject to (21.1) and } v(t) \overset{IID}{\sim} q. \end{aligned} \tag{21.2}$$

While we refer the reader to [278] for a more complete discussion about how to solve problem (21.2), and how to set out the learning via prediction error methods, here we summarize a number of simplifications that lead us to a simple closed-form procedure.

In (21.2) we have not made any assumption on the class of filters $\phi(\cdot)$, and there are many ways in which one can choose them. However, in texture analysis the dimension of the signal is huge (tens of thousands components) and there is a lot of redundancy. Therefore, we view the choice of filters as a dimensionality reduction step and seek for a decomposition of the image in the simple (linear) form $I(t) = \sum_{i=1}^{n} x_i(t)\theta_i \doteq Cx(t)$, where $C = [\theta_1, \dots, \theta_n] \in \mathbb{R}^{m \times n}$, $m \gg n$, and $\{\theta_i\}$ can be an orthonormal basis of $L^2$, a set of principal components, or a wavelet filter bank. Note that the inference method depends also upon what type of representation we choose for $q$. In principle, the unknown driving distribution belongs to an infinite-dimensional space. In this exposition we assume the simplest parametric class of densities, which is Gaussian $v(t) \overset{IID}{\sim} \mathcal{N}(0, Q)$, and $Q \in \mathbb{R}^{n \times n}$ is a symmetric positive-definite matrix. We assume a similar distribution for the measurement noise $w(t) \overset{IID}{\sim} \mathcal{N}(0, R)$, $R \in \mathbb{R}^{m \times m}$. Under these hypotheses model (21.1) reduces to the following linear Gauss-Markov model

$$\begin{cases} x(t+1) = Ax(t) + v(t) \,, & v(t) \sim \mathcal{N}(0, Q) \,, \quad x(0) = x_0 \,, \\ y(t) = Cx(t) + w(t) \,, & w(t) \sim \mathcal{N}(0, R) \,, \end{cases} \tag{21.3}$$

and the system identification problem consists in estimating the parameters $A, C, Q, R$ from the measurements $y(1), \dots, y(\tau)$. It is well known that this model can capture the second-order properties of a generic stationary stochastic process [524].

The first observation concerning model (21.3) is that the choice of matrices $A, C, Q$ is not unique, in the sense that there are infinitely many such matrices that give rise to exactly the same sample paths $y(t)$ starting from suitable initial conditions. This is immediately seen by substituting $A$ with $TAT^{-1}$, $C$ with $CT^{-1}$ and $Q$ with $TQT^T$, and choosing the initial condition $Tx_0$, where $T \in GL(n)$ is any invertible $n \times n$ matrix. In other words, the basis of the state-space is arbitrary, and any given process has *not* a unique model, but an *equivalence class* of models $\mathcal{R} \doteq \{[A] = TAT^{-1}, [C] = CT^{-1}, [Q] = TQT^T, \mid T \in GL(n)\}$. In order to identify a unique model of the type (21.3) from a sample path $y(t)$, it is necessary to choose a representative of each equivalence class: such a representative is called a *canonical model realization*, in the sense that it does not depend on the choice of basis of the state space (because it has been fixed).

While there are many possible choices of canonical models (see for instance [446]), we will make the assumption that $\text{rank}(C) = n$ and choose the canonical model that makes the columns of $C$ orthonormal: $C^T C = I_n$, where $I_n$ is the identity matrix of dimension $n \times n$. As we will see shortly, this assumption allows to infer a unique model that is tailored to the data in the sense of defining a basis of the state space such that its covariance $P \doteq \lim_{t\to\infty} E[x(t)x^T(t)]$ is asymptotically diagonal (see Equation (21.7)).

With the above simplifications one might use *subspace identification* techniques [524] to learn model parameters in closed-form in the maximum-likelihood sense, for instance with the well known N4SID algorithm [832]. Unfortunately this is not possible. In fact, given the dimensionality of our data, the requirements in terms of computation and memory storage of standard system identification techniques are far beyond the capabilities of the current state-of-the-art workstations. For this reason, following [278], we describe a closed-form sub-optimal solution of the learning problem, that takes few seconds to run on a current low-end PC when $m = 170 \times 110$ and $\tau = 120$.

## 21.3.1 Closed-form solution

Let $Y_1^\tau \doteq [y(1), \ldots, y(\tau)] \in \mathbb{R}^{m\times\tau}$ with $\tau > n$, and similarly for $X_1^\tau \doteq [x(1), \ldots, x(\tau)] \in \mathbb{R}^{n\times\tau}$ and $W_1^\tau \doteq [w(1), \ldots, w(\tau)] \in \mathbb{R}^{m\times\tau}$, and notice that

$$Y_1^\tau = C X_1^\tau + W_1^\tau \,. \tag{21.4}$$

Now let $Y_1^\tau = U\Sigma V^T$; $U \in \mathbb{R}^{m\times n}$; $U^T U = I$; $V \in \mathbb{R}^{\tau\times n}$, $V^T V = I$ be the singular value decomposition (SVD) [352] with $\Sigma = \text{diag}\{\sigma_1, \ldots, \sigma_n\}$, and $\{\sigma_i\}$ be the singular values, and consider the problem of finding the best estimate of $C$ in the sense of Frobenius: $\hat{C}(\tau), \hat{X}(\tau) = \arg\min_{C, X_1^\tau} \|W_1^\tau\|_F$ subject to (21.4). It follows immediately from the fixed rank approximation property of the SVD [352] that the unique solution is given by

$$\hat{C}(\tau) = U \,, \quad \hat{X}(\tau) = \Sigma V^T \,, \tag{21.5}$$

$\hat{A}$ can be determined uniquely, again in the sense of Frobenius, by solving the following linear problem: $\hat{A}(\tau) = \arg\min_A \|X_1^\tau - A X_0^{\tau-1}\|_F$, where $X_0^{\tau-1} \doteq [x(0), \ldots, x(\tau-1)] \in \mathbb{R}^{n\times\tau}$ which is trivially done in closed-form using the state estimated from (21.5):

$$\hat{A}(\tau) = \Sigma V^T D_1 V (V^T D_2 V)^{-1} \Sigma^{-1} \,, \tag{21.6}$$

where $D_1 = \begin{bmatrix} 0 & 0 \\ I_{\tau-1} & 0 \end{bmatrix}$ and $D_2 = \begin{bmatrix} I_{\tau-1} & 0 \\ 0 & 0 \end{bmatrix}$. Notice that $\hat{C}(\tau)$ is uniquely determined up to a change of sign of the components of $C$ and $x$. Also note that

$$E[\hat{x}(t)\hat{x}^T(t)] \equiv \lim_{\tau\to\infty} \frac{1}{\tau} \sum_{k=1}^{\tau} \hat{x}(t+k)\hat{x}^T(t+k) = \Sigma V^T V \Sigma = \Sigma^2 \,, \tag{21.7}$$

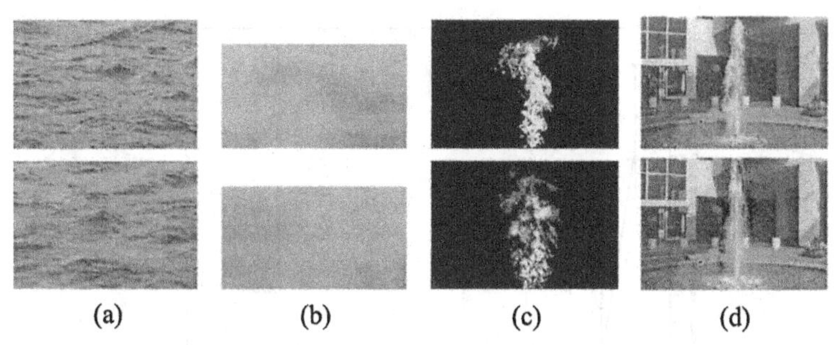

(a) (b) (c) (d)

Figure 21.1. Samples of four training sequences (top row) and four synthesized frames with the corresponding models (bottom row). (a) River sequence ($\tau = 120$ training images of $m = 170 \times 115$ pixels). Simulation is performed with a model of state dimension $n = 50$. (b) Steam sequence ($\tau = 120$, $m = 176 \times 96$), $n = 30$. (c) Fire sequence ($\tau = 150$, $m = 360 \times 243$), $n = 50$. (d) Fountain sequence ($\tau = 150$, $m = 320 \times 220$), $n = 50$. The river and steam sequences have been borrowed from the MIT Temporal Texture database, whereas the fire sequence comes from the Artbeats Digital Film Library. In all these cases the state dimension $n$ was given as input parameter. The movies are available on-line at `http://www.cs.ucla.edu/~doretto/projects/dynamic-textures.html`.

which is diagonal as mentioned in the first part of Section 21.7. Finally, the sample input noise covariance $Q$ can be estimated from

$$\hat{Q}(\tau) = \frac{1}{\tau}\sum_{i=1}^{\tau} \hat{v}(i)\hat{v}^T(i) \,, \tag{21.8}$$

where $\hat{v}(t) \doteq \hat{x}(t+1) - \hat{A}(\tau)\hat{x}(t)$. Should $\hat{Q}$ not be full rank, its dimensionality can be further reduced by computing the SVD $\hat{Q} = U_Q \Sigma_Q U_Q^T$ where $\Sigma_Q = \mathrm{diag}\{\sigma_{Q,1}, \ldots, \sigma_{Q,n_v}\}$ with $n_v \leq n$, and one can set $v(t) \doteq B\eta(t)$, with $\eta(t) \sim \mathcal{N}(0, I_{n_v})$, and $\hat{B}$ such that $\hat{B}\hat{B}^T = \hat{Q}$.

In the algorithm above we have assumed that the order of the model $n$ was given. In practice, this needs to be inferred from the data. Following [278], one can determine the model order empirically from the singular values $\sigma_1, \sigma_2, \ldots,$ by choosing $n$ as the cutoff where the singular values drop below a threshold. A threshold can also be imposed on the difference between adjacent singular values.

## 21.4 Model validation

One of the most compelling validations for a dynamic texture model is to simulate it to evaluate to what extent the synthesis captures the essential perceptual features of the original data. Given a typical training sequence of about one hundred frames, using the procedure described in Section 21.3.1 one can learn model parameters in a few seconds, and then synthesize a potentially infinite number of new images by simulating (21.3). To generate a new image one needs to draw a sample $v(t)$ from a Gaussian distribution with covariance $Q$, update the state

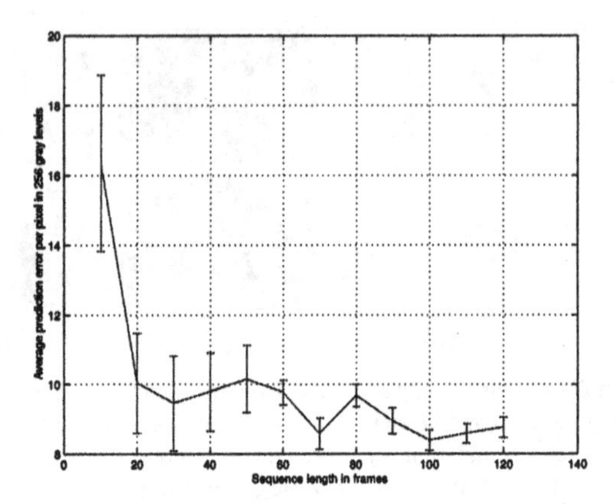


Figure 21.2. Error-bar plot of the average prediction error and standard deviation (for 100 trials), per pixel as a function of the length of the steam training sequence, expressed in gray levels (the range of pixel values is [0,255]). The state dimension is set to $n = 20$.

$x(t+1) = Ax(t) + v(t)$, and compute the image $I(t) = Cx(t)$. This can be done in real-time. Even though the result is best shown with movies, Figure 21.1 provides some examples of the kind of output that one can get (see [278] for more results). The simple model (21.3), that captures only the second-order temporal statistics of a video sequence, is able to represent most of the perceptual features of sequences of images of natural phenomena, such as fire, smoke, water, flowers or foliage in wind, etc., and even dynamic textures that are periodic signals in time [277].

An important question is how long should the input sequence be in order to capture the dynamics of the process. To answer this question experimentally, for a fixed state dimension, we consider the prediction error as a function of the length $\tau$, of the input (training) sequence. This means that for each length $\tau$, we predict the frame $\tau + 1$ (not part of the training set) and compute the prediction error per pixel in gray levels. We do so many times in order to infer the statistics of the prediction error, i.e. mean and variance at each $\tau$. Using one criterion for learning (the procedure in Section 21.3.1), and another one for validation (prediction error) is informative for challenging the model. Figure 21.2 shows an error-bar plot including mean and standard deviation of the prediction error per pixel for the steam sequence. The average error decreases and becomes stable after approximately 70 frames. The plot of Figure 21.2 validates *a-posteriori* the model (21.3) inferred with the procedure described in Section 21.3.1. Other dynamic textures have similar prediction error plots [278].

## 21.5 Recognition

According to the model (21.3) a dynamic texture is characterized by a linear dynamic system with multiple-input and multiple-output (MIMO) driven by white noise (which is also a vector autoregressive moving average (ARMA) model). Therefore, following [698], in order to build a recognition system able to categorize dynamic textures, one needs to first define a base measure in the space of vector ARMA models, and then to characterize probability distributions in that space. Defining an appropriate base measure in the space of vector ARMA models is not trivial, since each model entails a combination of an input density and state and output transition matrices that have a very particular Riemannian structure (they are *not* a linear space), and this problem remains unsolved to this day.

>From a pattern recognition viewpoint [288], constructing a probability density is not necessary to solve problems such as classification, clustering or grouping. For instance, the $k$-nearest neighbor algorithm only requires a distance to be implemented. This approach can be applied to the space of models, that will be endowed by a probability structure induced by the notion of distance that we have defined. More precisely, suppose a set of model samples $M_1$, $\cdots$, $M_N$, is given, where each model is labeled with $\lambda_j$, which is one out of $c$ classes. Given a new model sample $M$, the label $\lambda_m$ is chosen by taking a vote among the $k$ nearest model samples. That is, $\lambda_m$ is selected if the majority of the $k$ nearest neighbors have label $\lambda_m$. For $c = 2$ this happens with probability $\sum_{i=(k+1)/2}^{k} \binom{k}{i} P(\lambda_m|M)^i(1 - P(\lambda_m|M))^{k-i}$. It can be shown [288] that if $k$ is odd, $N \gg c$, and $c = 2$, the error rate is bounded above by the smallest concave function of $P^*$ (the optimal error rate) greater than $\sum_{i=0}^{(k+1)/2} \binom{k}{i} (P^{*i+1}(1-P^*)^{ki} + P^{*k-i}(1-P^*)^{i+1})$. Note that the analysis holds for $k$ fixed as $N \to \infty$, and that the rule approaches the minimum error rate for $k \to \infty$.

We assume that a model $M$ is given by the couple $(A, C)$. That is, we do not consider the covariance of the measurement noise $R$, since that carries no information on the underlying process. Moreover, we consider processes with different input noise covariance as equivalent, which means that we ignore $Q$.

### *21.5.1 Distances between dynamic texture models*

One of the difficulties in defining a distance between ARMA models is that each model $M$ is described not only by the parameters $(A, C)$, but by an equivalence class of such parameters, as pointed out in Section 21.7. Therefore, a suitable discrepancy measure has to compare *not* the parameters directly, but their equivalence classes. One technique for doing so has been proposed in [251]. It consists of building infinite observability matrices, whose columns span the space generated by the measurements $y(t)$ of the model (21.3), which is an $n$-dimensional subspace of the infinite-dimensional space of all possible measurements. Then one can compute the geometric angles between such subspaces through their embedding.

More formally, let $S \in \mathbb{R}^{m \times n}$ and $T \in \mathbb{R}^{m \times n}$ be two matrices with full column rank. The $n$ *principal angles* $\theta_k \in \left[0, \frac{\pi}{2}\right]$ between range($S$) and range($T$) are recursively defined for $k = 1, \ldots, n$ as

$$\begin{cases} \cos\theta_1 = \max\limits_{x,y \in \mathbb{R}^n} \frac{|x^T S^T T y|}{\|Sx\|_2 \|Ty\|_2} = \frac{|x_1^T S^T T y_1|}{\|Sx_1\|_2 \|Ty_1\|_2}, \\ \cos\theta_k = \max\limits_{x,y \in \mathbb{R}^n} \frac{|x^T S^T T y|}{\|Sx\|_2 \|Ty\|_2} = \frac{|x_k^T S^T T y_k|}{\|Sx_k\|_2 \|Ty_k\|_2}, \text{ for } k = 2, \ldots, n \\ \text{subject to } x_i^T S^T S x = 0 \text{ and } y_i^T T^T T y = 0, \text{ for } i = 1, 2, \ldots, k-1. \end{cases}$$

Now, let $M_1 \doteq (A_1, C_1)$ and $M_2 \doteq (A_2, C_2)$ be two models with the same output dimensionality. Their *infinite observability matrices* $\mathcal{O}_i$, for $i = 1, 2$, are defined as $\mathcal{O}_i \doteq \begin{bmatrix} C_i^T & A_i^T C_i^T & \ldots & (A_i^T)^n C_i^T & \ldots \end{bmatrix}^T \in \mathbb{R}^{\infty \times n}$, and we refer to the principal angles between the ranges of $\mathcal{O}_1$, and $\mathcal{O}_2$ as *subspace angles*. They can be computed in closed-form with a procedure described in [251].

For the case of minimum-phase single-input single-output (SISO) state space models that correspond to autoregressive (AR) models, one can use the subspace angles to define the so-called *Martin distance*:

$$d_M(M_1, M_2)^2 = \ln \prod_{i=1}^{n} \frac{1}{\cos^2\theta_i}, \tag{21.9}$$

which was originally proposed in [548] as function of the cepstrum coefficients[3] of the model, whereas the expression of the Martin distance as function of the subspace angles was introduced in [251]. It is also possible to define another distance, that uses only the biggest subspace angle, i.e. $d_F = \theta_n$. Geometrically $d_F$ is the *Finsler distance* between two subspaces viewed as two elements in the Grassman manifold $G(\infty, n)$ [874]. Roughly speaking, the difference between the Martin and Finsler distance is that $d_M^2$ is an $L^2$-norm but $d_F$ is an $L^\infty$-norm between linear systems.

Unfortunately, the generalization of $d_M^2$ and $d_F$ to the case of MIMO linear dynamic systems is not possible. For instance, it is not even guaranteed that the Martin distance be non-negative. Nevertheless, we used the idea of comparing two models by computing their subspace angles, and tested the ability of the Martin distance, Finsler distance, and the naïve Frobenius norm between model parameters, to classify dynamic textures within a $k$-nearest neighbor scheme.

### *21.5.2 Performance of the nearest neighbor classifier*

We illustrate tests of the distances proposed in the previous section against a database of 50 categories of dynamic textures, each of which represented by

[3] The cepstrum of a discrete-time process is the inverse Fourier transform of the logarithm of the power spectrum.

four models, following [698]. The models have state dimension $n = 20$, and have been extracted from sequences of length $\tau = 75$ frames of $m = 48 \times 48$ pixels, using the procedure illustrated in Section 21.3.1. The sequences capture natural phenomena like ocean waves, smoke, steam, fire, and plants. Included in the database are similar sequences with different dynamics. For example, there are water streams recorded from different angles, with flows moving in different orientations and at different speeds.

Between each pair of models of the database we computed the Frobenius norm, and the Martin and Finsler distance. Figure 21.3 shows a gray-level representation of the confusion matrix for a subset of 10 categories (40 sequences out of the entire database), for the case of Frobenius norm (top) and Martin distance (bottom). Moving along the horizontal axis, we marked the first (with an "o") and second (with an "x") nearest neighbors. For example, with reference to the results using Martin distance (bottom), the closest dynamic texture to Smoke1 (in the vertical axis) is Smoke2 (in the horizontal axis). Similarly, the second closest dynamic texture to Smoke1 is Water-Fall-b1.

If we define a hit when the first nearest neighbor of a sequence is one of the other three sequences in the same category, Figure 21.3 already highlights the differences between the Frobenius norm and the Martin distance. In the latter case most of the first nearest neighbors lie on the diagonal blocks (meaning that there are a lot of hits), whereas in the former they are almost randomly spread all over the blocks. In particular, if we count the number of correct hits for the whole database, in the case of Frobenius norm we obtain a hit ratio of 5.5%. This poor result is not unexpected since, even though ARMA models are linear, the space of model parameters is nonlinear and the Frobenius norm assumes linearity. On the other hand, the hit ratio of the Martin distance is 89.5%, whereas the Finsler distance is less efficient with a hit ratio of 24.5%.

The encouraging results obtained using the Martin distance suggest that, in principle, a comprehensive database of models of commonly occurring dynamic textures can be maintained, and a new sequence could be categorized, after learning its parameters, using the $k$-nearest neighbor rule.

## 21.6 Segmentation

Modeling the (global) spatio-temporal statistics of the entire video sequence can be a daunting task due to the complexity of natural scenes. An alternative consists of choosing a simple class of models, and then partition the scene into regions where the model fits the data within a specific accuracy. In this section, which follows [279], we discuss a simple model for partitioning the scene into regions where the spatio-temporal statistics, represented by a dynamic texture model, is constant. To perform this *segmentation* task we use a region-based approach pioneered in [591]. In particular, we revert to a level set framework of the Mumford-Shah functional introduced in [170, 809].

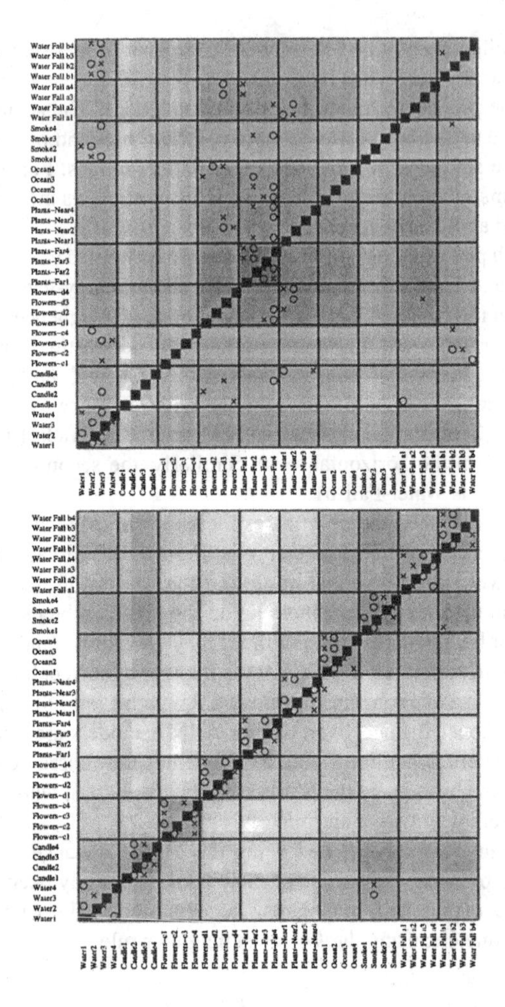


Figure 21.3. Gray-level representation of the confusion matrices for a subset of 10 dynamic texture categories (40 sequences out of 200 of the entire database), computed using the Frobenius norm (top) and the Martin distance (bottom).

Let $\Omega \subset \mathbb{R}^2$ be the domain of an image and $\{\Omega_i\}_{i=1,\dots,N}$ be a partition of $\Omega$ into $N$ (unknown) regions[4]. We assume that the intensities of the pixels $y_i(t)$, contained in the region $\Omega_i$, are a Gauss-Markov process that can be modeled with a dynamic texture model (21.3), with (unknown) parameters $A_i \in \mathbb{R}^{n_i \times n_i}$, $C_i \in \mathbb{R}^{m_i \times n_i}$, and $Q_i \in \mathbb{R}^{n_i \times n_i}$. Note that we allow the number of pixels $m_i$ to be different in each region, as long as $\sum_{i=1}^{N} m_i = m$, the size of the entire image,

[4]That is, $\Omega = \cup_{i=1}^{N} \Omega_i$ and $\Omega_i \cap \Omega_j = \emptyset, i \neq j$.

and that we require that neither the regions nor the parameters change over time, $\Omega_i, A_i, C_i, Q_i = \text{const}$. With this generative model, the segmentation problem can be formalized as follows: *Given a sequence of images* $\{y(t)\}_{t=1,\dots,\tau}$, $y(t) \in \mathbb{R}^m$, *with two or more distinct regions* $\Omega_i$, $i = 1, \dots, N \geq 2$ *that satisfy model (21.3), estimate both the regions* $\Omega_i$, *and model parameters of each region, namely the matrices* $A_i$, $C_i$, *and* $Q_i$.

If the regions $\Omega_i$, $i = 1, \dots, N$ were known, one would just be left with two problems. The first one is the learning of model parameters. This problem has already been solved in Section 21.3.1. Assuming that the parameters $A_i$, $C_i$, $Q_i$ have been inferred for each region, in order to set the stage for a segmentation procedure, one has to define a discrepancy measure among regions, i.e. between dynamic texture models. This problem have been approached in Section 21.7, and one can measure the discrepancy between different models by comparing either the subspace angles or their combination via the Martin distance (21.9).

On the other hand, if the dynamic texture associated with each pixel were known, then one could easily determine the regions by thresholding or other grouping or segmentation techniques. However, a dynamic texture associated with a certain pixel $\mathbf{x} \in \Omega$, as defined in Equation (21.3), depends on the whole region $\Omega_i$ containing $\mathbf{x}$. Therefore, we have a classic "chicken-and-egg" problem: If we knew the regions, we could easily identify the dynamical models, and if we knew the dynamical models we could easily segment the regions. Unfortunately, we know neither.

Since one can always explain the image with a few high-order models with large support regions (the entire image in the limit), or with many low-order models with small support regions (individual pixels in the limit), in order to render the chicken-and-egg problem well posed, a model complexity cost needs to be added, for instance the description length of model parameters and the boundaries of each region [675]. This significantly complicates the algorithms and the derivation. Following [279], we simplify the problem, and first associate a local *signature* $s(\mathbf{x})$ to each pixel $\mathbf{x} \in \Omega$, by integrating visual information on a fixed spatial neighborhood of that pixel $B(\mathbf{x}) \subset \Omega$; then we group together pixels with similar signatures in a region-based segmentation approach.

Each signature contains the cosines of the subspace angles between the local dynamic texture model corresponding to $\{y(\tilde{\mathbf{x}}, t)\}_{\tilde{\mathbf{x}} \in B(\mathbf{x}), t=1,\dots,\tau}$, and a reference dynamic texture model, for instance the one corresponding to a preselected spatio-temporal neighborhood centered at $x_0 \in \Omega$, i.e. $\{y(\tilde{\mathbf{x}}, t)\}_{\tilde{\mathbf{x}} \in B(\mathbf{x}_0), t=1,\dots,\tau}$. With this representation, a segmentation of the image plane $\Omega$ into a set of pairwise disjoint regions $\Omega_i$ of constant signature $s_i \in \mathbb{R}^n$ is obtained by minimizing the Mumford-Shah cost functional [591]:

$$E(\Gamma, \{s_i\}) = \sum_i \int_{\Omega_i} \left(s(\mathbf{x}) - s_i\right)^2 \mathrm{d}\mathbf{x} + \nu \left|\Gamma\right| , \tag{21.10}$$

simultaneously with respect to the region descriptors $\{s_i\}$, modeling the average signature of each region, and with respect to the boundary $\Gamma$, separating these

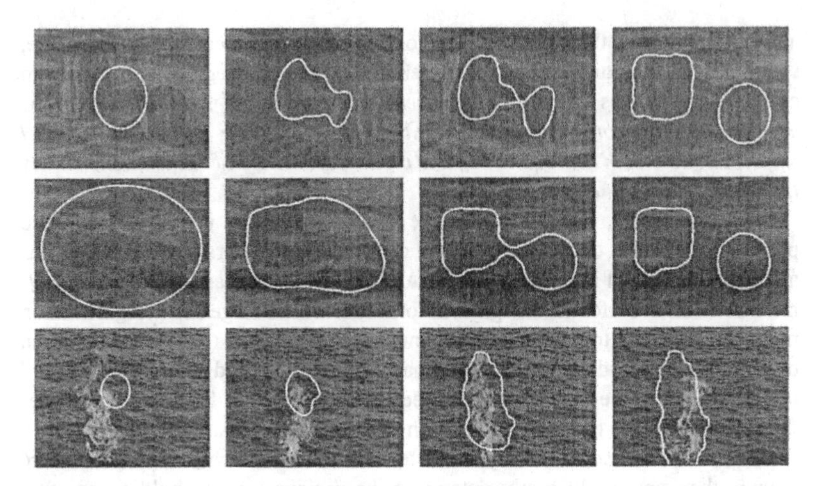

Figure 21.4. Top row: segmentation of two dynamic textures that share the same dynamics, and differ only for the "orientation" of their spatial statistics. Middle row: segmentation of two dynamic textures that are identical in appearance, and differ only in the dynamics. Bottom row: segmentation when the region boundaries (of the flame), are changing in time. In all the experiments, the local dynamic textures were defined on neighborhood of $11 \times 11$ pixels, whereas the state dimension was set to $n = 10$. The contour evolutions are available on-line at `http://www.cs.ucla.edu/~doretto/projects/dynamic-segmentation.html`.

regions. The first term in the functional (21.10) aims at maximizing the homogeneity with respect to the signatures in each region $\Omega_i$, whereas the second term aims at minimizing the length $|\Gamma|$ of the separating boundary.

Figure 21.4 demonstrates some aspects of dynamic texture segmentation, the reader is referred to [279] for a more complete account. The first row has a few snapshots of a sequence with ocean waves, where the portion of every frame within the square and the circle have been rotated by 90 degrees. The superimposed contour evolution shows that the segmentation system can partition the image plane based only on the spatial statistics of the images. On the other hand, the sequence in the second row has the square and circle filled with the same ocean waves of the background, but running at different speeds. In this case the algorithm segments based only on the dynamics of the regions. This ability is the most important aspect of this approach. The last row shows a sequence of fire combined with the ocean waves. Here the region boundaries are moving, against the initial hypothesis. The algorithm manage to estimates the "average" region where the flame is mostly present.

## 21.7 Discussion

This chapter, which draws on a series of works published recently [748, 278, 698, 279], illustrates that very simple statistical models can capture the phenomenology of very complex physical processes, such as water, smoke, fire etc. The fact that the synthesis from a very simple linear Gauss-Markov model is perceptually indistinguishable from the simulation of non-linear Navier-Stokes partial differential equations, such as those that govern fluid motion, is indication that such models may be sufficient to support detection and recognition tasks and, to a certain extent, even synthesis and animation [281]. This work shows that modeling image motion, i.e. deformations of the domain of the image, can be done through modeling image values, i.e. the range of the image. Depending on the statistical properties of the scene, this can be more or less efficient. Joint modeling of the variation in domain and range of the image can result in more efficient models, as it has been recently explored [282], pointing at a direction of new development.

### *Acknowledgements*

We would like to thank our precious collaborators. In particular, Section is based on material coauthored with A. Chiuso and Y. N. Wu [278]; Section is based on material coauthored with P. Saisan and Y. N. Wu [698]; and Section is based on material coauthored with D. Cremers and P. Favaro [279]. This work is supported by NSF ECS-0200511/IIS-0208197, ONR N00014-03-1-0850/N00014-02-1-0720, and AFOSR F49620-03-1-0095/E-16-V91-G2.

# Part V

# 3D from Images, Projective Geometry & Stereo Reconstruction

# Chapter22

# Differential Geometry from the Frenet Point of View: Boundary Detection, Stereo, Texture and Color

**S. Zucker**

Abstract

Frenet frames are a central construction in modern differential geometry, in which structure is described with respect to an object of interest rather than with respect to external coordinate systems. The Cartan moving frame model specifies how these frames adapt when they are transported along the object. We consider this as a model for integrating local information with information in a neighborhood for curve detection, stereo, texture, and color. These different objects results in a series of geometric compatibility constructions useful within a number of different optimization and probabilistic inference techniques.

## 22.1 Introduction

Many problems in computational vision that involve inferences over noisy, local measurements have been formulated with a geometrical component. Our goal in this Chapter is to organize a number of such problems according to their geometric content, to isolate a common thread between them that leads to differential geometry; and to introduce ideas from differential geometry to show how they can structure new approaches to seemingly unrelated computational vision problems. As described, the techniques can be used with a variety of different inference techniques, including relaxation labeling [414], belief propagation, graph cuts [113], Markov random fields, quadratic programming, and so on.

To prefigure the type of geometry we shall be concerned with, consider the problem of boundary detection. Starting with local "edge" operators that signal intensity differences in a small neighborhood around a point, the question is whether this intensity event is part of a boundary, or not. Since many objects have smooth boundaries, and since these boundaries project into the image as smooth curves, determining whether a putative boundary point continues through an image neighborhood containing that point is often key. Mathematically, since only a neighborhood is involved, the analysis is local. Computationally, since such questions can be asked around each point in the image, the local analysis must be applicable in a neighborhood around each point; i.e., it is parallel. Differential geometry is a mathematical abstraction of boundary completion that satisfies these requirements. It will lead, as we show, to connections between the local estimates that are specialized for each problem.

Expanding the above points, recall that the best linear approximation in an infinitesimal neighborhood to a smooth (boundary) curve is its tangent, and that this tangent approximation can be made around each point. Therefore the question becomes whether nearby tangents are consistently part of a single curve. To develop an intuition about what consistent might mean, recall the classical Gestalt demonstration of perceptual *good continuation* (Fig. 22.1). Observe how the "Figure 8" appears to continue across the crossing point; that is, how orientation is continued along the tangent direction. Many such demonstrations were developed in the early $20^{th}$ century ([483]).

Approximately a half century earlier a fundamental series of discoveries began concerning the differential geometry of curves, and they continued through the time period dominated by the Gestalt psychology movement. Frenet (in 1847) and, independently Serret (in 1851), introduced the idea of adapting a coordinate frame directly to a curve, rather than using extrinsic coordinates. The remarkable discovery was that changes in (derivatives of) this frame could be expressed directly in terms of the frame itself. The result is a beautiful expression of the theory of curves that fits precisely the requirements for perceptual organization above. The Frenet-Serret theory was extended by Darboux to surfaces a few decades later, and was then elaborated to the powerful *repère mobile*–the moving frame–by Élie Cartan. Moving frames are not slaves to any coordinate system; rather, they are adapted to the object under study, be it a curve, a surface (notice the texture flow in Fig. 22.1), a metric space or manifold. For computer vision applications, we shall adapt them to curves (in 2-D and in 3-D), to texture, and to color. Local approximations of how these frames move will provide the geometry of connections that can be used with the different inference techniques listed above.

There are many excellent texts describing this approach to differential geometry. We recommend [611, 753], which we have followed closely in preparing this Chapter. For related discussions see also [482]. This research was done in collaboration with Ohad Ben-Shahar, Lee Iverson, and Gang Li. I thank Pavel Dimitrov for illustrations and AFOSR, DARPA, NIH, and ONR for support.

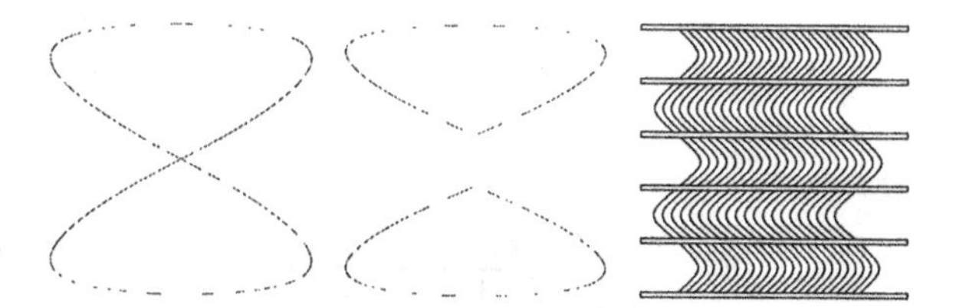

Figure 22.1. Perceptual organization is related to Gestalt notions of "good continuation." Observe how the "Figure 8" appears as a single curve, with smooth connections across the crossing point, and not as the non-generic arrangement of the two shapes in the middle. Such notions of orientation good continuation hold for textures as well; notice how this example appears to continue behind the occluders.

## 22.2 Introduction to Frenet-Serret

From a Newtonian perspective a curve can be thought of as the positions $\alpha(t) = (\alpha_1(t), \alpha_2(t), \alpha_3(t))$ in Euclidean 3-space swept out by a moving point $\alpha$ at parameter (time) $t$. Provided the coordinate functions $(\alpha_1, \alpha_2, \alpha_3)$ are differentiable, a *curve* can be defined as a differentiable map $\alpha : I \to \mathbb{E}^3$, from the open interval $I$ into $\mathbb{E}^3$. For now we shall assume the curve is simple, i.e., it does not cross itself, so the map is one-to-one and is an *immersion* of $I$ into $\mathbb{E}^3$.

The derivative of $\alpha$ gives the velocity or *tangent vector* of $\alpha$ at $t$

$$\alpha^{'}(t) = (\frac{d\alpha_1}{dt}(t), \frac{d\alpha_2}{dt}(t), \frac{d\alpha_3}{dt}(t), )_{\alpha(t)}$$

A curve is *regular* provided these derivatives are not zero simultaneously.

A reparameterization $s = s(t)$ yields the arc-length (unit speed) parameterization in which the length of each tangent vector is 1. We denote this unit speed curve by $\beta : I \to \mathbb{E}^3$ with $||\beta^{'}(s)|| = 1, s \in I$.

For simplicity, we work with $\beta$ for the remainder of this Section. We are interested in direction and, for non-straight lines, the rate at which the curve is bending. Intuition is helped by picturing the unit tangents as vectors in $\mathbb{E}^3$ attached to the points $\beta(s) \in \mathbb{E}^3$, that is, as a vector field along the curve. Euclidean coordinates for this vector field can again be differentiated:

$$\alpha^{''}(t) = (\frac{d^2\alpha_1}{dt^2}(t), \frac{d^2\alpha_2}{dt^2}(t), \frac{d^2\alpha_3}{dt^2}(t), )_{\alpha(t)}$$

to yield the acceleration, but geometrically the following construction will be more useful. (i) Denoting the unit tangent $T = \beta^{'}$, we obtain $T^{'} = \beta^{''}$, the curvature vector field. Observe $T^{'}$ is orthogonal to $T$ by differentiating $T \cdot T = 1$. The direction of the curvature vector is normal to $\beta$, and its length $\kappa(s) = ||T^{'}(s)||, s \in I$ is the *curvature*. (ii) The vector field $N = T^{'}/\kappa$ defines the *principal normal*, and (iii) the vector field $B = T \times N$ is the *binormal* vector field of $\beta$.

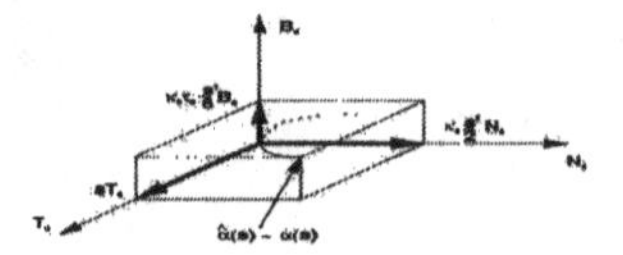

Figure 22.2. The Frenet frame attached to a point on a curve $\alpha(s)$ approximated to third order.

The *Frenet frame field* on $\beta$ is the triple $(T, N, B)$ such that $T \cdot T = N \cdot N = B \cdot B = 1$, all other dot products $= 0$, and the (i)–(iii) above hold (Fig. 22.2).

The remarkable property of this construction is that the derivatives of the frame can be expressed in terms of the frame itself. For $\kappa > 0$ and introducing the *torsion* $\tau$ we have:

$$\begin{pmatrix} T' \\ N' \\ B' \end{pmatrix} = \begin{bmatrix} 0 & \kappa & 0 \\ -\kappa & 0 & \tau \\ 0 & -\tau & 0 \end{bmatrix} \begin{pmatrix} T \\ N \\ B \end{pmatrix}. \tag{22.1}$$

These are the famous Frenet-Serret formulas. The torsion $\tau$ measures how rapidly the curve is twisting out of the (osculating) plane spanned by $(T, N)$. It is in this sense that the Frenet frame is adapted to the individual curve in a way that captures its essential (differential) geometric structure.

Basically all of information about the curve is contained in the Frenet-Serret formulas. The following theorem is fundamental in differential geometry: Let $\kappa, \tau : I \to \mathbb{R}$ be continuous ($\kappa(s) > 0, s \in I$). Then there is a curve $\beta : I \to \mathbb{E}^3$ with curvature function $\kappa(s)$ and torsion $\tau(s)$. Any two such curves differ only by a proper Euclidean motion.

Writing the Taylor approximation to the curve in the neighborhood of $\beta(0)$, and then substituting the Frenet formulas above and keeping only the dominant terms, we obtain:

$$\beta(s) \approx \beta(0) + s\beta'(0) + \frac{s^2}{2}\beta''(0) + \frac{s^3}{6}s\beta'''(0) \tag{22.2}$$

$$\approx \beta(0) + sT_0 + \kappa_0 \frac{s^2}{2} N_0 + \kappa_0 \tau_0 \frac{s^3}{6} B_0. \tag{22.3}$$

Thus the Frenet approximation shows how the tangent, curvature, and torsion effect the curve at each point (Fig.22.2).

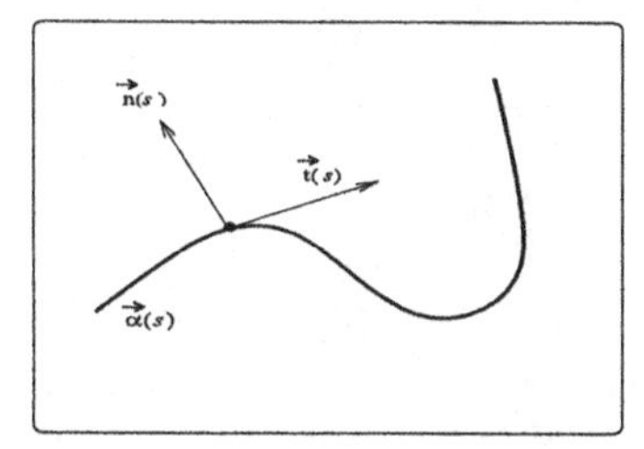


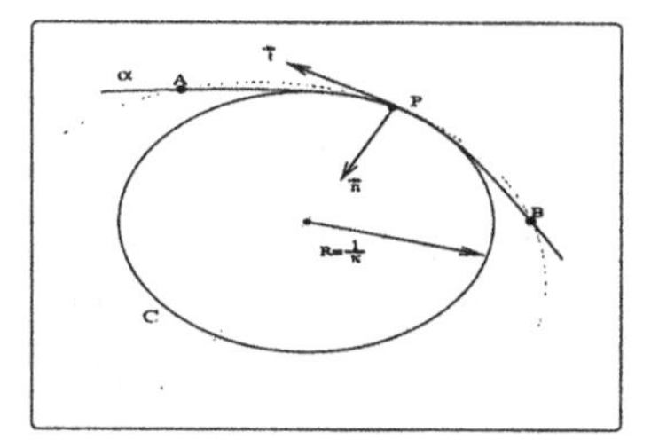


Figure 22.3. Two ways to think about the local structure of a curve in the plane. (left) The Frenet Frame is a (tangent, normal) coordinate frame that is adapted to the local structure of each point along a curve; and (right) the osculating circle is that circle with the largest contact with the curve among all circles tangent at that point.

## 22.3 Co-Circularity in $\mathbb{R}^2 \times S^1$

We now focus on curves in the plane $\mathbb{E}^2$. Observe that the first two terms in the Frenet approximation give the line in which the tangent (or best linear approximation) lies; the first three terms give the best quadratic approximation (a parabola) which, expressed in the (x,y) plane, has the shape $y = \kappa_0 x^2/2$ near $\beta(0)$.

The quadratic approximation around a point is determined by the curvature at that point, which can be defined in another way. Suppose the curve is not straight, and choose any three points on $\beta$ in the neighborhood of $\beta(0)$. Taking the limit as the three points approach $\beta(0)$, the *osculating circle* at that point is obtained. This is the unique circle tangent to the curve at that point such that its center lies on the normal and its radius is the inverse of the curvature (Fig.22.3).

The quadratic parabola is approximated by the *osculating circle* at that point, an observation introduced for the geometry of co-circularity [630][1]. The basic idea is illustrated in Fig. 22.4, which shows how local measurements of the tangent to a curve at an arbitrary point $q$ and at a nearby point in its neighborhood have different orientations. The geometry of consistency is given by Frenet: if the frame in the neighborhood of $q$ is transported along the curve to $q$, it should match the frame at $q$. If it does not, it is inconsistent.

However, the curve must be known before transport can be applied, but this is what we seek. The solution to this chicken-and-egg problem is to transport not along the actual curve, but along its approximation. We earlier showed that curvature dictates this approximation, and it can either be measured directly (which is what we think happens in neurobiology, [271]) or estimated by other means ([35]). In any case, once the system is discretized, the osculating circle and parabolic ap-

[1]Because of space limitations, references are very limited; we recommend that the original publications are consulted for additional references.

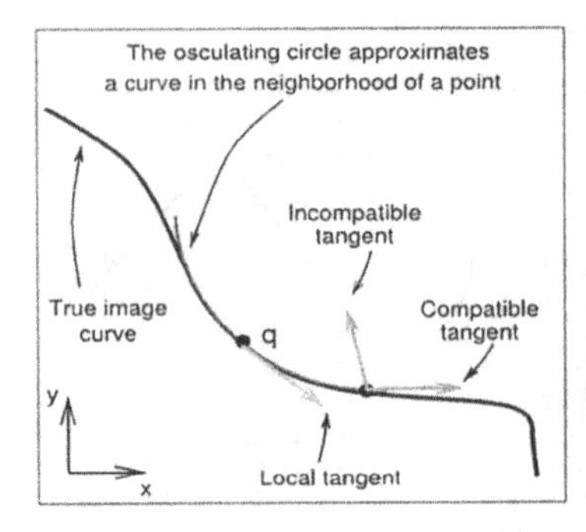


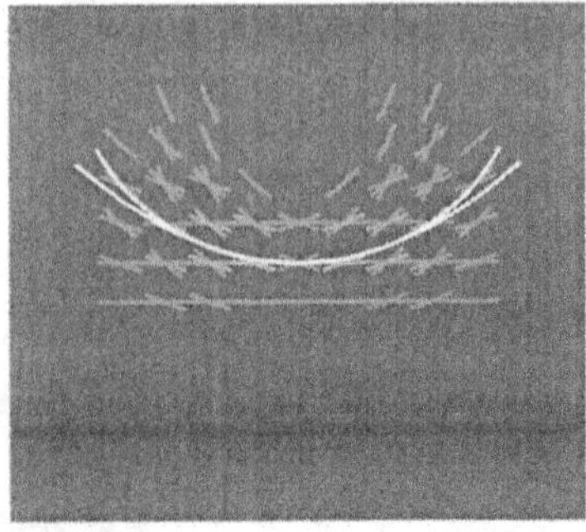

Figure 22.4. The geometry of co-circularity for curve detection in images. (left) Measurements of orientation differ at points along a curve. To determine whether they are consistent, nearby tangents are transported along the osculating circle approximation to the curve. If the transported tangents agree they are consistent; otherwise not. (right) To accomplish this transport operation in images, tangent position, orientation, and curvature must be discretized. This shows those nearby tangents that are consistent with a horizontal tangent at the center; that is, those tangent which, if transported along a (discretized) approximation to the osculating circle would support the central, horizontal tangent. (The width of the curve for this example is taken to be 3 pixels.) In the language of relaxation labeling, this is called an excitatory compatibility field. Note that the osculating circle and parabola approximations agree to within a fraction of a pixel over this neighborhood.

proximations agree to within a fraction of a pixel over the neighborhoods involved (Fig. 22.4); cf. [468]. Such geometric compatibility fields can be used with a number of different inference techniques, including relaxation labeling [414], belief propagation, and Bayes [480]. They are related to the forms that arise in elastica [589, 406]. For a different attempt to minimize a functional in curvature, see [732].

## 22.3.1 Multiple Orientations and Product Spaces

Thus far in this Chapter we have been concentrating exclusively on simple, regular curves. But the "figure 8" example in Fig. 22.1 is not simple, and it provided the motivation for the geometric approach. Which way should the curve be continued at the crossing point? For such examples, although $\beta(s_1) = \beta(s_2)$ for $s_1 \neq s_2$ at the crossing point, we have $\beta'(s_1) \neq \beta'(s_2)$, which provides a clue. Instead of assuming there is only one unique tangent per pixel, which is commonplace in computer vision [259], we shall allow more than one.

To allow multiple tangents at each position, it is natural to attach a copy of the space of all possible tangents to each position (Fig. 22.3.1). Since in principle tangent angle is distributed around the circle and position is a real number, the resultant space is $\mathbb{R}^2 \times S^1$. (Note differences from the classical coordinate representation.) This space is an example of another fundamental construct in modern differential geometry, the *unit tangent bundle* associated with a surface in $\mathbb{E}^3$. Intuitively one might think of a surface as being covered by (i.e., as a union of) all possible curves on that surface. More generally, the tangent bundle to a surface is the union of tangent spaces at all points. If the surface is 2-D, the tangent bundle is 4-D. The geometric compatibility fields can be applied in parallel to all tangents in this space. (We will be generalizing this construct in the next few Sections, and will show examples then.)

[929] discusses the relevance of this product construction for the neurobiology of vision.

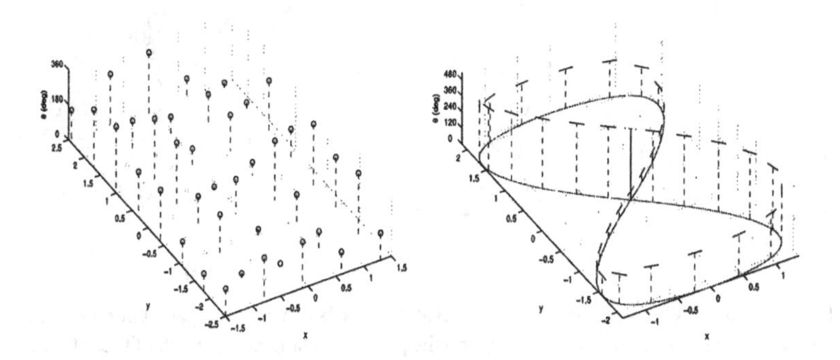


Figure 22.5. The need for higher-dimensional spaces than the image arises in representing non-simple or piecewise-regular curves. Since *a priori* a curve could be passing through any pixel at any orientation, it is natural to represent the (discretized) circle (the space of all unit vectors) $S^1$ at each (discretized) position (left). When the non-simple "figure 8" is lifted into the resultant space, the lift is a simple curve in $\mathbb{R}^2 \times S^1$ (right). The (position, orientation) space, which is abstract from the image, is sufficent to represent all possible curves in the image.

## 22.4 Stereo: Inferring Frenet 3-Frames from 2-Frames

We now move to 3-space, and consider the problem of inferring the structure of space curves from projection into two images. Earlier we showed that a curve in $\mathbb{R}^3$ has a tangent, normal, and binormal Frenet frame associated with every regular point along it. To sketch a geometric approach to stereo compatibility, for simplicity consider only the tangent in this frame and imagine it as an (infinitly) short line segment. This space tangent projects into a planar tangent in the left image and a planar tangent in the right image. Thus, space tangents project to pairs of image tangents. Now, consider the next point along the space curve; it too has a tangent, which projects to another pair of image tangents, one in the left image and one in the right image. Thus, in general, transport of the Frenet 3-frame in $\mathbb{R}^3$ from the second point back to the first has a correspondence in the left-right image pairs of 2-frames. [519] have developed this transport idea

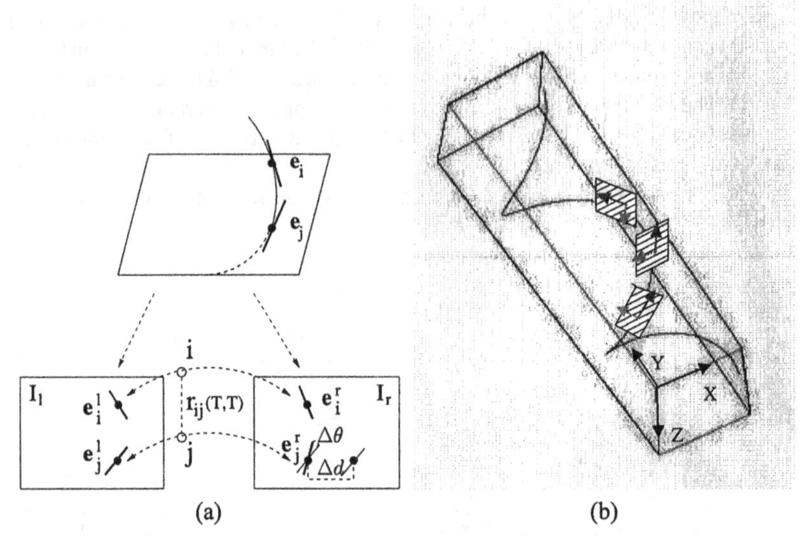


Figure 22.6. (a) Cartoon of the stereo relaxation process. A pair of space tangents associated with the Frenet approximation around the point with tangent $\mathbf{e}_j$. Each of these tangents projects to a (left,right) image tangent pair; compatibility between the space tangents thus corresponds to compatibility over (left,right) image tangent pairs. The projected tangents are shown as thick lines. One left image tangent is redrawn in the right image (as a thin line) to illustrate positional disparity ($\Delta d$)and orientation disparity ($\Delta\theta$). The compatibility between the tangent pair ($i$) and the pair ($j$) is denoted $r_{ij}$. Of course, for the full system the complete Frenet 2-frames are used to infer the Frenet 3-frame attached to the space curve. (b) Just as the osculating circle provided a local model for transport for image curves, a section of a helix provides a local model for a space curve. The $(T, N)$ components of the Frenet 3-frame define the osculating plane, which rotates as the frame is moved along the space curve.

to find corresponding pairs of image tangents such that their image properties match, as closely as the geometry can be approximated, the actual space tangents (Fig. 22.6). They show, in particular, that the stereo projection operator can be inverted to give the Frenet 3-frame and the curvature, but not the torsion. This builds on the related work of [267, 713, 653]

Two notions of disparity arise from the above transport model. First, the standard notion of positional disparity corresponds, through the camera model, to depth. Second, an orientation disparity is introduced if the space tangent is not in the epipolar plane. In the computational vision literature, orientation disparity is largely unexplored, but it is widely studied in visual psychophysics [410]. The geometric viewpoint shows how to use position and orientation disparities together. Typical reconstructions from this algorithm are shown in Fig. 22.7.

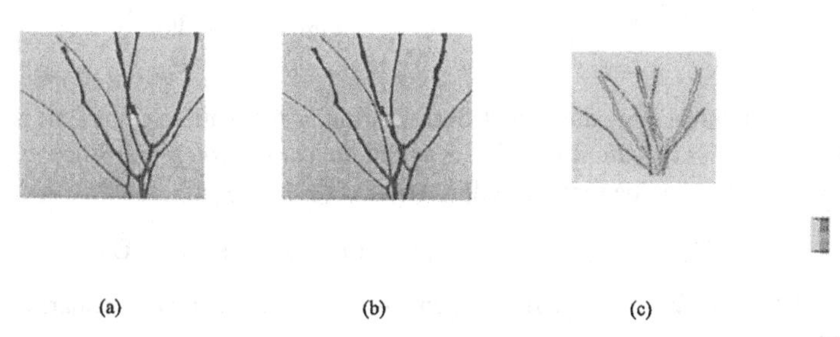

(a) (b) (c)

Figure 22.7. 3D reconstruction of a twig pair.(a) Left image (b) Right image; note in the highlighted region that subtleties in using the ordering constraint arise. Furthermore, occlusion of branches gives rise to discontinuities in orientation. Representing such discontinuities as multiple tangents facilitates proper matching. (c) Reconstruction. Depth scale is shown at right (units: meters).

## 22.5 Covariant Derivatives, Oriented Textures, and Color

We now denote orientation in the plane as a unit length tangent vector $\hat{\mathbf{E}}(\mathbf{q})$ attached to point $\mathbf{q} = (x, y) \in \mathbb{R}^2$. With such tangent vectors attached to every point of an oriented texture results in a unit length vector field, which creates a need to generalize the notion of transport: the frame can now be moved in any direction in the texture, rather than just along the curve.

Assuming good continuation as in the Introduction, a small translation $\mathbf{V}$ from the point q should rotate the vector $\hat{\mathbf{E}}(\mathbf{q})$ a small amount. Following the Frenet model, the frame $\{\hat{\mathbf{E}}_T, \hat{\mathbf{E}}_N\}$ is placed at the point $\mathbf{q}$ and the basis vector $\hat{\mathbf{E}}_T$ is identified with $\hat{\mathbf{E}}(\mathbf{q})$ – the tangent vector at $\mathbf{q}$ (Fig. 22.5). Note that $\hat{\mathbf{E}}_T$ is drawn at an angle $\theta$ – the local orientation measured relative to the x-axis – such that $(\mathbf{q}, \theta) \in \mathbb{R}^2 \times S^1$. Nearby tangents are displaced both in position and orientation according to the *covariant derivatives*, a tensor object whose components are essentially the partial derivatives of the underlying pattern. (Covariant derivatives generalize the earlier derivatives which were taken only along the curve; i.e., with respect to the arc length parameter $s$.) For vector fields the covariant derivative is taken in a direction given by another vector field, and is a vector. Again following Frenet, we observe that such covariant derivatives, $\nabla_{\mathbf{V}}\hat{\mathbf{E}}_T$ and $\nabla_{\mathbf{V}}\hat{\mathbf{E}}_N$, are naturally represented as vectors in the basis $\{\hat{\mathbf{E}}_T, \hat{\mathbf{E}}_N\}$ itself:

$$\begin{pmatrix} \nabla_{\mathbf{V}}\hat{\mathbf{E}}_T \\ \nabla_{\mathbf{V}}\hat{\mathbf{E}}_N \end{pmatrix} = \begin{bmatrix} w_{11}(\mathbf{V}) & w_{12}(\mathbf{V}) \\ w_{21}(\mathbf{V}) & w_{22}(\mathbf{V}) \end{bmatrix} \begin{pmatrix} \hat{\mathbf{E}}_T \\ \hat{\mathbf{E}}_N \end{pmatrix}. \tag{22.4}$$

The coefficients $w_{ij}(\mathbf{V})$ are *1-forms*, real-valued functions defined on tangent vectors. They are functions of the displacement direction vector $\mathbf{V}$, and since the basis $\{\hat{\mathbf{E}}_T, \hat{\mathbf{E}}_N\}$ is orthonormal, they are skew-symmetric $w_{ij}(\mathbf{V}) = -w_{ji}(\mathbf{V})$.

Thus $w_{11}(\mathbf{V}) = w_{22}(\mathbf{V}) = 0$ and the system reduces to:

$$\begin{pmatrix} \nabla_{\mathbf{V}}\hat{\mathbf{E}}_T \\ \nabla_{\mathbf{V}}\hat{\mathbf{E}}_N \end{pmatrix} = \begin{bmatrix} 0 & w_{12}(\mathbf{V}) \\ -w_{12}(\mathbf{V}) & 0 \end{bmatrix} \begin{pmatrix} \hat{\mathbf{E}}_T \\ \hat{\mathbf{E}}_N \end{pmatrix}. \quad (22.5)$$

This begins to resemble the Frenet-Serret formulas but is more general; it is *Cartan's connection equation*; $w_{12}(\mathbf{V})$ is called the *connection form*. Since $w_{12}(\mathbf{V})$ is linear in $\mathbf{V}$, it can be represented in terms of $\{\hat{\mathbf{E}}_T, \hat{\mathbf{E}}_N\}$:

$$w_{12}(\mathbf{V}) = w_{12}(a\,\hat{\mathbf{E}}_T + b\,\hat{\mathbf{E}}_N) = a\,w_{12}(\hat{\mathbf{E}}_T) + b\,w_{12}(\hat{\mathbf{E}}_N) \ .$$

The relationship between nearby tangents is thus governed by two scalars at each point.

$$\begin{aligned} \kappa_T &\triangleq w_{12}(\hat{\mathbf{E}}_T) \\ \kappa_N &\triangleq w_{12}(\hat{\mathbf{E}}_N) \end{aligned} \quad (22.6)$$

We interpret them as *tangential* ($\kappa_T$) and *normal* ($\kappa_N$) curvatures, since they represent a directional rate of change of orientation in the tangential and normal directions, respectively.

The connection equation describes the local behavior of orientation for the general two dimensional case, but is can be specialized to the one-dimensional case of curves developed earlier by considering only $\nabla_{\hat{\mathbf{E}}_T}$:

$$\begin{pmatrix} \nabla_{\hat{\mathbf{E}}_T}\hat{\mathbf{E}}_T \\ \nabla_{\hat{\mathbf{E}}_T}\hat{\mathbf{E}}_N \end{pmatrix} = \begin{bmatrix} 0 & w_{12}(\hat{\mathbf{E}}_T) \\ -w_{12}(\hat{\mathbf{E}}_T) & 0 \end{bmatrix} \begin{pmatrix} \hat{\mathbf{E}}_T \\ \hat{\mathbf{E}}_N \end{pmatrix}. \quad (22.7)$$

which, in our earlier notion, becomes:

$$\begin{pmatrix} T' \\ N' \end{pmatrix} = \begin{bmatrix} 0 & \kappa \\ -\kappa & 0 \end{bmatrix} \begin{pmatrix} T \\ N \end{pmatrix}. \quad (22.8)$$

We refer to $\kappa_T$ as the *tangential curvature* and $\kappa_N$ as the *normal curvature* - they represent the rate of change of the dominant orientation of the texture flow in the tangential and normal directions, respectively. In the language of frame fields, $\kappa_T$ and $\kappa_N$ are just the *coordinate functions* of $\nabla\theta$ with respect to $\{E_T, E_N\}$.

In the case of curves, the theory of frames is coupled to ordinary differential equations. For vector fields and texture flows, partial differential equations arise. In particular, since $E_T$ and $E_N$ are rigidly coupled, and we have

$$\begin{aligned} \kappa_T &= \nabla \times E_T \\ \kappa_N &= \nabla \cdot E_T \ . \end{aligned} \quad (22.9)$$

If $\kappa_T$ and $\kappa_N$ were known functions of position $q = (x, y)$, a PDE could be solved for the rotation angle $\theta(q)$. Thus $\kappa_T$ and $\kappa_N$ are not completely independent, and integrability conditions arise. In particular, unless $\kappa_T$ and $\kappa_N$ are both equal to zero, they cannot be constant simultaneously in a neighborhood around $q$, however small, or else the induced flow is nonintegrable. [72] show that, given any texture flow $\{E_T, E_N\}$, its curvature functions $\kappa_T$ and $\kappa_N$ must satisfy the

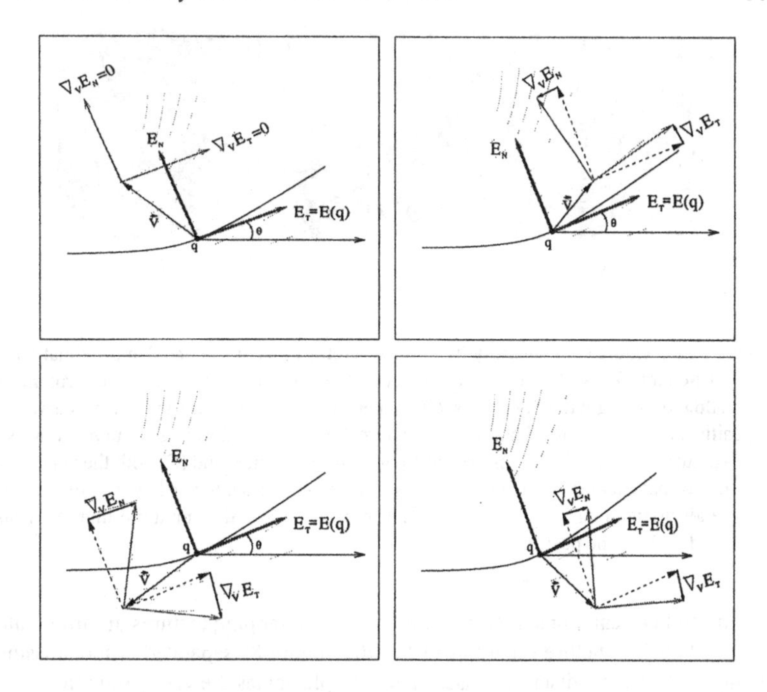


Figure 22.8. Displacement (transport) of a Frenet frame within a vector field or an oriented texture amounts to rotation, but differs for different displacements. The covariant derivative specifies the frame's initial rate of rotation for any direction vector **V**. The four different cases in this figure illustrate how this rotation depends on **V** both quantitatively (i.e,, different magnitudes of rotation) and qualitatively (i.e., clockwise, counter-clockwise, or zero rotation). A pure displacement in the tangential direction ($\hat{\mathbf{E}}_T$) specifies one rotation component (the tangential curvature) and a pure displacement in the normal direction ($\hat{\mathbf{E}}_N$) specifies the other (normal curvature) component.

relationship

$$\nabla\kappa_T \cdot E_N - \nabla\kappa_N \cdot E_T = \kappa_T^2 + \kappa_N^2$$

With osculating circles the natural local model for the geometry of regular planar curves, and helices the natural model for regular space curves, [72] show that the natural local model for textures and flows is a helicoid in $\mathbb{R}^2 \times S^1$. This follows intuitively because each streamline or intergral curve through the flow can be locally approximated by a section of an osculating circle; this lifts to a section of a helix. The helicoid is a ruled surface built of these lifts. Local sections of the helicoid can be projected into the image and discretized to provide connection or compatibility fields for textures and flows (Fig. 22.9).

The result of applying this system to overlapping flows is shown in Fig. 22.10. Notice in particular how woven textures can be thought of as multiple threads, or curves, overlapping one another. This emerged from our discussion of represent-

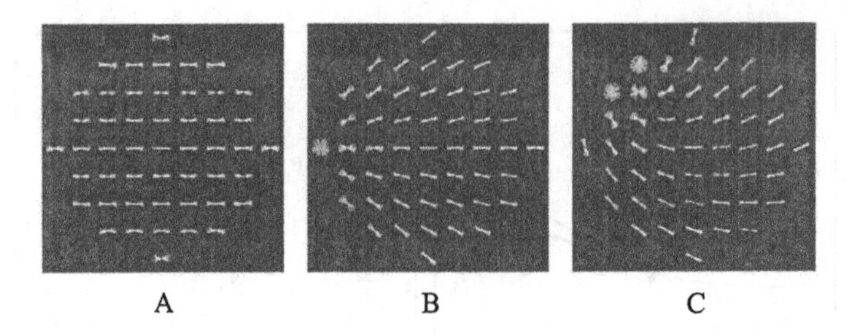


Figure 22.9. Texture compatability fields are discretizations of a helicoid approximation to a flow lifted into $\mathbb{R}^2 \times S^1$. Three examples are shown: **(A)** both curvatures are zero; this is the analog to a straight line for curves; **(B)** tangential curvature is zero and normal curvature is positive; this shows a local portion of a texture flow in which the integral curves converge to a (singular) point, as lines converge to a point in the distance; and **(C)** both the tangential and the normal curvatures are positive. This is the general case: notice how singular points (where all orientations are possible) arise. These are indicated as multiple line segments displayed at the same position.

ing multiple orientations at each point. When overlapping textures are lifted into $\mathbb{R}^2 \times S^1$ their structure separates just as the "figure 8" separated at the crossing point. But now, in a discrete sense, such multiple values are very common.

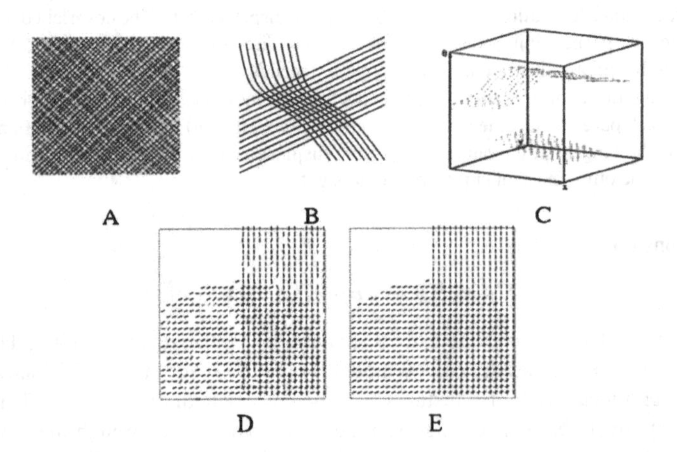


Figure 22.10. Examples of texture patterns rich in orientation. (A) A woven texture with two dominant orientations. This is an extension of (B) two overlapping textures, which are naturally separated when lifted into $\mathbb{R}^2 \times S^1$ in (C). The bottom panels illustrate how a noisy pattern (D) is refined using the geometric compatibilities in Fig. 22.9 to (E), thereby enforcing a Gestalt-like good contination of the flows.

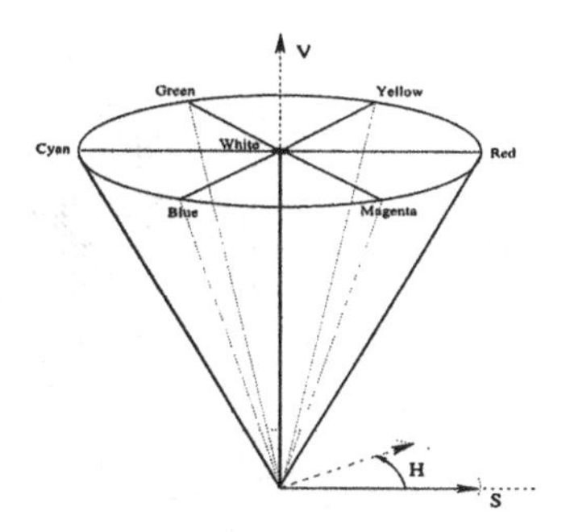


Figure 22.11. The HSV color representation in $\mathcal{S}^1 \times [0, 1]^2$ and the color wheel.

## 22.5.1 *Hue Flows*

While color is normally thought of as a point in (R,G,B)-space, it can also be represented in the psychophysically motivated HSV color space. Here a color image is a mapping $\mathcal{C} : \mathbb{R}^2 \rightarrow \mathcal{S}^1 \times [0, 1]^2$ (see Fig.22.11). The hue component across the image is a mapping $\mathcal{H} : \mathbb{R}^2 \rightarrow \mathcal{S}^1$ and thus can be represented as a unit length vector field over the image, which [72] called the *hue field.* Displays of the hue field reveal that it may vary greatly, albeit smoothly, even *within* perceptually coherent objects (see Fig 22.12.

Many color image enhancement algorithms are based on a form of anisotropic diffusion [2, 153], using either a vectorial representation or a manifold representation [787]. While diffusion in color space can work within very smooth regions, it does have the tendency to blur inappropriately.

Hue compatibility fields can be defined analogously to texture compatibility fields–see[73]. As expected, concepts of hue curvatures naturally arise, which express how the hue is flowing from one image position to those in its neighborhood. Just as with texture flows, a tangential and a normal hue curvature are required. Since the local behavior of the hue is characterized (up to Euclidean transformation) by this pair of curvatures, it is natural to conclude that nearby measurements of hue should relate to each other based on these curvatures. Or, put differently, measuring a particular curvature pair at a point should induce a field of coherent measurements, i.e., a hue function in its neighborhood. Coherence of hue to its spatial context can then be determined by examining how well it fits with those around it. Again, a helicoidal approximation in (position, hue) space arises.

Such flows are relevant to image denoising; for estimating mutual reflectance and color bleeding; for estimating smooth surface variations as separate from lighting variations (for lightness algorithms); and for separating cast shadow boundaries and highlights from other types of intensity edges.

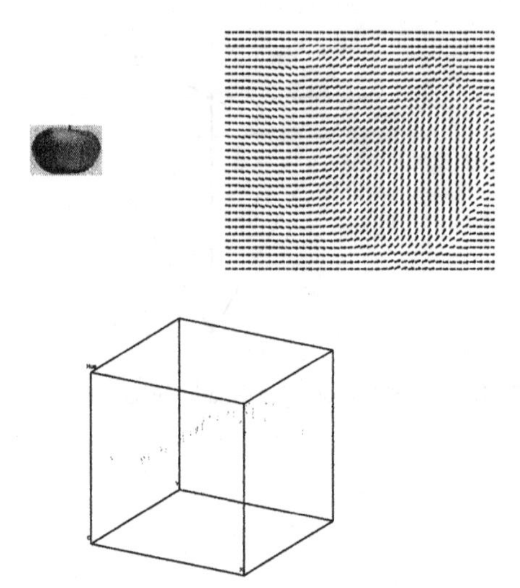

Figure 22.12. A flow perspective on color images is provided by their hue fields. These are typically piecewise smooth. Most importantly, hue can vary smoothly even within perceptually coherent objects. **(top)** A natural image of an apple with varying hue. Notice that the everyday expression of "red apple" is limited. The corresponding hue field changes smoothly across the image of the apple's surface. **(bottom)** A 3D representation of the hue filed, where hue is represented as height. Identifying the top face with the bottom (since hue is a circle) leads to the (position, hue) space.

## 22.6 Discussion

In this Chapter we co-developed ideas from modern differential geometry and problems in computer vision. The differential geometry was based on Frenet and Serret's ideas of attaching frames directly to curves, rather than expressing curve structure in terms of extrinsic coordinate functions. Such ideas were carried to a remarkable stage by Cartan, whose moving frame concept is now central in mathematics. The covariant derivative emerges for differential variation of frames in flows, as the normal derivative was useful for transporting a frame along a curve.

The moving frame concept provides a natural abstraction for perceptual organization problems, at least for those that can be defined over short distances. We considered curve detection in 2D and stereo as the projection of 3D curves to illustrate the power of this geometric abstraction. Techniques for integrating orientation disparity with positional disparity emerged. But the real power was seen for flows, in which textures and hues were considered.

Although the notion of tangent was introduced as the best linear approximation to a curve, modern definitions abstract via a limiting operation to an equivalence class of curves. Our discussion attempted to avoid any unnecessary abstraction, so that all concepts had a direct counterpart in computer vision terms.

Consideration of non-simple curves motivated an elaboration of the types of representations normally considered in computer vision from image-based ones to those that attach a space of possibilities at each point. It is commonplace to assume boundaries have a well-defined orientation at each point, but this holds for only a restricted class of curves. Local occlusion clues involving "T" junctions provide an important example of non-smooth curves, and our elaborated representation is capable of handling them as well.

The space of possible frames also has an important representation in differential geometry, and is related to fibre bundles. We just touched on such concepts in this Chapter, but fully expect them to be playing a much richer role in future applications of differential geometry to computational vision.

# Chapter23

# Shape From Shading


## E. Prados and O. Faugeras



Abstract

Shape From Shading is the process of computing the three-dimensional shape of a surface from one image of that surface. Contrary to most of the other three-dimensional reconstruction problems (for example, stereo and photometric stereo), in the Shape From Shading problem, data are minimal (we use a single image!). As a consequence, this inverse problem is intrinsically a difficult one. In this chapter we describe the main difficulties of the problem and the most recent theoretical results. We also give some examples of realistic modelings and of rigorous numerical methods.


## 23.1 Introduction

The "Shape From Shading" problem (SFS) is to compute the three-dimen--sional shape of a surface from the *brightness* of *one* black and white image of that surface; see figure 23.1.

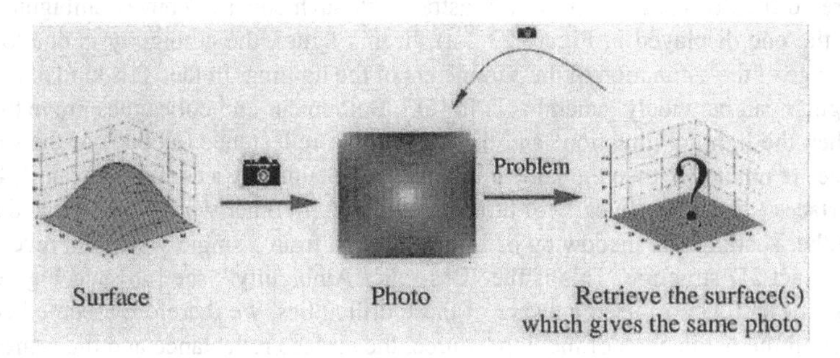


Figure 23.1. The "Shape-from-Shading" problem.

In the 70's, Horn [405] was the first to formulate the Shape From Shading problem simply and rigorously as that of finding the solution of a nonlinear first-order

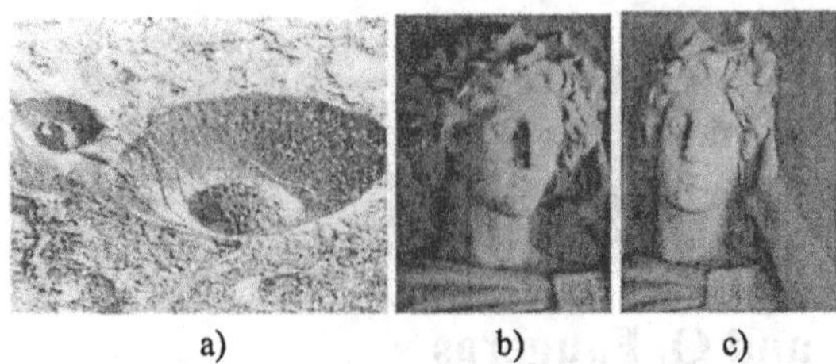


a) The crater illusion [638]: From the image we perceive two craters, a small and a big one. But we can turn these craters into volcanoes (although upside down) if we imagine the light source to be at the bottom of the picture rather than at the top. This picture is actually that of a pair of ash cones in the Hawaiian Island, not that of a pair of craters. b-c) "Bas-relief Ambiguity" [68]: Frontal and side views of a marble bas-relief sculpture. Notice how the frontal views appear to have full 3-dimensional depth, while the side view reveals the flattening. This demonstrates that the image b) can be produced by two surfaces: the three-dimensional surface we imagine by visualizing image b) and the actual bas-relief which is at the origin of the two photos b) and c).

Figure 23.2. Examples of Shape From Shading ambiguities.

Partial Differential Equation (PDE) called the brightness equation. In a first period (in the 80's) the authors focus on the computational part of the problem, trying to compute directly numerical solutions. Questions about the existence and uniqueness of solutions to the problem were simply not even posed at that time with the important exception of the work of Bruss and Brooks [134, 118]. Nevertheless, due to the poor quality of the results, these questions as well as those related to the convergence of numerical schemes for computing the solutions became central in the last decade of the 20th century. Today, the Shape From Shading problem is known to be an ill-posed problem. For example, a number of articles show that the solution is not unique [118, 608, 609, 690, 68, 292, 665, 663]. The encountered difficulties have often been illustrated by such concave/convex ambiguities as the one displayed in Figure 23.2-a). In this figure, the ambiguity is due to a change of the estimation of the parameters of the lighting. In fact, this kind of ambiguity can be widely generalized. In [68], Belhumeur and colleagues prove that when the lighting direction[1] and the Lambertian reflectance (albedo) of the surface are unknown, then the same image can be obtained by a continuous family of surfaces (depending linearly of three parameters). In other words, they show that neither shading nor shadowing of an object, seen from a single viewpoint reveals its exact 3D structure. This is the "Bas-relief Ambiguity", see [68] and Figures 23.2-b) and 23.2-c). Being aware of these difficulties, we therefore assume here that all the parameters of the light source, the surface reflectance and the camera are known.

[1] In the case of a distant light source.

As we have mentioned above, the modeling of the Shape From Shading problem introduced by Horn leads to a PDE: the brightness equation. This equation arises from the following

$$I(x_1, x_2) = R(\mathbf{n}(x_1, x_2)),$$

$(x_1, x_2)$ are the coordinates of a point $x$ in the image. The brightness equation connects the reflectance map ($R$) to the brightness image ($I$). At the exception of an extremely small number of papers, for example [44, 506, 667], almost all the Shape From Shading methods assume that the scene is Lambertian. In this case, the reflectance map is the cosine of the angle between the light vector $\mathbf{L}(x)$ and the normal vector $\mathbf{n}(x)$ to the surface:

$$R = \cos(\mathbf{L}, \mathbf{n}) = \frac{\mathbf{L}}{|\mathbf{L}|} \cdot \frac{\mathbf{n}}{|\mathbf{n}|}, \qquad (23.1)$$

(where $R$, $\mathbf{L}$ and $\mathbf{n}$ depend on $(x_1, x_2)$).

## 23.2 Mathematical formulation of the SFS problem

In this section, we formulate the SFS problem as that of solving some explicit PDEs. These explicit equations arise from equations (23.1).
Let $\Omega$ be an open subset of $\mathbb{R}^2$ representing the image domain (e.g. the rectangle $]0, X[\times]0, Y[$). We represent the scene by a three-dimensional surface $\mathbf{S} = \{S(x);\ x \in \overline{\Omega}\}$, which can be explicitly parameterized by using the function $S$ defined on the closure $\overline{\Omega}$ into $\mathbb{R}^3$. The particular type of parameterization is irrelevant here but may vary according to the camera type (orthographic versus pinhole) and to mathematical convenience. In this work, we assume that the light source is unique and punctual. For $y \in \mathbb{R}^3$, we denote $\mathbf{L}(y)$ the <u>*unit*</u> vector representing the light source direction at the point $y$. If the light source is located at infinity then the light vector field is uniform (i.e. constant). In this case, we denote by $\mathbf{L} = (\alpha, \beta, \gamma)$ with $\gamma > 0$, and $\mathbf{l} = (\alpha, \beta)$. If the light source is located at the optical center, then $\mathbf{L}(S(x)) = S(x)/|S(x)|$.

### *23.2.1 "Orthographic SFS" with a far light source*

This is the traditional setup for the SFS problem. Here, we assume in particular that the camera performs an orthographic projection of the scene. For such a modeling, it is natural to denote by $u$ the *distance* of the points in the scene to the camera; in other words, $\mathbf{S}$ is parameterized by $S : x \mapsto (x, u(x))$. For such a parameterization, a normal vector $\mathbf{n}(x)$ at the point $S(x)$ is given by[2]

[2] The two columns of the Jacobian $DS(x)$ are tangent vectors to $\mathbf{S}$ at the point $S(x)$. Their cross product is a normal vector.

$\mathbf{n}(x) = (-\nabla u, 1)$. The SFS problem is then, given $I$ and $\mathbf{L}$, to find a function $u : \overline{\Omega} \longrightarrow \mathbb{R}$ satisfying the brightness equation:

$$\forall x \in \Omega, \quad I(x) = (-\nabla u(x) \cdot \mathbf{l} + \gamma)/\sqrt{1 + |\nabla u(x)|^2}.$$

In the SFS literature, this equation is rewritten in a variety of ways as $H(x, p) = 0$, where $p = \nabla u$. For example, Rouy and Tourin [690] introduce

$$H_{R/T}(x, p) = I(x)\sqrt{1 + |p|^2} + p \cdot \mathbf{l} - \gamma.$$

In [290], Dupuis and Oliensis consider

$$H_{D/O}(x, p) = I(x)\sqrt{1 + |p|^2 - 2p \cdot \mathbf{l}} + p \cdot \mathbf{l} - 1.$$

In the case where $\mathbf{L} = (0, 0, 1)$, Lions et al. [522] deal with:

$$H_{Eiko}(x, p) = |p| - \sqrt{\frac{1}{I(x)^2} - 1}. \qquad \text{(called the Eikonal equation).}$$

The function $H$ is called the *Hamiltonian.*

### *23.2.2 "Perspective SFS" with a far light source*

"Perspective SFS" assumes that the camera performs a perspective projection of the scene. We therefore assume that $\mathbf{S}$ can be explicitly parameterized by the *depth modulation* function $u$ defined on $\overline{\Omega}$. In other words, we choose $S(x) = u(x).(x, -\mathrm{f})$, $\forall x \in \overline{\Omega}$, where f denotes the focal length. For such a parameterization, a normal vector $\mathbf{n}(x)$ at the point $S(x)$ is given by[2] $\mathbf{n}(x) = (\mathrm{f}\nabla u(x), u(x) + x \cdot \nabla u(x))$. Combining the expression of $\mathbf{n}(x)$ and the change of variables[3] $v = ln(u)$, we obtain from the irradiance equation (23.1) the following Hamiltonian [663, 788, 228]:

$$H_{P/F}(x, p) = I(x)\sqrt{\mathrm{f}^2|p|^2 + (x \cdot p + 1)^2} - (f\,\mathbf{l} + \gamma x) \cdot p - \gamma;$$

### *23.2.3 "Perspective SFS" with a point light source at the optical center*

Here, we parameterize $\mathbf{S}$ by $S(x) = u(x)\frac{f}{\sqrt{|x|^2 + f^2}}(x, -f)$, $\forall x \in \overline{\Omega}$. In this case, we can choose[2] $\mathbf{n}(x) = (f\nabla u - \frac{f u(x)}{|x|^2 + \mathrm{f}^2}\, x\,,\ \nabla u \cdot x + \frac{f u(x)}{|x|^2 + \mathrm{f}^2}\, f)$. Combining the expression of $\mathbf{n}(x)$ and the change of variables[3] $v = ln(u)$, we obtain from equation (23.1) the following Hamiltonian [664, 662]:

$$H_{OptC}(x, p) = I(x)\sqrt{\mathrm{f}^2|p|^2 + (p \cdot x)^2 + Q(x)^2} - Q(x).$$

[3]We assume that the surface is visible (in front of the retinal plane) hence $u > 0$.

### 23.2.4 A generic Hamiltonian

In [664, 662], Prados and Faugeras proved that all the previous SFS Hamiltonians are special cases of the following "generic" Hamiltonian:

$$H_g(x,p) = \kappa_x \sqrt{|A_x p + \mathbf{v}_x|^2 + K_x^2} + \mathbf{w}_x \cdot p + c_x,$$

with $\kappa_x, K_x \geq 0$, $c_x \in \mathbb{R}$, $\mathbf{v}_x, \mathbf{w}_x \in \mathbb{R}^2$ and $A_x \in \mathcal{M}_2(\mathbb{R})$, the set of $2 \times 2$ real matrices. They also showed that this "generic" Hamiltonian can be rewritten as a supremum:

$$H_g(x,p) = \sup_{a \in \bar{B}_2(0,1)} \{-f_g(x,a) \cdot p - l_g(x,a)\};$$

see [662] for the detailed expressions of $f_g$ and $l_g$. This generic formulation considerably simplifies the analysis of the problem. Theorems about the characterization and the approximation of the solutions are proved as much as possible for this generic SFS Hamiltonian. In particular, this formulation *unifies the orthographic and perspective* SFS problems. Also, from a practical point of view, *a single algorithm can be used to numerically solve these various problems.*

## 23.3 Mathematical study of the SFS problem

### 23.3.1 Related work

It is well-known that the SFS problem is an ill-posed problem even when we assume complete control of the experimental setup. For example, the previous SFS PDEs do not have a unique solution: several surfaces can yield the same image [292]. Before computing a numerical solution, it is therefore very important to answer the following questions. Does there exist a solution? If yes, in what sense is it a solution (classical or weak)? Is the solution unique? The various approaches for providing answers to these questions can be classified in two categories. **First**, Dupuis and Oliensis [290] and Kozera [493] deal with smooth (classical) solutions. More precisely, Dupuis and Oliensis [290] prove the uniqueness of some constrainted $C^2$ solutions, and they characterize some $C^1$ solutions. Kozera works with hemi-spheres and planes [493]. Nevertheless, we can design smooth images "without (smooth) shape" [119, 479]; also, because of noise, of errors on parameters (focal length, light position, etc) and of incorrect modeling (interrecflections, extended light source, nonlambertian reflectance...) there never exist in practice such smooth solutions with real images. In other respects, this also explains why the global methods (e.g. [290, 470, 606]) which are completely based on such regularity assumptions are somewhat disappointing with real images. This leads to consider the problem in a weaker framework. **Second**, in the 90s, Lions, Rouy and Tourin [690, 522] propose to solve the SFS problem by using the notion of viscosity solutions. Recently, their approach has been extended by Prados and Faugeras [665, 663] and by Falcone [144]. The theory of viscosity solutions is interesting for a variety of reasons: 1) it ensures the existence of weak solutions as soon as the intensity image is (Lipschitz) continuous; 2) it allows to characterize all solutions; 3) any particular solution is effectively computable. Nevertheless, the work of Lions et al., Prados and Faugeras, Falcone et al. [690, 522, 665, 663, 144] has a very important weakness: the characterization of a viscosity solution and its computation *require* in particular *the knowledge of its values on the boundary of the image*. This is *quite unrealistic* because in practice such values are not known. At the opposite of the work based on the viscosity solutions, Dupuis and Oliensis [290] characterize some $C^1$ solutions with much less data. In particular, they do not specify the values of the solution on the boundary of the image. Considering the advantages and the drawbacks of all these methods, Prados et al. [664, 661] propose a *new* class of *weak* solutions which guarantees the existence of a solution[4] in a large class of situations including some where there do not exist smooth solutions. They call these new solutions: "Singular Discontinuous Viscosity Solutions" (SDVS). The notion of SDVS allows to unify the mathematical frameworks proposed in the SFS literature and to generalize the previous main theoretical results.

### 23.3.2 *Nonuniqueness and characterization of a solution*

The results presented in this section are based on the notion of SDVS [664]. Let us recall that the viscosity solutions are solutions in a weak sense and that the classical (differentiable) solutions are particular viscosity solutions. For more details about this notion of weak solutions, we refer the reader to [52]. For an intuitive approach connected to computer vision, see for example [664] and references therein.

Since the CCD sensors have finite size, we assume that $\Omega$ is *bounded.* In this case, it is well known that the Hamilton-Jacobi equations of the form $H(x, \nabla u(x)) = 0, \forall x \in \Omega$, (and so the SFS equations considered here) do not have a unique viscosity solution [52]. It follows that for characterizing (and for computing) a solution, we need to impose additional constraints. In [664] (but also implicitly, in [290]) it is shown that the idea of state contraints (also called "Soner conditions") provides a more convenient notion of boundary condition than Dirichlet's[5] or Neumann's[6]. The "state contraint" is a boundary condition which is reduced to

$$H(x, u(x), \nabla u(x)) \geq 0 \qquad \text{on } \partial\Omega,$$

in the viscosity sense (see for example [52]). This constraint corresponds to the Dirichlet conditions

$$\forall x \in \partial\Omega, \quad u(x) = \varphi(x) \quad \text{with } \varphi(x) = +\infty,$$

[4]Corresponding to Dupuis and Oliensis' solution, if one exists.

[5]Dirichlet conditions consists in fixing the values of the solutions.

[6]Neumann conditions consists in fixing the values of the derivatives the solutions.

in the viscosity sense. In a sense, completing an equation with state constraints consists in choosing the highest viscosity solution. The interest of the notion of state constraints is twofold: 1) in contrast with the Dirichlet and Neumann boundary conditions, the state constraints do not require any data[7]. 2) the notion of state constraints can be approximately expressed as "$u(x)$ increases when $x$ tends to $\partial\Omega$"; see [664]. So the addition of this constraint provides a relevant solution as soon as the original surface verifies this basic assumption. Let us emphasize that this constraint is in fact not a strong one since, for example, the condition is satisfied as soon as the image to be processed contains an object of interest in front of a background.

The main difficulty encountered when one attempt to solve the SFS equations (described in section 23.2) is due to the fact that even if we impose Dirichlet or Soner (state constraints) boundary conditions all over the boundary of the image, these constraints are not sufficient for obtaining the uniqueness of the solution. **For characterizing a weak solution (SDVS) or a classical solution ($C^1$), it is necessary and sufficient to impose (in addition) *Dirichlet constraints* at the *singular points* which are *local "minima"*[8]**; at the other points, we just impose state constraints [661]. Let us remind the reader that the set of the *singular points* is $\mathcal{S} = \{ x \in \Omega \mid I(x) = 1 \}$. These points are those of maximal intensity[9] and correspond with the points for which the surface normal coincides with the light direction.

Therefore, in practice, to be able to recover the original surface[10], we need to know what are the singular points which are local minima and the height of the surface at all these particular points. In the cases where we do not have this knowledge (unfortunately, we do not have it in practice!), we are unable to recover the exact original surface. Nevertheless, let us note that Prados and Faugeras' framework allows to understand exactly what we compute, namely the SDVS (which coincides with the value function considered in particular by Dupuis and Oliensis [290]). In practice, we fix the height of the solution at the singular points and on the boundary of the image, when we know it, and we "send" these values to infinity when this information is not available (i.e., we impose a state constraint). Finally Prados and his coworkers prove that, with such constraints, there *exists* a *unique* SDVS of the SFS equations[11].

---

[7] Dirichlet (respectively, Neumann) boundary conditions require the knowledge of the exact values of the solution (respectively, the exact values of $\nabla u(x) \cdot n(x)$, where $n(x)$ is the unit inward normal vector to $\partial\Omega$ at the point $x$) on the boundary of the image. In the SFS problem, we rarely have such data at our disposal.

[8] More precisely, the minima of $u - \varphi$, where $\varphi$ is the adequate subsolution [661].

[9] Let us recall that we have assumed that $I(x) = cos(\mathbf{n}, \mathbf{L})$.

[10] i.e., in SFS, the photographed surface.

[11] With some weak adequate assumptions; see [661].

## 23.4 Numerical solutions by "Propagation and PDEs methods"

In section 23.2, we have shown that the SFS problem can be considered as that of solving a first order PDE. In this section, we consider the numerical SFS methods consisting in solving *directly* the *exact* SFS PDE. We call them "propagation and PDEs methods". These numerical methods do not make any linearizations (at the opposite of the linear methods; see [291] for a recent state of the art). Moreover, they do not introduce any biases in the equations contrary to the variational methods which, for example, add regularization or integrability terms. For more details about variational approaches in Shape From Shading, we refer the reader to Horn and Brooks' book [408] and to the survey of Durou and his coworkers [291] (and references therein).

### *23.4.1 Related work*

The propagation and PDEs methods can be subdivided into two classes. The *"single-pass" methods* and the *iterative methods*. The main single-pass methods are: the method of *characteristic strips* (introduced by Horn [405]), the method of *propagation of the equal-height contours* (introduced by Bruckstein [125] and improved by Kimmel and Bruckstein [471]), the *fast marching* method (proposed by Sethian and Kimmel [728, 474]). Amongst the iterative methods let us cite in particular: the algorithm introduced by Rouy and Tourin [690] and its extensions by Prados and Faugeras [665, 663], the algorithms of Dupuis and Oliensis [290] based on the control theory and differential games, the algorithms of Falcone et al. [144] based on finite elements. Let us note that, at the exception of the work of Prados and Faugeras [663, 664], all these methods deal only with the Eikonal equation [405, 125, 690, 471, 728] or with the orthographic SFS with oblique light source [290, 144, 474, 665].

In spite of the multiplicity of these methods, we can prove that **they all compute approximations of the same solution.** In particular, the initial equal-height contours method of Bruckstein [125] is a variant of the method of the characteristic strips of Horn [405]. In [125], Bruckstein assumes that the initial curve is an equal-height contour. By imposing such special Dirichlet boundary conditions, he drops the Neumann boundary conditions required by the basic method of the characteristic strips (see [479] for a nice and rigorous study of these methods). Basically both above methods are Lagrangian methods that suffer from unstability and topological problems, see for example [618]. To alleviate these problems Kimmel and Bruckstein [471] propose to upgrade Bruckstein's method by using a Eulerian formulation of the problem. In other respects, the connection between the front propagation problems and the Hamilton Jacobi equations are well known. In particular, roughly speaking, it has beeb proved that the viscosity solution of the Hamilton Jacobi equation associated with a front propagation corresponds with the evolution of the initial contour defined by Huygens' principle; see for example [302]. In the same way, the other methods we cite above (Sethian's, Rouy-Tourin's, Dupuis-Oliensis', Falcone's and Prados-Faugeras' methods) compute some approximations of the viscosity solutions of the SFS equations. In particular in [731], Sethian and Vladimirsky prove that the numerical solutions computed by the fast marching/ordered upwind methods converge toward the continuous viscosity solution (with Dirichlet boundary data on the boundary of the image). In [664], Prados and Faugeras generalize and unify the results proved in [690, 290, 144, 665, 663]. More precisely, they show that in all cases, the authors compute approximations of the SDVS. Basically, the difference between the work [690, 290, 144, 665, 663] is based on the choice of the boundary conditions; see [661]. In a general manner, all propagation and PDE methods require additional constraints: in particular, Dirichlet, Neumann or Soner boundary conditions. In other words, the computed solutions are characterized by the boundary conditions. These boundary conditions must contain enough information. Also, this information is thereby propagated "along" the solutions. Let us note that except for Horn's [405] and Bruckstein and Kimmel's method [125, 471], all the previous methods can deal with various Dirichlet/Soner boundary conditions. More precisely, the algorithms of Rouy and Tourin [690], Dupuis and Oliensis [290], Sethian [728] and Prados and Faugeras [665, 663] can use Dirichlet and/or Soner conditions on the boundary of the image $\partial\Omega$ at all the singular points $\mathcal{S}$ and on any other part of the image (for example, on an equal-height contour...). For instance, when we do not know the values of the solution at any points of the image, we can impose state constraints (i.e. Soner conditions) on $\partial\Omega \cup \mathcal{S}$ except for one point where we must impose a Dirichlet boundary condition. Contrary to these methods, let us note that Horn's [405] requires Dirichlet and Neumann boundary conditions and that Bruckstein's [125, 471] require the knowledge of an equal-height contour. This last constraint is a very specific Dirichlet condition and is much stronger than the previous ones. Note that implicitly, Bruckstein methods [125, 471] also impose state constraints on $\partial\Omega \cup \mathcal{S}$.

Finally, from a more numerical point of view, we can also remark that the approximation scheme considered by Sethian [728] is the one designed by Rouy and Tourin in [690]. Moreover, Prados and Faugeras' schemes are extensions of the Rouy and Tourin's scheme and their solutions coincide with those of Oliensis' schemes.

### *23.4.2 An example of provably convergent numerical method: Prados and Faugeras' method*

In this section, we present the provably convergent numerical method of Prados and Faugeras [662]. Let us recall that this method unifies in particular the iterative methods of Rouy and Tourin [690], Prados et al. [665, 663] and Dupuis and Oliensis [290].

We consider here a finite difference approximation scheme. The reader unfamiliar with the notion of approximation schemes can refer to [53] or [662]. Let us just recall that, following [53], an approximation scheme is a functional equation of the form

$$S(\rho, x, u(x), u) = 0 \qquad \forall x \in \overline{\Omega},$$

which "approximates" the considered PDE. $S$ is defined on $\mathcal{M} \times \overline{\Omega} \times \mathbb{R} \times B(\overline{\Omega})$ into $\mathbb{R}$, $\mathcal{M} = \mathbb{R}^+ \times \mathbb{R}^+$ and $\rho = (h_1, h_2) \in \mathcal{M}$ defines the size of the mesh that is used in the corresponding numerical algorithms. $B(D)$ is the space of bounded functions defined on a set $D$.

**Definition 1.** *We say that a scheme $S$ is stable*[12] *if for all fixed mesh size $\rho$ it has solutions and if all the solutions are bounded independently of $\rho$.*

For ensuring the stability of a scheme, it is globally sufficient that it is monotonous (i.e. the function $u \mapsto S(\rho, x, t, u)$ is nonincreasing) and that the function $t \mapsto S(\rho, x, t, u)$ is nondecreasing, see [662]. For obtaining such a scheme, Prados and Faugeras [662] approximate the generic Hamiltonian $H_g$ by

$$H_g(x, \nabla u(x)) \approx \sup_{a \in \overline{B}(0,1)} \left\{ \sum_{i=1}^{2} (-f_i(x,a)) \frac{u(x) - u(x + s_i(x,a) h_i \overrightarrow{e_i})}{-s_i(x,a) h_i} - l_g(x,a) \right\}$$

where $f_i(x,a)$ is the $i^{th}$ component of $f_g(x,a)$ and $s_i(x,a)$ is its sign. Thus, they obtain the approximation scheme $S_{impl}(\rho, x, u(x), u) = 0$ with $S_{impl}$ defined by:

$$S_{impl}(\rho, x, t, u) = \sup_{a \in \overline{B}(0,1)} \left\{ \sum_{i=1}^{2} (-f_i(x,a)) \frac{t - u(x + s_i(x,a) h_i \overrightarrow{e_i})}{-s_i(x,a) h_i} - l_g(x,a) \right\}.$$

By introducing a fictitious time $\Delta\tau$, they also transform this implicit scheme in a "semi-implicit" scheme (also monotonous):

$$S_{semi}(\rho, x, t, u) = t - \ ( u(x) + \Delta\tau \ S_{impl}(\rho, x, u(x), u) \ ),$$

where $\Delta\tau = (f_g(x, a_0) \cdot (1/h_1, 1/h_2))^{-1}$; $a_0$ being the optimal control associated with $S_{impl}(\rho, x, u(x), u)$. Let us emphasize that these two schemes have exactly the same solutions and that they verify the previous monotonicity conditions (with respect to $t$ and $u$). Prados and Faugeras prove in [662] the stability of these two schemes.
By construction, these two schemes are consistent[12] with the SFS equations as soon as the brightness image $I$ is Lipschitz continuous; see [662]. Using the stability and the monotonicity of the schemes and some uniqueness results, it follows directly from [53] that the solutions of the approximation schemes $S_{impl}$ and $S_{semi}$ converge towards the unique viscosity solution of the considered equation (complemented with the adequate boundary conditions) when the mesh size vanishes; see [662].

We now describe an *iterative algorithm* that computes numerical approximations of the solutions of a scheme $S(\rho, x, u(x), u) = 0$ for all fixed $\rho = (h_1, h_2)$. We denote, for $k \in \mathbb{Z}^2$, $x_k = (k_1 h_1, k_2 h_2)$, and $Q := \{k \in \mathbb{Z}^2 \text{ s.t. } x_k \in \overline{\Omega}\}$. We

[12] Following Barles and Souganidis definitions [53].

call "pixel" a point $x_k$ in $\overline{\Omega}$. Since $\overline{\Omega}$ is bounded the number of pixels is finite. The following algorithm computes for all $k \in Q$ a sequence of approximations $U_k^n$ of $u(x_k)$:

**Algorithm:**

1. *Initialisation ($n = 0$): $\forall k \in Q$, $U_k^0 = u_0(x_k)$;*
2. *Choice of a pixel $x_k$ and modification (step $n+1$) of $U_k^n$: we choose $U^{n+1}$ such that*

$$\begin{cases} U_l^{n+1} = U_l^n & \text{if} \quad l \neq k, \\ S(\rho, x_k, U_k^{n+1}, U^n) = 0; \end{cases}$$

3. *Choose the next pixel $x_k$ (using alternating raster scans [243]) and go back to 2.*

In [662], Prados and Faugeras prove that if $u_0$ is a supersolution of the SFS scheme $S_{impl}$ (respectively, $S_{semi}$) then step 2 of the algorithm has always a unique solution and that the computed numerical solutions converge (when $n \to +\infty$) toward the solutions of the scheme. Many details about the implementation of the algorithm can be found in [662].

## 23.5 Examples of numerical results

In this section, we show some examples of numerical results on real images. In these experiments, we test the implicit generic SFS algorithm of Prados and Faugeras. At the same time, we suggest some *applications* of the SFS methods hoping that the results will convince the reader of the *applicability of this method to real problems*.

Let us recall that we have assumed that the camera is geometrically and photometrically calibrated. In the experiments of sections 23.5.1 and 23.5.2 we know the focal length (5.8 mm) and approximately the pixel size (0.0045 mm; CCD size = 1/2.7") of the digital camera (Pentax Optio 330GS). In section 23.5.3, we choose some arbitrary reasonable parameters. Let us note that in these tests, we also make some educated guesses for gamma correction (when the photometric properties of the images seem incorrect).

### 23.5.1 Document restoration using SFS

In this section, we consider a reprographic system to remove the geometric and photometric distortions generated by the classical photocopy of a bulky book. Note that several solutions have been proposed in the SFS literature. Let us cite in particular the work of Wada et al. [850], Cho et al. [193] and Courteille et al. [228]. Here, the acquisition process we use is a classical camera. The book is illuminated by a single light source located at infinity or close to the optical center (following the models described in section 23.2). The acquired images are then processed using Prados and Faugeras' SFS method to obtain the shape of the photographed page. Let us emphasize that, for obtaining a compact experimental system, the camera must be located relatively close to the book. Therefore the *perspective model is especially relevant* for this application. Also, the distortion due to the perspective clearly appears in the image a) of figure 23.4. In this SFS method we assume that the albedo is constant. In this application, this does

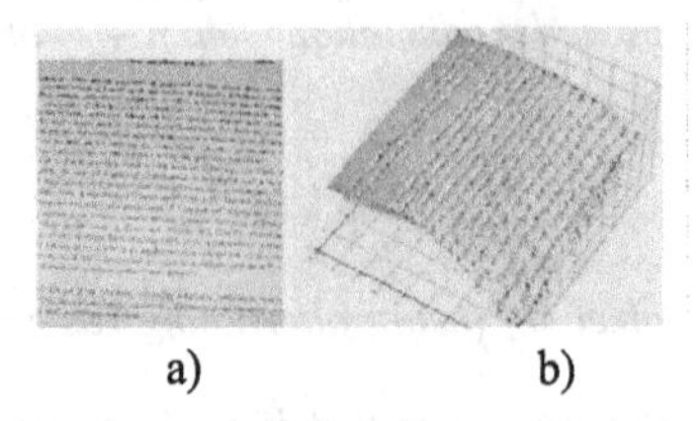

a) b)

Figure 23.3. a) Real image of a page of text [size $\simeq 800 \times 800$]; b) Surface recovered from a) by Prados and Faugeras' generic algorithm (without removing the printed parts of a)),

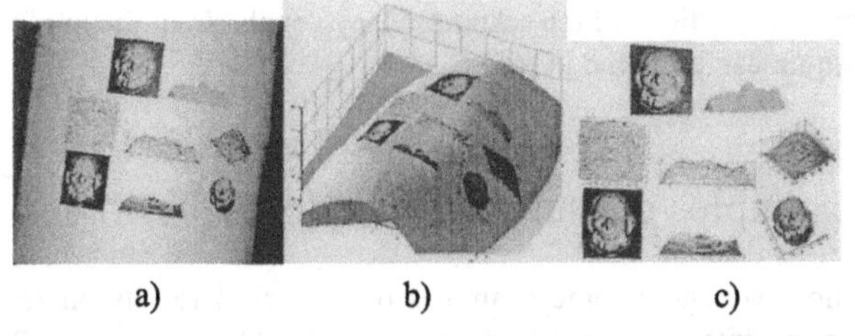

a) b) c)

Figure 23.4. a) real image of a page containing pictures and graphics [size $\simeq 2000 \times 1500$], b) surface (textured by the printed parts of a)) recovered from a) by Prados and Faugeras' generic algorithm (after having removed and inpainted the ink parts of a)). c) An orthographic projection of the surface b): the geometric (and photometric) distorsions are significantly reduced.

not hold because of the printed parts. Before recovering the surface of the page, we therefore localize the printed parts by using image statistic (similar to Cho's [193]) and we erase them automatically by using an inpainting algorithm. This step can produce an important pixel noise. Nevertheless, this is not a problem for us because, as figure 23.3-b) shows, *Prados and Faugeras' SFS method is extremely robust to pixel noise*: figure 23.3-b) displays the result produced by this algorithm (after 10 iterations) using the image of a text page with its pigmented parts, Fig.23.3-a). In this test, characters are considered as noise. Once we have recovered the three-dimensional shape of the page, we can then flatten the surface. Note that at each step of this restoration process we can keep the correspondences with the pixels in the image. Thus, at the final step, we can restore the printed parts.

To prove the applicability of this method, we have tested it on a page wrapped on a cylindrical surface[13] (we have used a cheap camera and flash in an approxi-

[13]For emphasizing the perspective effect.

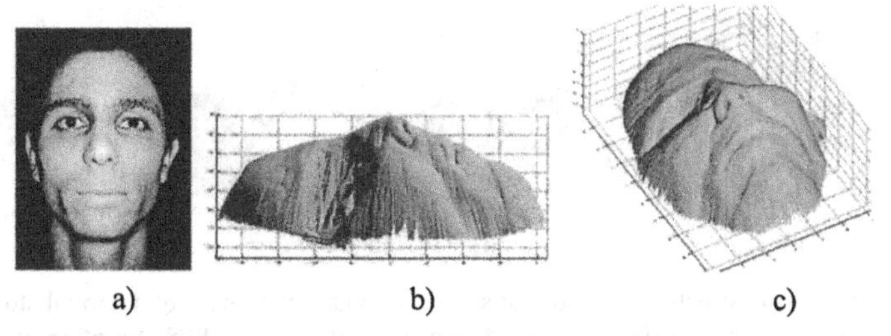

a) b) c)

Figure 23.5. a) Real face image [size $\simeq$ 450 $\times$ 600]; b) surface recovered from a) by the generic SFS algorithm with the perspective model with the light source located at the optical center; c) surface recovered by the generic SFS algorithm with the same modeling hypotheses as for b) after the inpainting process.

mately dark room). Figure 23.4 shows the original image in a), the reconstructed surface (after 10 iterations) (textured by the ink parts of a)) in b) and an orthographic projection of the reconstructed surface, in e). Figure 23.4-c) indicates that this method allows to remove the perspective and photometric distortions.

### *23.5.2 Face reconstruction from SFS*

In this section we propose a very simple protocol based on SFS for face reconstruction. We use one camera equiped with a basic flash in an approximately dark place. We have tested the implicit generic SFS algorithm on a real image of a face (using a small amount of make-up to make it more Lambertian) located at $\simeq$700 mm of the camera in an approximately dark place (see Fig.23.5-a)). Figure 23.5-b) shows the surface recovered by the generic algorithm with the perspective model with a point light source at the optical center. As in the previous application, the albedo is not constant over the whole image. Therefore we removed[14] the eyes and the eyebrows in the image by using an inpainting algorithm. Figure 23.5 shows in c) the surface recovered from the image obtained after the inpainting process.

### *23.5.3 Potential applications to medical images*

In this section, we are interested in applying the SFS method to some medical images. Our interest is motivated, for example, by the work of Craine et al. [232] (who use SFS for correcting some errors on the quantitative measurement of areas in the cervix, from colonoscopy images). We have applied Prados and Faugeras' algorithm to an endoscopic image of a normal stomach[15] (see figure 23.6-a)). For producing such an image, the light source must be very close to the camera, because of space constraints. So the adequate modeling is that of the "perspective SFS" with the light source located at the optical center. In figure 23.6-b), we show the result

[14] Can be automated by matching the image to a model image already segmented.

[15] Suggested by Tankus and Sochen [789]; http://www.gastrolab.net/

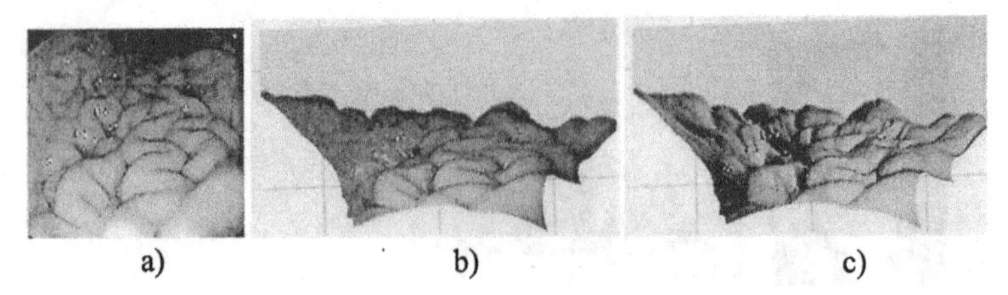


Figure 23.6. Reconstruction of a normal stomach. a) Original image of a normal stomach [size$\simeq$ 200 $\times$ 200]; b) surface recovered from a) by the generic SFS algorithm with the perspective model with the light source located at the optical center; c) surface b) visualized with a different illumination.

obtained. To further show the quality of the reconstruction, we display the surface b) with a different illumination. Finally, notice that the stomach wall is not perfectly Lambertian (see Fig.23.6-a)). This suggests the robustness of this SFS method to departures from the Lambertian hypothesis.

## 23.6 Conclusion

After having presented the SFS problem, we have described its main difficulties: in practice, the classical SFS equations are ill-posed. In a second time, we have focused on the numerical methods. We have considered the propagation and PDEs methods; in particular Prados and Faugeras' methods. We have demonstrated the applicability of the SFS methods by displaying some experimental results with real images. Finally, we have suggested that SFS may be useful in a number of real-life applications.

# Chapter24

# 3D from Image Sequences: Calibration, Motion and Shape Recovery

**M. Pollefeys**

Abstract

In this chapter we discuss how to recover the motion and calibration of a camera and the shape of a static object from an image sequence. The problem can be split into four subproblems: (1) computing the geometric relation between neighboring images, (2) estimating the motion and calibration of the camera, (3) computing a dense set of correspondences between neighboring images, (4) reconstruction of the 3D object shape. The approach we present here is fully automatic and can deal with photo or video sequences acquired with an uncalibrated hand-held camera. The different algorithms can also be used to provide solutions for other applications.

## 24.1 Introduction

In recent years a lot of progress has been made in the area of 3D reconstruction from images. Computer vision researchers have obtained a deep theoretical understanding of the geometric relations between multiple views of a scene [389, 311]. This has resulted in the development of robust algorithms to compute those geometric relations automatically and has made it possible to work with uncalibrated imagery. The adoption of bundle adjustment algorithms from photogrammetry has resulted in significant accuracy improvements for calibration, motion and shape recovery algorithms. During the same period significant progress was also made in the area of stereo matching and 3D surface reconstruction. By building on those advances it is now possible to implement a processing pipeline that automatically obtains a detailed 3D model from an image sequence acquired with an uncalibrated hand-held camera. In the remainder of this chapter we will discuss how this can be done. First, we introduce some notations and background, then we discuss the solution to the different subproblems. Section 24.2 deals with computing the relation between neighboring views based on tracked or matched feature points. Section 24.3 explains how the 3D structure of the feature points and the motion and calibration of the camera can be computed. Section 24.4 deals with dense correspondence matching and depth estimation and Section 24.5 describes how the computed information can be combined to construct a 3D surface model. An overview of the presented approach is shown in Fig. 24.1.

### *24.1.1 Notations and background*

In this section we briefly introduce some of the geometric concepts and notations used throughout this chapter. A more in depth description of these geometric concepts can be found in [389, 311]. A *perspective camera* is modeled through the projection equation

$$\lambda \mathtt{x} = \mathbf{P}\mathtt{X} \tag{24.1}$$

where $\lambda$ represents a non-zero scale factor, $\mathtt{X}$ is a 4-vector that represents 3D world point in homogeneous coordinates, $\mathtt{x}$ is a 3-vector that represents a corresponding 2D image point and $\mathbf{P}$ is a $3 \times 4$ projection matrix. In a metric or Euclidean coordinate frame $\mathbf{P}$ can be factorized as follows

$$\mathbf{P} = \mathbf{K}\mathbf{R}^\top[\mathbf{I}|\text{-}\mathtt{t}] \text{ where } \mathbf{K} = \begin{bmatrix} f & s & u \\ & rf & v \\ & & 1 \end{bmatrix} \tag{24.2}$$

contains the intrinsic camera parameters, $\mathbf{R}$ is a rotation matrix representing the orientation and $\mathtt{t}$ is a 3-vector representing the position of the camera. The intrinsic camera parameter $f$ represents the focal length measured in width of pixels, $r$ is the aspect ratio of pixels, $(u, v)$ represent the coordinates of the principal point and $s$ is a term accounting for the skew. In general, $s$ can be assumed zero. In practice, the principal point is often close to the center of the image, and the aspect ratio $r$ close to 1. In many cases the camera does not perfectly satisfy the perspective projection model and distortions have to be taken into account, the most important being radial distortion. In practice, when the amount of radial distortion is limited it is sufficient to model the radial distortion as follows:

$$\lambda \mathtt{x} \sim \mathbf{P}(\mathtt{X}) = \mathbf{K}\mathcal{R}(\mathbf{R}^\top[\mathbf{I}|\text{-}\mathtt{t}]\mathtt{X}) \text{ with } \mathcal{R}(\mathtt{x}) = (1+\kappa_1(x^2+y^2))[x\, y\, 0]^\top+[0\,0\,1]^\top \tag{24.3}$$

where $\kappa_1$ indicates the amount of radial distortion that is present in the image. For high accuracy applications, higher-order terms also have to be used. In this chapter the notation $d(.,.)$ will be used to indicate the Euclidean distance between entities in the images.

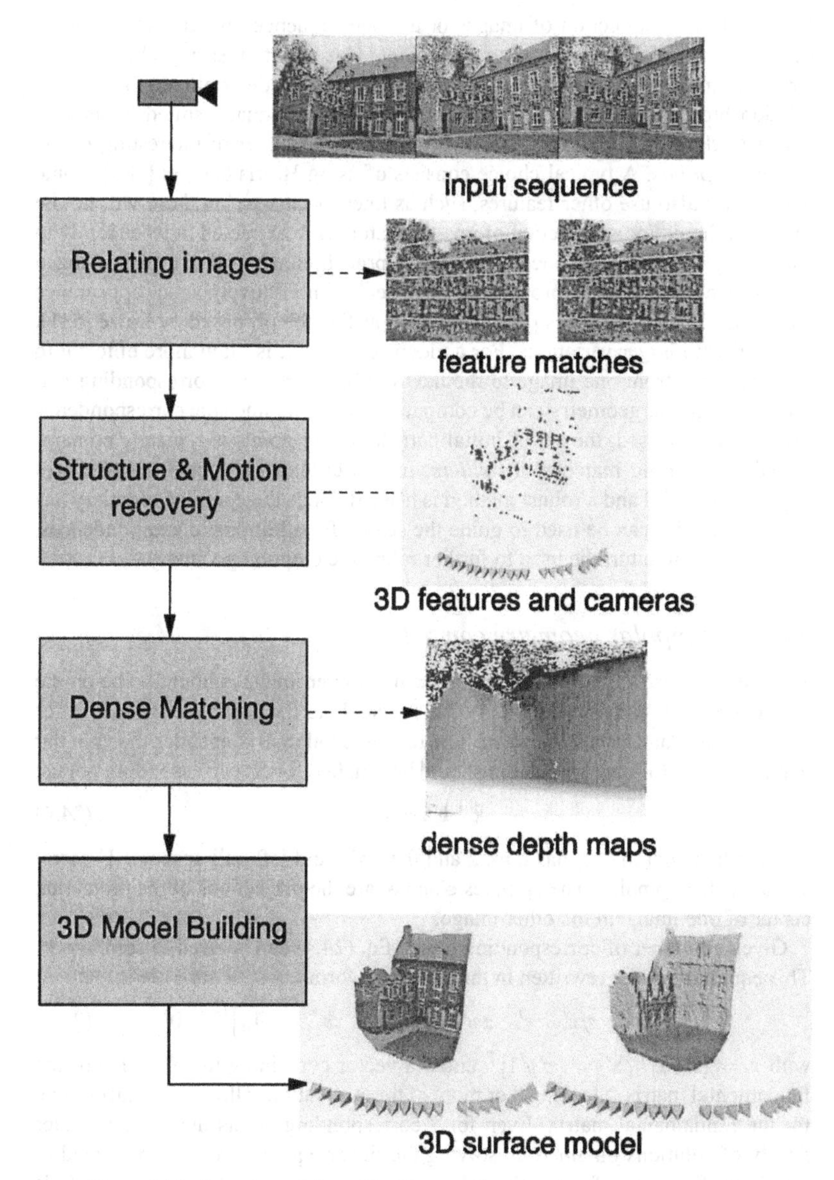


Figure 24.1. Overview of the calibration, motion and shape recovery pipeline.

## 24.2 Relating images

Starting from a collection of images or a video sequence, the first step consists of relating the different images to each other. This is not an easy problem. A restricted number of corresponding points is sufficient to determine the geometric relationship between images. Since not all points are equally suited for matching or tracking, the first step consist of selecting a number of interesting points or *feature points*. A typical choice consists of using Harris corners [385]. Some approaches also use other features, such as lines or curves, but these will not be discussed here. For a collection of images features are extracted independently in each image and then matched. A simple approach consists of using normalized cross-correlation. This approach can only deal with relatively small appearance variations. More advanced approaches, such as the one proposed by Lowe [531], can deal with larger variations. For a video sequence it is often more efficient to track features from one image to the next [738]. From these corresponding features the epipolar geometry can be computed. However, since the correspondence problem is ill-posed, the set of initial corresponding points is typically contaminated with wrong matches or *outliers*. In this case, a traditional least-squares approach will fail and a robust method is needed. Once the epipolar geometry has been obtained it can be used to guide the search for additional correspondences. These can then in turn be used to further refine the epipolar geometry.

### 24.2.1 *Epipolar geometry computation*

The point $\mathrm{x}'$ corresponding to the point $\mathrm{x}$ in another image is bound to be on the projection of its line of sight $\mathrm{l}' \sim \mathbf{F}\mathrm{x}$ where $\mathbf{F}$ is the *fundamental matrix* for the two views under consideration. The fundamental matrix encodes the epipolar geometry. The following equation should be satisfied for all corresponding points:

$$\mathrm{x}'^{\top}\mathbf{F}\mathrm{x} = 0 \ . \tag{24.4}$$

The fundamental matrix has rank 2 and the right and left null-space of $\mathbf{F}$ corresponds to the epipoles. The epipoles $\mathrm{e}$ and $\mathrm{e}'$ are the projections of the projection center of one image in the other image.

Given a number of corresponding points Eq. (24.4) can be used to compute $\mathbf{F}$. This equation can be rewritten in the following form:

$$\begin{bmatrix} xx' & yx' & x' & xy' & yy' & y' & x & y & 1 \end{bmatrix} \mathbf{f} = 0 \tag{24.5}$$

with $\mathrm{x} = [x\, y\, 1]^{\top}, \mathrm{x}' = [x' y' 1]^{\top}$ and $\mathbf{f}$ a vector containing the elements of the fundamental matrix. Stacking 8 or more of these equations allows to linearly solve for the fundamental matrix. Even for 7 corresponding points the one parameter family of solutions obtained by solving the linear equations can be restricted to 1 or 3 solutions by enforcing the cubic rank-2 constraint $\det(\mathbf{F}_1 + \lambda\mathbf{F}_2) = 0$. If the camera calibration is known an algorithm using only 5 corresponding points can be used [603]. As pointed out by Hartley [386] it is important to normalize the image coordinates before solving the linear equations. Otherwise the columns of Eq. (24.5) would differ by several orders of magnitude and the error would concentrate on the coefficients corresponding to the smaller columns. If feature points are well spread over the image the following normalization is appropriate:

$$\mathbf{x}_N = \mathbf{K}^{-1}\mathbf{x} \text{ with } \mathbf{K}_N = \begin{bmatrix} \frac{w+h}{2} & 0 & \frac{w}{2} \\ & \frac{w+h}{2} & \frac{h}{2} \\ & & 1 \end{bmatrix} \tag{24.6}$$

with $w$ and $h$ the width and height of the image. As we will see later this normalization will also be useful for other processing steps.

The initial set of correspondences can contain large number of outliers which will cause least-squares approaches to fail. To deal with this problem we use the RANSAC approach proposed by Fischler and Bolles [316]. A hypothesis for the fundamental matrix is obtained from a randomly selected minimal subset of data, and is used to classify each correspondence as an inlier or an outlier w.r.t the hypothesis under consideration. If the initial data sample contains no outliers, it can be expected that a large number of inliers will support the solution, otherwise the initial subset is probably contaminated with outliers. This procedure is repeated until the probability of having selected at least one outlier-free sample is in excess of 99%. The expression for this probability is $\Gamma = 1 - (1 - \gamma^p)^m$ with $\gamma$ the maximal fraction of inliers that has been observed, and $p$ the number of features in each sample ($p = 7$ for a fundamental matrix) and $m$ the number of trials. Once the epipolar geometry has been computed, it can be used to guide the matching process towards additional matches along the epipolar line.

In the case of a video sequence, consecutive frames are very close together and the computation of the epipolar geometry is ill conditioned. To avoid this problem we propose to only consider properly selected key-frames for the structure and motion recovery. Using appropriately spaced key-frames is also important for further steps such as the dense stereo matching. If it is important to compute the motion for all frames, such as for insertion of virtual objects in a video sequence [224], the pose for in-between frames can be computed afterwards. We propose to use model selection [800] to select the next key-frame only once the epipolar geometry model explains the tracked features better than the simpler homography model.

## 24.3 Structure and motion recovery

In the previous section it was seen how different views could be related to each other. In this section we will build on this to retrieve the structure of the scene and the motion of the camera, as well as the calibration.

At first two images are selected and used to set up and projective coordinate frame for the reconstruction. Then, the pose of the camera for the other views is determined in this frame and each time the existing reconstruction is refined and extended with newly observed features. In this way the pose estimation of views that do not share features with the two initial views also becomes possible. Typically, a view is only matched with its predecessor in the sequence. In most cases this works fine, but in some cases (e.g. when the camera moves back and forth) it can be advantageous to also relate a new view to a number of additional views. Once the structure and motion has been determined for the whole sequence, the results can be refined through a projective bundle adjustment. Then, the ambiguity of the reconstruction can be restricted to a similarity transformation through self-calibration. Finally, a metric bundle adjustment can be carried out to obtain an optimal estimation of the structure and motion.

### *24.3.1 Initial structure and motion*

Two images of the sequence are used to set up a projective reference frame. The world frame is aligned with the first camera. The second camera is chosen so that the epipolar geometry corresponds to the computed $\mathbf{F}_{12}$:

$$\begin{array}{lcl} \mathbf{P}_1 & = & [\ \mathbf{I}_{3\times 3} \mid \mathtt{0}_3\ ] \\ \mathbf{P}_2 & = & [\ [\mathtt{e}_{12}]_\times \mathbf{F}_{12} + \mathtt{e}_{12}\mathtt{a}^\top \mid \sigma \mathtt{e}_{12}\ ] \end{array} \qquad (24.7)$$

It was shown [306, 387] that this is sufficient to guarantee that the reconstruction differs from the recorded scene by no more than a projective transformation. Eq. (24.7) is not completely determined by the epipolar geometry, but has 4 more degrees of freedom. The 3-vector a determines the position of the reference plane (i.e. the plane at infinity in an affine or metric frame) and $\sigma$ determines the global scale of the reconstruction. The parameter $\sigma$ can simply be put to one. If care is taken to perform all computations in a projectively invariant way (by performing measurements in image space and by using homogeneous coordinates for 3D entities), one can simply choose $\mathtt{a} = [0\,0\,0]^\top$.

Once the two initial projection matrices have been fully determined the correspondences between the two views can be reconstructed through triangulation. Due to noise the lines of sight will not exactly intersect. As mentioned before, with a projective basis it is important to minimize an image distance, not a 3D distance. In this case the distance between the reprojected 3D point and the image points is appropriate:

$$d(\mathtt{x}_1, \mathbf{P}_1\mathtt{X})^2 + d(\mathtt{x}_2, \mathbf{P}_2\mathtt{X})^2 \qquad (24.8)$$

It was noted in [388] that the only important choice is to select in which epipolar plane the point is reconstructed. Once this choice is made it is trivial to select the optimal point in the plane. A bundle of epipolar planes has only one parameter. Minimizing the following equation is thus equivalent to minimizing equation (24.8).

$$d(\mathtt{x}_1, \mathtt{l}_1(\alpha))^2 + d(\mathtt{x}_2, \mathtt{l}_2(\alpha))^2 \qquad (24.9)$$

with $\mathtt{l}_1(\alpha)$ and $\mathtt{l}_2(\alpha)$ the epipolar lines obtained in function of the parameter $\alpha$ describing the bundle of epipolar planes. It turns out (see [388]) that this equation is a polynomial of degree 6 in $\alpha$. The global minimum of equation (24.9) can thus easily be computed. In both images the point on the epipolar line $\mathtt{l}_1(\alpha)$ and $\mathtt{l}_2(\alpha)$ closest to the points $\mathtt{x}_1$ and $\mathtt{x}_2$ respectively is selected. Since these points are in epipolar correspondence their lines of sight meet at a 3D point.

### *24.3.2 Updating the structure and motion*

The previous section dealt with obtaining an initial reconstruction from two views. This section discusses how to add a view to an existing reconstruction. First the pose of the camera is determined, then the structure is updated based on the added view and finally new points are initialized.

For every additional view the pose towards the pre-existing reconstruction is determined, then the reconstruction is updated. The first step consists of finding the epipolar geometry as described in Section 24.2.1. Then the matches which correspond to already reconstructed points are used to infer correspondences between 2D and 3D. Based on these the projection matrix $\mathbf{P}$ is computed using a robust RANSAC procedure. Eliminating $\lambda$ from Eq. (24.1) yields two linear equations per correspondence:

$$\begin{bmatrix} \mathtt{X}^\top & \mathtt{0} & -x\mathtt{X}^\top \\ \mathtt{0} & \mathtt{X}^\top & -y\mathtt{X}^\top \end{bmatrix} \mathbf{p} = \begin{bmatrix} 0 \\ 0 \end{bmatrix} \tag{24.10}$$

with $\mathbf{p}$ a 12-vector containing the coefficients of the projection matrix. In this case a minimal sample of 6 matches is needed to compute $\mathbf{P}$. A point is considered an inlier if there exists a 3D point that projects sufficiently close to all associated image points. We propose to verify this by attempting to refine the previous solution for $\mathbf{X}$ based on all observations, including the one in the new view. Because this verification step is computationally expensive (as this has to be done for each generated hypothesis), it is advised to use a modified version of RANSAC with early termination of unpromising hypotheses [200].

One important problem is that the computation of the camera pose for an uncalibrated camera remains ambiguous when all points are located on a plane. One possible solution to this consists of using model selection to detect this case and to delay the computation of the corresponding camera projection matrices until after self-calibration since at this point the ambiguity can be resolved [656].

Once the pose for a new view has been determined, the 3D reconstruction of feature points is refined. This can be done using an iteratively reweighted least-squares algorithm for each point. Eq. (24.1) can now be rewritten to become linear in $\mathtt{X}$:

$$\begin{bmatrix} \mathtt{P}_3 x - \mathtt{P}_1 \\ \mathtt{P}_3 y - \mathtt{P}_2 \end{bmatrix} \mathtt{X} = \begin{bmatrix} 0 \\ 0 \end{bmatrix} \tag{24.11}$$

with $\mathtt{P}_i$ the $i$-th row of $\mathbf{P}$ and $(x, y)$ being the image coordinates of the point. An estimate of $\mathtt{X}$ is computed by solving the system of linear equations obtained from all views where a corresponding image point is available. To obtain a better solution the criterion $\sum d(\mathbf{P}\mathtt{X}, \mathtt{x})$ should be minimized. This can be approximately obtained by weighting Eq.(24.11) with $\frac{1}{\mathtt{P}_3\tilde{\mathtt{X}}}$ where $\tilde{\mathtt{X}}$ correspond to the previous solution for $\mathtt{X}$. This procedure can be repeated one or two times. By solving this system of equations through SVD a normalized homogeneous point is automatically obtained. If a 3D point is not observed the position is not updated. In this case one can check if the point was seen in a sufficient number of views to be kept in the final reconstruction. We recommend not to use points seen in less than 3 views. This avoids having an important number of outliers due to spurious matches.

Of course, in an image sequence some new features will appear in every new image. If point matches are available that were not related to an existing point in the structure, then a new point can be initialized as described in Section 24.3.1.

### *24.3.3 Refining structure and motion*

Once the structure and motion has been obtained for the whole sequence, it is recommended to refine it through a global minimization step. A maximum likelihood estimation can be obtained through *bundle adjustment* [804]. The goal is to find the parameters of the camera view $\mathbf{P}_j$ and the 3D points $\mathtt{X}_j$ for which the mean squared distances between the observed image points $\mathtt{x}_{ij}$ and the reprojected image points $\mathbf{P}_i(\mathtt{X}_j)$ is minimized. The camera projection model should also take radial distortion into account. For $m$ views and $n$ points the following criterion should be minimized:

$$\min_{\mathbf{P}_i,\mathtt{X}_j} \sum_{i=1}^{m} \sum_{j=1}^{n} d(\mathtt{x}_{ij}, \mathbf{P}_i(\mathtt{X}_j))^2 \tag{24.12}$$

If the errors on the localization of image features are independent and satisfy a zero-mean Gaussian distribution then it can be shown that bundle adjustment corresponds to a maximum likelihood estimator. This minimization problem is huge, but the special structure of the problem can be exploited to solve the problem much more efficiently [389, 804]. The key reason for this is that a specific residual is only dependent on one point and one camera, which results in a very sparse Jacobian.

### *24.3.4 Upgrading from projective to metric*

The reconstruction obtained as described in the previous sections is only determined up to an arbitrary projective transformation which is insufficient for visualization and to perform measurements. Therefore, we need to upgrade our reconstruction by restricting the ambiguity to at most a similarity transformation. In recent years many *self-calibration* approaches have been proposed to achieve this. The first self-calibration algorithms were concerned with unknown but constant intrinsic camera parameters (e.g. [310]). Later algorithms for varying intrinsic camera parameters have also been proposed (e.g. [654]). An important issue with self-calibration is that in some cases the motion of the camera is not general enough to allow for self-calibration to recover the calibration uniquely [772].

One of the most important concepts for self-calibration is the *absolute conic* and its projection in the images. The simplest way to represent the absolute conic is through the dual absolute quadric $\Omega^*$ [803]. In a Euclidean coordinate frame $\Omega^* = \mathrm{diag}(1,1,1,0)$ and one can easily verify that it is invariant to similarity transformations. Inversely, it can also be shown that a transformation that leaves the dual quadric $\Omega^*$ unchanged is a similarity transformation. For a projective reconstruction $\Omega^*$ can be represented by a $4 \times 4$ rank-3 symmetric positive semi-definite matrix. According to the properties mentioned above a transformation that transforms $\Omega^* \rightarrow \mathrm{diag}(1,1,1,0)$ will bring the reconstruction within a similarity transformation of the original scene. Our goal is thus to compute the location of $\Omega^*$ in our projective reference frame.

The projection of the dual absolute quadric in the image is described by the following equation:

$$\lambda\omega^* = \mathbf{P}\Omega^*\mathbf{P}^\top \ . \tag{24.13}$$

It can be easily verified that in a Euclidean coordinate frame the image of the absolute quadric is directly related to the intrinsic camera parameters:

$$\omega^* = \mathbf{K}\mathbf{K}^\top \tag{24.14}$$

Since the images are independent of the projective basis of the reconstruction, Eq. (24.14) is also valid for a projective reconstruction and constraints on the intrinsic camera parameters can be translated to constraints on the location of $\Omega^*$.

If the self-calibration constraints on the camera intrinsics yield linear constraints on $\omega^*$, a linear self-calibration algorithm is thus easily obtained [654]. If the images have been normalized using Eq. (24.6), a focal length of a 60mm lens corresponds to 1 and thus focal lengths in the range of 20mm to 180mm would end up in the range $[1/3, 3]$. The principal point should be mapped close to the origin. The aspect ratio is typically also around 1 and the skew can be assumed 0 for all practical purposes. Making this prior knowledge more explicit and estimating reasonable standard deviations one could assume for example $f \approx rf \approx 1 \pm 3$, $u \approx v \approx 0 \pm 0.1$, $r \approx 1 \pm 0.1$ and $s = 0$ which approximately corresponds to $\frac{\omega^*_{11}}{\omega^*_{33}} \approx 1 \pm 9$, $\frac{\omega^*_{22}}{\omega^*_{33}} \approx 1 \pm 9$, $\frac{\omega^*_{22}}{\omega^*_{11}} \approx 1 \pm 0.2$, $\omega^*_{12} \approx 0 \pm 0.01$, $\omega^*_{13} \approx \omega^*_{23} \approx 0 \pm 0.1$. The constraints on the left-hand side of Eq. (24.13) should also be verified on the right-hand side. The uncertainty can be taken into account by weighting the equations.

$$\begin{array}{rclrcl}
\frac{1}{9\lambda}\left(P_1\Omega^*P_1{}^\top - P_3\Omega^*P_3{}^\top\right) & = & 0 & \frac{1}{0.01\lambda}\left(P_1\Omega^*P_2{}^\top\right) & = & 0 \\
\frac{1}{9\lambda}\left(P_2\Omega^*P_2{}^\top - P_3\Omega^*P_3{}^\top\right) & = & 0 & \frac{1}{0.1\lambda}\left(P_1\Omega^*P_3{}^\top\right) & = & 0 \\
\frac{1}{0.2\lambda}\left(P_1\Omega^*P_1{}^\top - P_2\Omega^*P_2{}^\top\right) & = & 0 & \frac{1}{0.1\lambda}\left(P_2\Omega^*P_3{}^\top\right) & = & 0
\end{array} \tag{24.15}$$

with $P_i$ the $i$th row of $\mathbf{P}$ and $\lambda$ a scale factor that is initially set to 1 and later on to $P_3\tilde{\Omega}^*P_3{}^\top$ with $\tilde{\Omega}^*$ the result of the previous iteration. Since $\Omega^*$ is a symmetric $4 \times 4$ matrix it is parametrized through 10 coefficients. An estimate of the dual absolute quadric $\Omega^*$ can be obtained by solving the above set of equations for all views through linear least-squares. The rank-3 constraint should be imposed by forcing the smallest singular value to zero. This scheme can be iterated until the $\lambda$ factors converge (typically after a few iterations). Although the equations related to the focal length are very much down-weighted, they can be important to regularize the solution when the camera performs a (quasi-)critical motion sequence [656]. The upgrading transformation $\mathbf{T}$ can be obtained from diag $(1, 1, 1, 0) = \mathbf{T}\Omega^*\mathbf{T}^\top$ by eigenvalue decomposition of $\Omega^*$. The metric structure and motion is then obtained from

$$\mathbf{P}_M = \mathbf{P}\mathbf{T}^{-1} \text{ and } \mathtt{X}_M = \mathbf{T}\mathtt{X}. \tag{24.16}$$

This initial metric reconstruction can then further be refined through a metric bundle adjustment. In this bundle adjustment the constraints on the camera intrinsics have to be enforced. These constraints can both be hard constraints (typically imposed through parameterization) or soft constraints (imposed by including an additional term in the minimization criterion). A good choice of constraints for a photo camera consists of imposing a constant focal length (if no zoom was used), a constant principal point, a constant radial distortion, an aspect ratio of one and the absence of skew. For a camcorder/video camera it is important to also estimate the (constant) aspect ratio as this can significantly differ from 1.

## 24.4 Dense surface estimation

Once the camera motion and calibration have been computed, multi-view reconstruction algorithms can be used to compute the surface of the recorded scene. A multitude of approaches have been proposed in the computer vision literature, e.g. [607, 308, 498]. Here we present a pragmatic approach well suited for images acquired with a hand-held camera. This approach combines two-view stereo matching with a multi-view correspondence linking process [481]. This combines the advantages of small-baseline matching with wide-baseline triangulation. In addition, this scheme is much less sensitive to inaccurate geometric and photometric calibration and avoids most problems with occlusions. First, stereo rectification is performed on neighboring images and a stereo matching algorithm is used to obtain a dense set of correspondences. Then, a dense depth map is computed by combining results from multiple stereo pairs.

### *24.4.1 Rectification and stereo matching*

Since the calibration between successive image pairs has been computed, the known epipolar geometry constrains the correspondence search to one dimension. To simplify the matching process the images are warped so that the corresponding epipolar lines become corresponding scanlines. This process is called image pair *rectification.* Stereo matching can be performed more efficiently on rectified image pairs because image regions do not have to be warped separately for each

Figure 24.2. Rectified image pair (left,right) and computed disparity map (center).

disparity evaluation, a simple image shift is sufficient. Most stereo algorithms expect image pairs to be rectified.

For some motions (i.e. when the epipole is located in the image) standard rectification based on planar homographies is not possible and a more advanced procedure should be used. We propose to use an approach that works for all possible motions and guarantees minimal image sizes (without losing information) [655]. The key idea is to use polar coordinates with the epipole as origin. Corresponding lines are given through the epipolar geometry. By taking the orientation into account the matching ambiguity can be reduced to half epipolar lines. A minimal image size is achieved by computing the angle between two consecutive epipolar lines so that the worst case pixel on the line preserve its area. To avoid image degradation, both correction of radial distortion and rectification can be performed in a single resampling step.

Stereo algorithms take a rectified image pair as input and compute a disparity map which encodes the horizontal displacement between corresponding pixels (see Figure 24.2). The correspondence search is typically limited to a specific disparity range. This range depends on the depth of the observed scene and the camera configuration and can be computed from tracked/matched features. The simplest stereo algorithms minimize the matching cost for each pixel separately. More advanced algorithms, such as the one we have used to compute the examples shown in this chapter [831], perform an optimization over a complete scanline that trades off matching cost with horizontal continuity. The most advanced –but also computational most expensive– algorithms perform an optimization over the whole image trading off matching cost with horizontal and vertical continuity. A complete taxonomy of stereo algorithms can be found in [711] and a more in-depth discussion of some algorithms can be found in other chapters of this book.

### 24.4.2 *Multi-view linking*

The pairwise disparity estimation allows us to compute image to image correspondence between adjacent rectified image pairs, and independent depth estimates for each camera viewpoint. An optimal joint estimate is achieved by fusing all independent estimates into a common depth map. The fusion can be performed in an economical way through controlled correspondence linking (see Figure 24.3). A

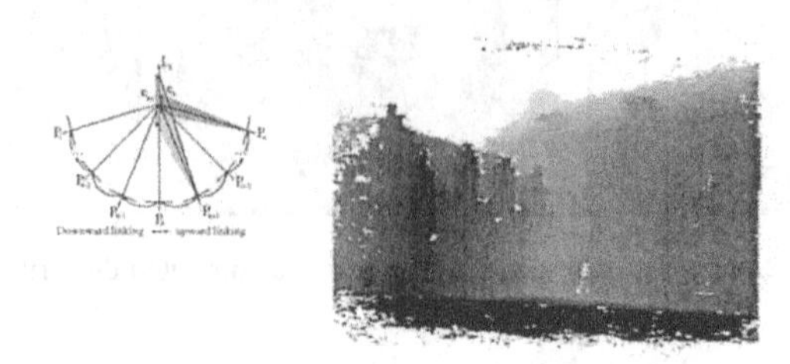

Figure 24.3. Depth fusion and uncertainty reduction from correspondence linking (left) and depth map (right).

point is transferred from one image to the next as follows:

$$x' = R'^{-1}(R(x) + D(R(x))) \quad (24.17)$$

with $R(.)$ and $R'(.)$ functions that map points from the original image into the rectified image and $D(.)$ a function that corresponds to the disparity map. When the depth obtained from the new image point $x'$ is outside the confidence interval the linking is stopped, otherwise the result is fused with the previous values through a Kalman filter. The variance provided by the Kalman filter can also be stored for later use. More details on this approach can be found in [481]. This approach combines the advantages of small baseline matching and wide baseline triangulation. It can provide a very dense depth map by avoiding most occlusions (by linking to all its direct neighbors independently). The depth resolution is increased through the combination of multiple viewpoints and a large global baseline while the matching is simplified through the small local baselines. Due to multiple observations of a single surface point the texture can be enhanced and noise and highlights can be removed. By only comparing image pixels between neighboring images, this approach is also robust to small errors in calibration and can deal with some view-dependent variations and limited changes in lighting or exposure between images.

## 24.5 3D surface reconstruction

In the previous sections a dense structure and motion recovery approach was given. This yields all the necessary information to build detailed 3D surface models. In practice, the 3D surface is approximated by a triangular mesh to reduce geometric complexity and to tailor the model to the requirements of computer graphics visualization systems. A simple approach consists of overlaying a 2D triangular mesh on top of one of the images and then building a corresponding 3D mesh by warping the vertices of the triangles in 3D space according to the values found in the corresponding depth map. To reduce noise it is recommended to first smooth the depth image (the kernel can be chosen of the same size as the mesh triangles). The image itself can be used as texture map (the texture coordinates are trivially obtained as the 2D coordinates of the vertices).

It can happen that for some vertices no depth value is available. In these cases the corresponding triangles are not reconstructed. The same happens when triangles are placed over discontinuities. This is achieved by selecting a maximum angle between the normal of a triangle and the line-of-sight through its center (e.g. 85 degrees). This simple approach works very well on the dense depth maps as obtained through multi-view linking.

To reconstruct more complex shapes it is necessary to combine results from multiple depth maps. The simplest approach consists of generating separate models independently and then loading them together in the graphics system. Since all depth-maps can be located in a single metric frame, registration is not an issue. When necessary, a volumetric depth map integration approach [240] can be used to obtain a single 3D consensus surface. For optimal results the variance, obtained from the Kalman filter in the multi-view linking step, should be used.

Figure 24.4. Reconstruction of ancient Medusa head: video frame and recovered structure and motion for key-frames (top), textured and shaded view of 3D reconstruction (bottom).

The example shown in Fig. 24.4 was recorded using a consumer camcorder (Sony TRV900). A 20 second shot was made of a Medusa head located on the entablature of a monumental fountain in the ancient city of Sagalassos (Turkey). The recorded object is about $1m$ across. Using progressive-scan frames of 720 ×

576 are obtained. Key-frames are automatically selected and the structure of the tracked features and the motion and calibration of the camera is computed, see upper-right of Fig. 24.4. It is interesting to notice that for this camera the aspect ratio is actually not 1, but around 1.09 which can be observed by comparing the upper-left and the lower-left image in Fig. 24.4 (notice that it is the real picture that is unnaturally stretched vertically). The next stage consisted of computing a dense surface representation. To this effect stereo matching was performed for all pairs of consecutive key-frames. Using our multi-view linking approach a dense depth map was computed for a central frame and the corresponding image was applied as a texture. Several views of the resulting model are shown in Fig. 24.4. The shaded view allows to observe the high-quality of the recovered geometry. We have also performed a quantitative evaluation of the results. The accuracy of the reconstruction was considered at two levels. Errors in the camera motion and calibration computations result in a global bias on the reconstruction. From the results of the bundle adjustment we have estimated this error to be of the order of $3mm$ for points on the reconstruction. The depth computations indicate that 90% of the reconstructed points have a relative error of less than $2mm$. Note that the stereo correlation uses a $7\times7$ window which corresponds to a size of $5mm\times5mm$ on the object and therefore the measured depth will typically correspond to the dominant visual feature within that patch.

## 24.6 Conclusion

In this chapter we have presented the steps needed to automatically compute 3D models from image sequences. In the presented system we have attempted to extract as much information as possible from the video sequence itself to make our approach as flexible as possible. However, this can also lead to some degeneracies such as critical motion sequences or pose estimation from planes which require additional measures. Therefore, when this information can be obtained easily from some other source (e.g. pre-calibration) this might benefit efficiency and robustness. However, when the only available information source are the images themselves, it is critical to be able to extract all the necessary information from them. Here we have focussed on the acquisition of photo-realistic 3D models of objects from images recorded with an uncalibrated hand-held camera, but many of the algorithms and solutions presented here can also be used to solve different problems. We have for example re-used many of the presented algorithms to implement the software for a pan-tilt stereo-head designed to reconstruct the 3D terrain model around a Mars lander. A key element in the success of this project was the ability to calibrate from images. Another example is automatic matchmoving. The first part of the presented processing pipeline can be used to compute the camera motion so that virtual objects can be correctly aligned with the real objects in video sequences. Some commercial products, such as 2D3's Boujou and RealViz' MatchMover, use algorithms similar to the ones described in this chapter. Also, besides explicit 3D models, it is possible to build alternative visual representations of the scene such as lightfields. We have adapted the presented pipeline to efficiently capture unstructured lightfields by waving the camera over the scene of interest. Our unstructured lightfield approach avoids the need for a single consistent 3D representation and renders view-dependent effects such as highlights.

## *Acknowledgement*

Many of the results shown in this chapter where obtained using software developed at the K.U.Leuven and we are grateful to Luc Van Gool, Reinhard Koch, Maarten Vergauwen, Frank Verbiest, Jan Tops, Kurt Cornelis, Geert Van Meerbergen and Jason Repko for their contributions to the presented work. The partial support of the FWO project G.0223.01 and NSF grants IIS-0237533 and IIS-0313047 are gratefully acknowledged.

# Chapter25

# Multi-view Reconstruction of Static and Dynamic Scenes

**M. Agrawal, A. Mittal and L. Davis**

Abstract

We explore the reconstruction of a three-dimensional scene from multiple images captured from far away viewpoints (wide-baseline camera arrangement). Such an arrangement is required for complex scenes where the visibility from any one viewpoint is not sufficient to adequately reconstruct the entire scene. Also, such an arrangement reduces the error in triangulation of the features, thereby improving the accuracy of reconstruction. Our emphasis is on algorithms that recover a volumetric model of the scene from calibrated cameras by explicitly modeling and detecting occlusions. We present a brief overview of the state of the art in such methods for multi-view reconstruction. In particular, algorithms based on a probabilistic framework have become quite popular and produce very accurate models. Several such probabilistic volume reconstruction methods will be described. For the dynamic parts of the scene, where an online reconstruction is needed, simpler methods are required. An important case of such scenes is that of walking people in a surveillance scenario. For this case, we present fast online algorithms that recover approximate shape and appearance models and 3D trajectories of people as they walk in the scene. Finally, we address the problem of optimal placement of cameras in order to acquire the best possible image data for the reconstruction of a given scene according to the particular task specification.

## 25.1 Introduction

Reconstruction of surfaces from multiple images has been a central research problem in computer vision for a long time. Early work in this area focused on developing stereo algorithms for binocular camera configurations. There is a volume of literature on binocular stereo with a number of algorithms that work well on many types of images. More recently, however, due to significant advances in computational power, vision systems using multiple cameras are becoming increasingly feasible and practical. Example of multi-view vision systems include the 3D room developed by Saito et al. [699] and the KECK Laboratory by Davis et al. [250]. These systems are able to capture multiple synchronized images of indoor scenes. This has generated a renewed interest in the computer vision community to develop efficient, scalable, and robust algorithms for 3D reconstruction from multiple images.

Going from binocular to multiple views has the advantage of potentially increasing the stability and accuracy of the reconstruction, as the baseline is increased. However, in order to fully exploit this potential, the algorithm must be able to handle occlusions, especially if the views are widely separated. The disparity map representation, which is widely used in binocular stereo, is unable to represent partially occluded background regions (due to the fact that only a single disparity value is assigned to each pixel in the reference image). Therefore, most multi-view algorithms use an explicit representation of the 3D volume of the scene (in Section 25.2.8, we will present an algorithm that uses an alternative representation, multiple depth and visibility maps to represent the scene). The goal of reconstruction is to find volume elements (voxels) that lie on the surface of the objects in the scene. In this chapter, we explore the problem of building a three dimensional model of a static as well as dynamic scene from multiple images captured from far away fully calibrated viewpoints (wide-baseline camera arrangement). Section 25.2 discusses reconstruction of static scenes followed by reconstruction of dynamic scenes in Section 25.3. Finally, the problem of optimal placement of sensors for multi-view systems is discussed in Section 25.4.

## 25.2 Reconstruction of Static Scenes

Reconstruction of static scenes from multiple images is an intensely researched area. The main challenge for the wide-baseline camera arrangement is to detect and handle occlusions. Therefore, we will restrict our discussion to approaches that model and detect occlusions explicitly. One of the simplest ways to build three-dimensional models is from multiple silhouette images of an object. The visual hull algorithm for reconstruction from silhouette images will be described in Section 25.2.1. The visual hull algorithm does not take into account the photometric properties of the scene. The voxel coloring framework discussed in Section 25.2.2 utilizes these photometric constraints to build photo-consistent models. This algorithm, however, works for only a special arrangement of cameras. The space carving algorithm presented in Section 25.2.3 is a generalization of voxel coloring that works for arbitrary placement of cameras. More recently, probabilistic approaches have become quite popular as they take into consideration alternative hypotheses that better explain all the images. Several such probabilistic algorithms will be described in Sections 25.2.5 through 25.2.8. In particular, Section 25.2.7 presents our probabilistic surface reconstruction algorithm. Here, the problem is formulated as one of estimating the probability that a 3D point in the scene lies on the object's surface. An iterative scheme is presented that updates this probability based on the visibility constraints that exist in the images.

### 25.2.1 Visual Hull

A silhouette image is a binary image with the value at a point indicating whether that image point is part of the background or the object. The binary silhouette images can be obtained by background subtraction algorithms or by segmentation. When the cameras are calibrated, each point in a silhouette image defines a ray in scene space that intersects the object at some unknown depth along this ray. The entire silhouette can thus be extruded for each camera, creating a cone-like volume that bounds the extent of the object. The volumetric representation of the object can then be obtained by intersecting these volumes.

The volume obtained by the intersection of the generalized cones associated with a set of cameras is only an approximation of the true 3D shape. Laurentini [502] characterized the best approximation obtainable in the limit by the infinite number of silhouettes captured from all viewpoints outside the convex hull of the object as the *visual hull*. The visual hull is guaranteed to enclose the object, but since it does not capture concavities, it might not be the same as the object. In practice, only a finite number of silhouettes is used resulting in an approximation of this visual hull.

The reconstructed 3D volume is efficiently represented by using an octree [775]. An octree is a tree-structured representation that can be used to describe a bounded volume. The octree is constructed by recursively subdividing each cube into eight sub cubes, starting at the root node (a single large cube). The current voxel is projected into all the images and tested to determine if it intersects the silhouette in each image. If the projected voxel does not intersect the silhouette in at least one image, the voxel is carved out, that is, marked transparent. If the projected voxel intersects only silhouette pixels in every image, the voxel is marked opaque. Otherwise, the voxel intersects both background and silhouette points in the images and is termed ambiguous. This ambiguous voxel is then subdivided into octants, and each sub voxel is processed recursively. The process is terminated when either the desired octree resolution is attained or the voxel projects to sub pixel area within the images.

A distributed version of this algorithm was implemented by Borovikov and Davis in [104]. Figure 25.1 illustrates the volumetric model obtained by using sixteen silhouette images in an indoor setting.

### 25.2.2 Voxel Coloring

An approach along different lines is the photo-consistent voxel coloring algorithm by Seitz and Dyer [721]. The voxel coloring problem is to assign colors (radiances) to voxels (points) in a 3D volume so as to maximize photo-consistency

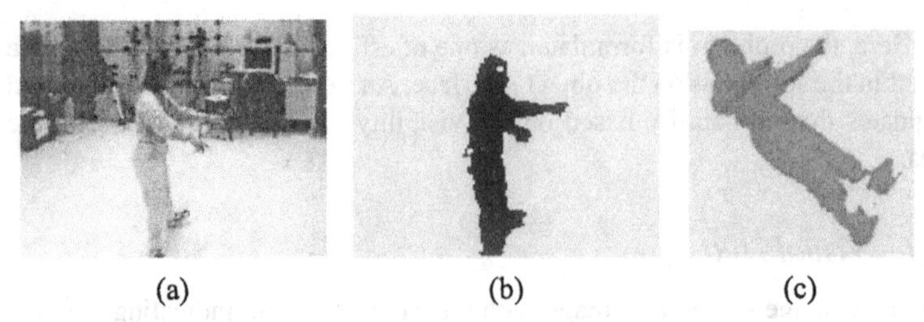

(a) (b) (c)

Figure 25.1. Illustration of visual hull algorithm: (a) observed scene, (b) silhouette image obtained by background subtraction, (c) voxelated VRML model

with a set of input images. That is, rendering the colored voxels from each input viewpoint should reproduce the original image as closely as possible. Using the notion of photo-consistency, voxels that are not on the surface are automatically carved out in the process.

For a Lambertian scene, photo-consistency implies that a voxel must project to similar colors in all views in which it is visible. Therefore, without noise or quantization effects, a photo-consistent voxel should project to a set of pixels with equal color values. The consistency of a set of colors can be defined as their standard deviation or, alternatively, the maximum of the $L_1$, $L_2$, or $L_\infty$ norm between all pairs of the projected colors in which it is visible. The voxel is considered to be on a surface if the measure is less than some threshold. For points *not* on the surface of the scene, the colors need not be similar, as illustrated in Figure 25.2(a).

The catch here is the fact that photo-consistency should be applied to only those views in which a voxel is visible. Therefore, occlusions must be detected before applying photo-consistency. However, to detect these occlusions, we must know the scene geometry first. Therefore, this becomes the chicken-and-egg problem. Seitz and Dyer solved this problem by imposing what they called the ordinal visibility constraint on the camera locations. This constraint requires that the cameras be placed such that no scene point should be contained within the convex hull of the camera centers. This placement provides a depth ordering of points in the scene so that all the voxels can be visited in a single scan in near-to-far order relative to every camera. Typically, this condition is met by placing all the cameras on one side of the scene and scanning voxels in planes that are successively further from the cameras. Hence, the problem of detecting occlusions is solved by the scene traversal ordering used in the algorithm; the order is such that if voxel $V$ occludes $V_0$ then $V$ is visited before $V_0$. This traversal greatly simplifies the computation of voxel visibility and allows a scene to be reconstructed in a single scan of the voxels.

The voxel coloring algorithm begins with a reconstruction volume of initially opaque voxels that encompasses the scene to be reconstructed. Voxels are traversed in the order of increasing distance from the camera volume. Each opaque voxel is projected in the images and tested for photo-consistency. Those that are found to be inconsistent are carved away, that is, made transparent. The consistency test is governed by a threshold on the color variation in the projected

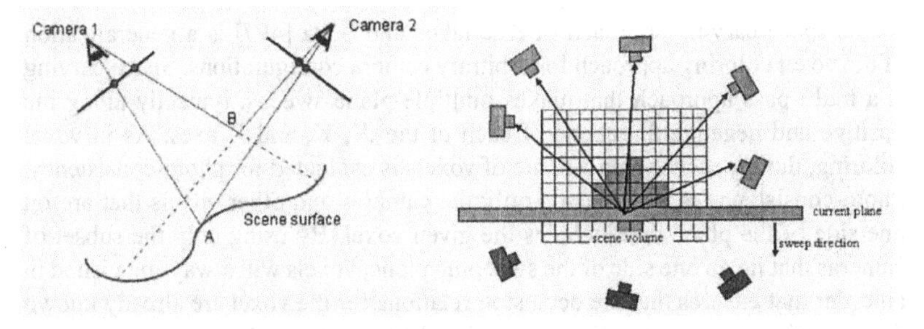


(a) Illustration of photo-consistency (b) Illustration of space carving

Figure 25.2. Illustration of photo-consistency and space carving. (a) Two cameras see consistent colors for the point A on the surface. For point B, not on the surface, the cameras see inconsistent colors. (b) For the current sweeping plane, only cameras that are above it (green) are used for the consistency check. Cameras that lie on the other side of the sweeping plane (red) are ignored.

images. The threshold corresponds to the maximum allowable correlation error. An overly conservative (small) value of the threshold results in an accurate but incomplete reconstruction. On the other hand, a large threshold yields a more complete reconstruction, but one that includes some erroneous voxels. In practice, the threshold should be chosen according to the desired characteristics of the reconstructed model, in terms of accuracy vs. completeness. The algorithm stops when all the remaining opaque voxels are photo-consistent. When these final voxels are assigned the colors they project to in the input images, they form a model that closely resembles the scene.

The voxel coloring approach yields excellent volumetric reconstruction provided the scene is colorful enough. However, the approach reconstructs only one of the potentially numerous scenes consistent with the input images. Consequently, it is susceptible to aperture problems caused by image regions of near-uniform color. These regions cause cusps in the reconstruction, since voxel coloring yields the reconstruction closest to the camera volume. Thus, reconstruction of regions with similar colors is biased.

## *25.2.3 Space Carving*

The ordinal visibility constraint imposes a significant limitation on allowable camera configurations in the voxel coloring approach. In particular, a widely used configuration in which the cameras surround the scene is not handled by this algorithm, as such an arrangement will not yield a near-to-far ordering of voxels relative to the camera volume. In the absence of such an ordering of voxels, there is no guarantee that the visibility of a voxel will not change after it has been checked for photo-consistency once. Therefore, algorithms that allow arbitrary camera placements must allow for multiple passes through the voxels.

The space carving approach of Kutulakos and Seitz [497] is a generalization of the voxel coloring approach for arbitrary camera configurations. Space carving is a multi pass approach that makes multiple-plane sweeps, typically along the positive and negative directions of each of the $X$, $Y$, and $Z$ axes. As in voxel coloring, during each sweep a plane of voxels is evaluated for photo-consistency. Photo-consistency is tested using only the cameras and other voxels that are on one side of the plane that includes the given voxel. By using only the subset of cameras that lie on one side of the sweeping plane, voxels will always be visited in an order that ensures that the occlusion relations for the voxel are already known for that subset of cameras. Figure 25.2(b) illustrates this, wherein for the current sweeping plane, only cameras that lie above it are used in the consistency check. Thus, when a voxel is evaluated, the transparencies of voxels that might occlude it from the cameras currently being used is known. Therefore, its photo-consistency may be easily evaluated from the set of cameras in which it is visible. If the voxel is inconsistent, it is carved out. Multiple iterations of the plane sweeps are necessary until no non-photo consistent voxels can be found on the surface of the carved volume and the process is terminated.

At each iteration of the multi plane-sweep traversal of voxels, only those voxels that are photo-inconsistent in the subset of selected cameras are removed. If the test for photo-consistency is such that if a voxel is photo-inconsistent in a subset of views, then it will also be photo-inconsistent in the entire set of views, then the algorithm will never carve out voxels that would be photo-consistent in the final model. Because carving is conservative, the set of uncarved voxels produced by the algorithm is a superset of any other photo-consistent model. This superset of all photo-consistent volumes of the set of images is termed the *photo hull* and this encloses the true shape of the object.

An alternative to this multi sweep approach used by space carving is the generalized voxel coloring algorithm of Culbertson et al. [239]. This algorithm simply iterates over all the boundary voxels of the scene, checking for their photo-consistency and removing those voxels that are not photo-consistent. Iterations are carried out until no change occurs in a complete pass. The main difference is that no specific plane is being swept in the scene. Without the plane sweep constraint the test of visibility of a voxel is more complicated. To improve its efficiency, the algorithm maintains a data structure, that stores, for every pixel in the image, the surface voxel that is visible along the pixel's visual ray. Thus, visibility may easily be determined by checking to see if this visible voxel matches the projection of the voxel in question. When a voxel is carved out, this data structure needs to be updated as the visibility of other voxels will also change. This update operation will result in correct visibility but will be considerably slower. Fortunately, because carving is conservative, the data structure can be updated less frequently, and the resulting out-of-date visibility can still be used for carving at the possible cost of additional iterations.

### *25.2.4 Probabilistic Approaches*

A common problem underlying the approaches discussed until now is that they make hard decisions in carving away voxels. Therefore, if a voxel is carved away in error, there is no way to recover this voxel at a later step and this leads to a cascading effect, thereby generating large errors in reconstruction. This is manifested in the final 3D model as large holes. Also, space carving requires the user to specify a global variance threshold for performing the photo-consistency check of the projected voxels (a voxel is consistent if the variance of its projected colors is less than this global threshold). A small threshold leads to incomplete reconstructions, whereas larger thresholds result in more errors. These two shortcomings of the space carving algorithm can be addressed in a probabilistic framework that does not make hard decisions.

### *25.2.5 Probabilistic Space Carving*

Broadhurst et al. [117] have proposed a probabilistic extension of the space carving algorithm for the case where the cameras satisfy the ordinal visibility constraint and therefore the images can be processed in a single sweep. Most of the ideas from the original space carving framework are retained but, significantly, the existence of a voxel is not a binary function anymore. Instead, each voxel is assigned a probability that it belongs to the true 3D surface. The voxel array is processed using the plane sweep algorithm, starting with the plane closest to the cameras. The probabilities of the planes prior to the current plane are used to determine visibility for voxels in the current plane. This is then used to compute the probability of the current plane of voxels by comparing the likelihoods for the voxel being opaque and transparent using the Bayesian framework.

The algorithm uses two models to describe a voxel. The first model describes what the projection of a voxel looks like in the image, and the second model describes what an image looks like when a voxel is removed. When a voxel exists, its projections in all the images are modeled by a spherical Gaussian distribution in RGB space, provided, of course the voxel is visible in that view. The probability of a voxel being visible in a particular view is computed using the probabilities of all the voxels in the line of sight. The second possibility is that the voxel is transparent. In this case when the transparent voxel is projected into each of the images, the image samples will have actually arisen from different voxels. In this case, it is assumed that each sample is locally independent and a transparent voxel is represented by a set of independent models (one for each image).

The entire scene is traversed in a single plane sweep. During the sweep, for each voxel, the probability of a voxel being opaque is determined using Bayes' theorem. The voxel and independent models are used to compute the likelihoods of the data given the models and Bayes' rule is used to compute the probabilities of the voxel.

### *25.2.6 Roxels: Responsibility Weighted Voxels*

The framework developed by Debonet and Viola [253] is also probabilistic. They use a probabilistic framework to represent voxels with partial opacity. Consequently, this approach is able to reconstruct opaque as well as transparent objects. In addition, this algorithm allows arbitrary camera placement and is therefore more general than the previous algorithm.

The Roxel algorithm assigns colors and opacities to a uniform voxel space. The key observation of this algorithm is the fact that the observed pixel intensity is a weighted linear combination of the colors along the ray, and the weights are a function of the voxel transparencies. These weights are termed as the responsibility of a voxel for the observation at that pixel. The Roxel algorithm alternates between estimation of the colors, estimation of responsibilities, and estimation of opacities. The voxel colors can be computed from the images and the voxel responsibilities by inverting this linear system. Symmetrically, the voxel colors and images can be used to compute the responsibilities. Finally, the responsibilities can be used to compute the opacities and vice versa.

It is assumed that initially each voxel along a cast ray is equally responsible for that pixel. The entire procedure is repeated until the global opacity estimate converges. At convergence, global color and transparency are extracted and combined to form the final semi transparent voxelated space that accurately reflects the constraints provided by the input image viewpoints, the positions and shape of both solid and transparent objects, and the uncertainty that remains.

### *25.2.7 Probabilistic Surface Reconstruction*

The probabilistic space carving algorithm presented in Section 25.2.5 can be applied to image sequences that can be processed in a single sweep. In other words, the cameras must satisfy the ordinal visibility constraint, which is too restrictive. The Roxel algorithm allows arbitrary placement of the cameras and represents transparency with uncertainties. However, in real situations we generally encounter opaque objects. Our probabilistic surface reconstruction algorithm [9] reconstructs an opaque scene from an arbitrary set of cameras. Our algorithm is iterative and estimates the probability that a scene point lies on the true 3D surface. This is done by explicitly estimating the probabilities that a 3D scene point is visible in a particular view.

The key idea behind this algorithm is the visibility constraint that is inherent in the scene. For a consistent viewing of an opaque scene, the following two properties must be satisfied for all points in the scene.

1. If a scene point is occluded from one view, then there must be another surface point along the ray joining that scene point to the camera center of that view.

2. Conversely, if a surface point is visible in a view, then there cannot be another surface point along the ray joining the camera center to that surface point.

The algorithm uses these two visibility constraints in all the views simultaneously to refine the probability that a scene point lies on the true surface, in an iterative manner.

To apply these viewing constraints, we must determine the visibility of a voxel. We use the pixel intensities of the projected images to obtain estimates for whether or not a surface point is occluded in a particular view. Under the assumption that the scene to be reconstructed is approximately Lambertian, if the $3D$ point $X$ is not occluded in views $i$ and $j$, then the pixel intensities at the projections must match, that is, $I_i(x_i) \approx I_j(x_j)$. Interpreted differently, if the absolute value of the difference in pixel intensities $\delta = \|I_i(x_i) - I_j(x_j)\|$ is large, then it is highly probable that $X$ is occluded in one of the views $i$ and $j$. The converse, however, is not true, unless of course each 3D point is uniquely colored. That is to say, even if $\|I_i(x_i) - I_j(x_j)\|$ is 0, it is possible that $X$ is occluded in one of the views. Therefore, the probability of a voxel being visible in two views is a function $f$ of the pixel intensity difference $\delta$, where $f$ must satisfy the following two properties.

1. $f$ should be high for small values of $\delta$ and should decrease as $\delta$ increases.

2. For small values of $\delta$, the value of $f$ should reflect the uncertainty that exists on whether or not $X$ is visible if $\delta$ is small.

For a voxel to be visible in a subset of the views, it must be visible in all such pairs of views. Under the assumption of independence, the probability of this event can be obtained by multiplying the probabilities of visibility of each such pair. Furthermore, we assume that each point in the scene is visible in at least $V$ views. Knowledge about the camera placement can be used to obtain a conservative estimate of $V$ . In the worst case, $V$ can be safely assigned a value of two. This reflects the fact that only those scene points visible in at least two images can be reconstructed. Therefore, the probability that a voxel is visible in a particular view can be determined from the subset of $V$ views that includes that particular view and has the maximum probability of visibility.

The algorithm is iterative. Starting with a distribution of probability that is high for many scene points, including those which are not surface points, our algorithm uses the visibility constraints to reduce the probabilities of the non-surface points, and, at the same time boost the probabilities of the true surface points. So in the end, the non-surface points are "carved" away. A probabilistic measure $R(X)$ is introduced that measures how well $X$ satisfies the visibility constraints, which is then used to update the probabilities for the next iteration. By constraint 1, all points farther away from the camera along the line of sight than the current location should be either a non surface point or a surface point that is not visible along the viewing ray. The probability of this event is given by the sum of the probabilities of these two events Similarly, by constraint 2, all points along the

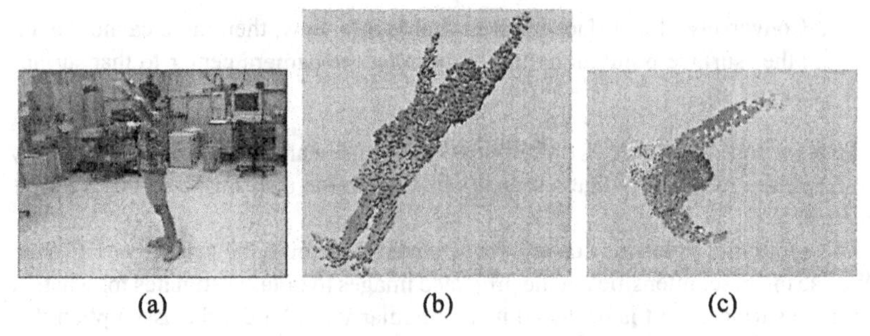

Figure 25.3. Results of probabilistic surface reconstruction algorithm: (a) one of the 14 input images, (b) and (c) two rendered views of the VRML model

viewing ray that are closer to the camera than $X$ must be carved away, provided $X$ is visible in that view.

These two constraints are then translated into probabilities. Both the constraints must be simultaneously satisfied for a surface point to be a visible surface point in a particular view. Therefore, the evidence for $X$ being a visible surface point in view $i$, $E_i(X)$ is simply the product of these two probabilities of satisfying the constraints. Once the evidence $E_i(X)$ for each voxel and viewing direction are computed, they are then scaled so that the maximum $E_i(X)$ along each viewing direction is 1. This converts the absolute probabilities $E_i(X)$ to relative probabilities $R_i(X)$ and accounts for the fact that along each viewing ray, there must be one surface point that is visible along that ray. Since a voxel is visible in at least $V$ images, we can sort the $R_i(X)$ and multiply the $V$ largest values to obtain the overall uncertainty $R(X)$.

Using these relative probabilities $R(X)$ the probabilities of a voxel being a surface point are updated using Bayes' rule, wherein the probabilities of the previous iteration are taken as the prior. The iterations are initialized from probabilities of visibility computed with the assumption of visibility in a minimum of $V$ views. The subset chosen is one with the highest probability of visibility. The algorithm converges in about 20 to 30 iterations. At each iteration of the algorithm, voxels with maximum probabilities along each viewing ray represent the reconstructed surface for that iteration, and the color for each voxel is determined as the average color of its projection in all the viewing directions in which it is not occluded. Most surfaces in the real world are smooth almost everywhere, except at surface discontinuities. At each iteration, we take this into account by considering a small $3D$ window centered at each voxel and then replacing the probability by the average probability in that window. Figure 25.3 shows the results of applying this algorithm on fourteen views of a human subject captured using synchronized color cameras placed on the four walls of a room.

### *25.2.8 Probabilistic Image-Based Stereo*

The probabilistic approaches described until now reconstruct a volumetric model of the scene. These approaches use a discretized volume and reconstruct the volume model to a predefined accuracy; for example, the number of octree levels is fixed from the start. However, the space requirement in a volume-based representation is cubic. Thus, for large images, a volume-based representation may not be suitable because of speed and memory requirements. Stretcha et al. [768] present a probabilistic stereo algorithm that estimates the depth map representation for each input image. These depth maps are relative to the positions and view directions of the cameras, and can later be integrated into a single model.

Occlusions are handled through a set of visibility maps. For each pair of views, the visibility map for the first image indicates whether a pixel in the first view is visible in the second image and likewise for the visibility map of the second image. Each input image is regarded as a noisy measurement of an unknown image irradiance or 'true image'. One of the views is taken as a reference view, and the depth map is computed in the reference map attached to this camera. The problem is formulated as one of estimating the true image and the depth map. This is solved in a Bayesian framework, wherein the visibility maps are regarded as the hidden or unobservable variables that must also be solved for during the course of the optimization.

In the Bayesian framework, this is accomplished by maximizing the posterior probability of the unknown quantities given the images. This requires that the unknown variables be integrated over all possible values of the visibility map, which is computationally intractable. Instead, it is assumed that the probability density function for the visibility is centered about a single value, which is then estimated iteratively from the current values of the unknown variables. This leads to an Estimation-Maximization (EM) based solution, which iterates between (i) estimating values for the visibility maps, given the current estimate of the true image, its noise, and the depth map, and (ii) maximizing the posterior probability of the unknown variables, given the current estimate of the depth maps. For the maximization step, given the visibility maps, it is straightforward to update the true image and its noise from visible views. The depth map is updated by performing a gradient descent on an energy that assumes locally smooth prior depths. In the Expectation step, the visibility maps are computed photometrically, given the current estimates for the depth maps, color model, and noise. The depth map is used to find the corresponding pixel in the other image. The color model and its variance is then used to estimate the likelihood that a pixel is occluded or visible in the other view based on the observed color difference between the projected pixel and the voxel color. This is very similar to the visibility determination in the probabilistic surface reconstruction algorithm discussed in the previous section.

These two steps of the EM algorithm are carried out until convergence, at which we obtain the true image and the depth estimates. The algorithm has few free parameters, displays a stable convergence behavior, and generates accurate depth estimates. The algorithm has been applied to several real-world wide-baseline image pairs, producing impressive results. As it is an image-space approach, the true image, and the depth map can be used to perform view interpolation, resulting in a high-quality walkthrough of the scene.

## 25.3 Reconstruction of Dynamic Scenes

Dynamic scenes may be considered at different levels of complexity. One naive approach would simply involve performing full 3D reconstruction for each new frame. Such an approach, however, can be quite time-consuming, which might make the method impractical for real-time applications and does not utilize the temporal information available in the data. In this section, we will mostly describe methods that utilize such temporal continuity in order to improve the reconstruction results. These methods may also be categorized according to the precision of the desired reconstruction and the models and assumptions used. Such considerations affect the quality of the reconstruction and the computational time taken. Some methods attempt to obtain very accurate 3D reconstructions at the expense of computational time, while others obtain approximate reconstructions for the sake of real-time performance.

### *25.3.1 Visual Hull Algorithms*

Algorithms describing the visual hull reconstruction of static scenes have already been described in Section 25.2.1. One may obtain the visual hull of dynamic scenes by application of this visual hull reconstruction algorithm at each time instant separately. For real-time applications, this is feasible if the visual hull algorithm is fast enough. Borovikov and Davis [104] describe methods to achieve real-time voxel reconstruction for dynamic scenes using a distributed architecture. The local memory and network bandwidth requirements are reduced by hierarchical flow of data in the multi processor system. They utilize an efficient method to process octrees using a depth-first-search (DFS) order of traversal through the nodes. These features make the system extremely efficient for real-time volume reconstruction, and hence the visual hull algorithm can be applied to reconstruct dynamic scenes.

### *25.3.2 Approximate 3D Localization of Targets for Surveillance*

In the context of surveillance applications, one is often not required to obtain very detailed 3D reconstruction of objects in a scene. Approximate localization of targets along with extraction of simple appearance models for matching and tracking across time are generally sufficient. This task simplification is also necessitated by the requirement of real-time performance and often the coverage of a large area reducing the image resolution available for each target.

Different requirements can be considered for such multi sensor surveillance systems. Some of the trade-offs involved include visibility in a crowded scene, accuracy of object localization, running time of the algorithm, and coverage area. Depending on the system requirements, different sensor configurations can been considered.

One scenario of using multiple cameras involves a wide area with relatively sparse objects and little inter-object occlusions [215, 758]. In this case, one camera is typically sufficient for observing one part of the scene and an overlap is required for hand-off between the detections of different cameras. Multiple cameras may also be employed for the same part of the scene in order to deal with occlusions from static obstacles such as trees and buildings. The system developed at CMU under the DARPA VSAM project [215] is an example of such a system. The objective in this system was to cover a large area and detect and track objects over large distances as the object moves in the scene. Pan/tilt/zoom cameras are then utilized to follow and zoom onto detected objects. Each camera detects objects in its field of view. Such objects are then classified using neural networks and linear discriminant analysis.

Such detections are then brought into a common 3-D coordinate system by the use of geodetic coordinates that utilize the latitude, longitude, and elevation with respect to the WGS84 datum (so-called "GPS coordinates"). This allows easy integration of all the detections in a common frame of reference. To determine 3D locations of objects, wide-baseline triangulation is utilized when the views of multiple sensors overlap. When there is no overlap between the views, domain constraints are needed. If the assumption is used that the object is in contact with the domain, one can determine the contact location by passing a viewing ray through the bottom of the object in the image and intersecting it with a model representing the terrain. If a scene plane is available, it can be utilized. However, large outdoor scenes may contain significantly varied terrain. To handle these situations, geolocation is performed using ray intersection with a full terrain model provided by a georeferenced digital elevation map (DEM). Tracking is performed by generating a hypothesis for object location and comparing such a 3D hypothesis to the detections from each camera, using a variety of factors including geometrical proximity, object classification and color-histogram-based appearance models. The best matched detection is then assigned to a tracked object, and split/merge/enter/exit schemes are used to alter the number of objects being tracked. The tracking methodology is very similar to the original paper on Monte Carlo-based tracking by Isard and Blake [420].

When the cameras look at the same scene from different viewpoints, two different approaches can be considered. The approach that many such systems [245] take is to sacrifice visibility for matching accuracy by using "stereo" pairs of sensors. Stereo matching is performed within each pair of cameras, and matched points are reconstructed in 3D by triangulation. Then, the 3D information is integrated across such stereo pairs in a global coordinate system. Some systems perform such integration in 3D space by clustering 3D triangulated points into people-shaped blobs. However, most systems [245, 578] assume that objects are upright and moving on a ground plane and perform such clustering in an orthographic vertical projection The *plan-view* image thus created simplifies and speeds up the correspondence in time since only a 2D search is required. Detection and tracking are then performed in this plan-view image. The tracking in these systems is often facilitated by the use of automatically developed appearance models consisting of color histograms obtained from the detected blobs.

The alternate approach is to utilize wide-baseline cameras for increasing visibility of the cameras. When occlusion is moderate, one can consider an approach where detections are performed independently in each view and such detections are simply merged across views in a consistent manner without regard to any appearance constraints [461]. Geometric and temporal constraints are used to match the trajectories of such detections.

When occlusion is significant from any given viewpoint and one wishes to maximize the visibility from a given number of sensors by placing sensors as far away as possible, one requires more sophisticated reasoning that combines the visual and geometric constraints in a unified framework. Our work on this topic [578] addresses this scenario. In particular, we have addressed the problem of automatically segmenting, detecting, and tracking multiple people in multi perspective video where the scene being viewed is sufficiently "crowded" that one cannot assume that any or all of the people in the scene would be visually isolated from any vantage point. Figure 25.4 shows images of a typical scenario captured from 6 views (only 2 views are shown in the figure due to space limitations).

The system handles the case of partial occlusions by explicitly segmenting the foreground region belonging to different people. Bayesian classification is used and a probabilistic scheme is used for setting priors in such a procedure. The scheme, which assumes knowledge of approximate shape and location of objects, dynamically assigns priors for different objects at each pixel so that occlusion information is encoded in the priors.

The image segmentations thus obtained are utilized by a region-based stereo algorithm that is capable of finding 3D points inside an object if the regions belonging to the object in two views are known. No exact point matching is required. This is especially useful in wide-baseline camera systems where exact matching is very difficult due to self-occlusion and a substantial change in viewpoint.

Rather than performing inference in a single view, the system combines the evidence gathered from different camera pairs using occlusion analysis so as to obtain a globally optimum detection and tracking of objects. Higher weight is given to those pairs having a clearer view of a location than those whose view is potentially obstructed by some objects. The weight is also determined dynamically and uses approximate shape features to give a probabilistic answer for the level of occlusion.

Good segmentation of people in a crowded scene is facilitated by models of the people being viewed. Unfortunately, the problem of detecting and finding the positions of the people requires accurate image segmentation in the face of occlusions. Therefore, we take a unified approach to the problem and solve both of them simultaneously. The algorithm uses segmentation results to find people's ground

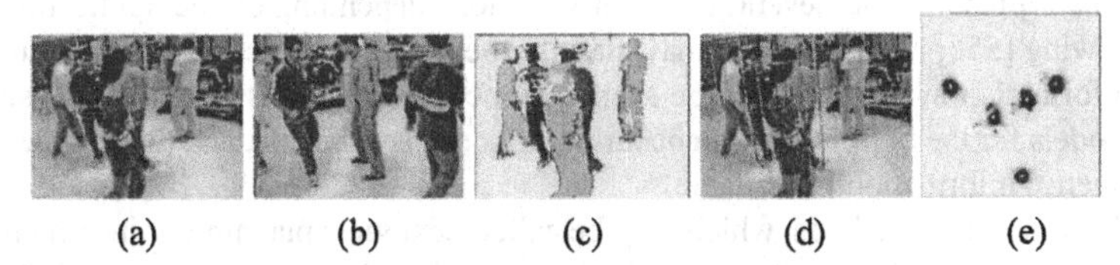


Figure 25.4. Results of our algorithm [578] in a crowded scene: (a) and (b) images from a 6-perspective sequence at a particular time instant, (c) result of segmentation of image (a) using our system, (d) result of detection and tracking as seen from image (a), (e) plan-view likelihood map obtained at this instant

plane positions and then uses the ground plane positions thus obtained to obtain segmentations; the process is iterated until the results are stable. This helps to obtain both good segmentations and ground plane position estimates simultaneously. Some results from this system are shown in Figure 25.4.

Finally, there are methods that try to achieve detection and tracking of multiple occluding objects from a single view or a few views [423, 668, 920]. The idea is to develop appearance and motion models of objects while they are visible in order to predict their trajectories when they are not visible from any view. When the objects become visible again, they are matched to the lost objects based on their appearance and motion characteristics. Such trajectory matching is often performed in a Bayesian sense by utilizing Monte Carlo samples to estimate the distribution of the state of the system at any given time. Although some significant progress has been made in this area, the problem is quite hard due to missing data. Thus, all such algorithms inevitably give inaccurate results when the object density is high, leading to inter-object occlusions and cases where the objects appear to be very close to each other.

## 25.4 Sensor Planning

Until now, we have described algorithms for 3D reconstruction of both static and dynamic scenes for varying model complexity. Another important, although relatively less researched factor, that affects the performance of any reconstruction method is the placement of the sensors for acquiring the best possible data suitable for the method. In this section, we address the problem of optimal placement of sensors for such systems.

As described in the previous section, different systems have different requirements and may vary according to several characteristics: wide-baseline cameras for better triangulation and visibility vs. short-baseline cameras for better matching, accuracy of reconstruction vs. the coverage of any object, non overlapping cameras for maximum coverage vs. overlapping cameras for high-density areas, and so on. Optimal sensor planning is a requirement for all of such systems.

Sensor planning has several different variations depending on the application. Following [556] and [791], one may classify these methods based on the amount of information available about the scene: (1) no information is available, (2) set of models for the objects that can occur in the scene is available, and (3) complete geometric information is available.

The first set of methods, which may be called next view planning or incremental scene reconstruction, attempts to build a model of the scene incrementally by successively sensing the unknown world from effective sensor configurations using the information acquired about the world up to this point [648, 556, 496]. The sensors are controlled based on several criteria such as occlusions, ability to view the largest unexplored region, and ability to perform good stereo matching. Such constraints are translated into constraints on the camera positions, and satisfaction of these constraints guarantees optimum and stable acquisition. The second set of methods assumes knowledge about the objects that can be present in the scene. The task, then, is to develop sensing strategies for model-based object recognition and localization [906].

The third set of methods assumes that complete geometric information is available and determines the location of static cameras so as to obtain the best views of a scene. The objective is either to detect the dynamic objects in the scene or to recover the appearance characteristics of the static parts. This problem was originally posed in the computational geometry literature as the "art-gallery problem" [614]. The traditional formulation of such problem requires only one camera to view any part of the scene and utilizes the simple assumption that two points are called visible if the straight line segment between them lies entirely inside the polygon. Even with such simple definition of visibility, the problem is NP-complete. The reader is referred to [614] for a survey of work done in this area.

Several recent papers have incorporated more complicated constraints such as incidence angle and range into the problem and obtain an approximate solution to the resultant NP-complete problem via randomized algorithms [354]. Several others [230, 671, 792, 556] have studied and incorporated more complex constraints based on factors such as resolution, focus, field of view, visibility, view angle, and prohibited regions. The set of possible sensor configurations satisfying all such constraints for all the features in the scene is then determined.

In addition to the "static" constraints that have been considered so far, there are additional constraints that arise when dynamic obstacles are present. Our work in this area has focused on analyzing visibility constraints in the presence of random dynamic obstacles, and maximization of system performance given task specification.

The visibility analysis probabilistically determines the visibility rate of objects at different locations, given that visibility from even one or a few (two in the case of stereo matching) sensors may be sufficient. Stated differently, the probability that the object is visible from at least one (or two for stereo matching) sensor is

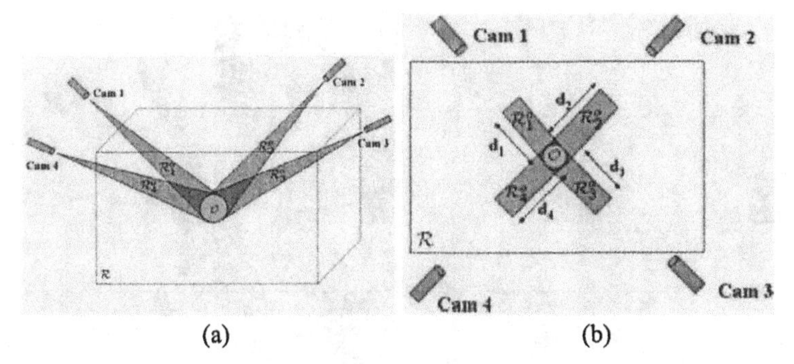


Figure 25.5. Scene geometry: (a) 3D case, (b) 2.5D case, where the sensors have finite heights

evaluated at all possible locations. Such analysis assumes a random occurrence of objects in a region according to a density function and then evaluates the probability that no such object appears in a region of occlusion where the occurrence of another object would cause the target object to be occluded from a given camera, as shown in Figure 25.5.

The multi-view probabilistic analysis is then combined with several other static constraints such as image resolution, stereo matching, field of view, and background scene. An inherent difficulty in the integration of such constraints is the trade-off that is typically involved between different constraints. For instance, a reduction in the distance from the camera enhances resolution, but might increase the viewing angle from the camera and cause difficulties in stereo matching, or may cause a part of the object to go out of the camera field of view. We have proposed a generic formulation that integrates a variety of such constraints and trade-offs in a single quality measure according to user requirements and also utilizes the multi-view visibility constraints in a natural way. Integration of such a quality measure over a given region of interest leads to the development of a cost function that can then be minimized for efficient sensor planning. Since exact optimization of such criteria is an NP-hard problem, methods are proposed that yield "good" configurations for most cases. Customization of the method for a given system allows the method to be utilized for a variety of different tasks and applications. Figure 25.6 illustrates the result of such sensor planning for some example scenes.

## 25.5 Conclusion

Reconstruction of scenes from multiple cameras has made significant progress over the past few years. We presented a brief overview of the state of the art in multi-view reconstruction of static and dynamic scenes. We focused on algorithms that utilized the wide-baseline camera arrangement and modeled occlusions explicitly. Such algorithms can be either based on pure geometric intersections

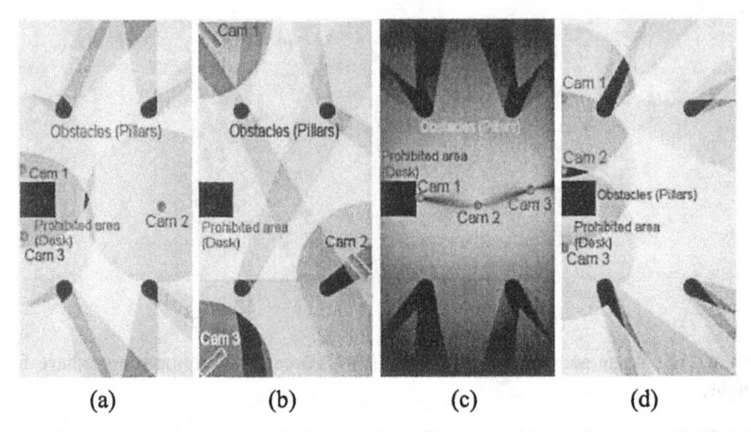


Figure 25.6. Sensor planning results. Optimum configuration: (a) omni-cameras (360° field of view), (b) field of view restricted to 90°, (c) a stereo requirement with omni-cameras, (d) no visibility with the left wall as background (object color matches the left wall)

(visual hull algorithm) or on the concept of photo-consistency (voxel coloring and space carving). In particular, algorithms based on a probabilistic framework are receiving more attention these days as they tend to produce superior reconstruction results. Several such probabilistic algorithms were described in detail.

Next, we considered the reconstruction of dynamic scenes. The main improvement over frame-by-frame reconstruction is the utilization of the temporal continuity constraints existing in such scenes. In this context, we described efficient visual hull methods for detailed reconstruction as well as fast but approximate target localization methods for multi-view surveillance applications. Finally, we addressed the problem of planning the placement of the sensors so that the data is acquired in an optimal manner for a given reconstruction task.

# Chapter26

# Graph Cut Algorithms for Binocular Stereo with Occlusions

**V. Kolmogorov and R. Zabih**

Abstract

Most binocular stereo algorithms assume that all scene elements are visible from both cameras. Scene elements that are visible from only one camera, known as occlusions, pose an important challenge for stereo. Occlusions are important for segmentation, because they appear near discontinuities. However, stereo algorithms tend to ignore occlusions because of their difficulty. One reason is that occlusions require the input images to be treated symmetrically, which complicates the problem formulation. Worse, certain depth maps imply physically impossible scene configurations, and must be excluded from the output. In this chapter we approach the problem of binocular stereo with occlusions from an energy minimization viewpoint. We begin by reviewing traditional stereo methods that do not handle occlusions. If occlusions are ignored, it is easy to formulate the stereo problem as a pixel labeling problem, which leads to an energy function that is common in early vision. This kind of energy function can be minimized using graph cuts, which is a combinatorial optimization technique that has proven to be very effective for low-level vision problems. Motivated by this, we have designed two graph cut stereo algorithms that are designed to handle occlusions. These algorithms produce promising experimental results on real data with ground truth.

## 26.1 Traditional stereo methods

Computing stereo depth is a traditional problem in computer vision, and has been the focus of a great deal of work (see [120, 711] for recent surveys). Given a pair of images taken at the same time, two pixels are said to correspond if they show the same scene element. The goal of stereo is to compute correspondences between pixels, which then determines depth. The binocular stereo problem is typically formulated as follows:

For every pixel in one image, find the corresponding pixel in the other image.

We will refer to this as the *traditional stereo problem.*

The problem formulation above has many advantages. It easily fits within a class of problems that arise in early vision called *pixel labeling problems,* where the goal is to assign each pixel $p = (p_x, p_y) \in \mathcal{P}$ a label from some set $\mathcal{L}$. The label set $\mathcal{L}$ depends upon the particular problem; for example, in image denoising, $\mathcal{L}$ is intensities. In stereo, $\mathcal{L}$ consists of disparities.

Pixel labeling problems have been widely studied in computer vision. The problem is naturally formulated in terms of energy minimization, where the goal is to find the labeling $f = (f_1, \ldots, f_p, \ldots, f_{|\mathcal{P}|})$ that minimizes

$$E(f) = \sum_p D_p(f_p) + \sum_{\{p,q\}\in\mathcal{N}} V(f_p, f_q). \tag{26.1}$$

Here $D_p$ is the penalty for assigning a label to the pixel $p$; $\mathcal{N}$ is a set of pairs of adjacent pixels, representing a neighborhood system; and $V$ is the penalty for assigning a pair of labels to adjacent pixels. The first term of equation 26.1 gives a data cost for $f$, which requires $f$ to respect the observed data, while the second term imposes spatial smoothness. Note that this energy function has an elegant connection to the probabilistic framework provided by Markov Random Fields [520], where the first term comes from the likelihood and the second comes from the prior.

The traditional stereo problem can be easily formulated as a pixel labeling problem. We will assign the label $f_p$ to the pixel $p$ when the pixel $p$ in one image $I$ corresponds to the pixel $p + f_p$ in the other image $I'$. (Note that the set $\mathcal{P}$ consists of pixels in $I$.) The *matching penalty* $D_p$ will enforce photoconsistency, which is the tendency of corresponding pixels to have similar intensities. The natural form of $D_p$ is $D_p(f_p) = ||I(p) - I'(p + f_p)||^2$.

The smoothness penalty $V$ will depend on what kind of scene geometry we expect. If $V$ gives too large a penalty for very different $f_p, f_q$, the solution will tend to oversmooth. With fronto-parallel scenes, the natural choice is $V(f_p, f_q) = \lambda \cdot T[f_p \neq f_q]$, where the indicator function $T[\cdot]$ is 1 if its argument is true and otherwise 0. This choice of $V$ is referred to as the Potts model. There are also more complex forms of $V$ that naturally handle slanted or curved surfaces [88, 113, 521] (surprisingly, these often rely on the Potts model).

The terms $D$ and $V$ can be easily visualized as tables, which are $|\mathcal{L}| \times 1$ or $|\mathcal{L}| \times |\mathcal{L}|$, respectively. For stereo with the Potts model, they are

$$D_p = \begin{array}{|c|}\hline (I(p_x,p_y) - I'(p_x,p_y))^2 \\ \hline (I(p_x,p_y) - I'(p_x - 1,p_y))^2 \\ \hline (I(p_x,p_y) - I'(p_x - 2,p_y))^2 \\ \hline \vdots \\ \hline \end{array} \qquad V = \begin{array}{|c|c|c|c|}\hline 0 & \lambda & \cdots & \lambda \\ \hline \lambda & 0 & \cdots & \lambda \\ \hline \vdots & \vdots & \ddots & \vdots \\ \hline \lambda & \lambda & \cdots & 0 \\ \hline \end{array}$$

This visualization will prove useful when we describe how to minimize the energy function.

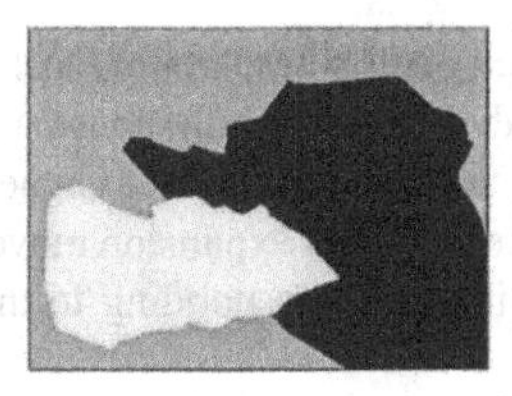 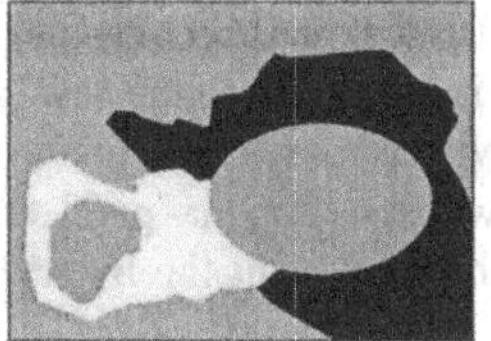 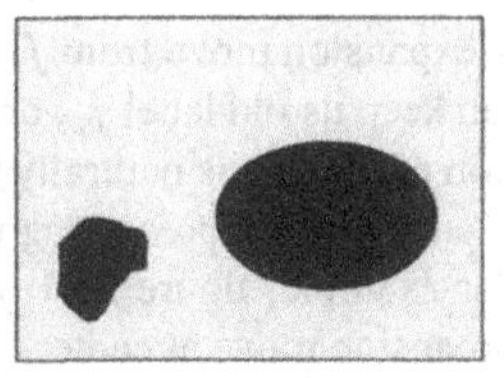

Figure 26.1. Expansion move example. The input labeling is shown at left. An expansion move is shown in the middle, and the corresponding binary labeling is shown at right.

### *26.1.1 Energy minimization via graph cuts*

A major advantage of pixel labeling problems is that they can now be rapidly solved by powerful optimization algorithms such as graph cuts [113, 424]. If the label set $\mathcal{L}$ consists of contiguous integers and if $V$ is a convex function of $f_p - f_q$, then the global minimum of $E$ can be rapidly computed in a single graph cut [424]. However, if $V$ is convex it will give a large penalty for very different $f_p, f_q$, and hence will oversmooth. Any class of smoothness terms that includes the Potts model is NP-hard to minimize [113], so a good local minimum is the best that we can hope to achieve.

If $V$ is a metric on labels, then it is possible to efficiently minimize $E$ using the expansion move algorithm. The Potts model is a metric, as are some other popular choices of $V$ that do not oversmooth [113]. The expansion move algorithm computes a strong local minimum, in a sense that we will describe with more precision shortly. Given a label $\alpha$ and a labeling $f$, another labeling $f'$ is defined to be an $\alpha$-expansion move from $f$ if for every pixel $p$

$$f'(p) \neq f(p) \implies f'(p) = \alpha.$$

Intuitively, $f'$ is obtained from $f$ by assigning the label $\alpha$ to an arbitrary set of pixels. An example of an expansion move is shown in figure 26.1, with $f$ at the left and $f'$ in the middle.

The expansion move algorithm cycles through the labels in some order (fixed or random). For a particular label $\alpha$, it computes the lowest energy expansion move from the current labeling, and moves to that labeling if its energy is lower. This is obviously a greedy algorithm, and terminates with a labeling that is a local minimum with respect to expansion moves. More precisely, when it terminates with a labeling $\hat{f}$ there is no $\alpha$-expansion move from $\hat{f}$ whose energy is lower than $E(\hat{f})$, for any label $\alpha$.

The number of expansion moves from a given labeling is $O(|\mathcal{L}| \cdot 2^{|\mathcal{P}|})$ (recall that $|\mathcal{P}|$ is the number of pixels). It is possible to prove that the energy of a local minimum with respect to expansion moves lies within a fixed multiplicative factor of the energy of the global minimum. The factor is at least 2, and depends on the exact form of $V$ (see [113] for details).

The key challenge in the expansion move algorithm lies in solving the following subproblem: given a labeling $f$ and a label $\alpha$, find the lowest energy

$\alpha$-expansion move from $f$. In an expansion move, each pixel $p$ has two options: it can keep its old label $f_p$, or it can switch to the new label $\alpha$. As a result, an expansion move can be naturally viewed as a binary image; there is a single bit assigned to each pixel, representing which option that pixel selects in this expansion move. For example, figure 26.1 shows at right the binary image corresponding to the expansion move at center.

We can thus view the problem of finding the lowest energy expansion move as an energy minimization problem over binary images. To formalize this, consider a binary image $\mathbf{x} = \{x_p \mid p \in \mathcal{P}\}$. The labeling associated with $\mathbf{x}$, given an initial labeling $f$ and a label $\alpha$, will be $\alpha$ at pixels where $\mathbf{x}$ is 1, and the same as $f$ elsewhere. We will write this labeling as $f^\alpha[\mathbf{x}]$. The problem of finding the lowest energy expansion move is to find the $\mathbf{x}$ that minimizes $E(f^\alpha[\mathbf{x}])$, given $f$ and $\alpha$.

We can now rewrite the energy $E$ as a new energy function $\mathcal{E}(\mathbf{x})$, where $\mathcal{E}(\mathbf{x}) = E(f^\alpha[\mathbf{x}])$. The new energy function is defined on binary images, and is given by

$$\mathcal{E}(\mathbf{x}) = \sum_p \mathcal{E}_p(x_p) + \sum_{p,q} \mathcal{E}_{p,q}(x_p, x_q)$$

Just as before, the two terms can be visualized as tables, where

$$\mathcal{E}_p = \begin{array}{|c|}\hline D_p(f_p) \\ \hline D_p(\alpha) \\ \hline\end{array} \qquad \mathcal{E}_{p,q} = \begin{array}{|c|c|}\hline V(f_p, f_q) & V(f_p, \alpha) \\ \hline V(\alpha, f_q) & V(\alpha, \alpha) \\ \hline\end{array}$$

The problem of minimizing $\mathcal{E}(\mathbf{x})$ can be solved exactly with a single graph cut as long as $\mathcal{E}_{p,q}$ has a property called *regularity*, introduced in [491]. $\mathcal{E}_{p,q}$ is regular if the sum of its diagonal elements is less than or equal to the sum of its off-diagonal elements; so a sufficient condition is

$$V(\alpha, \alpha) + V(l, l') \leq V(l, \alpha) + V(\alpha, l') \tag{26.2}$$

for any labels $l, l', \alpha$. As long as this condition is met, the general-purpose construction given in [491] can be used to minimize $\mathcal{E}$, and hence to find the lowest energy expansion move. Note that if $V$ is a metric, it clearly satisfies this condition since $V(\alpha, \alpha) = 0$ and so equation 26.2 is just the triangle inequality.

In summary, the traditional stereo problem is a pixel labeling problem. With the appopriate choices of $D_p$ and $V$ it can be formulated as an energy minimization problem. When $V$ is a metric, a strong local minimum can be computed using the expansion move algorithm. This stereo algorithm, due to [113], yields very good experimental results. For example, the majority of the top-ranked methods on the Middlebury stereo database rely on graph cuts [711].

## 26.2 Stereo with occlusions

The traditional stereo problem formulation, however, has some serious disadvantages. First, note that the problem formulation treats the input images asymmetrically, which is unnatural. The pixels to be labeled $\mathcal{P}$ come from the

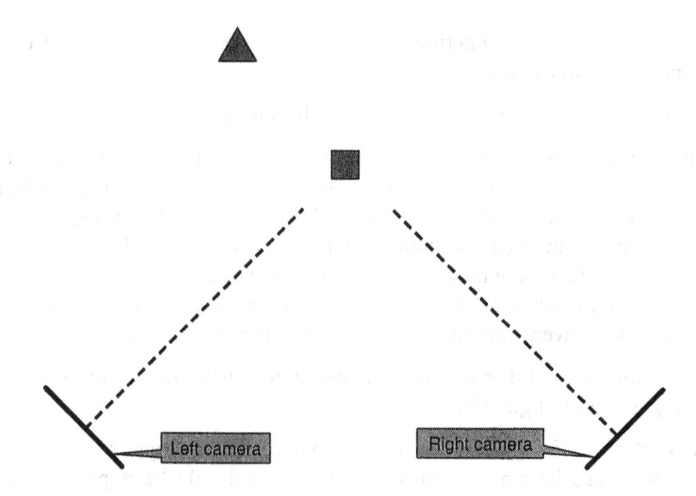


Figure 26.2. It is impossible for the left camera to see the square and the right camera to see the triangle at the same time.

primary image $I$, while $I'$ appears only in the data term $D_p$. Second, by assigning every pixel in $I$ a label, we assume that every scene element is visible in both images. This cannot be true if the scene has more than one depth.[1]

Worst of all, however, certain labelings $f$ imply physically impossible 3D scenes, and hence must be excluded from consideration. This results from the geometry of the imaging process. An example showing this constraint is shown in figure 26.2. A general-purpose pixel labeling algorithm will almost invariably generate solutions that violate these geometric constraints. For binocular stereo, these geometric constraints center on occlusions, which are scene elements that are only visible from one camera.

In this chapter we describe two binocular stereo algorithms that handle occlusions. We take an energy minimization approach, and rely on the expansion move algorithm to minimize the energy. One key challenge is that graph cuts perform *unconstrained* energy minimization [491], while binocular stereo with occlusions requires a *constrained* energy minimization algorithm.

The energy minimization approach to binocular stereo with occlusions consists of following three steps:

- Pick a representation for the problem. In other words, we need to choose the space of valid (physically possible) configurations $C_{valid}$ and define the correspondence between configurations and real scenes.

[1] While it is possible to augment the label set by adding a label that means "this pixel is occluded", this approach does not address the other difficulties of the traditional problem formulation.

- Design an energy function $E : \mathcal{C}_{valid} \rightarrow \mathbb{R}$ that captures the desired properties of a solution.
- Develop an algorithm for minimizing this energy.

Note that these steps are strongly interconnected. With a poor choice of representation, it may be hard or impossible to impose the correct problem constraints. Even if an energy function does captures all the desirable properties, computing a good minimum may be computationally intractable. We will get an effective algorithm only if all three issues are properly addressed.

Ideally, a representation for the stereo problem should have the following properties: for a given configuration it should be easy to determine

(P1) whether it is valid or not (i.e. whether there exists a real scene corresponding to this configuration); and

(P2) what pixels in the left and in the right image correspond to each other. This is crucial since photoconsistency should only be imposed between corresponding pixels.

There are two obvious types of representations for stereo: voxel-style representations, and representations based on labeling pixels. Voxel-style representations rely on an explicit representation of the 3D space that the scene may occupy. Such representations have been used in many approaches, including voxel coloring [722], space carving [498] and silhouette intersection [549]. Pixel labeling approaches include all the standard stereo methods, such as those surveyed in [120, 711].

### *26.2.1 Notation*

We will redefine $\mathcal{P}$ to now be the set of pixels in the left and in the right images (the previous definition was asymmetric). Let $\mathcal{V}$ be the set of (unordered) pairs of pixels that may potentially correspond. For simplicity we assume that images are rectified; then we have

$$\mathcal{V} = \{\langle p, q \rangle \mid p_y = q_y \text{ and } q_x - p_x \in \mathcal{L}\}$$

where $\mathcal{L}$ is the set of possible disparities: $\mathcal{L} = \{0, -1, \ldots, -d_{max}\}$. (We assume that disparities lie in some limited range, so each pixel in the left image can potentially correspond to one of $|\mathcal{L}|$ possible pixels in the right image, and vice versa). We call a pair $v = \langle p, q \rangle \in \mathcal{V}$ a *voxel*. Its disparity is denoted as $d(v)$ (i.e. $d(v) = q_x - p_x \in \mathcal{L}$).

Note that each voxel $v \in \mathcal{V}$ corresponds to a point in 3D space, as shown in figure 26.3. The disparity $d(v)$ directly depends on the depth of this point, i.e. its distance to the cameras. If the cameras are parallel then $-d(v)$ is inversely proportional to the depth. In a more general situation the relationship can be more complicated. In this chapter we assume that disparity is a monotonically increasing function of the depth. In other words, the farther a point from the cameras the larger the disparity.

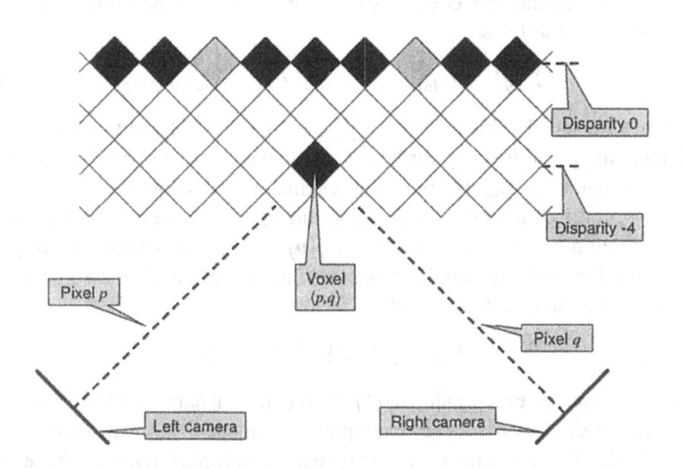


Figure 26.3. Voxel labeling algorithm. Voxels are given a binary label (active or not); dark shaded voxels are labeled as active. The disparity of the voxel $\langle p, q\rangle$ is $d(\langle p, q\rangle) = -4$. To simplify the drawing, orthographic projection is assumed. Note that the two gray-shaded voxels cannot be active if $\langle p, q\rangle$ is active.

For a voxel $v = \langle p, q\rangle$ we can compute the matching penalty $M(v)$ describing how photoconsistent the intensity of pixel $p$ is with the intensity of pixel $q$. The simplest function is the squared difference of intensities: $M(v) = ||I(p) - I'(q)||^2$; however, more elaborate functions (for example, [87]) tend to give better results.

## 26.3 Voxel labeling algorithm

Our first approach, which first appeared in [490], is directly inspired by property P2. A configuration will just be a labeling $g : \mathcal{V} \to \{0, 1\}$ such that $g(v)$ is 1 if the pixels $p$ and $q$ in voxel $v = \langle p, q\rangle$ correspond to each other, and 0 otherwise. In other words $g(v) = 1$ if and only if the the 3D point corresponding to voxel $v$ is present in the scene and is visible from both cameras. If this is the case we will say that $v$ is *active*.

The set of all configurations is $\mathcal{C} = \{0, 1\}^{\mathcal{V}}$. However, not all configurations in $\mathcal{C}$ are valid. Some of them violate the *uniqueness* constraint which says that a pixel in one image can correspond to at most one pixel in the other image. Let us define the set $\mathcal{C}_{valid}$ as follows: the configuration $g \in \mathcal{C}$ is valid if for any two distinct voxels $v$, $v'$ involving the same pixel (i.e. $v = \langle p, q\rangle$, $v' = \langle p, q'\rangle$ with $q \neq q'$) at least one of them has label 0: $g(v) = 0$ or $g(v') = 0$.

Now let us define the energy we will minimize. It has three terms: data, occlusion and smoothness:

$$E(g) = E_{data}(g) + E_{occ}(g) + E_{smooth}(g) \tag{26.3}$$

The data term will be $E_{data}(g) = \sum_{v \in \mathcal{V}} g(v) \cdot M(v)$. Note that this sum contains matching penalties only for voxels $v$ which are active in configuration $g$ (i.e. $g(v) = 1$). The term $E_{occ}(g)$ penalizes occlusions: it is equal to $C_{occ} \cdot |\mathcal{P}_{occ}(g)|$ where $C_{occ}$ is the penalty for an occlusion and $\mathcal{P}_{occ}(g)$ is the set of pixels occluded in configuration $g$ (i.e. pixels $p$ such that $g(v) = 0$ for all voxels $v = \langle p, q \rangle \in \mathcal{V}$).

The smoothness term involves a notion of neighborhood; we assume that there is a neighborhood system on voxels

$$\mathcal{N}_{\mathcal{V}} \subset \{\{v, v'\} \mid v, v' \in \mathcal{V}\}.$$

We require that for every pair $\{v, v'\} \in \mathcal{N}_{\mathcal{V}}$ the disparities of voxels $v$ and $v'$ are the same: $d(v) = d(v')$. For example, we can specify $\mathcal{N}_{\mathcal{V}}$ as follows: voxels $\langle p, q \rangle$,$\langle p', q' \rangle$ with the same disparity are neighbors if pixels $p, p'$ in the left image are 4-neighbors. Now the smoothness term can be written:

$$E_{smooth}(g) = \sum_{\{v,v'\} \in \mathcal{N}_{\mathcal{V}}} \lambda \cdot T[g(v) \neq g(v')].$$

To summarize, the voxel labeling stereo algorithm solves the constrained minimization problem:

$$g^* = \arg \min_{g \in \mathcal{C}_{valid}} E(g), \tag{26.4}$$

where $E(g)$ is defined in equation 26.3.

## 26.4 Pixel labeling algorithm

The representation discussed above might seem natural for stereo correspondence problem since it allows to identify corresponding pixels easily. However, it has several drawbacks. First, the smoothness term involved is rather restrictive — basically, it is the Potts model on voxels (see [492] for more details). Second, the set $\mathcal{C}_{valid}$ contains configurations which do not correspond to any physical scene. Consider, for example, the configuration $g$ with $g(a) = 0$ for every voxel $v \in \mathcal{V}$. Every pixel is occluded in this configuration; thus, the configuration contains "holes". From a practical point of view, we can ensure that we will not get such a configuration by setting penalty for occlusion to a sufficiently large value.

We now describe a different approach, first published in [490], which uses a representation proposed by [451]. We know that each pixel sees some element of the scene (even though this element may not be seen from the other camera). Our goal will be to compute the depth of this pixel (or, rather, its disparity). Thus, a configuration is a mapping $f : \mathcal{P} \rightarrow \mathcal{L}$. The set of all configurations is $\mathcal{C} = \mathcal{L}^{\mathcal{P}}$.

As in the previous case, not all configurations are valid. Formally, the configuration $f \in \mathcal{C}$ is valid if for every voxel $v = \langle p, q \rangle$ the following property holds: if $f(p) = d(v)$ then $f(q) \leq d(v)$. This can be understood intuitively in terms of figure 26.2; if the left camera sees the square, the right camera cannot see any scene element that is behind the square.

Our energy function will be

$$E(f) = E_{data}(f) + E_{smooth}(f). \tag{26.5}$$

Similar to the previous case, we would like the data term to be a sum only over *active* voxels. Let us discuss how we can identify such voxels in this representation. The voxel $v = \langle p, q \rangle$ is active if the corresponding 3D point is present in the scene and is visible from both cameras. This means that $f(p) = f(q) = d(v)$. This in turn motivates the following data term:

$$E_{data}(f) = \sum_{v=\langle p,q \rangle \in \mathcal{V}} [f(p) = f(q) = d(v)] \cdot D(v)$$

where $D(v)$ measures how similar intensities of pixels $p$ and $q$ are. For technical reasons explained in section 26.5 we need the term $D(v)$ to be non-positive. We set $D(v) = \min\{M(v) - K, 0\}$ where $K$ is a positive constant.

The smoothness term is very similar to that of tradional stereo problem, except that it is enforced for both images rather than just the left image:

$$E_{smooth}(f) = \sum_{\{p,q\} \in \mathcal{N}} V(f_p, f_q)$$

where $V$ can be, for example, the Potts model: $V(f_p, f_q) = \lambda \cdot T[f_p \neq f_q]$.

We thus obtain the following constrained minimization problem:

$$f^* = \arg \min_{f \in \mathcal{C}_{valid}} E(f), \tag{26.6}$$

where $E(f)$ is defined in equation 26.5.

## 26.5 Minimizing the energy

In this section we sketch how we solve the constrained minimization problems given in equations 26.4 and 26.6. First, we convert our constrained minimization problems into unconstrained ones. We add a hard constraint term $E_{valid}$ which is zero if a configuration is valid, and infinite otherwise. In the case of the pixel labeling algorithm, for example, the energy becomes

$$E(f) = E_{data}(f) + E_{smooth}(f) + E_{valid}(f).$$

All terms of this energy (including $E_{valid}$) can be written as a sum over pairs of pixels. In other words, the energy has the same functional form as in equation 26.1, only the neighborhood system $\mathcal{N}$ is different and terms $V$ are replaced by some other functions. Moreover, the representation of our problem resembles that of traditional stereo problem (section 26.1). As in traditional stereo, our goal is to assign disparities to pixels; the only change in representation is to consider pixels in both images. Thus, it is easy to adapt the expansion move algorithm described in section 26.1 to our minimization problem. We just need to ensure that for each $\alpha$-expansion the corresponding binary energy function is regular. We show in [490] that this condition holds assuming that terms $D(v)$ are non-positive.

In order to apply the expansion move algorithm to the voxel labeling problem, we need to modify the definition of $\alpha$-expansion. Indeed, the definition given in section 26.1 applies to multi-label variables, while our problem has binary labels. We say that configuration $g'$ is within a single $\alpha$-expansion move from configuration $g$ if voxels which are inactive in $g$ and whose disparity is different from $\alpha$ are also inactive in $g'$. Then for every valid configuration $g$ and disparity $\alpha$ it is possible to compute an optimal $\alpha$-expansion move using graph cuts (see [490] for details).

## 26.6 Experimental results

### 26.6.1 *Implementational details*

**Expansion move algorithm** We selected disparities $\alpha \in \mathcal{L}$ in random order, and kept this order for all iterations. We performed three iterations. (The number of iterations until convergence was at most five but the result was practically the same). The voxel labeling algorithm was initialized with a configuration where every voxel was inactive; the pixel labeling algorithm was initialized with all pixels having disparity zero.

**Matching penalty** For our matching penalty $M$ we made use of the method of [87] to handle sampling artifacts, with a slight variation: we compute intensity intervals for each band (R,G,B) using four neighbors, and then take the average data penalty. (We used color images; results for grayscale images are slightly worse).

**Smoothness terms** We used a Potts model for both algorithms (in one case this is the Potts model on voxels, while in the other the Potts model on pixels). This model is controlled by one parameter characterizing the penalty for a pair of neighboring voxels or pixels. This parameter, however, can depend on the pair. We can use this property to discourage discontinuities between adjacent pixels with very similar intensities. This trick is referred to as "static cues" in [113] and is quite useful for stereo.

For the pixel labeling algorithm we set

$$V_{p,p'}(f_p, f_{p'}) = \lambda_{p,p'} \cdot T[f_p \neq f_{p'}]$$

where $\lambda_{p,p'}$ was implemented as the following empirically selected decreasing function of $\Delta I(p, p')$ (the $L_\infty$ norm of the intensity difference between $p$ and $p'$):

$$\lambda_{p,p'} = \begin{cases} 3\lambda & \text{if } \Delta I(p,p') < 5, \\ \lambda & \text{otherwise.} \end{cases}$$

For the voxel labeling algorithm we used a similar expression:

$$\lambda_{v,v'} = \begin{cases} 3\lambda & \text{if } \max(\Delta I(p,p'), \Delta I(q,q')) < 8, \\ \lambda & \text{otherwise,} \end{cases}$$

where $v = \langle p, q \rangle$, $v' = \langle p', q' \rangle$ and $p$ and $p'$ are pixels in the same image, as well as $q$ and $q'$.

**Choice of parameters** The energy function for the voxel labeling algorithm as defined above depends on two numbers: occlusion penalty $C_{occ}$ and smoothness interaction strength $\lambda$. Similarly, the pixel labeling algorithm depends on the parameters $K$ and $\lambda$. These parameters should be tuned for for different datasets to reflect our prior knowledge about the scene geometry, amount of noise in the images and other factors. Selecting the parameters automatically, however, is a very challenging task.

We set $K$ in the pixel labeling algorithm using a simple heuristic which tries to estimate the amount of noise in the images. Details are given in [488]. It can be shown [488, 492] that $K/2$ approximately corresponds to the occlusion penalty, so for the voxel labeling algorithm we set $C_{occ} = K/2$. Finally, the parameter $\lambda$ was chosen to be proportional to $K$: $\lambda = K/5$.

### 26.6.2 *Algorithm performance*

We have compared three algorithms: our voxel and pixel labeling algorithms with occlusions ("[KZ '01]" and "[KZ '02]") and a traditional stereo algorithm proposed in [113] ("[BVZ]"). The latter technique was found to be the best algorithm for stereo according to [782]. In addition, we tested the algorithms in two modes: with reporting occlusions (some of the pixels in the left image are marked as occluded) and without reporting occlusions (all pixels in the left image are labeled with some disparity).

Determining occluded areas in the voxel and pixel labeling algorithms is easy since they output what pixels correspond to each other. The information produced by [BVZ], however, is not sufficient to determine where occlusions are. To produce occlusions, we have augmented the algorithm: we introduced a new label "occluded" with some fixed penalty.

Note that our voxel labeling algorithm does not produce depths for all pixels. We have filled occluded regions using some postprocessing: we have assigned to occluded pixels the depth label of the closest non-occluded left neighbor lying in the same scanline.

We primarily experimented with images from [711]; output is shown in figures 26.4–26.6. The running times below were obtained on 450MHz UltraSPARC

II processor. We used the max flow algorithm of [110], which is specifically designed for the kinds of graphs that arise in vision.

| stereo pair | image size | number of labels | running times [KZ '01] | [KZ '02] | [BVZ] |
|---|---|---|---|---|---|
| Tsukuba | 384 x 288 | 16 | 69 secs | 80 secs | 35 secs |
| Sawtooth | 434 x 380 | 20 | 115 secs | 141 secs | 66 secs |
| Venus | 434 x 383 | 22 | 145 secs | 159 secs | 85 secs |

First we evaluated the three algorithms in the mode without reporting occlusions. Error statistics using the ground truth from [711] are as follows:

| stereo pair | [KZ '01] | [KZ '02] | [BVZ] |
|---|---|---|---|
| Tsukuba | 5.82 (1.18) | 5.91 (1.86) | 7.17 (1.93) |
| Sawtooth | 12.13 (0.71) | 11.77 (0.67) | 11.86 (0.62) |
| Venus | 15.40 (1.07) | 13.19 (0.69) | 16.90 (0.75) |

We determined the percentage of the pixels where the algorithm did not compute the correct disparity ("errors" — the first number), or a disparity within $\pm 1$ of the correct disparity ("gross errors" — the second number). We counted only pixels that are not occluded according to the ground truth since depth labels of such pixels cannot be determined from the photoconsistency constraint.

We have also computed error statistics for the Tsukuba stereo pair in the mode with reporting occlusions.

| algorithm | Errors | Gross errors | False negatives | False positives |
|---|---|---|---|---|
| [KZ '01] | 6.56% | 2.17% | 41.33% | 1.33% |
| [KZ '02] | 6.51% | 2.66% | 44.16% | 1.03% |
| [BVZ] | 7.28% | 2.14% | 77.59% | 0.62% |

The first two columns count only pixels that are not occluded according to the ground truth. We considered labeling a pixel as occluded to be a gross error. The last two columns show error rates for occlusions.

## 26.7 Conclusions

We have presented two stereo algorithms that handle occlusions. The pixel labeling algorithm can be viewed as an improvement over the voxel labeling algorithm for two reasons. First, unlike voxel labeling, pixel labeling explicitly prohibits "holes" in the scene. In other words, it takes into account the fact that for any real scene the layer with disparity 0 (corresponding to the plane at infinity) is filled. Second, our pixel labeling method allows not only Potts interactions, but other useful smoothness terms (for example, truncated linear terms).

The major limitation of our approach lies in its bias towards fronto-parallel surfaces. With a sloped surface, our methods yield occlusions at discontinuities resulting from discretizing disparities. These occlusions are treated in the same

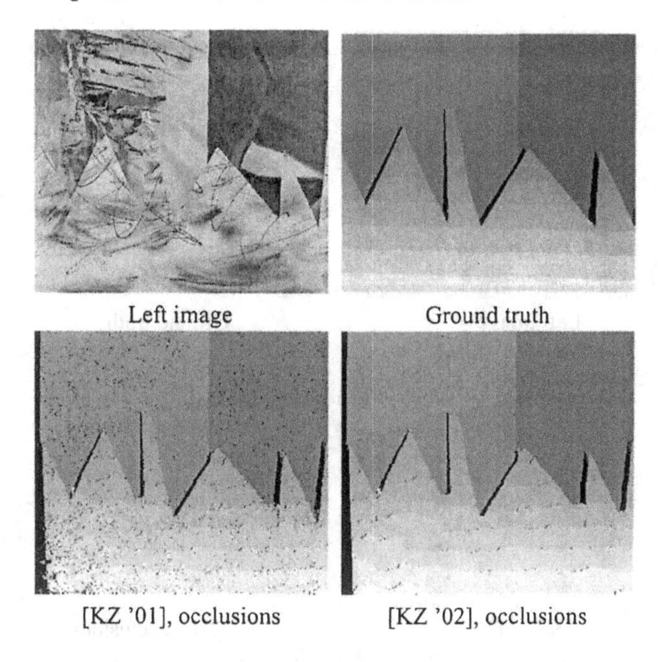


Figure 26.4. Sawtooth results (occlusions are shown in black).

way as real occlusions at object boundaries. Note that in the pixel labeling algorithm the problem can be alleviated by using a truncated linear smoothness term instead of Potts model.

It is possible to extend our algorithms to handle multiple cameras [488, 490, 492]. However, no scene point can lie inside the convex hull of the camera centers. This is the same class of camera configurations where voxel coloring [722] can be used, and includes many situations of practical interest.

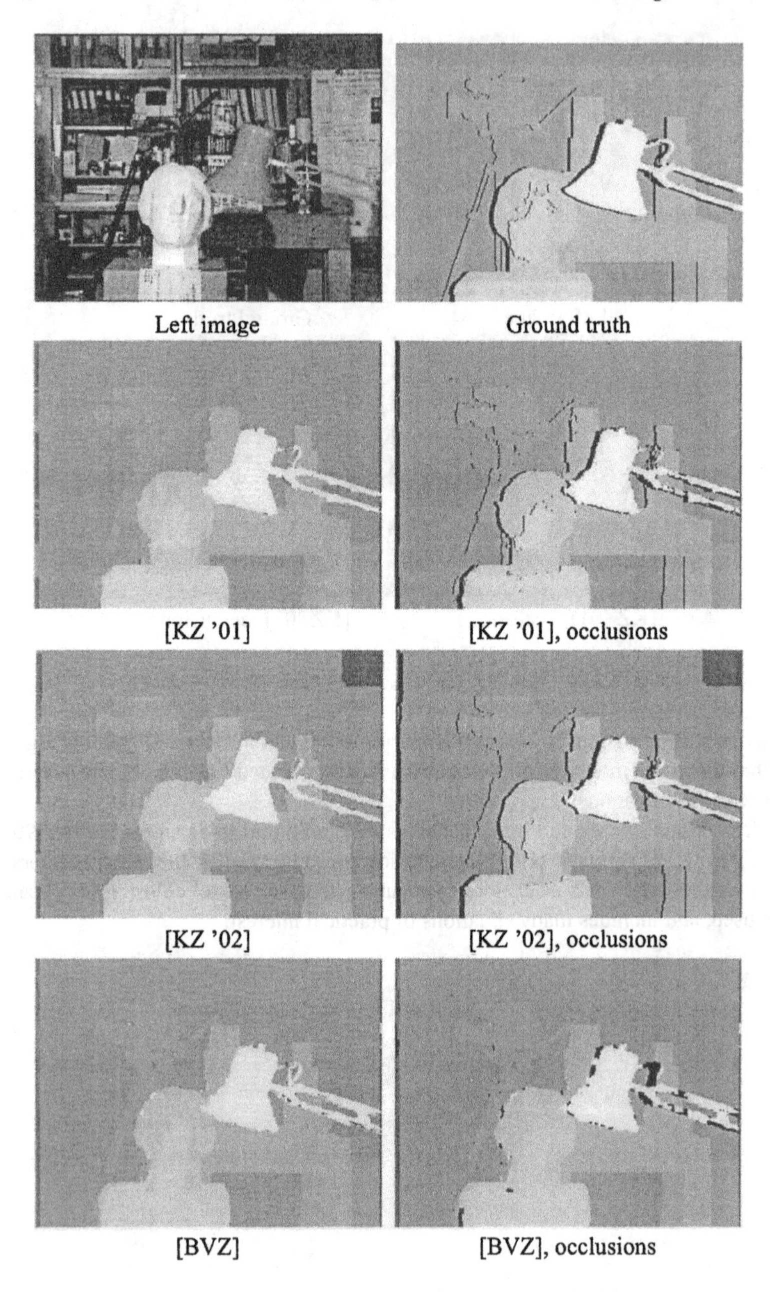


Figure 26.5. Tsukuba results (occlusions are shown in black).

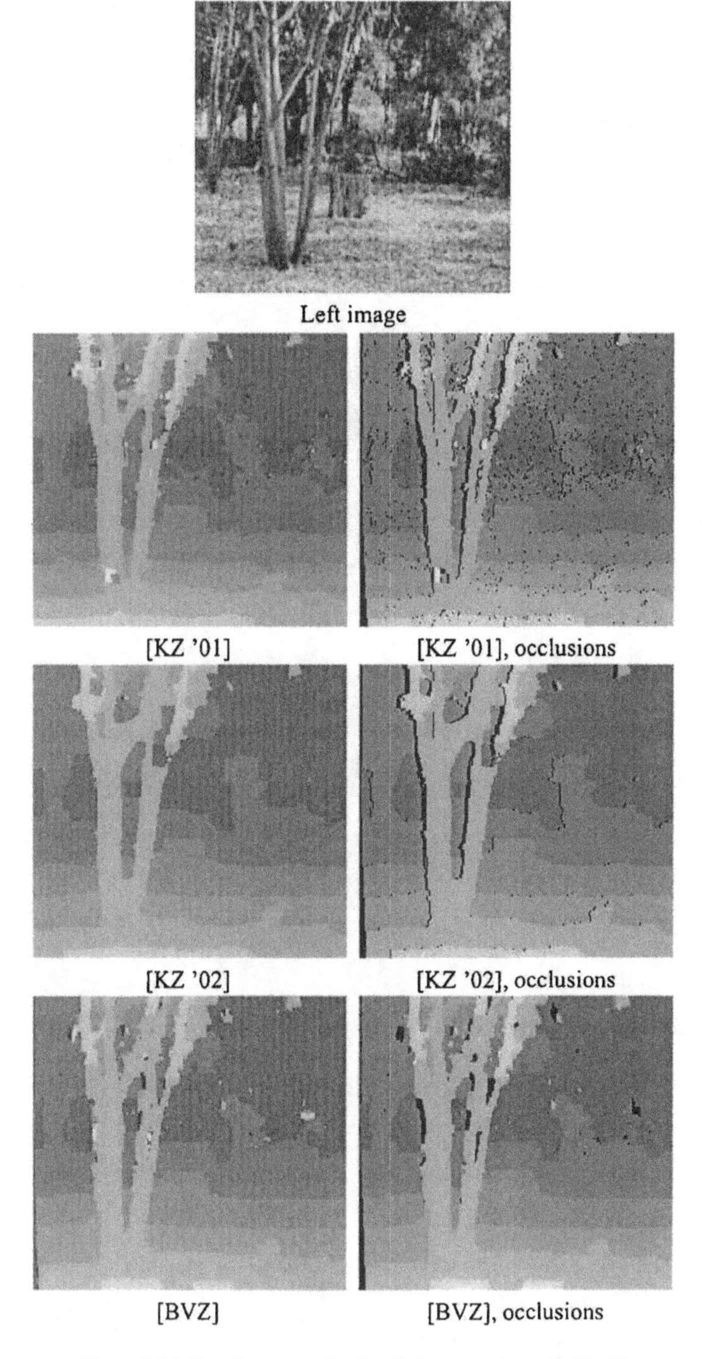


Figure 26.6. Tree image results (occlusions are shown in black).

# Chapter27

# Modelling Non-Rigid Dynamic Scenes from Multi-View Image Sequences

**J.-P. Pons, R. Keriven and O. Faugeras**

Abstract

This chapter focuses on the problem of obtaining a complete spatio-temporal description of some objects undergoing a non-rigid motion, given several calibrated and synchronized videos of the scene. Using stereovision and scene flow methods in conjunction, the three-dimensional shape and the non-rigid three-dimensional motion field of the objects can be recovered. We review the unrealistic photometric and geometric assumptions which plague existing methods. A novel method based on deformable surfaces is proposed to alleviate some of these limitations.

## 27.1 Introduction

Recovering the geometry of a scene from several images taken from different viewpoints, namely *stereovision*, is one of the oldest problems in computer vision. More recently, some authors have considered estimating the dense non-rigid three-dimensional motion field of a scene, often called *scene flow* [1] [835], from multiple video sequences. In this case, the input data are a two-dimensional array of images, in which each row is a multi-view stereovision dataset for a given time instant, and each column is a video sequence captured by a given camera.

Combining stereovision and scene flow allows to build a spatio-temporal model of a dynamic event. Once such a model is available, some novel virtual views of the event can be generated by interpolation across space and time [834].

[1] The scene flow should not be confused with the optical flow, which is the two-dimensional motion field of points in an image. Tbe optical flow is the projection of the scene flow in the image plane of a camera.

Stereovision and scene flow estimation both require to match different images of the same scene, in other words to find points in different cameras and in different frames corresponding to a same physical point. Once the correspondence problem is solved, the shape and the three-dimensional motion of the scene can be recovered easily by triangulation. Unfortunately, this problem is a very difficult task in computer vision because a scene patch generally has different shapes and appearances when seen from different points of view and at different times.

In Section 27.2, we report some important works on multi-view stereovision, scene flow estimation, and their integration. In particular, we show that, in order to solve the correspondence problem, most existing stereovision and scene flow algorithms rely on unrealistic simplifying assumptions that disregard either/both shape/appearance changes between different images of the scene.

In Section 27.3, we propose a new method that overcomes some of these limitations. Our method uses the prediction error [777] as a metric for shape and motion estimation. Both problems then translate into a generic image registration task. The latter is entrusted to a similarity measure chosen depending on imaging conditions and scene properties. In particular, our method can be made robust to appearance changes due to non-Lambertian materials and illumination changes. Our method results in a simpler, more flexible, and more efficient implementation than other deformable surfaces approaches. The computation time on large datasets does not exceed thirty minutes. Moreover, our method is compliant with a hardware implementation with graphics processor units.

Finally, in Section 27.4, we show some experimental results. Our stereovision algorithm yields very good results on a variety of datasets including specularities and translucency. We have successfully tested our scene flow algorithm on a challenging multi-view video sequence of a non-rigid event.

## 27.2 Previous Work

### *27.2.1 Multi-view complete stereovision*

Doing a complete review of the stereovision area is out of the scope of this chapter. Here, we are particularly interested in obtaining a complete scene reconstruction from a high number of input views. So we discard the methods in which the geometry of the scene is represented by depth maps or disparity maps. Indeed, these methods compute several partial models which have to be fused at post-processing. Moreover, they cannot handle visibility globally and consistently since no complete model of the scene is available during the estimation. However, let us mention two important works in this category: the graph cuts method of [490] and the PDE-based method of [769]. The interested reader should also refer to [711] for a good taxonomy of dense two-frame rectified stereo correspondence algorithms.

Thus, in the following, we focus on multi-view complete stereovision methods. These methods fall into two categories: the *space carving* framework and the *deformable surfaces* framework.

In the space carving framework [498], the scene is represented by a three-dimensional array of voxels. Each voxel can be labeled empty or occupied. When the algorithm starts, all voxels are occupied. Then the volume is traversed in an adequate order. If a voxel is not consistent with all the input images, it is relabeled empty. The order of the traversal is important because the visibility of the voxels is taken into account in the consistency test. In an older method called voxel coloring [722], there was a constraint on the placement of the cameras, and the algorithm needed a single pass. Space carving handles arbitrary camera configurations but is a little more expensive computationally.

The space carving framework suffers from several important limitations. First, these methods make hard decisions. Once a voxel is carved away, it cannot be recovered. And if one voxel is removed in error, further voxels can be erroneously removed in a cascade effect. This limitation is partially alleviated by the probabilistic space carving method [117].

Second, in the original space carving algorithm, the photo-consistency test derives from a brightness constancy constraint: corresponding points are asssumed to have the same color. This is a very naive assumption on the photometric properties of the scene. It requires a precise photometric calibration of the different cameras and only applies to strictly Lambertian scenes. In other words, this measure cannot cope with appearance changes between different images. Moreover, the choice of the global threshold on the color variance is often problematic. Recently, there have been some attempts to relax these photometric constraints [802, 904].

Third, the voxel-based representation disregards the continuity of shape, which makes it very hard to enforce any kind of spatial coherence. As a result, space carving is very sensitive to noise and outliers, and typically yields very noisy reconstructions.

We now turn to a review of stereovision methods based on deformable surfaces. These methods inherit from the active contour method pioneered in [455]. Here, contrarily to the space carving framework, the formulation is continuous and has a geometric interpretation. The unknown scene is modelled by a two-dimensional surface, and scene reconstruction is stated in terms of an energy minimization. An initial surface, positioned by the user, is driven by a partial differential equation minimizing an energy functional.

The most prominent work in this category is the level set stereovision method of [309]. In this work, the stereovision problem is formulated as a minimal surface approach, in the spirit of the geodesic active contours method [155]. In other words, the energy functional is written as the integral on the unknown surface of a data fidelity criterion. This criterion is the normalized cross correlation between image pairs.

The surface evolution is implemented in the level set framework [618]. On the one hand, the implicit representation offers numerical stability and the ability to handle topological changes automatically. On the other hand, it is quite expensive computationally, even with a narrow band approach. So, some authors have proposed an implementation with meshes [285] including a tangential smoothing operator to preserve the quality of the mesh and a merging/splitting procedure to handle topological changes.

Recently, some authors have proposed a new stereovision method to cope with non-Lambertian scenes [439]. Their method can estimate both the shape and the non-Lambertian reflectance of the scene. The surface deformation is driven by the minimization of the rank of a radiance tensor. This method outputs a geometric and photometric model which allows to predict the appearance of novel views.

Interestingly, in [309, 439], the geometric interpretation allows to agregate neighborhood information during the matching process, for a better robustness to noise and to realistic imaging conditions. But in return, these methods have to handle the geometric distortion between the different views. If fixed matching windows are used, the underlying assumption is the fronto parallel hypothesis: camera retinal planes are identical and the scene is an assembly of planes parallel to them. This assumption can still be found in recent work. In [517], the authors disregard projective distortion and attempt to minimize its impact by computing the stereo discrepancy of a scene patch with its two most front-facing cameras only. However, this approach is valid only for a high number of spatially well-distributed cameras.

In [309, 439], projective distortion is handled at least partially by taking into account the tangent plane to the object. For example, in [439], the radiance tensors are computed by sampling image intensities on a tesselation of the tangent plane. Thus, the matching score depends not only on the position of the surface but also on its orientation. Unfortunately, this first-order shape approximation results in a very complex minimizing flow involving second-order derivatives of the matching score. The computation of these terms is tricky, time-consuming and unstable, and, to our knowledge, all authors have resigned to drop them.

### *27.2.2 Scene flow estimation*

Three-dimensional motion estimation from multiple video sequences has long been limited to rigid or piecewise-rigid scenes or parametric models.

The problem of computing a dense non-rigid three-dimensional motion field from multiple video sequences has been addressed only recently. Two types of methods prevail in the scene flow literature.

The first family of methods [915, 149, 599] relies on the spatio-temporal derivatives of the input images. As pointed out in [835], estimating the scene flow from these derivatives without regularization is an ill-posed problem. Indeed, the associated normal flow equations only constrain the scene flow vector to lie on a line parallel to the iso-brightness contour on the object. This is nothing but a 3D version of the aperture problem for optical flow. In [149, 599], several samples of the spatio-temporal derivatives are combined in order to overconstrain the scene flow, whereas in [915], the aperture problem is solved by combining the normal flow constraint with a Tikhonov smoothness term. However, due to the underlying brightness constancy assumption, and to the local relevance of spatio-temporal derivatives, these differential methods apply mainly to slowly-moving lambertian scenes under constant illumination.

In the second family of methods [835, 915], scene flow is constructed from previously computed optical flows in all the input images. However, the latter may be noisy and/or physically inconsistent through cameras. The heuristic spatial smoothness constraints applied to optical flow may also alter the recovered scene flow.

### *27.2.3 Shape-motion integration*

Shape and motion estimations are linked. Indeed, the knowlegde of the shape is required to compute the scene flow. Conversely, the motion in the different cameras constrains the shape of the scene. This suggests that more robustness and more precision can be expected when properly fusing stereovision and scene flow estimation. More precisely, the correspondences across cameras and over time satisfy a round-about compatibility constraint that can be used to disambiguate the matching process.

There have been a few attempts to perform this integration [599, 915, 836]. But due to their increased computational cost and their modelling complexity, these techniques have not gained a significant popularity.

For example, in [836], shape and scene flow are estimated simultaneously using a plane-sweep carving algorithm in a 6D space. But this approach has a very high computational and memory cost, and is unable to enforce the smoothness of the recovered motion. As a result, in [834], the same authors renounce to the fusion: the shape-motion consistency is enforced by modifying the voxel representation and the scene flow at post-processing.

## 27.3 The Prediction Error as a New Metric for Stereovision and Scene Flow Estimation

We propose a common variational framework for complete stereovision and scene flow estimation which correctly handles projective distortion without any approximation of shape and motion and which can be made robust to appearance changes.

The metric used in our framework is the ability to predict the other input views from one input view and the estimated shape or motion. This is related to the methodology proposed in [777] for evaluating the quality of motion estimation and stereo correspondence algorithms. But in our method, the prediction error is used for the estimation itself rather than for evaluation purposes.

Our method consists in maximizing, with respect to shape and motion, the similarity between each input view and the predicted images coming from the other views. We adequatly warp the input images to compute the predicted images, which simultaneously removes projective distortion. For example, in the case of stereovision, we reproject the image taken by one camera onto the hypothetical surface, then we predict the appearance of the scene in the other views by projecting this texture-mapped surface in the other cameras. If the estimation of geometry is perfect, the predicted images coincide exactly with the corresponding input images, modulo noise, calibration errors, appearance changes and semi-occluded areas. This motivates our approach: we seek a shape or a motion maximizing the quality of the prediction.

Interestingly, this can be formulated as a generic image registration task. The latter is entrusted to a measure of image similarity chosen depending on imaging conditions and scene properties. This measure is basically a function mapping two images to a scalar value. The more similar the two images are, the lower the value of the measure is. Consequently, our formulation is completely decoupled from the nature of the image similarity measure used to assess the quality of the prediction. It can be the normalized cross correlation, some statistical measures such as the correlation ratio or the mutual information [398], or any other application-specific measure. Through this choice, we can make the estimation robust to camera spectral sensitivity differences, non-Lambertian materials and illumination changes.

Furthermore, contrarily to [309, 439, 285, 517], our method is not a minimal surface approach, i.e. our energy functional is not written as the integral on the unknown surface of a data fidelity criterion. In this approach, the data attachment term and the regularizing term are mixed whereas we may have to control them separately. As a consequence, to design non trivial regularity constraints, one has to twist the metric. A good discussion of this topic can be found in [749]. The authors show in some numerical experiments that better results can be achieved by integrating the similarity on the images rather than on the surface.

Consequently, in our method, the energy is defined as the sum of a matching term computed in the images and of a regularity constraint. The latter is required to make the problem well-posed. It is application-specific. For example, it could be designed to preserve shape or motion discontinuities. Here we focus on the design of the matching term and we settle for a straightforward regularization for each problem.

To minimize our energy functionals, we perform a gradient descent. We use a multi-resolution coarse-to-fine strategy to decrease the probability of getting stuck in irrelevant local minima.

Our method for scene flow estimation neither needs previous optical flow computations nor makes use of ambiguous spatio-temporal image derivatives. It directly evolves a 3D vector field to register the input images captured at different times. It can recover large displacements thanks to the multi-resolution strategy and can be made robust to illumination changes through the design of the similarity measure.

Our method processes entire images from which projective distortion has been removed, thereby avoiding the complex machinery usually needed to match win-

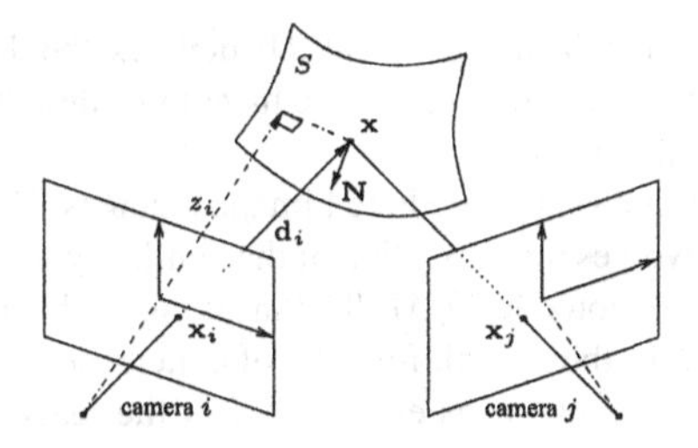


Figure 27.1. The camera setup and our notations.

dows of different shapes. Moreover, its minimizing flow is much simpler than in [309, 439]. This results in elegant and efficient algorithms.

### 27.3.1 Stereovision

In the following, let a surface $S \subset \mathbb{R}^3$ model the shape of the scene. We note $I_i : \Omega_i \subset \mathbb{R}^2 \to \mathbb{R}^d$ the image captured by camera $i$. The perspective projection performed by the latter is denoted by $\Pi_i : \mathbb{R}^3 \to \mathbb{R}^2$. Our method takes into account the visibility of the surface points. In the sequel, we will refer to $S_i$ as the part of $S$ visible in image $i$. The reprojection from camera $i$ onto the surface is denoted by $\Pi_{i,S}^{-1} : \Pi_i(S) \to S_i$. With this notation in hand, the reprojection of image $j$ in camera $i$ via the surface writes $I_j \circ \Pi_j \circ \Pi_{i,S}^{-1} : \Pi_i(S_j) \to \mathbb{R}^d$. We note $M$ a generic measure of similarity between two images.

The matching term $\mathcal{M}$ is the sum of the dissimilarity between each input view and the predicted images coming from all the other cameras. Thus, for each oriented pair of cameras $(i, j)$, we compute the similarity between $I_i$ and the reprojection of $I_j$ in camera $i$ via $S$, on the domain where both are defined, i.e. $\Omega_i \cap \Pi_i(S_j)$, in other words after discarding semi-occluded regions:

$$\mathcal{M}(S) = \sum_i \sum_{j \neq i} \mathcal{M}_{ij}(S) , \tag{27.1}$$

$$\mathcal{M}_{ij}(S) = M|_{\Omega_i \cap \Pi_i(S_j)} \left( I_i \, , \, I_j \circ \Pi_j \circ \Pi_{i,S}^{-1} \right) . \tag{27.2}$$

We now compute the variation of the matching term with respect to an infinitesimal vector displacement $\delta S$ of the surface. Figure 27.1 displays the camera setup and our notations. We neglect the variation related to visibility changes. This technical assumption is commonly used in the stereovision literature [309, 439, 285, 517]. Using the chain rule, we get that

$$\left. \frac{\partial \mathcal{M}_{ij}(S + \epsilon\, \delta S)}{\partial \epsilon} \right|_{\epsilon=0} = \int_{\Omega_i \cap \Pi_i(S_j)} \underbrace{\partial_2 M(\mathbf{x}_i)}_{1 \times d} \underbrace{DI_j(\mathbf{x}_j)}_{d \times 2} \underbrace{D\Pi_j(\mathbf{x})}_{2 \times 3} \underbrace{\left. \frac{\partial \Pi_{i,S+\epsilon\,\delta S}^{-1}(\mathbf{x}_i)}{\partial \epsilon} \right|_{\epsilon=0}}_{3 \times 1} d\mathbf{x}_i \, ,$$

where $\mathbf{x}_i$ is the position in image $i$ and $D\cdot$ denotes the Jacobian matrix of a function. For convenience to the reader, we have indicated the dimensions of the different matrices in the product.

When the surface moves, the predicted image changes. Hence the variation of the matching term involves the derivative of the similarity measure with respect to its second argument, denoted by $\partial_2 M$. The meaning of this derivative is detailed in Subsection 27.3.3. In the sequel, for sake of conciceness, we have omitted the images for which this derivative is evaluated. But the reader must be aware that the predicted images, as well as the domains where the similarity measures are computed, change along the minimizing flow.

We then use a relation between the motion of the surface and the displacement of the reprojected surface point $\mathbf{x} = \Pi_{i,S}^{-1}(\mathbf{x}_i)$:

$$\left.\frac{\partial \Pi_{i,S+\epsilon\,\delta S}^{-1}(\mathbf{x}_i)}{\partial \epsilon}\right|_{\epsilon=0} = \frac{\mathbf{N}^T \delta S(\mathbf{x})}{\mathbf{N}^T \mathbf{d}_i}\,\mathbf{d}_i\,,$$

where $\mathbf{d}_i$ is the vector joining the center of camera $i$ and $\mathbf{x}$, and $\mathbf{N}$ is the outward surface normal at this point.

Finally, we rewrite the integral in the image as an integral on the surface by the change of variable $d\mathbf{x}_i = -\frac{\mathbf{N}^T\mathbf{d}_i}{z_i^3}\,d\mathbf{x}$ , where $z_i$ is the depth of $\mathbf{x}$ in camera $i$, and we obtain that the gradient of the matching term is

$$\nabla \mathcal{M}_{ij}(S)(\mathbf{x}) = -\delta_{S_i\cap S_j}(\mathbf{x})\left[\partial_2 M(\mathbf{x}_i) DI_j(\mathbf{x}_j) D\Pi_j(\mathbf{x})\frac{\mathbf{d}_i}{z_i^3}\right]\mathbf{N}\,, \tag{27.3}$$

where $\delta_.$ is the Kronecker symbol. As expected, the gradient is zero in the regions not visible from both cameras. The reader should also note that the term between square brackets is a scalar function.

The regularization term is typically the area of the surface, and the associated minimizing flow is a mean curvature motion. The evolution of the surface is then driven by

$$\frac{\partial S}{\partial t} = \left[-\lambda H + \sum_i \sum_{j\neq i} \delta_{S_i\cap S_j}\, \partial_2 M\; DI_j\; D\Pi_j \frac{\mathbf{d}_i}{z_i^3}\right]\mathbf{N}\,, \tag{27.4}$$

where $H$ denotes the mean curvature of $S$, and $\lambda$ is a positive weighting factor.

## 27.3.2 *Scene flow*

Let now $S^t$ model the shape of the scene and $I_i^t$ be the image captured by camera $i$ at time $t$. Let $\mathbf{v}^t : S^t \to \mathbb{R}^3$ be a 3D vector field representing the motion of the scene between $t$ and $t+1$. The matching term $\mathcal{F}$ is the sum over all cameras of the dissimilarity between the images at time $t$ and the corresponding images at $t+1$ warped back in time using the scene flow.

$$\mathcal{F}(\mathbf{v}^t) = \sum_i \mathcal{F}_i(\mathbf{v}^t)\,, \tag{27.5}$$

$$\mathcal{F}_i(\mathbf{v}^t) = M\left(I_i^t\,,\; I_i^{t+1} \circ \Pi_i \circ (\Pi_{i,S^t}^{-1} + \mathbf{v}^t)\right)\,. \tag{27.6}$$

As the reader can check easily, its gradient writes

$$\nabla^T \mathcal{F}_i(\mathbf{v}^t) = -\delta_{S_i^t} \frac{\mathbf{N}^T \mathbf{d}_i}{z_i^3} \partial_2 M \, DI_i^{t+1} \, D\Pi_i \, . \tag{27.7}$$

In this case, the regularization term is typically the harmonic energy of the flow over the surface, and the corresponding minimizing flow is an intrinsic heat equation [78]. Then, the evolution of the scene flow is driven by

$$\frac{\partial \mathbf{v}^t}{\partial \tau} = \mu \, \Delta_{S^t} \mathbf{v}^t + \sum_i \delta_{S_i^t} \frac{\mathbf{N}^T \mathbf{d}_i}{z_i^3} \left[ \partial_2 M \, DI_i^{t+1} \, D\Pi_i \right]^T , \tag{27.8}$$

where $\tau$ is the fictious time of the minimization, $\Delta_{S^t}$ denotes the Laplace-Beltrami operator on the surface, and $\mu$ is a positive weighting factor.

### 27.3.3 Some similarity measures

For sake of completeness, we present two similarity measures than can be used in our framework: cross correlation and mutual information. Cross correlation assumes a local affine dependency between the intensities of the two images, whereas mutual information can cope with general statistical dependencies. We have picked these two measures among a broader family of statistical criteria proposed in [398] for multimodal image registration.

In the following, we consider two scalar images $I_1, I_2 : \Omega \subset \mathbb{R}^2 \to \mathbb{R}$. The measures below can be extended to vector (e.g. color) images by summing over the different components.

The minimizing flows given in Subsections 27.3.1 and 27.3.2 involve the derivative of the similarity measure with respect to the second image, denoted by $\partial_2 M$. The meaning of this derivative is the following: given two images $I_1, I_2 : \Omega \to \mathbb{R}^d$, we note $\partial_2 M(I_1, I_2)$ the function mapping $\Omega$ to the row vectors of $\mathbb{R}^d$, verifying for any image variation $\delta I$:

$$\left. \frac{\partial M(I_1, I_2 + \epsilon \, \delta I)}{\partial \epsilon} \right|_{\epsilon=0} = \int_\Omega \partial_2 M(I_1, I_2)(\mathbf{x}) \, \delta I(\mathbf{x}) \, d\mathbf{x} \, . \tag{27.9}$$

Cross correlation is still the most popular matching measure in the stereovision area. Most methods still use fixed square or rectangular matching windows. In this case, the choice of the window size is a difficult trade-off between match reliability and oversmoothing of depth discontinuities due to projective distortion [711]. Some authors alleviate this problem by using adaptative windows.

In our method, since we match distortion-free images, the size of the matching window is not related to a shape approximation. The matter here is in how big a neighborhood the assumption of affine dependency is valid. Typically, non-Lambertian scenes require to reduce the size of the correlation window, making the estimation less robust to noise and outliers.

In our implementation, we use smooth Gaussian windows with an infinite support instead of hard windows. Gaussian windows are more elegant as regards the continuous formulation of our problem and can be implemented efficiently with fast recursive filtering. Thus, we gather neighborhood information using convolutions by a Gaussian kernel. As this is the only difference with the traditional definition of normalized cross correlation, we do not give the full expression of the measure here. Moreover, due to space limitations, we invite the reader to refer to our technical report [657] for the expression of $\partial_2 M$ in this case.

Mutual information is based on the joint probability distribution of the two images, estimated by the Parzen window method with a Gaussian of standard deviation $\beta$:

$$P(i_1, i_2) = \frac{1}{|\Omega|} \int_\Omega G_\beta \left( I_1(\mathbf{x}) - i_1 \,,\, I_2(\mathbf{x}) - i_2 \right) d\mathbf{x} \,. \tag{27.10}$$

We note $P_1, P_2$ the marginals. Our measure is the opposite of the mutual information of the two images:

$$M^{MI}(I_1, I_2) = -\int_{\mathbb{R}^2} P(i_1, i_2) \log \frac{P(i_1, i_2)}{P_1(i_1) P_2(i_2)} \, di_1 \, di_2 \,. \tag{27.11}$$

Its derivative with respect to the second image writes

$$\begin{aligned} \partial_2 M^{MI}(I_1, I_2)(\mathbf{x}) &= \zeta(I_1(\mathbf{x}), I_2(\mathbf{x})) \,, \\ \zeta(i_1, i_2) &= \frac{1}{|\Omega|} G_\beta \star \left( \frac{\partial_2 P}{P} - \frac{P_2'}{P_2} \right) (i_1, i_2) \,. \end{aligned} \tag{27.12}$$

In practice, along the minimizing flow, the $\zeta$ function changes slowly relative to $I_1$ and $I_2$. So, in our implementation, we update it only every ten iterations.

## 27.4 Experimental Results

We have implemented our method in the level set framework [618], motivated by its numerical stability and its ability to handle topological changes automatically. However, our method is not specific to a particular surface model. Thus, an implementation with meshes would be straightforward.

The predicted images can be computed very efficiently thanks to graphics card hardware-accelerated rasterizing capabilities. In our implementation, we determine the visibility of surface points in all cameras using OpenGL depth buffering, we compute the reprojection of an image to another camera via the surface using projective texture mapping, and we discard semi-occluded areas using shadow-mapping [720].

The bottleneck in our current implementation is the computation of the similarity measure. Since it only involves homogeneous operations on entire images, we could probably resort to a graphics processor unit based implementation with fragment shaders (see *http://www.gpgpu.org*).

| Name | #Images | Image size | Level set size | Time (sec.) |
|---|---|---|---|---|
| Buddha | 25 | $500 \times 500$ | $128^3$ | 530 |
| Bust | 24 | $300 \times 600$ | $128 \times 128 \times 256$ | 1831 |

Table 27.1. Description of the stereovision datasets used in our experiments.

### 27.4.1 Stereovision

Table 27.1 describes the two challenging stereovision datasets used in our experiments. These datasets are publicly available from the OpenLF software (LFM project, Intel). "Buddha" is a synthetic scene simulating a translucent material and "Bust" includes strong specularities. However, cross correlation with a small matching window (standard deviation of 2 pixels) yields very good results.

Using all possible camera pairs is quite expensive computationally. Moreover, it is often not necessary since, when two cameras are far apart, no or little part of the scene is visible in both views. Consequently, in practice, we only pick pairs of neighboring cameras.

Our method is very efficient. The computation time does not exceed 30 minutes on a 2 GHz Pentium IV PC under linux. The number of iterations is 600 for both datasets. However, in practice, the convergence is often attained earlier. Hence the computation time could be reduced using an appropriate stopping criterion. In our experiments, the regularizer is a mean curvature motion.

We show our results in Figures 27.2 and 27.3. For each dataset, we display some of the input images, the ground truth, then some views of the estimated shape.

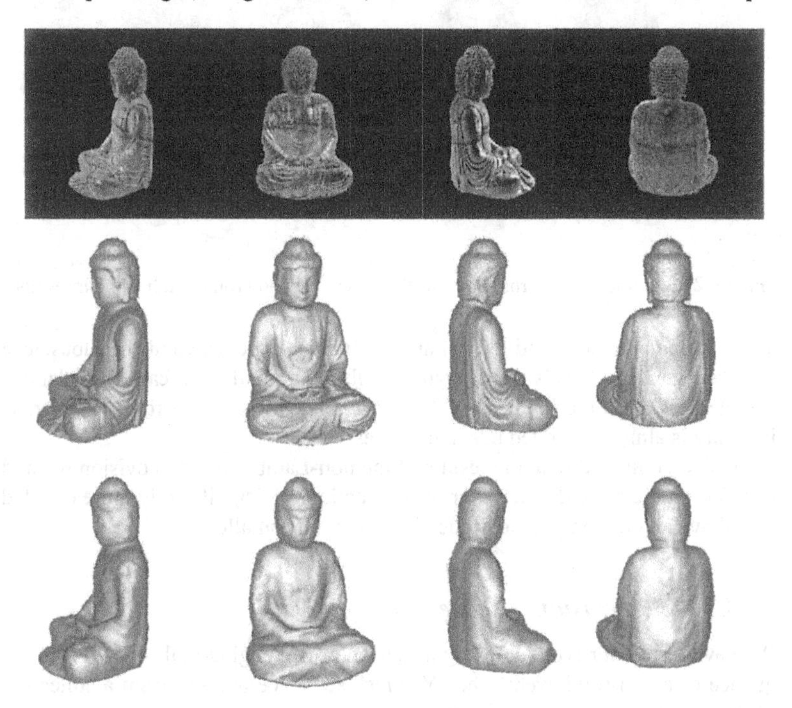

Figure 27.2. Some images from the "Buddha" dataset, ground truth and our results.

The overall shape of the objects is successfully recovered, and a lot of details

Figure 27.3. Some images from the "Bust" dataset, pseudo ground truth and our results.

are captured: the nose and the collar of "Buddha", the ears and the moustache of "Bust". A few defects are of course visible. Some of them can be explained. The depression in the forehead of "Bust" is related to a very strong specularity: intensity is almost saturated in some images.

Finally, compared with the results of the non-Lambertian stereovision method of [439] on the same datasets, our reconstructions are significantly more detailed and above all our computation time is considerably smaller.

### *27.4.2 Stereovision + scene flow*

We have tested our scene flow algorithm on a challenging multi-view video sequence of a non-rigid event. The "Yiannis" sequence is taken from a collection of datasets that were made available to the community by P. Baker and J. Neumann (University of Maryland) for benchmark purposes. This sequence shows a character talking while rotating his head. It was captured by 22 cameras at 54 fps plus 8 high-resolution cameras at 6 fps. Here we focus on the 30 synchronized se-

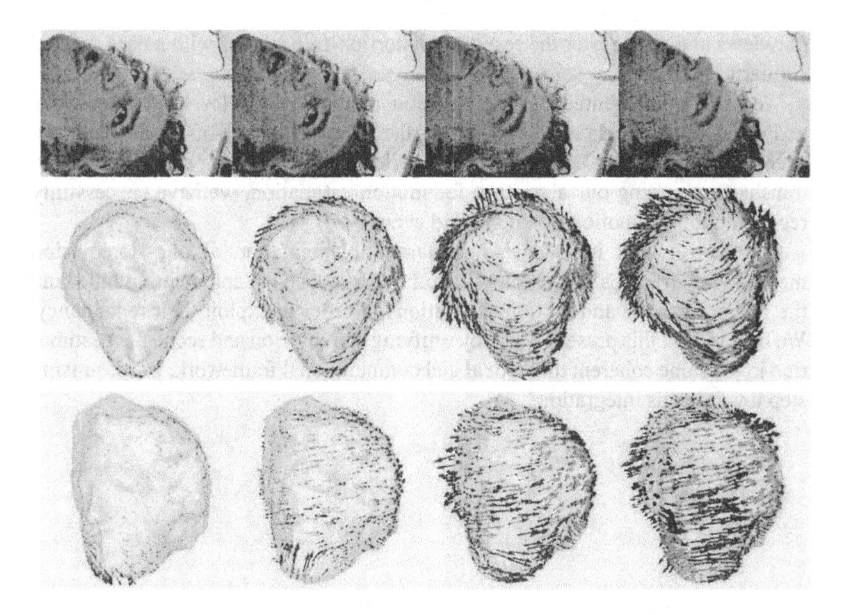

Figure 27.4. First images of one sequence of the "Yiannis" dataset and our results.

quences at the lower frame rate to demonstrate that our method can handle large displacements.

We have applied successively our stereovision and scene flow algorithms: once we know the shape $S^t$, we compute the 3D motion $\mathbf{v}^t$ with our scene flow algorithm. Since $S^t + \mathbf{v}^t$ is a very good estimate of $S^{t+1}$, we use it as the initial condition in our stereovision algorithm and we perform a handful of iterations to refine it. This is mush faster than restarting the optimization from scratch.

Figure 27.4 displays the first four frames of one of the input sequence and our estimation of shape and 3D motion at corresponding times. We successfully recover the opening then closing of the mouth, followed by the rotation of the head while the mouth opens again. Moreover, we capture displacements of more than twenty pixels. Our results can be used to generate time-interpolated 3D sequences of the scene. See the *Odyssée Lab* web page for more results.

## 27.5 Conclusion and Future Work

We have presented a novel method for multi-view stereovision and scene flow estimation which minimizes the prediction error. Our method correctly handles projective distortion without any approximation of shape and motion, and can be made robust to appearance changes. To achieve this, we adequatly warp the input views and we register the resulting distortion-free images with a user-defined similarity measure.

We have implemented our stereovision method in the level set framework and we have obtained results comparing favorably with state-of-the-art methods, even on complex non-Lambertian real-world images including specularities and translucency. Using our algorithm for motion estimation, we have successfully recovered the 3D motion of a non-rigid event.

Our future work includes a hardware implementation of our stereovision method with graphics processor units to further reduce the computation time, and the fusion of shape and motion estimations in order to exploit their redundancy. We believe that this present work, by unifying stereovision and scene flow estimation in the same coherent theoretical and computational framework, is a promising step towards this integration.

# Part VI

# Applications: Medical Image Analysis

# Chapter28

## Interactive Graph-Based Segmentation Methods in Cardiovascular Imaging

**L. Grady, Y. Sun and J. Williams**

Abstract

We examine the use of three techniques, graph cuts, isoperimetric minimization and random-walk partitioning for the interactive segmentation of cardiovascular medical images. These methods can often be used effectively without heavy reliance on learned or explicitly encoded priors. We illustrate, through the use of a toy problem, the basic difference in the performance characteristics of the methods. Subsequently, the suitability of each method to a particular segmentation application in the cardiovascular imaging domain is demonstrated.

## 28.1 Introduction

Isolation and quantification of structures in medical images is a continuous and varied source of segmentation problems. Segmentation methods which rely heavily on learned or explicit prior information often require significant customization before they can be applied to a specific problem. It is often preferable to use methods which can be quickly tested on the problem and then later enhanced with priors to improve accuracy. Graph partitioning algorithms are one such family of methods. In particular, we look at three segmentation techniques based on graph partitioning that at first glance may appear similar, but on closer inspection demonstrate unique behaviors. It is the distinct nature of these behaviors that can make one preferable over another for a specific application.

Although the graph cuts algorithm [362, 899] has been successfully employed in many applications, it is fundamentally a two-label algorithm. In fact, finding a minimal cut separating multiple terminals is an NP problem, although [113]

| Algorithm | Functional | Field | Constraints |
|---|---|---|---|
| Graph cuts | $Q(x) = x^T L x$ | $x = \{0, 1\}$ | Seeds fixed to $\{0, 1\}$ |
| Random walker | $Q(x) = x^T L x$ | $0 \leq x \leq 1$ | Seeds fixed to $\{0, 1\}$ |
| Isoperimetric | $Q(x) = \frac{x^T L x}{x^T d}$ | $0 \leq x$ | Seeds fixed to $\{0\}$ |

Table 28.1. A tabulated comparison of the three algorithms. See text for details.

provided an algorithm for getting within a bound of the optimal solution. The algorithm finds the smallest cut between two seed groups. In cases of weak object boundaries or small seed groups there is a tendency to find the cut that minimally encloses the seeds. The random walker algorithm proposed in [357] has a similar user interface (i.e., user "painting"), but does not suffer from the "small cut" problem and extends naturally to an arbitrary number of labels. The practical cost of this computation is currently higher than that of performing a binary graph cut on a similarly sized image graph. As will be discussed later, this algorithm also has a formal relationship to the graph cuts algorithm.

Both graph cuts and the random walker algorithm require specification of seed points for each output label in the resulting segmentation. In the case where a foreground/background segmentation is desired, a user is often interested in specifying only a few pixels in the foreground region instead of labeling pixels in both the foreground and background. Additionally, if one wants to apply one of these algorithms by specifying seeds automatically, it is easier to automatically specify foreground seeds than both foreground and background seeds. The isoperimetric algorithm of [356] naturally extends the random walker algorithm to a situation where only foreground labels are provided. Given a foreground-labeled pixel (or pixel group), the isoperimetric algorithm may be derived by starting a random walker at each unlabeled pixel and calculating the expected number of steps before the walker reaches a labeled seed. As with the random walker algorithm, these probabilities may be calculated analytically with simulation of a random walk. The expected number of steps may be converted into a foreground/background segmentation by finding a threshold that produces the minimal *isoperimetric ratio*, from which this algorithm was originally derived [356]. Not surprisingly, there is a formal relationship between the random walker and the isoperimetric algorithm.

## 28.2 Characteristic Behaviors of the Algorithms

Table 28.1 illustrates differences between these three algorithms. These differences may appear to be subtle compared with the similarities. From a practical standpoint, one might wonder if these algorithms return similar results or whether we can expect essentially identical behavior. It is certainly true that applying each technique to a simple segmentation task (e.g., a black circle in a white back-

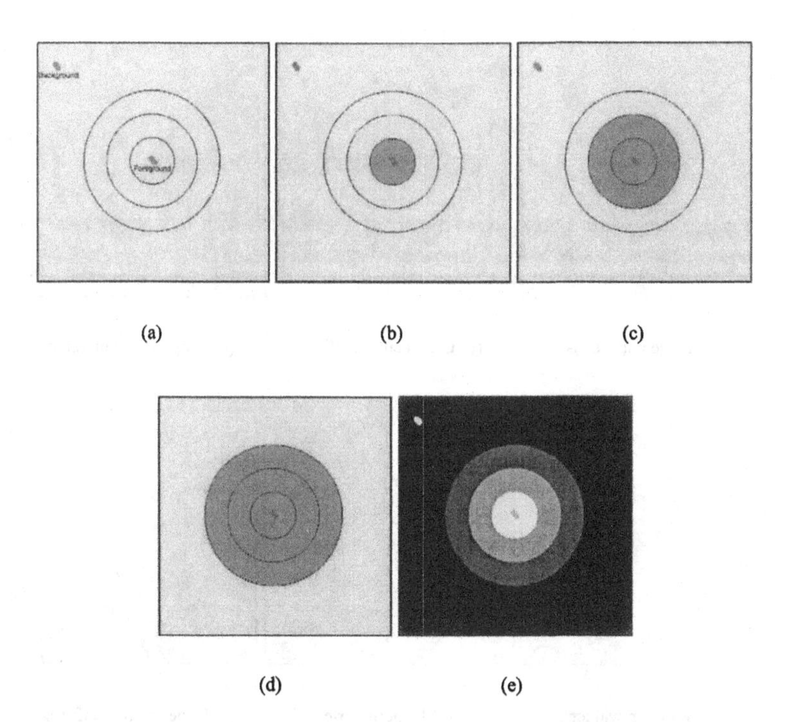

Figure 28.1. Different "personalities" of the three algorithms. (a) Input image with user-specified foreground and background seeds. (b) Graph cuts finds the innermost circle because it represents the smallest cut between seeds. (c) Random walk partitioning finds the middle circle because it is the most "equal" boundary between the two seeds. (d) The isoperimetric algorithm finds the outermost circle, since it minimizes the isoperimetric ratio in (28.8). Note that no background seeds were used when applying the isoperimetric algorithm. (e) Both the random walker and isoperimetric algorithms give a soft segmentation that is converted into a hard segmentation. For this image, each produce s the same soft segmentation (up to a scaling constant) that may be interpreted as a probability that a pixel lies in the foreground segment.

ground) will produce the same segmentation. However, Figures 28.1 and 28.2 are intended to illustrate the different "personalities" of each algorithm.

Figure 28.1 shows three concentric circles, with a foreground seed in the innermost circle and a background seed outside of the outermost circle. Given this user input, it is unclear how the "true" segmentation should be defined. Note that there is no ambiguity in the boundaries or difference in statistics of the regions. The real issue to be addressed by an algorithm that is given these seeds is: What does the user want? Depending on the user (or the goal), there are three valid outputs: The innermost object (small circle), the middle boundary between the foreground/background seeds (middle circle), or the entire group of objects (the

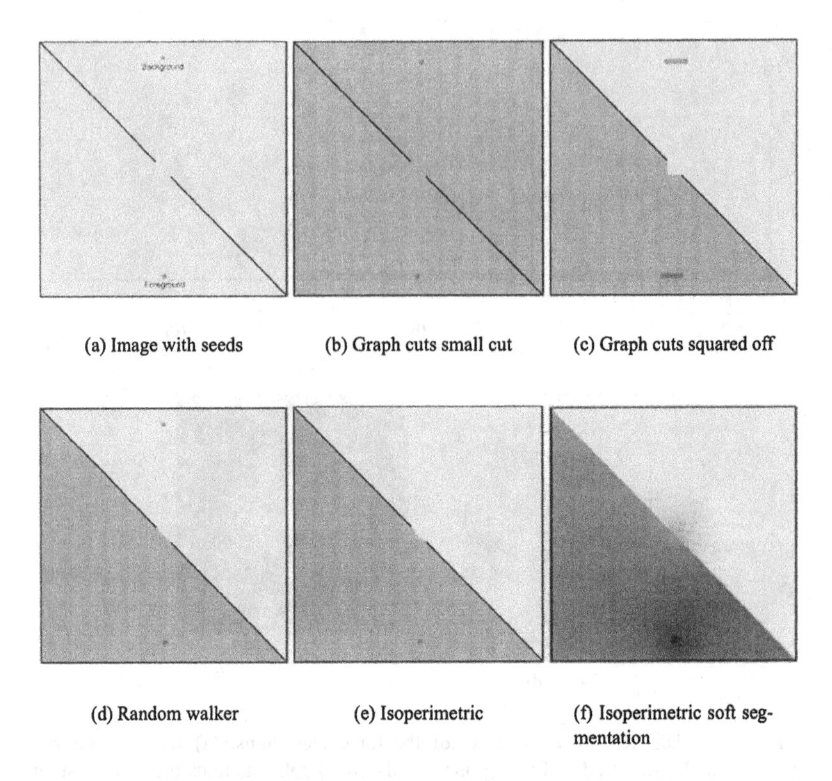

(a) Image with seeds (b) Graph cuts small cut (c) Graph cuts squared off

(d) Random walker (e) Isoperimetric (f) Isoperimetric soft segmentation

Figure 28.2. Weak edge behavior of the three algorithms. (a) Diagonal line image with user-specified seeds. (b) "Small cut" behavior of graph cuts. (c) More seeds overcome small cut issue, but yields a "blocky" segmentation (see text). (d) Random walker. (e) The isoperimetric algorithm finds the desired cut, but without background seeds. (f) Solution to (28.11) thresholded to produce a hard segmentation of (e) having a minimum isoperimetric ratio.

outermost circle). These results are exactly those given by the three algorithms respectively. Note that, regardless of the number of concentric circles, graph cuts will always choose the smallest, random walker will choose the middle and isoperimetric will choose the largest.

Graph cuts chooses the innermost circle because it will find the cut having the smallest cost. Each black/white transition of a ring will bear the same cost to cut, so the circle with smallest circumference will be preferred. This property is valuable because it finds the smallest object "unit" surrounding a foreground seed. The downside to this behavior is that a larger object (especially a textured object) requires many more seeds to locate. Additionally, this property is also the source of the "small cut" problem that can result in a return of the trivial boundary that minimally encloses the seeds.

A random walker starting from a given pixel will be more likely to reach the seed that requires crossing a smaller number of circles. Therefore, for an odd number of concentric circles, the random walker algorithm will always choose the middle circle. Note, however, that the probabilities form a "wedding cake" between the inner and outer circle that could be used (using another rule, e.g., minimum cut) to find any of the intermediate circles as seen in Figure 28.1. This behavior of the random walker algorithm is beneficial because the "small cut" problem is avoided and the most "equal" boundary between the seeds is found. However, this behavior can also result in the user having to place background seeds close to the foreground object in order to get the desired segmentation. Additionally, if an *even* number of concentric circles are present, the random walker algorithm will, in its neutrality between the two seeds, return a circle that threads the middle two circles instead of "snapping" the segmentation to the nearest circle.

The isoperimetric algorithm will also produce a "wedding cake" distribution , where each level corresponds to a circle. Note that the background seed is not employed in the isoperimetric algorithm. Given this solution, the isoperimetric algorithm looks for a threshold that produces a cut minimizing the isoperimetric ratio, defined as $h = \frac{x^T L x}{x^T d}$ (see below for detail). This ratio may be thought of intuitively as the ratio of the surface area (i.e., dual to the cut) to the volume. In a continuum setting, the isoperimetric ratio of a circle is $h = \frac{2\pi r}{\pi r^2} = \frac{2}{r}$ which will get smaller (and thus preferred) for a larger radius. Since the isoperimetric algorithm chooses the threshold that minimizes the isoperimetric ratio, the largest circle will be returned out of the "wedding cake" distribution produced by the solution.

Figure 28.2 illustrates another aspect of the personality of the algorithms. Here, a (broken) black line was drawn on a white image. All three algorithms exhibit the ability to locate the weak boundary even though there is no intensity cue at the gap and the statistics of both regions are identical. However, with graph cuts there are two issues. First, we initially see the "small cut" problem of when small seed groups are placed. However, even when the seed groups are made large enough, the algorithm finds a "squared off" cut that is unappealing. The reason for this squared off cut is because a 4-connected lattice is employed and, therefore, the squared off cut has the same cut cost as the diagonal cut, so one of these cuts is simply returned by the algorithm. This issue may be ameliorated by using a lattice with increased connectivity (e.g., 8-connected) at the cost of increased memory consumption. However, even for a 4-connected lattice, the random walker and isoperimetric algorithm neither exhibit a "squared off" solution nor suffer from the "small cut" problem.

## 28.3 Applications on CT Cardiovascular data

Computed tomography (CT) imaging has, in the last 5 years, undergone a revolution in resolution. Premium multi-slice scanners now have between 16 and 256 detector rows with gantry rotational latencies of less than half a second. These advances have meant not only an increase in spatial, but also in temporal resolution. CT angiography (CTA) uses injected contrast to opacify the cardiovascular system for high-resolution imaging. CTA is now the modality of choice for imaging 3D cardiovascular morphology. Due to the huge amount of data produced by these scanners (2GB volumes are now not uncommon), automated an semi-automated post-processing techniques are no longer a curiosity, they are indispensable tools for the radiologist.

### *28.3.1 Segmenting Individual Heart Chambers using Graph Cuts*

Electrophysiological ablation procedures, like pulmonary vein isolation for curing atrial fibrillation, are today guided by a combination of electrophysiological and morphological criteria. Therefore it is helpful for the electrophysiologist to have 3D visualizations of the cardiac chamber which is subject to RF ablation available for pre-procedural planning, intra-procedural catheter guidance, and post-procedural follow-up. This requires the tools for heart chamber segmentation from CTA images.

The requirements of the segmentation tools are:

1. Accuracy: Segmentation shall be as close as possible to the ground truth provided by the user.
2. Easy to use: The tool shall need minimal user input.
3. Performance: The algorithm shall be fast and memory efficient.

Since fully automatic segmentation inherently has the problem of reliability and repeatability, an interactive segmentation is more attractive. Interactive methods take advantage of the user knowledge of the anatomy, and increase the overall procedure efficiency.

Even with contrast, accurate chamber segmentation with minimal user interaction is still a challenging problem. The difficulty is largely due to the weak boundaries between chambers. For example, the left atrium and left ventricle often have similar intensity due to direct blood pool connection through the mitral valve. Image noise and different imaging protocols across various sites also pose a challenge for the robustness of the segmentation algorithm.

### *28.3.2 Multi-Resolution Banded Graph Cuts*

Boykov and Jolly [108] describe an interactive graph cuts algorithm. The algorithm assumes that some voxels have been identified as object or background seeds based on *a priori* knowledge from the anatomy. It computes a globally optimal binary segmentation that completely separates the object seeds and the background seeds.

Despite the power of finding a globally optimal solution, the major difficulty of the graph cuts algorithm lies in the enormous computational costs and memory consumption. Typical CT scans generate a 3D volume of hundreds of slices of images. For example, a volume of 512×512×300 has 75M voxels. A graph that stores the nodes and edges can easily consume over 1GB memory. Performing a graph cut segmentation on this volume using the max-flow algorithm [345] can take several minutes.

The approach we use to make the graph cuts algorithm practical in this context is a multi-resolution banded formulation. With the prior knowledge of the rough size of the chamber, the background seeds can be automatically detected after the user specifies an object seed inside the chamber. This makes it possible to achieve one-click-segmentation. What makes this possible is the compact shape of the heart chamber and the relatively homogeneous intensity in the chamber due to the contrast agent injection.

The idea of this algorithm is first to get a rough segmentation using graph cuts on a reduced resolution graph. The low resolution estimate is then used to guide a high resolution banded cut.

The algorithm contains five steps:

1. Apply a seeded region growing [4] from the object seeds in the low resolution volume. The growing stops when reaching a predefined maximal distance that depends on the *a priori* knowledge of the typical chamber size. The result may contain outliers due to leaking to left ventricle, the pulmonary arteries, and the bones.
2. Dilate a layer from the boundary of the region growing. The outer layer of the dilation is marked as background seeds.
3. Apply graph cuts in low resolution to get a rough segmentation of the left atrium. It is fast to solve because there are significantly fewer nodes for the low resolution graph.
4. Dilate a layer from the boundary of the rough segmentation to form a band whose inner boundary is marked as object seeds and outer boundary is marked as background seeds.
5. In high resolution, apply the graph cuts to get an accurate segmentation result. It is also fast to solve because the graph is built on the narrow band that only contains a few layers of voxels.

### *28.3.3 Empirical Results*

Using the multi-resolution and banded graph cuts, the segmentation of the heart chambers can be achieved with a single mouse click inside the left atrium.

Fig. 28.3(a) shows the one click segmentation of the left atrium using the multi-resolution and banded graph cuts algorithm. It is segmented from a CT volume with the resolution of 512×512×370 and it takes less than 15 seconds to segment the left atrium on a Pentium 4 2.4GHz computer and uses less than 200MB memory. Fig. 28.3(b) shows the separate heart chambers and vessels, including left atrium, left ventricle, right atrium, right ventricle, and aorta. These chambers are segmented individually using the multi-resolution, banded graph cuts algorithm.

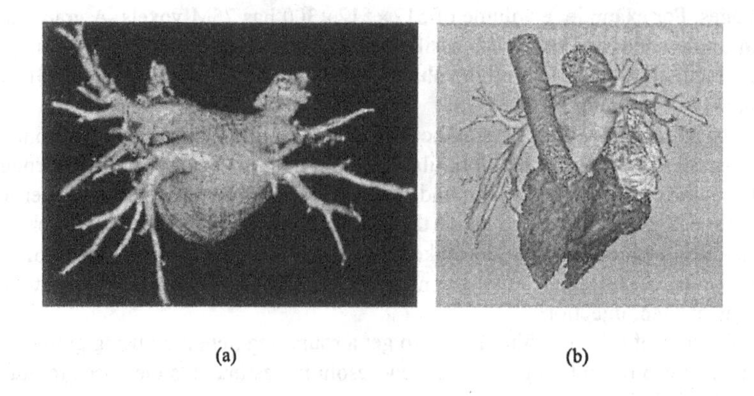

(a) (b)

Figure 28.3. Renderings of cardiac segmentations. (a) Left atrium. (b) Multiple chambers.

### 28.3.4 Random Walks for Simultaneous Chamber Segmentation

Where graph cuts are well suited to extraction of a single foreground object from a general background, extension to parallel multi-label segmentation does not follow naturally. Alternatively, the random walker algorithm can segment multiple regions in a single interactive step.

Assume that the user has provided $K$ labeled pixels (hereafter referred to as **seed points** or **seeds**). For each unlabeled pixel, we ask: Given a random walker starting at this location, what is the probability that it first reaches each of the $K$ seed points? It will be shown that this calculation may be performed exactly without the simulation of a random walk. By performing this calculation, we assign a $K$-tuple vector to each pixel that specifies the probability that a random walker starting from each unlabeled pixel will first reach each of the $K$ seed points. A final segmentation may be derived from these $K$-tuples by selecting for each pixel the most probable seed destination for a random walker. In this approach, we treat an image (or volume) as a purely discrete object — a graph with a fixed number of vertices and edges. Each edge is assigned a real-valued weight corresponding to the likelihood that a random walker will cross that edge (e.g., a weight of zero means that the walker may not move along that edge).

It has been established [450, 283] that the probability of a random walker first reaching a seed point exactly equals the solution to the Dirichlet problem [227] with boundary conditions at the locations of the seed points and the seed point in question fixed to unity, while the others are set to zero. A steady-state, DC circuit analogy is also given in [357]. Using the principle of superposition from circuit theory, it can be easily shown that the probabilities at each node sum to unity (as expected). For this reason, we need only solve $K - 1$ systems, given $K$ labels, since the remaining system is known via the unity constraint (i.e., at each node subtract the sum of the $K - 1$ solutions at that node from unity to recover the solution to the remaining system).

### *28.3.5 The Random Walker Algorithm*

We begin by defining a precise notion for a graph. A **graph** [382] consists of a pair $G = (V, E)$ with **vertices (nodes)** $v \in V$ and **edges** $e \in E \subseteq V \times V$. An edge, $e$, spanning two vertices, $v_i$ and $v_j$, is denoted by $e_{ij}$. A **weighted graph** assigns a value to each edge called a **weight**. The weight of an edge, $e_{ij}$, is denoted by $w(e_{ij})$ or $w_{ij}$. The **degree** of a vertex is $d_i = \sum w(e_{ij})$ for all edges $e_{ij}$ incident on $v_i$. Given a set of nonnegative weights, the probability that a random walker at node $v_i$ transitions to node $v_j$ is given by $p_{ij} = \frac{w_{ij}}{d_i}$. The following will also assume that our graph is connected.

In order to represent the image structure (given at the pixels) by random walker biases (i.e., edge weights), one must define a function that maps a change in image intensities to weights. Since this is a common feature of graph based algorithms for image analysis, several weighting functions are commonly used in the literature [737, 112, 356]. Additionally, it was proposed in [923] to use a function that maximizes the entropy of the resulting weights. In this work we have preferred (for empirical reasons) the typical Gaussian weighting function given by

$$w_{ij} = \mathrm{e}^{-\beta(g_i - g_j)^2}, \tag{28.1}$$

where $g_i$ indicates the image intensity at pixel $i$. The value of $\beta$ represents the only free parameter in this algorithm. In practice, we employ

$$w_{ij} = \mathrm{e}^{-\frac{\beta}{\rho}(g_i - g_j)^2} + \epsilon, \tag{28.2}$$

where $\epsilon$ is a small constant (we take $\epsilon = 10^{-6}$) and $\rho$ is a normalizing constant $\rho = \max(g_i - g_j), \forall i, j$. The purpose of (28.2) is to keep the choice of $\beta$ relevant to images of different quantization and contrast, as well as make sure that none of the weights go identically to zero (resulting in a possible disconnection).

The discrete Dirichlet problem has been discussed thoroughly in the literature [85, 283] and a convenient form for the solution is given in [358]. We will now review the method of solution.

Define the discrete Laplacian matrix [561] as

$$L_{v_i v_j} = \begin{cases} d_i & \text{if } i = j, \\ -w_{ij} & \text{if } v_i \text{ and } v_j \text{ are adjacent nodes,} \\ 0 & \text{otherwise,} \end{cases} \tag{28.3}$$

where $L_{v_i v_j}$ is used to indicate that the matrix $L$ is indexed by vertices $v_i$ and $v_j$.

Partition the vertices into two sets, $V_M$ (marked/seed nodes) and $V_U$ (unmarked nodes) such that $V_M \cup V_U = V$ and $V_M \cap V_U = \emptyset$. Note that $V_M$ contains all seed points, regardless of their label. Then, we may reorder the matrix $L$ to reflect the subsets

$$L = \begin{bmatrix} L_M & B \\ B^T & L_U \end{bmatrix}. \tag{28.4}$$

Denote the probability assumed at each node, $v_i$, for each label, $s$, by $x_i^s$. Define the set of labels for the seed points as a function $\phi(v_j) = s, \ \forall v_j \in V_M$, where

$s \in \mathbb{Z}, 0 < s \leq K$. Define the $|V_M| \times 1$ (where $|\cdot|$ denotes cardinality) marked vector for each label, $s$, at node $v_j \in V_M$ as

$$m_j^s = \begin{cases} 1 & \text{if } \phi(v_j) = s, \\ 0 & \text{if } \phi(v_j) \neq s. \end{cases} \tag{28.5}$$

As demonstrated in [358], the solution to the combinatorial Dirichlet problem may be found by solving

$$L_U x^s = -B m^s, \tag{28.6}$$

which is just a sparse, symmetric, positive-definite, system of linear equations with $|V_U|$ number of equations and the number of nonzero entries bounded from above by $2|E| + |V|$. Since $L_U$ is guaranteed to be nonsingular for a connected graph [84], the solution, $x^s$, is guaranteed to exist and be unique. Therefore, the potentials for all the labels may be found by solving the system

$$L_U X = -BM, \tag{28.7}$$

where $X$ has columns taken by each $x^s$ and $M$ has columns given by each $m^s$. As mentioned above, one must solve only $K - 1$ systems, given $K$ labels, since the probabilities at each node must sum to unity.

### *28.3.6 Numerical solution*

Many good methods exist for solving large, sparse, symmetric, linear systems of equations (e.g., [352, 666]). A direct method, such as $LU$ decomposition with partial pivoting has the advantage that the computation necessary to solve (28.7) is only negligibly increased over the amount of work required to solve (28.6). Unfortunately, current medical data volumes frequently exceed $256 \times 256 \times 256 \approx 16M$ voxels, and hence require the solution of an equal number of equations. Furthermore, there is no reason to believe that the resolution will not continue to increase. Most contemporary computers simply do not have enough memory to allow an $LU$ decomposition with that number of equations.

The standard alternative to the class of direct solvers for large, sparse systems is the class of iterative solvers [374]. These solvers have the advantages of a small memory requirement and the ability to represent the matrix-vector multiplication as a function. In particular since, for a lattice, the matrix $L_U$ has a circulant nonzero structure (although the coefficients are changing), one may avoid storing the matrix entirely. Instead, a vector of weights may be stored (or computed on the fly, if memory is at a premium) and the operation $L_U x_U^s$ may be performed very cheaply. Furthermore, sparse matrix operations (like those required for conjugate gradients) may be efficiently parallelized [273, 363](e.g., for use on a GPU). Because of the relationship of (28.6) to a finite differences approach to solving the Dirichlet problem on a hypercube domain, the techniques of numerical solution to PDEs may also be applied. Most notably, the algebraic multigrid method [734, 258] achieves near-optimal performance for the solution to equations like (28.6).

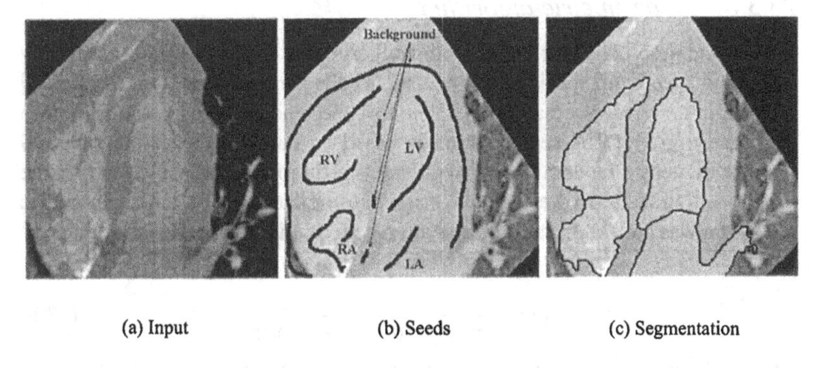


(a) Input (b) Seeds (c) Segmentation

Figure 28.4. Random walker algorithm applied to a four-chamber slice of a cardiac CTA volume. (a) Original four-chamber slice. (b) User-specified seeds of each chamber (i.e., left ventricle, left atrium, right ventricle, right atrium, background). (c) Resulting segmentation boundaries.

We have implemented the standard conjugate gradients algorithm with a modified incomplete Cholesky preconditioning [56], representing the matrix-vector multiplication implicitly, as described above on an Intel Xeon 2.4 GHz dual-processor with 1GB of RAM. Solution of (28.6) using conjugate gradients (tolerance $= 10^{-4}$, sufficient for the algorithm) for a $256 \times 256$ image with two randomly placed seed points required approximately 3 seconds.

To summarize, the steps of the random walker algorithm are:

1. Obtain a set, $V_M$, of marked pixels (seeds) with $K$ labels from the user.
2. Using (28.1), map the image intensities to edge weights in the lattice.
3. Solve (28.7) outright for the probabilities or solve (28.6) for each label except the final one, $f$ (for computational efficiency). Set $x_i^f = 1 - \sum_{s<f} x_i^s$.
4. Obtain a final segmentation by assigning to each node, $v_i$, the label corresponding to $\max_s (x_i^s)$.

## 28.3.7 Empirical Results

Using (28.1), we transformed a four-chamber view of a CTA heart volume into a weighted graph and applied the random walker algorithm. Results are displayed in Figure 28.3.7. The random walker algorithm was chosen for this problem because five labels are required to segment the four chambers of the heart (each chamber plus the background). Segmentation of this $256 \times 256$ image required approximately 20 seconds of computation time. We have found this algorithm reliable on a large variety of CTA cardiac data with varying levels of noise.

### 28.3.8 Isoperimetric algorithm

Neither the graph cuts nor the random walker algorithm can be used when only foreground seeds are specified by the user. The isoperimetric algorithm of [356] may be interpreted as a natural extension of the random walker algorithm to a single seed group. The segmentation is based on computation of the expected number of steps a random walker will take, starting from each pixel, to find the user-specified seeds. However, in the original formulation [356], the isoperimetric algorithm was derived from a segmentation goal of minimizing the **isoperimetric ratio**

$$Q(x) = \frac{x^T Lx}{\min(x^T d, (1^T - x^T)d)}, \tag{28.8}$$

where $1^T$ is the vector of all ones and $d$ is the vector of node degrees. The indicator vector, $x$, is defined as

$$x_i = \begin{cases} 0 & \text{if } v_i \in \overline{S}, \\ 1 & \text{if } v_i \in S, \end{cases} \tag{28.9}$$

where $S$ indicates the set of foreground nodes. Unfortunately, a combinatorial minimization of this problem is NP-Hard [581]. Consequently, the vector $x$ was relaxed to take real values and the "volume" (represented combinatorially by the denominator) was fixed to a constant, i.e., $x^T d = k$ (see [356] for a full exposition).

Using a Lagrange multiplier to perform a constrained minimization gives the energy as

$$Q(x) = x^T Lx - \lambda(x^T d - k), \tag{28.10}$$

and the resulting minimum as

$$Lx = \frac{1}{2}\lambda d. \tag{28.11}$$

Since we are only concerned with relative values of the solution, and in order that (28.11) represents the expected number of steps required to find a seed, we ignore the scalar factor $\frac{\lambda}{2}$, setting $\frac{\lambda}{2} = 1$.

Although the Laplacian matrix in (28.11) is singular, the incorporation of user-specified seeds, i.e., $x_i = 0, \forall v_i \in V_M$ removes the singularity. The solution, $x_i$, at node, $v_i$, obtained through solution of (28.11) gives the expected number of steps that a random walker would take to find a seed node (see [794] for justification of this interpretation). Indeed, if one were to solve (28.11) for two seeds, $v_1$ and $v_n$, then

$$Lx^1 = \begin{bmatrix} -1^T d_U - d_n \\ d_U \\ d_n \end{bmatrix}, \tag{28.12}$$

where $1^T$ represents the vector of all ones, $d_U$ represents the vector containing the degrees of unlabeled nodes. The reason that (28.12) holds is because premultiplication of both sides by $1^T$ produces zero on the left hand side, so the right hand side must be balanced. Then,

$$L(x^2 - x^1) = \begin{bmatrix} d_1 \\ d_U \\ -1^T d_U - d_1 \end{bmatrix} - \begin{bmatrix} -1^T d_U - d_n \\ d_U \\ d_n \end{bmatrix} = \begin{bmatrix} 1^T d_U + d_n + d_1 \\ 0 \\ -1^T d_U - d_n - d_1 \end{bmatrix}. \tag{28.13}$$

Since it is known [85] that multiplication of the solution to the random walker problem, $x_{\text{RW}}$, (given the same two seeds as above) by the Laplacian results in

$$Lx_{\text{RW}} = \begin{bmatrix} \rho \\ 0 \\ -\rho \end{bmatrix}, \tag{28.14}$$

where $\rho$ represents the **effective conductance** between nodes $v_1$ and $v_n$, the solution to two (or more) isoperimetric systems (28.11) also yields the random walker probabilities (up to a scaling and shift). It is intuitive that this should be true, since we would expect that a random walker having fewer expected steps to reach one seed over another would also be most likely to reach that seed first.

Computation of a solution to (28.11) yields a notion of how "far away" a given node is to a seed point, but it does not give a hard segmentation. Therefore, in accordance with [356], we convert the solution to (28.11) to a hard segmentation by thresholding the solution, $x$, at the value that produces a hard segmentation minimizing (28.8). Only $n$ thresholds must be evaluated (i.e., one for each node) and the values of (28.8) may be evaluated quickly, leading to a fast production of a hard segmentation (see [356] for more details). Producing a hard segmentation in this way guarantees that all nodes belonging to the foreground segment are connected or, if more than one seed group is present, each group of foreground pixels is connected to a seed [356]. Note that this procedure for converting a soft segmentation into a hard segmentation is very similar to what is performed in the NCuts algorithm [737].

To summarize, the steps of the isoperimetric algorithm are:

1. Obtain a set, $V_M$, of marked pixels (seeds) indicating foreground.
2. Using (28.1), map the image intensities to edge weights in the lattice.
3. Solve (28.11) for the expected number of steps taken by a random walker starting from each pixel to reach a node in $V_M$.
4. Obtain a hard segmentation by trying $n$ thresholds, $\alpha$, of $x$ and choosing the segmentation that produces the smallest ratio given in (28.8). Assign each node, $v_i$, to foreground if $x_i \leq \alpha$ and to background if $x_i > \alpha$.

### *28.3.9 Bone-Vessel Separation*

In CTA scans, contrast enhanced and calcified blood vessels can appear with the same intensity profile as bones. Further confounding an automatic bone/vessel segmentation is the fact that bones and vessels often touch each other and partial volume effects produce a gradual, diffuse boundary. Due to the difficulty of this

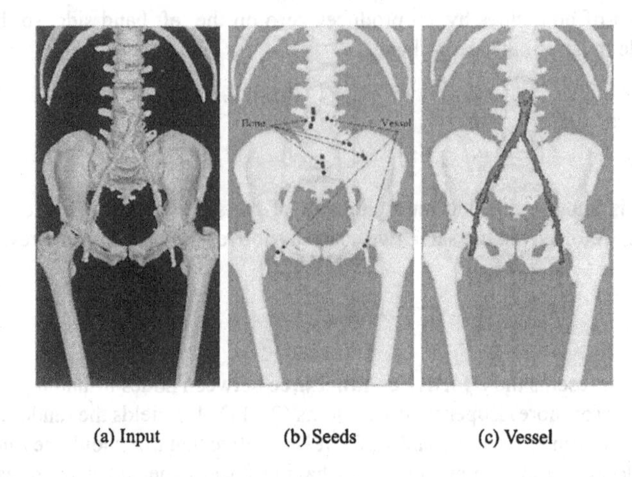


(a) Input (b) Seeds (c) Vessel

Figure 28.5. Isoperimetric algorithm applied to vessel/bone separation. This problem is difficult because bone and vessel have similar intensities and touch each other (i.e., have a weak boundary) at several areas. (a) Original 3D image. (b) User-specified seeds of vessel (foreground) and bone (background). See text for meaning of background seeds in this application. (c) Resulting segmentation vessel segmentation.

situation, a user-guided mode is required to accurately separate bones and blood vessels.

The ideal user interface for this application is for a user to provide a single click on a blood vessel, without taking the time to label the bone. We first used an automatic segmentation algorithm to produce an initial bone/vessel segmentation. After this initial stage was completed, the user was able to click on a blood vessel and the isoperimetric algorithm was run on the subgraph of the original volume defined by the initial, automatic, segmentation. Application of the isoperimetric algorithm to this problem often gives the correct answer after placement of one or two vessel seed points. However, there were cases in which the vessel segmentation "bled" into the bone. To handle these circumstances, we allowed the user to enter "bone" seeds as well. Inclusion of a bone seed at node, $v_i$, had two effects:

1. Scaling the $d_i$ entry on the right hand side of (28.11) by a large factor, $C >> 1$. This scaling has the effect of "pushing back" the solution to (28.11) from node $v_i$. Alternatively, the scaling may also be interpreted as an additional injection of current at $v_i$ in the circuit analogy of the isoperimetric algorithm presented in [356].
2. Limiting the threshold, $\alpha$, to $\alpha < x_i$. With this limit, the resulting hard segmentation never encompasses the bone seed.

The performance of the isoperimetric algorithm is data dependent in this semi-automatic application. However, the algorithm supported interactive performance (sub-second) for the $64 \times 64 \times 64$ sub-volumes used in our application.

## 28.4 Conclusions

The three algorithms described in this section exhibit distinct behaviors that make each suitable for a given problem. In practice, once a particular method is found to work well on a given problem without priors, priors are then encoded or learned from examples to improve overall performance. It remains a challenge to improve the computational efficiency of these techniques. Already we have seen performance improvements of one to two orders of magnitude by mapping the core solvers for these problems into commodity graphics hardware (GPU).

# Chapter29

## 3D Active Shape and Appearance Models in Cardiac Image Analysis

**B. Lelieveldt, A. Frangi, S. Mitchell, H. van Assen, S. Ordas, J. Reiber and M. Sonka**

Abstract

This chapter introduces statistical shape- and appearance models and their biomedical applications. Three- and four-dimensional extension of these models are the main focus. Approaches leading to automated landmark definition are introduced and discussed. The applicability is underlined by presenting practical examples of 3D medical image segmentation.
Keywords: Statistical shape models, point distribution models, active shape models, active appearance models, multi-view models, landmarking, cardiac segmentation, magnetic resonance.

## 29.1 Introduction

Three-dimensional diagnostic organ imaging is now possible with X-ray computed tomography, magnetic resonance, positron emission tomography, single photon emission tomography, and ultrasound to name just the main imaging modalities. While imaging modalities are developing rapidly, the images are mostly analyzed visually and therefore qualitatively. The ability to quantitatively analyze the acquired image data is still not sufficiently available in routine clinical care. The wealth of information buried in these acquired data is not fully exploited because of the tedious and time-consuming character of manual analyses. This is even more so when dynamic three-dimensional image data need to be processed and analyzed.

Much effort has been described on data driven approaches to automate the segmentation of medical images, however there are three main reasons why these frequently exhibit lower success rate in comparison with human expert observers, especially when applied to clinical-quality images – data driven methods do not incorporate a sufficient amount of a priori knowledge about the segmentation problem; do not consider three-dimensional or temporal context as an integral part of their functionality; and position the segmentation boundaries at locations of the strongest local image features not considering true anatomical boundary locations and shape constraints. For these reasons, model driven image analysis has received considerable attention over the last decade. Especially statistical models of shape and appearance have found widespread application in biomedical segmentation problems. In this chapter, we will briefly introduce statistical shape- and appearance models and their biomedical applications, discuss issues inherent to 3D extension of these models, and focus on application examples of segmentation of 3D medical volume data.

### *29.1.1 Background*

In general, statistical models capture the mean shape and shape variations from a training set. Building on the principles explored by Kendall [458], and Dryden and Mardia [284], Cootes and Taylor developed a statistical point distribution model (PDM), originally for shape analysis. This approach has helped to gain insight into typically occurring anatomical variations [220, 223]. Point Distribution Models describe populations of shapes using statistics of sets of corresponding landmarks of the shape instances [220, 223, 752]. By aligning $N$ shape samples (consisting of $n$ landmark points) and applying a principal component analysis (PCA) on the sample distribution, any sample $\mathbf{x}$ within the distribution can be expressed as an average shape $\overline{\mathbf{x}}$ with a linear combination of eigenvectors $P$ superimposed

$$\mathbf{x} = \overline{\mathbf{x}} + P\mathbf{b} \ . \tag{29.1}$$

In two-dimensional models, $p = \min(2n, N-1)$ eigenvectors $P$ form the principal basis functions, while in a three-dimensional model; $p = \min(3n, N-1)$ eigenvectors are formed [1]. In both cases the corresponding eigenvalues provide a measure for compactness of the distribution along each axis. By selecting the largest $q$ eigenvalues, the number of eigenvectors can be reduced, where a proportion $k$ of the total variance, $\mathcal{V}_T$, is described such that

$$\sum_{i=1}^{q} \lambda_i \geq k \cdot \mathcal{V}_T \quad \text{where} \quad \mathcal{V}_T = \sum_{i=1}^{p} \lambda_i \ . \tag{29.2}$$

One of the primary contributions of PDMs was an ease of automated learning of the model parameters from sets of corresponding points.

Apart from shape analysis, the learned PDM eigenvariations can be applied to image segmentation and motion tracking. This PDM extension is known as Active Shape Model (ASM), and consists of an iterative image matching scheme designed to fit the model to image data, while constraining the allowed model deformations within the trained statistically plausible limits. ASMs may use a

[1] The minimum operator is needed since we frequently have more corresponding shape points than training set samples.

gray-level model of scan lines perpendicular to the model contour or surface to estimate new update positions for each landmark points. Alternatively, update points can be generated by an edge detector or a (neural or fuzzy) pixel classifier. The differences between the cloud of candidate image feature points and the model points drive the model alignment and deformation in each iteration. The model deformation is statistically constrained to lie within the subspace spanned by the selected modes of variation of the PDM. Early applications of Active Shape Models address segmentation of for instance echocardiographic data [223] and deep neuroanatomical structures from MR images of the brain [293]. In recent literature, a diversity of other, mainly 2D biomedical applications have been described for a range of imaging modalities and organs.

The third type of landmark based model is the Active Appearance Model (AAM) introduced by Cootes [219, 222]. AAMs are an extension of PDMs with a statistical intensity model of a complete volumetric image patch, as opposed to merely scan lines near the landmarks in the ASM matching. An AAM is built by warping a mesh tessellating the training shapes to the mesh of the average shape. Obviously, this requires a consistent mesh node localization in all shapes of the training set. After intensity normalization to zero mean and unit variance, the intensity average and principal components are computed. A subsequent combined PCA on the shape and intensity model parameters yields a set of components that simultaneously capture shape and texture variability. AAM matching is based on minimizing a criterion expressing the difference between model intensities and the target image. This enables a rapid search for the correct model location during the matching stage of AAMs, while utilizing precalculated derivative images for the optimizable parameters. The sum of squares of the difference between the model-generated patch and the underlying image may serve as a simple criterion for matching quality.

AAMs have shown to be highly robust in the segmentation of routinely acquired single-phase, single slice cardiac MR [575] and echocardiographic images [105], because they exploit prior knowledge about the cardiac shape, image appearance and observer preference in a generic way. For a detailed background on 2D Active Appearance Shape and Appearance Models and their application to image segmentation, the reader is referred to [218, 759].

This chapter focuses on dimensional extension of landmark based models, and is mainly limited to the well-established PDMs, ASMs and AAMs. For completeness, we also mention a few proposed interesting alternatives to landmark-based statistical modeling that also enable 3D statistically constrained segmentation:

- Statistical Deformation Models [696, 529] are constructed by registering several training sets using multi-level free-form deformations. These free-form deformations are parameterized using a control point grid, and statistical analysis using PCA is performed on the control point sets, yielding an average deformation and principal components.

- Probabilistic Atlases have been widely applied to 3D shape modeling, see for instance [528]. These are typically constructed by rigidly registering a set of 3D manual segmentations, and probability maps are generated by blurring the segmented structure for each image, and averaging over all subjects. These models can be applied to segmentation using expectation maximization algorithms.

- A multiscale 3D shape modeling approach called M-reps was developed by Pizer et al. [649]. M-reps support a coarse-to-fine hierarchy and model shape variations via probabilistically described boundary positions with a width- and scale-proportional tolerances. Points on the surface are expressed in a local object coordinate frame spanned by a medial skeleton, yielding object specific point correspondence.

### 29.1.2 *Issues inherent to 3D extension*

Initially, most PDM, ASM and AAM models were applied to 2D modeling and matching problems. However, because many modern imaging modalities deliver (dynamic) 3D image data, extension to higher dimensions is desired. A critical issue to achieve extension of PDMs to 3 and higher dimensions is point correspondence: the landmarks have to be placed in a consistent way over a large database of training shapes, otherwise an incorrect parameterization of the object class would result. In a 2D case, the most straightforward definition of point correspondence is by identifying evenly spaced sampling points on a boundary from one characteristic landmark to the next, although this may lead to a suboptimal sampling. In a 3D case, the problem of defining a unique sampling of the 3D object surfaces is more complex, and far from trivial. Because of this, 3D point correspondence has recently been intensively researched, and three main approaches can be distinguished:

- Correspondence by parameterization: this has mainly been applied to relatively simple geometries that can be described using a spherical or cylindrical coordinate system, in combination with a few well-defined landmarks to fixate the coordinate frame. Applying this coordinate definition on all the samples yields parametrically corresponding landmarks, as will be exemplified in Section 29.2.3.2.

- Correspondence by registration or fitting by mapping a 3D surface tesselation of one sample to all the other samples. Lorenz et al. [527] for instance propose a 3D deformable surface that is matched to binary segmentations of new samples. By projecting the tesselation of the matched template to the new sample, correspondence for the new sample is achieved. Alternatively, non-rigid volumetric registration can be applied to define dense correspondences between training samples, as will be detailed later on in Section 29.2.1. These approaches have the advantage that topologically more complex shapes can be handled.

- Correspondence by optimal encoding: Davies et al. [247] has applied for instance a Minimum Description Length (MDL) criterion to evaluate the quality of correspondence in terms of the ability to encode the whole training set for a given landmark distribution. Stegmann [759] have shown that these MDL encoded models optimize model properties such as compactness and specificity. Davies [246] also developed a 3D MDL approach for 3D objects, however to our knowledge this model has only been applied to shape analysis, and not for segmentation. In this chapter, no further applications are given for these MDL based approaches.

In the next section, we introduce a number of recently proposed 3D extensions of the PDMs, ASMs and AAMs. First, the issue of automatically defining a dense correspondence over the training set is addressed, detailing an example approach based on non-rigid registration. The achieved point correspondence is then incorporated into an Active Shape Model that can be applied to multi-modal and multi-planar sparse data. Subsequently, several higher dimensional extensions of Active Appearance Models are discussed.

## 29.2 Methods

### *29.2.1 3D Point Distribution Models*

Frangi et al. [327] have described a methodology for the construction of three-dimensional (3D) statistical shape models of the heart, from a large image database of dynamic MRI studies. Non-rigid registration is employed for the automatic establishing of landmark correspondences across populations of healthy and diseased hearts. The general layout of the method is to align all the images of the training set to an atlas that can be interpreted as a mean shape. Once all the necessary transformations are obtained, they are inverted and used to propagate any number of arbitrarily sampled landmarks on the atlas, to the coordinate system of each subject. In this way, while it is still necessary to manually draw the contours in each training image, this technique relieves from manual landmark definition for establishing the point correspondence across the training set. The method can easily be set to build either 1- or 2-chamber heart models. Moreover, its generality allows for using it with other modalities (e.g., SPECT, CT) and organs with shape variability close to that of the heart (e.g., liver, kidneys). A detailed description of the method can be found in [327], and can be summarized as follows:

1. The manually drawn contours in the training set are converted into labeled shapes by flood-filling each (closed) sub-part with a different scalar value.
2. The labeled shapes are aligned through a global transformation (rigid registration with nine degrees of freedom: translation, rotation, and anisotropic scaling) to a Reference Sample (RS) randomly chosen from the training set. The RS is therefore considered as the first atlas estimate.

3. A new atlas is constructed by shape-based averaging of the aligned shapes. This is performed by averaging the images in their distance transform domain, and defining a new labeled shape by considering the zero iso-surface of each sub-part separately.

4. To minimize the bias introduced by the choice of the RS, steps 2 and 3 are repeated until the atlas becomes stable. At this point, the atlas is said to be in a Reference Coordinate System (RCS).

5. Subsequently, each rigidly aligned shape is locally deformed (using non-rigid registration) in order to accommodate to the RCS atlas.

6. The obtained local transformations are averaged and the resulting transformation is applied to the RCS atlas. The new atlas is said to be in a Natural Coordinate System (NCS) and diminishes the influence of the RS selection.

7. A new set of global and local transformations are recalculated in the same way as in steps 2 and 5 (Fig. 29.1(a)).

8. Finally, any automatically generated landmarks in the NCS atlas can be propagated to the training shapes through the transformations in step 7 (Fig. 29.1(b)).

9. In order to build the statistical shape models, the autolandmarked shapes are normalized with respect to a reference coordinate frame, eliminating differences across objects due to rotation, translation and size.

Once the shape samples are aligned, the remaining differences are solely shape related, and PCA can be performed. In Fig. 29.2, the first 4 eigenmodes of the obtained model are displayed. The main characteristic variations consist of size difference, twisting, rotation and their combinations. The 1st mode describes the size differences of the hearts. The 2nd mode indicates the large variation of the right ventricle. The 3rd and 4th modes describe the bending and twisting of the left ventricle. Higher modes combine vertical bends and less global deformations, but with decreasing impact to the total shape.

### *29.2.2 3D Active Shape Models*

The bi-ventricular model described above was extended with a matching algorithm to apply it to image segmentation [827]. A key design criterion behind this matching approach was applicability to data acquired with arbitrary image slice orientations, from different modalities (MR and CT), and even to sparsely sampled data with arbitrary image slice orientations. This implies that:

- only 2D image data may be used for updating the 3D model, to ensure applicability to arbitrarily oriented sparse data
- generation of update points is executed based on relative intensity difference to remove dependence on training-based gray-level models.

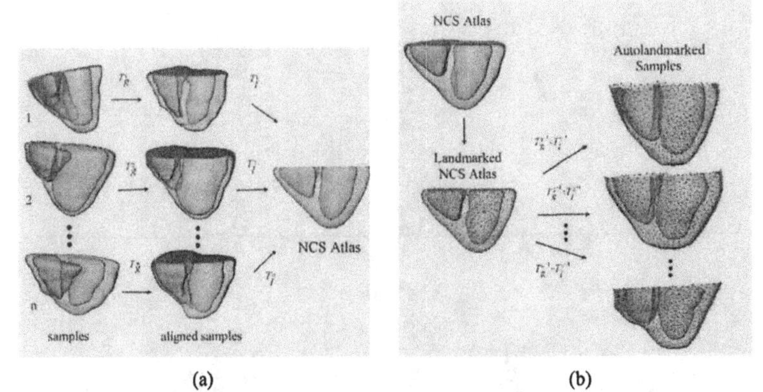


(a) (b)

Figure 29.1. (a) Final transformations. A set of final global ($T_g$) and local ($T_l$) transformations can take any sample shape of the training set, to the NCS atlas coordinate system. (b) Landmark propagation. Once the final global and local transformations are obtained, they are inverted and used to propagate any number of arbitrarily sampled landmarks on the NCS atlas, to the coordinate system of the original samples.

To accomplish this, the landmark points are embedded in a surface triangular mesh. During the matching, this mesh is intersected by the image planes, generating 2D contours spanned by the intersections of the mesh triangles. To remove dependencies on image orientation or limited resolution, model update information is represented by 2D point-displacement vectors. The 2D update vectors located at the intersections of the mesh with the image slices are first propagated to the nodes of the mesh, and projected to the local surface normals. Multiple contributions from different mesh intersections to a single mesh node are averaged to yield a single 3D update vector per node. Scaling, rotation, and translation differences between the current state of the model and the point cloud representing the candidate updates are eliminated by alignment. The current mesh state is aligned with the candidate model state using the Iterative Closest Point algorithm [82]. Successively, the parameter vector $\boldsymbol{b}$ controlling model deformation is calculated. An adjustment to $\boldsymbol{b}$ with respect to the previous iteration is computed, using both $\boldsymbol{x}_{proposed}$ and $\boldsymbol{x}_{current}$

$$\boldsymbol{b}' = \boldsymbol{b}_{current} + \Delta\boldsymbol{b} = \boldsymbol{b}_{current} + \Phi^T(\boldsymbol{x}_{proposed} - \boldsymbol{x}_{current}) \tag{29.3}$$

with $\boldsymbol{x}_{current}$ representing the aligned current state of the mesh, and $\boldsymbol{b}_{current}$ representing the parameter vector describing the current shape of the model within the statistical context.

In the classic ASM[223], model updates were generated using a (multi-resolution) statistical gray level model (GLM) in each sample point: this requires a modality-dependent training stage. To enable application to different modalities without retraining, a Fuzzy Inference System (FIS) was selected instead, which determines the 2D point-displacement vectors by pixel classification based on

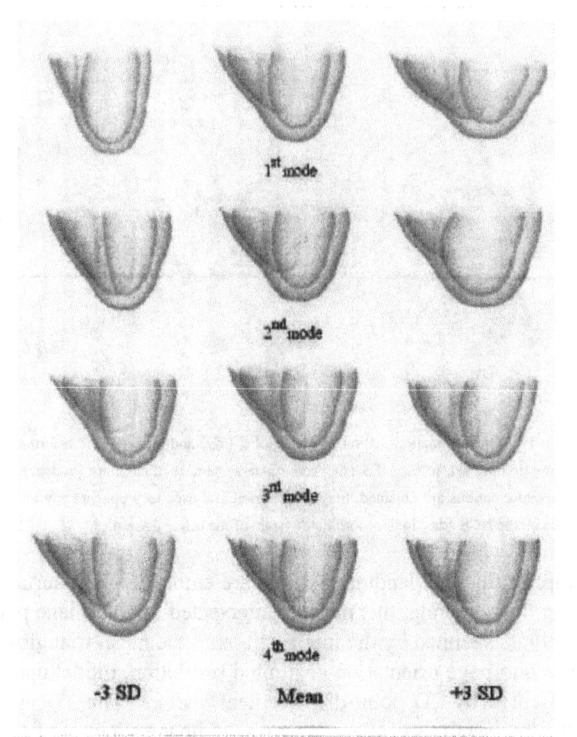


Figure 29.2. The 2-chamber PDM principal modes of variation for end-diastolic phase. The shapes are generated by varying a single model parameter ($b_i$), fixing all others at zero standard deviations (SD=$\sqrt{\lambda_i}$) from the mean shape.

relative intensity differences between tissues in the sampled data. This pixel classification is realized by fuzzy clustering of the intensity strips in the vicinity of the surface. Based on the resulting membership functions, different tissue transitions from blood to muscle and muscle to air can be inferred, which form the update points for the model matching.

The combined active shape model has been applied to cardiac MR data, where the left and right ventricle are segmented (see Figure 29.3(a)). Modality independence of the model has been shown in [827] , where the model is applied to MR and CT images without retraining. Independence of planar orientation is illustrated in Figure 29.3(b)

Alternatively, Kaus et al. [457] describe an ASM-based approach, where the matching mechanism is embedded in the internal energy term of an elastically deformable model. Training samples are manual segmentations expressed as binary volumes, and point correspondence is achieved by fitting a template mesh with a fixed point topology to each binary training sample. Contrary to van Assen et al. [827], they model the endo- and epicardial shapes separately. However, a coupling is realized by integrating connecting vertices between both surfaces and adding a

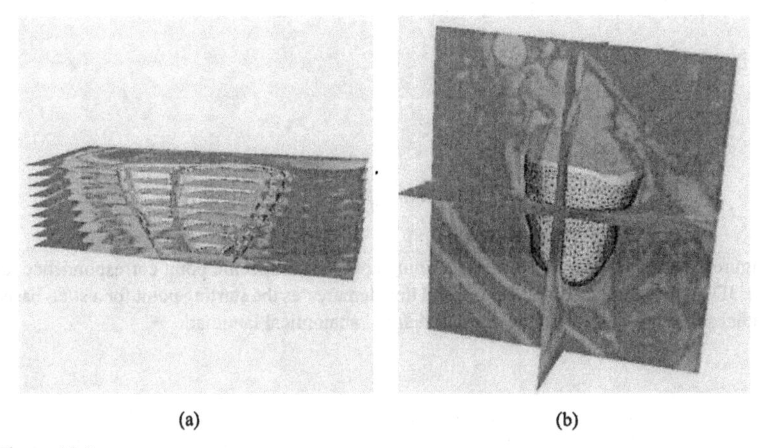

(a) (b)

Figure 29.3. (a) Application of the bi-ventricular 3D ASM to a cardiac MR image volume. (b) Matching of the model to 3 sparse, arbitrarily oriented MR image planes

connection term to the internal energy. In addition, they adopt a spatially varying feature model for each landmark. This approach has the advantage that statistical shape constraints are imposed on the allowed elastic mesh deformation, while allowing for some flexibility to deviate from the trained shapes to accommodate for untrained shape variability.

### *29.2.3 3D and 4D Active Appearance Models*

#### 29.2.3.1 2D + time Active Appearance Models

Segmentation of sequences of 2D images, such as echocardiographic or cardiac MR slices is often hampered by the fact that segmentation results are not time-continuous. Especially, sequential application of 2D segmentation techniques in subsequent frames may yield spatial and temporal discontinuities. To resolve that, an extension to 2D + time modeling has been proposed in [105, 828], where the temporal dimension is encoded into the model. In addition to spatial correspondence, time-correspondence is defined by defining "landmark time frames" such as end-diastole and end-systole. The shapes are interpolated to a fixed number of frames using a nearest neighbor interpolation. This time-correspondence allows the shape- and intensity vectors to be simply concatenated over the whole sequence and treated as 2D images, and the 2D AAM machinery described earlier can be applied unaltered. Though strictly speaking, this is not a fully 3D model, this way the segmentation is performed on all frames simultaneously, yielding time-continuous results. This approach has been validated on echocardiographic image sequences [105] and slice-based cardiac MR image sequences [828].

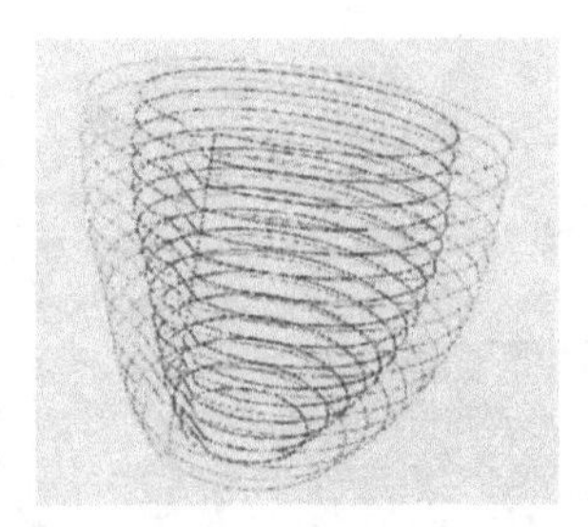

Figure 29.4. Line parameterization, defining application specific point correspondence for the 3D cardiac left ventricle. The vertical line demarcates the starting point for a slice-based radial sampling, and is based on a well defined anatomical landmark

### 29.2.3.2 3D Active Appearance Models: Modeling Volume Appearance

As discussed in the 3D ASM section, Active Shape Models are updated based on local intensity models in the vicinity of the landmarks. AAMs however differ in the sense that a complete intensity volume is modeled along with the shape, and model matching is based on trying to "blend in" the model in the target image. Mitchell et al. [574] have developed a 3D extension, where point correspondence is based on an application specific coordinate system (see Figure 29.4).

To create such an appearance model of a full volume, all the sample volumes are warped to the average shape to eliminate shape variation and yield voxel-wise correspondence across all the training samples. The voxel intensities can be represented as a shape-free vector of intensity values. Warping an image $\mathbf{I}$ to a new image $\mathbf{I}'$ involves creating a function which maps control points $\mathbf{x}_i$ to $\mathbf{x}'_i$ as well as the intermediate points in between. For the 2D case, either piecewise affine warping or thin-plate spline warping is adequate and landmark points are used to construct the shape area as a set of triangles.

In 3D models, piecewise affine warping is extended to tetrahedra with four corners, $\mathbf{x}_1$, $\mathbf{x}_2$, $\mathbf{x}_3$, and $\mathbf{x}_4$. Any point within the tetrahedron is represented as $\mathbf{x} = \alpha\mathbf{x}_1 + \beta\mathbf{x}_2 + \gamma\mathbf{x}_3 + \delta\mathbf{x}_4$. In a general case creating a tetrahedral representation of volume is solved using a 3D Delaunay Triangulation algorithm. Because all volumes are warped to the average volume, barycentric coordinates, $\alpha, \beta, \gamma, \delta$ are precomputed for each fixed voxel point eliminating the time consuming process of searching for the enclosing tetrahedron for each voxel point during the matching.

After the warping phase, the shape-free intensity vectors are normalized to an average intensity of zero and an average variance of one as described above. Next, PCA is applied to the shape-free intensity vectors to create an intensity model. In agreement with the AAM principle, shape information and intensity information are combined into a single active appearance model. Lastly, another PCA is applied to the coefficients of the shape and intensity models to form a combined appearance model [222].

In the equations below, the subscript $s$ corresponds to shape parameters while the subscript $g$ represents intensity (gray-level) parameters. To summarize, the 3D AAM is created as follows:

1. Let $\mathbf{x}_i$ denote a vector of 3D landmark points for a given sample $i$. Compute a 3D PDM and approximate each shape sample as a linear combination of eigenvectors, where $\mathbf{b}_s = P_s^T(\mathbf{x} - \overline{\mathbf{x}})$ represents the sample shape parameters.

2. Warp each image to the mean shape using a warping such as piecewise affine or thin plate spline warping to create shape-free intensity vectors.

3. Normalize each intensity vector, applying a global intensity transform with parameters $\mathbf{h}_i$, to match the average intensity vector $\overline{\mathbf{g}}$.

4. Perform a PCA on the normalized intensity images.

5. Express each intensity sample as a linear combination of eigenvectors, where $\mathbf{b}_g = P_g^T(\mathbf{g} - \overline{\mathbf{g}})$ represents the sample shape parameters.

6. Concatenate the shape vectors $\mathbf{b}_s$ and gray-level intensity vectors $\mathbf{b}_g$ in the following manner

$$\mathbf{b} = \begin{pmatrix} W\mathbf{b}_s \\ \mathbf{b}_g \end{pmatrix} = \begin{pmatrix} WP_s^T(\mathbf{x} - \overline{\mathbf{x}}) \\ P_g^T(\mathbf{g} - \overline{\mathbf{g}}) \end{pmatrix} , \tag{29.4}$$

   the weighting matrix $W$ is a diagonal matrix relating the different units of shape and intensity coefficients.

7. Apply a PCA to the sample set of all $\mathbf{b}$ vectors, yielding the appearance model

$$\mathbf{b} = Q\mathbf{c} . \tag{29.5}$$

#### 29.2.3.3 3D Active Appearance Models: Matching

Matching an appearance model to image data involves minimizing e.g. the root-mean-square intensity difference between the image data and appearance model instance by modifying the affine transformation, global intensity parameters, and the appearance coefficients. A gradient descent method is used that employs the relation between model coefficient changes and changes in the voxel intensity difference between the target image and the synthesized model [222].

Gradient descent optimization requires the partial derivatives of the error function defined by the intensity of the target and synthesized model volume. While it is not possible to create such a function analytically, these derivatives may be approximated using fixed matrices computed by randomly perturbing model coefficients for a set of known training images and observing the resulting difference in error images [222].

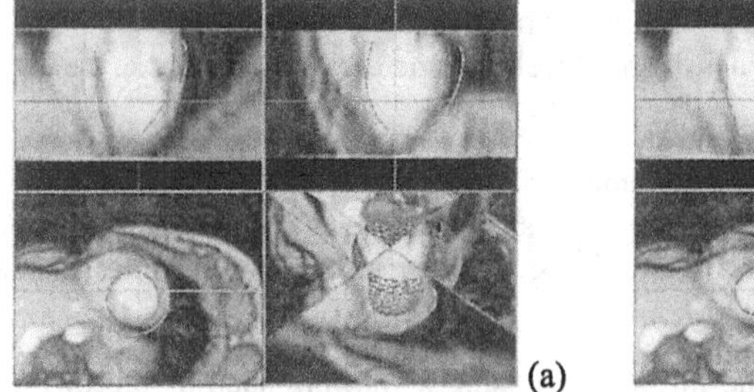


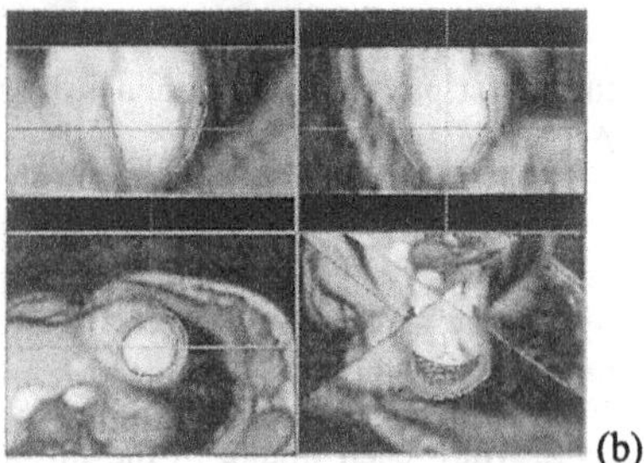


Figure 29.5. 3D AAM matching process. (a) The initial position of the model in the volumetric data set. (b) Final match result. The straight lines show the position of frames in the other two cutting planes.

Fig. 29.5 demonstrates the model matching process, from initial model position to the final fit. The 3D AAM was validated on cardiac MR and echocardiographic data in [574].

29.2.3.4 Multi-view Active Appearance Models

The 3D and 2D + time AAMs described above have mainly been designed to segment a single image set at a time, whereas cardiac MR patient examinations typically consist of a number of standardized acquisitions depicting different geometrical or functional features of the heart. For instance, the short-axis, long-axis, perfusion, rest-stress and delayed enhancement images provide complementary information about different aspects of cardiac function of the same heart. Because it involves views of the same heart, the shape features and image appearance in the different views are highly correlated: for example, an apical LV infarction may exhibit wall thinning in the apical regions in both a 4-chamber and a 2-chamber view. So far, such existing correlations between different parts of an integral patient examination have not been integrated into segmentation algorithms. To accomplish such behavior, the so-called Multi-View Active Appearance Model (AAM) was developed: an AAM extension that captures the coherence and correlation between multiple parts of a patient examination. Model training and matching are performed on multiple 2D views simultaneously, combining information from all views to yield a segmentation result. The Multi-View model is constructed by aligning the training shapes for different views separately, and concatenating the aligned shape vectors $x_i$ for each of the $N$ views. A shape vector for $N$ frames is defined as:

$$x = (x_1^T, x_2^T, x_3^T, \ldots) \ . \tag{29.6}$$

By applying a PCA on the sample covariance matrix of the combined shapes, a shape model is computed for all frames simultaneously. The principal model components represent shape variations, which are intrinsically coupled for all views. For the intensity model, the same applies: an image patch is warped on

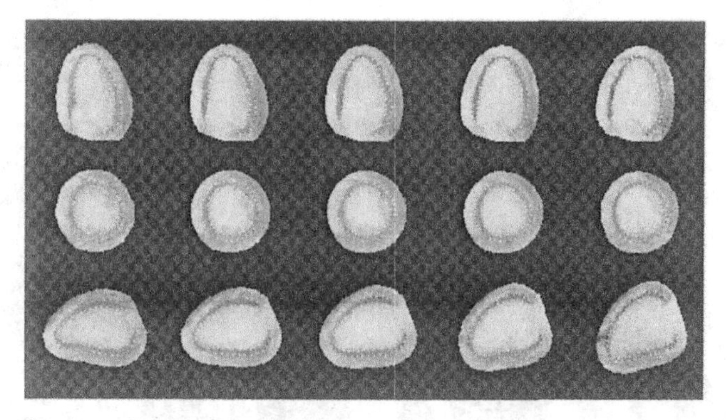

Figure 29.6. The most significant eigenmode for the multi-view AAM, varying from -2 (left) to +2 (right) standard deviations. Note that the appearance simultaneously changes in the 4-chamber (top row), short-axis (middle row) and 2-chamber views.

the average shape for view $i$ and sampled into an intensity vector $g_i$, the intensity vectors for each single frame are normalized to zero mean and unit variance, and concatenated:

$$g = (g_1^T, g_2^T, g_3^T, \ldots) \ . \tag{29.7}$$

Analogous to the other AAMs, a PCA is applied to the sample covariance matrices of the concatenated intensity sample vectors. Subsequently, each training sample is expressed as a set of shape and appearance coefficients. A combined model is computed from the combined shape–intensity sample vectors. In the combined model, the shape and appearance of both views are strongly interrelated, as is illustrated in Figure 29.6.

Like in all AAMs, estimation of the gradient matrices for computing parameter updates during image matching is performed by applying perturbations on the model, pose, and texture parameters, and measuring their effect on the residual images. Because of the correlations between views in the model, a disturbance in an individual model parameter yields residual images in all views simultaneously. The pose parameters however, are perturbed for each view separately: the model is trained to accommodate for trivial differences in object pose in each view, whereas the shape and intensity gradients are correlated for all views. In the matching procedure, the pose transformation for each view is also applied separately, whereas the model coefficients intrinsically influence multiple frames at once. Hence, the allowed shape and intensity deformations are coupled for all frames, while the pose parameter vectors for each view are optimized independently.

Multi-view AAMs have been successfully applied to segmentation of long-axis cardiac MR views and left-ventricular angiograms [613]. In Figure 29.7, examples of matching results are given for combined long- and short-axis cardiac MR scans.

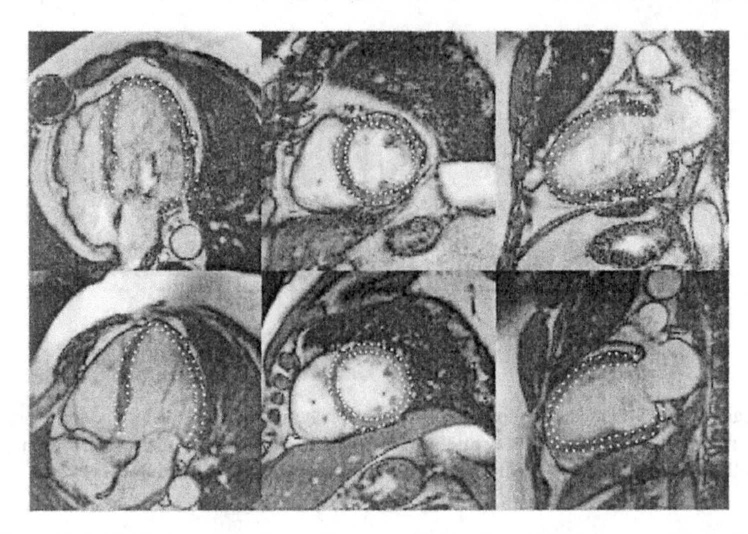

Figure 29.7. Multi-view AAM detected contours (white dotted lines) for two patients (top and bottom row) in a 4-chamber (left), short-axis(middle) and 2-chamber view (right).

#### 29.2.3.5 3D + time Active Appearance Models

Applying 3D AAM segmentation to the full cardiac cycle would require multiple models for different phases because any temporal knowledge of the interrelationship between frames would be lost. To extend the 3D AAM framework to 3D + time, Stegmann [759] has proposed to incorporate a time element to the model by phase–normalizing objects to a common time correspondence and concatenating shape and texture vectors of individual phases into a single shape and texture vector. In essence, this is similar to applying a multi-view AAM to different 3D time frames. Also, Stegmann has greatly improved the matching performance of 3D AAMs compared to the technique of Mitchell [574], by for instance introducing the "whiskers AAM": AAMs augmented with ASM-like scan line profiles that increase the model context awareness and lock-in range. In addition, the decreased computation time in his implementation (0.4 s for simultaneously segmenting the end diastolic and end systolic frames) enables an exhaustive search of several model initializations, rendering manual initialization unnecessary.

## 29.3 Discussion and Conclusion

Active Shape and Appearance models are being employed in medical image analysis more and more frequently. As most diagnostic imaging modalities nowadays deliver a high resolution, three-dimensional depiction of organs (sometimes over time), this chapter focused on higher dimensional extensions of Active Shape and Appearance models. Both models utilize a Point Distribution Model principle that captures the shape of an object from a set of examples in a compact mathematical description. To extend from 2D to higher dimensional PDMs, the definition of point correspondence in 3D is the most critical issue. Correspondence by registration or fitting has shown great potential in clinical applications, and can be applied to topologically complex shapes. Combining this registration-based correspondence with correspondence by optimal encoding may further optimize model properties such as compactness and specificity.

Active shape models are matched to image data by locally updating the model based on image information in the vicinity of the landmarks: main challenges for extending ASMs to 3D lie in generating update points using a robust (preferably modality and training independent) classifier. In addition, the use of an intermediary mesh combined with local mesh updates enables application to sparse, arbitrarily oriented image planes; this is not possible with AAMs due to the requirement of a densely sampled intensity volume. For AAMs, the main extension to higher dimension lies in defining a robust volume tesselation in 3D. Extensions to 2D + time, 3D + time and multiple views mainly rely on concatenating shape and intensity vectors for multiple time instances or geometric views.

A major limitation of the approaches described in this chapter is the fact that all methods rely on a balanced and representative training set. In case of a too–limited–number of training samples, or when presented with unrepresentative cases, the shape models may be overconstraining the segmentation results towards the model. A solution lies in applying a constraint relaxation when the model is close to its final solution, as has been proposed by Kaus et al. [457].

In conclusion, the field of 3D statistical shape modeling is rapidly expanding, with several biomedical applications. The landmark-based approaches introduced in this chapter have demonstrated to be an important step towards automated segmentation of dynamic volume data, because they utilize shape and appearance knowledge in a principled manner.

# Acknowledgments

This work was supported in part by NIH grant R01-HL071809. A. F. Frangi holds a Ramon y Cajal Research Fellowship from the Spanish Ministry of Education and Science [MEyC]. His research is also supported by MEyC grants TIC2002-04495-C02 and FIT-070000-2003-585, by ISCIII grant IM3-G03/185, and by a grant from MAPFRE Medicine Foundation. S. Ordas is supported by the MEyC grant AP2002-3955. B. P. F. Lelieveldt is supported by the Science Foundation of the Netherlands, Vernieuwingsimpuls grant 016.026.017.

captures the shape of an object from a set of examples in a compact mathematical description. In [illegible] from [illegible] dimensional PDMs, the definition of point correspondence [illegible] correspondence [illegible] [illegible] [illegible] applications, and [illegible] topologically [illegible] shapes. [illegible] correspondence [illegible] [illegible] [illegible] compactness and [illegible].

Active [illegible] models are matched to image data to locally update the model [illegible] in the variation of [illegible] image [illegible] 3D [illegible] generating [illegible] robust [illegible] and [illegible] independently [illegible] In addition, the [illegible] intermediary [illegible] combined with local [illegible] search [illegible] [illegible] image [illegible] [illegible] For AAMs, the main [illegible] lies in defining a [illegible] representation [illegible] to 2D [illegible] 3D [illegible] [illegible] rely on [illegible] shape and [illegible] for multiple [illegible] view.

[illegible] limitation of the approaches described in this chapter is [illegible] all [illegible] and representative training set. In case of a [illegible] when presented with [illegible] [illegible] [illegible] model [illegible] [illegible] the model [illegible] [illegible] et al. [illegible]

In [illegible] 3D [illegible] The [illegible] based [illegible] [illegible] knowledge [illegible]

## Acknowledgments

This work was supported in part by NIH grant [illegible] [illegible] Research [illegible] from the [illegible] Ministry of Education and [illegible] The research is also supported by [illegible] grants [illegible] and [illegible] [illegible] a grant from [illegible] [illegible] supported by the [illegible] [illegible] [illegible] grant [illegible] 6.0 [illegible]

# Chapter30

# Characterization of Diffusion Anisotropy in DWI


## Y. Chen



Abstract

Diffusion-weighted magnetic resonance imaging (DWI) is unrivaled in its ability to quantify changes in biological tissue microstructure noninvasively. The quantification is based on the anisotropy of water diffusion and fiber pathways determined from DWI measurements. This chapter is devoted to the study of the characterization of diffusion anisotropy. Two methods for characterizing diffusion anisotropy are introduced. One uses transition probability density function (PDF), and the other uses apparent diffusion coefficient (ADC) profiles. Techniques for estimating the PDF and ADC profiles from high angular resolution DWI are reviewed. In particular we presented a variational framework for the estimation of the PDF modeled as a mixture of two Gaussians. We also described a variational model for the estimation of the ADC profiles represented by a truncated spherical harmonic series, and the algorithm for the characterization of diffusion anisotropy using ADC profiles. These two models are distinguished by simultaneous smoothing and estimation. Experimental results indicate the effectiveness of these models in enhancing and revealing intravoxel information.


## 30.1 Introduction

Diffusion-weighted magnetic resonance imaging (DWI) adds to conventional MRI the capability of measuring the random motion of water molecules, referred to as water diffusion. The mobility of water molecules within tissue depends on the microstructure of the tissue. For instance, in most gray matter in the brain, the mobility of water molecules is the same in all directions and is termed isotropic diffusion. However, in fibrous tissues, such as cardiac muscle and brain white matter, water diffusion is with preferred direction along the dominant fiber orientation, and hindered to different degrees in different directions, causing diffusion anisotropy. DWI renders such complex information non-invasively and in vivo about how water diffuses into intricate 3-d representations of tissue. The anisotropy of water diffusion in tissue, and the sensitivity of water diffusion to the underlying tissue microstructure form the basis for the utilization of DWI to infer neural connectivity [67], and to probe tissue structures, compositions, architectures, and organizations [62, 67].

The goal of DWI data analysis is to characterize diffusion anisotropy and reconstruct fiber pathways. The changes in diffusion anisotropy or fiber pathways reflect the changes in underlying tissue properties, that can often be correlated with processes that occur in development, degeneration, disease, and aging ([660]). In this chapter we focus our attention on the characterization of diffusion anisotropy, that is to classify the diffusion as isotropic, anisotropic with one fiber or anisotropic with multiple fibers within a voxel. Two types of methods in the study of this problem will be introduced in the next two sections.

One of the methods uses the probability density function (PDF) $p$ on the displacement r of water diffusion over a period of time $t$. Since $p(\mathbf{r}, t)$ is largest in the directions of least hindrance to diffusion and smaller in other directions, the information about $p(\mathbf{r}, t)$ reveals fiber orientations and diffusion anisotropy. The standard methodology employed in most DWI experiments is the the Stejskal-tanner pulsed gradient spin echo method [762]. Two magnetic field gradient pulses of strength $\mathbf{G}$ and duration $\delta$ with a temporal separation of $\Delta$ between the onset of the pulses are applied to the simple spin-echo sequence. If the duration of the pulses $\delta$ is negligible comparing with $\Delta$, the attenuation of the MR signal $s(\mathbf{q})$ with respect to the diffusion sensitizing gradient $\mathbf{q}$ measures the Fourier transformation (FT) of the average PDF $p(\mathbf{r}, \Delta)$ on a spin displacement r over diffusion time $\Delta$ [64]:

$$s(\mathbf{q}) = s_0 \int p(\mathbf{r}, \Delta) e^{i\mathbf{q}\cdot\mathbf{r}} d\mathbf{r}, \tag{1.1}$$

where $\mathbf{q} = (2\pi)^{-1}\gamma\delta G$, $\gamma$ is gyromagnetic ratio of protons in water, and $s_0$ is the MR signal in the absence of any gradient.

The other method for the characterization of diffusion anisotropy utilizes the apparent diffusion coefficient (ADC). The ADC in DWI is defined as a function $d(\theta, \phi)$ in the Stejskal-tanner equation:

$$s(\mathbf{q}) = s_0 e^{-bd(\theta,\phi)}, \tag{1.2}$$

where $(\theta, \phi)$ $(0 \leq \theta < \pi, 0 \leq \phi < 2\pi)$ represents the direction of $\mathbf{q}$ in spherical coordinates, the $b$-factor is defined as $b = 4\pi^2|\mathbf{q}|^2(\Delta - \delta/3)$. For Gaussian diffusion, the PDF is a Gaussian:

$$p(\mathbf{r}, t) = \frac{1}{\sqrt{(4\pi t)^3 det(D)}} exp\{\frac{-\mathbf{r}^T D^{-1}\mathbf{r}}{4t}\}, \tag{1.3}$$

and the measurement $s(\mathbf{q})$ is related to the diffusion tensor $D$ by

$$s(\mathbf{q}) = s_0 e^{-b\mathbf{u}^T D\mathbf{u}}, \tag{1.4}$$

where $\mathbf{u}$ is the normalized $\mathbf{q}$. In this case the ADC is $\mathbf{u}^T D\mathbf{u}$, that is independent of the diffusion time $t$ and the magnitude $|\mathbf{q}|$ of the diffusion sensitizing gradient $\mathbf{q}$. For non-Gaussian diffusion equation (1.2) can be used to estimate $d(\theta,\phi)$ when the diffusion time $\Delta$ and diffusion gradient strength $G$ (hence $|\mathbf{q}|$) are fixed. The ADC profiles $d(\theta,\phi)$ for non-Gaussian diffusion is much more complicated. By high angular resolution acquisitions with larger $b$ value, it is possible to reveal the complex shape of the ADC profiles, which provides the information about the variance of diffusivities in different directions, and indicates the presence of multiple intravoxel fiber populations [11, 328, 818]. Recently the spherical harmonic approximation [13, 183, 329] and high rank tensors representation [620] of the ADC profiles have been used for characterizing diffusion anisotropy for non-Gaussian diffusion.

## 30.2 Estimation of PDF

From equation (1.1), the PDF $p(\mathbf{r},\Delta)$ can be estimated from the inverse FT of $s(\mathbf{q})/s_0$. However, it requires a large number of measurements of $s(\mathbf{q})$ over a wide range of $\mathbf{q}$ for a inverse FT. Recently, Tuch et al. [818] developed $q$-space imaging method to obtain high angular resolution diffusion (HARD) measurements. In [868] Wedeen et al. succeed in acquiring 512 measurements of $s(\mathbf{q})$ in each scan to perform a stable inverse FT. In related work Ozarslan et al. [621], estimated the PDF by taking a inverse FT on simulated DWI signals. The simulation considers the diffusion in a cylinder, when the applied diffusion gradient makes an angle $\theta$ with the direction of the cylinder. The signal attenuation $s(\mathbf{q})/s_0$ is given by the formula in [750].

A more common approach to estimate a transition PDF of diffusion over time $t$ from much sparser set of measurements $s(\mathbf{q})$ is assuming $p(\mathbf{r},t)$ to be a Gaussian. Under this assumption the measurement $s(\mathbf{q})$ is related to the diffusion tensor $D$ via (1.4). The diffusion tensor $D$ is a $3 \times 3$ positive definite matrix. By using model (1.3) the reconstruction of $p(\mathbf{r},t)$ can be posed as estimating the diffusion tensor $D$ via (1.4), which in principle requires only six independent diffusion-weighted measurements $s(\mathbf{q})$ plus $s_0$. This technique is known as diffusion tensor imaging (DTI). Based on the theory, that the principle eigenvector (PE) of $D$ parallels to the mean fiber orientation, it is possible to infer the orientation of the diffusion within a voxel. DTI is in particular useful for creating white matter fiber tracts [64, 190, 411].

However, it has been recognized that the single Gaussian model is inappropriate for assessing multiple fiber tract orientations, when complex tissue structure is found within a voxel [11, 63, 328, 329, 818, 868]. A simple extension to non-Gaussian diffusion is to assume that the multiple compartments within a voxel are in slow exchange and the diffusion within each compartment is a Gaussian [11, 328, 632, 817]. Under these assumption the diffusion can be modeled by a mixture of $n$ Gaussians:

$$p(\mathbf{r},t) = \sum_{i=1}^{n} f_i((4\pi t)^3 det(D_i))^{-1/2} e^{\frac{-\mathbf{r}^T D_i^{-1}\mathbf{r}}{4t}}, \tag{2.2}$$

where $f_i$ is the volume fraction of the voxel with the diffusion tensor $D_i$, $f_i \geq 0$, $\sum_i f_i = 1$, and $t$ is the diffusion time. Inserting (2.2) into equation (1.1) it yields

$$s(\mathbf{q}) = s_0 \sum_{i=1}^{n} f_i e^{-b\mathbf{u}^T D_i \mathbf{u}}. \tag{2.3}$$

To estimate $D_i$ and $f_i$, at least $7n - 1$ measurements $s(\mathbf{q})$ plus $s_0$ are required. In [817] Tuch et al. acquired HARD images with a large $b$-values, and extended the DTI to a mixture of two Gaussians to the voxels, where the signal $s(\mathbf{q})$ exhibited multiple local maxima. The $D_i$ and $f_i$ in the mixture problem (2.3) was solved with certain physiological constraints on the eigenvalues of $D_i$. Without constraints solving (2.3) is an ill posed problem. In [632] Parker et al. used the mixture model to estimate the PDF for the voxels where the Gaussian model fits the data poorly. Such voxels were identified by using the spherical harmonic representation of the ADC profiles. The Gaussian mixture problem (2.3) was solved by using Levenberg-Marquard algorithm [666].

To enhance the accuracy and stability in the estimation of biGaussian density function, recently, we developed [182] a variational framework to estimate $D_i$ and $f_i$ in (2.3). Different from the methods developed in [817] and [632], where the estimation was performed at each voxel independently, the model in [182] incorporated a smoothness constraint into the estimation. Then, the mixture problem solving is well-posed, and $D_i$ and $f_i$ $(i = 1, 2)$ were estimated over entire volume simultaneously by a joint smoothing and data fitting. This algorithm took two steps. The first step is to find the region where the diffusion is strongly isotropic or anisotropic with one fiber, (i.e. the location where the single Gaussian model fits well). This region was determined using the SHS representation of the ADC profiles $d(\theta, \phi)$. The detail of this method is given in the next section. Denote this region by $\Omega_1$. In the second step we solve the following minimization problem:

$$\min_{L_1,L_2,f} \int_\Omega (\sum_{i=1}^{2} |\nabla L_i|^{P_i(\mathbf{x})} + |\nabla f|^{P_f(\mathbf{x})}) d\mathbf{x} + \lambda_1 \int_{\Omega_1} (f-1)^2 d\mathbf{x}$$

$$+\lambda_2 \int_\Omega \int_0^{2\pi} \int_0^{\pi} |\sum_{i=1}^{2} f_i e^{-b\mathbf{u}^T L_i L_i^T \mathbf{u}} - e^{-bd}|^2 sin\theta d\theta d\phi d\mathbf{x}, \tag{2.4}$$

with the constraint $L_i^{m,m} > 0$. In (2.4) $f = f_1$ in (2.3), $L_i$ is a lower triangular matrix from the Cholesky factorization of $D_i$ $(D_i = L_i L_i^T)$ (see [862]). With the constraints on $L^{m,m}$ the factorization is unique, and $D$ is positive definite. In (2.4)

$|\nabla L_i|^p = \sum_{1 \le m,n \le 3} |\nabla L_i^{m,n}|^p$.

$$p_i(\mathbf{x}) = 1 + \frac{1}{1 + k|\nabla G_\sigma * L_i|^2}, \quad p_f(\mathbf{x}) = 1 + \frac{1}{1 + k|\nabla G_\sigma * f|^2},$$

with parameters $k$, $\sigma > 0$, and Gaussian kernel $G_\sigma$. Region $\Omega_1$ is given as a prior. The second term can help to skip local minima of (2.4). Finally the location where the model solution $f \equiv 1$ is adjusted from $\Omega_1$ by the smoothing term in (2.4). This property makes the model less sensitive to the preliminary choice of $\Omega_1$.

The smoothing terms in this model is featured by minimizing a nonstandard growth function, i.e. $p_i$ and $p_f$ are functions of $x$ rather than a constant as the standard $L^p$ norm. Similar idea has been applied in image restoration in [99, 184]. By the choice of $p_i(\mathbf{x})$ (also for $p_f$) the speed and direction of the smoothing governed by these terms at each point $\mathbf{x}$ varies according to the image gradient. At the locations where the magnitudes of the image gradients are high, $p_i(\mathbf{x}) \approx 1$, the diffusion at these locations is based on minimizing the total variation norm of the image gradient, and the direction of the diffusion is strictly tangential to the edges [98, 162, 695]. In homogeneous regions the image gradients are very small, $p_i(\mathbf{x}) \approx 2$, and the diffusion is essentially isotropic. At all other locations, the image gradient forces $1 < p_i < 2$, and the diffusion is between isotropic and total variation based, and varies depending on the local properties of the image. Therefore, the smoothing resulting from this model is very adaptive, and preserves the features in $L_i$ and $f$.

Model (2.4) has been applied to a set of HARD MRI human data. The raw HARD MR images were obtained on a GE 3.0 Tesla scanner with TR/TE=1000/85ms. The field of view =220 mm x 220 mm. 24 axial sections covering the entire brain with the slice thickness=3.8 mm and the intersection gap=1.2 mm. The diffusion-sensitizing gradient encoding is applied in 55 directions with $b = 1000s/mm^2$. Thus, a total of 56 diffusion-weighted images, with a matrix size of 256 x 256, were obtained for each slice section.

To accommodate the constraint on $L_i$ into the model in our numerical scheme we let $L^{m,m} = b_m^2$. By solving (2.4) we obtained the solutions $L_i$ and $f$, and consequently, $D_i = L_i L_i^T$ $(i = 1, 2)$. Fig. 1a shows the model solution $f$ in a health adult brain slice through the external capsule. Fig. 1b represents a color pie, which is implemented by relating the azimuthal angle ($\phi$) of the vector to color hue (H) and the polar angle ($\theta \geq \pi/2$) to the color saturation (S). We define $H = \phi/2\pi$, $S = 2(\pi - \theta)/\pi$, and $Value(V) = 1$ in SHV, so ($\phi$, $\theta$) is corresponding to a vector in the lower hemisphere. The upper hemisphere is just an antipodally symmetric copy of the lower one. The $xy$ plane is the plane of discontinuity. Figs. 1c and 1d show the color representation of the directions for the PE of $D_1(\mathbf{x})$ and $D_2(\mathbf{x})$, respectively, in the same slice as in Fig. 1a. By comparing the color-coding in Figs. 1c and 1d with the color pie shown in Fig. 1b, the directions of the PE's are uniquely determined.

To examine the accuracy of the model in recovering intravoxel information we selected a region inside the corpus callosum, where the diffusion is known as one-fiber diffusion. We computed the direction in which the ADC profiles $d$ is maximized. This direction field is shown in Fig. 2a. On the other hand we solved (2.4), and obtained $f \approx 1$ on this region. The direction field generated from the PE of $D_1$ is then shown in Fig. 2b. These two vector fields are comparable, and the one in Fig. 2b is more regularized due to the regularization terms in the model.

Most recently, we [373] proposed to replace the data fidelity term in (2.4) by

$$\int_{\Omega}\int_{0}^{2\pi}\int_{0}^{\pi} |\sum_{i=1}^{2} f_i s_0 e^{-b\mathbf{u}^T L_i L_i^T \mathbf{u}} - s(\mathbf{q})|^2 sin\theta d\theta d\phi d\mathbf{x}.$$

If we determine the strong Gaussian diffusion region $\Omega_1$ using the DWI signal $s(\mathbf{q})$ and $s_0$, this change enable us to solve the mixture problem without any prior knowledge on $d$.

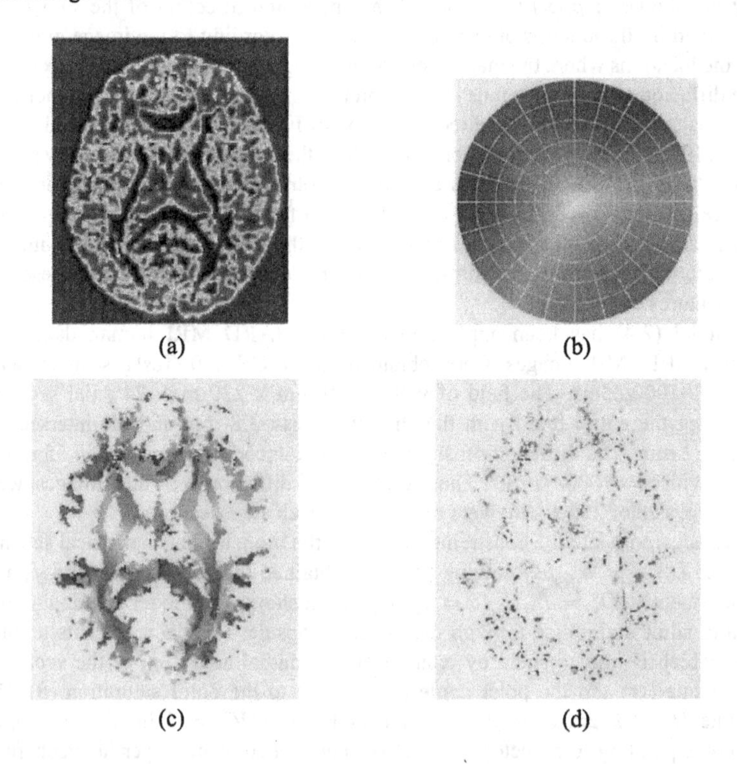

Figure 30.1. (a). Model solution $f$, (b). color pie, (c). color-coding of the 1st fiber direction mapping, (d). color-coding of the 2nd fiber direction mapping.

For sparsely distributed data it is difficult to get a desirable estimate for the PDF without a model, since a inverse FT requires the measurements $s(\mathbf{q})$ from a wide range of $\mathbf{q}$, while the $s(\mathbf{q})$'s with high $|\mathbf{q}|$ have very poor signal to noise ratio. One of the alternatives for characterizing diffusion anisotropy is to use the ADC profiles estimated from HARD MRI measurements.

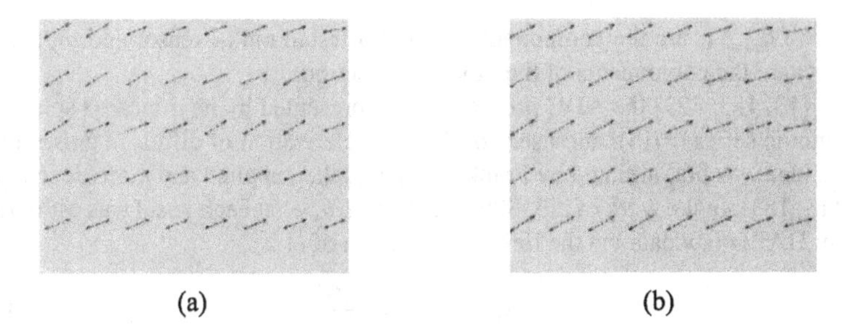

(a) (b)

Figure 30.2. Direction field obtained by (a) maximizing $d$, (b). the PE of $D_1$ (solution of (2.4))

## 30.3 Estimation of ADC profiles

In this section we discuss methods for estimating ADC profiles and characterizing diffusion anisotropy using the estimated ADC profiles.

The ADC profiles $d(\mathbf{x}, \theta, \phi)$ is related to the observed signal in DWI through the Stejskal-tanner equation (1.2). For Gaussian diffusion $d(\mathbf{u}) = b\mathbf{u}^T D\mathbf{u}$. Many models for smoothing and estimation the diffusion tensor $D$ in DTI model (1.4) having the ability of preserving the positive definite, anisotropy, or directional property of the PE of $D$ have been developed [178, 260, 313, 633, 660, 812, 813, 838, 864, 872].

The trace of $D$ provides a measure of the total diffusion within a voxel. The PE of $D$ indicates the direction of the diffusion. In particular the fractional anisotropy (FA), which is a measure of the orientational coherence of the diffusion compartments within a voxel [647], has become the most widely used measure of diffusion anisotropy. The FA is defined as

$$FA = \sqrt{\frac{3}{2}}\sqrt{\frac{(\lambda_1-\lambda_2)^2+(\lambda_2-\lambda_3)^+(\lambda_3-\lambda_1)^2)}{(\lambda_1+\lambda_2+\lambda_3)^2}}, \tag{3.1}$$

where $\lambda_i$ $(i = 1, 2, 3)$ are the eigenvalues of $D$. If fibers are strongly aligned within a voxel, the FA is high, and the diffusion is anisotropy at that voxel. If diffusion is isotropy, the FA is zero.

For non-Gaussian diffusion, ADC profiles are more complex. Tuch et al. [818] recognized that HARD imaging with high $b$-values is able to exhibit the variance of the signal as a function of diffusion gradient orientations. This admitted a generation of the concept of DTI to higher order tensors to characterize complex diffusion properties [13, 183, 329, 620].

To quantify diffusion anisotropy in [620] özarslan et al. extended DTI model (1.4) to the high rank tensor model:

$$log\frac{s(\mathbf{q})}{s_0} = -b\sum_{i_1=1}^{3}\sum_{i_2=1}^{3}\cdots\sum_{i_l=1}^{3} D_{i_1 i_2 \ldots i_l} g_{i_1} g_{i_2}\cdots g_{i_l},$$

where $D_{i_1 i_2 \ldots i_l}$ are the components of the Cartesian rank-$l$ tensor and $g_{i_k}$, $(1 \leq k \leq l)$ are the components of the gradient direction.

In [13, 183, 329] the ADC profiles were represented by its truncated spherical harmonic series (SHS), and used for the characterization of diffusion anisotropy. This idea was first initiated by Frank [329], and then applied and more developed in [13, 183]. In the work of [13, 329] the ADC $d(\theta, \phi)$ at each voxel was estimated from HARD raw data via the linearized version of (1.2):

$$d(\mathbf{q}) = -\frac{1}{b} log \frac{s(\mathbf{q})}{s_0}, \tag{3.2}$$

and then, approximated by its truncated SHS:

$$d(\theta, \phi) = \sum_{l=0}^{l_{max}} \sum_{m=-l}^{l} A_{l,m} Y_{l,m}(\theta, \phi), \tag{3.3}$$

where $Y_{l,m}(\theta, \phi)$: are the spherical harmonics, which are complex valued functions defined on the unit sphere. The odd-order terms in (3.3) are set to be zero, since the measurements are made by a series of 3-d rotation, and hence, $d(\theta, \phi)$ is antipodal symmetry. In [329] the $A_{l,m}$'s ($l$ is even) are determined by inverse spherical harmonic transform:

$$A_{l,m} = \int_0^{2\pi} \int_0^{\pi} -\frac{1}{b} log \frac{s(\mathbf{q})}{s_0} Y_{l,m}(\theta, \phi) sin\theta d\theta d\phi, \tag{3.4}$$

and in [13] they are estimated as the least-squares solutions of

$$-\frac{1}{b} log \frac{s(\mathbf{q})}{s_0} = \sum_{l=0}^{l_{max}} \sum_{m=-l}^{l} A_{l,m} Y_{l,m}(\theta, \phi). \tag{3.5}$$

Then, the coefficients $A_{l,m}$'s were used to characterize the diffusion anisotropy. In their algorithm the voxels with the significant 4th order ($l = 4$) components in SHS are characterized as anisotropic with two-fiber orientations (shorten as two-fibers), while voxels with the significant 2nd order ($l = 2$) but not the 4th order components are classified as anisotropic with single fiber orientation (shorten as one-fiber), which is equivalent to the DTI model. Voxels with the significant 0th order ($l = 0$) but not the 2nd and 4th order components are classified as isotropic. The truncated order is getting higher as the structure complexity increases. Their experimental results showed that non-Gaussian profiles arise consistently in various brain regions where complex tissue structure is known to exist. Fig.3 from [13] shows typical ADC profiles (left column) from each of three regions: pons (top), optic radiation (middle), and corona radiata (bottom), together with truncated SHS of orders 0, 2, 4, 6, and 8 (second from left to right columns). In each case, there is significant difference between the order 4 and order 2 models, which indicates significant non-Gaussian behavior. The models with order greater than 4 do not appear to change the overall profile shape significantly.

By this method the characterization of the diffusion anisotropy depends heavily on the coefficients in SHS (3.2). To improve the accuracy and stability in the

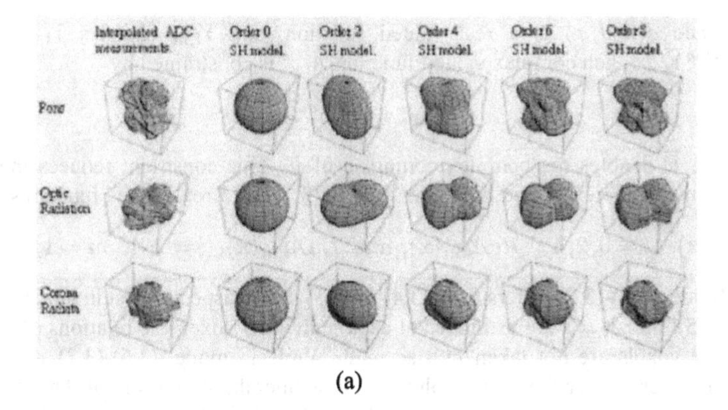


(a)

Figure 30.3. Typical ADC profiles (left) together with spherical harmonic models of orders 0, 2, 4, 6, and 8 (second from left to right) from each of the three regions (This Fig. is from [13]).

estimation of $A_{l,m}$ and enhance the characterization of anisotropy recently we [183] proposed a variational model that has the ability of simultaneously smoothing and estimating the ADC profiles from noisy HARD measurements $s(\mathbf{q})$, and preserving the relevant features, positiveness and antipodal symmetry properties of *d*. The basic idea of this approach is to approximate the ADC profiles at each voxel by a 4th order SHS (consider the case that the maximum number of fibers within a single voxel is two), whose coefficients are determined by solving the following constrained minimization problem:

$$\min_{A_{l,m}(\mathbf{x}),\tilde{s}_0(\mathbf{x})} \int_\Omega \{ \sum_{l=0,2,4} \sum_{m=-l}^{l} |\nabla A_{l,m}(\mathbf{x})|^{p_{l,m}(\mathbf{x})} + |\nabla \tilde{s}_0(\mathbf{x})|^{p(\mathbf{x})} \} d\mathbf{x}$$

$$+\frac{\lambda}{2}\int_\Omega \{ \int_0^{2\pi} \int_0^{\pi} |s(\mathbf{x},\mathbf{q}) - \tilde{s}_0(\mathbf{x}) e^{-bd(\mathbf{x},\theta,\phi)}|^2 sin\theta d\theta d\phi + |\tilde{s}_0 - s_0|^2 \} d\mathbf{x}, \quad (3.6)$$

with the constraint:

$$d(\mathbf{x},\theta,\phi) = \sum_{l=0,2,4} \sum_{m=-l}^{l} A_{l,m}(\mathbf{x}) Y_{l,m}(\theta,\phi) > 0, \quad (3.7)$$

where $\lambda > 0$ is a parameter,

$$p_{l,m}(\mathbf{x}) = 1 + \frac{1}{1 + k|\nabla G_\sigma * a_{l,m}|^2}, \; p(\mathbf{x}) = 1 + \frac{1}{1 + k|\nabla G_\sigma * s_0|^2}. \quad (3.8)$$

In (3.8) $k$, $\sigma > 0$ are parameters, $G_\sigma$ is the Gaussian kernel, and $a_{l,m}$ is the least-squares solution of (3.5).

Since $d(\mathbf{x}, \theta, \phi)$ is a real valued function, and $Y_{l,m}$ satisfies $Y_{l,-m} = (-1)^m \overline{Y_{l,m}}$, each complex valued function $A_{l,m}$ is constrained by

$$A_{l,-m} = (-1)^m \overline{A_{l,m}}$$

where $\overline{F}$ denotes the complex conjugate of $F$. This constraint reduces the 15 unknown complex valued functions $A_{l,m}$ in (3.7) to 15 real valued functions:

$$A_{l,0}(\mathbf{x}),\ (l = 0, 2, 4),\ \ ReA_{l,m}(\mathbf{x}),\ and\ ImA_{l,m}(\mathbf{x}), (l = 2, 4,\ \ m = 1, \ldots, l).$$

Model (3.6)-(3.7) differs from (3.4) or (3.5) in two aspects. First in model (3.4) or (3.5), the $A_{l,m}$'s are estimated at each individual voxel, the relations of $A_{l,m}$ across voxels are not taken into account. While in model (3.6)-(3.7) $A_{l,m}(\mathbf{x})$ are recovered over the entire volume with a smoothness constraint. Due to the present of the regularization term problem (3.6) is well-posed, and able to reveal the smooth change of diffusion anisotropy across voxels. Secondly, in (3.6) the estimation of $d$ is based on the original Stejskal-tanner equation (1.2) rather than its (log) linearized form (3.2). It has been observed in [864] that the original model provided better results in tensor field estimation from DTI. The smoothing terms in (3.6) are based on minimizing a nonstandard growth functional. As explained in the previous section by the choices of $p_{l,m}(\mathbf{x})$ and $p(\mathbf{x})$, the smoothing is isotropic in the homogeneous region, TV based along the edges, and varies in between isotropic and TV based in other regions depending on the image gradients at the location. Since the diffusion governed by this model is very adaptive so that the features in $A_{l,m}(\mathbf{x})$ and $\tilde{s}_0(\mathbf{x})$ are well preserved. The positiveness and antipodal symmetric properties of $d$ are constrained in (3.7).

The algorithm for the characterization of diffusion anisotropy in [183] is mainly based on the $A_{l,m}$'s in SHS estimated from HARD raw data, and the variance $\sigma$ of the ADC profiles $d(\theta, \phi)$ about its mean, which is as follows.

1. If

$$R_0(\mathbf{x}) =: \frac{|A_{0,0}|(\mathbf{x})}{\sum_{l=0,2,4} \sum_{m=-l}^{l} |A_{l,m}|(\mathbf{x})}, \tag{3.9}$$

is large, or the variance $\sigma(\mathbf{x})(\mathbf{x})$ of $d(\mathbf{x}, \theta, \phi)$ about its mean $\sigma(\mathbf{x})$ is small at a particular voxel $\mathbf{x}$, the diffusion at this voxel is classified as isotropic.

2. If

$$R_2(\mathbf{x}) =: \frac{\sum_{m=-2}^{m=2} |A_{2,m}|(\mathbf{x})}{\sum_{l=0,2,4} \sum_{m=-l}^{l} |A_{l,m}|(\mathbf{x})}, \tag{3.10}$$

is large at a voxel $\mathbf{x}$, the diffusion at this voxel is characterized as one-fiber diffusion.

3. For each uncharacterized voxel after the above two steps, search the directions $(\theta, \phi)$, where $d(\theta, \phi)$ attains its local maxima. If there is only one local maximum, $d$ is viewed as one-fiber diffusion. For the rest of the voxels that have more than one local maximum (say 3), the diffusion anisotropy is further

characterized by the weights:

$$W_i =: \frac{d(\theta_i, \phi_i) - d_{min}}{\sum_{i=1}^{3} d(\theta_i, \phi_i) - 3d_{min}},$$

where $(\theta_i, \phi_i)$ $(i = 1, 2, 3)$ are the directions in which $d$ attains 3 local maxima. If one of the weights is significant, it is considered as one fiber diffusion. If two weights are similar but much larger than the third one, it is viewed as two-fiber diffusion, if all three weights are similar, then higher order approximation of SHS for $d$ is required.

We applied model (3.6)-(3.7) to simulated data to test whether this model can efficiently reconstruct ADC profiles from noisy HARD measurements.

The simulated data was a set of $s_0$ and $A_{l,m}$ on a 3d lattice of dimension $6 \times 6 \times 5$. This volume consists of two homogeneous regions. In the region 1 $s_0(\mathbf{x}) = 562$, $A_{0,0} = 6.28 \times 10^{-3}$, $A_{2,0} = -8.81 \times 10^{-4}$, $A_{4,0} = 6.15 \times 10^{-5}$, $ReA_{2,1} = 5.22 \times 10^{-3}$, $ReA_{2,2} = 5.08 \times 10^{-4}$, $ReA_{4,1} = -8.47 \times 10^{-5}$, $ReA_{4,2} = 4.92 \times 10^{-5}$, $ReA_{4.3} = 3.10 \times 10^{-5}$, $ReA_{4.4} = -1.38 \times 10^{-4}$, $ImA_{2,1} = -1.82 \times 10^{-4}$, $ImA_{2,2} = -1.13 \times 10^{-3}$, $ImA_{4,1} = 9.62 \times 10^{-5}$, $ImA_{4,2} = 3.46 \times 10^{-5}$, $ImA_{4.3} = -3.58 \times 10^{-6}$, $ImA_{4.4} = 1.75 \times 10^{-5}$. In the region 2 $s_0(\mathbf{x}) = 378$, $A_{0,0} = 6.08 \times 10^{-3}$, $A_{2,0} = 2.04 \times 10^{-4}$, $A_{4,0} = 2.63 \times 10^{-4}$, $ReA_{2,1} = 6.63 \times 10^{-5}$, $ReA_{2,2} = -9.71 \times 10^{-5}$,$ReA_{4,1} = 1.27 \times 10^{-4}$, $ReA_{4,2} = 2.22 \times 10^{-4}$, $ReA_{4.3} = 1.24 \times 10^{-4}$, $ReA_{4.4} = 4.19 \times 10^{-5}$, $ImA_{2,1} = 5.77 \times 10^{-6}$, $ImA_{2,2} = 9.56 \times 10^{-6}$, $ImA_{4,1} = 6.51 \times 10^{-5}$, $ImA_{4,2} = 6.64 \times 10^{-5}$, $ImA_{4.3} = 7.52 \times 10^{-5}$, $ImA_{4.4} = 3.71 \times 10^{-5}$.

Fig. 4 shows the true, noise, and recovered ADC profiles $d(\mathbf{x}, \theta, \phi)$ for a particular slice of size 4 extracted from the volume $6 \times 6 \times 5$. The ADC profiles $d(\mathbf{x}, \theta, \phi)$ shown in Fig. 4a were computed by using (3.3) with the simulated data. Using this true $d$ the corresponding $s_{true}(\mathbf{x}, \theta, \phi)$ was constructed via (1.2) with $b = 1000s/mm^2$. Then we generated the noisy HARD MRI signal $s_n$ by adding a zero mean Gaussian noise with standard deviation $s = 0.5$. Using $s_n$ and simulated $s_0$ we estimated $A_{l,m}$ as the least-squares solutions of (3.5), and model solutions of (3.6)-(3.7), and then, obtained two corresponding $d$'s via (3.3) shown in Figs. 4b and 4c, respectively. Comparing these three figures, it is clear that the noisy measurements $s_n$ changed the original shapes of $d$ from Fig. 4a into Fig. 4b, while by applying model (3.6)-(3.7) to the noisy data to reconstruct the ADC profiles, the shapes of $d$ in Fig. 4a were recovered, as shown in Fig. 4c This experiment demonstrated that model (3.6)-(3.7) was effective in simultaneously regularizing and recovering ADC profiles.

Model (3.6)-(3.7) has also been applied to human HARD MR brain data for the estimation of ADC profiles and characterization of diffusion anisotropy.

The raw DWI data was acquired in the same way as in the experiment shown in Fig. 1. The diffusion-sensitizing gradient encoding is applied in fifty-five directions. Model (3.6)-(3.7) was applied to the raw data, and the coefficients $A_{l,m}$'s in SHS were obtained as the steady state solutions to the flow of the Euler-Lagrange equation associated with the energy function in (3.6) with constraint (3.7). The

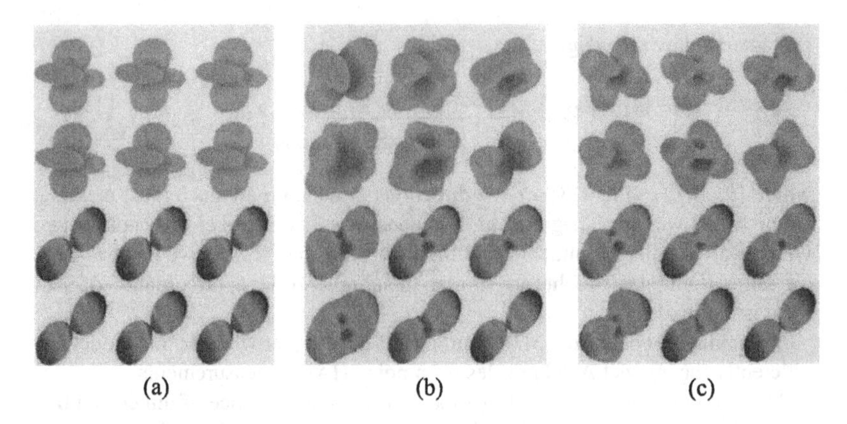


Figure 30.4. (a)-(c). The $d$ generated by (3.3), where the $A_{l,m}$'s are simulated data in (a), the least-squares solutions of (3.5) with the noisy measurement $s_n$ in (b), and the model solutions of (3.6)-(3.7) in (c).

initial $A_{l,m}$'s were chosen as the least squares solutions of (3.5). Then $d(\mathbf{x}, \theta, \phi)$ was computed via (3.3).

To characterize the diffusion anisotropy, we used the estimated $A_{l,m}(\mathbf{x})$'s to calculate $R_0(\mathbf{x})$ in (3.9), $R_2(\mathbf{x})$ in (3.10), and variance $\sigma(\mathbf{x})$ of $d(\mathbf{x}, \theta, \phi)$ about its mean: $\sigma(\mathbf{x}) = \int_0^{\pi} \int_0^{2\pi} (d(\mathbf{x}, \theta, \phi) - \sum_{i=1}^{55} d(\mathbf{x}, \theta_i, \phi_i)/55)^2 d\theta d\phi$. The diffusion at the voxels with $R_0(\mathbf{x}) > 0.856$, or $\sigma(\mathbf{x}) < 19.65$ were classified as isotropic. The diffusion at the voxels with $R_2(\mathbf{x}) > 0.75$ were considered as one-fiber diffusion. For the remaining voxels we further classified them by using the method in the third step of the characterization algorithm above. The thresholds used for $R_0$, $R_2$ and $\sigma$ were selected by using their histograms.

Fig. 5 shows the comparison of $R_2(\mathbf{x})$ maps and FA map in (3.1) for a particular slice in a health adult brain volume. Fig. 5a displays the FA image obtained by using advanced system software from GE. Fig. 5b-5d present the $R_2(\mathbf{x})$ images obtained by using (3.10), where the $A_{l,m}(\mathbf{x})$'s are estimated from three different models. The $A_{l,m}(\mathbf{x})$'s used to obtain $R_2(\mathbf{x})$ in Fig. 5b are directly computed from (3.4). Those used to obtain $R_2(\mathbf{x})$ in Figs. 5c and 5d are the least-squares solutions of (3.5) and the solutions of (3.6)-(3.7), respectively. In Figs. 5a-5d the voxels with high levels of intensities are characterized as one-fiber diffusion.

Although the image in Fig. 4a is obtained from a conventional DTI model (1.4), but it still comparable with the $R_2$ map, since single tensor diffusion characterized by SHS representation agrees with that characterized by the DTI model. However, in DTI a voxel with a low intensity of FA indicates isotropic diffusion, while using our algorithm, multi-fibers diffusion may occur at the location with the low value of $R_2$.

It is clearly evident that the ability to characterize anisotropic diffusion is enhanced, as shown in Figs. 5a-5d. Fig. 5b indicates again that the estimations of $A_{l,m}$ directly from the *log* signals usually are not good. Even the least-squares solution of (3.5) are not always effective. This can be seen by comparing the anatomic region inside the square of Figs. 5c and 5d, which are zoomed in Figs. 5e and 5f, respectively. There is a dark broken line showing on the map of the external capsule (arrow to the right on Fig. 5e), this same region was recovered by model (3.6)-(3.7) Our results also showed the connection in a cortical associative tract (arrow to the left in Fig. 5f), however, this connection was not mapped out on Fig. 5c or the zoomed image in Fig. 5e. In fact this connection was not mapped out on Figs 5a-5b either. Moreover, these two connection voxels are characterized by the third step in our algorithm as anisotropic diffusion with two-fibers (arrow to the right and left in Fig. 6b below). All these mapped connections are consistent with the known neuroanatomy. Combined together, this experimental result indicates that joint smoothing and estimation of the ADC profiles governed by model (3.6)-(3.7) has the advantage over the existing models in the enhancement of the ability to characterize diffusion anisotropy.

Fig. 6a shows a partition of isotropic diffusion, anisotropic diffusion with one-fiber, and two-fibers for the same slice displayed in figure 5. The two-fibers, one-fiber, and isotropic diffusion regions were further characterized by the white, gray, and black regions, respectively. The region inside the white square in Fig. 6a, which is the same one squared in Figs. 5c and 5d, is zoomed in Fig. 6b. It is clearly to see the two voxels directed by arrows in Fig. 6b are classified as diffusion with two-fibers. Fig. 6c represents the shapes of $d(\mathbf{x}, \theta, \phi)$ at three particular voxels (upper, middle and lower rows). The $d$'s in all three voxels are computed using (3.3). However, the $A_{l,m}(\mathbf{x})$ used in computing $d$ on the left column are the least-squares solutions of (3.5), while on the right column they are the model solutions of (3.6)-(3.7). The first and second rows show two voxels that can be characterized as isotropic diffusion using model (3.5), but as diffusion with two-fibers after applying model (3.6)-(3.7). These two voxels are the same voxels as in Fig. 5f (also Fig. 6b) directed by arrows. The lower row of Fig. 6c shows the one-fiber diffusion was enhanced after applying model (3.6)-(3.7).

Finally, we would like to point out that the number of the local maxima in ADC profiles indicates the number of the fibers through a voxel. However, for non-Gaussian diffusion the directions in which the ADC profiles attains the local maxima may not be the same as the fiber orientations.

## 30.4 Conclusion

The quantification of diffusion anisotropy in biological tissues are very complex. Two types of methods for the characterization of diffusion anisotropy were introduced in this chapter. One of them used the PDF of the diffusion of a mixture of $n$ Gaussians. The second method was based on the significant components in the SHS approximation of the ADC profiles. The methods for estimation of the mixture of Gaussians and SHS approximation of the ADC profiles were also presented in this chapter.

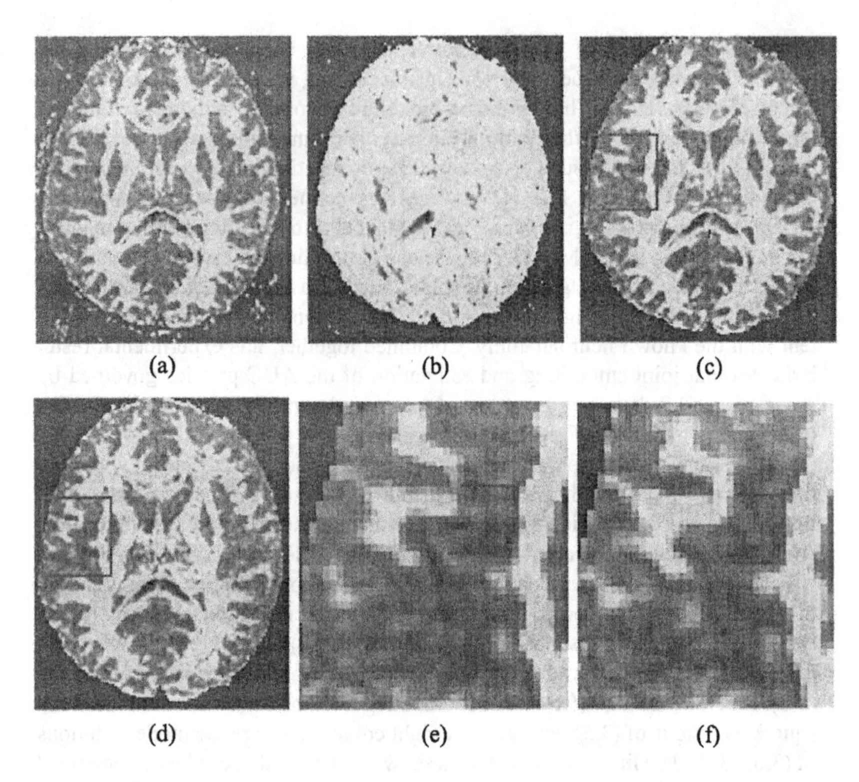


Figure 30.5. (a). FA from GE software, (b)-(d). $R_2$ with the $A_{l,m}$'s as the solutions of (1.7), least-squares solutions of (2.7), and model solutions, respectively. (e)-(f). Enlarged portions inside the squares in (c) and (d), respectively.

The second method of characterization did not require a prior knowledge for the PDF. However, the characterization might be less accurate, if the order of the significant components in the SHS was the only factor accounted for. In order to improve accuracy in [183] the variance, and the number of the local maxima of the ADC profiles was applied as the additional measurements in the characterization. But the extra measurements brought more parameters to be determined. Therefore, better models and methods to study diffusion processes inside the tissue are needed.

# Acknowledgments

I thank Qingguo Zeng and Weihong Guo for the implementation of models (2.4) and (3.6)-(3.7). I am grateful to Dr. Yijun Liu and Dr. Guojun He of Psychiatry and Neuroscience, for providing HARD images, and Prof. Baba Vemuri of CISE and Prof. Murali Rao of Mathematics, for helpful discussions.

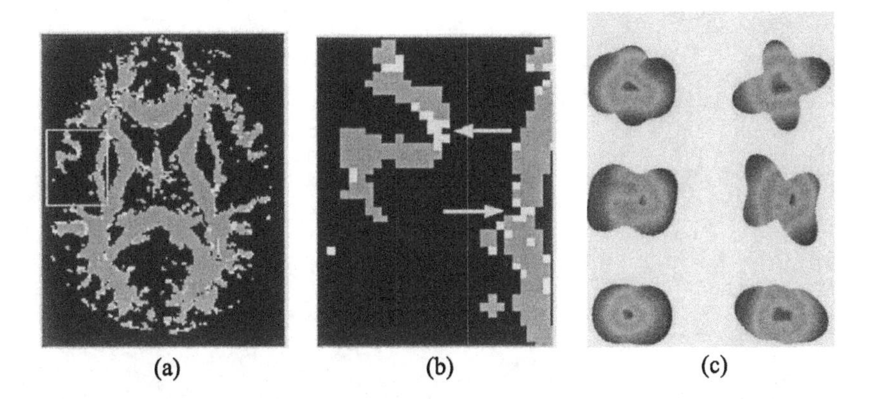


Figure 30.6. (a). Classification: White, gray, and black voxels are identified as two-fibers, one-fiber, and isotropic diffusion respectively. (b). Enlarged portions inside the white squares in (a). (c). Three typical *d*'s (upper, middle and lower rows) computed via (3.3). $A_{l,m}$ used in (3.3) on the left and right columns are the least-squares solutions of (3.5), and model solutions of (3.6)-(3.7), respectively.

# Chapter31

# Segmentation of Diffusion Tensor Images

**Z. Wang and B. Vemuri**

Abstract

Diffusion tensor images(DTI), which are matrix valued data sets, have recently attracted increased attention in the fields of medical imaging and visualization. In this chapter, we review the state of art in DTI segmentation and present some details of our recent approach to this problem.

## 31.1 Introduction

In their seminal work [63], Basser et al. introduced diffusion tensor magnetic resonance imaging (DT-MRI) as a new MRI modality from which anisotropic water diffusion can be inferred quantitatively. Since then, DT-MRI has became a powerful method to study the tissue microstructure e.g., white/gray matter connectivity in the brain or the spinal cord in vivo. DTI analysis consists of a variety of interesting problems: diffusion weighted image (DWI) acquisition, DTI restoration, DTI segmentation, DTI registration, fiber tracking and visualization. From all these, research on the DTI segmentation problem has only recently received much attention and will continue to do so in the near future. Segmentation is a fundamental problem in medical imaging and computer vision in general. DTI has the added advantage of providing directional as well as scalar information in one image as opposed to just the scalar (contrast) information present in standard MRI. In the following, we will present a brief overview of various techniques currently in vogue in segmenting DTI.

In the context of DTI segmentation in literature, there are three major approaches . The first approach is based on clustering techniques. To the best of our knowledge, the only published work using a clustering technique for DTI segmentation is due to Wiegell et al. [889]. They used k-means to achieve automatic segmentation of thalamic nuclei from DTI.

The second approach is based on the now widely popular geometric active contour model, which was independently pioneered in the computer vision community by Malladi et al. [536, 538] and in the applied mathematics community by Caselles et al. [152] and later refined in several approaches, leading to the geodesic active contours and the region-based active contours respectively in [156, 462, 174, 808, 625]. Recently, the geodesic active contour models have been used to handle DTI segmentation in [313, 646, 925]. Zhukov et al. [925] proposed a level set segmentation method that segments DTIs by segmenting scalar-valued images computed from the diffusion tensor. The scalar-valued images are functions of the eigen values of the diffusion tensors. However, this is not truly a matrix-valued image segmentation method since the direction information contained in the diffusion tensors are ignored. In [313], Feddern et al. extended the concept of image gradients to matrix-valued images for segmenting the same. The stopping criteria in the standard geodesic active contour is modified to a decreasing function of gradient magnitude of the matrix-valued image. In the same year, Pichon et al. [646] introduced an interesting diffusion flow by using an alignment penalty of the curve tangent to the dominant eigen vector (the eigen vector with the largest eigen value of the diffusion tensor) field as the conformal factor in a geodesic active contour. A group of curves can evolve in this diffusion flow and cluster together to yield a segmentation of the DTI.

The third approach is based on region-based geometric active contours and provides a much more interesting insight into the segmentation of symmetric positive definite (SPD) matrix-valued images, in particular DTIs [860, 858, 687, 508, 510]. In [860], Wang and Vemuri were the first to apply the Mumford-Shah functional [591] using an implementation involving the region-based active contours in a level-set framework to achieve matrix-valued image segmentation. This was done by incorporating a matrix distance based on the matrix Frobenius norm. Simultaneously, Rousson et al. [687] extended the geodesic active regions by incorporating region statistics of matrices for DTI segmentation. In both works [860, 687], a diffusion tensor is treated as a matrix wherein every component is independent and equally weighted. Still, they report realistic results for segmenting important subcortical structures like the corpus callosum from rat brains and human brains.

Each of the diffusion tensors in the DTI however can be viewed as the covariance matrix of a local diffusion process. In [858], Wang and Vemuri were the first to use this fact in the context of DTI segmentation. In particular, they proposed a novel diffusion tensor "distance" based on concepts grounded in information theory and incorporated it in active contour without edges model [174] for DTI segmentation. Soon after, the concepts presented in Wang and Vemuri [858] were extended by Lenglet et al. [508, 510] to the case of general probability density field segmentation using region statistics as grouping criteria. In particular, they use the Fisher information metric on the manifold of a family of probability density functions (pdf) as a distance for a family of pdfs, an idea that was also mentioned in [858] but not carried through. Very recently in [857, 859], Wang and Vemuri further extend their work in [858] to the case of the piece-wise smooth regions, based on the curve evolution implementation of the Mumford-Shah functional ([808, 175].

Rest of this chapter is organized as follows: in section 31.2, we briefly present the k-means algorithm for DTI segmentation. Then, in section 31.3, several methods applying or extending boundary based active contours are discussed. In section 31.4, we present our recent work on DTI segmentation in detail. Finally, in section 31.5 we present the conclusions.

## 31.2 K-means for DTI segmentation

The K-means algorithm has long been used for unsupervised clustering [288]. Recently, Wiegell et al. [889] applied this technique for segmentation of thalamic nuclei from DTI. Specifically, they set the number of clusters $n$ to be 14 based on visual inspection and the clustering measure between a voxel $j$ and a cluster $k$ is defined as:

$$E_{jk} = \|\mathbf{x}_j - \bar{\mathbf{x}}_k\|_{\mathbf{W}k} + \gamma\|\mathbf{D}_j - \bar{\mathbf{D}}_k\|_F \quad (31.1)$$

where $\mathbf{x}_j$ is the location of voxel $j$, $\bar{\mathbf{x}}_k$ is the center of cluster $k$, $\mathbf{D}_j$ is the diffusion tensor at voxel $j$, $\bar{\mathbf{D}}_k$ is the mean of the diffusion tensors in cluster $k$ which is simply a channel by channel mean. The first term is a Mahalanobis voxel distance defined as $\|\mathbf{x}\|_{\mathbf{W}k} = \sqrt{\mathbf{x}^T\mathbf{W}k^{-1}\mathbf{x}}$ where $\mathbf{W}k$ is the covariance matrix of the voxels in cluster $k$, and the second term is simply a Frobenius distance between two diffusion tensors defined as

$$\|\mathbf{D}_1 - \mathbf{D}_2\|_F = \sqrt{\sum_{ij}(D_{1,ij} - D_{2,ij})^2}$$

Note that due to its simplicity, the above form of diffusion tensor distance has also been used extensively in DTI restoration [871, 178] and DTI registration [12]. $\gamma$ is a weighting factor controls the tradeoff between voxel distance and diffusion tensor differences. The initialization of the algorithm is semi-automatic and the segmentation results are shown to agree with a histological atlas of the brain.

## 31.3 Boundary-based active contours for DTI segmentation

In their seminal work [455], Kass et al. introduced an elastically deformable contour dubbed the "snake" (a.k.a. "active contour") to find and link edges by evolving the "snake" in the image domain. However, this initial version had several limitations including the dependency on the parameterization and the inability to automatically change topology. A *geometric active contour* in a level-set framework was then proposed in the pioneering works of Malladi et al. [536, 537, 538] and Caselles et al. [152] to overcome these limitations. The level-set representation of evolving curves used in the works of Malladi et al. and Caselles et al. was first introduced by Dervieux and Thomasset in [261] and also independently developed and explored by Osher and Sethian [618] in fluid mechanics. Following the basic ideas of the geometric active contours in [536, 537, 538, 152], a variational formulation for the same was independently introduced in Caselles et al. [156] and Kichenassamy et al. [462], leading to the so called geodesic snakes. These models were then further developed to yield more general and stable models in [887, 740, 907, 189]. Since in these methods motion of the active contour is governed by the local image forces and local curve geometry defined along the boundary, it is appropriate to categorize them as boundary-based active contours models.

Lately, several authors have applied or extended some of these methods to DTI segmentation. Zhukov et al. [925] applied the following geometric active contour formulation to achieve DTI segmentation:

$$\frac{\partial \phi}{\partial t} = -\mathbf{F} \cdot \nabla \phi \tag{31.2}$$

where the $\phi$ is a function whose zero level set is the evolving curve $\mathbf{C}$ (or surface) and $\mathbf{F} = \mathbf{F}_{data} + \beta \mathbf{F}_{curv}$ is the speed of the evolving curve. The first term of the speed is data dependent and they use the gradient of grey scale features e.g., the gradient magnitude of some smoothed scalar volumes. Specifically, the scalar volumes they use are the trace of the diffusion tensor or a dimensionless anisotropy measure computed from DTI. The second term is the well known curvature-dependent smoothing term [618].

The works of Feddern et al. and Pichon et al. are both based on the following evolution equation of geodesic active contours:

$$\frac{\partial \phi}{\partial t} = g(.)|\nabla \phi| \nabla \cdot \frac{\nabla \phi}{|\nabla \phi|} + \nabla g(.) \cdot \nabla \phi \tag{31.3}$$

where $g(.)$ is a stopping function and depends on the local properties of the image such as the image gradient. In [313], Feddern et al. extended the gradient magnitude definition of vector-valued image to matrix-valued image using the following form:

$$gradMag(\mathbf{D}_\sigma) ::= \sqrt{\sum_{ij} |\nabla D_{\sigma,ij}|^2} \tag{31.4}$$

where $\mathbf{D}_\sigma$ is a channel by channel Gaussian smoothed DTI, $\mathbf{D}_{\sigma,ij}$ is the $ij-th$ component of $\mathbf{D}_\sigma$. They use $g(gradMag(\mathbf{D}_\sigma))$ in defining the stopping criterion function $g(.)$ and showed results for extracting the cortex from a 2D projection of a 3D human brain DTI.

Simultaneously, Pichon et al. proposed another modification of the geodesic active contours by rewriting the defining variational principle of the geodesic active contours. They define the weighted arc length $L$ of curve $\mathbf{C}$ by

$$L = \frac{1}{2}\int_0^L \|\mathbf{e}_1 - \mathbf{C}_s\|^2 F_a ds \tag{31.5}$$

where $\mathbf{e}_1$ and $F_a$ are the dominant eigen vector and the fractional anisotropy [65] of the DTI respectively. $L$ is minimized when the curve $\mathbf{C}$ is maximally aligned with the dominant eigen vector at places with high anisotropy. Then, they derive a diffusion flow by minimizing $L$ defined in (31.5) leading to:

$$\mathbf{C}_t = F_a \mathbf{C}_{ss} - curl(\mathbf{e}_1) \times \mathbf{C}_s - \nabla F_a \tag{31.6}$$

Thus, the active contour evolves to get maximally aligned to the dominant eigen vectors of the DTI and a group of curves can evolve to form a fiber bundle that can potentially be used to cluster the diffusion direction information. Note that the 3D curves corresponding to the nerve fibers are represented by the intersection of two 3D surfaces with fixed end points.

## 31.4 Region-based active contour for DTI segmentation

Region-based active contours involve the use of quantities defined over the whole image domain to evolve the curves and surfaces and they are preferred over the boundary-based cousins (discussed above) in medical image segmentation due to their robustness to noise and relative insensitivity to initialization. Currently there are two representative segmentation approaches. The first one is based on the Mumford-Shah functional [591] and was developed by us in a series of papers [860, 858, 857, 859]. The second approach is based on geometric active regions [625] developed by Lenglet et al. [508, 510] and Rousson et al. [687]. The second approach has been described in detail in an earlier chapter (of this book) by Deriche et al. and hence, we will focus on presenting the first approach pertaining to our latest work that incorporates an information theoretic diffusion tensor "distance" in the Mumford-Shah functional for DTI segmentation.

### *31.4.1 An information theoretic diffusion tensor "distance"*

In the context of DTI, water molecule diffusion inside a human or animal being imaged may be characterized by a rank two tensor $\mathbf{D}$ which is symmetric positive definite. This $\mathbf{D}$ is related to the displacement $\mathbf{r}$ of water molecules at each lattice point in the volumetric data at time $t$ via $p(\mathbf{r}|t,\mathbf{D}) = exp(\frac{-\mathbf{r}^T\mathbf{D}^{-1}\mathbf{r}}{4t})/\sqrt{(2\pi)^n|2t\mathbf{D}|}$. Thus, it is natural to use the distance measure between Gaussian distributions to induce a distance between these diffusion tensors. The most widely used information theoretic "distance" measure is the Kullback-Leibler divergence defined as

$$KL(p\|q) = \int p(\mathbf{x}) log \frac{p(\mathbf{x})}{q(\mathbf{x})} d\mathbf{x} \tag{31.7}$$

for two given densities $p(\mathbf{x})$ and $q(\mathbf{x})$. The KL divergence is not symmetric and a popular way to symmetrize it is

$$J(p,q) = \frac{1}{2}\left[KL(p\|q) + KL(q\|p)\right] \tag{31.8}$$

which is called the J-divergence. An information theoretic diffusion tensor "distance" can now be defined as the square root of the J-divergence, i.e.

$$d(\mathbf{T}_1, \mathbf{T}_2) = \sqrt{J(p(\mathbf{r}|t, \mathbf{T}_1), p(\mathbf{r}|t, \mathbf{T}_2))} \tag{31.9}$$

It is known that twice the KL divergence and thus twice the J-divergence is the squared distance between two infinitesimally nearby points on a Riemannian manifold of parameterized distributions [17]. Thus, taking the square root in (31.9) is justified. Furthermore, equation (31.9) has a very simple closed form for the case of Gaussian distributions and is given by

$$d(\mathbf{T}_1, \mathbf{T}_2) = \frac{1}{2}\sqrt{tr(\mathbf{T}_1^{-1}\mathbf{T}_2 + \mathbf{T}_2^{-1}\mathbf{T}_1) - 2n} \tag{31.10}$$

where $tr(\cdot)$ is the matrix trace operator, $n$ is the size of the square matrix $\mathbf{T}_1$ and $\mathbf{T}_2$. Note that the "distance" defined in equation (31.10) is not a true distance as it does not satisfy the triangle inequality. Rao's distance [33] between the Gaussian distributions $p(\mathbf{r}|t, \mathbf{T}_1)$ and $p(\mathbf{r}|t, \mathbf{T}_2))$ can be used to define a true distance between $\mathbf{T}_1$ and $\mathbf{T}_2$. However, this distance for diffusion tensors poses a computational difficulty for DTI segmentation in that it does not yield a closed form expression for the mean value of the DTI required in the piecewise constant segmentation model. Instead, we choose the diffusion tensor "distance" defined in (31.10) as it approximates the Rao's distance between diffusion tensors and it is also computationally efficient for the purpose of segmentation. Note that (31.10) has been proposed in various other contexts (for example [816]), however, to the best of our knowledge, this form of "distance" was proposed in the DTI analysis literature for the first time by our work in [858].

When the domain of the DTI undergoes an affine transformation, the diffusion tensors will also be transformed but by a congruent transformation. If the affine domain transformation is represented by $\mathbf{y} = \mathbf{Ax} + \mathbf{b}$, then the vector $\mathbf{r}$ representing the displacement of a water molecule will be transformed according to $\hat{\mathbf{r}} = \mathbf{Ar}$. Since $\mathbf{r}$ has a Gaussian distribution with covariance matrix $2t\mathbf{T}$, the transformed displacement $\hat{\mathbf{r}}$ has a covariance matrix of $2t\mathbf{ATA}^T$. Thus, the transformed DTI is given by

$$\hat{\mathbf{T}}(\mathbf{y}) = \mathbf{AT}(\mathbf{x})\mathbf{A}^T, \quad \mathbf{y} = \mathbf{Ax} + \mathbf{b} \tag{31.11}$$

The information theoretic diffusion tensor "distance" is invariant to such affine domain transformations, i.e.

$$d(\mathbf{T}_1, \mathbf{T}_2) = d(\mathbf{AT}_1\mathbf{A}^T, \mathbf{AT}_2\mathbf{A}^T) \tag{31.12}$$

Although the transformation of the diffusion tensor is actually a congruent transformation, the above invariance however will be referred to as "affine" invariance because, the congruent transformation on the diffusion tensors is induced by the affine transformation of the domain on which they are defined. It is easy to show that Frobenius norm of the diffusion tensor difference used in earlier published work [12, 871, 178, 860] does not have this property.

In [858], we proved the following novel theorem that allows the analytical computation of the mean value of a DTI.

**Theorem 1.** *The mean value of a diffusion tensor field is defined as*

$$\bar{\mathbf{M}}(\mathbf{T}, R) = min_{\mathbf{M} \in SPD(n)} \int_R d^2 \left[\mathbf{M}, \mathbf{T}(\mathbf{x})\right] d\mathbf{x} \tag{31.13}$$

*and is given by*

$$\bar{\mathbf{M}} = \sqrt{\mathbf{B}^{-1}} \left[ \sqrt{\sqrt{\mathbf{B}}\mathbf{A}\sqrt{\mathbf{B}}} \right] \sqrt{\mathbf{B}^{-1}} \tag{31.14}$$

*where* $\mathbf{A} = \int_R \mathbf{T}(\mathbf{x}) d\mathbf{x}$, $\mathbf{B} = \int_R \mathbf{T}^{-1}(\mathbf{x}) d\mathbf{x}$ *and* $SPD(n)$ *denotes the set of symmetric positive definite matrices of size* $n$.

This theorem is essential for the piecewise constant Mumford-Shah model used in the segmentation algorithm, wherein the DTI is modeled by piece-wise constant regions and the constant is the mean value taken over the region.

### *31.4.2 The DTI Segmentation Model*

In [857, 859], DTI segmentation in $\mathbb{R}^2$ was posed as a minimization of the following variational principle based on the Mumford-Shah functional [591]:

$$E(\mathbf{T}, \mathbf{C}) = \int_\Omega d^2(\mathbf{T}(\mathbf{x}), \mathbf{T}_0(\mathbf{x})) d\mathbf{x} + \alpha \int_{\Omega/\mathbf{C}} p(\mathbf{T})(\mathbf{x}) d\mathbf{x} + \beta |\mathbf{C}| \tag{31.15}$$

where the curve $\mathbf{C}$ is the boundary of the desired unknown segmentation, $\Omega \subset \mathbb{R}^2$ is the image domain, $\mathbf{T}_0$ is the given noisy DTI, $\mathbf{T}$ is a piecewise smooth approximation of $\mathbf{T}_0$ with discontinuities only along $\mathbf{C}$, $|\mathbf{C}|$ is the arc length of the curve $\mathbf{C}$, $\alpha$ and $\beta$ are control parameters, $d(.,.)$ is a measure of the distance between two diffusion tensors. The second term uses the Dirichlet integral [394] of the DTI $\mathbf{T}$ that is a map from $\mathbb{R}^2$ to $S$, where $S$ is a Riemannian manifold of SPD matrices of size $m$ with a metric $g$ induced by the Rao's distance for matrices. As there are $m(m+1)/2$ independent components in SPD matrices of size $m$, the dimension of $S$ is $m(m+1)/2$. Let the local coordinates of a neighborhood of $T(\mathbf{x})$ on $S$ be given by $\mathbf{u} = (u_1, ..., u_{m(m+1)/2})$, then

$$p(\mathbf{T})(\mathbf{x}) = \sum_{1 \le k \le n} \sum_{1 \le i,j \le m(m+1)/2} g_{ij}(u) \frac{\partial u^i}{\partial x_k} \frac{\partial u^j}{\partial x_k} \tag{31.16}$$

where $n = 2$ for 2D segmentation. The extension of the Mumford-Shah functional to 3D is straight forward and can be achieved simply by replacing the curve $\mathbf{C}$ with a surface $\mathbf{S}$ and the implementation in 3D is similar to that in 2D.

### *31.4.3 The Piecewise Constant Model for DTI Segmentation*

The variational principle in equation (31.15) will capture piecewise smooth regions while maintaining a smooth boundary, the balance between the smoothness of the DTI in each region and the boundaries is controlled by $\alpha$ and $\beta$. When $\alpha$ is extremely large, equation (31.15) is reduced to a simplified form which aims to capture piecewise constant regions of two types i.e., binary segmentation:

$$E(\mathbf{C}, \mathbf{T}_1, \mathbf{T}_2) = \int_R d^2(\mathbf{T}(\mathbf{x}), \mathbf{T}_1)d\mathbf{x} + \int_{R^c} d^2(\mathbf{T}(\mathbf{x}), \mathbf{T}_2)d\mathbf{x} + \beta|\mathbf{C}| \quad (31.17)$$

where $R$ is the region enclosed by $\mathbf{C}$ and $R^c$ is the region outside $\mathbf{C}$, $\mathbf{T}_1$ and $\mathbf{T}_2$ are the mean values of the DTI in region $R$ and $R^c$ respectively.

The above model can be viewed as a modification of the active contour model without edges for scalar valued images by Chan and Vese [174]. It can segment DTIs with two types of regions with different mean (constant) values (each region type however can have disconnected parts) in a very efficient way. In [858], we incorporated the information theoretic diffusion tensor "distance" (31.10) in this active contour model to achieve DTI segmentation.

The Euler Lagrange equation for the variational principle (31.17) is

$$\left[\beta k - d^2(\mathbf{T}, \mathbf{T}_1) + d^2(\mathbf{T}, \mathbf{T}_2)\right] \mathbf{N} = 0$$
$$\mathbf{T}_1 = \mathbf{M}(\mathbf{T}, R), \quad \mathbf{T}_2 = \mathbf{M}(\mathbf{T}, R^c)$$

where $k$ is the curvature of the curve $\mathbf{C}$, $\mathbf{N}$ is the outward normal to the curve. When $\mathbf{T}_1$ and $\mathbf{T}_2$ are fixed as $\mathbf{T}_1 = \mathbf{M}(\mathbf{T}, R)$ and $\mathbf{T}_2 = \mathbf{M}(\mathbf{T}, R^c)$, we have the following curve evolution for the above equation:

$$\frac{\partial \mathbf{C}}{\partial t} = -\left[\beta k - d^2(\mathbf{T}, \mathbf{T}_1(t)) + d^2(\mathbf{T}, \mathbf{T}_2(t))\right] \mathbf{N}$$

The curve evolution equation (31.18) can be easily written out in a level set framework leading to,

$$\frac{\partial \phi}{\partial t} = \left[\beta \nabla \cdot \frac{\nabla \phi}{|\nabla \phi|} - d^2(\mathbf{T}, \mathbf{T}_1) + d^2(\mathbf{T}, \mathbf{T}_2)\right] |\nabla \phi| \quad (31.18)$$

where $\phi$ is the signed distance function of $\mathbf{C}$.

We then developed a modified version of the Chan and Vese [174] implementation. Similar to [174], we used a two stage implementation in which the first stage involves evolving the embedding function $\phi$ according to equation (31.18) for a fixed $\mathbf{T}_1$ and $\mathbf{T}_2$. The second stage involves computing the mean values $\mathbf{T}_1$ and $\mathbf{T}_2$ for a fixed $\phi$. What is different here from [174] is the computation of $\mathbf{T}_1$ and $\mathbf{T}_2$ using (31.14). The major step in (31.14) is the computation of the square root of an SPD matrix and can be achieved by matrix diagonalization [857].

Equation (31.18) can be easily discretized using an explicit Euler scheme. Updating according to equation (31.18) on the whole domain $\Omega$ has a complexity of $O(|\Omega|)$ and will be rather slow when the the domain is large. Since we are only interested in the evolving the zero level set, updating only a narrow band around the zero level set will suffice and this can be achieved using the narrow band method described in [3, 539]. In order to maintain $\phi$ as a signed distance function of $\mathbf{C}$, it is necessary to reinitialize $\phi$ and can also be done only within a narrow band. There are also several other efficient numerical schemes that one may employ for example the multi-grid scheme as was done in Tsai et al. [808]. In our work, an explicit Euler scheme with the narrow band method yielded reasonably fast solutions (3-5secs. for the 2D synthetic data examples and 2-10 minutes for the 3D real DTI examples on a 1Ghz Pentium-3 CPU).

### *31.4.4 The Piecewise Smooth DTI Segmentation Model*

In certain cases, the piecewise constant assumption will be violated and the piecewise smooth model (31.15) has to be employed in such cases. In [857, 859], we extend our work to accommodate such cases. Following the curve evolution implementation of the Mumford-Shah functional by Tsai et al. [808] and Chan et al. [175], we use a two-stage scheme. In the smoothing stage, the curve is fixed and a smoothing inside the curve and outside the curve are done by preserving the discontinuity across the curve. In the curve evolution stage, the inside and outside of the smoothed DTI are fixed while the curve is allowed to move.

*Discontinuity Preserving Smoothing*
When the curve is fixed, we have the following energy functional:

$$E_{\mathbf{C}}(\mathbf{T}) = \int_{\Omega} d^2(\mathbf{T}(\mathbf{x}), \mathbf{T}_0(\mathbf{x}))d\mathbf{x} + \alpha \int_{\Omega/\mathbf{C}} p(\mathbf{T})(\mathbf{x})d\mathbf{x} \tag{31.19}$$

As we have

$$d^2(T(\mathbf{x} + h\mathbf{dx}_k), T(\mathbf{x})) = \sum_{1 \leq i,j \leq m(m+1)/2} g_{ij}(u)\frac{\partial u^i}{\partial x_k}\frac{\partial u^j}{\partial x_k}h^2 \tag{31.20}$$

where $d(.,.)$ represents Rao's distance between diffusion tensors. Since our diffusion tensor "distance" approximates Rao's distance between infinitesimally close diffusion tensors and is computationally sound, the above energy functional can be discretized as follows where we use $h = 1$:

$$E_{\mathbf{C}}(\mathbf{T}) = \sum_{\mathbf{x}} d^2(\mathbf{T}(\mathbf{x}, \mathbf{T}_0(\mathbf{x})) + \alpha \sum_{(\mathbf{x},\mathbf{y}) \in N_C} d^2(\mathbf{T}(\mathbf{x}), \mathbf{T}(\mathbf{y})) \tag{31.21}$$

where $N_C$ defines a collection of neighboring pixels. If a pair $(\mathbf{x}, \mathbf{y})$ cuts across the boundary, it is excluded from $N_C$.

We then have an energy functional of a DTI on a discrete grid and we can therefore compute its gradient with respect to this discrete DTI. A straight forward way to do this is to treat all the independent components of the diffusion tensors as the components of a vector and compute the gradient of this energy function with respect to this vector. However, the form of the gradient will not be compact. Instead, we use the derivative of a matrix function $f(\mathbf{A})$ with respect to its matrix variable $\mathbf{A}$ i.e., the components $A_{ij}$ as follows:

$$\frac{\partial f(\mathbf{A})}{\partial \mathbf{A}} = \left[\frac{\partial f(\mathbf{A})}{\partial A_{ij}}\right] = \left[lim_{dt\to 0}\frac{f(\mathbf{A}+dt\mathbf{E}_{ij})-f(\mathbf{A})}{dt}\right] \tag{31.22}$$

where $\mathbf{E}_{ij}$ is a matrix with a 1 at location $(i,j)$ and 0 elsewhere.

The directional variation of $E$ with respect to a perturbation $\mathbf{V}$ on $\mathbf{T(x)}$ is given by

$$E_{\mathbf{C}}(\mathbf{T}(\mathbf{x})+\mathbf{V}) - E_{\mathbf{C}}(\mathbf{T}(\mathbf{x})) = \frac{1}{4}tr\left[(\mathbf{B}-\mathbf{T}^{-1}(\mathbf{x})\mathbf{A}\mathbf{T}^{-1}(\mathbf{x}))\mathbf{V}\right] \tag{31.23}$$

where $\mathbf{A} = \alpha\sum_{\mathbf{y}\in N_C(\mathbf{x})}\mathbf{T}^{-1}(\mathbf{y}) + \mathbf{T}_0^{-1}(\mathbf{x})$ and $\mathbf{B} = \alpha\sum_{\mathbf{y}\in N_C(\mathbf{x})}\mathbf{T}(\mathbf{y}) + \mathbf{T}_0(\mathbf{x})$. In particular, let $\mathbf{V} = dt\mathbf{E}_{ij}$, and $\mathbf{K} = \left[\mathbf{B}-\mathbf{T}^{-1}(\mathbf{x})\mathbf{A}\mathbf{T}^{-1}(\mathbf{x})\right]$, we have:

$$E_{\mathbf{C}}(\mathbf{T}(\mathbf{x})+dt\mathbf{E}_{ij}) - E_{\mathbf{C}}(\mathbf{T}(\mathbf{x})) = \frac{1}{4}tr(dt\mathbf{K}\mathbf{E}_{ij}) = \frac{1}{4}dtK_{ij}$$

then the gradient of $E_{\mathbf{C}}$ can be derived from equation (31.22) as:

$$\frac{\partial E_C}{\partial \mathbf{T}(\mathbf{x})} = \frac{1}{4}\mathbf{K} = \frac{1}{4}\left[\mathbf{B}-\mathbf{T}^{-1}(\mathbf{x})\mathbf{A}\mathbf{T}^{-1}(\mathbf{x})\right] \tag{31.24}$$

So the minimizer of the discrete variational principle (31.21) satisfies

$$\mathbf{B} = \mathbf{T}^{-1}(\mathbf{x})\mathbf{A}\mathbf{T}^{-1}(\mathbf{x}) \tag{31.25}$$

Note that the above analytical and compact equation (31.25) is a byproduct of our choice of diffusion tensor "distance" in the form of (31.10). It is not plausible to derive such a nice form using the exact information theoretic diffusion tensor distance as given in [508, 510].

*Curve Evolution Equation*

Once the discontinuity preserving smoothing of the DTI is achieved, the DTI is fixed and the curve $\mathbf{C}$ evolves for several steps in accordance with the minimization of the following energy functional:

$$\begin{aligned} E_{\mathbf{T}}(\mathbf{C}) &= \int_R d^2(\mathbf{T}_R(\mathbf{x}),\mathbf{T}_0(\mathbf{x}))d\mathbf{x} + \int_{R^c} d^2(\mathbf{T}_{R^c}(\mathbf{x}),\mathbf{T}_0(\mathbf{x}))d\mathbf{x} \\ &\quad +\alpha\int_R p(\mathbf{T}_R)(\mathbf{x})d\mathbf{x} + \alpha\int_{R^c} p(\mathbf{T}_{R^c})(\mathbf{x})d\mathbf{x} + \beta|\mathbf{C}| \end{aligned} \tag{31.26}$$

The gradient descent of the above energy functional is given by,

$$\frac{\partial \mathbf{C}}{\partial t} = \left\{-\beta k + \left[d^2(\mathbf{T}_R,\mathbf{T}_0) - d^2(\mathbf{T}_{R^c},\mathbf{T}_0)\right] + \alpha\left[p(\mathbf{T}_{R^c}) - p(\mathbf{T}_R)\right]\right\}\mathbf{N}$$

Again for implementation, we have

$$\begin{aligned} \frac{\partial \mathbf{C}}{\partial t} &= -\beta k\mathbf{N} + \left[d^2(\mathbf{T}_R,\mathbf{T}_0) - d^2(\mathbf{T}_{R^c},\mathbf{T}_0)\right]\mathbf{N} \\ &\quad +\alpha\left[\sum_{\mathbf{y}\in N_{R^c}(\mathbf{x})} d^2(\mathbf{T}_{R^c},\mathbf{T}_{R^c}(\mathbf{y})) - \sum_{\mathbf{y}\in N_R(\mathbf{x})} d^2(\mathbf{T}_R,\mathbf{T}_R(\mathbf{y}))\right]\mathbf{N} \end{aligned} \tag{31.27}$$

The level set form of (31.27) can be easily derived and implemented similarly as in the piecewise constant case [808, 175]. The major difference here lies in the computation of $\mathbf{T}_R$ and $\mathbf{T}_{R^c}$ instead of the simple mean diffusion tensor values. Since the gradient can be computed as in (31.24), it is easy to design efficient numerical algorithm to achieve the discontinuity preserving smoothing. In [857, 859], we use gradient descent with adaptive step size due to its simplicity however, more sophisticated techniques such the implicit Euler with preconditioned conjugate gradient can be applied and will be the focus of our future research.

### *31.4.5 Experimental Results*

In [858, 857, 859], we presented several sets of experiments on the application of our DTI segmentation algorithm. We will present excerpts of these results here for the purposes of illustration. The first one is on 2D synthetic data sets, the second one is on single slices of a real DTI and the last one is on a 3D real DTI. In these experiments, if not explicitly stated, the segmentation model used is the piecewise constant model in equation (31.17).

The purpose of the synthetic data experiments is to demonstrate the need to use the full information contained in the diffusion tensors for segmentation purposes as opposed to using scalar maps computed from the diffusion tensors. To this end, we synthesize two 2D diffusion tensor fields, both are $2 \times 2$ symmetric positive definite matrix valued images on a $128 \times 128$ lattice and have two homogeneous regions. The two regions in the first diffusion tensor field differ only in the orientations while the two regions in the second diffusion tensor field only differ in the scales. These two fields are visualized as ellipses at each lattice point, as shown in Fig. 31.1 top and bottom row respectively. Each ellipse's axes correspond to the eigenvector directions of the diffusion tensor and are scaled by the corresponding eigenvalues. With an arbitrary initialization, our model yields desired segmentation results as show in Fig. 31.1. The evolving boundaries of the segmentation are shown as curves in red. Note that the first diffusion tensor field can not be segmented by using only the scalar anisotropic properties of diffusion tensors as in [925] and the second diffusion tensor field can not be segmented by using only the dominant eigenvectors of the diffusion tensors. These two examples show that one must use the full information contained in diffusion tensors to achieve quality segmentation.

For the case of 2D slices of a 3D DTI from a normal rat brain, Fig. 31.2 depicts the segmentation procedure applied to extract the corpus callosum with the evolving segmentation boundary curve in red superimposed on the ellipsoid visualization of the DTI. In the final step, the essential part of the corpus callosum is captured by our piecewise constant segmentation model. To further get the horns of the corpus callosum, we use the segmentation results of the piecewise constant model as initialization and apply the piecewise smooth region model (see equation 31.15). The result is shown in Fig. 31.3, which depicts a significant refinement over the segmentation achieved using the piecewise constant region model in Fig.

31.2. In all the above experiments, the region corresponding to water surrounding the rat brain was excluded as it is of no significance in the biological context.

Finally we demonstrate 3D segmentation results for a normal rat brain DTI of size 114 × 108 × 12. First row of Fig. 31.4 depicts the initialization, intermediate and the final stages of the segmentation algorithm in order to segment the corpus callosum. In addition, intersections of the final 3D segmentation with different slices of the $D_{xx}$ component of the DTI are shown in the bottom row of Fig. 31.4. As seen from the overlays in these images, the segmentation of the corpus callosum is visually correct. It is evident that a significant part of the corpus callosum inside this volume is captured.

Validation of the segmentations in the real DTI case for three dimensions is a hard problem since developing methods for obtaining ground truth segmentations by manually segmenting DTI data sets is nontrivial. We will focus our future efforts in this research direction.

## 31.5 Conclusion

We reviewed several approaches in DTI segmentation ranging from the clustering method to region-based active contour models. In particular, we present our recent approach in detail. Our *novel DTI segmentation algorithm* incorporates an information theoretic diffusion tensor "distance" into the popular region-based active contour models [174, 175, 808]. The particular information theoretic discriminant we employed offers several advantages: It naturally follows from the physical phenomena of diffusion, is affine invariant and is computationally tractable. The computational tractability is facilitated by a novel theorem that we proved which allows for the computation of, the mean of the diffusion tensor field in closed form, and an analytical form of the discontinuity preserving smoothing of the diffusion tensor field. By using a discriminant on diffusion tensors, as opposed to either the eigen values or the eigen vectors of these diffusion tensors, we make full use of all the information contained in the diffusion tensors. Our approach was applied to synthetic and real DTI segmentation yielding very promising results. In situations where the data does not contain sufficient information for the algorithm to yield desired segmentations, one may resort to use of shape priors built using the DTI data sets and this is one of our current research foci.

**Acknowledgment**

Authors would like to thank Dr. T. Mareci and E. Özarslan for providing the DT-MRI data. We also would like to thank Dr. R. Deriche and Dr. M. Rousson for their useful suggestions on this chapter. This work was supported in part by the grant NIH RO1 NS42075.

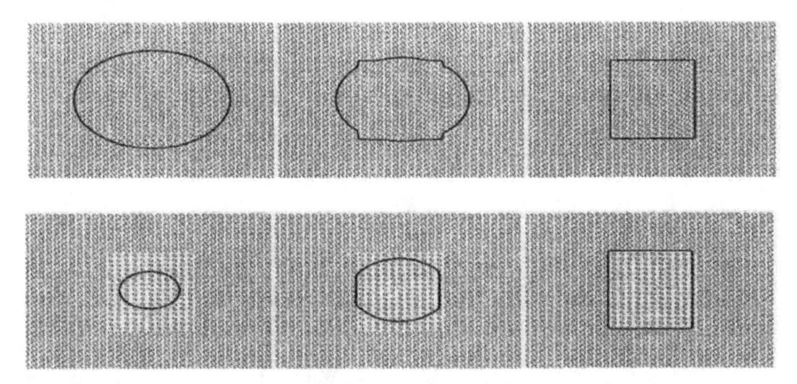

Figure 31.1. Segmentation of synthetic diffusion tensor fields with two regions. Top tow: Two regions are homogeneous and differ only in the orientations. Bottom row: two regions are homogeneous and differ only in scale. Left to right are the initial, intermediate and final steps of the curve evolution process for segmentation.

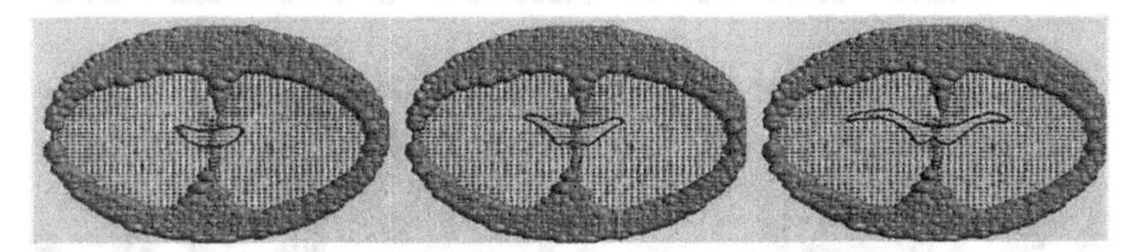

Figure 31.2. Segmentation of the corpus callosum from a real DTI slice. Left to right: initial, intermediate and final steps in segmenting the corpus callosum.

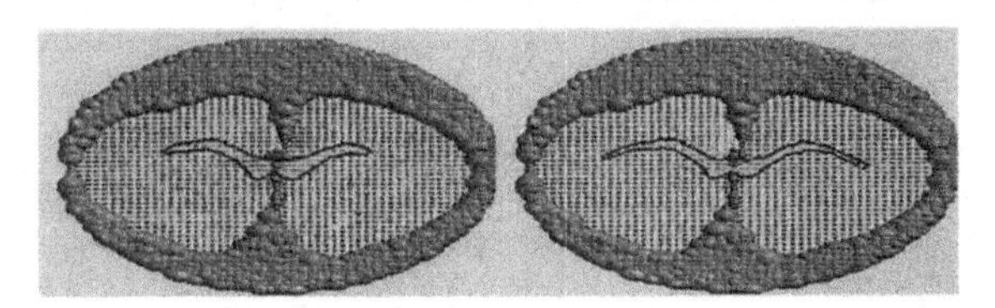

Figure 31.3. Segmentation of the corpus callosum from a real DTI slice using the piecewise smooth model. Left to right: initial and final steps in separating the corpus callosum .

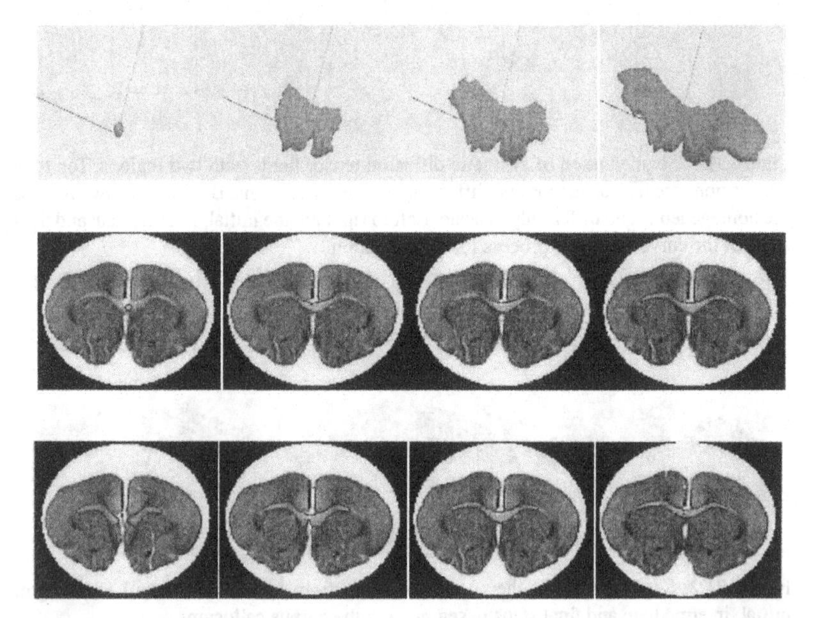

Figure 31.4. 3D Segmentation of the corpus callosum from the DTI of a normal rat brain. First row, left to right.: initial, intermediate and final steps in separating the corpus callosum. Second row, left to right: a 2D slice of the corresponding evolving 3D segmentation in the first row superimposed on the $D_{xx}$ component. Third row, left to right: different 2D slices of the final segmentation superimposed on the $D_{xx}$ component.

# Chapter32

# Variational Approaches to the Estimation, Regularization and Segmentation of Diffusion Tensor Images

**R. Deriche, D. Tschumperlé, C. Lenglet and M. Rousson**

Abstract

Diffusion magnetic resonance imaging probes and quantifies the anisotropic diffusion of water molecules in biological tissues, making it possible to non-invasively infer the architecture of the underlying structures. In this chapter, we present a set of new techniques for the robust estimation and regularization of diffusion tensor images (DTI) as well as a novel statistical framework for the segmentation of cerebral white matter structures from this type of dataset. Numerical experiments conducted on real diffusion weighted MRI illustrate the techniques and exhibit promising results.

## 32.1 Introduction

Diffusion magnetic resonance imaging is a relatively new modality [505] that acquires, at each voxel, data allowing the reconstruction of a probability density function characterizing the average motion of water molecules. As of today, it is the only non-invasive method that allows to distinguish the anatomical structures of the cerebral white matter. Well-known examples are the corpus callosum, the arcuate fasciculus or the corona radiata. These are commissural, associative and projective neural pathways, the three main types of fiber bundles, respectively connecting the two hemispheres, regions of a given hemisphere or the cerebral cortex with subcortical areas. Diffusion MRI is particularly relevant to a wide range of clinical applications related to pathologies such as acute brain ischemia, stroke, Alzheimer's disease or schizophrenia. It is also extremely useful in order to identify the neural connectivity patterns of the human brain [507] and references therein.

In 1994, Basser et al [63] proposed to model the probability density function (*pdf*) of the three-dimensional molecular motion $r$, at each voxel of a diffusion MR image, by a Gaussian distribution whose covariance matrix is given by the diffusion tensor. Diffusion tensor imaging (DTI) thus produces a volumic image containing, at each voxel, a $3 \times 3$ symmetric positive-definite matrix. The estimation of these tensors requires the acquisition of diffusion weighted images in several non-collinear sampling directions as well as a $T2$-weighted image. Numerous algorithms have been proposed to perform a robust estimation and regularization of these tensors fields [811], [839], [884], [541], [872], [863], [864], [226], [814], [124], [179], [557],[679]. Among all these works, it is worth pointing out that [864] was the first to use the original Stejskal-Tanner equation, and not the linearized form, in the data term. The authors showed the importance of this model and relied on the Cholesky decomposition to estimate the symmetric, positive-definite tensors. In sections 32.2 and 32.3, we will tackle the estimation and regularization tasks within a common variational framework while taking into account the symmetry and positive definiteness constraints.

Moreover, it is well-known that normal brain functions require specific cortical regions to communicate through fiber pathways. Based on DTI, most of the existing techniques addressing the issue of the anatomical connectivity mapping work on a fiber-wise basis. In other words, they do not take into account the global coherence that exists among fibers of a given tract. Recent work by Corouge et al [225] has proposed to cluster and align fibers by local shape parameterization so that a statistical analysis of the tract geometrical and physiological properties can be carried out. This work relies on the extraction of a set of streamlines from diffusion tensor images by the method proposed in [583] which is known to be sensitive to noise and unreliable in areas of fibers crossings.

For these reasons, we propose, in section 32.4, to directly perform the segmentation of diffusion tensor images in order to extract neural fibers bundles. Contrary to the methods proposed in [925], [889], [314],[861], [858] and [441], our approach is grounded on the expression of statistics in the space of multivariate Gaussian distributions [687], [508], [509]. We use this information in a level-set and region-based framework to evolve a surface while maximizing the likelihood of the region to extract. The central point in the developments of section 32.4 will be the choice of the probability metric, e.g. the dissimilarity measure used to compare any two probability density functions.

## 32.2 Estimation of Diffusion Tensor Images

### *32.2.1 Data acquisition*

Our dataset consists of 30 diffusion weighted images $S_k : \Omega \subset \mathbb{R}^3 \to \mathbb{R}$, $k = 1, ..., 30$ as well as a single image $S_0$ corresponding to the signal intensity in the absence of any diffusion-sensitizing gradient. They were obtained on a GE 1.5 $T$ Signa Echospeed with standard 22 $mT/m$ gradient field. The echoplanar images were acquired on 56 evenly spaced axial planes with $128 \times 128$ pixels in each slice. Voxel size is 1.875 $mm \times 1.875$ $mm \times 2.8$ $mm$. 6 gradient directions $\mathbf{g}_k$, each with 5 different $b$-factors and 4 repetitions were used. Imaging parameters were: $b$-factors between 0 and 1000 $s.mm^{-2}$, $TR = 2.5$ $s$, $TE = 84.4$ $ms$ and a square field of view of 24 $cm$ [659][1].

### 32.2.2 Linear estimation

We recall that the estimation of a field of $3\times3$ symmetric positive definite matrices $\mathbf{D}$ is performed by using the Stejskal-Tanner equation 32.1 [762] for anisotropic diffusion.

$$S_k(x) = S_0(x)\exp\left(-b\mathbf{g}_k^T\mathbf{D}(x)\mathbf{g}_k\right) \quad \forall x \in \Omega \tag{32.1}$$

where $\mathbf{g}_k$ are the normalized non-collinear gradient directions and $b$ the diffusion weighting factor. Many approaches have been derived to estimate the tensor field $\mathbf{D}$.

If we effectively restrict ourselves to 6 gradient directions, Westin et al. derived in [884] a compact analytical solution to equation 32.1 and, by doing so, eliminated the need to solve it for every single data point. The idea relies on the introduction of a dual tensor basis $\tilde{\mathbf{B}}_k$, computed from the tensor basis $\mathbf{B}_k = \mathbf{g}_k\mathbf{g}_k^T$, and which can be used to decompose any given tensor $\mathbf{D}(x)$. We then end up with the closed-form solution:

$$\mathbf{D} = \sum_{k=1}^{6} \frac{1}{b}\ln\left(\frac{S_0}{S_k}\right)\tilde{\mathbf{B}}_k \tag{32.2}$$

This method turns out to be sensitive to noise and easily influenced by potential outliers. This is due to the low number of measurements intrinsically used by this approach and by the choice of the minimization function (see [541] where the Geman-McLure M-estimator is used in order to reduce outlier-related artifacts). Moreover resulting tensors may not be positive definite, which requires a subsequent reprojection step [814].

### 32.2.3 Variational estimation

In order to deal with a more complete estimation approach, we propose to incorporate some important priors such as tensor positivity and regularity into a variational formulation of the estimation problem by minimizing the following energy on the manifold of real $3 \times 3$ symmetric positive-definite matrices

[1] Data courtesy of J.F. Mangin and J.B Poline, CEA/SHFJ, Orsay, France

$S^+(3,\mathbb{R})$:

$$\underset{\mathbf{D}(x)\in S^+(3,\mathbb{R})}{\operatorname{argmin}} \int_\Omega \sum_{k=1}^{n} \psi\left(\left|\ln\left(\frac{S_0(x)}{S_k(x)}\right) - b\mathbf{g}_k^T\mathbf{D}(x)\mathbf{g}_k\right|\right) + \alpha\rho(|\nabla\mathbf{D}(x)|)dx \tag{32.3}$$

where $\psi$ controls the robust estimation and the Lagrange multiplier $\alpha$, together with the scalar function $\rho$, drives the anisotropic regularity of the solution. Minimizing this criterion, in the constrained tensor space, leads to the following evolution equation:

$$\begin{cases} \mathbf{D}_{(t=0)} = \mathbf{Id} \\ \frac{\partial \mathbf{D}}{\partial t} = (\mathbf{G}+\mathbf{G}^T)\mathbf{D}^2 + \mathbf{D}^2(\mathbf{G}+\mathbf{G}^T) \end{cases}$$

where $\mathbf{G}$ corresponds to the gradient of the unconstrained criterion. defined as $G_{ij} = \sum_{k=1}^{n} \psi'(|v_k|)\text{sign}(v_k)\left(\mathbf{g}_k\mathbf{g}_k^T\right)_{ij} + \alpha\text{div}\left(\frac{\rho'(|\nabla\mathbf{D}|)}{|\nabla\mathbf{D}|}\nabla D_{ij}\right)$ with $v_k = \ln(S_0/S_k) - b\mathbf{g}_k^T\mathbf{D}\mathbf{g}_k$.
Note that if $\psi(v) = v^2$ and $\alpha = 0$, the criterion reduces to a simple multilinear regression by least square that generalizes the linear estimation method of Westin et al [884] and provides a positive definite solution since the minimization is done in the constrained space $S^+(3,\mathbb{R})$. This variational method converges to a much more consistent solution thanks to its global behavior. Concerning the implementation part, a carefully designed numerical scheme, based on manifold integration, to ensure that the estimate stays on $S^+(3,\mathbb{R})$ at each step of the gradient descent, is used to solve the associated Euler-Lagrange equations:

$$\mathbf{D}_{(t+dt)} = \mathbf{A}^T\mathbf{D}_{(t)}\mathbf{A} \text{ with } \mathbf{A} = \exp\left(\mathbf{D}_{(t)}(\mathbf{G}+\mathbf{G}^T)dt\right)$$

Our iterative method starts from a field of isotropic tensors that are evolving in $S^+(3,\mathbb{R})$ and are morphing until their shapes fit the measured data $S_0, S_k$. Enforcing the positiveness and regularity constraints has a large interest for DTI estimation, and leads to more accurate results than with classical methods. For more details, we refer the interested readers to the article [814].

## 32.3 Regularization of Diffusion Tensor Images

The variational estimation method naturally brings some spatial coherence and smoothness into the generated tensor field. However, the fundamental properties of diffusion tensors, like diffusivities and principal orientations, are contained in their spectral features. It can then be interesting to regularize the tensor field with regard to those spectral elements. This will bring more coherence into the tensor structural information and thus improve any subsequent processing such as the tracking of neural fibers.

### *32.3.1 On some non-spectral methods and their limitations*

Non-spectral methods are based on a direct anisotropic smoothing of the diffusion weighted data $S_k$ or consider each tensor as 6 independent scalar components $\mathbf{D}(x)_{ij}$ (by symmetry) with possible coupling. We thus evolve each $\mathbf{D}(x)_{ij}$ by minimizing the following quantity:

$$E(\mathbf{D}) = \int_\Omega \frac{\alpha}{2}|\mathbf{D}(x) - \mathbf{D}_0(x)|^2 + \rho(|\nabla\mathbf{D}(x)|)dx \tag{32.4}$$

where $\mathbf{D}_0$ designates the initial noisy tensor field and the field gradient norm $|\nabla\mathbf{D}|$ behaves as a coupling term between the tensors components. However, eigenvalues tend to diffuse faster than eigenvectors, resulting in a *swelling* effect on the tensors.

Spectral methods separately consider the eigen-elements of the tensors. Eigenvalues smoothing is typically performed by a vector-valued anisotropic PDE ([702] and references therein) satisfying the maximum principle in order to preserve the positiveness. The three orthonormal eigenvectors define a matrix of $O(3)$ which can be regularized by acting only on the principal eigenvector $\mathbf{u}^1$ and then reconstructing the associated tensor [226]. The field of orthonormal matrices can also be evolved under a scheme preserving the eigenvectors norms and angles [811]. This boils down to solving a system of coupled and constrained PDEs. However, all these approaches require a time-consuming step of eigenvectors realignment since a given vector and its opposite are both solution of the same singular value decomposition and thus yield artificially discontinuous vectors fields.

### *32.3.2 A fast isospectral method*

In [179], we proposed an efficient alternative to the previous spectral techniques, which does not require any spectral decomposition, by building flows acting on a given submanifold of the linear set of matrix-valued functions and preserving some constraints. We showed that this amounts to characterizing the velocity of the flows (ie. the tangent space of the submanifold) at each point of that submanifold. Actually, the relevant constraints (orthogonality, eigenvalues conservation ...) can be expressed by simply working with the proper Lie group or homogeneous space. For example, an isospectral flow acts on a field of real symmetric matrices and preserves their eigenvalues. Moreover, its velocity is directly derived from the matrices field gradient, hence no need for realignment. If $[X, Y]$ denotes the Lie bracket of $X$ and $Y$, e.g. $XY - YX$, the general form for our isospectral flow is given by:

$$\frac{\partial \mathbf{D}}{\partial t} = [\mathbf{D}, [\mathbf{D}, (\mathbf{G} + \mathbf{G}^T)]] \tag{32.5}$$

where $\mathbf{G}$ prescribes the desired regularization process, such as

$$G_{ij} = \operatorname{div}\left(\frac{\rho'(|\nabla\mathbf{D}|)}{|\nabla\mathbf{D}|}\nabla D_{ij}\right)$$

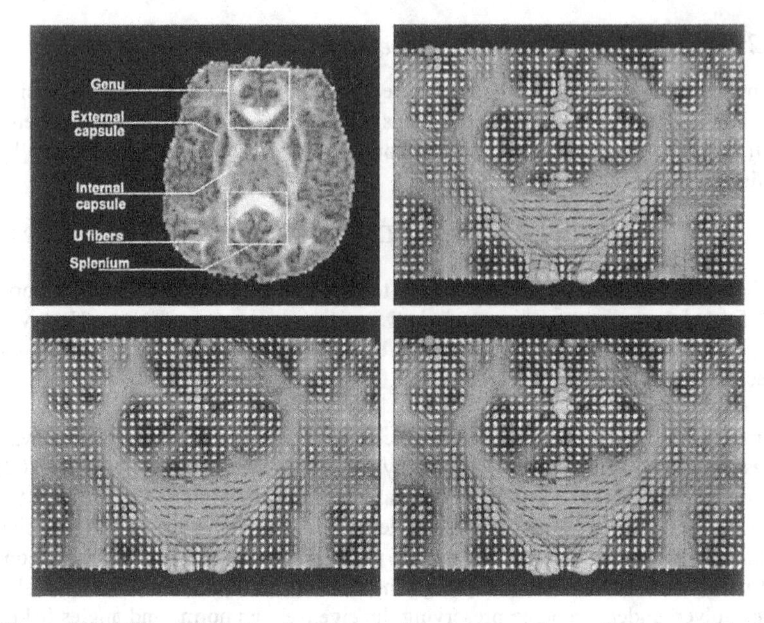


Figure 32.1. DTI regularization in the genu of the corpus callosum ([TOP LEFT]: Annotated fractional anisotropy axial slice, [TOP RIGHT]: Original tensors, [BOTTOM LEFT]: Non-spectral regularization, [BOTTOM RIGHT]: Isospectral flow)

$\rho$ denotes the same scalar function as in section 32.2.3 and preserves important structures of the tensor field. A specific reprojection-free scheme based on the exponential map can also be used to implement the PDE (32.5):

$$\mathbf{D}_{(t+dt)} = \mathbf{A}^T \mathbf{D}_{(t)} \mathbf{A} \text{ with } \mathbf{A} = \exp\left(dt[\mathbf{G} + \mathbf{G}^T, \mathbf{D}_{(t)}]\right)$$

Results of non-spectral smoothing and isospectral flow on diffusion tensors estimated in the genu of the corpus callosum are presented in figure 32.1.

## 32.4 Segmentation of Diffusion Tensor Images

The previous sections described algorithms for the estimation and the regularization of diffusion tensor images. We now focus on the segmentation of these tensor-valued images, seen as fields of Gaussian probability density functions. We first set up the level-set and region-based surface evolution framework that will be used throughout this section. We then progressively introduce the various statistical parameters associated with the probability metrics derived from the Euclidean distance, the Kullback-Leibler divergence and finally, the geodesic distance between probability density functions.

### 32.4.1 Level-set and region-based surface evolution

Our ultimate goal is to compute the optimal 3D surface separating an anatomical structure of interest from the rest of a diffusion tensor image. The region-based front evolution, as developed in [684], is an efficient and well-suited framework for our segmentation problem. We hereafter summarize the basic notions of this technique.

Let $\mathbf{s}$ be the optimal boundary between the object to extract $\Omega_1$ and the background $\Omega_2$. We introduce the level-set [261], [262] and [618] function $\phi : \Omega \rightarrow \mathbb{R}$, defined as follows:

$$\begin{cases} \phi(x) = 0, & \text{if } x \in \mathbf{s} \\ \phi(x) = \mathcal{D}_{Eucl}(x, \mathbf{s}), & \text{if } x \in \Omega_1 \\ \phi(x) = -\mathcal{D}_{Eucl}(x, \mathbf{s}), & \text{if } x \in \Omega_2 \end{cases}$$

where $\mathcal{D}_{Eucl}(x, \mathbf{s})$ stands for the Euclidean distance between $x$ and $\mathbf{s}$ and $\Omega = \Omega_1 \cup \Omega_2$. Furthermore, let $H_\epsilon(.)$ and $\delta_\epsilon(.)$ be regularized versions of the Heaviside and Dirac functions as defined in [174].
Let $q(x, r)$ be the probability density function of our random vector $r$ of $\mathbb{R}^3$ describing the water molecules average motion at a given voxel $x$ of a DTI dataset. We also denote by $p_1$ and $p_2$ the probability distributions of the *pdfs* $q(x, .)$ respectively in $\Omega_1$ or $\Omega_2$. Then, according to the Geodesic Active Regions model [625], and by adding a regularity constraint on the interface, the optimal partitioning of $\Omega$ in two regions $\Omega_1$ and $\Omega_2$ is obtained by minimizing:

$$\begin{aligned} E(\phi, p_1, p_2) = \nu \int_\Omega |\nabla H_\epsilon(\phi)| dx - \int_\Omega H_\epsilon(\phi) \log p_1(q(x, .)) dx \\ - \int_\Omega (1 - H_\epsilon(\phi)) \log p_2(q(x, .)) dx \end{aligned} \tag{32.6}$$

We have reached the point where we need to express $p_1$ and $p_2$, e.g. the probability distributions in the space of probability density functions $q(., r)$. This is the purpose of the next sections.

### 32.4.2 Multivariate Gaussian distributions as a linear space

When dealing with diffusion tensor images, we recall that the molecular motion is assumed to follow a Gaussian law of zero mean. The diffusion tensor can indeed be interpreted as the covariance matrix of the underlying Brownian motion. As proposed in [687], we start by considering the parameters space of three-dimensional Gaussian *pdfs* $q(., r)$ as linear, which boils down to reducing a diffusion tensor image to a vector-valued volume, each voxel being assigned with the 6-dimensional vector of the variances and covariances, and the probability metric being Euclidean.

Let $u(x)$ be the vector representation of a tensor $\mathbf{D}(x)$, the probability distributions of $u(x)$ in the regions $s = 1, 2$ are defined as:

$$p_s(u|\overline{u}_s, \Lambda_s) = \frac{1}{(2\pi)^3|\Lambda_s|^{1/2}} e^{-\frac{1}{2}(u-\overline{u}_s)^T \Lambda_s^{-1}(u-\overline{u}_s)}$$

The Euclidean mean vectors $\overline{u}_s$ and covariance matrices $\Lambda_s$ have to be estimated. They can simply be introduced as unknown in (32.6) and optimized for during the front evolution process. Our objective function 32.6 then becomes:

$$E(\phi, \{\overline{u}_{1,2}, \Lambda_{1,2}\}) = \nu \int_\Omega |\nabla H_\epsilon(\phi)| dx - \int_\Omega H_\epsilon(\phi) \log p_1(u(x)|\overline{u}_1, \Lambda_1) dx$$
$$- \int_\Omega (1 - H_\epsilon(\phi)) \log p_2(u(x)|\overline{u}_2, \Lambda_2) dx$$

This type of energy was studied in [684], [686], the Euler-Lagrange equations for $\phi$ yield the following evolution equation for the level-set function $\phi(x)$ $\forall x \in \Omega$:

$$\phi_t(x) = \delta_\epsilon(\phi)\Big(\nu \operatorname{div}\frac{\nabla\phi}{|\nabla\phi|} + \frac{1}{2}\log\frac{|\Lambda_2|}{|\Lambda_1|} - \frac{1}{2}(u(x) - \overline{u}_1)^T \Lambda_1^{-1}(u(x) - \overline{u}_1)$$
$$+ \frac{1}{2}(u(x) - \overline{u}_2)^T \Lambda_2^{-1}(u(x) - \overline{u}_2)\Big)$$

while it can be shown that the statistical parameters must be updated by their empirical estimates [687]. Adequate implementation schemes for this type of optimization can be found in [174]. If we restrict the covariance matrices to the identity, these equations simplify and the likelihoods in equation (32.6) simply become the Euclidean distance between the vectors $u$ and $\overline{u}_{s=1,2}$, which is equivalent to the Frobenius norm of the difference between the corresponding tensors, as studied in [861].
Figure 32.2 illustrates this method on a synthetic dataset where the Y-shape region to be segmented only differs from the background by the orientation of its tensors. A crossing area with low fractional anisotropy was created and Gaussian noise was separately added on the eigenvalues and eigenvectors to stress the algorithm.

Motivated by the method proposed by Wang and Vemuri in [858], we now derive the statistics and the associated evolution equation based on a more natural and widely used measure of dissimilarity between *pdf*s, known as the Kullback-Leibler divergence or relative entropy.

### 32.4.3 *Information-theoretic statistics between distributions*

We will show that this approach is not only more natural, in the sense that it is strongly rooted and used in the information theory community, but also more versatile since it enables the segmentation algorithm to work on fields of Gaussian densities as well as on non-parametric densities [508].
We consider a general probability density function $q(x, r)$ of the random vector $r$ of $\mathbb{R}^3$. The symmetrized Kullback-Leibler divergence can be used to express

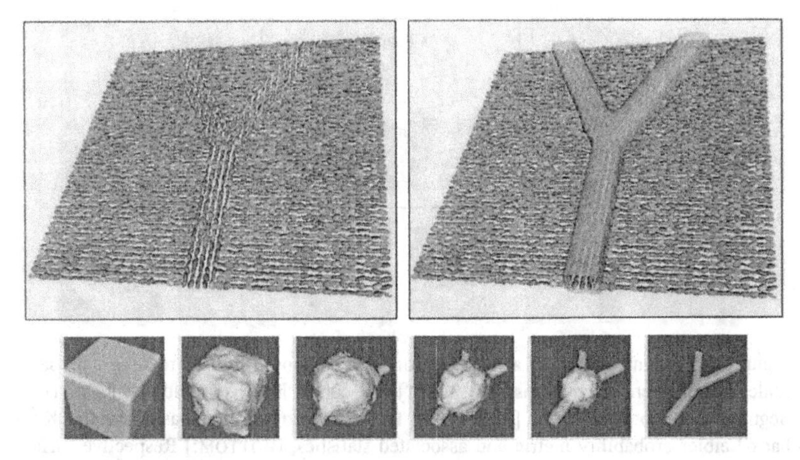

Figure 32.2. Segmentation (with Euclidean probability metric) of a noisy tensor field composed by two regions with same scales but different orientations ([TOP LEFT]: 2D-cut of the tensor field, [TOP RIGHT]: Final segmentation, [BOTTOM]: Surface evolution)

the dissimilarity between diffusion processes at different locations of $\Omega$. With $q(x,.), q(y,.)$ $\forall x, y \in \Omega$ two probability density functions from $\mathbb{R}^3$ onto $\mathbb{R}^+$, their symmetrized Kullback-Leibler divergence is given by:

$$\mathcal{D}_{kl}(q(x,.), q(y,.)) = \frac{1}{2}\int_{\mathbb{R}^3}\left(q(x,r)\log\frac{q(x,r)}{q(y,r)} + q(y,r)\log\frac{q(y,r)}{q(x,r)}\right) dr \tag{32.7}$$

We denote by $\overline{q}_1$ and $\overline{q}_2$ the mean probability density functions over $\Omega_1$ and $\Omega_2$ verifying equation 32.10. In this section, we make the assumption that the *pdfs* in $\Omega_1$ and $\Omega_2$ have respective Gaussian distributions $p_1^{kl}, p_2^{kl}$ with means $\overline{q}_1, \overline{q}_2$ and variances $\sigma_1^2, \sigma_2^2$:

$$p_{s=1,2}^{kl}(q|\overline{q}_s, \sigma_s^2) = \frac{1}{\sqrt{2\pi\sigma_s^2}}\exp\frac{-\mathcal{D}_{kl}^2(q, \overline{q}_s)}{2\sigma_s^2}$$

We can then rewrite our objective function 32.6 as follows:

$$E(\phi, \{\overline{q}_{1,2}, \sigma_{1,2}^2\}) = \nu\int_\Omega |\nabla H_\epsilon(\phi)|dx - \int_\Omega H_\epsilon(\phi)\log p_1^{kl}(q(x)|\overline{q}_1, \sigma_1^2)dx$$
$$- \int_\Omega (1 - H_\epsilon(\phi))\log p_2^{kl}(q(x)|\overline{q}_2, \sigma_2^2)dx \tag{32.8}$$

In the case where the $\sigma_s^2$ are equal to 1, this energy is equivalent to the one proposed in [858]. As for the Euclidean probability metric, the Euler-Lagrange equations yield the following evolution equation:

$$\phi_t(x) = \delta_\epsilon(\phi)\left(\nu\text{div}\frac{\nabla\phi}{|\nabla\phi|} + \log\frac{p_2^{kl}(q(x)|\overline{q}_2, \sigma_2^2)}{p_1^{kl}(q(x)|\overline{q}_1, \sigma_1^2)}\right) \tag{32.9}$$

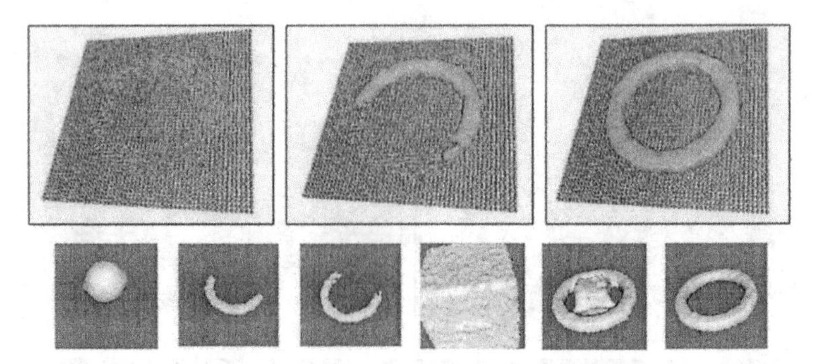

Figure 32.3. Segmentation of a noisy tensor field composed by two regions with same scales but different orientations ([TOP LEFT]: 2D-cut of the tensor field, [TOP CENTER]: Segmentation obtained from [858], [TOP RIGHT]: Segmentation based on the Kullback-Leibler probability metric and associated statistics, [BOTTOM:] Respective surface evolutions)

Moreover, it can be shown that the variance must be updated by its empirical estimation with respect to the Kullback-Leibler divergence, whereas some more work is needed for $\overline{q}_s$, defined as:

$$\overline{q}_{s=1,2} = \underset{q_s}{\text{argmin}} \frac{1}{|\Omega_s|} \int_{\Omega_s} \mathcal{D}_{kl}^2(q(x), q_s)dx \tag{32.10}$$

Indeed, for a general probability density function $q(., r)$, the variance is easily computed as in [686] but the estimation of the $\overline{q}_s$ might require the use of numerical approximation techniques if no closed-form expression is available. It turns out that, for Gaussian *pdfs* $q(., r)$, the energy 32.8 simplifies as follows:

$$\begin{aligned} &E(\phi, \{\overline{q}_{1,2}, \sigma_{1,2}^2\}) = \\ &\nu \int_\Omega |\nabla H_\epsilon(\phi)|dx + \frac{1}{2}\int_\Omega H_\epsilon(\phi)(\log(2\pi\sigma_1^2) + \mathcal{D}_{kl}^2(q(x), \overline{q}_1)\sigma_1^{-2})dx \\ &+\frac{1}{2}\int_\Omega (1 - H_\epsilon(\phi))(\log(2\pi\sigma_2^2) + \mathcal{D}_{kl}^2(q(x), \overline{q}_2)\sigma_2^{-2})dx \end{aligned} \tag{32.11}$$

Using the closed-form expressions provided in [858] for the symmetrized Kullback-Leibler divergence between two Gaussian *pdfs* and for the associated mean density $\overline{q}_s$ parameterized by the mean diffusion tensor $\overline{\mathbf{D}}_s$, the Euler-Lagrange equations for our energy yield (the dependence on $x$ is omitted for the sake of clarity):

$$\begin{aligned} \phi_t = \delta_\epsilon(\phi)\Bigg(&\nu \text{div}\frac{\nabla\phi}{|\nabla\phi|} + \frac{1}{2}\left(\frac{3(\sigma_1^2 - \sigma_2^2)}{2(\sigma_1^2\sigma_2^2)} + \log\frac{\sigma_2^2}{\sigma_1^2}\right) + \\ &\frac{1}{8}\left(\text{tr}\left(\mathbf{D}^{-1}\overline{\mathbf{D}}_2 + \overline{\mathbf{D}}_2^{-1}\mathbf{D}\right)\sigma_2^{-2} - \text{tr}\left(\mathbf{D}^{-1}\overline{\mathbf{D}}_1 + \overline{\mathbf{D}}_1^{-1}\mathbf{D}\right)\sigma_1^{-2}\right)\Bigg) \end{aligned} \tag{32.12}$$

Notice that we obtain additional terms (the $\sigma_s^2$ coefficients) in equation 32.12 if compared to the Euler-Lagrange equations proposed in [858].
Figure 32.3 illustrates the importance of the variance in our model.

The symmetrized Kullback-Leibler divergence, although it does not satisfy the triangle inequality, has many useful properties and is widely used to measure dissimilarities between *pdfs*. However, for particular densities like multivariate Gaussian distributions of fixed mean, better probability metrics are available. In the next section, we show how a Riemannian metric can be associated with the 6-dimensional parameters space of these densities using the Fisher information matrix. The geodesic distance, intrinsic mean and covariance matrix of multivariate Gaussian distributions, as well as curvature information, can be efficiently computed to yield a generalized Gaussian distribution of multivariate Gaussian densities. This generalized distribution can then be used in our segmentation framework.

### *32.4.4 A Riemannian approach to DTI segmentation*

We now consider the Riemannian manifold $\mathcal{M}$ of the family of three-dimensional Gaussian probability density functions parameterized by the 6 components of their covariance matrix $\Sigma$ (in other words, the diffusion tensor $\mathbf{D}$). Following the work by Rao [670] and Burbea-Rao [138], where a Riemannian metric was introduced in term of the Fisher information matrix, we wish to define the notion of geodesic distance and intrinsic statistics on this 6-dimensional manifold whose coordinate system, in some local chart, is given by a real vector parameter $\theta = (\theta_1, ..., \theta_6) \in \mathbb{R}^6$ such that for all random vector $r \in \mathbb{R}^3$, $\mathcal{M} = \{q(r|\theta),\ \theta \in \mathbb{R}^6\}$. In the following, we first show the main limitation of the Kullback-Leibler divergence together with its impact on the segmentation process. Then, we present the closed-form expression of the geodesic distance as well as original computational methods to approximate a generalized Gaussian distribution of multivariate Gaussian densities with common mean.

**The Fisher information matrix:** The manifold $(\mathcal{M}, g)$ equipped with the Fisher information matrix $g = g_{ij},\ i,j = 1, ..., 6$ has the structure of a Riemannian manifold [670], [746] when $g$ is non-degenerate. We recall that $g$ is defined as follows:

$$g_{ij} = \int_{\mathbb{R}^3} \frac{\partial \log q(r|\theta)}{\partial \theta_i} \frac{\partial \log q(r|\theta)}{\partial \theta_j} q(r|\theta) dr \tag{32.13}$$

By plugging the definition of a Gaussian *pdf* into equation 32.13, the $6 \times 6$ metric tensor, as presented in [509], can be expressed in terms of the parameters $\theta_i,\ i = 1, ..., 6$ used to describe the *pdfs*. Thus, instead of considering the parameterized *pdfs* as living in the linear space $\mathbb{R}^6$, we do take into account the Riemannian structure of the underlying manifold. Moreover, the Kullback-Leibler divergence $\mathcal{D}_{kl}$ turns out to be a Taylor approximation of the geodesic distance between two

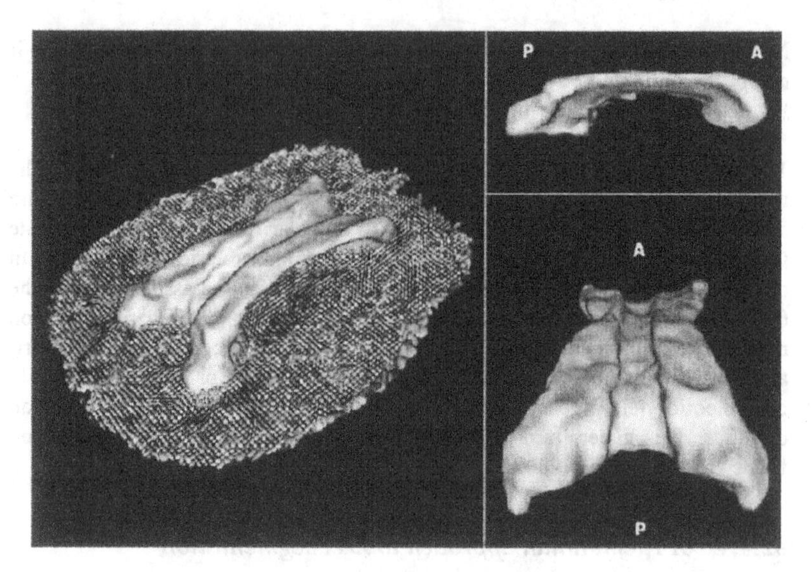


Figure 32.4. Segmentation of the corpus callosum with the Riemannian probability metric ([LEFT]: 3D view with an axial slice of diffusion tensors, [RIGHT]: A: Anterior, P: Posterior)

nearby distributions $q(r|\theta)$ and $q(r|\theta + d\theta)$, given suitable technical conditions. Indeed, as summarized in [45], it can be shown that:

$$\mathcal{D}_{kl}(\theta, \theta + d\theta) = \frac{1}{2}\mathbb{E}\left[\frac{\partial \log q(r|\theta)}{\partial \theta_i}\frac{\partial \log q(r|\theta)}{\partial \theta_j}\right] d\theta_i d\theta_j$$

This means that the infinitesimal squared geodesic distance $g_\theta(d\theta, d\theta)$ is twice the Kullback-Leibler divergence (this is also true for its symmetrized form). In other words, the method presented in the previous section assumes that we always compute distances between nearby elements of $\mathcal{M}$, which, in general, does not hold. For general *pdf*s, we may have no other choice but, in the more particular case of multivariate Gaussian densities with common mean, a closed-form solution of the geodesic distance is available, thus allowing the comparison of any two of these distributions. We now introduce this geodesic distance and derive the associated intrinsic statistical parameters.

**Geodesic distance and intrinsic statistics:** We recall that $S^+(m, \mathbb{R})$ denotes the set of $m \times m$ real symmetric positive-definite matrices $\Sigma$ (here $m = 3$). A detailed study on the definition of a statistical model on this nonlinear space was presented by the authors in [509]. Another recent work by Pennec et al [636] relies on a comparable approach to derive tensor fields filtering techniques. Following [509], [746], [137], [143], [325], [579] and [323], $S^+(m, \mathbb{R})$ can be characterized as an affine symmetric space for which the geodesic distance $\mathcal{D}_g$ between any two elements $\Sigma_1$ and $\Sigma_2$ was derived by Jensen.

**Theorem 32.4.1.** *(S.T. Jensen, 1976 [33])*
*Consider the family of multivariate Gaussian distributions with common mean vector but different covariance matrices. The geodesic distance between two members of the family with covariance matrices $\Sigma_1$ and $\Sigma_2$ is:*

$$\mathcal{D}_g(\Sigma_1, \Sigma_2) = \sqrt{\frac{1}{2}\mathrm{tr}(\log^2(\Sigma_1^{-1/2}\Sigma_2\Sigma_1^{-1/2}))} = \sqrt{\frac{1}{2}\sum_{i=1}^{m}\log^2(\lambda_i)}$$

*where the $\lambda_i$ are the roots of the determinantal equation $|\lambda\Sigma_1 - \Sigma_2| = 0$.*

We now explain how to estimate the empirical mean, as proposed by Fréchet [332], Karcher [454] and Pennec [635], as well as the empirical covariance matrix.

**Definition 32.4.1.** *The Gaussian distribution parameterized by $\overline{\Sigma} \in S^+(m, \mathbb{R})$ and defined as the empirical mean of N distributions $\Sigma_k$, $k = 1, ..., N$, achieves a local minimum of the function $\sigma^2 : S^+(m, \mathbb{R}) \to \mathbb{R}^+$ known as the empirical variance and defined as:*

$$\sigma^2(\Sigma_1, ..., \Sigma_N) = \frac{1}{N-1}\sum_{k=1}^{N}\mathcal{D}_g^2(\Sigma_k, \overline{\Sigma}) = \mathbb{E}[\mathcal{D}_g^2(\Sigma_k, \overline{\Sigma})]$$

Karcher proved in [454] that such a mean exists and is unique for manifolds of non-positive sectional curvature. This was shown to be the case for $S^+(m, \mathbb{R})$ in [746]. A closed-form expression of the mean cannot be obtained [579] but a gradient descent algorithm was proposed in [509]. A flow is derived from an initial guess $\overline{\Sigma}_0$ toward the mean of a subset of $S^+(m, \mathbb{R})$. The following evolution was obtained:

$$\overline{\Sigma}_{t+1} = \overline{\Sigma}_t^{1/2}\exp(-\frac{1}{N}\overline{\Sigma}_t^{1/2}\sum_{k=1}^{N}\log(\Sigma_k^{-1}\overline{\Sigma}_t)\overline{\Sigma}_t^{-1/2})\overline{\Sigma}_t^{1/2} \qquad (32.14)$$

The empirical covariance matrix $\Lambda^g$ relative to the mean $\overline{\Sigma}$ is defined as:

**Definition 32.4.2.** *Given N elements of $S^+(m, \mathbb{R})$ and a mean value $\overline{\Sigma}$, the empirical covariance matrix relative to $\overline{\Sigma}$ is defined as:*

$$\Lambda^g = \frac{1}{N-1}\sum_{k=1}^{N}\beta_k\beta_k^T$$

*where $\beta_k = \overline{\Sigma}\log(\Sigma_k^{-1}\overline{\Sigma})$ is the gradient of the squared geodesic distance $\nabla\mathcal{D}_g^2(\Sigma_k, \overline{\Sigma})$ in vector form.*

Finally, as detailed in [509], the Ricci curvature tensor $\mathcal{R}$ can be computed at the mean $\overline{\Sigma}$. Putting everything together and following Theorem 4 of [635], we have:

**Theorem 32.4.2.** *The generalized Gaussian distribution in $S^+(m,\mathbb{R})$ for a covariance matrix $\Lambda^g$ of small variance $\sigma^2 = tr(\Lambda^g)$ is of the form:*

$$p^g(\Sigma|\overline{\Sigma},\Lambda^g) = \frac{1+O(\sigma^3)+\epsilon(\sigma/\xi)}{\sqrt{(2\pi)^{m(m+1)/2}|\Lambda^g|}} \exp\frac{-\beta^T\gamma\beta}{2} \quad \forall\Sigma\in S^+(m,\mathbb{R})$$

*where $\beta = \overline{\Sigma}\log(\Sigma^{-1}\overline{\Sigma})$ is expressed in vector form and the concentration matrix is $\gamma = (\Lambda^g)^{-1} - \mathcal{R}/3 + O(\sigma) + \epsilon(\sigma/\xi)$. $\xi$ is the injection radius at $\overline{\Sigma}$ and $\epsilon$ is such that* $\lim_{0^+} x^{-\beta}\epsilon(x) = 0\ \forall\beta\in\mathbb{R}^+$.

**Implementation:** We can use the very same variational framework as the one described in section 32.4.2 in order to maximize the likelihoods of the diffusion tensors distributions in $\Omega_1$ and $\Omega_2$. This can now be achieved with respect to the geodesic distance by using $p^g_{s=1,2}(\Sigma|\overline{\Sigma}_s,\Lambda^g_s)$ and by accordingly evolving the level-set function $\phi$ toward the optimal segmentation. Figure 32.4 illustrates how well this approach performs on a real diffusion tensor image.

## 32.5 Conclusion

Diffusion magnetic resonance imaging gives a direct insight into the microstructure of biological tissues through the measurement of hindered molecular motion. In this chapter, we have described efficient and versatile numerical methods for the estimation and the regularization of the diffusion tensor images. We have also presented a novel statistical and geometric approach to the segmentation of DTI data. The central point of this front evolution framework relies on the definition of dissimilarity measures and statistics between diffusion tensors, seen as the covariance matrices of Gaussian probability density functions. The major contribution of this set of techniques is related to the robust extraction of anatomical structures in the brain white matter.

# Chapter33

# An Introduction to Statistical Methods of Medical Image Registration

**L. Zöllei, J. Fisher and W. Wells**

Abstract

After defining the medical image registration problem, we provide a short introduction to a select group of multi-modal image alignment approaches. More precisely, we choose four widely-used statistical methods applied in registration scenarios for analysis and comparison. We clarify the implicit and explicit assumptions made by each, aiming to yield a better understanding of their relative strengths and weaknesses. We also introduce a figural representation of the methods in order to provide an intuitive way of illustrating their similarities and differences.

## 33.1 Introduction

Registration of medical image data sets is the problem of identifying a set of geometric transformations which map the coordinate system of one data set to that of the others. Depending on the nature of the input modalities, we distinguish between uni-modal and multi-modal cases, according to whether the images being registered are of the same type. The multi-modal registration scenario is more challenging as corresponding anatomical structures will have differing intensity properties. In our analysis, we focus on the multi-modal case.

When designing a registration framework, one needs to decide on the nature of the transformations that will be used to bring images into agreement. For example, rigid transformations are generally sufficient in the case of bony structures while non-rigid mappings are mainly utilized for soft tissue matching. One must also evaluate the quality of alignment given an estimate of the aligning transformation. *Objective functions* or *similarity measures* are special-purpose functions that are designed to provide these essential numerical scores. The goal of a registration problem can then be interpreted as the optimization of such functions over the set of possible transformations. In general, these problems correspond to multi-dimensional non-convex optimization problems where we cannot automatically bracket the solution (as we would in the case of a 1D line-search). Thus an initial estimate of the aligning transformation is needed before the search begins.

In the past few decades there have been numerous types of objective functions proposed for solving the registration problem. Among these, there exist a variety of methods that are based on sound statistical principles. These include various maximum likelihood [512, 798], maximum mutual information [534, 883], minimum Kullback-Leibler divergence [201], minimum joint entropy [771] and maximum correlation ratio [678] methods. We are primarily interested in these, and in our discussion we select four of these registration approaches for further analysis. We explore the relative strengths and weaknesses of the selected methods, we clarify the type of explicit and implicit assumptions they make and demonstrate their use of prior information. By such an analysis and some graphical representations of the solution manifold for each method, we hope to facilitate a deeper and more intuitive understanding of these formulations.

In the past, similar or more detailed overview studies of the registration problem have been reported. Roche et al. [678], for example, have described the modeling assumptions in *uni-modal* registration applications and a general maximum likelihood framework for a certain set of multi-modal registration approaches, and we have described a *unified information theoretic* framework for analyzing multi-modal registration algorithms [927, 928].

## 33.2 The Similarity Measures

In our analysis, we discuss four objective criteria that rely on clear statistical principles: maximum likelihood (ML), approximate maximum likelihood (MLa), Kullback-Leibler divergence (KL) and mutual information (MI). While not an exhaustive list, these similarity measures are representative of a significant group of currently used registration algorithms. Many registration approaches either directly employ or approximate one of these measures.

While the analysis presented here carries straightforwardly to registration of multiple data sets, for simplicity, we focus on the case of two *registered* data sets, $u(x)$ and $v(x)$ sampled on $x \in \mathbb{R}^M$. These data sets represent, for example, two imaging modalities of the same underlying anatomy in an M-dimensional space. In practice, we observe $u(x)$ and $v_o(x)$ where the latter is related to $v(x)$ by

$$v_o(x) = v(T^*(x)) \quad \text{or} \quad v(x) = v_o\left(\left(T^*\right)^{-1}(x)\right), \tag{33.1}$$

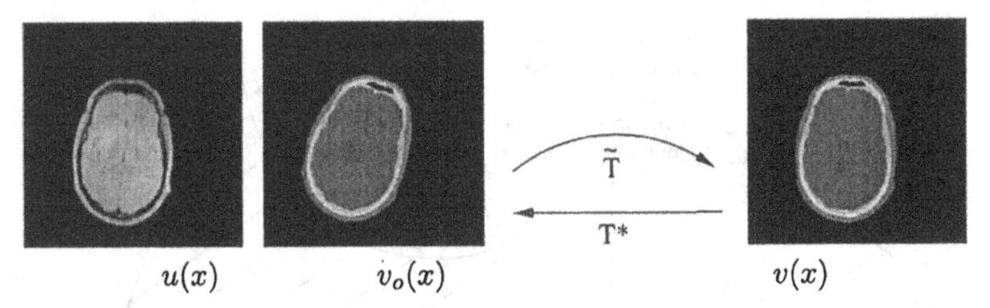


Figure 33.1. A 2D example of the registration problem. The *observed* input images are $u(x)$, an MRI slice, and $v_o(x)$, a CT slice. $v(x)$ is the CT slice that is in correct alignment with the MRI slice. The unknown transformation that relates the observed data to the aligned image is $T^*$. The goal of the registration algorithm is to make $\tilde{T}$ be the best estimate of $(T^*)^{-1}$.

where $T^* : \mathbb{R}^M \to \mathbb{R}^M$ is a bijective mapping corresponding to an unknown *relative* transformation. The goal of registration is to find an estimate of an *aligning* transformation $\tilde{T} \approx (T^*)^{-1}$ which optimizes some objective function of the observed data sets.[1] Figure 33.1 demonstrates the key components of the registration problem via a 2D example.

Throughout our analysis (and consistent with practice) spatial samples $x_i$ are modeled as independent random draws of a uniformly distributed random variable $X$ whose support is the domain of $u(x)$. Consequently, all the analyzed methods assume that

(IID-i) observed intensities $v_o(x_i)$ and $u(x_i)$ can be viewed as independent and identically distributed (*i.i.d.*) random variables, despite spatial dependencies present within the data.

This is a simple consequence of the property that a *function* of an *i.i.d.* random variable is itself an *i.i.d.* random variable under very general conditions.

## *33.2.1 Maximum Likelihood*

The maximum likelihood (ML) method of parameter estimation has served as the basis for many registration algorithms. Its popularity in parameter estimation can be explained by the fact that as the sample size increases, ML becomes the smallest variance unbiased estimator. As we will see, practical issues generally preclude a direct ML approach. Analysis of the method is however useful for comparison purposes. Given that the input images are related by an unknown transformation $T^*$ (see Figure 33.1), we parameterize the observed data samples

[1]Technically speaking, $u(x)$ may have undergone some transformation as well, but without loss of generality we assume it has not. If there were some canonical coordinate frame (e.g. an anatomical atlas) by which to register the data sets one might consider transformations on $u(x)$ as well.

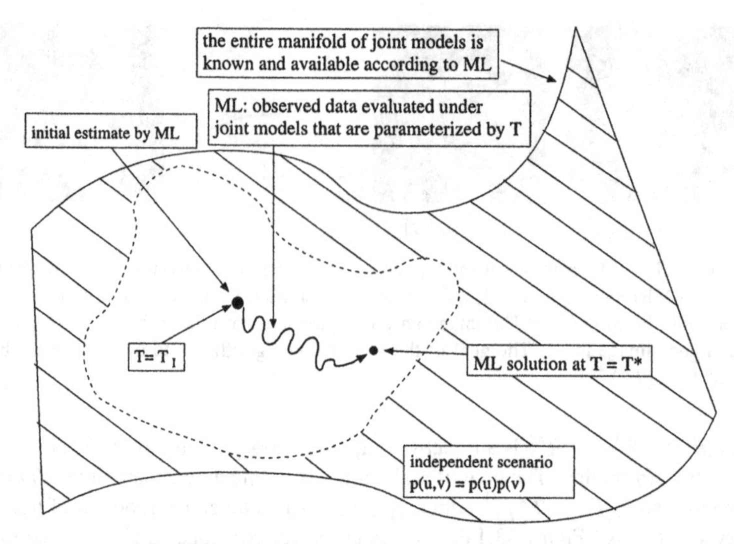


Figure 33.2. Joint density manifold of the registration search space parameterized by T. According to the classical ML approach, the entire manifold of joint models is known and available for the optimization task. The solution is defined at the location which maximizes the likelihood of the observed sample pairs. Here $T_I$ has been chosen as an initial estimate for the search.

(a sequence of joint measurements drawn *i.i.d*) as

$$\begin{aligned} \mathcal{Y}_{T^*} &= \{[u, v_{T^*}]_1, \ldots, [u, v_{T^*}]_N\} \\ &= \{[u(x_1), v(T^*(x_1))], \ldots, [u(x_N), v(T^*(x_N))]\} \\ &= \{[u(x_1), v_o(x_1)], \ldots, [u(x_N), v_o(x_N)]\}. \end{aligned}$$

According to the ML criterion, we obtain estimates by varying some parameters of a probabilistic model that is being evaluated on a set of observed data. In the case of our registration problem, the optimal geometrical transformation that *explains* the observations according to the ML criterion satisfies the (normalized) log-likelihood criterion:

$$T_{\mathrm{ML}} = \arg\max_T \mathcal{L}_T(\mathcal{Y}_{T^*}) \tag{33.2}$$

$$= \arg\max_T \frac{1}{N} \sum_i \log(p([u, v_{T^*}]_i; T)). \tag{33.3}$$

$\mathcal{L}_T(.)$ in Equation (33.2) indicates that we are evaluating a model parameterized by the transformation $T$.

This formulation of the registration problem implicitly assumes that

(ML-i) as $T$ approaches $T^*$, Equation (33.3) is non-decreasing.

An important distinction between currently used registration methods and the classical ML approach is that the former optimize the objective criterion by transforming the joint observations ($[u, v_{T^*}]_i$). In contrast, a classical ML approach optimizes the objective function by changing the parameters of the joint *density* model under which we evaluate the observations (as a function of transformation $T$), leaving the observations static throughout the search process. Below, we will indicate these differences via notional graphs of the solution paths of the selected methods. In Figure 33.2, according to the ML approach, the entire search space of joint models (parameterized by transformation $T$) is considered to be known and available. We let the initial estimate of this example be $T = T_I$ (the identity transformation). The solution lies at transformation $T^*$ that maximizes the likelihood function with respect to the currently observed images. Thus the initial guess by ML is modified in order to satisfy the criterion.

This framework highlights two practical obstacles to a direct ML approach. The optimization of Equation (33.2) requires the solution of a system of non-linear equations for which no direct global solution typically exists. Finding a globally optimal solution would likely require that $p(u, v; T)$ be pre-computed over all relative transformations $T$ (see Figure 33.2). An alternative is to use an optimization procedure that searches for a local optimum, which would require the ability to produce $p(u, v; T)$ on demand, as we search. The first approach may be impractical due to computational and memory limitations. While the second approach may be feasible, as far as we know, it has not been tested or used. The second obstacle is that there are configurations of the data for which a considerable set of transformations form an equivalence class under the ML criterion. As the relative transformations away from the solution $T = T^*$ become large, we observe empirically that the joint models tend toward statistical independence. In addition, they may tend towards the same independent model (more on this appears in Section 33.2.4, below). In this situation, the ML criterion will lose traction for such large transformations. (In Figure 33.2, such models are located outside of the dashed outline.) As we shall see, MI-based approaches can be interpreted as moving away from these models.

### 33.2.2 Approximate Maximum Likelihood

As mentioned above, the optimization of Equation (33.2) is generally a very difficult problem. Suppose, however, that we have a model of the joint density of our data sets at one particular parameter setting, specifically when the multimodal images are registered. We can estimate this model from other *registered* data sets and evaluate new observations under the resulting model. This idea was first suggested by Leventon and Grimson and we refer to it as an approximate maximum likelihood registration approach (MLa) [512]. (A similar approach has been discussed more recently in [924].) The approach makes two strong modeling assumptions:

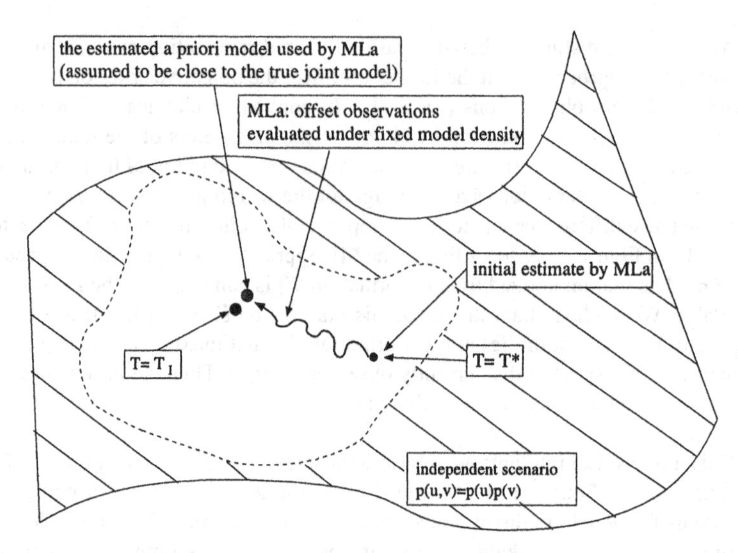


Figure 33.3. The approximate ML method (MLa) searches over the set of joint data sets offset by T. The goal is to maximize a criterion that is similar to likelihood with respect to a fixed model.

(MLa-i) It is feasible to estimate or learn a joint probability model over the data modalities of interest at the correct alignment[2], and

(MLa-ii) the resulting model accurately captures the statistical properties of other unseen image pairs (of the same anatomy and with the same modality pairing as the training set).

We denote the estimated joint density model as

$$p^{\circ}(u,v) \approx p(u,v;T_I).$$

As with all of the remaining methods, the MLa approach transforms the observations prior to evaluating the objective criterion. We denote the transformed observations as

$$\begin{aligned}
\mathcal{Y}_T &= \left\{\left[u(x_1), v_\circ(\hat{T}(x_1))\right], \ldots, \left[u(x_N), v_\circ(\hat{T}(x_N))\right]\right\} \\
&= \left\{\left[u(x_1), v(T^* \circ \hat{T}(x_1))\right], \ldots, \left[u(x_N), v(T^* \circ \hat{T}(x_N))\right]\right\} \\
&= \{[u(x_1), v(T(x_1))], \ldots, [u(x_N), v(T(x_N))]\} \\
&= \{[u, v_T]_1, \ldots, [u, v_T]_N\}. \qquad (33.4)
\end{aligned}$$

[2] Assuming manual or other types of ground truth results are available from previous registration experiments.

We emphasize that the transformation $T = \left(T^* \circ \hat{T}\right)$ in this particular notation refers to the relative transformation on $v(x)$ rather than on the observed image of $v_\circ(x)$. In practice, it is $\hat{T}$ that we apply to the observed image, so optimization is performed over $\hat{T}$ through $v_\circ(\hat{T}(x))$. This is equivalent to implicit optimization over $T$ through the relation $v(T(x)) = v_\circ(T^* \circ \hat{T}(x))$. While we express results on the implicit transformation, there are simple relationships which allow results to be expressed in terms of either $T$ or $\hat{T}$.

The MLa approach estimates $T$ to be the transformation that maximizes a criterion that is similar to the likelihood criterion:

$$T_{\text{MLa}} = \arg\max_T \mathcal{L}_{T_I}(\mathcal{Y}_T) \tag{33.5}$$

$$= \arg\max_T \frac{1}{N} \sum_i \log\left(p\left([u, v_T]_i ; T_I\right)\right). \tag{33.6}$$

Notice that, according to this approach the joint observations $([u, v_T]_i)$ are varied as a function of T and the model density $p^\circ$ is held static. It is under this particular fixed probability model that all the transformed inputs are evaluated. In Figure 33.3, we indicate the path of the MLa approach by tracing a sample search path. Beginning with the initial estimate, the algorithm searches over transformations to maximize the likelihood-like criterion with respect to the previously constructed, static density model.

The MLa method also makes an implicit assumption when solving the registration problem. It assumes that:

(MLa-iii) as $\hat{T}$ approaches $(T^*)^{-1}$, or equivalently as $(T^* \circ \hat{T})$ approaches $T_I$, Equation (33.6) is non-decreasing.

In general, one cannot guarantee the validity of this assumption. Theoretically, there might exist some counter-intuitive scenarios for which this implicit hypothesis would fail. The existence of these is explained by the information theoretic phenomenon of *typicality* [229]. A more detailed discussion of this issue is not in the scope of this chapter; it is described in an information-theoretic framework in [928].

This obstacle, in the context of multi-modal registration, may explain some shortcomings of the MLa approach that were observed empirically by Chung *et al.* [201]. It motivates their registration approach, which is described in the next section.

### *33.2.3 Kullback-Leibler Divergence*

Chung *et al.* suggested the use of KL divergence as a registration measure in order to align digital-subtraction angiography (DSA) and MR angiography (MRA) data sets [201]. Using the same modelling assumption as in MLa (i.e. a model

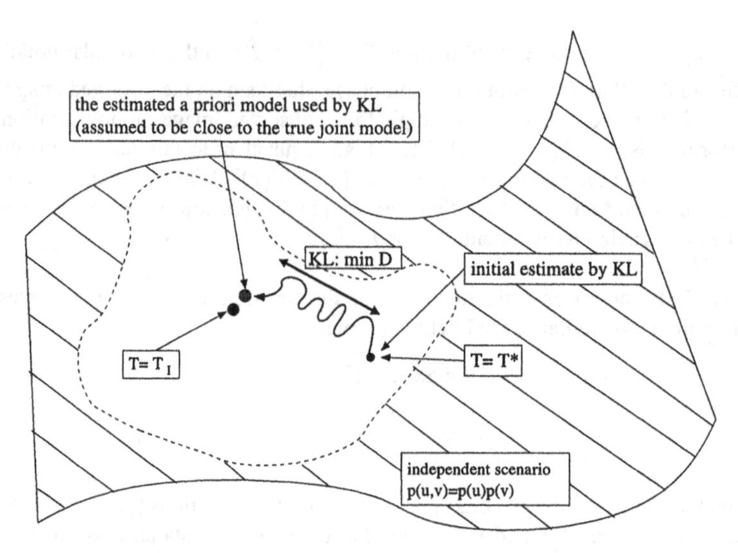


Figure 33.4. According to the KL framework, at each point on the manifold, a joint density is estimated from the offset data pairs. The aligning transformation is located where the KL distance (D) is minimized between that current estimate and a previously defined fixed model.

of the joint intensity data can be estimated from a set of registered data sets), they optimize an objective function based on a KL divergence term. That is, the distance between the joint density at the current transformation estimate and the fixed model is to be minimized:

$$T_{\text{KL}} \approx \arg\min_{T} D\left(\hat{p}(u,v;T) \| p^{\circ}(u,v)\right),$$

where $p^o$ is constructed as in the MLa approach from correctly registered data sets and $\hat{p}(u,v;T)$ is a probability model estimated from the transformed sets of observed pixel intensities $\{u(x_i), v(T(x_i))\}$ (or $\{u(x_i), v_o(\hat{T}(x_i))\}$ as discussed above). Whereas the previous methods utilize a likelihood function of the observed data sets, here numerical or Monte Carlo integration is used in order to calculate the KL divergence terms directly.

Consequently, in addition to assumptions MLa-i and MLa-ii, this approach makes the following hypothesis:

(KL-i) There is a reliable method for estimating $p(u,v;T)$ from transformed observations, and

(KL-ii) the KL divergence $D\left(p(u,v;T) \| p^{\circ}(u,v)\right)$ can be accurately estimated via numerical or Monte Carlo integration of

$$\int\int \hat{p}\left(u,v;T\right) \log\left(\frac{\hat{p}\left(u,v;T\right)}{p^{\circ}\left(u,v\right)}\right) dudv \qquad (33.7)$$

by substituting $\hat{p}\left(u,v;T\right)$ for $p\left(u,v;T\right)$ in the KL divergence integral.

The KL method has been demonstrated to be more robust with respect to, or less dependent on, the size of the sampling region (the area from which the joint sample pairs are drawn from) than the MLa (or the MI) approaches [201]. This robustness is demonstrated empirically [201] and can be partly explained by *typicality*, as discussed in the preceding section (Section 33.2.2).

Provided that both of the KL assumptions are valid (the density estimate and the integration methods are accurate), the KL divergence estimate is non-increasing as $\hat{T}$ approaches $(T^*)^{-1}$. This is supported by empirical comparisons in which KL did not exhibit some of the undesirable local extrema encountered in the MLa method[201]. Additionally, the authors emphasize that even though the estimated models represent a strong assumption, sufficient model distributions can be constructed even if manual alignment is unavailable. For example, the joint probability distribution could be estimated from segmented data for corresponding structures.

In relation to the previous methods, both the samples ($[u, v_T]_i$) *and* the evaluation density ($\hat{p}(u, v; T)$) are being varied as a function of the transformation $T$, while the algorithm approaches the static joint probability density model ($p^\circ(u, v)$) constructed prior to the alignment procedure. Instead of evaluating the joint characteristics of the transformed input data sets under the model distribution, the KL approach re-estimates the joint model ($\hat{p}(u, v; T)$) at every iteration and uses that when evaluating the observations. In Figure 33.4, the KL method is shown to approach the solution by minimizing the KL distance between the model and the current estimate.

### *33.2.4 Mutual Information and Joint Entropy*

As has been amply documented in the literature [534, 651, 652, 883], Mutual Information (MI) is a popular information theoretic objective criterion. It estimates the transformation parameter $T$ by maximizing the mutual information (or the statistical dependence) between the input image data sets:

$$T_{\mathrm{MI}} = \arg\max_T I\left(u; v_T\right).$$

One way to define the MI term is to use marginal and joint entropy measures. By definition, given random variables $A$ and $B$, mutual information is the sum of their marginal entropies minus their joint:

$$I(A, B) = H(A) + H(B) - H(A, B).$$

In the multi-modal alignment scenario that translates to

$$I(u; v_T) = H(p(u)) + H(p(v; T)) - H(p(u, v; T)). \tag{33.8}$$

If $T$ is restricted to the class of symplectic transformations (i.e. volume preserving), then $H(p(u))$ and $H(p(v; T))$ are invariant to $T$. In that case, maximization

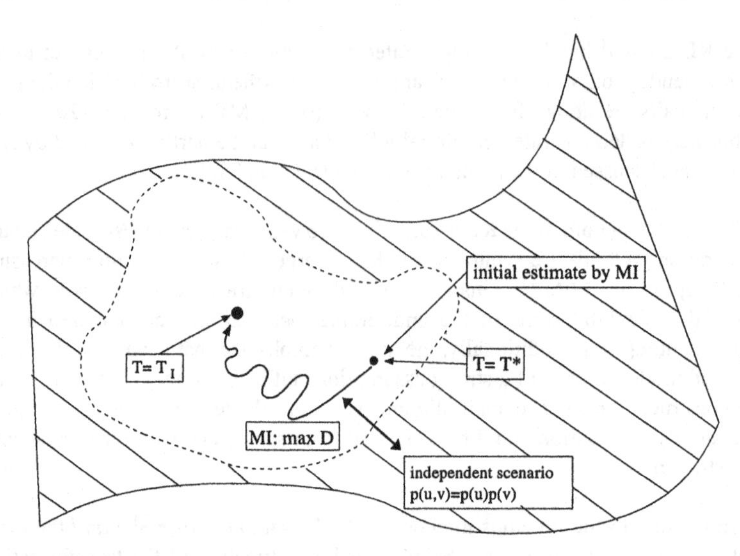


Figure 33.5. According to MI, the solution is located maximum KL distance away from the worst-case, independent scenario, where the joint density is defined as the product of its marginals: $p(u, v; T) = p(u)p(v; T)$.

of MI is equivalent to minimization of the joint entropy term, $H(p(u, v; T))$, the presumption being that this quantity is minimized when $\hat{T} = (T^*)^{-1}$. The minimization of the joint entropy term has also been widely used in the registration community.

MI can also be expressed as a KL divergence measure [494] as

$$I(u, v_T) = D\left(p(u, v; T) \| p(u)p(v; T)\right).$$

That is, mutual information is the KL divergence between the observed joint density term and the product of its marginals. Accordingly, the implicit assumption of MI-based methods is that:

(MI-i) as $(T^* \circ \hat{T})$ diverges from $T_I$ (as we are getting farther away from the ideal registration pose) the joint intensities look less statistically dependent, tending towards statistical independence.

This allows us to write the MI optimization problem as maximizing the divergence from the current density estimate to the scenario where the images are completely independent:

$$T_{\text{MI}} \approx \arg\max_T D\left(\hat{p}(u, v; T) \| \hat{p}(u)\hat{p}(v; T)\right).$$

As in the KL divergence alignment approach, both the samples and the evaluation densities are being simultaneously varied as a function of the transformation $T$. However, instead of approaching a known model point according to KL distance, the aim is to move farthest away from the condition of statistical independence among the images, in the KL sense. This behavior is illustrated in Figure 33.5.

Numerous variations on the mutual information metric have been introduced; for instance, one making it invariant to image overlap (normalized mutual information [771]) and another enhancing its robustness using additional image gradient information (gradient-augmented mutual information [651]). In this report, we do not list and analyze these, given that they operate with similar underlying principles.

## 33.3 Conclusion

We have provided a brief comparison of four well-known and widely used multimodal image registration methods. We illustrated the underlying assumptions which distinguish them, and specifically, we clarified the assumed behavior of joint intensity statistics as a function of transformation parameters. Considering the collection of approaches discussed, we see that the ML approach has not actually been used, in practice. The related MLa method and the KL divergence method exploit prior information in the form of static joint density estimates over previously registered data. Subsequently, both make similar implicit assumptions regarding the behavior of joint intensity statistics as the transformation estimate approaches the ideal alignment. In contrast, the MI approach makes no use of specific prior joint statistics – instead, it simply moves away from the general class of statistically independent models. Figure 33.6 serves as a visual guide to summarize how the different methods approach the solution.

## Acknowledgment

This work has been supported by NIH grant #R21CA89449 and #5P41RR13218 by NSF ERC grant (JHU EEC #9731748), by the Whiteman Fellowship and The Harvard Center for Neurodegeneration and Repair.

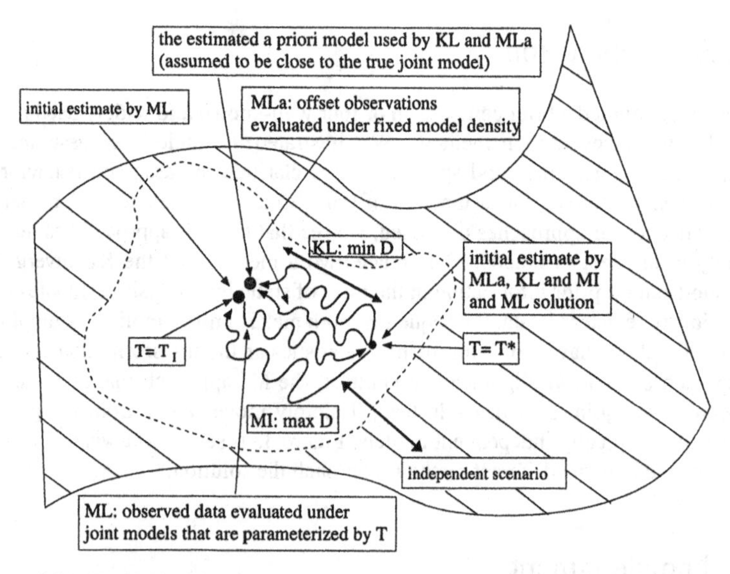


Figure 33.6. Manifold of the registration search space parameterized by transformation T. The illustration shows how each of the examined methods (ML, MLa, KL and MI) search through the settings in order to obtain the best estimate of the aligning transformation. Note that the ML method transforms the model to agree with the observed data, while the rest of the methods operate by transforming the observed data.

# Bibliography

## Bibliography


[1] K. Abter, W. Snyder, H. Burkhardt, and G. Hirzinger.Application of Affine Invariant Fourier Descriptors to Recognition of 3D Objects.*IEEE Transactions on Pattern Analysis and Machine Intelligence*, 12(7):640–647, 1990.

[2] S. Acton.Multigrid anisotropic diffusion.*IEEE Transactions on Image Processing*, 7(3):280–291, 1998.

[3] D. Adalsteinsson and J. Sethian.A Fast Level Set Method for Propagating Interfaces.*Journal of Computational Physics*, 118:269–277, 1995.

[4] R. Adams and L. Bischof.Seeded Region Growing.*IEEE Transactions on Pattern Analysis and Machine Intelligence*, 16:641–647, 1994.

[5] E. Adelson and J. Bergen.Spatiotemporal energy models for the perception of motion.*Journal of the Optical Society of America A*, 2:284–299, 1985.

[6] E. Adelson and J. Movshon.Phonemenal coherence of moving plaid patterns.*Nature*, 300(5892):523–525, 1982.

[7] A. Agarwala, M. Dontcheva, M. Agrawala, S. Drucker, A. Colburn, B. Curless, D. Salesin, and M. Cohen.Interactive digital photomontage.*ACM Transactions on Graphics*, 23(3):292–300, August 2004.

[8] J. Aggarwal and Q. Cai.Human Motion Analysis: A Review.*Computer Vision and Image Understanding*, 73(3):428–440, 1999.

[9] M. Agrawal and L. Davis.A probabilistic framework for surface reconstruction from multiple images.In *IEEE International Conference on Computer Vision and Pattern Recognition*, volume 2, pages 470–476, December 2001.

[10] M. Agrawal and L. Davis.Window-based, discontinuity preserving stereo.In *IEEE Conference on Computer Vision and Pattern Recognition*, pages I: 66–73, 2004.

[11] A. Alexander, K. Hasan, M. Lazar, J. Tsuruda, and D. Parker.Analysis of partial volume effects in diffusion-tensor MRI.*Magnetic Resononance Medicine*, 45:770–780, 2001.

[12] C. Alexander, C. Gee, and R. Bajcsy.Similarity Measure for Matching Diffusion Tensor images.In *British Machine Vision Conference*, pages 93–102, 1999.

[13] D. Alexander, G. Barker, and S. Arridge.Detection and modeling of non-gaussian apparent diffusion coefficient profiles in human brain data.*Magn. Reson. Med.*, 48:331–340, 2002.

[14] S. Alliney.A Property of the Minimum Vectors of a Regularizing Functional Defined by Means of the Absolute Norm.*IEEE Transactions on Signal Processing*, 45:913–917, 1997.

[15] L. Alvarez, P-L. Lions, and J-M. Morel.Image selective smoothing and edge detection by nonlinear diffusion.*SIAM Journal of Numerical Analysis*, 29:845–866, 1992.

[16] O. Amadieu, E. Debreuve, M. Barlaud, and G. Aubert.Inward and Outward Curve Evolution Using Level Set Method.In *IEEE International Conference on Image Processing*, volume III, pages 188–192, 1999.

[17] S. Amari.Information Geometry on Hierarchy of Probability Distributions.*IEEE Transaction on Information Theory*, 47(5):1701–1711, 2001.

[18] L. Ambrosio, V. Caselles, S. Masnou, and J-M. Morel.Connected Components of Sets of Finite Perimeter and Applications to Image Processing.*Journal of European Mathematical Society*, 3:39–92, 2001.

[19] L. Ambrosio, N. Fusco, and D. Pallara.*Functions of Bounded Variation and Free Discontinuity Problems*.Oxford University Press, 2000.

[20] L. Ambrosio and S. Masnou.A Direct Variational Approach to a Problem Arising in Image Reconstruction.*Interfaces and Free Boundaries*, 5:63–81, 2003.

[21] N. Amenta, M. Bern, and M. Kamvysselis.A New Voronoi-Based Surface Reconstruction Algorithm.In *Proc. SIGGRAPH 98, ACM*, pages 415–421, 1998.

[22] A. Amini, Y. Chen, M. Elayyadi, and P. Radeva.Tag surface reconstruction and tracking of myocardial beads from SPAMM-MRI with parametric b-spline surfaces.*IEEE Transactions on Medical Imaging*, 20(2):94–103, 2001.

[23] A. Amini, S. Tehrani, and T. Weymouth.Using dynamic programming for minimizing the energy of active contours in the presence of hard constraints.In *IEEE International Conference on Computer Vision*, pages 95–99, 1988.

[24] Amira.Amira Visualization and Modeling System.http://www.AmiraVis.com, 2004.

[25] P. Anandan.A computational framework and an algorithm for the measurement of visual motion.*International Journal of Computer Vision*, 2:283–310, 1989.

[26] P. Andresen and M. Nielsen.Non-rigid registration by geometry-constrined diffusion.*Medical Image Analysis*, 6:81–88, 2000.

[27] F. Andreu, V. Caselles, J. Diaz, and J. Mazón.Qualitative properties of the total variation flow.*Journal of Functional Analysis*, 188(2):516–547, February 2002.

[28] B. Appleton and H. Talbot.Globally optimal surfaces by continuous maximal flows.In *Digital Image Computing: Techniques and Applications, Proc. VIIth APRS*, volume 1, pages 623–632, December 2003.

[29] P. Arbelaez and L. Cohen.Energy partitions and image segmentation.*Journal of Mathematical Imaging and Vision*, 20(1-2):43–57, January - March 2004.

[30] R. Ardon and L. Cohen.Fast constrained surface extraction by minimal paths.In *Proc. IEEE Workshop on Variational and Level Set Methods in Computer Vision*, Nice, France, September 2003.

[31] G. Aronsson.On the Partial Differential Equation $u_x^2 u_{xx} + 2u_x u_y u_{xy} + u_y^2 u_{yy} = 0$.*Ark. for Math.*, 7:395–425, 1968.

[32] M. Ashikhmin.Synthesizing Natural Textures.In *ACM Symposium on Interactive 3D Graphics*, pages 217–226, 2001.

[33] C. Atkinson and A. Mitchell.Rao's distance measure.*Sankhya: The Indian Journal of Statistics*, 43(A):345–365, 1981.

[34] G. Aubert, M. Barlaud, O. Faugeras, and S. Jehan-Besson.Image segmentation using active contours: Calculus of variations or shape gradients ?*SIAM Applied Mathematics*, 63(6):2128–2154, 2003.

[35] J. August and S. Zucker.Sketches with curvature: The curve indicator random field and markov processes.*IEEE Transactions on Pattern Analysis and Machine Intelligence*, 25:387–401, 2003.

[36] J.-F. Aujol, G. Aubert, L. Blanc-Feraud, and A. Chambolle.Image Decomposition: Applications to Textured Images and SAR Images.Technical Report 4704, INRIA, France, 2002.

[37] S. Avidan.Support vector tracking.In *IEEE Conference on Computer Vision and Pattern Recognition*, 2001.

[38] S. Ayer and H. Sawhney.Layered Representation of Motion Video Using Robust Maximum-Likelihood Estimation of Mixture Models and MDL Encoding.In *IEEE International Conference in Computer Vision*, pages 777–784, Caibridge, USA, 1995.

[39] F. Azuola.*Error in the Representation of Anthropometric Data By Human Figure Models*.PhD thesis, University of Pennsylvania, Philadelphia, PA, 1996.

[40] A. Bab-Hadiashar and D. Suter.Robust optical flow computation.*International Journal of Computer Vision*, 29:59–77, 1998.

[41] N. Badler, C. Phillips, and B. Webber.*Simulating humans: Computer Graphics Animation and Control*.Oxford University Press, New York, NY, 1993.

[42] C. Bajaj, F. Bernardini, and G. Xu.Automatic Reconstruction of Surfaces and Scalar Fields from 3D Scans.In *Proc. SIGGRAPH 95, ACM*, pages 109–118, 1995.

[43] S. Baker and I. Matthews.Lucas-kanade 20 years on: A unifying framework: Part 1: The quantity approximated, the warp update rule, and the gradient descent approximation.*International Journal of Computer Vision*, 56(3):221–255, March 2004.

[44] S. Bakshi and Y. Yang.Shape from shading for non-lambertian surfaces.In *IEEE International Conference on Image Processing*, volume 94, pages 130–134, 1994.

[45] V. Balasubramanian.A geometric formulation of Occam's razor for inference of parametric distributions.PUPT 1588, Princeton University, December 2001.

[46] C. Ballester, M. Bertalmio, V. Caselles, G. Sapiro, and J. Verdera.Filling-in by Joint Interpolation of Vector Fields and Grey Levels.*IEEE Transactions on Image Processing*, 10:1200–1211, 2001.

[47] C. Ballester, V. Caselles, and J. Verdera.Disocclusion by Joint Interpolation of Vector Fields and Gray Levels.*Multiscale Modeling and Simulation*, 2:80–123, 2003.

[48] C. Ballester, V. Caselles, J. Verdera, M. Bertalmio, and G. Sapiro.A Variational Model for Filling-in Gray level and Color Images.In *Proc. Eighth International Conference on Comp. Vision ICCV*, volume 1, pages 10–16, Vancouver, Canada, 2001.

[49] Y. Bao and H. Krim.Towards bridging scale-space and multiscale frame analyses.In A. Petrosian and F. Meyer, editors, *Wavelets in Signal and Image Analysis*,

volume 19 of *Computational Imaging and Vision*, chapter 6. Kluwer, Dordrecht, 2001.

[50] Z. Bar-Joseph, R. El-Yaniv, D. Lischinski, and M. Werman.Texture mixing and texture movie synthesis using statistical learning.*IEEE Transactions on Visualization and Computer Graphics*, 7(2):120–135, 2001.

[51] E. Bardinet, L. Cohen, and N. Ayache.A parametric deformable model to fit unstructured 3D data.*Computer Vision and Image Understanding*, 71(1):39–54, 1998.

[52] G. Barles.*Solutions de Viscosité des Equations de Hamilton–Jacobi*.Springer–Verlag, 1994.

[53] G. Barles and P. Souganidis.Convergence of Approximation Schemes for Fully Non-Linear Second Order Equations.*Asymptotic Analysis*, 4:271–283, 1991.

[54] E. Barret, P. Payton, N. Haag, and M. Brill.General Methods for Determining Projective Invariants in Imagery.*CVGIP: Image Understanding*, 53:46–65, 1991.

[55] E. Barrett, P. Payton, and M. Brill.Contributions to the Theory of Projective Invariants for Curves in Two or Three Dimensions.In *DARPA/ESPRIT Workshop on the use of Invariants in Computer Vision, Reykjavik, Iceland*, 1991.

[56] R. Barrett, M. Berry, J. Chan, T. Demmel, J. Donato, J. Dongarra, V. Eijkhout, R. Pozo, C. Romine, and H. Van der Vorst.*Templates for the Solution of Linear Systems: Building Blocks for Iterative Methods*.Number 43 in Miscellaneous Titles in Applied Mathematics Series. SIAM, November 1993.

[57] C. Barrón and I. Kakadiaris.Estimating Anthropometry and Pose from a Single Image.*Computer Vision And Image Understanding*, 81(3):269–284, March 2001.

[58] C. Barrón and I. Kakadiaris.On the improvement of anthropometry and pose estimation from a single uncalibrated image.*Machine Vision and Applications*, 14(4):229–236, 2003.

[59] J. Barron, D. Fleet, and S. Beauchemin.Performance of Optical Flow Techniques.*International Journal of Computer Vision*, 12:43–77, 1994.

[60] B. Bascle and R. Deriche.Region tracking through image sequences.In *IEEE International Conference on Computer Vision*, pages 302–307, 1995.

[61] R. Basri, L. Costa, D. Geiger, and D. Jacobs.Determining the similarity of deformable shapes.*Vision Research*, 38:2365–2385, 1998.

[62] P. Basser.Inferring microstructural features and the physiological state of tissues from diffusion-weighted images.*NMR Biomd.*, 8:333–344, 1995.

[63] P. Basser, J. Mattiello, and D. Le Bihan.MR diffusion tensor spectroscopy and imaging.*Biophysica*, 66:259–267, 1994.

[64] P. Basser, J. Mattiello, and D. Lebihan.Estimation of the effective self-diffusion tensor from the NMR.*Spin Echo. J. Magn. Reson.*, series B 103:247–254, 1994.

[65] P. Basser and C. Pierpaoli.Microstructural and Physiological Features of Tissue Eelucidated by Quantitative Diffusion-tensor MRI.*J. Magn. Reson.*, 111(3):209–219, 1996.

[66] A. Baumberg and D. Hogg.Learning flexible models from image sequences.In J-O. Eklundh, editor, *European Conference on Computer Vision*, pages 299–308. Springer-Verlag, 1994.

[67] C. Beaulieu.The basis of anisotropic water diffusion in the nervous system - a technical review.*NMR Biomed.*, 15:435–455, 2002.

[68] P. Belhumeur, D. Kriegman, and A. Yuille.The bas-relief ambiguity.*IJCV*, 35(1):33–44, 1999.

[69] G. Bellettini, V. Caselles, and M. Novaga.The Total Variation Flow in $\mathbb{R}^n$.*J. Differtial. Equations*, 184:475–525, 2002.

[70] G. Bellettini, G. Dal Maso, and M. Paolini.Semicontinuity and Relaxation Properties of a Curvature Depending Functional in 2d.*Ann. Scuola Normale Sup. di Pisa, Cl. Sci.*, 20:247–297, 1993.

[71] G. Ben-Arous, A. Tannenbaum, and O. Zeitouni.Stochastic approximations to curve shortening flows via particle systems.*Journal Diff. Equations*, 195:119–142, 2003.

[72] O. Ben-Shahar and S. Zucker.The perceptual organization of texture flow: A contextual inference approach.*IEEE Transactions on Pattern Analysis and Machine Intelligence*, 25:401–417, 2003.

[73] O. Ben-Shahar and S. Zucker.Hue geometry and horizontal connections.*Neural Networks*, 17:753–771, 2004.

[74] R. Benosman and S.-B. Kang, editors.*Panoramic Vision: Sensors, Theory, and Applications*, New York, 2001. Springer.

[75] J. Bergen, P. Anandan, K. Hanna, and R. Hingorani.Hierarchical Model-Based Motion Estimation.In *European Conference on Computer Vision*, pages 237–252, 1992.

[76] F. Bernardini, J. Mittleman, H. Rushmeier, C. Silva, and G. Taubin.The Ball-Pivoting Algorithm for Surface Reconstruction.*IEEE Transactions on Visualization and Computer Graphics*, 5:349–359, 1999.

[77] M. Bertalmio, A. Bertozzi, and G. Sapiro.Navier-Stokes, fluid-dynamics, and image and video inpainting.In *Proc. of IEEE-CVPR*, pages 355–362, 2001.

[78] M. Bertalmio, L. Cheng, S. Osher, and G. Sapiro.Variational Problems and Partial Differential Equations on Implicit Surfaces.*Journal of Computational Physics*, 174(2):759–780, 2001.

[79] M. Bertalmio, G. Sapiro, L.-T. Cheng, and S. Osher.Image Inpainting.In *ACM SIGGRAPH*, pages 417–424, 2000.

[80] M. Bertalmio, L. Vese, G. Sapiro, and S. Osher.Simultaneous structure and texture image inpainting.*IEEE-TIP*, 12(8):882–889, 2003.

[81] J. Besag.On the statistical analysis of dirty images.*Journal of Royal Statistics Society*, 48:259–302, 1986.

[82] P. Besl and N. McKay.A Method for Registration of 3-D Shapes.*IEEE Transactions on Pattern Analysis and Machine Intelligence*, 14:239–256, 1992.

[83] K. Bhat, S. Seitz, J. Hodgins, and P. Khosla.Flow-based video synthesis and editing.In *Proceedings of SIGGRAPH*, pages 360–363, 2004.

[84] N. Biggs.*Algebraic Graph Theory*.Number 67 in Cambridge Tracts in Mathematics. Cambridge University Press, 1974.

[85] N. Biggs.Algebraic Potential Theory on Graphs.*Bulletin of London Mathematics Society*, 29:641–682, 1997.

[86] J. Bigun, G. Granlund, and J. Wiklund.Multidimensional orientation estimation with applications to texture analysis and optical flow.*IEEE Transactions on Pattern Analysis and Machine Intelligence*, 13(8):775–790, 1991.

[87] S. Birchfield and C. Tomasi.A pixel dissimilarity measure that is insensitive to image sampling.*IEEE Transactions on Pattern Analysis and Machine Intelligence*, 20(4):401–406, April 1998.

[88] S. Birchfield and C. Tomasi.Multiway cut for stereo and motion with slanted surfaces.In *IEEE International Conference on Computer Vision*, pages 489–495, 1999.

[89] M. Black and P. Anandan.The robust estimation of multiple motions: Parametric and piecewise-smooth flow fields.*Computer Vision and Image Understanding*, 63:75–104, 1996.

[90] M. Black and A. Jepson.Eigentracking: Robust matching and tracking of articulated objects using a view-based representation.In *European Conference on Computer Vision*, pages 329–342, 1996.

[91] M. Black and A. Jepson.EigenTracking: Robust matching and tracking of articulated objects using a view-based representation.*International Journal of Computer Vision*, 26(1):63–84, 1998.

[92] A. Blake.Visual tracking: a longer research roadmap.Internal Report MSR-TR-2005-??, Microsoft Research, 2005.

[93] A. Blake, R. Curwen, and A. Zisserman.A framework for spatio-temporal control in the tracking of visual contours.*International Journal Computer Vision*, 11(2):127–145, 1993.

[94] A. Blake and M. Isard.*Active contours*.Springer, 1998.

[95] A. Blake, M. Isard, and D. Reynard.Learning to track the visual motion of contours.*Journal of Artificial Intelligence*, 78:101–134, 1995.

[96] A. Blake, C. Rother, M. Brown, P. Perez, and P. Torr.Interactive image segmentation using an adaptive gmmrf model.In *8th European Conference on Computer Vision*, volume I of *LNCS 3021*, pages 428–441, Prague, Czech Republic, May 2004. Springer-Verlag.

[97] W. Blaschke.Vorlesungen uber Differential Geometrie II.Berlin, Germany, 1923. Verlag von Julius Springer.

[98] P. Blomgren and T. Chan.Color TV : Total Variation Methods for Restoration of Vector Valued Images.*IEEE Transactions on Image Processing*, 7(3):304–309, 1998.

[99] P. Blomgren, T. Chan, P. Mulet, and C. Wong.Total Variation Image Restoration: Numerical Methods and Extensions.In *IEEE International Conference on Image Processing*, pages III, 384–387, 1997.

[100] H. Blum and R. Nagel.Shape Description Using Symmetric Axis Features.*Pattern Recognition*, 10:167–180, 1978.

[101] F. Bookstein.Principal Warps: Thin-plate Splines and the Decomposition of Deformations.*IEEE Trans. Patt. Anal. Mach. Intell.*, 11(6):567–585, June 1989.

[102] F. Bookstein.*Morphometric Tools for Landmark Data: Geometry and Biology*.Cambridge University Press, 1991.

[103] G. Borgefors.Distance Transformations in Digital Images.*Computer Vision, Graphics, and Image Processing*, 34:344–371, 1986.

[104] E. Borovikov and L. Davis.A distributed system for real-time volume reconstruction.In *Proc. Fifth IEEE International Workshop on Computer Architectures for Machine Perception*, 2000.

[105] J. Bosch, S. Mitchell, B. Lelieveldt, F. Nijland, O. Kamp, M. Sonka, and J. Reiber.Automatic segmentation of echocardiographic sequences by active appearance models.*IEEE Trans. Med. Imaging*, 21:1374–1383, 2002.

[106] C. Bowyer, K. and Dyer.Aspect graphs: An introduction and survey of recent results.*Int. J. Imaging Systems and Technology*, 2:315–328, 1990.

[107] Y. Boykov and G. Funka-Lea.Optimal Object Extraction via Constrained Graph-Cuts.*International Journal of Computer Vision (IJCV)*, 2005, to appear. (Earlier version is in ICCV'01, vol. I, pp. 105-112, July 2001).

[108] Y. Boykov and M-P. Jolly.Interactive Organ Segmentation using Graph Cuts.In *Medical Image Computing and Computer-Assisted Intervention*, pages 276–286, Pittsburgh, PA, October 2000.

[109] Y. Boykov and V. Kolmogorov.Computing geodesics and minimal surfaces via graph cuts.In *International Conference on Computer Vision*, volume I, pages 26–33, 2003.

[110] Y. Boykov and V. Kolmogorov.An experimental comparison of min-cut/max-flow algorithms for energy minimization in vision.*IEEE Transactions on Pattern Analysis and Machine Intelligence*, 26(9):1124–1137, September 2004.

[111] Y. Boykov, O. Veksler, and R. Zabih.Markov random fields with efficient approximations.In *IEEE Conference on Computer Vision and Pattern Recognition*, pages 648–655, 1998.

[112] Y. Boykov, O. Veksler, and R. Zabih.A New Algorithm for Energy Minimization with Discontinuities.In Hancock-E-R; Pelillo-M., editor, *Energy Minimization Methods in Computer Vision and Pattern Recognition. Second International Workshop, EMMCVPR'99, York, UK, 26-29 July 1999.*, pages 205–220, 26-29 July 1999.

[113] Y. Boykov, O. Veksler, and R. Zabih.Fast Approximate Energy Minimization via Graph Cuts.*IEEE Transactions on Pattern Analysis and Machine Intelligence*, 23:1222–1239, 2001.

[114] K. Bredies, D. Lorenz, P. Maass, and G. Teschke.A partial differential equation for continuous non-linear shrinkage filtering and its application for analyzing MMG data.In F. Truchetet, editor, *Wavelet Applications in Industrial Processing*, volume 5266 of *Proceedings of SPIE*, pages 84–93. SPIE Press, Bellingham, 2004.

[115] C. Bregler and J. Malik.Tracking people with twists and exponential maps.In *Proc. CVPR*, 1998.

[116] X. Bresson, P. Vandergheynst, and J. Thiran.A Priori Information in Image Segmentation: Energy Functional based on Shape Statistical Model and Image Information.In *IEEE International Conference on Image Processing*, volume 3, pages 428–428, Barcelona, Spain, 2003.

[117] A. Broadhurst, T. Drummond, and R. Cipolla.A Probabilistic Framework for Space Carving.In *IEEE International Conference on Computer Vision*, pages 388–393, 2001.

[118] M. Brooks.Two results concerning ambiguity in shape from shading.In *AAAI-83*, pages 36–39, 1983.

[119] M. Brooks, W. Chojnacki, and R. Kozera.Shading without shape.*Quarterly of Applied Mathematics*, 50(1):27–38, 1992.

[120] M. Brown, D. Burschka, and G. Hager.Advances in computational stereo.*IEEE Transactions on Pattern Analysis and Machine Intelligence*, 25(8):993–1008, August 2003.

[121] M. Brown and D. Lowe.Recognising Panoramas.In *IEEE International Conference in Computer Vision*, pages 1218–1225, 2003.

[122] T. Brox, A. Bruhn, N. Papenberg, and J. Weickert.High accuracy optical flow estimation based on a theory for warping.In *ECCV04*, pages Vol IV: 25–36, 2004.

[123] T. Brox, M. Rousson, R. Deriche, and J. Weickert.Unsupervised segmentation incorporating colour, texture, and motion.*Computer Analysis of Images and Patterns. Lecture Notes in Computer Science*, 2756:353–360, 2003.

[124] T. Brox, J. Weickert, B. Burgeth, and P. Mrázek.Nonlinear structure tensors.Technical Report 113, Department of Mathematics, Saarland University, Saarbrücken, Germany, October 2004.

[125] A. Bruckstein.On shape fron shading.*Computer Vision Graphics Image Process*, 44:139–154, 1988.

[126] A. Bruckstein, R. Holt, A. Netravali, and T. Richardson.Invariant Signatures for Planar Shape Recognition under Partial Occlusion.*CVGIP: Image Understanding*, 58(1):49–65, 1993.

[127] A. Bruckstein, N. Kazir, M. Lindenbaum, and M. Porat.Invariant Signatures for Partially Occluded Planar Shapes.*International Journal of Computer Vision*, 7(3):271–285, 1992.

[128] A. Bruckstein and A. Netravali.On Differential Invariants of Planar Curves and Recognizing Partially Occluded Planar Shapes.*Annals of Mathematics and Artificial Itelligence*, 13(3-4):227–250, 1995.

[129] A. Bruckstein, E. Rivlin, and I. Weiss.Scale Space Semi-local Invariants.*Image and Vision Computing*, 15(5):335–344, May 1997.

[130] A. Bruckstein, G. Sapiro, and D. Shaked.Afine Invariant Evolutions of Planar Polygons.*International Journal of Pattern Recognition and Artificial Intelligence*, 9(6):991–1014, 1996.

[131] A. Bruckstein and D. Shaked.On Projective Invariant Smoothing and Evolutions of Planar Curves and Polygons.*Journal of Mathematical Imaging and Vision*, 7(3):225–240, Jun 1997.

[132] A. Bruckstein and D. Shaked.Skew-Symmetry Detection via Invariant Signatures.*Pattern Recognition*, 31(2):181–192, Feb 1998.

[133] A. Bruhn, J. Weickert, and C. Schnorr.Lucas/Kanade meets Horn/Schunck: Combining local and global optic flow methods.*International Journal of Computer Vision*, 61(3):211–231, 2005.

[134] A. Bruss.The eikonal equation: Some results applicable to computer vision.*Journal of Mathematical Physics*, 23(5):890–896, 1982.

[135] K. Bubna and C. Stewart.Model Selection Techniques and Merging Rules for Range Data Segmentation Algorithms.*Computer Vision and Image Understanding*, 80:215–245, 2000.

[136] S. Buchin.Affine Differential Geometry.In *Gordon and Breach Science Publishers New York*. Science Press, Beijing, China, 1983.

[137] J. Burbea.Informative geometry of probability spaces.*Expositiones Mathematica*, 4:347–378, 1986.

[138] J. Burbea and C. Rao.Entropy differential metric, distance and divergence measures in probability spaces: A unified approach.*Journal of Multivariate Analysis*, 12:575–596, 1982.

[139] P. Burt and E. Adelson.A multiresolution spline with applications to image mosaics.*ACM Transactions on Graphics*, 2(4):217–236, October 1983.

[140] H. Buseman.*Convex Surfaces*.Interscience Publ., 1958.

[141] E. Calabi, P. Olver, C. Shakiban, and et al.Defferential and Numerically Invariant Signature Curves Applied to Object Recognition.*International Journal of Computer Vision*, 26(2):107–135, Feb 1998.

[142] L. Calabi and W. Hartnett.Shape Recognition, Prairie Fires, Convex Deficiences and Skeletons.*Amer. Math. Mon*, 75:335–342, 1968.

[143] M. Calvo and J. Oller.An explicit solution of information geodesic equations for the multivariate normal model.*Statistics and Decisions*, 9:119–138, 1991.

[144] F. Camilli and M. Falcone.An approximation scheme for the maximal solution of the shape-from-shading model.In *ICIP'96*, pages 49–52, 1996.

[145] V. Camion and L. Younes.Geodesic Interpolating Splines.In *Energy Minimization Methods for Computer Vision and Pattern Recognition*, pages 513–527. Springer, New York, 2001.

[146] A. Can, H. Shen, J. Turner, H. Tanenbaum, and B. Roysam.Rapid Automated Tracing and Feature Extraction from Live High-Resolution Retinal Fundus Images Using Direct Exploratory Algorithms.*IEEE Transactions on Information Technology in Biomedicine*, 3:125–138, 1999.

[147] A. Can, C. Stewart, B. Roysam, and H. Tanenbaum.A Feature-Based, Robust, Hierarchical Algorithm for Registering Pairs of Images of the Curved Human Retina.*IEEE Transactions on Pattern Analysis and Machine Intelligence*, 24:347–364, 2002.

[148] D. Capel and A. Zisserman.Automated mosaicing with super-resolution zoom.In *IEEE Computer Society Conference on Computer Vision and Pattern Recognition*, pages 885–891, Santa Barbara, June 1998.

[149] R. Carceroni and K. Kutulakos.Multi-View Scene Capture by Surfel Sampling: From Video Streams to Non-Rigid 3D Motion, Shape and Reflectance.*The International Journal of Computer Vision*, 49:175–214, 2002.

[150] J. Carr, R. Beatson, J. Cherrie, T. Mitchell, W. Fright, B. McCallum, and T. Evans.Reconstruction and Representation of 3D Objects with Radial Basis Functions.In *ACM SIGGRAPH*, pages 67–76, 2001.

[151] J. Carter.*Dual Methods for Total Variation-based Image Restoration*.PhD thesis, UCLA, Los Angeles, CA, 2001.

[152] V. Caselles, F. Catté, B. Coll, and F. Dibos.A geometric model for active contours in image processing.*Numerische Mathematik*, 66(1):1–31, 1993.

[153] V. Caselles, B. Coll, and J-M. Morel.Geometry and color in natural images.*Journal Mathematical Imaging and Vision*, 16:89–105, 2002.

[154] V. Caselles, B. Coll, and J.M. Morel.Topographic Maps and Local Contrast Changes in Natural Images.*Int. J. of Computer Vision*, 33:5–27, 1999.

[155] V. Caselles, R. Kimmel, and G. Sapiro.Geodesic Active Contours.In *IEEE International Conference in Computer Vision*, pages 694–699, 1995.

[156] V. Caselles, R. Kimmel, and G. Sapiro.Geodesic Active Contours.*International Journal of Computer Vision*, 22:61–79, 1997.

[157] V. Caselles, R. Kimmel, G. Sapiro, and C. Sbert.Minimal surfaces based object segmentation.*IEEE Transactions on Pattern Analysis and Machine Intelligence*, 19:394–398, 1997.

[158] V. Caselles, J.-M. Morel, G. Sapiro, and A. Tannenbaum.Introduction to the special issue on PDEs and geometry-driven diffusion in image processing and analysis.*IEEE Transactions on Image Processing*, 7(3):269–274, 1998.

[159] A. Chakraborty, L. Staib, and J. Duncan.Deformable boundary finding in medical images by integrating gradient and region information.*IEEE Transactions on Medical Imaging*, 15:859–870, december 1996.

[160] A. Chambolle.An Algorithm for Total Variation Minimization and Applications.*J. Math. Imaging Vision*, 20:89–97, 2004.

[161] A. Chambolle, R. DeVore, N. Lee, and B. Lucier.Nonlinear wavelet image processing: variational problems, compression, and noise removal through wavelet shrinkage.*IEEE Transactions on Image Processing*, 7(3):319–335, March 1998.

[162] A. Chambolle and P.-L. Lions.Image Recovery via Total Variation Minimization and Related Problems.*Numerische Mathematik*, 76:167–188, 1997.

[163] A. Chambolle and B. Lucier.Interpreting translationally-invariant wavelet shrinkage as a new image smoothing scale space.*IEEE Transactions on Image Processing*, 10(7):993–1000, 2001.

[164] G. Champleboux, S. Lavallee, R. Szeliski, and L. Brunie.From Accurate Range Imaging Sensor Calibration to Accurate Model-Based 3-D Object Localization.In *IEEE Conference on Computer Vision and Pattern Recognition*, pages 83–89, 1992.

[165] T. Chan and S. Esedoglu.Aspects of Total Variation Regularized $L^1$ Function Approximation.*UCLA CAM Report*, 04-07, 2004.Accepted for publication in SIAM J. Appl. Math.

[166] T. Chan, G. Golub, and P. Mulet.A Nonlinear Primal-dual Method for Total Variation-based Image Restoration.*SIAM J. Sci. Comp.*, 20:1964–1977, 1999.

[167] T. Chan, S. Kang, and J. Shen.Euler's Elastica and Curvature-based Image Inpainting.*SIAM J. Appl. Math.*, 63(2):564–592, 2002.

[168] T. Chan, M. Marquina, and P. Mulet.High-order Total Variation-based Image Restoration.*SIAM J. Sci. Comput.*, 22:503–516, 2000.

[169] T. Chan and F. Park.Data Dependent Multiscale Total Variation Based Image Decomposition and Contrast Preserving Denoising.*UCLA CAM Report*, 04-15, 2004.

[170] T. Chan, B. Sandberg, and L. Vese.Active contours without edges for vector-valued images.*Journal of Visual Communications and Image Representation*, 11(2):130–141, 2000.

[171] T. Chan and J. Shen.Mathematical Models of Local Non-texture Inpaintings.*SIAM J. Appl. Math.*, 62:1019–1043, 2001.

[172] T. Chan and J. Shen.Non-texture Inpaintings by Curvature-driven Diffusions (CDD).*J. Visual Comm. Image Rep.*, 12(4):436–449, 2001.

[173] T. Chan and L. Vese.An Active Contour Model without Edges.In *International Conference on Scale-Space Theories in Computer Vision*, pages 141–151, 1999.

[174] T. Chan and L. Vese.Active Contours without Edges.*IEEE Transactions on Image Processing*, 10:266–277, 2001.

[175] T. Chan and L. Vese.A Level Set Algorithm for Minimizing the Mumford–Shah Functional in Image Processing.In *IEEE Workshop on Variational and Level Set Methods in Comp. Vis.*, pages 161–168, 2001.

[176] T. Chan and C. Wong.Total Variation Blind Deconvolution.*IEEE Trans. Image Process.*, 7:370–375, 1998.

[177] G. Charpiat, O. Faugeras, and R. Keriven.Approximations of shape metrics and application to shape warping and empirical shape statistics.*Journal of Foundations Of Computational Mathematics*, 2004.in press.

[178] C. Chefd'hotel, O. Faugeras, D. Tschumperlé, and R. Deriche.Constrained flows of matrix-valued functions: application to diffusion tensor regularization.In *ECCV*, pages 251–265, 2002.

[179] C. Chefd'hotel, D. Tschumperlé, R. Deriche, and O. Faugeras.Regularizing flows for constrained matrix-valued images.*Journal of Mathematical Imaging and Vision*, 20(1-2):147–162, 2004.

[180] S. Chen.QuickTime VR – an image-based approach to virtual environment navigation.*Computer Graphics (SIGGRAPH'95)*, pages 29–38, August 1995.

[181] T. Chen and D. Metaxas.Gibbs prior models, marching cubes, and deformable models: a hybrid segmentation framework for medical images.In *Medical Imaging Copmuting and Computer-Assisted Intervention*, volume 2, pages 703–710, 2003.

[182] Y. Chen, W. Guo, Q. Zeng, G. He, B. Vemuri, and Y. Liu.Recovery of intra-voxel structure from hard MRI.In *IEEE International Symposium on Biomedical Imaging*, pages 1028–1031, Arlington, Virginia, 2004.

[183] Y. Chen, W. Guo, Q. Zeng, X. Yan, F. Huang, H. Zhang, G. He, B. Vemuri, and Y. Liu.Estimation, smoothing, and characterization of apparent diffusion coefficient profiles from high angular resolution DWI.In *IEEE Int. Conf. in Computer Vision and Pattern Recognition (CVPR)*, pages 588–593, Washington, D.C., 2004.

[184] Y. Chen, S. Levine, and M. Rao.Functionals with $p(x)$-growth in image restoration.*submitted to SIAM*, 2004.

[185] Y. Chen and G. Medioni.Object Modeling by Registration of Multiple Range Images.*Image and Vision Computing*, 10:145–155, 1992.

[186] Y. Chen and G. Medioni.Description of Complex Objects from Multiple Range Images Using an Inflating Balloon Model.*Computer Vision and Image Understanding*, 61:325–334, 1995.

[187] Y. Chen, H. Tagare, S. Thiruvenkadam, F. Huang, D. Wilson, K. Gopinath, R. Briggs, and E. Geiser.Using prior shapes in geometric active contours in a variational framework.*International Journal on Computer Vision*, 50(3):315–328, 2002.

[188] Y. Chen, H. Thiruvenkadam, H. Tagare, F. Huang, and D. Wilson.On the Incorporation of Shape Priors int Geometric Active Contours.In *IEEE Workshop in Variational and Level Set Methods*, pages 145–152, 2001.

[189] Y. Chen, B. C. Vemuri, and L. Wang.Image Denoising and Segmentation via Nonlinear Diffusion.*Computers and Mathematics with Applications*, 39(5/6):131–149, 2000.

[190] T. Chenevert, J. Brunberg, and J. Pipe.Anisotropic diffusion in human white matter: Demonstration with MR techniques in vivo.*Radiology*, 177:401–405, 1990.

[191] Y. Cheng.Mean Shift, Mode Seeking, and Clustering.*IEEE Transactions on Pattern Analysis and Machine Intelligence*, 17:790–799, 1995.

[192] C. Chesnaud, P. Réfrégier, and V. Boulet.Statistical region snake-based segmentation adapted to different physical noise models.*IEEE Transactions on Pattern Analysis and Machine Intelligence*, 21:1145–1156, November 1999.

[193] S. Cho and H. Saito.A Divide-and-Conquer Strategy in Shape from Shading problem.In *CVRP'97*, 1997.

[194] D. Chopp.Computing Minimal Surfaces via Level Set Curvature Flow.*Journal of Computational Physics*, 106:77–91, 1993.

[195] A. Chorin and J. Marsden.*A Mathematical Introduction to Fluid Mechanics*.Springer-Verlag, Third Edition, 1993.

[196] G. Christensen.*Deformable shape models for anatomy*.PhD thesis, Electrical engineering, Washington University, St. Louis, Missouri, August 1994.

[197] G. Christensen.Consistent Linear-elastic Transformations for Image Matching.In *Proceedings of Information Processing in Medical Imaging—IPMI 99*, pages 224–237. Springer–Verlag, 1999.

[198] H. Chui and A. Rangarajan.A New Point Matching Algorithm for Non-rigid Registration.*Computer Vision and Image Understanding*, 89:114–141, 2003.

[199] H. Chui, L. Win, J. Duncan, R. Schultz, and A. Rangarajan.A Unified Non-rigid Feature Registration Method for Brain Mapping.*Medical Image Analysis*, 7:112–130, 2003.

[200] O. Chum and J. Matas.Randomized ransac with td,d test.In *Proc. BMVC'02, Vol. 2*, pages 448–457, 2002.

[201] A. Chung, W. Wells III, A. Norbash, and W. Grimson.Multi-modal Image Registration by Minimizing Kullback-Leibler Distance.In *International Conference on Medical Image Computing and Computer-Assisted Intervention*, volume 2 of *Lecture Notes in Computer Science*, pages 525–532. Springer, 2002.

[202] U. Clarenz, U. Diewald, G. Dziuk, Rumpf. M., and R. Rusu.A Finite Element Method for Surface Restoration with Smooth Boundary Conditions.*Preprint*, 2003.

[203] I. Cohen and L. Cohen.A hybrid hyperquadric model for 2-D and 3-D data fitting.*Computer Vision and Image Understanding*, 63(3):527–541, May 1996.

[204] L. Cohen.On active contour models and balloons.*CVGIP: Image Understanding*, 53:211–218, 1991.

[205] L. Cohen.Avoiding local minima for deformable curves in image analysis.In *Curves and Surfaces with Applications in CAGD*, pages 77–84. A. Le Méhauté, C. Rabut, and L. L. Schumaker (eds.), 1997.

[206] L. Cohen.Multiple contour finding and perceptual grouping using minimal paths.*Jour. of Mathematical Imaging and Vision*, 14(3), 2001.CEREMADE TR 0101, Jan 2001.

[207] L. Cohen, E. Bardinet, and N. Ayache.Surface reconstruction using active contour models.In *SPIE Conference on Geometric Methods in Computer Vision*, pages 38–50, San Diego, CA, 1993.

[208] L. Cohen and I. Cohen.Finite-element methods for active contour models and balloons for 2-D and 3-D images.*IEEE Transactions on Pattern Analysis and Machine Intelligence*, 15:1131–1147, 1993.

[209] L. Cohen and R. Kimmel.Fast marching the global minimum of active contours.In *IEEE International Conference on Image Processing (ICIP'96)*, pages I:473–476, Lausanne, Suisse, September 1996.

[210] L. Cohen and R. Kimmel.Global minimum for active contour models: A minimal path approach.*International Journal of Computer Vision*, 24(1):57–78, August 1997.

[211] T. Cohignac.*Reconnaissance de Formes Planes*.PhD thesis, Universite Paris Dauphine, 1994.

[212] T. Cohignac, C. Lopez, and J.M. Morel.Integral and Local Affine Invariant Parameters and Application to Shape Recognition.In D. Dori and A. Bruckstein, editors, *Shape Structure and Pattern Recognition*. World Scientific Publishing, 1995.

[213] R. Coifman and D. Donoho.Translation invariant denoising.In A. Antoine and G. Oppenheim, editors, *Wavelets in Statistics*, pages 125–150. Springer, New York, 1995.

[214] R. Coifman and A. Sowa.New methods of controlled total variation reduction for digital functions.*SIAM Journal on Numerical Analysis*, 39(2):480–498, 2001.

[215] R. Collins, A. Lipton, H. Fujiyoshi, and T. Kanade.Algorithms for cooperative multi-sensor surveillance.*Proceedings of the IEEE*, 89(10):1456–1477, October 2001.

[216] D. Comaniciu, V. Ramesh, and P. Meer.Real-time tracking of non-rigid objects using mean shift.In *IEEE Conference on Computer Vision and Pattern Recognition*, pages 142–151, 2000.

[217] W. Cook, W. Cunningham, W. Pulleyblank, and A. Schrijver.*Combinatorial Optimization*.John Wiley & Sons, 1998.

[218] T. Cootes.Statistical models of appearance for computer vision.Technical report, University of Manchester, 2004.

[219] T. Cootes, C. Beeston, G. Edwards, and C. Taylor.Unified Framework for Atlas Matching Using Active Appearance Models.In *Information Processing in Medical Imaging*, pages 322–333, 1999.

[220] T. Cootes, D. Cooper, C. Taylor, and J. Graham.Trainable method of parametric shape description.*Image and Vision Computing*, 10(5):289–294, 1992.

[221] T. Cootes, G. Edwards, and C. Taylor.Active appearance models.In *European Conference on Computer Vision*, volume 1407, pages 484–500, 1998.

[222] T. Cootes, G. Edwards, and C. Taylor.Active appearance models.*IEEE Trans. Pattern Anal. and Machine Intelligence*, 23:681–685, 2001.

[223] T. Cootes, C. Taylor, D. Cooper, and J. Graham.Active shape models - their training and application.*Computer Vision and Image Understanding*, 61:38–59, 1995.

[224] K. Cornelis, M. Pollefeys, M. Vergauwen, and L. Van Gool.Augmented reality from uncalibrated video sequences.In *3D Structure from Images - SMILE 2000, LNCS, Vol.2018*, pages 150–167. Springer-Verlag, 2001.

[225] I. Corouge, S. Gouttard, and G. Gerig.A statistical shape model of individual fiber tracts extracted from diffusion tensor MRI.In *MICCAI*, pages 671–679, 2004.

[226] O. Coulon, D. Alexander, and S. Arridge.Diffusion tensor magnetic resonance image regularisation.*Medical Image Analysis*, 8(1):47–67, 2004.

[227] R. Courant and D. Hilbert.*Methods of Mathematical Physics*, volume 2.John Wiley and Sons, 1989.

[228] F. Courteille, A. Crouzil, J-D. Durou, and P. Gurdjos.Towards shape from shading under realistic photographic conditions.In *ICPR'04*, 2004.

[229] T. Cover and J. Thomas.*Elements of Information Theory*.Wiley-Interscience, 1991.

[230] C. Cowan and P. Kovesi.Automatic sensor placement from vision task requirements.*IEEE Transactions on Pattern Analysis and Machine Intelligence*, 10(3):407–416, May 1988.

[231] T. Cox and M. Cox.*Multidimensional Scaling*.Chapman & Hall, London, 2001.

[232] B. Craine, Craine E., C. O'Toole, and Q. Ji.Digital imaging colposcopy: Corrected area measurements using Shape-from-Shading.*IEEE Transactions on Medical Imaging*, 17(6):1003–1010, 1998.

[233] D. Cremers, T. Kohlberger, and C. Schnörr.Shape statistics in kernel space for variational image segmentation.*Pattern Recognition*, 36(9):1929–1943, September 2003.

[234] D. Cremers and S. Soatto.Motion Competition: A Variational Framework for Piecewise Parametric Motion Segmentation.*International Journal of Computer Vision*.to appear.

[235] D. Cremers and S. Soatto.A pseudo distance for shape priors in level set segmentation.In *IEEE Workshop on Variational, Geometric and Level Set Methods (VLSM)*, pages 169–176, Nice, 2003.

[236] D. Cremers, N. Sochen, and C. Schnorr.Multiphase Dynamic Labeling for Variational Recognition-driven Image Segmentation.In *European Conference on Computer Vision*, pages 74–86, Prague, Chech Republic, 2004.

[237] D. Cremers, F. Tischhäuser, J. Weickert, and C. Schnörr.Diffusion Snakes: Introducing statistical shape knowledge into the Mumford–Shah functional.*International Journal on Computer Vision*, 50(3):295–313, 2002.

[238] A. Criminisi, P. Perez, and K. Toyama.Region filling and object removal by exemplar-based image inpainting.*IEEE-TIP*, 13(9):1200–1212, 2004.

[239] W. Culbertson, T. Malzbender, and G. Slabaugh.Generalized voxel coloring.In B. Triggs, A. Zisserman, and R. Szeliski, editors, *Vision Algorithms: Theory and Practice (Proc. Int. Workshop on Vision Algorithms)*, volume 1883 of *Lecture Notes in Computer Science*, pages 100–115. Springer-Verlag, 2000.

[240] B. Curless and M. Levoy.A volumetric method for building complex models from range images.In *ACM SIGGRAPH*, pages 303–312, 1996.

[241] R. Curwen and A. Blake.Dynamic contours: real-time active splines.In A. Blake and A. Yuille, editors, *Active Vision*, pages 39–58. MIT, 1992.

[242] D. Cygansky, J. Orr, T. Cott, and R. Dodson.An Affine Transform Invariant Curvature Function.In *Proccedings of the First ICCV*, pages 496–500, London, England, 1987.

[243] P. Danielsson.Euclidean Distance Mapping.*Computer Graphics and Image Processing*, 14:227–248, 1980.

[244] M. Darboux.Sur Un Probleme de Geometrie Elementarie.*Bull. Sci. Math*, 2:298–304, 1878.

[245] T. Darrell, D. Demirdjian, N. Checka, and P. Felzenszwalb.Plan-view trajectory estimation with dense stereo background models.In *IEEE International Conference on Computer Vision*, pages II: 628–635, Vancouver, Canada, June 2001.

[246] R. Davies, C. Twining, T. Cootes, J. Waterton, and C. Taylor.3D statistical shape models using direct optimisation of description length.In *European Conference on Computer Vision – ECCV*, pages 3–21, 2002.

[247] R. Davies, C. Twining, T. Cootes, J. Waterton, and C. Taylor.A minumum description length approach to statistical shape modelling.*IEEE Transactions on Medical Imaging*, 21(5), May 2002.525–537.

[248] J. Davis.Mosaics of scenes with moving objects.In *IEEE Computer Society Conference on Computer Vision and Pattern Recognition*, pages 354–360, Santa Barbara, June 1998.

[249] J. Davis, S. Marschner, M. Garr, and M. Levoy.Filling Holes in Complex Surfaces Using Volumetric Diffusion.In *Proc. First International Symposium on 3D Data Processing, Visualization, and Transmission*, 2002.

[250] L. Davis, E. Borovikov, R. Cutler, D. Harwood, and T. Horprasert.Multi-perspective analysis of human action.In *Proc. Third International Workshop on Cooperative Distributed Vision*, November 1999.

[251] K. De Cock and B. De Moor.Subspace angles between linear stochastic models.In *IEEE Conference on Decision and Control*, volume 2, pages 1561–1566, 2000.

[252] P. Debevec and J. Malik.Recovering high dynamic range radiance maps from photographs.*Proceedings of SIGGRAPH 97*, pages 369–378, August 1997.

[253] J. Debonet and P. Viola.Roxels: Responsibility weighted $3d$ volume reconstruction.In *IEEE International Conference on Computer Vision*, volume 1, pages 415–425, 1999.

[254] E. Debreuve, M. Barlaud, G. Aubert, and J. Darcourt.Space time segmentation using level set active contours applied to myocardial gated SPECT.*IEEE Transactions on Medical Imaging*, 20(7):643–659, July 2001.

[255] D. Decarlo and D. Metaxas.Blended deformable models.*IEEE Transactions on Pattern Analysis and Machine Intelligence*, 18(4):443–448, 1996.

[256] M. Delfour and J. Zolésio.*Shape and geometries*.Advances in Design and Control, SIAM, 2001.

[257] A. Dempster, N. Laird, and D. Rubin.Maximum likelihood from incomplete data via the EM algorithm.*Journal Royal Statistical Society Series B*, 39:1–38, 1977.

[258] J. Dendy.Black Box Multigrid.*Journal of Computational Physics*, 48:366–386, 1982.

[259] R. Deriche.Using Canny's Criteria to Derive a Recursively Implemented Optimal Edge Detector.*International Journal of Computer Vision*, 1:167–187, 1987.

[260] R. Deriche, D. Tschumperlé, and C. Lenglet.DT-MRI estimation, regularization and fiber tractography.In *Proc. of the 2nd ISBI*, pages 9–12, Washington D.C, 2004.

[261] A. Dervieux and F. Thomasset.A finite element method for the simulation of rayleigh-taylor instability.*Lecture Notes in Mathematics*, 771:145–159, 1979.

[262] A. Dervieux and F. Thomasset.Multifluid incompressible flows by a finite element method.In W. Reynolds and R.W. MacCormack, editors, *Seventh International Conference on Numerical Methods in Fluid Dynamics*, volume 141 of *Lecture Notes in Physics*, pages 158–163, June 1980.

[263] T. Deschamps.*Curve and Shape Extraction with Minimal Path and Level-Sets techniques - Applications to 3D Medical Imaging*.PhD thesis, Université Paris-IX Dauphine, Paris, December 2001.

[264] T. Deschamps and L. Cohen.Fast extraction of minimal paths in 3D images and applications to virtual endoscopy.*Medical Image Analysis*, 5(4):281–299, December 2001.Video in the web version of the journal.

[265] T. Deschamps and L. Cohen.Fast extraction of tubular and tree 3D surfaces with front propagation methods.In *Proc. IEEE ICPR'02*, Quebec, Canada, August 2002.

[266] T. Deschamps and L. Cohen.Grouping connected components using minimal path techniques.In Springer, editor, *Geometrical Method in Biomedical image processing*. R. Malladi (ed.), 2002.

[267] F. Devernay and O. Faugeras.Computing differential properties of 3-d shapes from stereoscopic images without 3-d models.In *Proc. IEEE Conf. on Computer Vision and Pattern Recognition*, 1994.

[268] E. Dickmanns and V. Graefe.Applications of dynamic monocular machine vision.*Machine Vision and Applications*, 1:241–261, 1988.

[269] E. Dijkstra.A note on two problems in connection with graphs.*Numerische Mathematic*, 1:269–271, 1959.

[270] H. Dinh, G. Turk, and G. Slabaugh.Reconstructing Surfaces Using Anisotropic Basis Functions.In *Proceedings IEEE Int. Conference on Computer Vision 2001*, pages 606–613, 2001.

[271] A. Dobbins, S. Zucker, and M. Cynader.Endstopping and curvature.*Vision Research*, 29:1371–1387, 1989.

[272] D. Dobson and C. Vogel.Global Total Variation Minimization.*SIAM Journal on Numerical Analysis*, 37:646–664, 2000.

[273] J. Dongarra, I. Duff, D. Sorenson, and H. van der Vorst.*Solving Linear Systems on Vector and Shared Memory Computers*.Society for Industrial and Applied Mathematics, Philadelphia, 1991.

[274] D. Donoho.De-noising by soft thresholding.*IEEE Transactions on Information Theory*, 41:613–627, 1995.

[275] D. Donoho and I. Johnstone.Ideal spatial adaptation by wavelet shrinkage.*Biometrica*, 81(3):425–455, 1994.

[276] C. Dorai, J. Weng, and A. Jain.Optimal Registration of Object Views Using Range Data.*IEEE Transactions on Pattern Analysis and Machine Intelligence*, 19:1131–1138, 1997.

[277] G. Doretto.*DYNAMIC TEXTURES: modeling, learning, synthesis, animation, segmentation, and recognition*.PhD thesis, University of California, Los Angeles, CA, March 2005.

[278] G. Doretto, A. Chiuso, Y. Wu, and S. Soatto.Dynamic textures.*International Journal of Computer Vision*, 51(2):91–109, 2003.

[279] G. Doretto, D. Cremers, P. Favaro, and S. Soatto.Dynamic texture segmentation.In *IEEE International Conference in Computer Vision*, volume 2, pages 1236–1242, 2003.

[280] G. Doretto, E. Jones, and S. Soatto.Spatially homogeneous dynamic textures.In *European Conference on Computer Vision*, volume 2, pages 591–602, 2004.

[281] G. Doretto and S. Soatto.Editable dynamic textures.In *IEEE Conference on Computer Vision and Pattern Recognition*, volume 2, pages 137–142, 2003.

[282] G. Doretto and S. Soatto.Modeling dynamic scenes with active appearance.Technical report, UCLA Computer Science Department, December 2004.

[283] P. Doyle and L. Snell.*Random Walks and Electric Networks*.Number 22 in Carus mathematical monographs. Mathematical Association of America, Washington, D.C., 1984.

[284] I. Dryden and K. Mardia.*Statistical Shape Analysis*.John Wiley & Son, 1998.

[285] Y. Duan, L. Yang, H. Qin, and D. Samaras.Shape Reconstruction from 3D and 2D Data using PDE-Based Deformable Surfaces.In *European Conference on Computer Vision*, pages 238–251, 2004.

[286] J. Duchon.*Splines Minimizing Rotation-invariant Semi-norms in Sobolev Spaces, in Constructive Theory of Functions of Several Variables*, pages 85–100.Springer–Verlag, 1977.

[287] R. Duda and P. Hart.*Pattern Classification and Scene Analysis*.John Wiley & Sons, 1973.

[288] R. Duda, P. Hart, and P. Stork.*Pattern Classification*.Wiley Interscience, 2000.

[289] P. Dupuis, U. Grenander, and M. Miller.Variational Problems on Flows of Diffeomorphisms for Image Matching.*Quarterly of Applied Math.*, 56:587–600, 1998.

[290] P. Dupuis and J. Oliensis.An optimal control formulation and related numerical methods for a problem in shape reconstruction.*The Annals of Applied Probability*, 4(2):287–346, 1994.

[291] J-D. Durou, M. Falcone, and M. Sagona.A survey of numerical methods for shape from shading.Technical Report 2004-2-R, IRIT, 2004.

[292] J.-D. Durou and D. Piau.Ambiguous shape from shading with critical points.*JMIV*, 12(2):99–108, 2000.

[293] N. Duta and M. Sonka.Segmentation and interpretation of MR brain images: An improved active shape model.*IEEE Trans. Med. Imaging*, 17:1049–1062, 1998.

[294] H. Edelsbrunner and E. Mücke.Three-dimensional Alpha Shapes.*ACM Transanctions on Graphics*, 13:43–72, 1994.

[295] B. Efron and R. Tibshirani.*An Introduction to the Bootstrap*.Number 57 in Monographs on Statistics and Applied Probability. Chapman and Hall, 1993.

[296] A. Efros and W. Freeman.Image Quilting for Texture Synthesis and Transfer.In *Proc. SIGGRAPH 01, ACM Press*, pages 341–346, 2001.

[297] A. Efros and T. Leung.Texture synthesis by non-parametric sampling.In *IEEE International Conference on Computer Vision*, pages 1033–1038, 1999.

[298] G. Emile-Male.*The Restorer's Handbook of Easel Painting*.Van Nostrand Reinhold, New York, 1976.

[299] S. Esedoglu and S. Osher.Decomposition of Images by the Anisotropic Rudin-Osher-Fatemi Model.*Comm. Pure Appl. Math.*, 57:1609–1626, 2004.

[300] S. Esedoglu and J. Shen.Digital Inpainting Based on the Mumford-Shah-Euler Image Model.*European J. Appl. Math.*, 13:353–370, 2002.

[301] L. Evans and R. Gariepy.*Measure theory and fine properties of functions*.CRC Press, 1992.

[302] L. Evans and P. Souganidis.Differential games and representation formulas for solutions of hamilton-jacobi-isaacs equations.*Indiana Univ. Math. J.*, 33:773–797, 1984.

[303] H. Farid and E. Simoncelli.Differentiation of discrete multi-dimensional signals.*IEEE Transactions on Image Processing*, 13(4):496–508, 2004.

[304] G. Farin.*Curves and Surfaces for Computer–Aided Geometric Design*.Academic Press, San Diego, 1997.

[305] G. Farneback.Very high accuracy velocity estimation using orientation tensors, parametric motion models, and simultaneous segmentation of the motion field.In *IEEE International Conference on Computer Vision*, volume 1, pages 171–177, Vancouver, 2001.

[306] O. Faugeras.What can be seen in three dimensions with an uncalibrated stereo rig?In *Computer Vision – ECCV'92, LNCS, Vol.588*, pages 563–578. Springer-Verlag, 1992.

[307] O. Faugeras.Cartan's Moving Frame Method and its Application to the Geometry and Evolution of Curves in the Euclidian, Affine and Projective Planes.In J. Mundy, A. Zisserman, and D. Forsyth, editors, *Applications of Invariance in Computer Vision*, chapter Foundations, pages 11–46. Springer Verlag, 1993.

[308] O. Faugeras and R. Keriven.Complete Dense Stereovision Using Level Set Methods.In *European Conference on Computer Vision*, pages 379–393, 1998.

[309] O. Faugeras and R. Keriven.Variational Principles, Surface Evolution, PDEs, Level Set Methods, and the Stereo Problem.*IEEE Transactions on Image Processing*, 7:336–344, 1998.

[310] O. Faugeras, Q.-T. Luong, and S. Maybank.Camera self-calibration: Theory and experiments.In *Computer Vision – ECCV'92, LNCS, Vol.588*, pages 321–334. Springer-Verlag, 1992.

[311] O. Faugeras, Q.-T. Luong, and T. Papadopoulo.The geometry of multiple images.MIT press, 2001.

[312] O. Faugeras, F. Lustman, and G. Toscani.Motion and structure from motion from point and line matches.In *IEEE International Conference on Computer Vision*, pages 25–34, 1987.

[313] C. Feddern, J. Weickert, and B. Burgeth.Level-set Methods for Tensor-valued Images.In *IEEE Workshop on Variational and Level Set Methods in Comp. Vis.*, pages 65–72, 2003.

[314] C. Feddern, J. Weickert, B. Burgeth, and M. Welk.Curvature-driven PDE methods for matrix-valued images.Technical Report 104, Department of Mathematics, Saarland University, Saarbrücken, Germany, April 2004.

[315] J. Feldmar and N. Ayache.Rigid, affine and locally affine registration of free-form surfaces.*International Journal of Computer Vision*, 18:99–119, 1996.

[316] M. Fischler and R. Bolles.Random sampling consensus: a paradigm for model fitting with application to image analysis and automated cartography.*Commun. Assoc. Comp. Mach.*, 24:381–395, 1981.

[317] M. Fischler and R. Elschlager.The representation and matching of pictorial structures.*IEEE. Trans. Computers*, C-22(1), 1973.

[318] A. Fitzgibbon.Stochastic rigidity: image registration for nowhere-static scenes.In *IEEE International Conference in Computer Vision*, volume 1, pages 662–669, 2001.

[319] D. Fleet, M. Black, Y. Yacoob, and A. Jepson.Design and use of linear models for image motion analysis.*International Journal of Computer Vision*, 36(3):169–191, 2000.

[320] D. Fleet and A. Jepson.Computation of component image velocity from local phase information.*International Journal of Computer Vision*, 5:77–104, 1990.

[321] D. Fleet and A. Jepson.Stability of phase information.*IEEE Transactions on Pattern Analysis and Machine Intelligence*, 15:1253–1268, 1993.

[322] J. Fleet.*Measurement of Image Velocity*.Kluwer, Norwell, MA, 1992.

[323] P. Fletcher and S. Joshi.Principal geodesic analysis on symmetric spaces: Statistics of diffusion tensors.In *CVAMIA-MMBIA*, pages 87–98, 2004.ECCV'04 workshop.

[324] L. Ford and D. Fulkerson.*Flows in Networks*.Princeton University Press, 1962.

[325] W. Förstner and B. Moonen.A metric for covariance matrices.Technical report, Dept. of Geodesy and Geoinformatics, Stuttgart University, 1999.

[326] A. Frangi, W. Niessen, K. Vincken, and M. Viergever.Multiscale vessel enhancement filtering.In *Proc. MICCAI'98, Cambridge*, pages 130–137, 1998.

[327] A. Frangi, D. Rueckert, J. Schnabel, and W. Niessen.Automatic construction of multiple-object three-dimensional statistical shape models: application to cardiac modeling.*IEEE Transactions on Medical Imaging*, 21(9):1151–66, 2002.

[328] L. Frank.Anisotropy in high angular resolution diffusion-weighted MRI.*Magn Reson Med*, 45:935–939, 2001.

[329] L. Frank.Characterization of hisotropy in high angular resolution diffusion weighted MRI.*Magn Reson Med*, 47:1083–1099, 2002.

[330] W. Freeman and E. Adelson.The design and use of steerable filters.*IEEE Transactions on Pattern Analysis and Machine Intelligence*, 13:891–906, 1991.

[331] K. Fritzsche, A. Can, H. Shen, C. Tsai, J. Turner, H. Tanenbaum, C. Stewart, and B. Roysam.Automated Model Based Segmentation, Tracing and Analysis of Retinal Vasculature from Digital Fundus Images.In J. S. Suri and S. Laxminarayan, editors, *State-of-The-Art Angiography, Applications and Plaque Imaging Using MR, CT, Ultrasound and X-rays*, pages 225–298. Academic Press, 2003.

[332] M. Fréchet.Les éléments aléatoires de nature quelconque dans un espace distancié.*Annales de l'Institut H. Poincaré*, 10(4):215–310, 1948.

[333] K. Fukunaga and L. Hostetler.The estimation of the gradient of a density function, with applications in pattern recognition.*IEEE Trans. Info. Theory*, IT-21:32–40, 1975.

[334] M. Gage and R. Hamilton.The heat equation shrinking convex plane curves.*Journal of Differential Geometry*, 23:69–96, 1986.

[335] M. Gastaud, M. Barlaud, and G. Aubert.Combining shape prior and statistical features for active contour segmentation.*IEEE Transactions on Circuits and Systems for Video Technology*, 14(5):726–734, May 2004.

[336] D. Gavrila.Multi-feature hierarchical template matching using distance transforms.In *Proc. of IEEE International Conference on Pattern Recognition*, pages 439–444. Brisbane, Australia, 1998.

[337] D. Gavrila.The Visual Analysis of Human Movement: A Survey.*Computer Vision and Image Understanding*, 73(1):82–98, 1999.

[338] D. Gavrila and V. Philomin.Real-time object detection for smart vehicles.In *IEEE International Conference on Computer Vision*, pages 87–93, 1999.

[339] N. Gelfand, L. Ikemoto, S. Rusinkiewicz, and M. Levoy.Geometrically Stable Sampling for the ICP Algorithm.In *Proceedings of the 4th International Conference on 3-D Digitial Imaging and Modeling*, pages 260–267, 2003.

[340] D. Geman and B. Jedynak.An active testing model for tracking roads in satellite images.*IEEE Trans. on Pattern Analysis and Machine Intelligence*, 18(1):1–14, 1996.

[341] S. Geman and D. Geman.Stochastic Relaxation, Gibbs Distributions, and the Bayesian Restoration of Images.*IEEE Transactions on Pattern Analysis and Machine Intelligence*, 6:721–741, 1984.

[342] S. Geman and D. McClure.Statistical methods for tomographic image reconstruction.*Bulletin of the International Statistical Institute*, LII-4:5–21, 1987.

[343] D. Gennery.Visual tracking of known three-dimensional objects.*International Journal Computer Vision*, 7(3):243–270, 1992.

[344] G. Gerig, O. Kübler, R. Kikinis, and F. Jolesz.Nonlinear anisotropic filtering of MRI data.*IEEE Transactions on Medical Imaging*, 11:221–232, 1992.

[345] A. Gibbons.*Algorithmic Graph Theory*.Cambridge University Press, 1989.

[346] J. Gibson.*The Perception of the Visual World*.Houghton Mifflin, Boston, 1950.

[347] F. Girosi, M. Jones, and T. Poggio.Regularization Theory and Neural Network Architectures.*Neural Computation*, 7:219–269, 1995.

[348] C. Glasbey and Mardia K.A penalized likelihood approach to image warping.*J. R. Statist. Soc. B*, 63:465–514, 2001.

[349] S. Gold and A. Rangarajan.Graph matching by graduated assignment.In *CVPR96*, pages 239–244, 1996.

[350] A. Goldberg and R. Tarjan.A new approach to the maximum-flow problem.*Journal of the Association for Computing Machinery*, 35(4):921–940, October 1988.

[351] R. Goldenberg, R. Kimmel, E. Rivlin, and M. Rudzsky.Fast Geodesic Active Contours.*IEEE Transactions on Image Processing*, 10:1467–1475, 2001.

[352] G. Golub and C. Van Loan.*Matrix Computations*.The Johns Hopkins University Press, 3rd edition, 1996.

[353] J. Gomes and O. Faugeras.Reconciling distance functions and level sets.*Journal of Visual Communication and Image Representation*, 11:209–223, 2000.

[354] H. González-Banos and J.C. Latombe.A randomized art-gallery algorithm for sensor placement.In *SCG*, Medford, MA, June 2001.

[355] N. Gordon, D. Salmond, and A. Smith.Novel approach to nonlinear/non-Gaussian Bayesian state estimation.*IEE Proc. F*, 140(2):107–113, 1993.

[356] L. Grady.*Space-Variant Computer Vision: A Graph-Theoretic Approach*.PhD thesis, Boston University, Boston, MA, 2004.

[357] L. Grady and G. Funka-Lea.Multi-Label Image Segmentation for Medical Applications Based on Graph-Theoretic Electrical Potentials.In *Computer Vision and Mathematical Methods in Medical and Biomedical Image Analysis, ECCV 2004 Workshops CVAMIA and MMBIA*, pages 230–245, Prague, Czech Republic, 2004.

[358] L. Grady and E. Schwartz.Anisotropic Interpolation on Graphs: The Combinatorial Dirichlet Problem.Technical Report CAS/CNS-TR-03-014, Department of Cognitive and Neural Systems, Boston University, Boston, MA, July 2003.

[359] R. Gray and L. Davisson.*An introduction to statistical signal processing*.electronic document, available on-line at http://ee.stanford.edu/~gray/sp.html, August 2004.

[360] R. Gray, J. Young, and A. Aiyer.Minimum discrimination information clustering: modeling and quantization with gauss mixtures.In *International Conference on Image Processing*, Thessaloniki, Greece, October 2001.

[361] M. Grayson.The heat equation shrinks embedded plane curves to round points.*Journal of Differential Geometry*, 26:285–314, 1987.

[362] D. Greig, B. Porteous, and A. Seheult.Exact Maximum a posteriori Estimation for Binary Images.*Journal of the Royal Statistical Society, Series B*, 51(2):271–279, 1989.

[363] K. Gremban.*Combinatorial Preconditioners for Sparse, Symmetric Diagonally Dominant Linear Systems*.PhD thesis, Carnegie Mellon University, Pittsburgh, PA, October 1996.

[364] U. Grenander.*General Pattern Theory*.Oxford University Press, 1993.

[365] U. Grenander, Y. Chow, and D.M. Keenan.*HANDS. A Pattern Theoretical Study of Biological Shapes*.Springer-Verlag. New York, 1991.

[366] U. Grenander and M. Miller.Computational anatomy: An emerging discipline.*Quarterly of Applied Mathematics*, LVI(4):617–694, 1998.

[367] M. Grimaud.*La Géodésie Numérique en Morphologie Mathématique. Application a la Détection Automatique de Microcalcifications en Mammographie Numérique*.PhD thesis, E.N.S. des Mines de Paris, 1991.

[368] M. Grimaud.New measure of contrast : dynamics.*Image Algebra and Morphological Processing III, San Diego CA, Proc. SPIE*, 1992.

[369] H. Grossauer and O. Scherzer.Using the Complex Ginzburg-Landau Equation for Digital Inpainting in 2D and 3D.In *Scale-Space '03. Lecture Notes in Computer Science*, pages 225–236. Springer Verlag, 2003.

[370] H. Guggenheimer.Differential Geometry.*McGraw-Hill*, New York, 1963.

[371] F. Guichard and J-M. Morel.*Image Iterative Smoothing and P.D.E.'s*.Book in preparation, 2003.

[372] H. Guo, A. Rangarajan, S. Joshi, and L. Younes.Non-rigid Registration of Shapes via Diffeomorphic Point Matching.*ISBI 2004*, 2004.

[373] W. Guo, Q. Zeng, Y. Chen, and Y. Liu.White matter fiber tracking using multi-directional vector field.page 2004, Preprint.

[374] W. Hackbusch.*Iterative Solution of Large Sparse Systems of Equations*.Springer-Verlag, 1994.

[375] G. Hager and P. Belhumeur.Efficient region tracking with parametric models of geometry and illumination.*IEEE Transactions on Pattern Analysis and Machine Intelligence*, 27(10):1025–1039, 1998.

[376] G. Hager and K. Toyama.Xvision: combining image warping and geometric constraints for fast tracking.In *European Conference on Computer Vision*, pages 507–517, 1996.

[377] J. Hallam.Resolving observer motion by object tracking.In *International Joint Conference on Artificial Intelligence*, volume 2, pages 792–798, 1983.

[378] Halphen.*Sur les Invariants Differentiels des Courbes Gauches*.PhD thesis, J. Ecole Polytechnique, XXVII, 1880.

[379] J. Hammersley and P. Clifford.*Markov fields on finite graphs and lattices*.Preprint University of California, Berkeley, 1971.

[380] F. Hampel, E. Ronchetti, P. Rousseeuw, and W. Stahel.*Robust Statistics: The Approach Based on Influence Functions*.Wiley, New York, 1986.

[381] R. Haralick.Digital step edges from zero crossing of second directional derivatives.*IEEE Transactions on Pattern Analysis and Machine Intelligence*, 6:58–68, 1984.

[382] F. Harary.*Graph Theory*.Addison-Wesley, 1994.

[383] C. Harris.Geometry from visual motion.In A. Blake and A. Yuille, editors, *Active Vision*, pages 263–284. MIT, 1992.

[384] C. Harris.Tracking with rigid models.In A. Blake and A. Yuille, editors, *Active Vision*, pages 59–74. MIT, 1992.

[385] C. Harris and M. Stephens.A combined corner and edge detector.In *Fourth Alvey Vision Conference*, pages 147–151, 1988.

[386] R. Hartley.In defense of the eight-point algorithm.*IEEE Trans. on Pattern Analysis and Machine Intell.*, 19(6):580–593, June 1997.

[387] R. Hartley, R. Gupta, and T. Chang.Stereo from uncalibrated cameras.In *Proc. CVPR'92*, pages 761–764, 1992.

[388] R. Hartley and P. Sturm.Triangulation.*Computer Vision and Image Understanding*, 68(2):146–157, 1997.

[389] R. Hartley and A. Zisserman.*Multiple View Geometry in Computer Vision*.Cambridge University Press, 2000.

[390] H. Haussecker and J. Fleet.Estimating optical flow with physical models of brightness variation.*IEEE Transactions on Pattern Analysis and Machine Intelligence*, 23(6):661–673, 2001.

[391] T. Heap and D. Hogg.Wormholes in shape space: Tracking through discontinuous changes in shape.In *IEEE International Conference on Computer Vision*, 1998.

[392] D. Heeger and J. Bergen.Pyramid-Based Texture Analysis/Synthesis.In *Proceedings of ACM SIGGRAPH 95, ACM Press*, pages 229–238, 1995.

[393] D. Heeger and A. Jepson.Subspace methods for recovering rigid motion i: Algorithms and implementation.*International Journal of Computer Vision*, 7(2):95–117, January 1992.

[394] F. Hèlein.*Harmonic Maps, Conservation Laws and Moving Frames*.Cambridge University Press, 2 edition, 2002.

[395] A. Herbulot, S. Jehan-Besson, M. Barlaud, and G. Aubert.Shape gradient for image segmentation using information theory.In *IEEE International Conference on Acoustics, Speech, and Signal Processing*, volume 3, pages 21–24, Montreal, May 2004.

[396] A. Herbulot, S. Jehan-Besson, M. Barlaud, and G. Aubert.Shape gradient for image segmentation using mutual information.In *International Conference on Image Processing*, Singapore, October 2004.

[397] G. Hermosillo, C. Chefd'Hotel, and O. Faugeras.A Variational Approach to Multi-Modal Image Matching.Technical Report RR-4117, INRIA, 2001.

[398] G. Hermosillo, C. Chefd'hotel, and O. Faugeras.Variational Methods for Multimodal Image Matching.*The International Journal of Computer Vision*, 50:329–343, 2002.

[399] M. Hintermüller and W. Ring.A second order shape optimization approach for image segmentation.*SIAM Journal on Applied Mathematics*, 64(2):442–467, 2003.

[400] A. Hirani and T. Totsuka.Combining Frequency and Spatial Domain Information for Fast Interactive Image Noise Removal.In *SIGGRAPH 96*, pages 269–276, 1996.

[401] T. Hofmann and J. Buhmann.Pairwise data clustering by deterministic annealing.*IEEE Transactions on Pattern Analysis and Machine Intelligence*, 19(1):1–14, 1997.

[402] D. Hogg.Model-based vision: a program to see a walking person.*Journal in Image and Vision Computing*, 1(1):5–20, 1983.

[403] L. Hong and G. Chen.Segment-based stereo matching using graph cuts.In *CVPR04*, pages I: 74–81, 2004.

[404] H. Hoppe, T. DeRose, T. Duchamp, J. McDonald, and W. Stuetzle.Surface reconstruction from unorganized points.In *ACM SIGGRAPH*, 1992.

[405] B. Horn.Obtaining shape from shading information.In P. Winston, editor, *The Psychology of Computer Vision*. McGraw-Hill, New York, 1975.

[406] B. Horn.The Curve of Least Energy.*ACM Transactions on Mathematical Software*, 9:441–460, 1982.

[407] B. Horn.*Robot Vision*.MIT Press, 1986.

[408] B. Horn and M. Brooks.*Shape from Shading*.MIT Press, 1989.

[409] B. Horn and B. Schunck.Determinating Optical Flow.*Artificial Intelligence*, 17:185–203, 1981.

[410] I. Howard and B. Rogers.*Binocular Vision and Stereopsis*.Oxford Univ. Press, 1995.

[411] E. Hsu and S. Mori.Analytical expression for the NMR apparent diffusion coefficients in an anisotropy system and a simplified method for determing fiber orientation.*Magn. Resòn. Med.*, 34:194–200, 1995.

[412] X. Huang, D. Metaxas, and T. Chen.Metamorphs: Deformable shape and texture models.In *IEEE Conference in Computer Vision and Pattern Recognition*, volume 1, pages 496–503, 2004.

[413] X. Huang, N. Paragios, and D. Metaxas.Registration of Structures in Arbitrary Dimensions: Implicit Representations, Mutual Information & Free-Form Deformations.Technical Report DCS-TR-0520, Division of Computer & Information Science, Rutgers University, 2003.

[414] R. Hummel and S. Zucker.On the foundations of relaxation labeling processes.*IEEE Transactions on Pattern Analysis and Machine Intelligence*, 6:267–287, 1983.

[415] D. Huttenlocher, J. Noh, and W. Rucklidge.Tracking non-rigid objects in complex scenes.In *IEEE International Conference on Computer Vision*, pages 93–101, 1993.

[416] H. Igehy and L. Pereira.Image Replacement through Texture Synthesis.*IEEE International conference on Image Processing*, 3:186–189, 1997.

[417] M. Irani and P. Anandan.Video indexing based on mosaic representations.*Proceedings of the IEEE*, 86(5):905–921, May 1998.

[418] M. Irani and P. Anandan.Factorization with uncertainty.In *European Conference on Computer Vision*, pages 539–553, Dublin, 2000.

[419] M. Isard and A. Blake.Visual tracking by stochastic propagation of conditional density.In *European Conference on Computer Vision*, pages 343–356, 1996.

[420] M. Isard and A. Blake.C-conditional density propagation for visual tracking.*Internat. Journal of Computer Vision*, 29(1):5–28, August 1998.

[421] M. Isard and A. Blake.ICondensation: Unifying low-level and high-level tracking in a stochastic framework.In *European Conference on Computer Vision*, pages 893–908, 1998.

[422] M. Isard and A. Blake.A mixed-state Condensation tracker with automatic model switching.In *IEEE International Conference on Computer Vision*, pages 107–112, 1998.

[423] M. Isard and J. MacCormick.Bramble: A bayesian multiple-blob tracker.In *IEEE International Conference on Computer Vision*, pages II: 34–41, Vancouver, Canada, July 2001.

[424] H. Ishikawa.Exact optimization for Markov Random Fields with convex priors.*IEEE Transactions on Pattern Analysis and Machine Intelligence*, 25(10):1333–1336, October 2003.

[425] H. Ishikawa and D. Geiger.Occlusions, discontinuities, and epipolar lines in stereo.In *5th European Conference on Computer Vision*, pages 232–248, 1998.

[426] H. Ishikawa and D. Geiger.Segmentation by grouping junctions.In *IEEE Conference on Computer Vision and Pattern Recognition*, pages 125–131, 1998.

[427] A. Jackson, B. Lautrup, P. Johansen, and M. Nielsen.Products of random matrices.*Phys. Rev. E*, 66:article 66124, 2002.

[428] B. Jähne, H. Haussecker, H. Spies, D. Schmundt, and U. Schurr.Study of dynamical processes with tensor-based spatiotemporal image processing techniques.In

H. Burkhardt and B. Neumann, editors, *European Conference on Computer Vision*, pages 322–335, Freiburg, 1998. Springer.

[429] A. Jain and R. Dubes.*Algorithms for Clustering Data*.Prentice-Hall, 1988.

[430] E. Jaynes.On the rationale of maximum-entropy methods.*Proc IEEE*, 70(939), 1982.

[431] S. Jehan-Besson, M. Barlaud, and G. Aubert.Video object segmentation using eulerian region-based active contours.In *International Conference on Computer Vision*, Vancouver, Canada, October 2001.

[432] S. Jehan-Besson, M. Barlaud, and G. Aubert.DREAM$^2$S: Deformable regions driven by an eulerian accurate minimization method for image and video segmentation.*International Journal of Computer Vision*, 53(1):45–70, 2003.

[433] S. Jehan-Besson, M. Barlaud, G. Aubert, and O. Faugeras.Shape gradients for histogram segmentation using active contours.In *International Conference on Computer Vision*, pages 408–415, Nice, France, October 2003.

[434] R. Jensen.Uniqueness of Lipschitz Extensions: Minimizing the Sup Norm of the Gradient.*Arch. Rat. Mech. Anal.*, 123:51–74, 1993.

[435] A. Jepson and M. Black.Mixture models for optical flow computation.In *IEEE Computer Vision and Pattern Recognition, CVPR-93*, pages 760–761, New York, June 1993.

[436] A. Jepson, D. Fleet, and T. El-Maraghi.Robust online appearance models for visual tracking.In *IEEE Conference on Computer Vision and Pattern Recognition*, pages 415–422, 2001.

[437] J. Jia and C. Kang.Inference of segmented color and texture description by tensor voting.*IEEE-TPAMI*, 26(6):771–786, 2004.

[438] J. Jia, T. Wu, Y. Tai, and C. Tang.Video repairing: Inference of foreground and background under severe occlusion.In *Proc. CVPR*, 2004.

[439] H. Jin, S. Soatto, and A. Yezzi.Multi-view Stereo Beyond Lambert.In *IEEE Conference on Computer Vision and Pattern Recognition*, pages 171–178, 2003.

[440] H. Johnson and G. Christensen.Consistent Landmark and Intensity-Based Image Registration.*IEEE Transactions on Medical Imaging*, 21:450–469, 2002.

[441] L. Jonasson, X. Bresson, P. Hagmann, O. Cuisenaire, R. Meuli, and J.P. Thiran.White matter fiber tract segmentation in DT-MRI using geometric flows.*Medical Image Analysis*, 2004.In press.

[442] S. Joshi and M. Miller.Landmark Matching via Large Deformation Diffeomorphisms.*IEEE Transactions on Image Processing*, 9:1357–1370, 2000.

[443] O. Juan, R. Kerivan, and G. Postelnicu.Stochastic mean curvature motion in computer vision: stochastic active contours.*INRIA Report*, 2004.

[444] B. Julesz.Visual pattern discrimination.*IEEE Transactions on Information Theory*, 8(2):84–92, 1962.

[445] P. Juutinen.Absolutely Minimizing Lipschitz Extensions on a Metric Space.*Annales Acadademiae Scientiarum Fennicae Mathematica*, 27:57–67, 2002.

[446] T. Kailath.*Linear systems*.Prentice Hall, Inc., 1980.

[447] I. Kakadiaris and D. Metaxas.3D Human body model acquisition from multiple views.*International Journal of Computer Vision*, 30(3):191–218, 1998.

[448] I. Kakadiaris and D. Metaxas.Model-based estimation of 3D human motion.*IEEE Transactions on Pattern Analysis and Machine intelligence*, 22(12):1453–1459, 2000.

[449] I. Kakadiaris, D. Metaxas, and R. Bajcsy.Inferring 2D object structure from the deformation of apparent contours.*Computer Vision and Image Understanding*, 65(2):129–147, 1997.

[450] S. Kakutani.Markov Processes and the Dirichlet Problem.*Proceeding of the Japenese Academy*, 21:227–233, 1945.

[451] S.-B. Kang, R. Szeliski, and J. Chai.Handling occlusions in dense multi-view stereo.In *IEEE Conference on Computer Vision and Pattern Recognition*, 2001.Expanded version available as MSR-TR-2001-80.

[452] S.-B. Kang, M. Uyttendaele, S. Winder, and R. Szeliski.High dynamic range video.*ACM Transactions on Graphics*, 22(3):319–325, July 2003.

[453] G. Kanizsa.*Gramática de la Visión*.Paidos, 1986.

[454] H. Karcher.Riemann center of mass and mollifier smoothing.*Communications on Pure and Applied Mathematics*, 30:509–541, 1977.

[455] M. Kass, A. Witkin, and D. Terzopoulos.Snakes: Active Contour Models.In *IEEE International Conference in Computer Vision*, pages 261–268, 1987.

[456] M. Kass, A. Witkin, and D. Terzopoulos.Snakes: Active Contour Models.*International Journal of Computer Vision*, 1:321–332, 1988.

[457] M. Kaus, J. von Berg, J. Weese, W. Niessen, and V. Pekar.Automated segmentation of the left ventricle in cardiac.*Medical Image Analysis*, 8(3):245–254, 9 2004.

[458] D. Kendall.Shape manifolds, procrustean metrics and complex projective spaces.*Bulletin of London Mathematics Society*, 16, 1984.

[459] D. Kendall, D. Barden, T. Carne, and H. Le.*Shape and shape theory*.Wiley, Chichester, 1999.

[460] C. Kenney and J. Langan.A new image processing primitive: reconstructing images from modified flow fields.Technical report, University of California Santa Barbara, 1999.

[461] S. Khan and M. Shah.Consistent labeling of tracked objects in multiple cameras with overlapping fields of view.*IEEE Transactions on Pattern Analysis and Machine Intelligence*, 25(10):1355–1360, October 2003.

[462] S. Kichenassamy, A. Kumar, P. Olver, A. Tannenbaum, and A. Yezzi.Gradient flows and geometric active contour models.In *IEEE International Conference in Computer Vision*, pages 810–815, 1995.

[463] S. Kichenesamy, A. Kumar, P. Olver, A. Tannenbaum, and A. Yezzi.Conformal curvature flows: from phase transitions to active contours.*Archive for Rational Mechanics and Analysis*, 134:275–301, 1996.

[464] J. Kim, J. Fisher, M. Cetin, A. Yezzi, and A. Willsky.Incorporating complex statistical information in active contour-based image segmentation.In *International Conference on Image Processing*, pages 655–658, 2003.

[465] J. Kim, J. Fisher, A. Yezzi, M. Cetin, and A. Willsky.Non-Parametric Methods for Image Segmentation using Information Theory and Curve Evolution.In *IEEE International Conference on Image Processing*, 2002.

[466] J. Kim and H. Kim.GA-based image restoration by isophote constraint optimization.In *Proceedings of EURASIP*, 2003.

[467] G. Kimeldorf and G. Wahba.Some Results on Tchebycheffian Spline Functions.*Journal of Mathematical Analysis and Applications*, 33(1):82–95, 1971.

[468] B. Kimia, I Frankel, and A.M. Popescu.Euler spiral for shape completion.In K. Boyer and S. Sarker, editors, *Perceptual Organization for Artificial Vision Systems*, pages 289–310. Kluwer Academic Publishers, 2000.

[469] R. Kimmel, A. Amir, and A. Bruckstein.Finding shortest paths on surfaces using level sets propagation.*IEEE Trans. on PAMI*, PAMI-17(6):635–640, June 1995.

[470] R. Kimmel and A. Bruckstein."Global shape-from-shading".*CVGIP: Image Understanding*, pages 360–369, 1995.

[471] R. Kimmel and A. Bruckstein.Tracking level sets by level sets: a method for solving the shape from shading problem.*Computer Vision, Graphics and Image Understanding*, 62:47–58, 1995.

[472] R. Kimmel, N. Kiryati, and A. Bruckstein.Distance maps and weighted distance transforms.*Journal of Mathematical Imaging and Vision*, 6:223–233, May 1996.Special Issue on Topology and Geometry in Computer Vision.

[473] R. Kimmel, R. Malladi, and N. Sochen.Images as Embedded Maps and Minimal Surfaces: Movies, Color, Texture, and Volumetric Medical Images.*IJCV*, 39:111–129, 2000.

[474] R. Kimmel and J. Sethian.Optimal algorithm for shape from shading and path planning.*JMIV*, 14(2):237–244, 2001.

[475] D. King.*The Commissar Vanishes*.Henry Holt and Company, 1997.

[476] C. Kipnis and C. Landim.*Scaling Limits of Interacting Particle Systems*.Springer-Verlag, 1999.

[477] D. Kirsanov and S. Gortler.A discrete global minimization algorithm for continuous variational problems.*Harvard Computer Science Technical Report*, TR-14-04, July 2004, (also submitted to a journal).

[478] E. Klassen, A. Srivastava, W. Mio, and S. Joshi.Analysis of planar shapes using geodesic paths on shape spaces.*IEEE Transactions on Pattetn Analysis and Machine Intelligence*, 26(3):372–383, 2004.

[479] R. Klette, R. Kozera, and K. Schlüns.Shape from shading and photometric stereo methods.Technical Report CITR-TR-20, University of Auckland, New Zealand, 1998.

[480] D-C. Knill and W. Richards.*Perception as Bayesian Inference*.Cambridge University Press, 1996.

[481] R. Koch, M. Pollefeys, and L. Van Gool.Multi viewpoint stereo from uncalibrated video sequences.In *Computer Vision – ECCV'98, LNCS, Vol.1406*, pages 55–71. Springer-Verlag, 1998.

[482] J. Koenderink.*Solid Shape*.MIT Press, 1990.

[483] K. Koffka.*Principles of Gestalt Psychology*.Harcourt, Brace, and World, New York, 1935.

[484] A. Kokaram.On missing data treatment for degraded video and film archives: a survey and a new bayesian approach.*IEEE Transactions on Image Processing*, 13:397–415, 2004.

[485] A. Kokaram, R. Morris, W. Fitzgerald, and P. Rayner.Detection of Missing Data in Image Sequences.*IEEE Transactions on Image Processing*, 11:1496–1508, 1995.

[486] A. Kokaram, R. Morris, W. Fitzgerald, and P. Rayner.Interpolation of Missing Data in Image Sequences.*IEEE Transactions on Image Processing*, 11:1509–1519, 1995.

[487] D. Koller, K. Daniilidis, and H.H. Nagel.Model-based object tracking in monocular image sequences of road traffic scenes.*International Journal Computer Vision*, 10(3):257–281, 1993.

[488] V. Kolmogorov.*Graph Based Algorithms for Scene Reconstruction from Two or More Views*.PhD thesis, Cornell University, September 2003.

[489] V. Kolmogorov and R. Zabih.Visual correspondence with occlusions using graph cuts.In *IEEE International Conference on Computer Vision*, pages 508–515, 2001.

[490] V. Kolmogorov and R. Zabih.Multi-camera Scene Reconstruction via Graph Cuts.In *European Conference on Computer Vision*, volume 3, pages 82–96, 2002.

[491] V. Kolmogorov and R. Zabih.What energy functions can be minimized via graph cuts?*IEEE Transactions on Pattern Analysis and Machine Intelligence*, 26(2):147–159, February 2004.

[492] V. Kolmogorov, R. Zabih, and S. Gortler.Generalized multi-camera scene reconstruction using graph cuts.In *International Workshop on Energy Minimization Methods in Computer Vision and Pattern Recognition*, July 2003.

[493] R. Kozera.Uniqueness in shape from shading revisited.*JMIV*, 7:123–138, 1997.

[494] S. Kullback.*Information Theory and Statistics*.John Wiley and Sons, New York, 1959.

[495] H. Kunita.*Stochastic Flows and Stochastic Differential Equations*.Cambridge University Press, 1990.

[496] K. Kutulakos and C. Dyer.Recovering shape by purposive viewpoint adjustment.*International Journal of Computer Vision*, 12(2-3):113–136, April 1994.

[497] K. Kutulakos and S. Seitz.A theory of shape by space carving.In *IEEE International Conference on Computer Vision*, pages 307–314, 1999.

[498] K. Kutulakos and S. Seitz.A Theory of Shape by Space Carving.*International Journal of Computer Vision*, 38:199–218, 2000.

[499] V. Kwatra, A. Schodl, I. Essa, and A. Bobick.GraphCut textures: image and video synthesis using graph cuts.In *ACM Transactions on Graphics (SIGGRAPH)*, volume 22, July 2003.

[500] O. Ladyzhenskaya.*The Mathematical Theory of Viscous Incompressible Flow*.Gordon and Breach Science Publishers, New York-London, 1963.

[501] E. Lane.A Treatise on Projective Differential Geometry.In *University of Chicago Press*, 1941.

[502] A. Laurentini.The Visual Hull Concept for Silhouette-Based Image Understanding.*IEEE Transactions on Pattern Analysis and Machine Intelligence*, 16:150–162, 1994.

[503] H. Le and D. Kendall.The Riemannian structure of Euclidean shape spaces: a novel environment for statistics.*The Annals of Statistics*, 4216:1225–1271, 1993.

[504] H. Le and A. Kume.The fréchet mean shape and the shape of the means.*Adv. Appl. Prob. (SGSA)*, 32:101–113, 2000.

[505] D. Le Bihan, E. Breton, D. Lallemand, P. Grenier, E. Cabanis, and M. Laval-Jeantet.MR imaging of intravoxel incoherent motions: Application to diffusion and perfusion in neurologic disorders.*Radiology*, 161:401–407, 1986.

[506] K. Lee and C. Kuo.Shape from shading with a generalized reflectance map model.*CVIU*, 67(2):143–160, 1997.

[507] C. Lenglet, R. Deriche, and O. Faugeras.Inferring white matter geometry from diffusion tensor MRI: Application to connectivity mapping.In *ECCV*, pages 127–140, 2004.

[508] C. Lenglet, M. Rousson, and R. Deriche.Segmentation of 3D probability density fields by surface evolution: Application to diffusion MRI.In *MICCAI*, pages 18–25, 2004.

[509] C. Lenglet, M. Rousson, R. Deriche, and O. Faugeras.Statistics on multivariate normal distributions: A geometric approach and its application to diffusion tensor MRI.Research Report 5442, INRIA, Sophia Antipolis, June 2004.

[510] C. Lenglet, M. Rousson, R. Deriche, and O. Faugeras.Toward Segmentation of 3D Probability Density Fields by Surface Evolution: Application to Diffusion MRI.Technical Report 5243, INRIA, 2004.

[511] T. Leung and J. Malik.Contour Continuity in Region-Based Image Segmentation.In H. Burkhardt and B. Neumann, editors, *Proc. Euro. Conf. Computer Vision, volume 1, Freiburg (Germany)*, pages 544–559. Springer-Verlag, 1998.

[512] M. Leventon and W. Grimson.Multi-modal Volume Registration Using Joint Intensity Distributions.In *First International Conference on Medical Image Computing and Computer-Assisted Intervention*, Lecture Notes in Computer Science. Springer, 1998.

[513] M. Leventon, W. Grimson, and O. Faugeras.Statistical Shape Influence in Geodesic Active Controus.In *IEEE Conference on Computer Vision and Pattern Recognition*, pages I:316–322, 2000.

[514] A. Levin, A. Zomet, S. Peleg, and Y. Weiss.Seamless image stitching in the gradient domain.In *Eighth European Conference on Computer Vision*, volume IV, pages 377–389, Prague, May 2004.

[515] A. Levin, A. Zomet, and Y. Weiss.Learning how to inpaint from global image statistics.In *Proc. of International Conference on Computer Vision*, pages 305–312, 2003.

[516] M. Levoy, K. Pulli, B. Curless, S. Rusinkiewicz, D. Koller, L. Pereira, M. Ginzton, S. Anderson, J. Davis, J. Ginsberg, J. Shade, and D. Fulk.The Digital Michelangelo Project: 3D Scanning of Large Statues.In *Computer Graphics (SIGGRAPH) 2000*, pages 269–276, 1996.

[517] M. Lhuillier and L. Quan.Surface Reconstruction by Integrating 3D and 2D Data of Multiple Views.In *IEEE International Conference on Computer Vision*, 2003.

[518] G. Li.Robust regression.In D. Hoaglin, F. Mosteller, and J. Tukey, editors, *Exploring Data Tables, Trends, and Shapes*. Wiley, 1985.

[519] G. Li and S. Zucker.A differential geometric approach to stereo correspondence.In *Second IEEE Workshop on Variational, Geometric, and Level Set Methods in Computer Vision*, 2003.

[520] S. Li.*Markov Random Field Modeling in Computer Vision*.Springer-Verlag, 1995.

[521] M. Lin and C. Tomasi.Surfaces with occlusions from layered stereo.*IEEE Transactions on Pattern Analysis and Machine Intelligence*, 26(8):1073–1078, August 2004.

[522] P.-L. Lions, E. Rouy, and A. Tourin.Shape-from-shading, viscosity solutions and edges.*Numer. Math.*, 64:323–353, 1993.

[523] J. Liu and R. Chen.Sequential Monte Carlo methods for dynamic systems.*Journal of the American Statistical Association*, 93(443):1032–1044, 1998.

[524] L. Ljung.*System identification: theory for the user*.Prentice-Hall, Inc., 2nd edition, 1999.

[525] H. Longuet-Higgins and K. Prazdny.The interpretation of a moving retinal image.*Proceedings of the Royal Society*, B-208:385–397, 1980.

[526] W. Lorensen and H. Cline.Marching cubes: a high resolution 3D surface construction algorithm.In *ACM SIGGRAPH*, volume 21, pages 163–170, 1987.

[527] C. Lorenz and N. Krahnstover.Generation of point-based 3D statistical shape models for anatomical objects.*Computer Vision and Image Understanding*, 77(2):175–191, February 2000.

[528] M. Lorenzo-Valdes, G. Sanchez-Ortiz, A. Elkington, R. Mohiaddin, and D. Rueckert.Segmentation of 4D cardiac MR images using a probabilistic atlas and the em algorithm.*Medical Image Analysis*, 8(3):255–265, 2004.

[529] J. Lötjönen, S. Kivistö, J. Koikkalainen, D. Smutek, and K. Lauerma.Statistical shape model of atria, ventricles and epicardium from short- and long-axis MR images.*Med Imag Anal*, 8(3):371–386, September 2004.

[530] D. Lowe.Robust model-based motion tracking through the integration of search and estimation.*International Journal Computer Vision*, 8(2):113–122, 1992.

[531] D. Lowe.Object recognition from local scale-invariant features.In *Proc. ICCV'99*, pages 1150–1157, 1999.

[532] D. Lowe.Distinctive image features from scale-invariant keypoints.*International Journal of Computer Vision*, 60(2):91–110, November 2004.

[533] B. Lucas and T. Kanade.An Iterative Image Registration Technique with an Application to Stereo Vision.In *International Joint Conference on Artificial Intelligence*, pages 674–679, 1981.

[534] F. Maes, A. Collignon, D. Vandermeulen, G. Marchal, and P. Suetens.Multimodality image registration by maximization of mutual information.*IEEE Transactions on Medical Imaging*, 16(2):187–198, 1997.

[535] R. Malladi and J. Sethian.A Real-Time Algorithm for Medical Shape Recovery.In *IEEE International Conference in Computer Vision*, pages 304–310, Bombay, India, 1998.

[536] R. Malladi, J. Sethian, and B. Vemuri.A Topology Independent Shape Modeling Scheme.In *Proc. Int'l Society for Optical Engineering*, volume 2031, pages 246–258, 1993.

[537] R. Malladi, J. Sethian, and B. Vemuri.Evolutionary fronts for topology independent shape modeling and recovery.In *European Conference on Computer Vision*, pages 1–13, 1994.

[538] R. Malladi, J. Sethian, and B. Vemuri.Shape Modeling with Front Propagation: A Level Set Approach.*IEEE Transactions on Pattern Analysis and Machine Intelligence*, 17:158–175, 1995.

[539] R. Malladi, J. Sethian, and B. Vemuri.A Fast Level Set Based Algorithm for Topology-independent Shape Modeling.*J. Math. Imaging and Vision*, 6(2/3):269–289, 1996.

[540] S. Mallat.*A Wavelet Tour of Signal Processing*.Academic Press, San Diego, second edition, 1999.

[541] J. Mangin, C. Poupon, C. Clark, D. Le Bihan, and I. Bloch.Distortion correction and robust tensor estimation for MR diffusion imaging.*Medical Image Analysis*, 6(3):191–198, 2002.

[542] R. Mann, A. Jepson, and J. Siskind.Computational perception of scene dynamics.*Computer Vision and Image Understanding*, 65(2):113–128, 1997.

[543] S. Mann and R. Picard.Virtual bellows: Constructing high-quality images from video.In *First IEEE International Conference on Image Processing*, volume I, pages 363–367, Austin, November 1994.

[544] R. March and M. Dozio.A Variational Method for the Recovery of Smooth Boundaries.*Image Vision Computing*, 15:705–712, 1997.

[545] B. Markussen.A statistical approach to large deformation diffeomorphisms.In *CVPR04*, pages available on the CD–rom (GMBV workshop), 2004.

[546] A. Marquina and S. Osher.Explicit Algorithms for a New Time Dependent Model Based on Level Set Motion for Nonlinear Deblurring and Noise Removal.*SIAM J. Sci. Comp.*, 22:387–405, 2000.

[547] P. Martin, P. Réfrégier, F. Goudail, and F. Guérault.Influence of the noise model on level set active contour segmentation.*IEEE Transactions on Pattern Analysis and Machine Intelligence*, 26(6):799–803, June 2004.

[548] R. Martin.A metric for ARMA processes.*IEEE Transactions on Signal Processing*, 48(4):1164–1170, 2000.

[549] W. Martin and J. Aggarwal.Volumetric descriptions of objects from multiple views.*IEEE Transactions on Pattern Analysis and Machine Intelligence*, 5(2):150–158, March 1983.

[550] G. Martinez, I. Kakadiaris, and D. Magruder.Teleoperating ROBONAUT: A case study.In *British Machine Vision Conference*, Cardiff, UK, September 2-5 2002.

[551] S. Masnou.*Filtrage et Desocclusion d'Images par Méthodes d'Ensembles de Niveau*.PhD thesis, Université Paris-Dauphine., 1998.

[552] S. Masnou.Disocclusion: a Variational Approach using Level Lines.*IEEE Transactions on Image Processing*, 11:68–76, 2002.

[553] S. Masnou and J. Morel.Level-lines based Disocclusion.In *Proc. 5th IEEE Int. Conf. on Image Process.*, pages 259–263, Chicago, IL, 1998.

[554] U. Massari.Frontiere Orientate di Curvatura Media Assegnata in $l^p$.*Rend. Sem. Mat. Univ. Padova*, 53:37–52, 1975.

[555] G. Matheron.Les nivellements.Technical report, Centre de Morphologie Mathématique, 1997.

[556] J. Maver and R. Bajcsy.Occlusions as a guide for planning the next view.*IEEE Transactions on Pattern Analysis and Machine Intelligence*, 15(5):417–433, May 1993.

[557] T. McGraw, B. Vemuri, Y. Chen, M. Rao, and T. Mareci.DT-MRI denoising and neuronal fiber tracking.*Medical Image Analysis*, 8:95–111, 2004.

[558] T. McInerney and D. Terzopoulos.A dynamic finite element surface model for segmentation and tracking in multidimensional medical images with application to cardiac 4d image analysis.*Computerized Medical Imaging and Graphics*, 19(1), 1995.

[559] J. Meehan.*Panoramic Photography*.Watson-Guptill, 1990.

[560] C.-H. Menq, H.-T. Yau, and G.-Y. Lai.Automated Precision Measurement of Surface Profile in CAD-Directed Inspection.*IEEE Transactions on Robotics and Automation*, 8:268–278, 1992.

[561] R. Merris.Laplacian Matrices of Graphs: A Survey.*Linear Algebra and its Applications*, 197,198:143–176, 1994.

[562] D. Metaxas.*Physics-Based Deformable Models*.Kluwer Academic Publishers, 1996.

[563] F. Meyer.Algorithmes à base de files d'attente hiérarchique.Technical Report NT-46/90/MM, Centre de Morphologie Mathématique, 1990.

[564] F. Meyer.From connected operators to levelings.In H. Heijmans and J. Roerdink, editors, *Mathematical Morphology and its Applications to Image and Signal Processing*, pages 191–199. Kluwer, 1998.

[565] F. Meyer.Levelings, image simplification filters for segmentation.*J. of Mathematical Imaging and Vision*, 20:59–72, 2004.

[566] F. Meyer and P. Maragos.Nonlinear scale-space representation with morphological levelings.*J. Visual Commun. and Image Representation*, 11:245–265, 2000.

[567] Y. Meyer.*Oscillating Patterns in Image Processing and in some Nonlinear Evolution Equations*.AMS, 2001.(Lewis Memorial Lectures).

[568] Y. Meyer.*Oscillating Patterns in Image Processing and Nonlinear Evolution Equations*, volume 22 of *University Lecture Series*.AMS, Providence, 2001.

[569] K. Mikolajczyk and C. Schmid.A performance evaluation of local descriptors.In *IEEE Computer Society Conference on Computer Vision and Pattern Recognition*, volume II, pages 257–263, Madison, WI, June 2003.

[570] D. Milgram.Computer methods for creating photomosaics.*IEEE Transactions on Computers*, C-24(11):1113–1119, November 1975.

[571] M. Miller, S. Joshi, and G. Christensen.*Brain Warping*, pages 131–155.Academic Press, 1998.

[572] W. Mio and A. Srivastava.Elastic string models for representation and analysis of planar shapes.In *Proc. of IEEE Computer Vision and Pattern Recognition*, 2004.

[573] W. Mio, A. Srivastava, and X. Liu.Learning and bayesian shape extraction for object recognition.In *European Conf. on Computer Vision*, volume 3024 of *LNCS*, pages 62–73, Prague, 2004. Springer.

[574] S. Mitchell, J. Bosch, B. Lelieveldt, R. van der Geest, J. Reiber, and M. Sonka.3-D active appearance models: segmentation of cardiac MR and ultrasound images.*IEEE Transactions on Medical Imaging*, 21(9):1167–78, September 2002.

[575] S. Mitchell, B. Lelieveldt, R. van der Geest, H. Bosch, J. Reiber, and M. Sonka.Cardiac segmentation using active appearance models.*IEEE Trans. Med. Imaging*, 20:415–423, 2001.

[576] N. Mitra, N. Gelfand, H. Pottmann, and L. Guibas.Registration of Point Cloud Data from a Geometric Optimization Perspective.In *Proceedings of the Eurographics Symposium on Geometry Processing*, 2004.

[577] T. Mitsunaga and S. Nayar.Radiometric self calibration.In *IEEE Computer Society Conference on Computer Vision and Pattern Recognition*, volume 1, pages 374–380, Fort Collins, June 1999.

[578] A. Mittal and L. Davis.$M_2$tracker: A multi-view approach to segmenting and tracking people in a cluttered scene.*International Journal of Computer Vision*, 51(3):189–203, February 2003.

[579] M. Moakher.A differential geometric approach to the geometric mean of symmetric positive-definite matrices.*SIAM Journal on Matrix Analysis and Applications*, 2004.In press.

[580] T. Moeslund and E. Granum.A Survey of Computer Vision-Based Human Motion Capture.*Computer Vision and Image Understanding*, 81(3):231–268, 2001.

[581] B. Mohar.Isoperimetric Numbers of Graphs.*Journal of Combinatorial Theory, Series B*, 47:274–291, 1989.

[582] L. Moisan.Affine Plane Curve Evolution: A Fully Consistent Scheme.*IEEE Transactions on Image Processing*, 7:411–420, 1998.

[583] S. Mori, B.J. Crain, V.P. Chacko, and P.C.M. Van Zijl.Three-dimensional tracking of axonal projections in the brain by magnetic resonance imaging.*Annals of Neurology*, 45(2):265–269, 1999.

[584] P. Mrázek and J. Weickert.Rotationally invariant wavelet shrinkage.In B. Michaelis and G. Krell, editors, *Pattern Recognition*, volume 2781 of *Lecture Notes in Computer Science*, pages 156–163, Berlin, 2003. Springer.

[585] P. Mrázek, J. Weickert, and G. Steidl.Correspondences between wavelet shrinkage and nonlinear diffusion.In L. D. Griffin and M. Lillholm, editors, *Scale-Space Methods in Computer Vision*, volume 2695 of *Lecture Notes in Computer Science*, pages 101–116, Berlin, 2003. Springer.

[586] P. Mrázek, J. Weickert, and G. Steidl.Diffusion-inspired shrinkage functions and stability results for wavelet denoising.Technical Report 96, Dept. of Mathematics, Saarland University, Saarbrücken, Germany, October 2003.Submitted to *International Journal of Computer Vision*.

[587] P. Mrázek, J. Weickert, G. Steidl, and M. Welk.On iterations and scales of nonlinear filters.In O. Drbohlav, editor, *Proc. Eighth Computer Vision Winter Workshop*, pages 61–66, Valtice, Czech Republic, February 2003. Czech Pattern Recognition Society.

[588] D. Mumford.Mathematical theories of shape: do they model perception?In *Geometric Methods in Computer Vision*, pages 2–10, 1991.

[589] D. Mumford.Elastica and Computer Vision.In C. Bajaj, editor, *Algebraic geometry and its applications*, pages 491–506. Springer-Verlag, 1994.

[590] D. Mumford and J. Shah.Boundary detection by minimizing functionals.In *IEEE Conference on Computer Vision and Pattern Recognition*, pages 22–26, 1985.

[591] D. Mumford and J. Shah.Optimal Approximation by Piecewise Smooth Functions and Associated Variational Problems.*Communications on Pure and Applied Mathematics*, 42:577–685, 1989.

[592] D. Murray and A. Basu.Motion tracking with an active camera.*IEEE Trans. on Pattern Analysis and Machine Intelligence*, 16(5):449–459, 1994.

[593] H. Nagel.On the estimation of optical flow: relations between different approaches and some new results.*Artificial Intelligence*, 33:299–324, 1987.

[594] National Aeronautics and Space Administration.Man systems integration standards.Technical report, National Aeronautics and Space Administration, 1987.

[595] S. Negahdaripour.Revised definition of optical flow: integration of radiometric and geometric clues for dynamic scene analysis.*IEEE Transactions on Pattern Analysis and Machine Intelligence*, 20(9):961–979, 1998.

[596] R. Nelson and R. Polana.Qualitative recognition of motion using temporal texture.*CVGIP Image Understanding*, 56(1):78–89, 1992.

[597] O. Nestares and D. Fleet.Likelihood functions for general error-in-variables problems.In *IEEE International Conference on Image Processing*, pages vol. III, pp. 77–80, Barcelona, Spain, 2003.

[598] O. Nestares, D. Fleet, and D. Heeger.Likelihood functions and confidence bounds for total-least-squares estimation.In *IEEE Conference on Computer Vision and Pattern Recognition*, volume 1, pages 523–530, Hilton Head, 2000.

[599] J. Neumann and Y. Aloimonos.Spatio-Temporal Stereo Using Multi-Resolution Subdivision Surfaces.*The International Journal of Computer Vision*, 47:181–193, 2002.

[600] M. Nielsen and P. Johansen.A pde solution of brownian warping.In *ECCV 2004, Volume IV*, pages 180–191. Springer verlag, 2004.

[601] M. Nielsen, P. Johansen, A. Jackson, and B. Lautrup.Brownian warps.In *MICCAI 02*, volume 2489 of *LNCS*. Springer, 2002.

[602] M. Nikolova.Minimizers of cost-functions involving nonsmooth data-fidelity terms. *SIAM J. Numer. Anal.*, 40:965–994, 2002.

[603] D. Nister.An efficient solution to the five-point relative pose problem.In *Proc. CVPR'03, Vol.2*, pages 195–202, 2003.

[604] M. Nitzberg, D. Mumford, and T. Shiota.*Filtering, Segmentation, and Depth*.Springer-Verlag, Berlin, 1993.

[605] J. Ogden, E. H. Adelson, J. Bergen, and P. Burt.Pyramid based computer graphics.*RCA Engineer*, 30:4–15, 1985.

[606] T. Okatani and K. Deguchi.On classification of singular points for global shape from shading.In *ACCV'98*, volume 1351, pages 48–55, 1998.

[607] M. Okutomi and T. Kanade.A multiple baseline stereo.*IEEE Trans. on Pattern Analysis and Machine Intell.*, 15(4):353–363, 1993.

[608] J. Oliensis.Shape from shading as a partially well–constrained problem.*CVGIP: Image Understanding*, 54(2):163–183, 1991.

[609] J. Oliensis.Uniqueness in shape from shading.*IJCV*, 2(6):75–104, 1991.

[610] P. Olver.Joint Invariant Signatures.*Foundations on Computational Mathematics*, 1(1):3–67, Feb 2001.

[611] B. O'Neill.*Elementary Differential Geometry*.Academic Press, London, 1966.

[612] E. Ong and M. Spann.Robust optical flow computation based on least-median-of-squares regression.*International Journal of Computer Vision*, 31:51–82, 1999.

[613] C. Oost, B. Lelieveldt, M. Uzumcu, H. Lamb, J. Reiber, and M. Sonka.Multi-view active appearance models: Application to X-ray LV angiography and cardiac MRI.In *Information Processing in Medical Imaging – IPMI*, number 2732 in Lecture Notes in Computer Science, pages 234–45, 2003.

[614] J. O'Rourke.*Art Gallery Theorems and Algorithms*.Oxford University Press, August 1987.

[615] S. Osher, M. Burger, D. Goldfarb, J. Xu, and W. Yin.An Iterative Regularization Method for Total Variation Based Image Restoration.*UCLA CAM Report*, 04-13, 2004.

[616] S. Osher and R. Fedkiw.*The Level Set Method and Dynamic Implicit Surfaces*.Springer, 2002.

[617] S. Osher and N. Paragios.*Geometric Level Set Methods in Imaging, Vision and Graphics*.Springer Verlag, 2003.

[618] S. Osher and J. Sethian.Fronts propagating with curvature-dependent speed : Algorithms based on the Hamilton-Jacobi formulation.*Journal of Computational Physics*, 79:12–49, 1988.

[619] S. Osher, A. Sole, and L. Vese.Image Decomposition and Restoration Using Total Variation Minimization and the $H^{-1}$ Norm.*Multiscale Model. Simul.*, 1:349–370, 2003.

[620] E. Özarslan and T. Mareci.Generalized diffusion tensor imaging and analytical relationships between diffusion tensor imaging and high angular resolution diffusion imaging.*Magnetic Resononance Medicine*, 50:955–965, 2003.

[621] E. Özarslan, B. Vemuri, and T. Mareci.Fiber orientation mapping using generalized diffusion tensor imaging.In *Proc. of IEEE International Symposium on Biomedical Imaging (ISBI)*, pages 1036–1039, 2004.

[622] N. Paragios.*Geodesic Active Regions and Level Set Methods: Contributions and Applications in Artificial Vision*.PhD thesis, I.N.R.I.A./University of Nice-Sophia Antipolis, 2000.http://www.inria.fr/RRRT/TU-0636.html.

[623] N. Paragios and R. Deriche.A PDE-based Level Set approach for Detection and Tracking of moving objects.In *IEEE International Conference in Computer Vision*, pages 1139–1145, 1998.

[624] N. Paragios and R. Deriche.Geodesic Active Contours and Level Sets for the Detection and Tracking of Moving Objects.*IEEE Transactions on Pattern Analysis and Machine Intelligence*, 22:266–280, 2000.

[625] N. Paragios and R. Deriche.Geodesic Active Regions: A New Framework to Deal with Frame Partition Problems in Computer Vision.*Journal of Visual Communication and Image Representation*, 13:249–268, 2002.

[626] N. Paragios and R. Deriche.Geodesic Active Regions Level Set Methods for Supervised Texture Segmentation.*International Journal of Computer Vision*, 46(3):223–247, 2002.

[627] N. Paragios, O. Mellina-Gottardo, and V. Ramesh.Gradient Vector Flow Fast Geodesic Active Contours.In *IEEE International Conference in Computer Vision*, pages I:67–73, 2001.

[628] N. Paragios and M. Rousson.Shape analysis towards model-based segmentation.In S. Osher and N. Paragios, editors, *Geometric Level Set Methods in Imaging Vision and Graphics*. Springer Verlag, 2003.

[629] N. Paragios, M. Rousson, and V. Ramesh.Distance Tranforms for Non-Rigid Registation.*Computer Vision and Image Understanding*, 23:142–165, 2003.

[630] P. Parent and S. Zucker.Trace inference, curvature consistency and curve detection.*IEEE Transactions on Pattern Analysis and Machine Intelligence*, 11(8):823–839, August 1989.

[631] J. Park, D. Metaxas, A. Young, and L. Axel.Deformable models with parameter functions for cardiac motion analysis.*IEEE Transactions on Medical Imaging*, 15(3), 1996.

[632] G. Parker and D. Alexander.Probabilistic monte carlo based mapping of cerebral connections utilising whole-brain crossing fiber information.In *Information Processing in Medical Imaging*, pages 684–696, Ambleside UK, 2003.

[633] G. Parker, J. Schnabel, M. Symms, D. Werring, and G. Baker.Nonlinear smoothing for reduction of systematic and random errors in diffusion tensor imaging.*Magn. Reson. Med.*, 11:702–710, 2000.

[634] K. Patwardhan and G. Sapiro.Projection based image and video inpainting using wavelets.In *Proceedings International Conference on Image Processing, ICIP*, pages 857–860, 2003.

[635] X. Pennec.Probabilities and statistics on Riemannian manifolds: A geometric approach.Research Report 5093, INRIA, Sophia Antipolis, January 2004.

[636] X. Pennec, P. Fillard, and N. Ayache.A Riemannian framework for tensor computing.Research Report 5255, INRIA, Sophia Antipolis, July 2004.

[637] X. Pennec and J.-P. Thirion.A Framework for Uncertainty and Validation of 3-D Registration Methods Based on Points and Frames.*International Journal of Computer Vision*, 25:203–229, 1997.

[638] A. Pentland.Local shading analysis.*PAMI*, 6:170–187, 1984.

[639] P. Perez, M. Gangnet, and A. Blake.Poisson image editing.In *Proc. SIGGRAPH 03, ACM*, pages 313–318, 2003.

[640] P. Perez, C. Hue, J. Vermaak, and M. Gangnet.Color-based probabilistic tracking.In *European Conference on Computer Vision*, 2002.

[641] P. Perez, J. Vermaak, and A. Blake.Data fusion for visual tracking with particles.*Proc. IEEE*, 92(3):495–513, 2004.

[642] P. Perona and J. Malik.Scale space and edge detection using anisotropic diffusion.*IEEE Transactions on Pattern Analysis and Machine Intelligence*, 12:629–639, 1990.

[643] G. Peyre and L. Cohen.Geodesic re-meshing and parameterization using front propagation.In *Proc. IEEE Workshop on Variational and Level Set Methods in Computer Vision*, Nice, France, October 2003.

[644] R. Peyret and T.D. Taylor.*Computational methods for fluid flow*.Springer Verlag, 1993.

[645] P. Phillips, S. Sarkar, I. Robledo, P. Grother, and K. Bowyer.The gait identification challenge problem: data sets and baseline algorithm.In *Proceedings of the 15th International Conference on Pattern Recognition*, pages I: 385–388, 2002.

[646] E. Pichon, G. Sapiro, and A. Tannenbaum.*Lecture Notes in Control and Information Sciences*, chapter Segmentation of Diffusion Tensor Imagery, pages 239–247.Springer-Verlag, 2003.

[647] C. Pierpaoli and P. Basser.Toward a quantitative assessment of diffusion anisotropy. *Magnetic Medicine*, 36:893–906, 1996.

[648] R. Pito.A solution to the next best view problem for automated surface acquisition.*IEEE Transactions on Pattern Analysis and Machine Intelligence*, 21(10):1016–1030, October 1999.

[649] S. Pizer, S. Joshi, T. Fletcher, M. Styner, G. Tracton, , and J. Chen.Segmentation of single-figure objects by deformable M-reps.In *Medical Image Computing & Computer Assisted Interventions – MICCAI*, number 2208 in Lecture Notes in Computer Science, pages 862–871, 2001.

[650] Z. Pizlo and A. Rosenfeld.Recognition of Planar Shapes from Perspective Images using Contour Based Invariants.*Center for Automation Research*, CAR-TR-528, December 1990.

[651] J. Pluim, J. Maintz, and M. Viergever.Image registration by maximization of combined mutual information and gradient information.In *Proceedings of MICCAI*, Lecture Notes in Computer Science, pages 567–578. Springer, 2000.

[652] J. Pluim, J. Maintz, and M. Viergever.Mutual-information-based registration of medical images: a survey.*MedImg*, 22(8):986–1004, August 2003.

[653] S. Pollard, J. Mayhew, and J. Frisby.PMF: A stereo correspondence algorithm using a disparity gradient limit.*Perception*, 14:449–470, 1985.

[654] M. Pollefeys, R. Koch, and L. Van Gool.Self-calibration and metric reconstruction in spite of varying and unknown internal camera parameters.In *Proc. ICCV'98*, pages 90–95, 1998.

[655] M. Pollefeys, R. Koch, and L. Van Gool.A simple and efficient rectification method for general motion.In *Proc.ICCV'99*, pages 496–501, 1999.

[656] M. Pollefeys, F. Verbiest, and L. Van Gool.Surviving dominant planes in uncalibrated structure and motion recovery.In *Computer Vision - ECCV 2002, LNCS, Vol.2351*, pages 837–851, 2002.

[657] J.-P. Pons, R. Keriven, and O. Faugeras.Modelling Dynamic Scenes by Registrating Multi-View Image Sequences.Technical Report 5321, INRIA, 2004.

[658] J. Portilla and E. Simoncelli.A parametric texture model based on joint statistics of complex wavelet coefficients.*International Journal of Computer Vision*, 40(1):49–71, 2000.

[659] C. Poupon.*Détection des faisceaux de fibres de la substance blanche pour l'étude de la connectivité anatomique cérébrale*.PhD thesis, Ecole Nationale Supérieure de Télécommunications de Paris, 1999.

[660] C. Poupon, J. Mangin, C. Clark, V. Frouin, J. Regis, D. LeBihan, and I. Block.Towards inference of human brain connectivity from MR diffusion tensor data.*Med. Image Anal.*, 5:1–15, 2001.

[661] E. Prados, F. Camilli, and O. Faugeras.A viscosity method for Shape-From-Shading without boundary data.Technical Report RR-5296, INRIA, 2004.

[662] E. Prados and O. Faugeras.A mathematical and algorithmic study of the lambertian SFS problem for orthographic and pinhole cameras.Technical Report RR-5005, INRIA, 2003.

[663] E. Prados and O. Faugeras."Perspective Shape from Shading" and viscosity solutions.In *ICCV'03*, volume 2, pages 826–831, 2003.

[664] E. Prados and O. Faugeras.Unifying approaches and removing unrealistic assumptions in Shape from Shading: Mathematics can help.In *ECCV'04*, 2004.

[665] E. Prados, O. Faugeras, and E. Rouy.Shape from shading and viscosity solutions.In *ECCV'02*, volume 2351, pages 790–804, 2002.

[666] W. H. Press, S. Teukolsky, W Vetterling, and B. Flannery.*Numerical Recipes in C: The Art of Scientific Computing*.Cambridge University Press, 2nd edition, 2002.

[667] H. Ragheb and E. Hancock.A probabilistic framework for specular shape-from-shading.*Pattern Recognition*, 36:407–427, 2003.

[668] A. Rahimi, B. Dunagan, and T. Darrell.Simultaneous calibration and tracking with a network of non-overlapping sensors.In *IEEE International Conference on Computer Vision and Pattern Recognition*, pages I: 187–194, 2004.

[669] S. Rane, M. Bertalmio, and G. Sapiro.Structure and Texture Filling-in of Missing Image Blocks in Wireless Transmission and Compression Applications.*IEEE Transactions on Image Processing*, 12:296 – 303, 2003.

[670] C. Rao.Information and accuracy attainable in the estimation of statistical parameters.*Bull. Calcutta Math. Soc.*, 37:81–91, 1945.

[671] M. Reed and P. Allen.Constraint-based sensor planning for scene modeling.*IEEE Transactions on Pattern Analysis and Machine Intelligence*, 22(12):1460–1467, December 2000.

[672] J. Rehg and T. Kanade.Visual tracking of high dof articulated structures: an application to human hand tracking.In J-O. Eklundh, editor, *European Conference on Computer Vision*, pages 35–46. Springer-Verlag, 1994.

[673] I. Reid and D. Murray.Tracking foveated corner clusters using affine structure.In *IEEE International Conference on Computer Vision*, pages 76–83, 1993.

[674] J. Rieger and K. Rohr.Semi-algebraic solids in 3-space: A survey of moldelling schemes and implications for view graphs.*Image and Vision Computing*, 12(7):395–410, 1994.

[675] J. Rissanen.Modeling by shortest data description.*Automatica*, 14:465–471, 1978.

[676] J. Rissanen.*Stochastic Complexity in Statistical Inquiry, (2nd edition)*.World Scientific Press, 1998.

[677] E. Rivlin and I. Weiss.Local Invariants for Recognition.*IEEE Transactions on Pattern Analysis and Machine Intelligence*, 17(3):226–238, March 1995.

[678] A. Roche, G. Malandain, X. Pennec, and N. Ayache.The correlation ratio as a new similarity measure for multimodal image registration.In *Proceedings of MICCAI*, volume 1496 of *Lecture Notes in Computer Science*, pages 1115–1124. Springer, 1998.

[679] M. Rodríguez Florido, C.-F. Westin, and J. Ruiz-Alzola.Dt-mri regularization using anisotropic tensor field filtering.In *IEEE 2004 International Symposium on Biomedical Imaging*, pages 15–18, Arlington, VA, EEUU, April 2004.

[680] R. Ronfard.Region-based strategies for active contour models.*International Journal of Computer Vision*, 13:229–251, 1994.

[681] J. Rosen.The Gradient Projection Method for Nonlinear Programming. Part I. Linear Constraints.*SIAM*, 8(1):181–217, March 1960.

[682] V. Roth, J. Laub, M. Kawanabe, and J. Buhmann.Optimal cluster preserving embedding of nonmetric proximity data.*IEEE Transactions on Pattetn Analysis and Machine Intelligence*, 25(12):1540–1551, Dezember 2003.

[683] C. Rother, V. Kolmogorov, and A. Blake.Grabcut - interactive foreground extraction using iterated graph cuts.In *ACM Transactions on Graphics (SIGGRAPH)*, August 2004.

[684] M. Rousson.*Cues integrations and front evolutions in image segmentation*.PhD thesis, Université de Nice-Sophia Antipolis, 2004.

[685] M. Rousson and R. Deriche.A Variational Framework for Active and Adaptative Segmentation of Vector Valued Images.Technical Report 4515, INRIA, France, 2002.

[686] M. Rousson and R. Deriche.A variational framework for active and adaptive segmentation of vector valued images.In *IEEE Workshop on Motion and Video Computing*, pages 56–61, Orlando, Florida, 2002.

[687] M. Rousson, C. Lenglet, and R. Deriche.Level Set and Region Based Surface Propagation for Diffusion Tensor MRI Segmentation.In *Computer Vision Approaches to Medical Image Analysis and Mathematical Methods in Biomedical Image Analysis Workshop*, pages 87–98, 2004.

[688] M. Rousson and N. Paragios.Shape Priors for Level Set Representations.In *European Conference on Computer Vision*, pages II:78–93, Copenhangen, Denmark, 2002.

[689] M. Rousson, N. Paragios, and R. Deriche.Implicit Active Shape Models for 3D Segmentation in MR Imaging.In *Medical Imaging Copmuting and Computer-Assisted Intervention*, 2004.

[690] E. Rouy and A. Tourin.A Viscosity Solutions Approach to Shape-from-Shading.*SIAM Journal on Numerical Analysis*, 29:867–884, 1992.

[691] S. Rowe and A. Blake.Statistical mosaics for tracking.*Journal in Image and Vision Computing*, 14:549–564, 1996.

[692] S. Roy.Stereo without epipolar lines: A maximum-flow formulation.*International Journal of Computer Vision*, 34(2/3):147–162, August 1999.

[693] S. Roy and I. Cox.A maximum-flow formulation of the n-camera stereo correspondence problem.In *IEEE Proc. of Int. Conference on Computer Vision*, pages 492–499, 1998.

[694] S. Roy and V. Govindu.MRF solutions for probabilistic optical flow formulations.In *International Conference on Pattern Recognition (ICPR)*, September 2000.

[695] L. Rudin, S. Osher, and E. Fatemi.Nonlinear Total Variation Based Noise Removal.*Physica D*, 60:259–268, 1992.

[696] D. Rueckert, A. Frangi, and J. Schnabel.Automatic construction of 3D statistical deformation models of the brain using non-rigid registration.*IEEE Transactions on Medical Imaging*, 22(8):1014–25, August 2003.

[697] S. Rusinkiewicz and M. Levoy.Efficient Variants of the ICP Algorithm.In *Proceedings of the 3rd International Conference on 3-D Digitial Imaging and Modeling*, pages 224–231, 2001.

[698] P. Saisan, G. Doretto, Y. Wu, and S. Soatto.Dynamic texture recognition.In *IEEE Conference on Computer Vision and Pattern Recognition*, volume 2, pages 58–63, 2001.

[699] H. Saito, S. Baba, M. Kimura, S. Vedula, and T. Kanade.Appearance-based virtual view generation of temporally varying events from multi-camera images in the 3d room.In *Proc. Second International Conference on 3-D Digital Imaging and Modeling (3DIM99)*, pages 516–525, October 1999.

[700] P. Salembier and J. Serra.Flat zones filtering, connected operators and filters by reconstruction.*IEEE Transactions on Image Processing*, 3(8):1153–1160, August 1995.

[701] C. Samson, L. Blanc-Feraud, G. Aubert, and J. Zerubia.A Level Set Model for Image Classification.*International Journal of Computer Vision*, 40:187–197, 2000.

[702] G. Sapiro.*Geometric Partial Differential Equations in Image Processing*.Cabridge University Press, 2001.

[703] G. Sapiro and A. Bruckstein.The Ubiquitous Ellipse.*Acta Applicandae Mathematicae*, 38(2):139–147, 1995.

[704] G. Sapiro and D. Ringarch.Anisotropic diffusion of multivalued images with applications to color filtering.*IEEE Transactions on Image Processing*, 5:1582–1585, 1996.

[705] G. Sapiro and A. Tannenbaum.On invariant curve evolution and image analysis.*Indiana Journal of Mathematics*, 42:51–87, 1993.

[706] G. Sapiro and A. Tannenbaum.On affine planar curve evolution.*J. Functl. Anal.*, 119:79–120, 1994.

[707] J. Sato and R. Cippola.Affine Integral Invariants for Extracting Symmetry Axes.*Image and Vision Computing*, 15(8):627–635, Aug 1997.

[708] J. Sato and R. Cippola.Quasi Invariant Parameterizations and Matching of Curves in Images.*International Journal of Computer Vision*, 28(2):117–136, June-July 1998.

[709] H. Sawhney and R. Kumar.True multi-image alignment and its application to mosaicing and lens distortion correction.*IEEE Transactions on Pattern Analysis and Machine Intelligence*, 21(3):235–243, March 1999.

[710] H. Sawhney, R. Kumar, G. Gendel, J. Bergen, D. Dixon, and V. Paragano.Videobrush: Experiences with consumer video mosaicing.In *IEEE Workshop on Applications of Computer Vision*, pages 56–62, Princeton, October 1998.

[711] D. Scharstein and R. Szeliski.A Taxonomy and Evaluation of Dense Two-Frame Stereo Correspondence Algorithms.*The International Journal of Computer Vision*, 47:7–42, 2002.

[712] C. Schmid, R. Mohr, and C. Bauckhage.Evaluation of interest point detectors.*International Journal of Computer Vision*, 37(2):151–172, June 2000.

[713] C. Schmid and A. Zisserman.The geometry and matching of lines and curves over multiple views.*International Journal of Computer Vision*, 40(3):199–233, 2000.

[714] C. Schnorr.Determining optical flow for irregular domains by minimising quadratic functionals of a certain class.*International Journal of Computer Vision*, 6(1):25–38, 1991.

[715] C. Schnörr.Computation of discontinuous optical flow by domain decomposition and shape optimization.*IJCV*, 8(2):153–165, 1992.

[716] A. Schödl, R. Szeliski, D. Salesin, and I. Essa.Video textures.In *Proceedings of SIGGRAPH*, pages 489–498, 2000.

[717] I. Schoenberg.The Finite Fourier Series and Elementary Geometry.*Amer. Math. Monthly*, 57:390–404, 1950.

[718] S. Sclaroff and J. Isidoro.Active blobs.In *IEEE International Conference on Computer Vision*, 1998.

[719] T. Sederberg and S. Parry.Free-form deformation of solid geometric models.In *Proceedings of the 13th Annual Conference on Computer Graphics*, pages 151–160, 1986.

[720] M. Segal, C. Korobkin, R. van Widenfelt, J. Foran, and P. Haeberli.Fast Shadows and Lighting Effects Using Texture Mapping.*Computer Graphics*, 26:249–252, 1992.

[721] S. Seitz and C. Dyer.Photorealistic scene reconstruction by voxel coloring.In *IEEE International Conference on Computer Vision and Pattern Recognition*, pages 1067–1073, June 1997.

[722] S. Seitz and C. Dyer.Photorealistic Scene Reconstruction by Voxel Coloring.*The International Journal of Computer Vision*, 35:151–173, 1999.

[723] J. Serra.*Image Analysis and Mathematical Morphology*.Academic Press, 1982.

[724] J. Serra.*Image Analysis and Mathematical Morphology. II: Theoretical Advances*. Academic Press, London, 1988.

[725] J. Serra.Set connections and discrete filtering.In M. Couprie G. Bertrand and L. Perroton, editors, *Discrete Geometry for Computer Imagery*, Lecture Notes in Computer Science 1568, pages 191–207. Springer, 1999.

[726] J. Serra and P. Salembier.Connected Operators and Pyramids.In *Proc. SPIE Image Algebra Math. Morphology, SPIE 2030*, pages 65–76, San Diego, CA, 1993.

[727] J. Sethian.A Review of the Theory, Algorithms, and Applications of Level Set Methods for Propagating Interfaces.*Cambridge University Press*, pages 487–499, 1995.

[728] J. Sethian.*Level Set Methods*.Cambridge University Press, 1996.

[729] J. Sethian.Fast Marching Methods.*SIAM Review*, 41:199–235, 1999.

[730] J. Sethian.*Level Set Methods and Fast Marching Methods*.Cambridge University Press, second edition, 1999.

[731] J. Sethian and A. Vladimirsky.Ordered upwind methods for static hamilton–jacobi equations: Theory and algorithms.*SIAM J. of Numerical Analysis*, 41(1):325–363, 2003.

[732] A. Shaashua and S. Ullman.Structural saliency: The detection of globally salient structures using a locally connected network.In *International Conference on Computer Vision*, pages 321–327, 1988.

[733] J. Shah.A Common Framework for Curve Evolution, Segmentation and Anisotropic Diffusion.In *IEEE Conference on Computer Vision and Pattern Recognition*, pages 136–142, 1996.

[734] Y. Shapira.*Matrix-Based Multigrid: Theory and Applications*, volume 2 of *Numerical Methods and Algorithms*.Kluwer Academic Publishers, 2003.

[735] E. Sharon, A. Brandt, and R. Basri.Completion Energies and Scale.In *In Proc. IEEE Conf. Comp. Vision and Pattern Recognition (CVPR '97)*, pages 884–890, San Juan, Puerto Rico, 1997.

[736] E. Sharon and D. Mumford.2d-shape analysis using conformal mapping.In *Proc. CVPR*, pages 350–357, Washington, D.C., 2004. IEEE Comp. Soc.

[737] J. Shi and J. Malik.Normalized Cuts and Image Segmentation.*IEEE Transactions on Pattern Analysis and Machine Intelligence*, 22:888–905, 2000.

[738] J. Shi and C. Tomasi.Good features to track.In *Proc. CVPR'94*, pages 593 – 600, 1994.

[739] H.-Y. Shum and R. Szeliski.Construction of panoramic mosaics with global and local alignment.*International Journal of Computer Vision*, 36(2):101–130, February 2000.Erratum published July 2002, 48(2):151-152.

[740] K. Siddiqi, Y.-B. Lauziere, A. Tannenbaum, and S. Zucker.Area and Length Minimizing Flow for Shape Segmentation.*IEEE Transactions on Image Processing*, 7:433–443, 1998.

[741] J. Simon.Differentiation with respect to the domain in boundary value problems.*Numer. Funct. Anal. Optimiz.*, 2:649–687, 1980.

[742] L. Simon.*Lectures on Geometric Measure Theory*.Proceedings of the Centre for Mathematical Analysis, Australian National University, 1983.

[743] E. Simoncelli.*Distributed representation and analysis of visual motion*.PhD thesis, Department of Electrical Engineering, MIT, 1993.

[744] E. Simoncelli, E. Adelson, and D. Heeger.Probability distributions of optical flow.In *IEEE Conference on Computer Vision and Pattern Recognition*, pages 310–315, Mauii, 1991.

[745] E. Simoncelli and J. Portilla.Texture Characterization via Joint Statistics of Wavelet Coefficient Magnitudes.In *Proc. 5th IEEE Int. Conf. on Image Processing*, pages 62–66, Chicago, IL., Oct 4-7, 1998.

[746] L. Skovgaard.A Riemannian geometry of the multivariate normal model.*Scandinavian Journal of Statistics*, 11:211–233, 1984.

[747] C. Small.*The Statistical Theory of Shape*.Springer, 1996.

[748] S. Soatto, G. Doretto, and Y. Wu.Dynamic textures.In *IEEE International Conference in Computer Vision*, volume 2, pages 439–446, 2001.

[749] S. Soatto, A. Yezzi, and H. Jin.Tales of Shape and Radiance in Multi-view Stereo.In *IEEE International Conference on Computer Vision*, pages 974–981, 2003.

[750] O. Söderman and B. Jönsson.Restricted diffusion in cylindrical geometry.*J. Magn. Reson. A*, 117:94–97, 1995.

[751] J. Sokolowski and J. Zolesio.*Introduction to shape optimization*, volume 16 of *Springer series in computational mathematics*.Springer-Verlag, 1992.

[752] M. Sonka, V. Hlavac, and R. Boyle.*Image Processing, Analysis, and Machine Vision*.PWS, Pacific Grove, CA, 2nd edition, 1998.

[753] M. Spivak.*A Comprehensive Introduction to Differential Geometry*.Publish or Perish, Houston, 1975.

[754] M. Srinivasan, S. Zhang, M. Altwein, and J. Tautz.Honeybee navigation: Nature and calibration of the odometer.*Science*, 287(5454):851–853, 2000.

[755] A. Srivastava, S. Joshi, W. Mio, and X. Liu.Statistical shape anlayss: Clustering, learning and testing.*IEEE Transactions on Pattern Analysis and Machine Intelligence*, to appear, 2005.

[756] L. Staib and S. Duncan.Boundary finding with parametrically deformable models.*IEEE Transactions on Pattern Analysis and Machine Intelligence*, 14:1061–1075, 1992.

[757] C. Stauffer and W. Grimson.Adaptive background mixture models for real-time tracking.In *IEEE Conference on Computer Vision and Pattern Recognition*, pages 246–252, 1999.

[758] C. Stauffer and W. Grimson.Learning patterns of activity using real-time tracking.*IEEE Transactions on Pattern Analysis and Machine Intelligence*, 22(8):747–757, August 2000.

[759] M. Stegmann.*Generative Interpretation of Medical Images*.PhD thesis, Informatics and Mathematical Modeling Institute, Technical University of Denmark, 2004.

[760] G. Steidl, J. Weickert, T. Brox, P. Mrázek, and M. Welk.On the equivalence of soft wavelet shrinkage, total variation diffusion, total variation regularization, and SIDEs.*SIAM Journal on Numerical Analysis*, 42(2):686–713, 2004.

[761] A. Steiner, R. Kimmel, and A. Bruckstein.Planar Shape Enhancement and Exaggeration.*Graph Moddels and Image Processing*, 60(2):112–124, March 1998.

[762] E. Stejskal and J. Tanner.Spin diffusion measurements: Spin echoes in the presence of a time-dependent field gradient.*Journal of Chemical Physics*, 42:288–292, 1965.

[763] B. Stenger, A. Thayananthan, P. Torr, and R. Cipolla.Filtering using a tree-based estimator.In *IEEE International Conference on Computer Vision*, 2003.

[764] C. Stewart.Robust Parameter Estimation in Computer Vision.*SIAM Reviews*, 41:513–537, 1999.

[765] C. Stewart, C.-L. Tsai, and B. Roysam.The Dual-Bootstrap Iterative Closest Point Algorithm with Application to Retinal Image Registration.*IEEE Transactions on Medical Imaging*, 22:1379–1394, 2003.

[766] A. Stoddart, S. Lemke, A. Hilton, and T. Renn.Estimating Pose Uncertainty for Surface Registration.*Image and Vision Computing*, 16:111–120, 1998.

[767] G Strang.Maximal flow through a domain.*Mathematical Programming*, 26:123–143, 1983.

[768] C. Strecha, R. Fransens, and L. Van Gool.Wide-baseline stereo from multiple views: A probabilistic account.In *IEEE International Conference on Computer Vision and Pattern Recognition*, volume 2, pages 552–559, 2004.

[769] C. Strecha, T. Tuytelaars, and L. Van Gool.Dense Matching of Multiple Wide-Baseline Views.In *IEEE International Conference on Computer Vision*, pages 1194–1201, 2003.

[770] D. Strong and T. Chan.Edge-preserving and Scale-dependent Properties of Total Variation Regularization.*Inv. Probl.*, 19:165–187, 2003.

[771] C. Studholme, D. Hill, and D. Hawkes.An overlap invariant entropy measure of 3d medical image alignment.*Pattern Recognition*, 32(1):71–86, 1999.

[772] P. Sturm.Critical motion sequences for monocular self-calibration and uncalibrated euclidean reconstruction.In *Proc. CVPR'97*, pages 1100–1105, 1997.

[773] J. Sullivan, A. Blake, M. Isard, and J. MacCormick.Object localisation by bayesian correlation.In *IEEE International Conference on Computer Vision*, pages 1068–1075, 1999.

[774] M. Sussman, P. Smereka, and S. Osher.A Level Set Method for Computing Solutions to Incomprenissible Two-Phase Flow.*Journal of Computational Physics*, 114:146–159, 1994.

[775] R. Szeliski.Rapid octree construction from image sequences.*Computer Vision, Graphics and Image Processing: Image Understanding*, 58(1):23–32, July 1993.

[776] R. Szeliski.Video mosaics for virtual environments.*IEEE Computer Graphics and Applications*, 16(2):22–30, March 1996.

[777] R. Szeliski.Prediction Error as a Quality Metric for Motion and Stereo.In *IEEE International Conference on Computer Vision*, pages 781–788, 1999.

[778] R. Szeliski.Image alignment and stitching: A tutorial.Technical Report MSR-TR-2004-92, Microsoft Research, December 2004.

[779] R. Szeliski and J. Coughlin.Spline-based image registration.*International Journal of Computer Vision*, 22(3):199–218, 1997.

[780] R. Szeliski and S.-B. Kang.Recovering 3D shape and motion from image streams using nonlinear least squares.*Journal of Visual Communication and Image Representation*, 5(1):10–28, March 1994.

[781] R. Szeliski and H. Shum.Creating full view panoramic image mosaics and texture-mapped models.*Computer Graphics (SIGGRAPH'97 Proceedings)*, pages 251–258, August 1997.

[782] R. Szeliski and R. Zabih.An experimental comparison of stereo algorithms.In B. Triggs, A. Zisserman, and R. Szeliski, editors, *Vision Algorithms: Theory and Practice*, number 1883 in LNCS, pages 1–19, Corfu, Greece, September 1999. Springer-Verlag.

[783] M. Szummer and R. Picard.Temporal texture modeling.In *IEEE International Conference on Image Processing*, volume 3, pages 823–826, 1996.

[784] E. Tadmor, S. Nezzar, and L. Vese.A Multiscale Image Representation Using Hierarchical $(BV, L^2)$ Decompositions.*Multiscale Model. Simul.*, 2:554–579, 2003.

[785] H. Tagare, D. O'shea, and D. Groisser.Non-rigid shape comparison of plane curves in images.*J. Math. Imaging Vis.*, 16(1):57–68, 2002.

[786] P. Tan, S. Lin, L. Quan, and H. Shum.Highlight removal by illumination-constrained inpainting.In *Proceedings International Conference on Computer Vision*, pages 164–169, 2003.

[787] B. Tang, G. Sapiro, and V. Caselles.Color image enhancement via chromaticity diffusion.*IEEE Transactions on Image Processing*, 10:701–707, 2001.

[788] A. Tankus, N. Sochen, and Y. Yeshurun.A new perspective [on] Shape-from-Shading.In *ICCV'03*, volume 2, pages 862–869, 2003.

[789] A. Tankus, N. Sochen, and Y. Yeshurun.Reconstruction of medical images by perspective Shape-from-Shading.In *ICPR'04*, 2004.

[790] M. Tappen and W. Freeman.Comparison of graph cuts with belief propagation for stereo, using identical mrf parameters.In *IEEE Intl. Conference on Computer Vision (ICCV)*, October 2003.

[791] K. Tarabanis, P. Allen, and R. Tsai.A survey of sensor planning in computer vision.*IEEE Transactions on Robotics and Automation*, 11(1):86–105, February 1995.

[792] K. Tarabanis, R. Tsai, and A. Kaul.Computing occlusion-free viewpoints.*IEEE Transactions on Pattern Analysis and Machine Intelligence*, 18(3):279–292, March 1996.

[793] D. Terzopoulos and R. Szeliski.Tracking with Kalman snakes.In A. Blake and A.L. Yuille, editors, *Active Vision*, pages 3–20. MIT, 1992.

[794] P. Tetali.Random Walks and the Effective Resistance of Networks.*Journal of Theoretical Probability*, 4(1):101–109, 1991.

[795] K. Thornber and L. Williams.Analytical solution of stochastic completion fields. *Biological Cybernetics*, 75:141–151, 1996.

[796] K. Thornber and L. Williams.Characterizing the Distribution of Completion Shapes with Corners Using a Mixture of Random Processes.*Pattern Recognition*, 33:543–553, 2000.

[797] A. Tikhonov.The regularization of ill-posed problems.*Dokl. Akad. Nauk.*, SSR 153(1):49–52, 1963.

[798] S. Timoner.*Compact Representations for Fast Nonrigid Registration of Medical Images*.PhD thesis, Massachusetts Institute of Technology, 2003.

[799] P. Torr.Bayesian Model Estimation and Selection for Epipolar Geometry and Generic Manifold Fitting.*International Journal of Computer Vision*, 50:271–300, 2002.

[800] P. Torr, A. Fitzgibbon, and A. Zisserman.Maintaining multiple motion model hypotheses through many views to recover matching and structure.In *Proc. ICCV'99*, pages 485–491, 1998.

[801] K. Toyama and A. Blake.Probabilistic tracking in a metric space.In *IEEE International Conference on Computer Vision*, pages 50–59, 2001.

[802] A. Treuille, A. Hertzmann, and S. Seitz.Exemple-Based Stereo with General BRDFs.In *European Conference on Computer Vision*, pages 457–469, 2004.

[803] B. Triggs.The absolute quadric.In *Proc. CVPR'97*, pages 609–614, 1997.

[804] B. Triggs, P. McLauchlan, R. Hartley, and Fiztgibbon A.Bundle adjustment – a modern synthesis.In *Vision Algorithms: Theory and Practice, LNCS, Vol. 1883*, pages 298–372. Springer-Verlag, 2000.

[805] B. Triggs, P. McLauchlan, R. Hartley, and A. Fitzgibbon.Bundle adjustment — a modern synthesis.In *International Workshop on Vision Algorithms*, pages 298–372, Kerkyra, Greece, September 1999.

[806] A. Trouveé and L. Younes.Diffeomorphic matching problems in one dimension: designing and minimizing matching functionals.In *Proc. 6th European Conf. Computer Vision*, LNCS, pages 573–587. Springer, 2000.

[807] A. Tsai, A. Willsky, and A. Yezzi.A statistical approach to snakes for bimodal and trimodal imagery.In *IEEE International Conference on Computer Vision*, volume 2, pages 898–903, 1999.

[808] A. Tsai, A. Yezzi, W. Wells, C. Tempany, D. Tucker, A. Fan, W. Grimson, and A. Willsky.Model-based Curve Evolution Technique for Image Segmentation.In *IEEE Conference on Computer Vision and Pattern Recognition*, volume I, pages 463–468, 2001.

[809] A. Tsai, A. Yezzi, and A. Willsky.Curve evolution implementation of the Mumford-Shah functional for image segmentation, denoising, interpolation, and magnification.*IEEE Transactions on Image Processing*, 10(8):1169–1186, 2001.

[810] C.-L. Tsai, A. Majerovics, C. Stewart, and B. Roysam.Disease-Oriented Evaluation of Dual-Bootstrap Retinal Image Registration.In *Proceedings of the 6th International Conference on Medical Image Computing and Computer-Assisted Intervention*, volume II, pages 754–761, 2003.

[811] D. Tschumperlé and R. Deriche.Diffusion tensor regularization with constraints preservation.In *CVPR*, pages 948–954, 2001.

[812] D. Tschumperlé and R. Deriche.Regularization of orthonormal vector sets regularization with PDE's and applications.*IJCV*, 50(3):237–252, 2002.

[813] D. Tschumperlé and R. Deriche.Tensor field visualization with pde's and application to DT-MRI fiber visualization.In *Proc. of IEEE Workshop on VLSM*, pages 256–26, Nice, 2003.

[814] D. Tschumperlé and R. Deriche.Variational frameworks for DT-MRI estimation, regularization and visualization.In *ICCV*, pages 116–122, 2003.

[815] J. Tsitsiklis.Efficient Algorithms for Globally Optimal Trajectories.In *33$^{rd}$ Conference on Decision and Control*, pages 1368–1373, 1994.

[816] k. Tsuda, S. Akaho, and K. Asai.The em algorithm for kernel matrix completion with auxiliary data.*J. Mach. Learn. Res.*, 4:67–81, 2003.

[817] D. Tuch, T. Reese, M. Wiegell, N. Makris, J. Belliveau, and V. Wedeen.High angular resolution diffusion imaging reveals intravoxel white matter fiber heterogeneity.*Magn. Reson. Med.*, 48:577–582, 2002.

[818] D. Tuch, R. Weisskoff, J. Belliveau, and V. J. Wedeen.High angular resolution diffusion imaging of the human brain.In *Proc. of the 7th ISMRM*, page 321, Philadelphia, 1999.

[819] G. Turk and M. Levoy.Zippered Polygon Meshes from Range Images.In *Proc. SIGGRAPH 94, ACM*, pages 311–318, 1994.

[820] A. Tversky.Features of similarity.*Psychological Review*, 84(4):327–352, 1977.

[821] S. Ullman.Filling-in the Gaps: the Shape of Subjective Contours and a Model for Their Generation.*Biological Cybernetics*, 75:1–6, 1976.

[822] S. Ullman.The interpretation of structure from motion.*Proceedings of the Royal Society*, B-203:405–426, 1979.

[823] S. Uras, F. Girosi, A. Verri, and V. Torre.A computational approach to motion perception.*Biological Cybernetics*, 60:79–97, 1989.

[824] T. Ushijima and S. Yazaki.Convergence of a crystalline algorithm for the motion of a closed convex curve by a power of curvature $v = k^{\alpha}$.*SIAM. J. Numer. Anal.*, 37:500–522, 2000.

[825] M. Uyttendaele, A. Eden, and R. Szeliski.Eliminating ghosting and exposure artifacts in image mosaics.In *IEEE Computer Society Conference on Computer Vision and Pattern Recognition*, volume II, pages 509–516, Kauai, Hawaii, December 2001.

[826] C. Vachier and L. Vincent.Valuation of image extrema using alternating filters by reconstruction.*Image Algebra and Morphological Processing, San Diego CA, Proc. SPIE*, Juil. 1995.

[827] H. van Assen, M. Danilouchkine, F. Behloul, H. Lamb, R. van der Geest, J. Reiber, and B. Lelieveldt.Cardiac LV segmentation using a 3D active shape model driven by fuzzy inference.In *Medical Image Computing & Computer Assisted Interventions – MICCAI*, volume 2878 of *Lecture Notes in Computer Science*, pages 535–540. Springer Verlag, Berlin, 2003.

[828] R. van der Geest, B. Lelieveldt, E. Angelie, M. Danilouchkine, M. Sonka, and J. Reiber.Evaluation of a new method for automated detection of left ventricular contours in time series of magnetic resonance images using an active appearance motion model.*Journal of Cardiovascular Magnetic Resonance*, 6(3):609–617, 2004.

[829] L. Van Gool, R. Kempenaers, and A. Oosterlinck.Recognition and Semi-differential Invariants.In *International Conference on Pattern Recognition*, pages 454–460, Maui, Hawaii, 1991.

[830] L. Van Gool, T. Moons, and A. Oosterlinck.Semi-differential Invariants.In J. Mundy and A. Zisserman, editors, *Active Vision*. The MIT Press, 1992.

[831] G. Van Meerbergen, M. Vergauwen, M. Pollefeys, and L. Van Gool.A hierarchical symmetric stereo algorithm using dynamic programming.*Int. J. Comput. Vision*, 47(1/2/3):275–285, 2002.

[832] P. Van Overschee and B. De Moor.N4SID: subspace algorithms for the identification of combined deterministic-stochastic systems.*Automatica*, 30(1):75–93, 1994.

[833] R. Vaz and D. Cygansky.Generation of Affine Invariant Local Contour Feature Data.*Pattern Recognition Letters*, 11:479–483, 1990.

[834] S. Vedula, S. Baker, and T. Kanade.Spatio-Temporal View Interpolation.In *ACM Eurographics Workshop on Rendering*, pages 1–11, 2002.

[835] S. Vedula, S. Baker, P. Rander, R. Collins, and T. Kanade.Three-Dimensional Scene Flow.In *IEEE International Conference on Computer Vision*, pages 722–729, 1999.

[836] S. Vedula, S. Baker, S. Seitz, and T. Kanade.Shape and Motion Carving in 6D.In *IEEE Conference on Computer Vision and Pattern Recognition*, pages 592–598, 2000.

[837] O. Veksler.*Efficient Graph-based Energy Minimization Methods in Computer Vision*.PhD thesis, Cornell University, Ithaca, NY, August 1999.

[838] B. Vemuri, Y. Chen, M. Rao, and T. Mareci.Automatic fiber tractograph from dti and its validation.In *Proc. of the 1st IEEE ISBI*, pages 505–508, 2002.

[839] B. Vemuri, Y. Chen, M. Rao, T. McGraw, Z. Wang, and T. Mareci.Fiber tract mapping in the CNS using DT-MRI.In *VLSM*, pages 81–88, 2001.

[840] J. Verdera, V. Caselles, M. Bertalmio, and G. Sapiro.Inpainting Surface Holes.In *Proc. IEEE International Conference on Image Processing (ICIP)*, volume II, pages 903–906, Barcelona, Spain, 2003.

[841] L. Vese and T. Chan.A Multiphase Level Set Framework for Image Segmentation Using the Mumford and Shah Model.*International Journal of Computer Vision*, 50:271–293, 2002.

[842] L. Vese and S. Osher.Modelling Textures with Total Variation Minimization and Oscillating Patterns in Image Processing.*J. Sci. Computing*, 19:553–572, 2003.

[843] L. Vese and S. Osher.Modeling Textures with Total Variation Minimization and Oscillating Patterns In Image Processing.*J. Math. Imaging Vision*, 20:7–18, 2004.

[844] E. Villeger, G. Aubert, and L. Blanc-Feraud.Image Disocclusion Using a Probabilistic Gradient Orientation.*Preprint*, 2004.

[845] L. Vincent.Morphological area openings and closings for grayscale images.*Shape in Picture, NATO Workshop, Driebergen*, Sept. 1992.

[846] L. Vincent.Morphological grayscale reconstruction in image analysis: Applications and efficient algorithms.*IEEE Trans. in Image Procesing*, pages 176–201, 1993.

[847] P. Viola and M. Jones.Rapid object detection using a boosted cascade of simple features.In *IEEE Conference on Computer Vision and Pattern Recognition*, 2001.

[848] P. Viola, M. Jones, and D. Snow.Detecting pedestrians using patterns of motion and appearance.In *IEEE International Conference on Computer Vision*, pages 734–741, 2003.

[849] C. Vogel and M. Oman.Iterative methods for total variation denoising.*SIAM Journal on Scientific Computing*, 17:227–238, 1996.

[850] T. Wada, H. Ukida, and T. Matsuyama.Shape from shading with interreflections under proximal light source-3D shape reconstruction of unfolded book surface from a scanner image.In *ICCV'95*, 1995.

[851] G. Wahba.*Spline Models for Observational Data*.SIAM, Philadelphia, PA, 1990.

[852] S. Walden.*The Ravished Image*.St. Martin's Press, New York, 1985.

[853] J. Wang and E. Adelson.Representing Moving Images with Layers.*IEEE Transactions on Image Processing*, 3:625–638, 1994.

[854] L. Wang, W. Hu, and T. Tan.Recent developments in human motion analysis.*Pattern Recognition*, 36(3):585–601, March 2003.

[855] Y. Wang and S.-C. Zhu.A generative method for textured motion: analysis and synthesis.In *European Conference on Computer Vision*, pages 583–598, 2002.

[856] Y. Wang and S.-C. Zhu.Modeling complex motion by tracking and editing hidden Markov graphs.In *IEEE Conference on Computer Vision and Pattern Recognition*, volume 1, pages 856–863, 2004.

[857] Z. Wang.*Diffusion Tensor Field Restoration and Segmentation*.PhD thesis, University of Florida, 2004.

[858] Z. Wang and B. Vemuri.An Affine Invariant Tensor Dissimilarity Measure and its Application to Tensor-valued Image Segmentation.In *CVPR*, pages 228–233, 2004.

[859] Z. Wang and B. Vemuri.DTI Segmentation using an Information Theoretic Tensor Dissimilarity Measure.*IEEE Transactions on Medical Imaging*, 2004.Submitted.

[860] Z. Wang and B. Vemuri.Tensor Field Segmentation Using Region Based Active Contour Model.In *Europe. Conf. Comp. Vis.(4)*, pages 304–315, 2004.

[861] Z. Wang and B. Vemuri.Tensor field segmentation using region based active contour model.In *ECCV*, pages 304–315, 2004.

[862] Z. Wang, B. Vemuri, Y. Chen, and T. Mareci.Simultaneous smoothing and estimation of the tensor field from diffusion tensor MRI.In *Proc. of CVPR*, volume 2, Madison, WI, 2003.

[863] Z. Wang, B. Vemuri, Y. Chen, and T. Mareci.Simultaneous smoothing and estimation of the tensor field from diffusion tensor MRI.In *CVPR*, pages 461–466, 2003.

[864] Z. Wang, B. Vemuri, Y. Chen, and T. Mareci.A constrained variational principle for direct estimation and smoothing of the diffusion tensor field from complex DWI.*IEEE Transactions on Medical Imaging*, 23(8):930–939, 2004.

[865] W. Warren.Self-motion: Visual perception and visual control.In *Handbook of Perception and Cognition*, volume 5: Perception of Space and Motion. Academic Press, New York, 1995.

[866] A. Watson and A. Ahumada.Model of human visual-motion sensing.*Journal of the Optical Society of America A*, 2:322–342, 1985.

[867] J. Weber and J. Malik.Robust computation of optical-flow in a multiscale differential framework.*International Journal of Computer Vision*, 14(1):67–81, 1995.

[868] V. Wedeen, T. Reese, Tuchand D., M. Weigel, J. Dou, R. Weisskoffand, and D. Chesler.Mapping fiber orientation spectra in cerebral white matter with fourier transform diffusion MRI.In *Proc. of the 8th ISMRM*, page 82, Denver, 2000.

[869] L. Wei and M. Levoy.Fast texture synthesis using tree-structured vector quantization.In *Proceedings of SIGGRAPH*, pages 479–488, 2000.

[870] J. Weickert.*Anisotropic Diffusion in Image Processing*.Teubner, Stuttgart, 1998.

[871] J. Weickert.Diffusion and Regularization Methods for Tensor-valued Images.In *First SIAM-EMS Conf. Appl. Math. in Our Changing World*, 2001.

[872] J. Weickert and T. Brox.Diffusion and regularization of vector- and matrix-valued images.*Inverse Problems, Image Analysis, and Medical Imaging. Contemporary Mathematics*, 313:251–268, 2002.

[873] J. Weickert and G. Kuhne.Fast Methods for Implicit Active Contours.In S. Osher and n. Paragios, editors, *Geometric Level Set Methods in Imaging, Vision and Graphics*, pages 43–58. Springer, 2003.

[874] A. Weinstein.Almost invariant submanifolds for compact group actions.*Journal of the European Mathematical Society*, 2(1):53–86, 2000.

[875] I. Weiss.Projective Invariants of Shapes.*Center for Automation Research, University of Mariland*, TR-339, 1988.

[876] I. Weiss.Noise Resistant Invariants of Curves.*Center for Automation Research, University of Mariland*, CAR-TR-537, 1991.

[877] I. Weiss.Local Projective and Affine Invariants.*Annals of Mathematics and Artificial Itelligence*, 13(3-4):203–225, 1995.

[878] Y. Weiss.Smoothness in layers: Motion segmentation using nonparametric mixture estimation.In *IEEE Conference on Computer Vision and Pattern Recognition*, pages 520–526, Puerto Rico, 1997.

[879] Y. Weiss and E. Adelson.A unified mixture framework for motion segmentation: Incorporating spatial coherence and estimating the number of models.In *IEEE Conference on Computer Vision and Pattern Recognition*, pages 321–326, San Francisco, 1996.

[880] Y. Weiss and D. Fleet.Velocity likelihoods in biological and machine vision.In R. Rao, B. Olshausen, and M. Lewicki, editors, *Probabilistic models of the brain.* MIT Press, 2002.

[881] Y. Weiss, E. Simoncelli, and E. Adelson.Motion illusions as optimal percepts.*Nature Neuroscience*, 5(6):598–604, June 2002.

[882] M. Welk, J. Weickert, and G. Steidl.A four-pixel scheme for singular differential equations.In R. Kimmel, N. Sochen, and J. Weickert, editors, *Scale-Space and PDE Methods in Computer Vision*, Lecture Notes in Computer Science, Berlin, 2005. Springer.To appear.

[883] W. Wells III, P. Viola, H. Atsumi, S. Nakajima, and R. Kikinis.Multi-modal volume registration by maximization of mutual information.*Medical Image Analysis*, 1:35–52, 1996.

[884] C-F. Westin, S. Maier, H. Mamata, A. Nabavi, F. Jolesz, and R. Kikinis.Processing and visualization for diffusion tensor MRI.*Medical Image Analysis*, 6(2):93–108, 2002.

[885] Y. Wexler, E. Shechtman, and M. Irani.Space-time video completion.In *Proc. CVPR*, 2004.

[886] M. Wheeler, Y. Sato, and K. Ikeuchi.Consensus Surfaces for Modelling 3D Objects from Multiple Range Images.In *Proc. IEEE Int. Conference on Computer Vision*, pages 917–924, 1998.

[887] R. Whitaker.Volumetric deformable models: active blobs.In *Visualization in Biomedical Computing*, pages 122–134, 1994.

[888] R. Whitaker.A Level-Set Approach to 3D Reconstruction from Range Data.*International Journal of Computer Vision*, 29:203–231, 1998.

[889] M. Wiegell, D. Tuch, H. Larson, and V. Wedeen.Automatic Segmentation of Thalamic Nuclei from Diffusion Tensor Magnetic Resonance Imaging.*NeuroImage*, 19:391–402, 2003.

[890] E. Wilczynski.Projective Differential Geometry of Curves and Ruled Surfaces.In *Teubner*, Leipzig, Germany, 1906.

[891] L. Williams and A. Hanson.Perceptual Completion of Occluded Surfaces.In *Proceedings of the IEEE Computer Vision and Pattern Recognition*, pages 104–112, Seattle, WA, 1994.

[892] L. Williams and D. Jacobs.Stochastic Completion Fields: A Neural Model of Illusory Contour Shape and Salience.In *Proceedings of the 5th International Conference on Computer Vision*, pages 408–415, Cambridge, Mass, 1995.

[893] L. Williams and D. Jacobs.Local Parallel Computation of Stochastic Completion Fields.In *Proc. IEEE Conf. on Computer Vision and Pattern Recognition CVPR '96*, pages 161–168, 1996.

[894] O. Williams, A. Blake, and R. Cipolla.A sparse probabilistic learning algorithm for real-time tracking.In *IEEE International Conference on Computer Vision*, 2003.

[895] J. Wills, S. Agarwal, and S. Belongie.What went where.In *CVPR03*, pages I: 37–44, 2003.

[896] D. Wilson, E. Geiser, and J. Larocca.Automated analysis of echocardiographic apical 4-chamber images.In *Proceedings of the International Society for Optical*

*Engineering in Mathematical Modeling, Estimation, and Imaging*, volume 4121, pages 128–139, 2000.

[897] M. Worring, A. Smeulders, L. Staib, and J. Duncan.Parameterized feasible boundaries in gradient vector fields.*Computer Vision and Image Understanding*, 63(1):135–144, 1996.

[898] C. Wren, A. Azarbayejani, T. Darrell, and A. Pentland.Pfinder: Real-time tracking of the human body.*IEEE Trans. on Pattern Analysis and Machine Intelligence*, 19(7):780–785, 1997.

[899] Z. Wu and R. Leahy.An Optimal Graph Theoretic Approach to Data Clustering: Theory and its Application to Image Segmentation.*IEEE Pattern Analysis and Machine Intelligence*, 11:1101–1113, 1993.

[900] J. Xiao and M. Shah.Motion layer extraction in the presence of occlusion using graph cut.In *CVPR04*, pages II: 972–979, 2004.

[901] C. Xu, D. Pham, M. Rettmann, D. Yu, and J. Prince.Reconstruction of the human cerebral cortex from magnetic resonance images.*IEEE Transactions on Medical Imaging*, 18(6):467–480, 1999.

[902] C. Xu and J. Prince.Generalized gradient vector flow external forces for active contours.*Signal Processing: An International Journal*, 71(2):131–139, 1998.

[903] N. Xu, R. Bansal, and N. Ahuja.Object segmentation using graph cuts based active contours.In *IEEE Conference on Computer Vision and Pattern Recognition*, volume II, pages 46–53, 2003.

[904] R. Yang, M. Pollefeys, and G. Welch.Dealing with Textureless Regions and Specular Highlights: A Progressive Space Carving Scheme Using a Novel Photo-consistency Measure.In *IEEE International Conference on Computer Vision*, pages 576–584, 2003.

[905] L. Yatziv, G. Sapiro, and M. Levoy.Light field completion.In *Proceedings International Conference on Image Processing, ICIP*, 2004.

[906] Y. Ye and J. Tsotsos.Sensor planning for 3d object search.*Computer Vision and Image Understanding*, 73(2):145–168, February 1999.

[907] A. Yezzi.Modified curvature motion for image smoothing and enhancement.*IEEE Transactions on Image Processing*, 7:345–352, 1998.

[908] A. Yezzi, S. Kichenssamy, A. Kumar, P. Olver, and A. Tannebaum.A geometric snake model for segmentation of medical imagery.*IEEE Transactions on Medical Imaging*, 16(2):199–209, 1997.

[909] A. Yezzi, A. Tsai, and A. Willsky.A Statistical Approach to Snakes for Bimodal and Trimodal Imagery.In *IEEE International Conference in Computer Vision*, pages 898–903, 1999.

[910] A. Yezzi, A. Tsai, and A. Willsky.A statistical approach to snakes for bimodal and trimodal imagery.In *International Conference on Image Processing*, Kobe Japan, 1999.

[911] Y. You and M. Kaveh.A Regularization Approach to Joint Blur Identification and Image Restoration.*IEEE Trans. Image Process.*, 5:416–427, 1996.

[912] L. Younes.Computable elastic distances between shapes.*SIAM Journal of Applied Mathematics*, 58:565–586, 1998.

[913] L. Yuan, F. Wen, C. Liu, and H. Shum.Synthesizing dynamic texture with closed-loop linear dynamic systems.In *European Conference on Computer Vision*, volume 2, pages 603–616, 2004.

[914] A. Yuille and P. Hallinan.Deformable templates.In A. Blake and A. Yuille, editors, *Active Vision*, pages 20–38. MIT, 1992.

[915] Y. Zhang and C. Kambhamettu.On 3D Scene Flow and Structure Estimation.In *IEEE Conference on Computer Vision and Pattern Recognition*, pages 778–785, 2001.

[916] Z. Zhang.Iterative Point Matching for Registration of Free-Form Curves and Surfaces.*International Journal of Computer Vision*, 13:119–152, 1994.

[917] H-K. Zhao, T. Chan, B. Merriman, and S. Osher.A variational Level Set Approach to Multiphase Motion.*Journal of Computational Physics*, 127:179–195, 1996.

[918] H.-K. Zhao, S. Osher, and R. Fedkiw.Fast Surface Reconstruction Using the Level Set Method.In *Proc. First IEEE Workshop on Variational and Level Set Methods, in conjunction with Proc. IEEE ICCV 2001*, pages 194–202, 2001.

[919] J. Zhao and N. Badler.Nonlinear programming for highly articulated figures.*ACM Transactions on Graphics*, 13(4):313 – 336, October 1994.

[920] T. Zhao and R. Nevatia.Tracking multiple humans in crowded environment.In *IEEE International Conference on Computer Vision and Pattern Recognition*, pages II: 406–413, 2004.

[921] W. Zhao, R. Chellappa, J. Phillips, and A. Rosenfeld.Face recognition: A literature survey.*ACM Computing Surveys*, pages 399–458, 2003.

[922] S. Zhu and A. Yuille.Region Competition: Unifying Snakes, Region Growing, and Bayes/MDL for Multiband Image Segmentation.*IEEE Transactions on Pattern Analysis and Machine Intelligence*, 18:884–900, 1996.

[923] X. Zhu, J. Lafferty, and Z. Ghahramani.Combining Active Learning and Semi-Supervised Learning Using Gaussian Fields and Harmonic Functions.In *Proceedings of the ICML 2003 workshop on The Continuum from Labeled to Unlabel Data in Machine Learning and Data Mining*, pages 58–65, 2003.

[924] Y. Zhu and S. Cochoff.Likelihood maximization approach to image registration.*IEEE Transactions on Image Processing*, 11(12):1417–1426, 2002.

[925] L. Zhukov, K. Museth, D. Breen, R. Whitaker, and A. Barr.Level set segmentation and modeling of DT-MRI human brain data.*Journal of Electronic Imaging*, 12(1):2003, 125-133.

[926] W. Ziemer.*Weakly Differentiable Functions*.Springer Verlag, GTM, 120, 1989.

[927] L. Zöllei, J. Fisher, and W. Wells.A unified statistical and information theoretic framework for multi-modal image registration.In *Proceedings of IPMI*, volume 2732 of *Lecture Notes in Computer Science*, pages 366–377. Springer, 2003.

[928] L. Zöllei, J. Fisher, and W. Wells.A unified statistical and information theoretic framework for multi-modal image registration.Technical report, MIT, 2004.

[929] S. Zucker.Which computation runs in visual cortical columns?In J. Leo van Hemmen and T.J. Sejnowski, editors, *Problems in Systems Neuroscience*. Oxford University Press, 2004.

# Index